1514934-1   11/21/07

# GLUTAMATE AND ADDICTION

Series Editors:
*Ralph Lydic and*
*Helen A. Baghdoyan*

**Genetics and Genomics of Neurobehavioral Disorders,** edited by *Gene S. Fisch,* 2003

**Sedation and Analgesia for Diagnostic and Therapeutic Procedures**, edited by *Shobha Malviya, Norah N. Naughton, and Kevin K. Tremper,* 2003

**Neural Mechanisms of Anesthesia,** edited by *Joseph F. Antognini, Earl E. Carstens, and Douglas E. Raines,* 2002

**Glutamate and Addiction**
edited by *Barbara Herman,* 2002

**Molecular Mechanisms of Neurodegenerative Diseases**
edited by *Marie-Françoise Chesselet,* 2000

# Contemporary Clinical Neuroscience

# GLUTAMATE AND ADDICTION

Edited by

## BARBARA H. HERMAN, PhD

*Clinical Medical Branch, Division of Treatment Research and Development
National Institute on Drug Abuse (NIDA), National Institutes of Health (NIH)
Bethesda, MD*

*Coeditors*

### Jerry Frankenheim, PhD

*Pharmacology, Integrative & Cellular Neurobiology Research Branch (PICNRB)
Division of Neuroscience & Behavioral Research (DNBR), National Institute on Drug Abuse (NIDA)
National Institutes of Health (NIH), Bethesda, MD*

### Raye Z. Litten, PhD

*Treatment Research Branch, Division of Clinical and Prevention Research
National Institute on Drug Abuse (NIAAA), National Institutes of Health (NIH), Bethesda, MD*

### Philip H. Sheridan, MD

*Division of Neuropharmacological Drug Products, Center for Drug Evaluation and Research
Office of Drug Evaluation I, Food and Drug Administration, Rockville, MD*

### Forrest F. Weight, MD

*Laboratory of Molecular and Cellular Neurobiology, Division of Intramural Clinical and Biological
Research, National Institute on Drug Abuse (NIAAA) National Institutes of Health (NIH)
Bethesda, MD*

### Steven R. Zukin, MD

*Division of Treatment Research and Development, National Institute on Drug Abuse (NIDA)
National Institutes of Health (NIH), Bethesda, MD*

HUMANA PRESS ✻ TOTOWA, NEW JERSEY

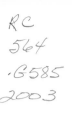

RC
564
.G585
2003

© 2003 Humana Press Inc.
999 Riverview Drive, Suite 208
Totowa, New Jersey 07512

www.humanapress.com

All rights reserved. No part of this book may be reproduced, stored in a retrieval system, or transmitted in any form or by any means, electronic, mechanical, photocopying, microfilming, recording, or otherwise without written permission from the Publisher.

The content and opinions expressed in this book are the sole work of the authors and editors, who have warranted due diligence in the creation and issuance of their work. The publisher, editors, and authors are not responsible for errors or omissions or for any consequences arising from the information or opinions presented in this book and make no warranty, express or implied, with respect to its contents.

Due diligence has been taken by the publishers, editors, and authors of this book to assure the accuracy of the information published and to describe generally accepted practices. The contributors herein have carefully checked to ensure that the drug selections and dosages set forth in this text are accurate and in accord with the standards accepted at the time of publication. Notwithstanding, as new research, changes in government regulations, and knowledge from clinical experience relating to drug therapy and drug reactions constantly occurs, the reader is advised to check the product information provided by the manufacturer of each drug for any change in dosages or for additional warnings and contraindications. This is of utmost importance when the recommended drug herein is a new or infrequently used drug. It is the responsibility of the treating physician to determine dosages and treatment strategies for individual patients. Further it is the responsibility of the health care provider to ascertain the Food and Drug Administration status of each drug or device used in their clinical practice. The publisher, editors, and authors are not responsible for errors or omissions or for any consequences from the application of the information presented in this book and make no warranty, express or implied, with respect to the contents in this publication.

Cover design by Patricia F. Cleary.

For additional copies, pricing for bulk purchases, and/or information about other Humana titles, contact Humana at the above address or at any of the following numbers: Tel.: 973-256-1699;Fax: 973-256-8341; E-mail: humana@humanapr.com or visit our website: http://humanapress.com

The opinions expressed herein are the views of the authors and may not necessarily reflect the official policy of the National Institute on Drug Abuse or any other parts of the US Department of Health and Human Services. The US Government does not endorse or favor any specific commercial product or company. Trade, proprietary, or company names appearing in this publication are used only because they are considered essential in the context of the studies reported herein.

This publication is printed on acid-free paper. ∞
ANSI Z39.48-1984 (American National Standards Institute) Permanence of Paper for Printed Library Materials.

**Photocopy Authorization Policy:**
Authorization to photocopy items for internal or personal use, or the internal or personal use of specific clients, is granted by Humana Press Inc., provided that the base fee of US $8.00 per copy, plus US $00.25 per page, is paid directly to the Copyright Clearance Center at 222 Rosewood Drive, Danvers, MA 01923. For those organizations that have been granted a photocopy license from the CCC, a separate system of payment has been arranged and is acceptable to Humana Press Inc. The fee code for users of the Transactional Reporting Service is: [0-89603-879-3/03 $10.00 + $00.25].

Printed in the United States of America. 10 9 8 7 6 5 4 3 2 1

Library of Congress Cataloging-in-Publication Data
Glutamate and addiction / edited by Barbara H. Herman; coeditors, Jerry Frankenheim...[et al.].
      p.;cm.–(Contemporary clinical neuroscience)
  Includes bibliographical references and index.
  ISBN 0-89603-879-3 (alk. paper)
  1. Substance abuse–Pathophysiology. 2. Glutamic acid–Physiological effect. I. Herman, Barbara H. II. Frankenheim, Jerry. III. Series.
  [DNLM: 1. Substance-Related Disorders–physiopathology. 2. Glutamates–pharmacology. 3. Receptors, Glutamates–physiology. WM 270 G567 2002]
  RC564.G585 2002
  616.86–dc21
                                                                                            2002190242

## *Acknowledgments*

This book is dedicated to our families and collaborators who contributed and supported this effort to characterize the role of glutamatergic systems in addiction disorders and to develop new technologies for the treatment of these brain disorders. This book is also dedicated to Dr. Marian Fischman who died on October 23, 2001 during the final review phase of this book. Dr. Fischman's contributions to the medications development of cocaine and opiate addiction were an outstanding influence in understanding the biology and treatment of addiction disorders. Finally, this book is dedicated to the physicians Lawrence Kelley, MD, Jacqueline R. Honig, MD, Marc S. Myerson, MD, and the countless other valued medical personnel who brought one of us back from a near death experience during its creation.

To Alexandra Samantha Herman, Robert H. Herman, Anita S. Herman
And our families
*BHH, JF, RL, FW, SZ*

# Preface

Assembling *Glutamate and Addiction* was a two-and-a-half year labor of love. As editors, we all had the same goal in mind and pursued this with a fierce dedication. We felt that it was now time for a volume clarifying for the first time the relationship between glutamatergic systems and addiction. The past decade has seen a steady and escalating progression of scientific advances that have implicated a pivotal role of glutamatergic systems in cocaine, opiate, and alcohol dependence—both the etiology of these disorders and their treatment. As editors, we met as a group several times a year to discuss the progress and the ever emerging direction of the book. As senior editor, I am personally indebted to the superb job of the coeditors attracting the very best scientists in this field to contribute their important papers to this book.

To Philip H. Sheridan, MD of the Food and Drug Administration (FDA), for his marvelous ability to attract internationally known scientists to contribute to the first section of the book on the basic physiology and pharmacology of glutamate. The five stellar chapters in this section include ones by Borges and Dingledine; Witkin, Kaminski and Rogawski; Choi and Snider; Sanchez and Jensen; and Kaul and Lipton. A special thank you to Michael A. Rogawski, MD, PhD, Epilepsy Research Section, NINDS, NIH for being an early and avid supporter of this effort and bringing to our attention valuable contributors to this book. It is our hope that these five introductory chapters will provide a level playing field for all readers of this book to upgrade their basic understanding of glutamate before proceeding to the other research chapters focused on the relationship between glutamate and various addictive disorders.

To Jerry Frankenheim, PhD of the National Institute on Drug Abuse, the National Institutes of Health (NIH) for his wonderful role, as senior editor of Section II, in illustrating the role of glutamatergic systems in stimulant drugs of abuse including cocaine, amphetamine, and methamphetamine. Dr. Frankenheim displayed considerable care in editing this section. In addition, I am personally indebted to Dr. Frankenheim for his seamless job in serving as Acting Senior Editor of this volume for a two-month period when I was unavailable for this task. Section II is a truly remarkable part of the book in its thoroughness in covering virtually every aspect of the role of glutamate in stimulant drugs of abuse, with outstanding chapters by Pert, Post, and Weiss; Karler, Thai, and Calder; Wolf; Baker, Cornish, and Kalivas; Wang, Mao, and Lau; Pulvirenti; Vezina and Suto; Cadet; Burrows and Yamamoto; Itzhak, Martin, and Ali; Matsumoto and Pouw; Bisaga and Fischman; and Epping-Jordan. As we state in our dedication of this book, this effort also coincided with the tragic death of one of our beloved colleagues in the addiction field, Marian Fischman, PhD of Columbia University School of Medicine. Dr. Fischman was a vibrant human being, and one of the most vital forces in the research field of addiction medicine. A personal thank you to Adam Bisaga, MD who took over the task of writing and editing this chapter with Dr. Fischman in an extremely gracious and responsible fashion in the face of tragic circumstances.

We are extremely grateful to the authors who contributed to the valued third section of the book on glutamate and opiate drugs of abuse including heroin. The world-renown scientists in this section included Mao; Trujillo; Popik; and Rasmussen. An overview of this important topic is provided by Jianren Mao, MD, PhD of Harvard University School of Medicine. It is of interest to note that the researchers in this section were some of the first to provide evidence of a relationship between glutamate and various aspects of the addiction process.

In the final section, the relationship between glutamate and alcohol abuse and alcoholism is explored. Our superb editors of Section IV are Forrest F. Weight, MD and Raye Litten, PhD, both of the National Institute of Alcohol Abuse and Alcoholism (NIAAA). Personally, I am particularly grateful for the continuous role provided by Dr. Litten, who managed to come to virtually every editorial meeting across building lines and to quickly get his section collated into a deliverable form to our publisher, Humana Press.

I would like to thank Craig Adams and Elyse O'Grady of Humana Press for their superb editorial and publishing skills and their tireless efforts in cheering this effort on to its completion. Craig and Elyse supported this effort from the beginning and until its completion, with a compassion and expertise that I will forever admire.

Finally, I would like to thank my institute, the National Institute on Drug Abuse, NIH, for being supremely generous in allowing me the time to pursue this effort for the last two and a half years. Particular thanks goes to Alan Leshner, Ph.D., former Director, NIDA, Glen R. Hanson, PhD, DDS, current and Acting Director, NIAA, Frank Vocci, PhD, Director, Division of Treatment Research and Development (DTR&D), NIDA and Ahmed Elkashef, MD, Chief, Clinical Medical Branch (CMB), DTR&D, NIDA for permitting this effort to occur. We also thank the institute directors of NIAAA, Enoch Gordis, MD (former director) and the present top official of the FDA Bernard A. Schwertz, DVM, PhD, Acting Principal Deputy Commissioner and the past commissioner of the FDA, Jane E. Henney, MD for enabling the participation of individuals from their respective institutions.

I am personally touched by the numerous cards, letters and flowers that I received from family, friends, professional colleagues, and folks from Humana while in the hospital.

Our interest in glutamatergic systems and drug abuse disorders stems back to at least 1991, when the first preclinical evidence was presented for a role of this system in the development of opiate tolerance and withdrawal (cf. 1, 2). Indeed, a few years earlier, research in the late 1980s suggested a role of glutamate in stimulant drug addiction (3). From there, we as a group launched several efforts to try to synthesize the knowledge base that was quickly accumulating in this exciting area. Thanks to the efforts of the National Institutes of Health (NIH) and the Food and Drug Administration (FDA), approaches to understanding the biological and behavioral basis of drug addiction and developing new modalities for the treatment of drug addiction are now attaining some level of consistency across the world. A highlight in this trend for unification in theory and practice, is illustrated by the conceptual writings of Alan I. Leshner, PhD former Director, National Institute on Drug Abuse, who has tirelessly pioneered to increase the research and scientific basis for understanding drug addiction as a disorder of the brain (e.g., 4, 5). A similar emphasis on drug abuse as a brain disorder is noted in the very basic preclinical research of Stephen E. Hyman, MD, former Director, National Institute on Mental Health (e.g., 6, 7). Similarly, in a monthly letter developed by the National Institute on Alcohol Abuse and Treatment (NIAAA), Enoch Gordis, MD, former Director, NIAAA has describe numerous scientific advances detailing the role of various biochemical systems in alcohol dependence and the role of medication treatment in alcohol dependence (cf., 8, 9). An esteemed partner in this endeavor is Jane Henney, MD, former Commissioner, FDA whose institute is responsible for making certain that the medications that are developed for this indication are both efficacious and safe. We very much value the superb contributions of the authors in Section IV on glutamate and alcohol, who include Peoples; Crew, Rudolph, and Chandler; Becker and Redmond; Krystal, Petrakis, D'Souza, Mason, and Trevisan; Zieglgänsberger, Rammes, Spanagel, Danysz, and Parsons; Pasternak and Kolesnikov; and Potgieter. We all work together with these institutes and with the creative and brilliant scientists who undertake both the preclinical and clinical research to develop a rigorous science of drug addiction. It is our hope that this research will result in innovative treatments for drug abuse and addiction, and for understanding the basis of these disorders in the central nervous system.

The job of characterizing the role of glutamatergic systems in addiction disorders is now off to a solid beginning. With the recent advance and approval of glutamatergic antagonists for the indication of alcohol abuse and addiction in a variety of European countries, we have already started to witness some clinical payoff for the superbly innovative and thorough research of both preclinical and clinical sciences. We hope that this effort will launch a new decade starting in the year 2001, that will see yet even further advances in the glutamatergic field, both in the etiology and treatment of addiction disorders.

*Barbara H. Herman*, PhD

*References*

1. Herman, B.H., Vocci, F., Bridge, P. The effects of NMDA receptor antagonists and nitric oxide synthase inhibitors on opioid tolerance and withdrawal. *Neuropsychopharmacology* **13**: 269–292, 1995.

2. Herman, B.H. and O'Brien, C.P. Clinical medications development for opiate addiction: focus on nonopioids and opioid antagonists for the amelioration of opiate withdrawal symptoms and relapse prevention. *Seminars in Neuroscience* **9**: 158–172, 1997.

3. Karler, R., et al., Blockade of "reverse tolerance" to cocaine and amphetamine by MK-801. *Life Sci* **45**: 599–606, 1989.

4. Leshner A.I., Koob G.F. Drugs of abuse and the brain. *Proc Assoc Am Physicians* **111**:99–108, 1999.

5. Leshner A.I. Addiction is a brain disease, and it matters. *Science* **278**: 45–47, 1997.

6. Hyman S.E., Hyman S.E., Malenka R.C. Addiction and the brain: the neurobiology of compulsion and its persistence. *Nat Rev Neurosci*.**2**: 695–703, 2001.

7. Berke J.D., Hyman S.E. Addiction, dopamine, and the molecular mechanisms of memory. *Neuron* **25**: 515–532, 2000.

8. Gordis E. Advances in research on alcoholism and what they promise for future treatment and prevention. *Med Health R I* **82**:121, 1999.

9. Gordis E. The neurobiology of alcohol abuse and alcoholism: building knowledge, creating hope. *Drug Alcohol Depend.* **51**:9–11, 1998.

# Contents

Preface .................................................................................................. vii
Contributors ......................................................................................... xv

## I. INTRODUCTION: *PHYSIOLOGY AND PHARMACOLOGY OF GLUTAMATE*

*Philip H. Sheridan, Forrest F. Weight, and Barbara H. Herman Section Editors*

    **1**   Molecular Pharmacology and Physiology of Glutamate Receptors ................................................... 3
*Karin Borges and Raymond Dingledine*

    **2**   Pharmacology of Glutamate Receptors ........................................ 23
*Jeffrey M. Witkin, Rafal Kaminski, and Michael A. Rogawski*

    **3**   Glutamate and Neurotoxicity ...................................................... 51
*B. Joy Snider and Dennis W. Choi*

    **4**   Maturational Regulation of Glutamate Receptors and Their Role in Neuroplasticity ......................................... 63
*Russell M. Sanchez and Frances E. Jensen*

    **5**   Role of the NMDA Receptor in Neuronal Apoptosis and HIV-Associated Dementia .................................. 71
*Marcus Kaul and Stuart A. Lipton*

## II. GLUTAMATE: *STIMULANT DRUGS OF ABUSE (COCAINE, AMPHETAMINE, METHAMPHETAMINE)*

*Jerry Frankenheim and Barbara H. Herman, Section Editors*

    **6**   Role of Glutamate and Nitric Oxide in the Acquisition and Expression of Cocaine-Induced Conditioned Increases in Locomotor Activity ............................................... 83
*Agu Pert, Robert M. Post, and Susan R. B. Weiss*

    **7**   Interactions of Dopamine, Glutamate, and GABA Systems in Mediating Amphetamine- and Cocaine-Induced Stereotypy and Behavioral Sensitization ................................. 107
*Ralph Karler, David K. Thai, and Larry D. Calder*

    **8**   Addiction and Glutamate-Dependent Plasticity ......................... 127
*Marina E. Wolf*

    **9**   Glutamate and Dopamine Interactions in the Motive Circuit: *Implications for Craving* ......................................................... 143
*David A. Baker, Jennifer L. Cornish, and Peter W. Kalivas*

    **10**   Glutamate Cascade from Metabotropic Glutamate Receptors to Gene Expression in Striatal Neurons: *Implications for Psychostimulant Dependence and Medication* ................... 157
*John Q. Wang, Limin Mao, and Yuen-Sum Lau*

| | | |
|---|---|---|
| 11 | Glutamate Neurotransmission in the Course of Cocaine Addiction .................................................................................. | 171 |
| | *Luigi Pulvirenti* | |
| 12 | Glutamate and the Self-Administration of Psychomotor-Stimulant Drugs ................................................... | 183 |
| | *Paul Vezina and Nobuyoshi Suto* | |
| 13 | Roles of Glutamate, Nitric Oxide, Oxidative Stress, and Apoptosis in the Neurotoxicity of Methamphetamine ............ | 201 |
| | *Jean Lud Cadet* | |
| 14 | Methamphetamine Toxicity: *Roles for Glutamate, Oxidative Processes, and Metabolic Stress* ............................................... | 211 |
| | *Kristan B. Burrows and Bryan K. Yamamoto* | |
| 15 | Nitric Oxide-Dependent Processes in the Action of Psychostimulants ....................................................................... | 229 |
| | *Yossef Itzhak, Julio L. Martin, and Syed F. Ali* | |
| 16 | Effects of Novel NMDA/Glycine-Site Antagonists on the Blockade of Cocaine-Induced Behavioral Toxicity in Mice ............................................................................. | 243 |
| | *Rae R. Matsumoto and Buddy Pouw* | |
| 17 | Clinical Studies Using NMDA Receptor Antagonists in Cocaine and Opioid Dependence ........................................ | 261 |
| | *Adam Bisaga and Marian W. Fischman* | |
| 18 | The Role of mGluR5 in the Effects of Cocaine: *Implications for Medication Development* ............................ | *271* |
| | *Mark P. Epping-Jordan* | |

## III. GLUTAMATE AND OPIATE DRUGS (HEROIN) OF ABUSE

### *Barbara H. Herman and Jerry Frankenheim, Section Editors*

| | | |
|---|---|---|
| 19 | Role of the Glutamatergic System in Opioid Tolerance and Dependence: *Effects of NMDA Receptor Antagonists* ...... | 281 |
| | *Jianren Mao* | |
| 20 | The Role of NMDA Receptors in Opiate Tolerance, Sensitization, and Physical Dependence: *A Review of the Research, A Cellular Model, and Implications for the Treatment of Pain and Addiction* ................................ | 295 |
| | *Keith A. Trujillo* | |
| 21 | Modification of Conditioned Reward by *N*-Methyl-D-aspartate Receptor Antagonists ........................... | 323 |
| | *Piotr Popik* | |
| 22 | Morphine Withdrawal as a State of Glutamate Hyperactivity: *The Effects of Glutamate Receptor Subtype Ligands on Morphine Withdrawal Symptoms* ........................................ | 329 |
| | *Kurt Rasmussen* | |

IV. GLUTAMATE AND ALCOHOL ABUSE AND ALCOHOLISM

*Forrest F. Weight and Raye Z. Litten, Section Editors*

- 23 Alcohol Actions on Glutamate Receptors ................................... 343
  **Robert W. Peoples**
- 24 Glutamate and Alcohol-Induced Neurotoxicity .......................... 357
  **Fulton T. Crews, Joseph G. Rudolph, and L. Judson Chandler**
- 25 Role of Glutamate in Alcohol Withdrawal Kindling ................... 375
  **Howard C. Becker and Nicole Redmond**
- 26 Alcohol and Glutamate Neurotransmission in Humans: *Implications for Reward, Dependence, and Treatment* ........... 389
  **John H. Krystal, Ismene L. Petrakis, D. Cyril D'Souza, Graeme Mason, and Louis Trevisan**
- 27 Mechanism of Action of Acamprosate Focusing on the Glutamatergic System ....................................................... 399
  **W. Zieglgänsberger, G. Rammes, R. Spanagel, W. Danysz, and Ch. Parsons**
- 28 The NMDA/Nitric Oxide Synthase Cascade in Opioid Analgesia and Tolerance ......................................................... 409
  **Gavril W. Pasternak and Yuri Kolesnikov**
- 29 Overview of Clinical Studies for Acamprosate .......................... 417
  **Adriaan S. Potgieter**

Index ................................................................................................... 427

# Contributors

SYED F. ALI, PhD • *Neurochemistry Laboratory, Division of Neurotoxicology, National Center for Toxicological Research, Food and Drug Administration, Jefferson, AR*
DAVID A. BAKER, PhD • *Department of Physiology and Neuroscience, Medical School of South Carolina, Charleston, SC*
HOWARD C. BECKER, PhD • *Charleston Alcohol Research Center, Center for Drug and Alcohol Programs, Department of Psychiatry and Behavioral Sciences, Physiology, and Neuroscience, Department of Veterans Affairs Medical Center, Medical University of South Carolina, Charleston, SC*
ADAM BISAGA, MD • *Division on Substance Abuse, Department of Psychiatry, Columbia University College of Physicians and Surgeons, New York, NY*
KARIN BORGES, PhD • *Department of Pharmacology, Emory University School of Medicine, Atlanta, GA*
KRISTAN B. BURROWS, PhD • *Program in Basic and Clinical Neuroscience, Department of Psychiatry, Case Western Reserve University Medical School, Cleveland, OH*
JEAN LUD CADET, MD • *Molecular Psychiatry Division, National Institute on Drug Abuse, National Institutes of Health, Baltimore, MD*
LARRY D. CALDER, PhD • *Department of Pharmacology, University of Utah School of Medicine, Salt Lake City, UT*
L. JUDSON CHANDLER, PhD • *Department of Physiology/Neuroscience and Psychiatry, Medical University of South Carolina, Charleston, SC*
DENNIS W. CHOI, MD, PhD • *Center for the Study of Nervous System Injury and Department of Neurology, Washington University School of Medicine, St. Louis, MO*
JENNIFER L. CORNISH, PhD • *National Institute on Drug Abuse, National Institutes of Health, Baltimore, MD*
FULTON T. CREWS, PhD • *Director, Center for Alcohol Studies, University of North Carolina at Chapel Hill, Chapel Hill, NC*
W. DANYSZ • *Merz Co., Frankfurt, Germany*
RAYMOND DINGLEDINE, PhD • *Department of Pharmacology, Emory University School of Medicine, Atlanta, GA*
D. CYRIL D'SOUZA, MD • *Department of Psychiatry, Yale University School of Medicine, New Haven, CT; Alcohol Research Center, VA Connecticut Healthcare System, West Haven, CT; and NIAAA Center for the Translational Neuroscience of Alcoholism, Ribicoff Research Facilities, Connecticut Mental Health Center, New Haven, CT*
MARK P. EPPING-JORDAN, PhD • *Addex Pharmaceuticals, Institut de Biologie Cellulaire et du Morphologie, Universit de Lausanne, Lausanne, Switzerland*
MARIAN W. FISCHMAN, PhD (deceased) • *Division on Substance Abuse, Department of Psychiatry, Columbia University College of Physicians and Surgeons, New York, NY*
JERRY FRANKENHEIM, PhD • *DNBR, National Institute on Drug Abuse (NIDA), Bethesda, MD*
BARBARA H. HERMAN, PhD • *Clinical Medical Branch, Division of Treatment Research and Devlopment, National Institute on Drug Abuse (NIDA), National Instiutes of Health (NIH), Bethesda, MD*

YOSSEF ITZHAK, PhD • *Department of Psychiatry and Behavioral Sciences, University of Miami School of Medicine, Miami, FL*
FRANCES E. JENSEN, MD • *Children's Hospital, Boston, MA, and Harvard Medical School, Boston, MA*
PETER W. KALIVAS, PhD • *Department of Physiology and Neuroscience, Medical School of South Carolina, Charleston, SC*
RAFAL KAMINSKI, MD, PhD • *Drug Development Group, Intramural Research Program, National Institute on Drug Abuse, National Institutes of Health, Bethesda, MD*
RALPH KARLER, PhD • *Department of Pharmacology, University of Utah School of Medicine, Salt Lake City, UT*
MARCUS KAUL, PhD • *Center for Neuroscience and Aging, The Burnham Institute, La Jolla, CA*
YURI KOLESNIKOV, MD, PhD • *The Laboratory of Molecular Neuropharmacology, Memorial Sloan-Kettering Cancer Center, New York, NY*
JOHN H. KRYSTAL, MD • *Department of Psychiatry, Yale University School of Medicine, New Haven, CT; Alcohol Research Center, VA Connecticut Healthcare System, West Haven, CT; and NIAAA Center for the Translational Neuroscience of Alcoholism, Ribicoff Research Facilities, Connecticut Mental Health Center, New Haven, CT*
YUEN-SUM LAU, PhD • *Division of Pharmacology, School of Pharmacy, University of Missouri–Kansas City, Kansas City, MO*
STUART A. LIPTON, MD, PhD • *Center for Neuroscience and Aging, The Burnham Institute, La Jolla, CA*
RAYE Z. LITTEN, PhD • *Division of Clinical and Prevention Research, National Institute on Alcohol Abuse and Alcoholism (NIAAA), Bethesda, MD*
JIANREN MAO, MD, PhD • *MGH Pain Center, Department of Anesthesia and Critical Care, Massachusetts General Hospital, Harvard Medical School, Boston, MA*
LIMIN MAO, MD • *Division of Pharmacology, School of Pharmacy, University of Missouri–Kansas City, Kansas City, MO*
JULIO L. MARTIN, PhD • *Department of Psychiatry and Behavioral Sciences, University of Miami School of Medicine, Miami, FL*
GRAEME MASON, PhD • *Department of Psychiatry, Yale University School of Medicine, New Haven, CT; Alcohol Research Center, VA Connecticut Healthcare System, West Haven, CT; and NIAAA Center for the Translational Neuroscience of Alcoholism, Ribicoff Research Facilities, Connecticut Mental Health Center, New Haven, CT*
RAE R. MATSUMOTO, PhD • *Department of Pharmaceutical Sciences, College of Pharmacy, University of Oklahoma Health Sciences Center, Oklahoma City, OK*
CH. PARSONS • *Merz Co., Frankfurt, Germany*
GAVRIL W. PASTERNAK, MD, PhD • *The Laboratory of Molecular Neuropharmacology, Department of Anesthesiology, Memorial Sloan-Kettering Cancer Center, New York, NY*
ROBERT W. PEOPLES, PhD • *Unit on Cellular Neuropharmacology, Laboratory of Molecular and Cellular Neurobiology, National Institute on Alcohol Abuse and Alcoholism, National Institutes of Health, Bethesda, MD*
AGU PERT, PhD • *Biological Psychiatry Branch, National Institute of Mental Health, National Institutes of Health, Bethesda, MD*

ISMENE L. PETRAKIS, MD • *Department of Psychiatry, Yale University School of Medicine, New Haven, CT; Alcohol Research Center, VA Connecticut Healthcare System, West Haven, CT; and NIAAA Center for the Translational Neuroscience of Alcoholism, Ribicoff Research Facilities, Connecticut Mental Health Center, New Haven, CT*
PIOTR POPIK, MD, PhD • *Institute of Pharmacology, Polish Academy of Sciences, Kraków, Poland*
ROBERT M. POST, PhD • *Biological Psychiatry Branch, National Institute of Mental Health, National Institutes of Health, Bethesda, MD*
ADRIAAN S. POTGIETER, MD • *Marketing and Business Development, European Society of Cardiology*
BUDDY POUW, MD • *Department of Pharmaceutical Sciences, College of Pharmacy, University of Oklahoma Health Sciences Center, Oklahoma City, OK*
LUIGI PULVIRENTI, MD • *Department of Neuropharmacology, The Scripps Research Institute, La Jolla, CA*
G. RAMMES • *Max-Planck Institute of Psychiatry, Munich, Germany*
KURT RASMUSSEN, PhD • *Lilly Research Laboratories, Eli Lilly & Co., Lilly Corporate Center, Indianapolis, IN*
NICOLE REDMOND • *Charleston Alcohol Research Center, Center for Drug and Alcohol Programs, Department of Psychiatry and Behavioral Sciences, Physiology, and Neuroscience, Department of Veterans Affairs Medical Center, Medical University of South Carolina, Charleston, SC*
MICHAEL A. ROGAWSKI, MD, PhD • *Epilepsy Research Section, National Institute of Neurological Disorders and Stroke, National Institutes of Health, Bethesda, MD*
JOSEPH G. RUDOLPH, PhD • *NIH/NIAAA/DICBR/LNG, Rockville, MD*
RUSSELL M. SANCHEZ, PhD • *Children's Hospital, Boston, MA, and Harvard Medical School, Boston, MA*
PHILIP H. SHERIDAN, MD • *Division of Neuropharmacological Drug Products, Center for Drug Evaluation and Research, Office of Drug Evaluation I, Food and Drug Administration, Rockville, MD*
B. JOY SNIDER, MD, PhD • *Center for the Study of Nervous System Injury and Department of Neurology, Washington University School of Medicine, St. Louis, MO*
R. SPANAGEL • *Central Institute of Mental Health, Mannheim, Germany*
NOBUYOSHI SUTO, MA • *Department of Psychiatry, The University of Chicago, Chicago, IL*
DAVID K. THAI, PhD • *Department of Pharmacology, University of Utah School of Medicine, Salt Lake City, UT*
LOUIS TREVISAN, MD • *Department of Psychiatry, Yale University School of Medicine, New Haven, CT; Alcohol Research Center, VA Connecticut Healthcare System, West Haven, CT*
KEITH A. TRUJILLO, PhD • *Department of Psychology, California State University San Marcos, San Marcos, CA*
PAUL VEZINA, PhD • *Department of Psychiatry, The University of Chicago, Chicago, IL*
JOHN Q. WANG, MD, PhD • *Division of Pharmacology, School of Pharmacy, University of Missouri–Kansas City, Kansas City, MO*
FORREST F. WEIGHT, MD • *Division of Intramural Clinical and Biological Research, National Institute on Alcohol Abuse and Alcoholism (NIAAA), Bethesda, MD*

SUSAN R. B. WEISS, PhD • *Biological Psychiatry Branch, National Institute of Mental Health, National Institutes of Health, Bethesda, MD*
JEFFREY M. WITKIN, PhD • *Neuroscience Discovery Research, Lilly Research Laboratories, Lilly Corporate Center, Indianapolis, IN*
MARINA E. WOLF, PhD • *Department of Neuroscience, FUHS/The Chicago Medical School, North Chicago, IL*
BRYAN K. YAMAMOTO, PhD • *Department of Pharmacology, Boston University School of Medicine, Boston, MA*
W. ZIEGLGÄNSBERGER, MD, PhD • *Max-Planck Institute of Psychiatry, Munich, Germany*
STEVEN R. ZUKIN, MD • *DTRD, National Institute on Drug Abuse, Bethesda, MD*

# I Introduction
*Physiology and Pharmacology of Glutamate*

*Section Editors*

Philip H. Sheridan
Forrest F. Weight
Barbara H. Herman

# 1
# Molecular Pharmacology and Physiology of Glutamate Receptors

### Karin Borges, PhD and Raymond Dingledine, PhD

## 1. INTRODUCTION

Glutamate receptors represent the main excitatory receptors in synaptic transmission in the brain and have been intensively studied over the last 15 yr. Although clinical settings involving glutamate receptor modulators or antagonists usually involve stroke, acute brain injury, epilepsy, and neuropathic pain, both metabotropic and ionotropic classes of glutamate receptor also appear to play a role in addiction and cognition. For example, sensitization to cocaine upon chronic exposure to this stimulant appears to be mediated in part by $Ca^{2+}$ influx through α-amino-3-hydroxy-5-methyl-4-isoxazole propionic acid (AMPA) receptors *(1)*, and an mGluR2 agonist attenuates the disruptive effects of phencyclidine on working memory *(2)*. We will provide an overview of the molecular and physiological properties of glutamate receptors and review their subunit-specific pharmacology. As much as possible, we will focus on features of glutamate receptor activation and desensitization that may be most relevant to addiction and cognitive processing. More extensive information on glutamate receptor pharmacology can be found in the literature *(3–5)*.

## 2. METABOTROPIC RECEPTORS

### 2.1. Introduction into mGluR Classifications and Their Classical G-Protein-Coupled Signaling Pathways

The metabotropic receptors all contain seven transmembrane domains (TM) and are coupled to G-proteins. They are classified into three groups according to their pharmacology (Table 1). Many excellent extensive reviews for the mGluRs are avaible (e.g., refs. *6–9*). Metabotropic glutamate receptors are widely expressed in the brain, except for mGluR6, which only occurs in the retina. Group II mGluRs are found in presynaptic membranes or extrasynaptically, group III receptors function as autoreceptors in the presynaptic terminal membrane, and group I mGluRs are often expressed perisynaptically, near the postsynaptic density *(10)*. Astrocytes can express mGluR3 and mGluR5 (reviewed in ref. *11*) and outside the brain, mGluRs occur, for example, in the heart *(12)*. Group I receptors are coupled to $G_q$-proteins, which, by activating phospholipase C, produce inositol triphosphate ($IP_3$), which then activates the endoplasmic $IP_3$ receptor and triggers the release of calcium from intracellular stores. Group I receptors also activate or inhibit voltage-gated ion channels. Group II and III receptors couple to $G_i/G_0$ proteins that either block adenylate cyclase or calcium channels or activate potassium channels. An example of the different signaling pathways as occurring in the CA1 area of the hippocampus is shown in Fig. 1. The figure also displays the interaction of mGluRs with other receptors and ion channels.

From: *Contemporary Clinical Neuroscience: Glutamate and Addiction*
Edited by: Barbara H. Herman et al. © Humana Press Inc., Totowa, NJ

# Table 1
## Established and Commonly Used Compounds That Can Distinguish Among mGluR Receptor Groups and Between mGluR1 and mGluR5

|  | Receptor | Effectors | Agonists | Antagonists |
|---|---|---|---|---|
| Group I | mGluR1 | Gq | DHPG | LY367385, CPCCOEt |
|  | mGluR5 | Gq | CHPG, CBPG, DHPG | MPEP |
| Group II | mGluR2 | $G_i/G_o$ | LY354740, APDC, DCG-IV | LY341495[a] |
|  | mGluR3 | $G_i/G_o$ |  |  |
| Group III | mGluR4 | $G_i/G_o$ | L-AP4, L-SOP, PPG | MAP4 |
|  | mGluR6 | $G_i/G_o$ |  |  |
|  | mGluR7 | $G_i/G_o$ |  |  |
|  | mGluR8 | $G_i/G_o$ |  |  |

*Note:* For further information, see, for example, 9. The full names of the abbreviated compounds in alphabetical order are as follows: L-AP4, L-(+)-amino-4-phosphonobutyric acid; APCD, 4-aminopyrrolidine-2,4-dicarboxylic acid; CBPG, (S)-(+)-2-(3′-dicarboxycyclopropyl(1.1.1)pentyl)-glycine; CHPG, (R,S)-2-chloro-5- hydroxyphenylglycine; CPCCOEt, cyclopropan[b]chromen-1a-carboxylate; DCG-VI, (2′S,2′R,3′R)-2-(2′3′-dicarboxycyclopropyl)glycine; DHPG, 3,5-dihydroxyphenylglycine; LY341495, 2S-2 amino-2-(1S,2S-2-carboxycyclopropan-1-yl)-3-xanth-9-yl)propanoic acid; LY354740,(1S,2S,5R,6S)-(+)-2-aminobicyclo[3.1.0]hexane-2,6-dicarboxylic acid; LY367385, (+)-2-methyl-4-carboxy-phenylglycine; MAP4, α-methyl-L-amino-4-phosphonobutyrate; PPG, (R,S)-4-phosphonophenylglycine; L-SOP, L-serine O-phosphate.

[a] Note that at high concentrations, the group II mGluR antagonist LY341495 can also block group III and I receptors.

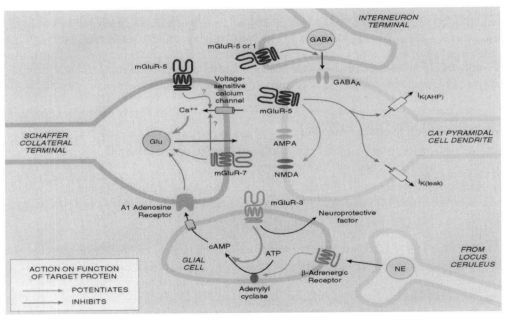

Fig. 1. Example of physiological roles of mGluRs at the Schaffer collateral synapse in CA1 in the hippocampus. mGluR5 and mGluR7 are located in the presynaptic terminal, where they inhibit glutamate release directly by effects on the release machinery or indirectly by inhibiton of voltage-gated calcium channels. mGluR5 is expressed at the postsynaptic terminal which increases pyramidal cell excitability by reducing potassium currents. Moreover, mGluR5 activation can potentiate NMDA receptor-mediated currents. Inhibitory GABA-ergic terminals express group I mGluRs, and by inhibiting GABA release, they can indirectly increase pyramidal cell excitability. Finally, glial cells express mGluR3. There is evidence that glial mGluR stimulation leads to release of a neuroprotective factor. Also, mGluR3 activation can potentiate β-adrenergic responses, leading to release of cAMP or adenosine, which stimulates A1 adenosine receptors and reduces glutamate release from the presynaptic terminal. (From ref. 13, with permission.) (Color illustration in insert following p. 142.)

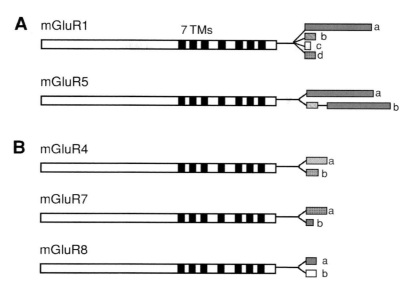

Fig. 2. Schematic representation of splice variants of mGluR proteins. Only translated regions are depicted; alternative splicing in untranslated regions is not shown. The seven transmembrane domains are shown in black. The different C-terminal domains are indicated by different patterns. (**A**) Group I mGluRs. The mGluR1a C-tail is homologous to those of mGluR5a and mGluR5b (all shown in gray). The C-terminal domains of mGluR5a and mGluR5b are the same. (**B**) Within the group III mGluRs, no homology between C-terminal domains is found. Adapted from ref. 7; new splice variants for mGluR7 and mGluR8 (14,15) are added.

Some mGluRs are alternatively spliced. Alternative splicing in translated regions occurs only at the C-termini in mGluR1, 4, 5, 7, and 8, as shown Fig. 2. In some cases, alternative splicing leads to different interactions with other signaling molecules (see Section 2.2.).

## 2.2. Association with Other Intracellular Signaling Proteins and Targeting Proteins

In addition to mGluR signaling via G-protein cascades, mGluR interactions with other signaling molecules are being discovered. For example, group I mGluRs with homologous C-termini (mGluR1a and mGluR5) couple to Homer proteins (16). Constitutively expressed Homer proteins (the long forms of Homer 1b, 1c, 2, and 3) physically link mGluR1a or mGluR5 to the endoplasmic IP$_3$ receptor. This signaling complex can be disrupted by the truncated Homer 1a, which is up-regulated as an immediate early gene after certain forms of long-term potentiation and after seizures. Similarly, the long Homer forms inhibited group I mGluR-mediated regulation of N-type calcium channels and M-type potassium channels, whereas the truncated forms did not (17). Moreover, Homer interacts with the scaffold protein Shank, which links Homer to many other cytoplasmic and membrane proteins. By virtue of its ability to bridge receptors and cytoplasmic proteins, Homer controls the trafficking of mGluR1a and mGluR5 into and out of synapses (18).

In addition to linking mGluRs to signaling molecules, the mGluR C-termini can be involved in receptor targeting, as observed in other receptors. For example, the last 60 amino acids target mGluR7 to axons and dendrites, whereas mGluR2 is excluded from axons (19). Calmodulin binds to C-terminal regions of mGluR5 and mGluR7, which are also phosphorylated by protein kinase C (PKC) (20–22). Calcium/calmodulin binding and PKC phosphorylation are mutually occlusive, similar to their roles at NMDA receptors (see Section 3.7.). Moreover, mGluR7 seems to be able to associate with the PKC α-subunit and protein interacting with C-kinase 1 (PICK1), because they can be coimmunoprecipitated from transfected COS cells and PICK1 can reduce phosphorylation of mGluR7a in

**Table 2**
**Glutamate Receptor Subunits and Their Genes**

| Group | Receptor Family | Subunit | Gene | Chromosome (human) | GenEmbl accession numbers | | |
|---|---|---|---|---|---|---|---|
| | | | | | Mouse | Rat | Human |
| 1 | AMPA | GluR1 | GRIA1 | 5q33 | X57497 | X17184 | 157354 |
| 1 | AMPA | GluR2 | GRIA2 | 4q32–33 | X57498 | M85035 | A46056 |
| 1 | AMPA | GluR3 | GRIA3 | Xq25–26 | | M85036 | X82068 |
| 1 | AMPA | GluR4 | GRIA4 | 11q22–23 | | M36421 | U16129 |
| 2 | Kainate | GluR5 | GRIK1 | 21q21.1–22.1 | X66118 | M83560 | U16125 |
| 2 | Kainate | GluR6 | GRIK2 | 6q16.3–q21 | D10054 | Z11715 | U16126 |
| 2 | Kainate | GluR7 | GRIK3 | 1p34–p33 | | M83552 | U16127 |
| 3 | Kainate | KA-1 | GRIK4 | 11q22.3 | | X59996 | S67803[a] |
| 3 | Kainate | KA-2 | GRIK5 | 19q13.2 | D10011 | Z11581 | S40369 |
| 4 | NMDA | NR1 | GRIN1 | 9q34.3 | D10028 | X63255 | X58633 |
| 5 | NMDA | NR2A | GRIN2A | 16p13.2 | D10217 | D13211 | U09002 |
| 5 | NMDA | NR2B | GRIN2B | 12p12 | D10651 | M91562 | U28861[a] |
| 5 | NMDA | NR2C | GRIN2C | 17q24–q25 | D10694 | D13212 | |
| 5 | NMDA | NR2D | GRIN2D | 19q13.1qter | D12822 | D13214 | U77783 |
| 6 | NMDA | NR3A | GRIN3A[b] | | | L34938 | |
| 7 | Orphan | $\delta 1$ | GRID1 | | D10171 | Z17238 | |
| 7 | Orphan | $\delta 2$ | GRID2 | 4q22 | D13266 | Z17239 | |

[a] Partial sequence.
[b] as proposed in ref. 3.
Source: ref. 3, with permission

vitro (23). PICK1 also interacts with AMPA receptors (see Section 3.6.). Thus, an extensive network of cytoplasmic proteins exists that serve to anchor, target, and modulate metabotropic glutamate receptors. As described in the following section, many of these proteins play similar roles for the ionotropic glutamate receptors.

The mitogen-activated protein (MAP) kinase ERK2 can be activated by mGluR stimulation, by a G-protein-mediated mechanism (24,25). However, a G-protein-independent mGluR1 signaling pathway appears to occur in CA3 pyramidal cells, because a transient activation of a cation conductance follows activation of a Src-family kinase (26).

## 3. IONOTROPIC RECEPTORS

### 3.1. Ionotropic Receptor Classes and Their Subunits

The main features of ionotropic glutamate receptors will be discussed here. Additional information can be found in more extensive reviews (e.g. refs. 3, 27, and 28). Studies on glutamate receptor knock-out and transgenic mice are summarized in other reviews (29,30) and information on the ionotropic glutamate receptor promoters can be found in ref. 30. The mammalian ionotropic glutamate receptors are divided into three classes according to their subunit composition and pharmacology. They are named after their high-affinity agonists: AMPA receptors containing the GluR1–4 or GluRA–D subunits; kainate receptors comprising GluR5-7, KA1, and KA2 subunits; and N-methyl-D-aspartate (NMDA) receptors with the subunits NR1, NR2A–D, and NR3A (Table 2). The two distantly related orphan receptors, $\delta 1$ and $\delta 2$, do not form functional homomeric channels. However, neurodegeneration in the Lurcher mouse is caused by a $\delta 2$ mutation that produces a constitutively active, $Ca^{2+}$-permeable channel resembling an AMPA or kainate receptor (31,32).

**Table 3**
**Subunits That Were Coimmunoprecipitated from Various Brain Regions**

| Location | Subunits | Reference | Ref. no. |
|---|---|---|---|
| Forebrain | NR1 + NR2A + NR2B | Chazot and Stephenson, 1997 | 43 |
| Adult rat cerebral | Major: NR1 + NR2A + NR2B | Luo et al., 1997 | 44 |
| | Minor: NR1 + NR2A | | |
| | Minor: NR1 + NR2B | | |
| Rat neocortex | NR1 + NR2A + NR2B | Sheng et al., 1994 | 45 |
| Adult rat cerebral | NR1 + NR2A + NR2D | Dunah et al., 1998 | 46 |
| | NR1 + NR2B + NR2D | | |
| CA1 hippocampus | Major GluR1 + GluR2 | Wenthold et al., 1996 | 47 |
| | Major GluR2 + GluR3 | | |
| | Minor GluR1 alone | | |
| | Minor GluR1 + GluR3 | | |
| Cerebellum | GluR1 + GluR4 | Ripellino et al., 1998 | 48 |
| | GluR6/7+KA2[a] | | |

[a] GluR6 and 7 could not be distinguished.

Ionotropic glutamate receptors are expressed in the spinal cord and in the brain, with high expression mainly in neurons. However, functional ionotropic glutamate receptors and their subunits are also found in astrocytes, oligodendrocytes, and glial precursor cells (reviewed in refs. *11* and *33*), and in microglia *(34)*. Moreover, ionotropic glutamate receptor expression has also been found outside the brain [e.g., in the heart *(35)* and in the male lower urogenital tract *(36)*]; receptors were shown to be functional in pancreatic islet cells *(37,38)*. Moreover, in the skin, glutamate receptors appear to play a role in pain perception *(39,* and references therein). Thus, drugs impermeable to the blood-brain barrier that are targeted to glutamate receptors might find uses in several clinical situations.

Although most ionotropic glutamate receptors are localized postsynaptically, recently NMDA receptor immunoreactivity has also been found on presynaptic terminals *(40,41)*.

### 3.2. Subunit Composition and Stoichiometry

Subunits of each class assemble into cation channels, which are mainly permeable to sodium, potassium, and, to a varying degree, calcium; GluR6 receptors containing R at the Q/R editing site (see Section 3.4.) are also permeable to chloride *(42)*. In native and recombinant receptors, to date only subunits within a given family have been shown to coassemble (e.g., AMPA receptor subunits assemble with AMPA but not kainate or NMDA receptor subunits). The properties of the receptor are determined by which subunits assemble and only certain recombinant subunit compositions have been found to form functional channels. Coimmunoprecipitation of two subunits by an antibody directed against one subunit has been used to suggest which subunits coassemble in native receptors (Table 3; also see review in ref. *49*). The very N-terminal regions of AMPA receptor subunits appear to determine the specificity of subunit assembly *(50,51)*.

Recombinant AMPA receptor subunits seem to assemble in all combinations as homomers or heteromers. Most cells in the brain express more than one AMPA receptor subunit mRNA, and immunoprecipitations from the hippocampal CA1 field revealed mainly heteromeric complexes *(47)*. Functional kainate receptors appear to require the expression of GluR5, GluR6, or GluR7; KA1 or KA2 have not been found to form functional channels by themselves.

In mammalian cells, functional NMDA receptors require the expression of NR1 subunits together with one or more NR2 subunits. The NR2 subunits influence many properties of the receptor, e.g., desensitization and deactivation rates (reviewed in ref. *3*).

It is still not clear whether ionotropic glutamate receptors are tetrameric or pentameric. At least two different NR1 splice variants can coexist in one receptor complex *(52,53)*, which implies that there can be more than one NR1 subunit in one complex. Hawkins et al. *(54)*, using FLAG- and c-myc-tagged NR2B subunits, showed that the NMDA receptor can contain three NR2 subunits: one NR2A and two NR2B subunits. Assuming that two NR1 subunits occur in the NMDA receptor complex, a pentamer is possible, although most biophysical studies are more consistent with a tetrameric structure (e.g., refs. *55–57,* but see refs. *58* and *59*). Convincing resolution of this issue will require large-scale receptor purification or perhaps high-resolution imaging of the receptors.

### 3.3. Topology and Crystal Structure

The transmembrane topology of the ionotropic glutamate receptor subunits with four domains buried in the membrane (M1–M4) is shown in Fig. 3A. Only three of the membrane-buried regions are transmembrane domains (M1, M3, and M4). The M2 domain is a re-entrant loop and contributes key amino acid residues to the inner wall of the open channel. With their M2 loop, GluRs resemble more the potassium channel structure than other receptor ion channels with four transmembrane domains (e.g., nicotinic acetylcholine receptors). Part of the extracellular N-terminus and the region between M3 and M4 form a clamshell-like agonist-binding site (Fig. 3B–D), which was crystallized and the structure solved for the GluR2 subunit by Armstrong et al. *(60)*. A conserved amino-acid-binding pocket is proposed to exist in all glutamate receptors. This pocket is formed from two globular domains (S1 and S2) drawn from the sequence adjacent to the M1 domain and the M3–M4 loop, respectively. One interesting conclusion from the crystal structure is that the ligand-binding pocket appears to be contained within a single subunit rather than lie at an interface between two or more subunits. This arrangement could therefore result in multiple agonist-binding sites in each functioning receptor. Kainate binds deep within the S1–S2 cleft, in the process contacting both lobes of the closed form of the clamshell structure. Agonist binding and subsequent closure of the clamshell structure could lead to channel opening by creating a mechanical force or torque on the receptor that is transmitted to the transmembrane region, which, in turn, could increase the likelihood that the channel structure itself undergoes a conformational change to the open state *(60,61)*.

### 3.4. Splice and Editing Variants

Several splice variants are known for most ionotropic glutamate receptor subunits (Fig. 4) and the different exons confer various properties. Among the AMPA receptors, the alternative flip or flop exons determine the desensitization kinetics of the receptor. The flip variants are typically more slowly desensitizing and are expressed in embryonic and adult animals, whereas flop variants appear later around postnatal d 8 in rats and continue to be expressed in adult animals. Similar to the metabotropic glutamate receptors, alternatively spliced C-termini occur for most of the AMPA and kainate receptor subunits with as yet largely unknown function (but see Sections 3.6. and 3.7.). Among NMDA receptor subunits, only the NR1 subunit is known to be alternatively spliced, namely, in exons 5, 21, and 22. This results in three NR1 C-terminal isoforms, named C1, C2, and C2′, and within the N-terminal NR1 domain, exon 5 (domain N) can be present or absent. So far, no splice variants among the NR2 subunits have been found. Relative to the NR1 subunit, NR2 subunits contain a much longer intracellular C-terminus.

Among the AMPA and kainate receptor subunits, GluR1-6 pre-mRNAs are subject to editing, a process that changes a single amino acid codon (Fig. 4A,B; reviewed in ref. *62*). In each case, editing involves an intronic sequence that forms a loop with the exonic region, in which a selected adenosine is recognized by an RNA-modifying enzyme. The editing enzyme deaminates adenosine to inosine, which base pairs like guanosine and changes the codon. Two of the known RNA-editing enzymes, ADAR1 and ADAR2, appear to edit glutamate receptor subunits with different substrate specificities *(62,63)*. In normal rodents or humans, GluR2 mRNAs are >99% edited at the Q/R site in M2, whereas

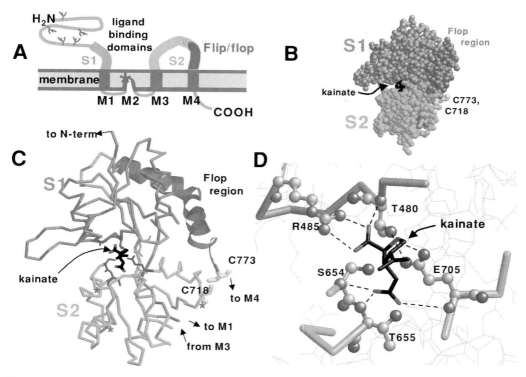

Fig. 3. Glutamate receptor topology and crystal structure of the agonist-binding pocket. (A) Schematic of an ionotropic glutamate receptor subunit with the two domains that contain agonist-binding residues colored in orange (S1) and turquoise (S2). The general topology with the M2 re- entrant loop is common to all ionotropic glutamate receptors. However, the Q/R editing site indicated by a red asterix in M2 occurs only in GluR2, 5, and 6. Moreover, the region preceding M4 in violet is only alternatively spliced in AMPA receptor subunits, resulting in the flip or flop exons. Note that the intracellular C-terminal region varies considerably in length between NR2 subunits and the other glutamate receptor subunits with a short C-tail. (B) Space-filled representation of the crystallized kainate-bound S1 and S2 domains joined by an 11-residue linker peptide, in the same colors as in (A). The helical flop region is located on a solvent-exposed face of the protein. The position of a single kainate agonist molecule (black) within a deep gorge of the protein is indicated; the two disulfide-bonded cysteines (C718 and C773) are shown in yellow. (C) Backbone representation of the subunit, with kainate (black) docked into its binding site. The kainate-binding residues are shown as stick figures in magenta, the two cysteines in yellow, and the flop helix structure in violet. The two green residues (E402 and T686) do not directly bind to kainate but, instead, interact with each other, helping to hold the clamshell in the closed conformation. Red asterisks mark the positions of S662 and S680 (lower left), which are important in GluR6 for PKA phosphorylation, and N721 (adjacent to the yellow C722), which controls agonist sensitivity in GluR5 and GluR6. (D) Close-up view of the ligand-binding pocket. The binding residues are in space-filled representation, with atoms colored conventionally (gray = carbon, light blue = nitrogen, red = oxygen). [From ref. *3* with original pdb files kindly provided by E. Gouaux *(60).*] (Color illustration in insert following p. 142.)

GluR5 and GluR6 are only partially edited at this site; in each case, editing increases the subunit's calcium permeability and (for GluR2) single-channel conductance (see Section 4). Editing at the other sites, R/G in GluR2–4, which accelerates recovery from desensitization, and I/V and Y/C in M1 of GluR5–6, which changes ion permeability in GluR6Q, changes during development. The importance of GluR2 editing is highlighted by the findings that both GluR2 editing-deficient mice *(64)* and ADAR2 knockout mice *(65)* are prone to seizures and die young. Moreover, the ADAR2 knockout

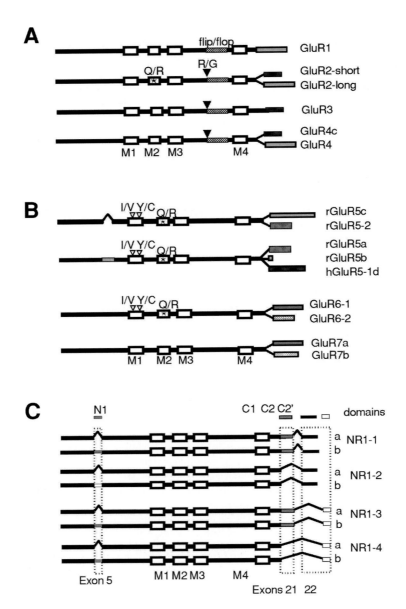

**Fig. 4.** Splice and editing variants of alternatively spliced ionotropic glutamate receptor subunits. The diagram shows the basic structures of ionotropic glutamate receptor subunits, the regions buried in the membrane (white boxes, M1–M4), and the alternatively spliced cassettes (boxes with different patterns). Within the AMPA, kainate, or NR1 subunit variants, the homologous alternatively spliced domains are indicated by the same pattern. **(A)** AMPA receptor subunits with flip/flop and C-terminal splice variants. The position of the Q/R and R/G editing sites are indicated. **(B)** Rat (prefix r) and human (prefix h) kainate receptor subunits with C-terminal and N-terminal splice variants. The editing sites I/V, Y/C, and Q/R in the M1 and M2 domains of kainate receptor subunits are shown. **(C)** Alternative splicing of NR1. Exons 5, 21, and 22 are alternatively spliced, giving rise to the cassettes N1, C1, C2, and C2′. No splice variants of the NR2 subunits are known at this time.

mice were rescued when both GluR2 alleles (Q) were modified to encode for the edited (R) version of GluR2, suggesting that GluR2 is the physiologically most important substrate of ADAR2 *(65)*.

## 3.5. Phosphorylation

Phosphorylation modulates the function of AMPA, kainate, and NMDA receptors (summarized in ref. *3*) and is associated with synaptic plasticity. For example, phosphorylation of GluR1 at Ser831 by calcium calmodulin kinase II (CAMKII) increases AMPA receptor-mediated current flow and is one of the mechanisms by which long-term potentiation (LTP) occurs. In contrast, during long-term depression, GluR1 is dephosphorylated at protein kinase A (PKA) phosphorylation site, Ser845, decreasing the opening probablility of AMPA receptors. The GluR2–4 subunits do not contain these two phosphorylatable serines. Instead, GluR2 can be phosphorylated at Ser880 by PKC, which is located within the sequence critical for PDZ domain binding. GluR2 phosphorylation at Ser880 decreases GRIP1 binding *(66,67;* see Section 3.6). Moreover, Ser842 of GluR4 can be phosphorylated by PKA, PKC, and CAMKII, and Thr830 is a potential PKC phosphorylation site *(68)*.

Phosphorylation of NMDA receptors at several sites increases ionic currents through activated receptors. PKC phosphorylation increases the opening probability of NMDA receptors *(69)*. However, it is not clear whether the identified PKC phosphorylation sites, Ser890, Ser896, and Thr879 in the alternatively spliced C1 domain of NR1, confer the effect *(70)*. Phosphorylation of Ser890 also inhibits receptor clustering. Little is known about NMDA receptor phosphorylation by PKA and CAMKII, although at least two sites can be phosphorylated: Ser897 in the C1 domain of NR1 by PKA and Ser1303 in NR2B by CAMKII. Tyrosine phosphorylation of NMDA receptors is another way to increase NMDA receptor currents. The endogenous tyrosine kinase Src phosphorylates three C-terminal tyrosines (Y1105, Y1267, and Y1387) in NR2A, which reduces $Zn^{2+}$ blockade, thereby potentiating the receptor *(71;* see Section 4.3.1.). The potentiation of NMDA receptor currents by Src is one mechanism leading to LTP in CA1 pyramidal cells *(72)*. Another tyrosine kinase that phosphorylates NR2A and NR2B is Fyn. Fyn knock-out mice are impaired in LTP and spatial learning *(73)*.

## 3.6. Association with Intracellular Proteins

The application of the yeast two-hybrid system resulted in the cloning of many intracellular proteins that associate with glutamate receptors, in most cases at their C-terminus. Many of these proteins contain three PDZ domains (named after the proteins PSD-95, Dlg, and ZO1, all of which contain the domain), one src homology domain 3 (SH3), and a guanylate kinase (GK) domain, and thus belong to the PSD-95 or synapse-associated protein (SAP) family (reviewed in ref. *74*). Other glutamate receptor-associated proteins do not contain PDZ domains or are signaling molecules. Most of these proteins seem to play a role in receptor membrane insertion [e.g., N-ethylmaleimide-sensitive protein (NSF)], anchoring to the cytoskeleton, clustering, localization, or forming signaling complexes with different signaling molecules.

The interactions of intracellular proteins with AMPA receptors are reviewed by Braithwaite et al. *(75)*. GluR2, GluR3, and GluR4c have a similar C-terminal sequence [IESV(V/I)KI] containing a PDZ-binding domain. This sequence interacts with three proteins containing seven PDZ domains, GRIP and AMPA receptor-binding protein (ABP or GRIP2), and a shorter splice variant of the latter containing six PDZ domains. Moreover, depending the phosphorylation state, they interact with PICK1, which seems to be involved in AMPA receptor clustering and also binds to mGluR7a. Recently, Hayashi et al. *(76)* proposed that a GluR1 C-terminal sequence (TGL) also interacts with PDZ domains. They showed delivery of GFP-tagged GluR1 into synapses mediated by CAMKII or LTP. The delivery was blocked by mutating the predicted C-terminal PDZ interaction site to AGL, but not by mutating the CAMKII phosphorylation site Ser831. Thus, LTP seemed to depend on GluR1–PDZ domain interactions.

GluR2 interacts with NSF, a protein without PDZ domains and involved in membrane-fusion events, such as exocytosis of synaptic vehicles. NSF seems to act as a chaperone for AMPA receptor (re-)insertion into the membrane (see also review in ref. *77*). Moreover, SAP-97 is one the proteins belonging to the SAP family that binds AMPA, kainate, and NMDA receptors, and GluR6 can bind to PSD-95.

NMDA receptor subunits bind to a vast number of SAP proteins, including PSD-95 (NR2A). NR2A, B, and C can also interact with PSD-93/chapsyn 110, SAP97, and SAP102 (e.g., reviewed in ref. *78*). These proteins associate to other proteins building a scaffold and bridging to the cytoskeleton, such as CRIPT – microtubuli, MAP1A, GKAP, SAP90/PSD-95-associated proteins (SAPAPs), or neuroligin (see section 3.8.). Proteins lacking PDZ domains but interacting with NMDA receptor subunits are neurofilament subunit L (NR1), yotiao (NR1), α-actinin (NR1, NR2B), and spectrin (NR2A and B). Additional proteins, with an as-yet unidentified function, were identified recently by mass spectrocsopy *(79)*.

### 3.7. Non-isotopic Signaling Cascades

Traditionally, ionotropic receptors signal by ion flux through their ion channels. However, recent studies reported new signaling mechanisms, such as association with G-proteins, protein tyrosine kinases, or calmodulin. For example, AMPA receptors have been reported to be associated with $G\alpha_{i1}$ *(80)*, and pertussis toxin-sensitive MAP kinase activation was observed after AMPA receptor activation *(81)*. Furthermore, in retinal ganglion cells, an AMPA-induced suppression of the cGMP-gated current could be blocked by pertussis toxin *(82)*. Moreover, the C-tail of GluR2 was found to associate with the Src-family kinase Lyn, and in cerebellar granule cell cultures, stimulation of AMPA receptors resulted in Lyn activation and subsequent MAP kinase activation *(83)*. Finally, kainate receptors may couple to $G_i/G_0$ proteins in the hippocampus and inhibit GABA release presynaptically *(84,85)*.

NMDA receptors coimmunoprecipitate with Src protein and phospholipase C, and the NR2B C-tail has a high affinity for autophosphorylated CAMKII. Moreover, the C-terminus of the NR2D subunit can associate with the SH3 domain of c-Abl, but not other tested tyrosine kinases, and inhibit c-Abl *(86)*. Furthermore, the calcium-dependent binding of calmodulin to the C1 cassette of NR1 can be weakened by PKC phosphorylation. Both PKC phosphorylation and calmodulin binding weakens the association of NR1 with spectrin.

### 3.8. Extracellular Proteins Binding Glutamate Receptors

A new concept is the bridging of receptors and ion channels across the extracellular space in the central nervous system (CNS) (reviewed in ref. *87*), which was first discovered at the neuromuscular junction. For example, PSD-95 can form a link between NMDA receptors and neuroligin. Neuroligin is a protein with a single transmembrane domain that is expressed at the postsynaptic membrane and can associate in the extracellular space with neurexin. Neurexin binds to the PDZ domain of CASK, which, via its SH3 domain, binds to calcium channels located in the presynaptic membrane. Neuroligin can induce presynaptic differentiation in neurons. Another secreted surface molecule that clusters GluR1-3 is Narp (neuronal activity-regulated pentraxin). Narp was cloned as an immediate early gene with a long half-life induced by seizures *(88)*. Narp is expressed presynaptically and postsynaptically at excitatory synapses in the hippocampus and the spinal cord and induces GluR1 aggregation *(89)*.

## 4. GENERAL FEATURES OF SYNAPTIC POTENTIALS MEDIATED BY GLUTAMATE RECEPTORS

Glutamate released from presynaptic terminals activates both presynaptic and postsynaptic metabotropic receptors to modulate synaptic transmission. Glutamate also activates all ionotropic receptors and mediates the vast majority of "fast" synaptic transmission in the CNS, whereas aspartate is an agonist at NMDA but not AMPA receptors.

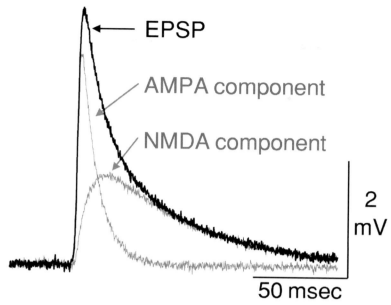

**Fig. 5.** Dual-component excitatory postsynaptic potential. Shown are simulated EPSPs based on recordings from hippocampal interneurons in the presence of bicuculline to block postsynaptic GABAergic inhibitory postsynaptes potential (IPSPs). The AMPA and NMDA receptor components and their algebraic sum are indicated.

In neurons, the kinetics, pharmacologic sensitivity, and $Ca^{2+}$ permeability of glutamatergic excitatory post synaptic potentials (EPSPs) are strongly influenced by which ionotropic receptors are activated and their subunit composition. Rapid desensitization of AMPA receptors coupled with a slow onset of NMDA receptor activation causes glutamatergic EPSPs in most brain regions to be biphasic (Fig. 5). The early component (lasting 10–20 ms) is dominated by AMPA receptors and the later component (up to several hundred milliseconds) by NMDA receptors.

Three classic features of NMDA receptor activation—glycine action, $Mg^{2+}$ block, and high $Ca^{2+}$ permeability—will be summarized briefly. The amino acid glycine was originally reported to potentiate NMDA receptor activation (90), but shortly thereafter was recognized to be an essential coagonist at NMDA receptors (91). NMDA receptors are thus the first and still the only neurotransmitter receptor known to require simultaneous activation by two agonists. $Mg^{2+}$ is a voltage-dependent channel blocker of NMDA receptors, the block being very strong at hyperpolarized potentials (–80 mV and below), but progressively relieved by depolarization. Synaptic plasticity mediated by NMDA receptors is thus associative in nature, dependent on relief of the $Mg^{2+}$ block by AMPA receptor-mediated depolarization. NMDA receptor-linked synaptic plasticity is mediated by a high $Ca^{2+}$ flux through open NMDA receptor channels.

The $Ca^{2+}$ permeability of AMPA and kainate receptors depends on the presence or absence of editing in the Q/R site, receptors containing exclusively unedited subunits exhibiting much higher $Ca^{2+}$ permeability than edited receptors (Figs. 3 and 4A,B). Among the AMPA receptors, only the GluR2 subunit mRNA is edited at the Q/R site, with the consequence that receptors lacking GluR2 are about four times more permeable to $Ca^{2+}$ than to $Na^+$ or $K^+$ (92). A glutamine (Q) or arginine (R) residing in the Q/R site of all known functional kainate receptor subunits also influences their $Ca^{2+}$ permeability. The homologous amino acid in NMDA receptor subunits is asparagine, which endows all NMDA receptors with high $Ca^{2+}$ permeability. Indeed, replacement by site-directed mutagenesis of this asparagine with an arginine produces NMDA receptors with very low $Ca^{2+}$ permeability, similar to that of GluR2-containing AMPA receptors (93). Likewise, replacement of the arginine in GluR2 with glutamine results in a receptor with high $Ca^{2+}$ permeability. From these findings and others in which the permeability of receptors containing mutations in the pore region has been evaluated (e.g., ref. 94) it is concluded that the cation "selectivity filter" is similar among all glutamate receptors.

The high $Ca^{2+}$ permeability of NMDA and certain AMPA and kainate receptor channels leads to the transient activation of several $Ca^{2+}$-activated enzymes, including $Ca^{2+}$/calmodulin-dependent protein kinase II, the phosphatase calcineurin, protein kinase C, phospholipase A2, phospholipase C, and nitric oxide synthase. Activation of one or more of these enzymes is thought to be at the root of synaptic plasticities mediated by $Ca^{2+}$ permeable NMDA and AMPA receptors; for example, CAMKII can induce LTP by phosporylating GluR1, leading to potentiation of AMPA receptor-mediated currents (section 3.5.).

### 4.1. Synaptic Functions of Metabotropic Receptors

Postsynaptically located metabotropic glutamate receptors modulate both transmitter-activated and voltage-gated ion channels and thereby influence the strength of synaptic transmission. Both L-type and N-type $Ca^{2+}$ channels are inhibited by activation of group I or II mGluR, and L-type channels are additionally inhibited by group III mGluR activation. Inhibition of $Ca^{2+}$ entry presumably contributes to the observed reduction of $Ca^{2+}$-dependent $K^+$ currents in many neurons, but in cerebellar granule cells, mGluR activation increases the activity of $Ca^{2+}$-dependent and inwardly rectifying $K^+$ channels. The net effect on excitability is thus difficult to predict. Many ligand-gated channels are also modulated by mGluRs including AMPA, NMDA, and $GABA_A$ receptors. Whether activation of mGluR acts to inhibit or potentiate a receptor is often cell-specific. For example, in hippocampal pyramidal cells, mGluR activation potentiates NMDA receptors (see Fig. 1), but in cerebellar granule cells, mGluR activation inhibits NMDA receptors. In both cases, the effect is reduced by protein kinase C inhibitors.

Some mGluRs are located presynaptically and serve as autoreceptors that limit transmitter release. For example, glutamate-mediated EPSPs are reduced by activation of group II mGluRs at mossy fiber terminals onto CA3 hippocampal pyramidal neurons and by activation of group III mGluRs on Schaffer collateral synapses made onto CA1 pyramidal cells (see Fig. 1). The mechanism may involve reduction of $Ca^{2+}$ entry into the presynaptic terminal.

### 4.2. Glial Glutamate Receptors

We mainly emphasize the synaptic roles of glutamate receptors expressed by neurons. However, there is increasing evidence that glial glutamate receptors also play physiological roles in the brain (see reviews in refs. 33 and 95). The most direct evidence for activation of glial glutamate receptors by neuronally released glutamate was obtained recently for oligodendrocyte precursor cells in the hippocampus of young and mature rats (96). Schaffer collateral/commissural fiber stimulation led to AMPA receptor-mediated EPSCs in oligodendrocyte precursor cells, which were identified by immunostaining and electron microscopy. Moreover, electron microscopy revealed glutamatergic synapses from boutons onto oligodendrocyte precursor cells that had been physiologically identified and filled with biocytin. The function of the glutamatergic input onto oligodendrocyte precursor cells in vivo is yet unknown. In vitro, glutamate receptor activation of oligodendrocyte precursor cell cultures inhibits their proliferation and maturation into oligodendrocytes (97)

Evidence for mGluR receptor activation by neuronally released glutamate in astrocytes *in situ* is accumulating. Several studies reported a rise of intracellular calcium concentration, calcium waves, or oscillations in astrocytes after electrical stimulations of adjacent nerve fibers (98–100). For example, Schaffer collateral stimulation led to intracellular calcium waves in astrocytes in the stratum radiatum *in situ* (99). Astrocytic calcium waves appeared to not be a consequence of transient potassium release from activated neurons, but rather to activation of glial glutamate receptors. First, astrocytic calcium waves could be blocked by an mGluR antagonist, which had a negligible effect on (neuronal) field potentials. Second, the calcium waves were not blocked by the ionotropic glutamate receptor antagonist kynurenate, which blocked the postsynaptic response. Moreover, work from Winder et al. *(101)* strongly suggests that in the rat hippocampal CA1 region, glia and not neurons contain the metabotropic group II receptor mGluR3 and that the coactivation of group II metabotropic glutamate and β-adrenergic receptors occurs in glia and not in CA1 pyramidal neurons (*see* Fig. 1).

### 4.3. Subunit-Dependent Modulation of Ionotropic Glutamate Receptors

The classical competitive blockers of the glutamate site on NMDA receptors are phosphono derivatives of short-chain (five to seven carbons) amino acids such as 2-amino-5-phosphonopentanoic acid. In contrast, the glycine site of NMDA receptors, as well as the glutamate sites of AMPA and kainate receptors, are competitively blocked by halogenated quinoxalinediones and kynurenic acid derivatives. However, numerous compounds have been identified that act on modulatory sites of NMDA receptors and, to a lesser extent, AMPA and kainate receptors. We will concisely review the pharmacology of these modulatory sites, particularly as they relate to cognitive or addictive processes.

#### 4.3.1. Noncompetitive Antagonists of NMDA Receptors

Several addictive or abused drugs inhibit NMDA receptors, often in a subunit-dependent fashion. Ethanol noncompetitively inhibits recombinant or native NMDA receptors at intoxicating concentrations, inhibition being more pronounced at NR2A- or NR2B-containing receptors than those with NR2C or NR2D *(102)*. Inhibition of NR1/NR2A (but not NR2B) receptors was reduced by expression of α-actinin protein *(103)*, which links the NR1 and NR2B subunits to actin and thus to the cytoskeleton *(104)*. Interestingly, treatment of ethanol-dependent mice with a selective NR2B-containing NMDA receptor antagonist, ifenprodil, reduced the intensity of ethanol withdrawal signs *(105)*. Moreover, chronic exposure of mice to ethanol increased expression of NR2B subunit protein in the limbic forebrain; the two findings together are suggestive of a role for NR2B upregulation in the development of physical dependence to ethanol *(105)*. At higher concentrations, ethanol also inhibits native AMPA receptors expressed by neurons of the medial septum/diagonal band *(106)*.

Nitrous oxide (laughing gas) at anesthetic concentrations and toluene, two inhaled drugs of abuse, are noncompetitive NMDA receptor blockers *(107,108)*. The effect of toluene is strongest at NR2B-containing receptors, but little is known about the nitrous oxide effect.

Protons inhibit NMDA receptors that lack exon 5 of the NR1 subunit, with half-inhibition near pH 7.4; exon 5 is a very basic short extracellular loop that is thought to act as a tethered modulator of NMDA receptor gating *(109)*. Ifenprodil is the exemplar of an important class of NR2B-selective NMDA receptor antagonists. These drugs inhibit receptor activation by potentiating proton inhibition of the receptors *(110)*. Inhibition by ifenprodil is occluded if the NR1 subunit contains exon 5. Zinc appears to play a similar role for NR2A-containing NMDA receptors *(111,112)*. Paoletti et al. *(113)* have proposed a structure near the N-terminus of NR2A that binds zinc in a clamshell-like cleft (Fig. 6); presumably, a similar structure in NR2B binds ifenprodil and related compounds.

#### 4.3.2. Modulators of Desensitization of AMPA and Kainate Receptors

A prominent molecular determinant of AMPA receptor desensitization is the flip/flop alternative exon, which is a helical region lying on a solvent-exposed surface of the subunit in the third extracellular loop (Fig. 3A–C). Receptors with subunits containing predominantly the flop exon have threefold to fivefold faster desensitization than those with the flip variants *(114)*. Several drug classes, as typified by cyclothiazide and the cognitive enhancer aniracetam, have been identified that modulate desensitization of AMPA receptors. The effects of cyclothiazide on AMPA receptors are strongly influenced by the Ser/Asn residue at position 750. Conversion of Ser750 in GluR1flip to glutamine, which is the

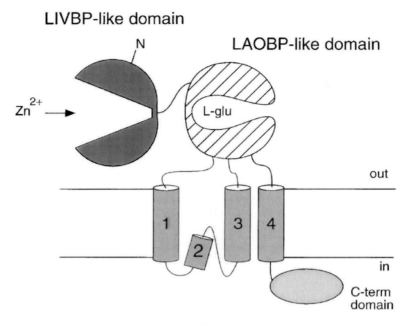

**Fig. 6.** Structure of the NR2A subunit as proposed by Paoletti et al. *(113)*. extracellular domains form two clamshell-like binding domains, both homologous to bacterial periplasmic amino-acid-binding proteins (LIVBP and LAOPB). The very N-terminal LIVBP-like domain acts as a $Zn^{2+}$ sensor and the LAOPB-like domain binds the agonist glutamate. (From ref. *113*, with permission of Elsevier Science and P. Paoletti.)

homologous residue found in the cyclothiazide-insensitive kainate receptors, abolishes cyclothiazide actions on AMPA receptors *(115)*. Conversely, the introduction of a serine residue into the homologous site on GluR6 imparts cyclothiazide sensitivity *(115)*.

Kainate receptor desensitization is insensitive to cyclothiazide, but it can be irreversibly reduced by concanavalin A, a lectin carbohydrate *(116)*. Concanavalin A is effective only for receptors containing GluR5 or GluR6, but not GluR7, suggesting different glycosylation patterns between GluR7 and GluR5/6 *(117)*.

## 5. CONCLUSIONS

The most prominent focus of glutamate receptor pharmacology has been in the fields of stroke, epilepsy and parkinsonism. However, the cognitive/memory disruption caused by certain NMDA receptor antagonists and the cognitive enhancement by AMPA receptor modulators or mGluR2 agonists remind us that both ionotropic and metabotropic glutamate receptors play important roles in normal mental processes. Moreover, there is increasing recognition that synaptic plasticities underlying components of addiction may be mediated by glutamate receptors whose subunit composition has been altered during chronic exposure to addictive drugs. In the past decade, much has been learned of the structure and function of the multiple classes of glutamate receptor. One looks forward to more precise and targeted study of the roles of particular glutamate receptors or subunits in the addictive or reward processes with the improved availability of genetically modified mice.

## ACKNOWLEDGMENT

We are grateful to Rosalyn Lightfoot for typing of the references.

# REFERENCES

1. Kelz, M. B., Chen, J., Carlezon, W. A., Jr., Whisler, K., Gilden, L., Beckmann, A. M., et al. (1999) Expression of the transcription factor deltaFosB in the brain controls sensitivity to cocaine. *Nature* **401,** 272–276.
2. Moghaddam, B. and Adams, B. W. (1998) Reversal of phencyclidine effects by a group II metabotropic glutamate receptor agonist in rats. *Science* **281,** 1349–1352.
3. Dingledine, R., Borges, K., Bowie, D., and Traynelis, S. F. (1999) The glutamate receptor ion channels. *Pharmacol. Rev.* **51,** 7–61.
4. Bräuner-Osborne, H., Egebjerg, J., Nielsen, E. O., Madsen, U., and Krogsgaard-Larsen, P. (2000) Ligands for glutamate receptors: design and therapeutic prospects. *J. Med. Chem.* **43,** 2609–2645.
5. Fletcher, E. J. and Lodge, D. (1996) New developments in the molecular pharmacology of α-amino-3-hydroxy-5-methyl-4-isoazole propionate and kainate receptors. *Pharmacol. Ther.* **70,** 65–89.
6. Pin, J., DeColle, C., Bessis, A.-S., and Ascher, F. (1999) New perspectives for the development of selective metabotropic glutamate receptor ligands. *Eur. J. Pharmacol.* **375,** 277–294.
7. Conn, P. J. and Pin, J. (1997) Pharmacology and functions of metabotropic glutamate receptors. *Annu. Rev. Pharmacol. Toxicol.* **37,** 205–237.
8. Schoepp, D. D., Jane, D. E., and Monn, J. A. (1999) Pharmacological agents acting at subtypes of metabotropic glutamate receptors. *Neuropharmacology* **38,** 1431–1476.
9. Cartmell, J. and Schoepp, D. D. (2000) Regulation of neurotransmitter release by metabotropic glutamate receptors. *J. Neurochem.* **75,** 889–907.
10. Takumi, Y., Matsubara, A., Rinvik, E., and Ottersen, O. P. (1999) The arrangment of glutamate receptors in excitatory synapses. *Ann. NY Acad. Sci.* **868,** 474–482.
11. Condorelli, D. F., Conti, F., Gallo, V., Kirchhoff, F., Seifert, G., Steinhäuser, C., et al. (1999) Expression and functional analysis of glutamate receptors in glial cells. *Adv. Exp. Med. Biol.* **468,** 49–67.
12. Gill, S. S., Pulido, O. M., Mueller, R. W. and McGuire, P. F. (1999) Immunochemical localization of the metabotropic glutamate receptors in the rat heart. *Brain Res. Bull.* **48,** 143–146.
13. Conn, P. J. (1999) Metabotropic glutamate receptors. *Sci. Med.* **6,** 28–37.
14. Flor, P. J., Van Der Putten, H., Rüegg, D., Lukic, S., Leonhardt, T., Bence, M., et al. (1997) A novel splice variant of a metabotropic glutamate receptor, human mGluR 7b. *Neuropharmacology* **36,** 153–159.
15. Corti, C., Restituito, S., Rimland, J. M., Brabet, I., Corsi, M., Pin, J. P., and Ferraguti, F. (1998) Cloning and characterization of alternative mRNA forms for the rat metabotropic glutamate receptors mGluR7 and mGluR8. *Eur. J. Neurosci.* **10,** 3629–3641.
16. Xiao, B., Tu, J. C., and Worley, P. F. (2000) Homer: a link between neural activity and glutamate receptor function. *Curr. Opin. Neurobiol.* **10,** 370–374.
17. Kammermeier, P. J., Xiao, B., Tu, J. C., Worley, P. F., and Ikeda, S. R. (2000) Homer proteins regulate coupling of group I metabotropic glutamate receptors to N-type calcium and M-type potassium channels. *J. Neurosci.* **20,** 7238–7245.
18. Roche, K. W., Tu, J. C., Petralia, R. S., Xiao, B., Wenthold, R. J., and Worley, P. F. (1999) Homer 1b regulates the trafficking of group I metabotropic glutamate receptors. *J. Biol. Chem.* **274,** 25,953–25,957.
19. Stowell, J. N. and Craig, MC. (1999) Axon/dendrite targeting of metabotropic glutamate receptors by their cytoplasmic carboxy-terminal domains. *Neuron* **22,** 525–536.
20. Minakami, R., Jinnai, N., and Sugiyama, H. (1997) Phosphorylation and calmodulin binding of the metabotropic glutamate receptor subtype 5 (mGluR5) are antagonistic *in vitro. J. Biol. Chem.* **272,** 20291–20298.
21. Nakajima, Y., Yamamoto, T., Nakayama, T., and Nakanishi, S. (1999) A relationship between protein kinase C phosphorylation and calmodulin binding to the metabotropic glutamate receptor subtype 7. *J. Biol. Chem.* **274,** 27,573–27,577.
22. O'Connor, V., El Far, O., Bofill-Cardona, E., Nanoff, C., Freissmuth, M., Karschin, A., et al. (1999) Calmodulin dependence of presynaptic metabotropic glutamate receptor signaling. *Science* **286,** 1180–1184.
23. Dev, K. K., Nakajima, Y., Kitano, J., Braithwaite, S. P., Henley, J. M., and Nakanishi, S. (2000) PICK1 interacts with and regulates PKC phosphorylation of mGluR7. *J. Neurosci.* **20,** 7252–7257.
24. Peavy, R. D. and Conn, P. J. (1998) Phosphorylation of mitogen-activated protein kinase in cultured rat cortical glia by stimulation of metabotropic glutamate receptors. *J. Neurochem.* **71,** 603–612.
25. Ferraguti, F., Baldani-Guerra, B., Corsi, M., Nakanishi, S. and Corti, C. (1999) Activation of the extracellular signal-regulated kinase 2 by metabotropic glutamate receptors. *Eur. J. Neurosci.* **11,** 2073–2082.
26. Heuss, C., Scanziani, M., Gähwiler, B. H., and Gerber, U. (1999) G-Protein-independent signaling mediated by metabotropic glutamate receptors. *Nat. Neurosci.* **2,** 1070–1077.
27. Ozawa, S., Kamiya, H., and Tsuzuki, K. (1998) Glutamate receptors in the mammalian central nervous system. *Prog. Neurobiol.* **54,** 581–618.
28. Hollmann, M. and Heinemann, S. (1994) Cloned glutamate receptors. *Annu. Rev. Neurosci.* **17,** 31–108.
29. Sprengel, R. and Single F. N. (1999) Mice with genetically modified NMDA and AMPA receptors. *Ann. NY Acad. Sci.* **868,** 494–501.

30. Myers, S. J., Dingledine, R., and Borges, K. (1999) Genetic regulation of glutamate receptors. *Annu. Rev. Pharmacol. Toxicol.* **39,** 221–241.
31. Zuo, J., DeJager, P. L., Takahashi, K. A., Jiang, W., Linden, D. J., and Heintz, N. (1997) Neurodegeneration in Lurcher mice caused by mutation in δ2 glutamate receptor gene. *Nature* **388,** 769–767.
32. Wollmuth, L. P., Kuner, T., Jatzke, C., Seeburg, P. H., Heintz, N., and Zuo, J. (2000) The Lurcher mutation identifies delta 2 as an AMPA/kainate receptor-like channel that is potentiated by $Ca^{(2+)}$. *J. Neurosci.* **20,** 5973–5980.
33. Verkhratsky, A. and Steinhäuser, C. (2000) Ion channels in glial cells. *Brain Res. Rev.* **32,** 380–412.
34. Noda, M., Nakanishi, H., Nabekura, J., and Akaike, N. (2000) AMPA—kainate subtypes of glutamate receptor in rat cerebral microglia. *J. Neurosci.* **20,** 251–258.
35. Gill, S. S., Pulido, O. M., Mueller, R. W., and McGuire, P. F. (1998) Molecular and immunochemical characterization of the ionotropic glutamate receptors in the rat heart. *Brain Res. Bull.* **46,** 429–434.
36. Gonzalez-Cadavid, N. F., Ryndin, l., Vernet, D., Magee, T. R., and Rajfer, J. (2000) Presence of NMDA receptor subunits in the male lower urogenital tract. *J. Androl.* **21,** 566–578.
37. Inagaki, N., Kuromi, H., Gonoi, T., Okamoto, Y., Ishida, H., Seino, Y., et al. (1995) Expression and role of ionotropic glutamate receptors in pancreatic islet cells. *FASEB J.* **9,** 686–691.
38. Weaver, C. D., Yao, T. L., Powers, A. C., and Verdoorn, T. A. (1996) Differential expression of glutamate receptor subtypes in rat pancreatic islets. *J. Biol. Chem.* **271,** 12,977–12,984.
39. Omote, K., Kawamata, T., Kawamata, M., and Namiki A. (1998) Formalin-induced release of excitatory amino acids in the skin of the rat hindpaw. *Brain Res.* **787,** 161–164.
40. Van Bockstaele, E. J. and Colago, E. E. (1996) Selective distribution of the NMDA-R1 glutamate receptor in astrocytes and presynaptic axon terminals in the nucleus locus coeruleus of the rat brain: an immunoelectron microscopic study. *J. Comp. Neurol.* **369,** 483–496.
41. Paquet, M. and Smith, Y. (2000) Presynaptic NMDA receptor subunit immunoreactivity in GABAergic terminals in rat brain. *J. Comp. Neurol.* **423,** 330–347.
42. Burnashev, N., Villarroel, A., and Sakmann, B. (1996) Dimensions and ion selectivity of recombinant AMPA and kainate receptor channels and their dependence on Q/R site residues. *J. Physiol.* **496,** 165–173.
43. Chazot, P. L. and Stephenson, F. A. (1997) Molecular dissection of native mammalian forebrain NMDA receptors containing the NR1 C2 exon: direct demonstration of NMDA receptors comprising NR1, NR2A, and NR2B subunits within the same complex. *J. Neurochem.* **69,** 2138–2144.
44. Luo, J., Wang, Y., Yasuda, R. P., Dunah, A. W., and Wolfe, B. B. (1997) The majority of *N*-methyl-D-aspartate receptor complexes in adult rat cerebral cortex contain at least three different subunits (NR1/NR2A/NR2B). *Mol. Pharmacol.* **51,** 79–86.
45. Sheng, M., Cummings, J., Roldan, L. A., Jan, Y. N., and Jan LY. (1994) Changing subunit composition of heteromeric NMDA receptors during development of rat cortex. *Nature* **368,** 144–147.
46. Dunah, A. W., Luo, J., Wang, Y. H., Yasuda, R. P., and Wolfe, B. B. (1998) Subunit composition of *N*-methyl-D-aspartate receptors in the central nervous system that contain the NR2D subunit. *Mol. Pharmacol.* **53,** 429–437.
47. Wenthold, R. J., Petralia, R. S., Blahos, J., II, and Niedzielski, A. S. (1996) Evidence for multiple AMPA receptor complexes in hippocampal CA1/CA2 neurons. *J. Neurosci.* **16,** 1982–1989.
48. Ripellino, J. A., Neve, R. L., and Howe J. R. (1998) Expression and heteromeric interactions of non-*N*-methyl-D-aspartate glutamate receptor subunits in the developing and adult cerebellum. *Neuroscience* **82,** 485–497.
49. Dunah, A. W., Yasuda, R. P., Luo, J., Wang, Y. H., Prybylowski, K. L., and Wolfe, B. B. (1999) Biochemical studies of the structure and function of the *N*-methyl-D-aspartate subtype of glutamate receptors. *Mol. Neurobiol.* **19,** 151–179.
50. Leuschner, W. D. and Hoch, W. (1999) Subtype-specific assembly of α-amino-3-hydroxy-5-methyl-4-isoxazole propionic acid receptor subunits is mediated by their N-terminal domains. *J. Biol. Chem.* **274,** 16,907–16,916.
51. Kuusinen. A., Abele, R., Madden, D. R., and Keinänen, K. (1999) Oligomerization and ligand-binding properties of the ectodomain of the α-amino-3-hydroxy-5methyl-4-isoxazole propionic acid receptor subunit GluRD. *J. Biol. Chem.* **274,** 28,937–28,943.
52. Chazot, P. L. and Stephenson, F. A (1997) Biochemical evidence for the existence of a pool of unassembled C2 exon-containing NR1 subunits of the mamalian forebrain NMDA receptor. *J. Neurochem.* **68,** 507–516.
53. Blahos, J, 2nd and Wenthold, R. J. (1996) Relationship between *N*-methyl-D-aspartate receptor NR1 splice variants and NR2 subunits. *J. Biol. Chem.* **271,** 15,669–15,674.
54. Hawkins, L. M., Chazot, P. L., and Stephenson, F. A. (1999) Biochemical evidence for the co-association of three *N*-methyl-D-aspartate (NMDA) R2 subunits in recombinant NMDA receptors. *J. Biol. Chem.* **274,** 27,211–27,218.
55. Laube, B., Kuhse, J. and Betz, H. (1998) Evidence for a tetrameric structure of recombinant NMDA receptors. *J. Neurosci.* **18,** 2954–2961.
56. Rosenmund, C., Stern-Bach, Y., and Stevens, C. F. (1998) The tetrameric structure of a glutamate receptor channel. *Science* **280,** 1596–1599.
57. Behe, P., Stern, P., Wyllie, D. J. A., Nassar, M., Schoepfer, R., and Colquhoun, D. (1995) Determination of NMDA NR1 subunit copy number in recombinant NMDA receptors. *Proc. Roy. Soc. (Lond.) B* **262,** 205–213.

58. Premkumar, L. S. and Auerbach, A. (1997) Stoichiometry of recombinant N-methyl-D-aspartate receptor channels inferred from single-channel current patterns. *J. Gen. Physiol.* **110**, 485–502.
59. Ferrer-Montiel, A. V. and Montal, M. (1996) Pentameric subunit stoichiometry of a neuronal glutamate receptor. *Proc. Natl. Acad. Sci. USA* **93**, 2741–2744.
60. Armstrong, N., Sun, Y., Chen, G., and Gouaux, E. (1998) Structure of a glutamate-receptor ligand-binding core in complex with kainate. *Nature* **395**, 913–917.
61. Abele, R, Keinänen, K., and Madden, D. R. (2000) Agonist-induced isomerization in a glutamate receptor ligand-binding domain. *J. Biol. Chem.* **275**, 21,355–21,363.
62. Seeburg, P. H., Higuchi, M., and Sprengel, R. (1998) RNA editing of brain glutamate receptor channels: mechanism and physiology. *Brain Res. Rev.* **26**, 217–229.
63. Bass, B. L., Nishidura, K., Keller, W., Seeburg, P. H., Emeson, R. B., O'Connell, M. A., et al. (1997) A standardized nomenclature for adenosine deaminases that act on RNA. *RNA* **3**, 947–949.
64. Brusa, R., Zimmermann, F., Koh, D. S., Feldmeyer, D., Gass, P., Seeburg, P. H, et al. (1995) Early-onset epilepsy and post-natal lethality associated with an editing-deficient GluR-B allele in mice. *Science* **270**, 1677–1680.
65. Higuchi, M., Maas, S., Single, F. N., Hartner, J., Rozov, A., Burnashev, N., et al. (2000) Point mutation in an AMPA receptor gene rescues lethality in mice deficient in the RNA-editing enzyme ADAR2. *Nature* **406**, 78–81.
66. Matsuda, S., Mikawa, S., and Hirai, H. (1999) Phosphorylation of serine-880 in GluR2 by protein kinase C prevents its C terminus from binding with glutamate receptor-interacting protein. *J. Neurochem.* **73**, 1765–8.
67. Chung. H. J., Xia, J., Scannevin, R. H., Zhang, X., and Huganir, R. L. (2000) Phosphorylation of the AMPA receptor subunit GluR2 differentially regulates its interaction with PDZ domain-containing proteins. *J. Neurosci.* **20**, 7258–7267.
68. Carvalho, A. L., Kameyama, K., and Huganir, R. L. (1999) Characterization of phosphorylation sites on the glutamate receptor 4 subunit of the AMPA receptors. *J. Neurosci.* **19**, 4748–4754.
69. Chen, L. and Huang, L,-Y. M. (1992) Protein kinase C reduces $Mg^{2+}$ block of NMDA-receptor channels as a mechanism of modulation. *Nature* **356**, 521–523.
70. Tingley, W. G., Ehlers, M. D., Kameyama, K., Doherty, C., Ptak, J. B., Riley, C. T., et al. (1997) Characterization of protein kinase A and protein kinase C phosphorylation of the N-methyl-D-aspartate receptor NR1 subunit using phosphorylation site-specific antibodies. *J. Biol. Chem.* **272**, 5157–5166.
71. Zheng, F., Gingrich, M. B., Traynelis, S. F., and Conn, P. J (1998) Tyrosine kinase potentiates NMDA receptor current by reducing tonic $Zn^{2+}$ inhibition. *Nat. Neurosci.* **1**, 185–191.
72. Lu, Y. M., Roder, J. C., Davidow, J., and Salter, M. W. (1998) Src activation in the induction of long-term potentiation in CA1 hippocampal neurons. *Science* **279**, 1363–1367.
73. Grant, S. G. N., O'Dell, T. J., Karl, K. A., Stein, P. L., Soriano, P., and Kandel, E. R. (1992) Impaired long-term potentiation, spatial learning, and hippocampal development in *Fyn* mutant mice. *Science* **258**, 1903–1910.
74. Fujita, A. and Kurachi, Y. (2000) SAP family proteins. *Biochem. Biophys. Res. Commun.* **269**, 1–6.
75. Braithwaite, S. P., Meyer, G., and Henley, J. H (2000) Interactions between AMPA receptors and intracellular proteins. *Neuropharmacology* **39**, 919–930.
76. Hayashi, Y., Shi, S.-H., Esteban, J. A., Piccini, A., Poncer, J.-C., and Malinow, R. (2000) Driving AMPA receptors into synapses by LTP and CaMKII: requirement for GluR1 and PDZ domain interaction. *Science* **287**, 2262–2267.
77. Lin, J. W. and Sheng, M. (1998) NSF and AMPA receptors get physical. *Neuron* **21**, 267–270.
78. Kennedy, M. B. (1998) Signal transduction molecules at the glutamatergic postsynaptic membrane. *Brain Res. Rev.* **26**, 243–257.
79. Husi, H., Ward, M. A., Choudhary, J. S., Blackstock, W. P., and Grant, S. G .N. (2000) Proteomic analysis of NMDA receptor-adhesion protein signaling complexes. *Nat. Neurosci.* 3,661–669.
80. Wang, Y., Small, D. L., Stanimirovic, D. B., Morley, P., and Durkin, J. P. (1997) AMPA receptor-mediated regulation of a $G_i$-protein in cortical neurons. *Nature* **389**, 502–504.
81. Wang, Y. and Durkin, J. P. (1995) α-Amino-3-hydroxy-5-methyl-4-isoxazolepropionic acid, but not N-methyl-D-aspartate, activates mitogen-activated protein kinase through G-protein βγ subunits in rat cortical neurons. *J. Biol. Chem.* **270**, 22,783–22,787.
82. Kawai, F. and Sterling, P. (1999) AMPA receptor activates a G-protein that suppresses a cGMP-gated current. *J. Neurosci.* **19**, 2954–2959.
83. Hayashi, Y., Umemori, H., Mishina, M., and Yamamoto, T. (1999) The AMPA receptor interacts with and signals through the protein tyrosine kinase Lyn. *Nature.* **397**, 72–76.
84. Rodriques-Moreno, A. and Lerma, J. (1998) Kainate receptor modulation of GABA release involves a metabotropic function. *Neuron* **20**, 1211–1218.
85. Cunha, R. A., Malva, J. O., and Ribeiro, J. A. (1999) Kainate receptors coupled to $G_i/G_o$ proteins in the rat hippocampus. *Mol. Pharmacol.* **56**, 429–433.
86. Glover, R. T., Angiolieri, M., Kelly, S., Monaghan, D. T., Wang, J. Y., Smithgall, T. E., et al. (1998) Interaction of the N-methyl-D-aspartic acid receptor NR2D subunit with the c-Abl tyrosine kinase. *J. Biol. Chem.* **275**, 12725–12729.

87. Rao, A., Harms, K. J., and Craig, A. M. (2000) Neuroligation: building synapses around the neurexin-neuroligin link. *Nat. Neurosci.* **3,** 747–749.
88. Tsui, C. C., Copeland, N. G., Gilbert, D. J., Jenkins, N. A., Barnes, C., and Worley, P. F. (1996) Narp, a novel member of the pentraxin family, promotes neurite outgrowth and is dynamically regulated by neuronal activity. *J. Neurosci.* **16,** 2463–2478.
89. O'Brien, R. J., Xu, D., Petralia, R. S., Steward, O., Huganir, R. L., and Worley, P. (1999) Synaptic clustering of AMPA receptors by the extracellular immediate-early gene product Narp. *Neuron* **23,** 309–323.
90. Johnson, J. W. and Ascher, P. (1987) Glycine potentiates the NMDA response in cultured mouse brain neurons. *Nature* **325,** 529–531.
91. Kleckner, N. W. and Dingledine, R. (1988) Requirement for glycine in activation of NMDA receptors expressed in *Xenopus* oocytes. *Science* **241,** 835–837.
92. Washburn, M. S., Numberger, M., Zhang, S., and Dingledine, R. (1997) Differential dependence of GluR2 expression of three characteristic features of AMPA receptors. *J. Neurosci.* **17,** 9393–9406.
93. Burnashev, N., Schoepfer, R. Monyer, H., Ruppersberg, J. P., Gunther, W., Seeburg, P. H., et al. (1992) Control by asparagine residues of calcium permeability and magnesium blockade in the NMDA receptor. *Science* **257,** 1415–1419.
94. Kuner, T., Wollmuth, L. P., Karlin, A., Seeburg, P. H., and Sakmann, B. (1996) Structure of the NMDA receptor channel M2 segment inferred from the accessibility of substituted cysteines. *Neuron* **17,** 343–352.
95. Bezzi, P., Vesce, S., Panzarasa, P., and Volterra, A. (1999) Astrocytes as active participants of glutamatergic function and regulatiors of its homeostasis. *Adv. Exp. Med. Biol.* **468,** 69–80.
96. Bergles, D. E., Roberts, J. D. B., Somogyi, P., and Jahr, C. E. (2000) Glutamatergic synapse on oligodendrocyte precursur cells in the hippocampus. *Nature* **405,** 187–191.
97. Gallo, V., Zhou, J. M., McBain, C. J., Wright, P., Knutson, P. L., and Armstrong, R. C. (1996) Oligodendrocyte progenitor cell proliferation and lineage progression are regulated by glutamate receptor-mediated $K^+$ channel block. *J. Neurosci.* **16,** 2659–2670.
98. Dani, J. W., Chernjavsky, A., and Smith S. J. (1992) Neuronal activity triggers calcium waves in hippocampal astrocyte networks. *Neuron.* **8,** 429–440.
99. Porter, J. T. and McCarthy, K. D. (1996) Hippocampal astrocytes *in situ* respond to glutamate released from synaptic terminals. *J Neurosci.* **16,** 5073–5081.
100. Pasti, L, Volterra, A., Pozzan, T., and Carmignoto, G. (1997) Intracellular calcium oscillations in astrocytes: a highly plastic, bidirectional form of communication between neurons and astrocytes *in situ. J. Neurosci.* **17,** 7817–7830.
101. Winder, D. G., Ritch, P. S., Gereau, R. W. IV, and Conn P. J. (1996) Novel glial–neuronal signalling by coactivation of metabotropic glutamate and beta-adrenergic receptors in rat hippocampus. *J. Physiol.* **494,** 743–755.
102. Peoples, R. W. and Weight, F. F. (1997) Anesthetic actions on excitatory amino acids receptors, in *Anesthesia: Biologic Foundation.* (Yaksh, T., ed.) Lippincott–Raven, philadelphia/pp. 239–258.
103. Anders, D. L., Blevins, T., Smothers, C. T., and Woodward, J. J. (2000) Reduced ethanol inhibition of *N*-methyl-D-aspartate receptors by deletion of the NR1 C0 domain or overexpression of alpha-actinin-2 proteins. *J. Biol. Chem.* **275,** 15,019–15,024.
104. Wyszynski, M., Kharazia, V., Shanghvi, R., Rao, A., Beggs, A. H., Craig, A. M., et al. (1998) Differential regional expression and ultrastructural localization of alpha-actinin-2, a putative NMDA receptor-anchoring protein, in rat brain. *J. Neurosci.* **18,** 1383–1392.
105. Narita, M., Soma, M., Mizoguchi, H., Tseng, L. F., and Suzuki, T. (2000) Implications of the NR2B subunit-containing NMDA receptor localized in mouse limbic forebrain in ethanol dependence. *Eur. J. Pharmacol.* **401,** 191–195.
106. Frye, G. D. and Fincher, A. (2000) Sustained ethanol inhibition of native AMPA receptors on medial septum/diagonal band (MS/DB) neurons. *Br. J. Pharmacol.* **129,** 87–94.
107. Cruz, S. L., Mirshahi, T., Thomas, B., Balster, R. L., and Woodward J. J. (1998) Effects of the abused solvent toluene on recombinant *N*-methyl-D-aspartate and non-*N*-methyl-D-aspartate receptors expressed in *Xenopus* oocytes. *J. Pharmacol. Exp. Ther.* **286,** 334–340.
108. Jevtovic-Todorovic, V., Todorovic, S. M., Mennerick, S., Powell, S., Dikranian, K., Benshoff, N., et al. (1998) Nitrous oxide (laughing gas) is an NMDA antagonist, neuroprotectant and neurotoxin. *Nat. Med.* **4,** 460–463.
109. Traynelis, S. F, Hartley, M., and Heinemann, S. F. (1995) Control of proton sensitivity of the NMDA receptor by RNA splicing and polyamines. *Science* **268,** 873–876.
110. Mott, D. D., Doherty, J. J., Zhang, S., Washburn, M. S., Fendley, M. J., Lyuboslavsky, P., et al. (1998) Enhancement of protein inhibition: a novel mechanism of inhibition of NMDA receptors by phenylethanolamines. *Nat. Neurosci.* **1,** 659–667.
111. Choi, Y. B. and Lipton, S. A. (1999) Identification and mechanism of action of two histidine residues underlying high-affinity $Zn^{2+}$ inhibition of the NMDA receptor. *Neuron* **23,** 171–180.
112. Low, C. M., Zheng, F., Lyuboslavsky, P., and Traynelis, S. F. (2000) Molecular determinants of coordinated proton and zinc inhibition of *N*-methyl-D-aspartate NR1/NR2A receptors. *Proc. Natl. Acad. Sci. USA* **97,** 11,062–11,067.

113. Paoletti, P., Perin-Dureau, F., Fayyazuddin, A., Le Goff, A., Callebaut, I., and Neyton, J. (2000) Molecular organization of a zinc binding N-terminal modulatory domain in a NMDA receptor subunit. *Neuron* **28,** 911–925.
114. Mosbacher, J., Schoepfer, R., Monyer, H., Burnashev, N., Seeburg, P. H., and Ruppersberg, J. P. (1994) A molecular determinant for submillisecond desensitiation in glutamate receptors. *Science* **266,** 1059–1061.
115. Partin, K. M., Bowie, D., and Mayer, M. L. (1995) Structural determinants of allosteric regulation in alternatively spliced AMPA receptors. *Neuron* **14,** 833–843.
116. Partin, K. M., Patneau, D. K., Winter, C. A., Mayer, M. L., and Buonanno, A. (1993) Selective modulation of desensitization at AMPA versus kainate receptors by cyclothiazide and concanavalin A. *Neuron* **11,** 1069–1082.
117. Schiffer, H. H., Swanson, G. T., and Heinemann, S. F. (1997) Rat GluR 7 and a carboxyterminal splice variant, GluR 7b, are functional kainate receptor subunits with a low sensitivity to glutamate. *Neuron* **19,** 1141–1146.

# 2
# Pharmacology of Glutamate Receptors

## Jeffrey M. Witkin, PhD, Rafal Kaminski, MD, PhD, and Michael A. Rogawski, MD, PhD

## 1. INTRODUCTION

Glutamate is the principal excitatory neurotransmitter in the central nervous system, playing an essential role in virtually every brain system and their associated behavioral functions. Modulation of glutamate-mediated neurotransmission is a major way in which the function of the nervous system is modified in response to experience (*see* Chapter 4). Drug addiction and the related behavioral and pharmacological states of sensitization, tolerance, and physical dependence appear to develop and be maintained in ways that may be functionally equivalent to other neural plasticity phenomena for which changes in the strength of glutamate-mediated synaptic transmission plays a principal role (cf. ref. *1*). As such, understanding the physiological roles of glutamatergic neurotransmission and its modulation by drugs of abuse is important in identifying the neurobiological substrates of addiction and relevant molecular targets for therapeutic intervention. Agents that target glutamate receptors are the main pharmacological tools available for investigating the functions of glutamatergic synapses and their roles in behavior. The present chapter is therefore devoted to glutamate-receptor pharmacology and provides the background necessary to appreciate the use of drugs that target glutamate receptors in the specific applications discussed in later Chapters. Here, we focus on the specificity of action of these agents, their particular advantages and disadvantages as tools for neurobiological inquiry, and their efficacies and side-effect profiles, including information from studies of human subjects where available. In addition, we briefly consider agents that modify glutamate neurotransmission through effects on glutamate release, by inhibition of NAALADase, and through effects on the downstream effector nitric oxide.

In addition to their potential therapeutic roles in a range of neurological and psychiatric disorders such as epilepsy, stroke, anxiety, depression, chronic pain, and cognitive impairments, drugs affecting glutamatergic neurotransmission are also potentially important therapeutic modalities in drug-dependence disorders. Basic information on glutamate pharmacology is necessary to guide drug discovery efforts. Moreover, there is an increasing body of information on the effects of drugs that modify glutamate neurotransmission in human subjects in health and disease. An appreciation of these clinical studies is crucial for the appropriate selection of drugs for clinical trials in drug dependence.

As noted in other chapters, compounds that modulate glutamatergic transmission can interact with several types of glutamate receptors, including the ionotropic *N*-methyl-D-aspartate (NMDA), α-amino-3-hydroxy-5-methyl-4-isoxazole propionic acid (AMPA), and kainate receptors as well as a diversity of metabotropic receptors, each with unique functional roles and distributions in the central nervous system. Thus, the potential range of biological effects of glutamate modulators is theoretically quite diverse and the biological evidence is in concordance with this prediction.

From: *Contemporary Clinical Neuroscience: Glutamate and Addiction*
Edited by: Barbara H. Herman et al. © Humana Press Inc., Totowa, NJ

## 2. NMDA RECEPTORS

The NMDA receptors are ionotropic receptors that function as cation channels mediating neuronal depolarization and excitation and play a special role in the induction of synaptic plasticity phenomena. The receptor has recognition sites for the coagonists glutamate and glycine, as well as for other modulatory agents whose binding can alter ion flux through the channel (see Chapter 1; Fig. 1). Agonists at the glutamate and glycine sites activate the NMDA receptor, leading to excitation. Antagonists at these sites have the opposite effect of blocking NMDA-receptor gating and dampening excitation. NMDA-receptor function can also be inhibited by agents that block the ionophore of the channel, that bind to diverse modulatory sites on the channel, or that alter the redox or phosphorylation state of the channel. Drugs that modulate the NMDA receptor in these various ways can have diverse behavioral effects and varying degrees of toxicity. Additional specificity can be determined by the selectivity of certain agents for NMDA receptors composed of different subunit combinations, each with distinct biophysical properties and distributions in the nervous system. Other and, in some cases, more exhaustive reviews of the molecular and behavioral pharmacology of glutamate-receptor systems, which include extensive discussions of structure–activity relationships, are available (2–5).

### 2.1. Uncompetitive Antagonists

NMDA receptor–channel blockers are uncompetitive antagonists because their binding and blocking action requires the receptor–channel to be gated in the open state. Occupancy of a binding site within the ionophore of the channel by these compounds prevents the flux of cations through the ion channel, thus producing a functional block of NMDA-receptor responses (6,7). Unlike competitive NMDA recognition site antagonists (see Section 2.2.), uncompetitive antagonists share with other non-competitive antagonists of the NMDA receptor the theoretical advantage that blockade would not be overcome by high synaptic levels of glutamate. An additional potential advantage of uncompetitive antagonists is use dependence, in which the inhibitory action may specifically be potentiated at sites of excessive receptor activation. Nonetheless, these potential advantages may, in some cases, come at the cost of undesirable side effects.

Blockade of the NMDA-receptor ion channel can be produced by a host of structurally diverse compounds, including $Mg^{2+}$, which serves as an endogenous modulator. Compounds that have been studied most extensively (including, in most cases, studies in humans) are listed in Table 1. Uncompetitive antagonists may be divided into two major classes, based on their binding affinities, kinetic characteristics, and side-effect profiles (8). Dissociative anestheticlike compounds, such as dizocilpine (MK-801), phencyclidine (PCP), and aptiganel, generally have blocking affinities below 100 n$M$, have slower rates of binding and receptor dissociation, and exhibit a high degree of trapping. Such compounds have a high behavioral toxicity and a propensity for inducing psychotomimetic effects. Ketamine is a notable exception to this rule in that it exhibits dissociative anestheticlike behavioral properties, but it has a binding affinity in the range of low-affinity channel-blocking compounds that are generally less toxic. Dissociative anestheticlike compounds vary in the degree to which they specifically interact with NMDA receptors to the exclusion of other targets. For example, PCP binds with high affinity not only to the NMDA-receptor ion channel but can also interact with the dopamine transporter (9,10) and voltage-activated potassium channels (11); N-allylnormetazocine (SKF 10,047) is a high-affinity ligand of both NMDA and sigma receptors (12). An understanding of the full spectrum of the pharmacological activities of these compounds is critical to assessing their behavioral effects, particularly because sigma receptors and the dopamine uptake carrier have been implicated as targets for drugs of abuse. In contrast to the complex pharmacology of some dissociative anestheticlike agents, dizocilpine is a highly selective uncompetitive NMDA-receptor antagonist and is not known to affect other receptors or ion channels at concentrations comparable to those that block NMDA receptors.

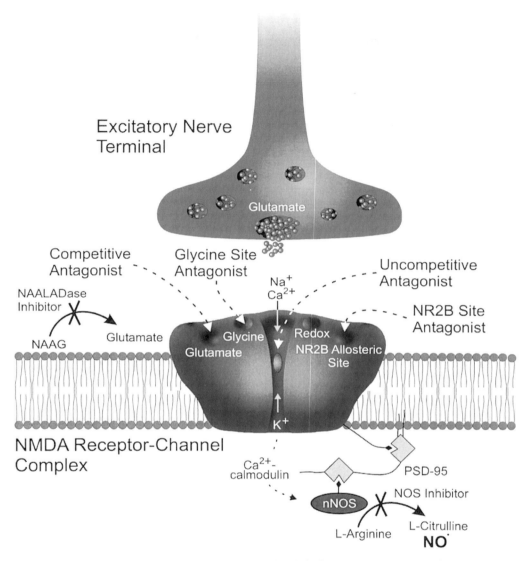

**Fig. 1.** A representation of the NMDA receptor with associated binding and regulatory domains. (Color illustration in insert following p. 142.)

Uncompetitive NMDA-receptor antagonists exhibit evidence of predicted clinical efficacy in a host of models of neurological and psychiatric disorders. For example, a series of uncompetitive antagonists dose-dependently protect against the convulsant effects of cocaine that are not otherwise robustly sensitive to standard anticonvulsant agents (Fig. 2A). At comparable doses, these compounds also produce neurological impairment in the inverted-screen test (Fig. 2B). The potencies of these NMDA antagonists to block the convulsant effects of cocaine are positively associated with their affinities for the ion channel (Fig. 3A). There is also a correlation between binding affinity and propensity to induce ataxia (Fig. 3B). In addition, some of the antagonists produce diverse behavioral effects characterized as dissociative anestheticlike, including locomotor stimulation

**Table 1**
**Some Uncompetitive (Channel-Blocking) NMDA Receptor Antagonists**

| Compound | Clinical status |
|---|---|
| Dissociative Anestheticlike | |
|   Ketamine | Marketed—anesthesia (human and veterinary use) |
|   Phencyclidine (PCP) | Clinical information available |
|   Dizocilpine [(+)-MK-801] | Prior investigation—epilepsy |
|   Aptiganel (CNS-1102, Cerestat) | Clinical investigation—stroke |
|   Dextrorphan | Prior investigation—stroke |
|   TCP | |
|   $N$-Allylnormetazocine (SKF 10,047) | |
| Low Affinity | |
|   Dextromethorphan | Marketed—cough suppression; Prior investigation—stroke |
|   Memantine | Marketed (Europe)—dementia |
|   Amantadine | Marketed—influenza, Parkinson's disease |
|   Ramacemide (FPL-12924) | Prior development—epilepsy, stroke |
| | In development—Huntington's disease, Parkinson's disease |
|   ARL-15896 (AR-R15896) | In development—stroke |
|   ADCI | Prior investigation—epilepsy |
|   Ibogaine | In development—drug dependence |

(Fig. 4), stereotypies, and disruptions in learning and memory (cf. ref. *15* and *16*). These compounds also produce subjective effects as assessed in drug discrimination paradigms that are similar to those of PCP. Such PCP-like discriminative stimulus effects generally correlate with affinity for the NMDA-receptor ion channel *(8,17–19)*. Importantly, the diverse behavioral actions of dissociative anestheticlike uncompetitive NMDA antagonists occur at doses that are predicted to produce clinical efficacy. Thus, such agents are expected to exhibit little separation between clinical efficacy and side effects.

Clinical experience with PCP, ketamine, dizocilpine, and aptiganal has indeed provided evidence of poor tolerability. These agents all induce troubling behavioral side effects, including hallucinations, poor concentration, ataxia, sedation, depression, and other symptoms resembling those of schizophrenia *(20,21)*. As PCP is a highly abused substance, drug abuse is an additional concern with respect to the clinical use of drugs in this class. In addition, uncompetitive blockers of the NMDA receptor may cause cellular toxicity, including neuronal vacuolization at low doses *(22)* and necrosis at higher doses *(23)*. However, the generality of these findings to primates may be limited and it is now apparent that some NMDA antagonists produce, if anything, only transient vacuolization without necrosis.

A second major class of channel-blocking NMDA receptor antagonists are the low-affinity uncompetitive antagonists that are in clinical practice, in advanced stages of clinical trials, or in other phases of development for a host of neurological indications *(24)*. (Table 1). ADCI and memantine are two members of this class of drugs, which generally have more favorable side-effect profiles in comparison with dissociative anestheticlike agents *(25–27)*. For example, ADCI exhibits a large separation (18-fold) between its potency to block cocaine-induced seizures and its potency to produce neurological impairment, which contrasts dramatically with the situation for dizocilpine in which impairment occurs at 50% of the dose required for seizure protection *(13)*. ADCI also suppressed withdrawal seizures in mice physically dependent on ethanol and produces anticataleptic effects at doses that do not produce ataxia, stereotypy, or impairment of locomotion *(27,28)*. Similarly, memantine blocks NMDA-induced convulsions at doses that are 8- to 18-fold lower than those producing ataxia or effects on locomotion; in contrast, the anticonvulsant activity of dizocilpine is evident only at behaviorally

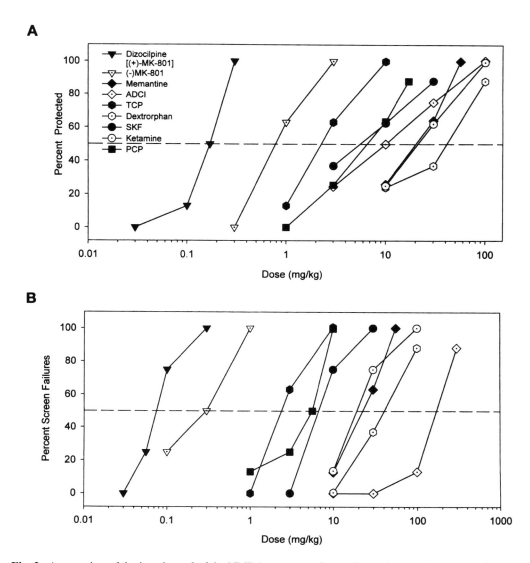

**Fig. 2.** Antagonists of the ion channel of the NMDA protect against anticonvulsant-resistant convulsant effects of cocaine in mice (**A**) but produce behavioral side effects at comparable doses (**B**). (From ref. *13* with permission of the publisher.)

disrupting doses *(29)*. In addition, there is general agreement that low-affinity channel-blocking NMDA-receptor antagonists like ADCI do not replicate the discriminative stimulus effects of dizocilpine or PCP *(18)*. However, some of these agents, notably memantine, may, in some cases, substitute *(18,29,30)*. In another model predictive of PCP-like side effects *(14)*, memantine produced ataxia like that of high-affinity ligands but did not produce locomotor stimulation typical of dissociative anestheticlike agents *(29)*. Memantine has been used in Europe for the treatment of dementia, Parkinson's disease, drug-induced extrapyramidal syndromes, neuroleptic malignant syndrome, and spasticity *(2,31)*. Although side effects have been reported *(32–34)*, memantine appears to carry a far lower risk of inducing psychotomimetic symptoms and neurological impairment than high-affinity

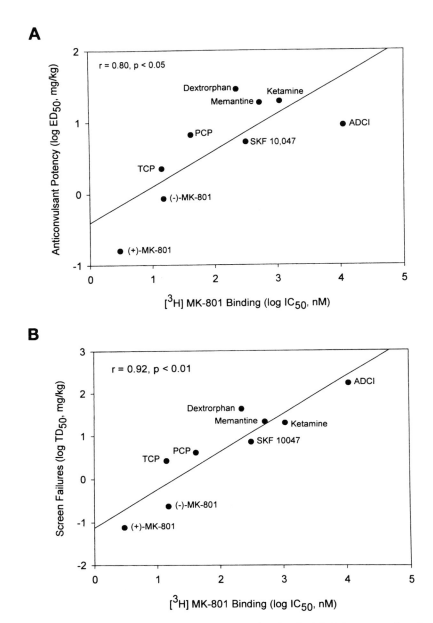

**Fig. 3.** Affinities of uncompetitive NMDA receptor antagonists for the NMDA receptor ion channel correlate with their potencies to protect against cocaine-induced convulsions (**A**) and to produce behavioral side effects as measured by the inverted screen test (**B**). (From ref. *13*, with permission from the publisher.)

antagonists *(2)*. The drug is currently under development for the treatment of neuropathic pain and AIDS dementia. Preclinical and clinical data suggest that amantadine may have similar therapeutic activities *(2,31)*. However, amantadine has very weak activity as an NMDA-receptor antagonist in vitro and does not protect against NMDA-induced convulsions in vivo *(29)*. Whether it acts as a low-affinity NMDA-receptor antagonist at clinically relevant doses is uncertain. NPS 1506 is a moderate-affinity uncompetitive NMDA-receptor antagonist with neuroprotective activity in rodent models of

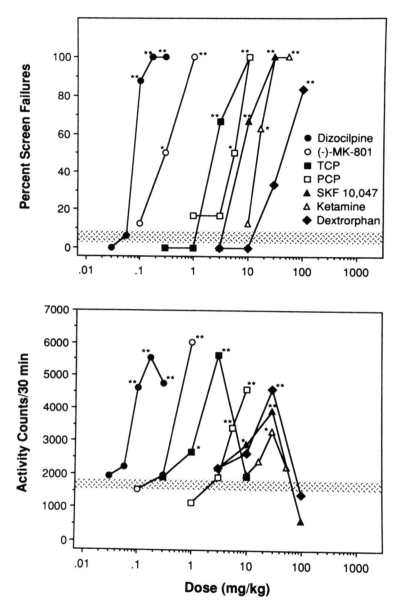

**Fig. 4.** High-affinity NMDA receptor antagonists of the ion channel produce both a unique combination of sedativelike and stimulantlike behavioral effects in mice. (From ref. *14*, with permission from the publisher.)

stroke and head trauma *(35)*. Early clinical trials have revealed good tolerability at doses in excess of those that confer neuroprotection in rodents.

The basis for the improved toxicity profiles of low-affinity channel-blocking NMDA antagonists is not fully understood, but may relate to a variety of factors, including faster blocking kinetics, partial trapping, reduced agonist-independent (closed-channel) block, subunit selectivity (leading to regional differences in action), and combined block at allosteric sites on the NMDA-receptor complex apart from the channel-blocking site *(8,36,37)*. In addition, the low affinity of these compounds for the

**Table 2**
**Some Competitive NMDA Receptor Antagonists**

| Compound | Clinical status |
|---|---|
| Selfotel (CGS 19755) | Prior development—stroke |
| D-CPP-ene (SDZ EAA-494) | In development—traumatic brain injury |
| MDL 100,453 | In development—epilepsy, stroke |
| NPC 17742 | Preclinical—stroke |
| (+)-CPP | |
| CGP 37849 | |
| NPC 12626 | |
| LY 274614 | |
| LY 235959 | |
| LY 233536 | |

NMDA-receptor-associated ion channel also makes the possibility that they will have promiscuous actions at sites other than NMDA receptors *(8,38)*. For example, felbamate, in addition to its actions on NMDA receptors, is also a low-efficacy positive allosteric modulator of GABA receptors *(39)* and remacemide, its active *des*-glycine metabolite ARL 12495AA, and ADCI all act as state-dependent blockers of voltage-activated Na$^+$ channels *(see* ref. *8)*. Such actions on multiple receptor targets may produce additive or synergistic therapeutic activities, but toxicities at each target may be distinct and nonadditive, resulting in superior therapeutic activity and reduced toxicity in comparison with agents that target NMDA receptors alone.

A body of preclinical *(see* ref. *40;* subsequent chapters of this book) and clinical (cf. ref. *41)* evidence attests to the potential therapeutic utility of ibogaine for the treatment of drug dependence. Although this compound has multiple pharmacological actions, ibogaine is well recognised as a low-affinity channel blocker of the NMDA receptor, an effect that may, at least in part, be responsible for its antiaddictive activity *(40,42,43)*. Ibogaine produces striking behavioral disruptions at doses claimed to be effective in drug-dependence treatment, and at high doses, it produces hallucinations. In addition, the drug exhibits PCP-like subjective effects as determined in drug discrimination experiments *(42)*. Analogs of ibogaine and of ibogamine have recently been patented with claims of antiaddictive efficacy in rats *(see* ref. *44)*.

## 2.2. Competitive NMDA Recognition-Site Antagonists

Competitive NMDA-receptor antagonists inhibit activation of NMDA receptors by occupying the glutamate recognition site of the receptor–channel complex and preventing binding of the neurotransmitter glutamate *(see* Fig. 1). Like uncompetitive antagonists, competitive NMDA-receptor antagonists have a diversity of potential clinical applications, including the treatment of drug dependence (see subsequent chapters of this book). A large, structurally diverse group of glutamate recognition-site antagonists is now available with good bioavailability and favorable pharmacokinetic properties (Table 2). Many of these agents have long-lasting activity after oral administration (e.g., CGS 19755, D-CPPene, CGP 37849, LY 274614).

Although studies in animals suggested that competitive NMDA recognition-site antagonists would have reduced liability for producing dissociative anestheticlike motor and subjective side effects *(see* refs. *2,3, 45,* and *46)*, neurobehavioral toxicities have occurred in humans that were not fully anticipated. For example, in clinical trials for stroke and traumatic head injury, D-CPPene and CGS 19755 demonstrated side effects reminiscent of those observed with dizocilpine and PCP (cf. refs. *2,47,* and *48)*. Indeed, competitive antagonists can produce prominent dissociative anestheticlike motor disturbances such as head weaving, body rolling, hyperlocomotion, and ataxia in rodents *(15)*. Moreover,

Carter *(49)* observed a positive correlation between the potencies of both competitive and uncompetitive NMDA-receptor antagonists for inducing motor incoordination in the rotorod test and their potencies to inhibit NMDA-induced lethality (an in vivo measure of NMDA-receptor blockade). At high doses, the competitive antagonists CGS 19755, NPC 17742, (±)-CPP, and LY 233536 all produced full or nearly full substitution for the discriminative stimulus effects of dizocilpine in mice *(19)*. Although competitive antagonists may not have the same propensity for producing certain behavioral side effects such as stereotypies and disturbances of locomotion that are common with uncompetitive NMDA antagonists, these effects do occur and may be especially prominent in sensitized states such as in amygdaloid kindled animals *(50,51)*. Furthermore, as is the case with uncompetitive antagonists, competitive antagonists can also produce neuronal vacuolization and morphological damage in certain brain regions *(22)*. Tolerance does not seem to develop to either the therapeutic activity or the side effects *(52)*. On the whole, the preclinical and clinical data have led many investigators to lose enthusiasm for the clinical potential of competitive antagonists.

Despite the potential drawbacks of this class of agents, selected competitive antagonists may have reduced liability for side effects in comparison with other members of the class. For example, although the protective indicies (ratio of $TD_{50}$ value for induction of toxicity and $ED_{50}$ for seizure protection) of many competitive antagonists is near 1 when tested in mice against cocaine-induced convulsions *(13)*, at least one competitive blocker, LY 233536, has a protective index as high as 7. This compound has previously been reported to display a less debilitating behavioral profile than that of other competitive antagonists *(14)*. On the other hand, the competitive antagonist LY 274614 and its active isomer, LY 235959, produced profound and potent suppression of behavior *(14,19,53)*. The lack of substitution of these latter two compounds in mice trained to discriminate dizocilpine from saline may have been due to these behavioral effects that made it impossible to test higher doses. The recognition that there can be large differences in the behavioral toxicities of competitive NMDA recognition-site antagonists suggests that it may be premature to conclude that this class of agent is not clinically viable.

## 2.3. Glycine-Site Ligands

Activation of NMDA receptors by glutamate requires the presence of a coagonist at a distinct recognition domain on the receptor complex referred to as the glycine site *(54,55)* (Fig. 1). More recently, the term glycine$_B$ site has been used to distinguish the regulatory coagonist site on NMDA receptors from the inhibitory strychnine-sensitive glycine receptor, a $Cl^-$ channel. Although glycine was originally believed to be the endogenous coagonist, recent evidence indicates that D-serine may serve this role, at least at some synapses *(56)*. Competitive antagonists at the glycine site can inhibit NMDA receptors in a manner similar to that of antagonists at the glutamate recognition site. In fact, a diverse group of glycine site ligands are available (Table 3) whose efficacies range from full agonism (D-serine) to partial agonism (approx 90% with ACPC), to weak partial agonism [approx 10% with (+)-HA-966], to full antagonism (7-Cl-kynurenic acid). These compounds exhibit diverse activities in animal models of anxiety, depression, memory *(57)*, and drug dependence (see subsequent chapters this book; also refs. *2* and *49*).

In contrast to other classes of NMDA antagonist, glycine-site antagonists (full antagonists and partial agonists) generally do not produce dissociative anestheticlike motor or subjective side effects in preclinical models at therapeutically relevant doses (cf. refs. *3,16,* and *58–60*), although the drugs can cause muscle relaxation and motor impairment under some circumstances (cf. ref. *2*). For example, in studies of cocaine-induced convulsions, glycine-site partial agonists exhibit protective activity at doses that do not produce motor impairment *(13)* and ACPC prevents the acquisition of cocaine-induced place preference without producing reinforcing effects of its own *(61)*.

The basis for the improved toxicity profiles of glycine-site antagonists in comparison with competitive and uncompetitive NMDA-receptor antagonists is not well understood. Many of the glycine-site ligands display poor bioavailability after systemic administration. Therefore, it has been

**Table 3**
**Some NMDA Receptor Strychnine-Insensitive Glycine$_B$-Site Partial Agonists and Antagonists**

| Compound | Clinical status |
|---|---|
| Partial Agonists | |
| ACPC | Prior development—stroke, depression |
| D-Cycloserine | Prior development—dementia |
| (+)-HA-966 | |
| Antagonists | |
| GV-150526 | In development—stroke |
| Licostinel (ACEA-1021) | In development—stoke, head injury |
| ZD-9379 | In development—stroke, pain |
| 7-Cl-kynurenic acid | Preclinical |
| 5,7-diCl-kynurenic acid | |
| MDL-100,748 | Preclinical—epilepsy |
| MRZ-2/570 | Preclinical—drug dependence |
| MRZ-2/571 | Preclinical—drug dependence |
| MRZ-2/576 | Preclinical—drug dependence |
| NMCQX | |
| NMDX | |
| L-698,544 | |

suggested that the low incidence of side effects could be the result of poor central accessibility (e.g., ref. 2). However, this explanation is not fully satisfactory given the host of effects produced by these agents upon systemic administration in animal models of neuropsychiatric disease. Several newer compounds have been developed with high potency and good central activity after systemic administration (e.g., NMCQX, NMDX, and L-698,544). It remains to been seen if these agents also display favorable side-effect profiles. Another possible explanation for the improved tolerability of glycine-site antagonists is that they could selectively interact with specific NMDA receptor subtypes that are expressed in distinct neuronal populations in a regionally specific fashion (see ref. 2). For example, the glycine-site antagonist CGP 61594 appears to selectively target NMDA receptors containing the NR2B subunit *(62)*. Moreover, agents with partial agonist activity may have better tolerability than full antagonists.

Because at least some glycine-site ligands have actions at other sites, these additional actions may contribute to their pharmacological effects. For example, quinoxaline ligands such as NMCQX and NMDX also have affinity for AMPA receptors, albeit at slightly higher micromolar concentrations and L-698,544 exhibits good systemic activity (cf. ref. *63*), but this compound also interacts with the glutamate ($K_b$, 6.7 µ$M$) and AMPA ($K_b$, 9.2 µ$M$) recognition sites with somewhat lower affinity than for the glycine site (IC$_{50}$, 0.41 µ$M$). Therefore, the efficacy of such agents in vivo might be due to competitive blockade of NMDA or AMPA receptors. Note, however, that the additional actions of compounds like these may provide a therapeutic advantage.

Like the neurotransmitter glutamate, glycine-site agonists enhance the opening of NMDA receptors *(64)*. Such agonists would therefore oppose the actions of NMDA antagonists and may specifically promote the dissociation of use-dependent blockers such as uncompetitive antagonists *(65–67)*. Consequently, glycine-site agonists can theoretically modulate responses to NMDA-receptor blockers, and in behavioral studies, glycine and glycine-site agonists D-serine and D-alanine have been shown to antagonize the stereotypies, ataxia, and locomotor stimulation produced by PCP and dizocilpine *(68–72)*. However, D-serine does not block the subjective effects of PCP as measured in rats discriminating PCP from saline *(60)*.

**Table 4**
**Some Selective Antagonists of NMDA Receptor Isoforms Containing NR2B Subunits**

| Compound | Clinical status |
| --- | --- |
| Ifenprodil | Marketed (outside US)—peripheral vascular disease |
| Eliprodil (SL 82.0715) | Prior development—stroke, traumatic brain injury |
| CP-101,606 | In development—stroke |
| Ro 25-6981 | Preclinical—stroke |
| Conantokin G | In development—epilepsy |
| Felbamate | Marketed—epilepsy |
| ADCI | Prior investigation—epilepsy |

## 2.4. NR2B-Selective Ligands

NMDA receptors in mammalian neurons are believed to be heterooligomers formed by the coassembly of an obligatory NR1 subunit and at least one type of NR2 subunit (see Chapter 1). Four NR2 subunits have been identified, each of which exhibits a distinct regional distribution and developmental expression pattern. The four distinct NR2 subunits confer unique pharmacological and biophysical properties upon the NMDA receptors from which they are assembled. In addition, the NR1 subunit exists in multiple alternatively spliced forms that can provide further diversity to NMDA receptors. Selective targeting of specific NMDA-receptor subtypes is a reasonable approach for the development of NMDA-receptor antagonists with improved therapeutic activity and reduced toxicity. Many antagonists show small potency differences at NMDA receptors composed of different subunits. For example, the competitive NMDA recognition-site antagonist (±)-CPP appears to have modestly higher affinity for NMDA receptors composed of a form of the NR1 subunit containing an insert generated by the inclusion of 21 amino acids coded by the alternatively spliced *exon 5* (see refs. 2 and 73). To date, however, the only NMDA-receptor isoforms that can be targeted in a therapeutically relevant way are those containing the NR2B subunit (Fig. 1). NR2B selectivity was first described for the phenylethylamine ifenprodil, and, subsequently, the related compounds eliprodil, CP101,606, Ro 25-6981, and nylidrin were found to have similar selectivity *(74–77)*. (see Table 4) and actions governed by their ability to regulate proton inhibition (cf. ref. 78). Despite their relatively potent NMDA-receptor-blocking activity, ifenprodil and eliprodil generally display more favorable toxicity profiles than other NMDA receptor antagonists *(79)*. At neuroprotective doses, they do not produce neurological impairment, do not substitute for PCP in drug-discrimination studies, nor do they induce amnestic effects in passive avoidance testing *(2,3)*. In addition, ifenprodil and eliprodil block cocaine-induced seizures at doses that do not produce failures in the inverted screen test *(13)*. Moreover, clinical trials have indicated that these compounds do not produce psychostimulant, amnesic, or psychotomimetic effects in humans. Finally, eliprodil does not appear to produce the pathomorphological changes that have been observed with other NMDA receptor antagonists *(80)*. These NR2B-selective antagonists may also have additive or synergistic effects when given in conjunction with other NMDA-receptor antagonists (cf. ref. 81).

The anticonvulsants felbamate and ADCI also exhibit moderate NR2B selectivity and this may in part account for their favorable neurobehavioral tolerability in comparison with other NMDA-receptor antagonists *(18,82,83)*. Conantokin G, a component of a cone snail toxin that is currently in development as an anticonvulsant, appears to act as a competitive NMDA recognition-site antagonist with selectivity for the NR2B subunit *(84)*. The butyrophenones haloperidol, droperidol, and spiperone also exhibit NR2B-selective antagonist activity, but this is not of practical utility in behavioral studies because these compounds have a diversity of other pharmacological actions, the most prominent of which is dopamine receptor blockade *(75,85,86).*

Although the basis for the improved neurobehavioral toxicity of NR2B-selective NMDA-receptor antagonists is not well understood, it could relate to the more restricted distribution of NR2B-containing NMDA receptors. Thus, unlike the NR2A subunit that is distributed ubiquitously in the central nervous system, expression of the NR2B subunit in adults is largely restricted to the forebrain *(87).*

## 3. AMPA AND KAINATE RECEPTORS

AMPA receptors mediate a large proportion of fast excitatory neurotransmission in the central nervous system. Although the role of kainate receptors is not as well understood, emerging evidence indicates that these receptors contribute to fast excitatory neurotransmission at selected synapses in the brain and spinal cord *(88),* and may also play a role in regulating inhibitory neurotransmission *(89,90).* In the past several years, highly selective ligands for both AMPA and kainate receptors have been discovered (Table 5).

### *3.1. AMPA Receptors*

AMPA receptor antagonists may modulate AMPA receptor function by acting at the glutamate recognition site (competitive antagonists) or at a distinct allosteric regulatory site for 2,3-benzodiazepines (negative allosteric modulators). In addition, $Ca^{2+}$-permeable AMPA receptors can be antagonized by a diverse group of cationic channel blockers, including polyamines, spider and wasp toxins such as Joro spider toxin (JSTX), and IEM-1460. Recent and detailed reviews of AMPA-receptor structure, function, and pharmacology are available *(7,91–93).*

#### *3.1.1. Competitive Antagonists*

Competitive antagonists acting at the glutamate recognition site of AMPA receptors have shown promising neuroprotective, anticonvulsant, analgesic, and anxiolytic activities in animal models *(92,93).* Competitive antagonists also display beneficial effects upon certain stages of drug dependence (see subsequent chapters, this book). competitive AMPA-receptor antagonists include quinoxalinediones such as CNQX (6-cyano-7-nitroquinoxaline-2,3-dione), NBQX [2,3-dihydroxy-6-nitro-7-sulfamoylbenzo (F)quinoxaline], and YM90K [6-(1*H*-imidazol-l-yl)-7-nitro-2,3-(1*H*-4*H*)-quinoxalinedione], as well as several structurally novel compounds including isatin oximes such as NS 257 {(1,2,3,6,7,8-hexahydro-3-(hydroxyimino)- *N,N*-[$^3$H]dimethyl-7-methyl-2-oxobenzo[2,1-b:3,4-c′]dipyrrole-5-sulfonamide} and decahydroisoquinolines such as LY293558 {(3S,4aR,6R,8aR)-6-[2-(1(2)H3-tetrazole-5-yl)ethyl] decahydroisoquinoline-3-carboxylic acid}. Most of these antagonists do not discriminate well between AMPA and kainate receptors *(94).* This is also the case for the water-soluble and systemically available isatin oxime NS 257, which not only blocks AMPA and kainate receptors but also interacts with NMDA receptors *(95,96).* NBQX, however, exhibits modest selectivity for AMPA-preferring receptors.

NBQX, like most other AMPA-receptor antagonists, generally produces motor and memory impairments in experimental animals at neuroprotective or anticonvulsant doses. However, in some cases therapeutic effects can be produced with little toxicity. For example, there is a good separation between the doses of NBQX that protect against cocaine-induced convulsions and those that induce motor impairment in mice *(13).* Moreover, NBQX attenuated both cocaine- and methamphetamine-induced locomotor stimulation in mice at doses that did not influence spontaneous locomotor activity *(97).* In addition, NBQX prevented the induction of sensitization to locomotor stimulant effects of cocaine *(98)* and CNQX blocked the reinstatement of cocaine self-administration induced by intra-accumbens AMPA or dopamine *(99).* NBQX did not substitute for the discriminative stimulus effects of dizocilpine, which may suggest that it is devoid of behavioral effects that are associated with NMDA-receptor antagonists *(19).*

**Table 5**
**Some AMPA and Kainate Receptor Ligands**

| Compound | Clinical status |
|---|---|
| Competitive AMPA Recognition-Site Antagonists | |
| CNQX | |
| NBQX | Prior investigation—healthy volunteers (crystaluria) |
| YM90K (YM900) | Prior investigation—stroke |
| YM872 | Clinical investigation—stroke |
| ZK200775 (Fanapanel) | Clinical investigation—stroke |
| NS-257 | |
| Ro 48-8587 | |
| LU 115455 | |
| LU 136541 | |
| Negative Allosteric AMPA Receptor Modulators | |
| GYKI 52466 | |
| LY 300164 (GYKI 53773; talampanel) | Prior investigation—epilepsy, amyotrophic lateral sclerosis |
| LY 300168 (GYKI 53655) | |
| LY 303070 | |
| GYKI 47261 | |
| SYM 2206 | |
| SYM 2189 | |
| Positive Allosteric AMPA Receptor Modulators | |
| Aniracetam | In development—cognitive enhancer |
| Piracetam | Marketed (Europe)—cognitive enhancer |
| CX516 | In development—cognitive enhancer, schizophrenia |
| Cyclothiazide | |
| Diazoxide | |
| S18986 | |
| IDRA-21 | |
| Kainate Receptor Antagonists | |
| NonSelective | |
| NBQX, CNQX, DNQX | |
| NS-102 | |
| GluR5 Selective | |
| LY 293558 (enantiomer of LY 215490) | Clinical investigation—pain |
| LY 377770 (enantiomer of LY 294486) | |
| LY 382884 (enantiomer of LY 307130) | |

(Above GluR5-selective decahydroisoquinolines also block AMPA receptors LY 293558 >LY294486>LY 382884)

In humans NBQX, at low doses, was without significant effect on the electroencephalogram, body temperature, or cardiovascular function *(100)*. In animal studies, YM90K demonstrates an improved pharmacokinetic profile in comparison with NBQX and is a potent neuroprotective and anticonvulsant agent *(101)*. YM90K was tested in human healthy volunteers and appeared to be well tolerated, but its low solubility raises concerns *(102)*. A series of compounds structurally related to YM90K but with improved selectivity for AMPA receptors have been synthesized *(103,104)*. YM872 {2.3-dioxo-7-(1*H*-imidazol-1-yl)6-nitro-1,2,3,4-tetrahydro-1-quinoxal inyl]acetic acid} has improved solubility and a selectivity for AMPA receptors close to that of NBQX or YM90K *(105)*. This compound displays remarkably potent neuroprotective effects after experimental ischemia in various species, is effective even when injected several hours following ischemia *(106,107)*, and is also an effective analgesic *(108)*.

LY293558 is a structurally distinct competitive antagonist that in addition to AMPA-receptor-blocking activity also has high affinity for GluR5 kainate receptors *(109,110)*. LY 293558 has demonstrated neuroprotective, antinociceptive, and anticonvulsant activity (cf. ref. *111*). In a number of experimental studies, LY 293558 showed beneficial effects on aspects of morphine addiction, including the development of sensitization *(112)* and tolerance *(113)* and on morphine-withdrawal signs *(114)*. In clinical evaluation for experimental pain, LY 293558 had selective antihyperalgesic activity without significant effects on nociceptive stimuli in normal skin. LY 293558 did not produce neurological or cognitive deficits; however, mild and transient visual obscuration was observed after administration of a high dose of this compound *(115)*.

New generations of competitive AMPA antagonists are now available, although chracterization of their effects in vivo is limited. For instance, Ro 48-8587 is a highly potent and selective AMPA antagonist *(116)*. Compounds in a pyrrolyl-quinoxalinedione series (e.g., LU 115455, LU 136541) display more than a 1000-fold higher affinity at AMPA receptors than at kainate receptors and have shown efficacy against amygdaloid-kindled seizures at doses without motor impairing effects in rats *(117)*. Another novel quinoxalinedione derivative, ZK200775, has greatly improved solubility and had a wide therapeutic window as a neuroprotective agent *(118)*.

The overall efficacy of AMPA-receptor blockade (e.g., by NBQX) may be augmented synergistically by low doses of NMDA-receptor antagonists without increasing adverse effects *(119–121)*. LU 73068, which acts as an antagonist at the glycine site of NMDA receptors as well as an antagonist of AMPA receptors, is also an effective anticonvulsant agent in kindled rats with a more favorable therapeutic profile than either the AMPA antagonist NBQX or the glycine-site antagonist L-701,324 given separately *(121)*.

### 3.1.2. Negative Allosteric Modulators

2,3-Benzodiazepines such as GYKI 52466 and its 3*N*-acetyl analog GYKI 53405 and 3*N*-methylcarbamyl analog GYKI 53655 (LY300168) act as selective noncompetitive AMPA receptor antagonists *(25,122–124)* through negative allosteric modulation of AMPA receptor function *(125,126)*. Like competitive AMPA receptor antagonists, these compounds posses a broad therapeutic potential in the treatment of conditions associated with excessive glutamate release, such as epilepsy, stroke, traumatic brain injury, and certain neurodegenerative disorders. *(92,127)*. As noncompetitive antagonists, 2,3-benzodiazepines have the theoretical advantage that blockade would not be overcome by high synaptic glutamate levels *(128)*. Interactions of 2,3-benzodiazepine derivatives with AMPA receptors are stereoselective, with activity residing in the minus isomers (e.g., LY 300164 and LY 303070), whereas the respective (+) isomers (e.g., LY 300165 and LY 303071) are devoid of AMPA receptor antagonistic properties *(129)*. LY 300164 (GYKI 53773, talampanel) is orally available and has undergone clinical evaluation. It exhibited preliminary evidence of efficacy in epilepsy trials (phase II) and good tolerability, although high plasma concentrations were associated with ataxia, diplopia, sedation, and euphoria *(92,93)*.

Recently, a variety of analogs of the original 2,3-benzodiazepines have been described that exhibit AMPA receptor-blocking activity and the possibility of reduced side effects. for example, several novel heterocyclic condensed and halogen-substituted 2,3-benzodiazepine derivatives such as GYKI 47261 have been shown to have good potency and broad anticonvulsant and neuroprotective activity comparable to GYKI 52466 and LY 300164 *(130)*. In addition, a class of substituted 6,7-methylenedioxy 1,2-dihydrophtalazines (SYM 2206, SYM 2207) that are closely related structurally to the 2,3-benzodiazepines also act as noncompetitive AMPA receptor antagonists with anticonvulsant and neuroprotective activity *(131,132)*. Several 7-deoxygenated analogs have recently been reported to have similar AMPA-receptor-blocking activity, but one of these compounds, the 2-*N*-propylcarbamoyl SYM 2189, is said to have a higher protective index and longer duration of action than the other members of this class, although its overall in vivo potency is low in comparison with 2,3-benzodiazepines

*(132)*. In contrast, a SYM 2207 analog having a ketone substituent [4-(4-aminophenyl)-6,7-methylene-dioxy-phthalazin-1(2H)-one] has 10-fold greater anticonvulsant potency than GYKI 52466, but a reduced toxicity profile and a longer duration of action *(133)*. In in vitro studies, this analog is less potent than GYKI 52466 and GYKI 53655, suggesting that greater bioavailability accounts for its improved anticonvulsant potency.

Derivatives of methaqualone have also recently been shown to have very potent AMPA-receptor antagonist activity, although the selectivity and biological characterization of these compounds remains to be reported *(134)*.

### 3.1.3. Positive Allosteric Modulators

Compounds that decrease the rate of desensitization or deactivation of AMPA receptors may enhance excitation mediated by the receptor and have novel therapeutic uses (e.g., in the treatment of cognitive disorders) *(see refs. 91 and 135)*. Two main classes of positive modulators of the AMPA receptor can be distinguished: pyrrolidones (aniracetam, piracetam, and CX516) and benzothiadiazides (cyclothiazide, diazoxide, S18986, and IDRA-21). Many of these AMPA receptor modulators including IDRA-21 *(136,137)* and CX516 *(138,139)* have been shown to improve performance of experimental animals in memory tasks. In both healthy volunteers *(140)* and elderly subjects, CX516 had significant beneficial effects on memory *(144)* without significant central or peripheral side effects. Positive allosteric modulators also have predicted therapeutic potential in schizophrenia based on the observation that schizophrenic patients have a reduced brain density of AMPA receptors in postmortem evaluation (see the review in ref. *142*). Synergistic effects of CX516 and antipsychotic drugs were observed against methamphetamine-induced behavioral effects in rats, suggesting that this compound may be a useful adjuvant in treating schizophrenia *(143)*; clinical trials are currently underway. However, given the potential for excitotoxicity or seizures with enhancement of glutamatergic neurotransmission, the use of these agents will need to be approached cautiously.

Thiocyanate ion is a negative allosteric modulator of AMPA receptors that acts by enhancing desensitization of the receptor *(144)*. Recently, thiocyanate has been found to interact with the same site as the positive modulator cyclothiazide, and it can thus be considered an "inverse agonist" at this site *(145)*. It should be noted that thiocyanate and cyclothiazide do not directly interact with the negative modulatory site for 2,3-benzodiazepines. Enhancement of AMPA-receptor desensitization is an intriguing potential therapeutic strategy that may permit inhibitory modulation of excessive glutamatergic neurotransmission with reduced side effects *(92)*. Thiocyanate could provide clues to the development of such inhibitory modulators.

## 3.2. Kainate Receptors

Kainate receptors are ionotropic glutamate receptors that, like NMDA and AMPA receptors, are expressed widely in the central nervous system where they may play a role in fast excitatory transmission *(146,147)*. Five kainate receptor subunits are known that have approx 40% sequence homology with AMPA receptor subunits GluR1–4. Three of these subunits (GluR5–7) are believed to be principal subunits—they can form functional receptors by themselves—and two (KA1–2) may be auxiliary subunits that form heteromeric assemblies with the principal subunits and modify their behavior, but do not form functional receptors by themselves.

Until recently, the functional roles of kainate receptors was poorly understood because of the lack of specific pharmacological tools. However, the availability of several more or less selective kainate receptor agonists and antagonists—most notably a series of decahydroisoquinolines that specifically block kainate receptors containing the GluR5 subunit—has begun to allow insight into the roles of kainate receptors in neurotransmission. In addition to their well-known AMPA receptor-blocking activity, quinoxalinediones also interact with kainate receptors with lower affinity; CNQX and DNQX are slightly more potent against AMPA than kainate receptors, whereas NBQX may be up to 30-fold more potent against AMPA receptors *(94)*. The first selective kainate receptor ligand to be described was the oxime derivative NS-102 (6,7,8,9-tetrahydro-5-nitro-1*H*-benz[*g*]indole-2,3-dione-3-oxime) *(148,149)*. This compound interacts with both GluR5 and GluR6 receptors and also with AMPA receptors with approximately 10-fold lower affinity. However, poor solubility considerably limits the utility of this antagonist. The decahydroisoquinolines, which act as competitive antagonists of the glutamate recognition site on kainate receptors, have proven to be very useful in both in vitro and in vivo studies of kainate-receptor function. These agents preferentially interact with GluR5 kainate receptors and do not have measurable activity at GluR6 kainate receptors. The decahydroisoquinoline LY293558 was originally described as an AMPA-receptor antagonist and only later recognized to block GluR5 kainate receptors with similar affinity. Such mixed AMPA- and GluR5-selective antagonists can be used to uncover the function of GluR5 kainate receptors when studied in conjunction with selective AMPA-receptor antagonists such as 2,3-benzodiazepines that only block kainate receptors at high concentrations *(150)*. More recently, decahydroisoquinolines, including LY377770 and LY382884, have become available that preferentially interact with GluR5 kainate receptors and bind to AMPA receptors only at much higher concentrations *(151,152)*.

Using the decahydroisoquinolines, it has been possible to demonstrate that GluR5 kainate receptors participate in excitatory synaptic transmission in certain brain regions including the hippocampus *(153)* and amygdala *(88)*. This class of kainate receptors also appears to play a prominent role in regulating inhibitory GABA-mediated neurotransmission in the hippocampus *(89,154)*. In addition, kainate receptors probably contribute to excitatory transmission in the cerebellum *(155)* and spinal cord *(156)*. Indeed, recent pharmacological evidence supports a role for GluR5 kainate receptors in nociceptive responses *(157)* although this has been questioned on the basis of studies in which the GluR5 expression was reduced through gene targeting *(158)*. From the perspective of drug abuse, the existence of kainate receptors in the amygdala is of particular significance, given the role of this brain region in fear, anxiety, and memory. Importantly, GluR5 kainate receptors seem to mediate a novel form of heterosynaptic plasticity that could potentially be a mechanism underlying long-term adaptations to drugs of abuse *(159)*.

## 4. METABOTROPIC RECEPTORS

Metabotropic glutamate receptors (mGluRs) are seven transmembrane domain G-protein-coupled receptors that are activated by glutamate and that regulate neuronal excitability by a multiplicity of postsynaptic and presynaptic mechanisms. Eight distinct mGluRs have been identified by molecular cloning, some of which exist in multiple alternatively spliced forms. The family of mGluRs can be divided into two major groups based on the second-messenger systems to which they are coupled: those that increase phosphatidylinositol (PI) hydrolysis through activation of phospholipase C (mGluR1, mGluR5) and those that are negatively coupled to cyclic AMP formation via inhibition of adenyl cyclase (mGluR2–4, mGluR6–8). On the basis of sequence homology, the second-messenger mechanisms to which they are coupled, and pharmacological criteria, it has been useful to further divide mGluRs into three major groups. Group I corresponds to the mGluRs linked to PI turnover, whereas groups II (mGluR2, mGluR3) and III (mGluR4, mGluR6-8) encompass mGluRs linked to adenyl cyclase inhibition. In functional terms, the changes in second messengers induced by activation of metabotropic glutamate receptors can modulate the activity of $Ca^{2+}$ and $K^+$ channels and also directly affect the neurotransmitter release machinery, leading to changes in neuronal excitability and in the release of neurotransmitters at both glutamatergic and nonglutamatergic synapses. Generally, activation of group II mGluRs leads to presynaptic depression and consequent dampening of stimulus-evoked glutamate release, whereas activation of group III mGluRs inhibits the release of both GABA and glutamate. Of particular relevance to drug abuse, mGluRs can also modulate the release of dopamine, which has been most directly implicated in the rewarding aspects of many

**Table 6**
**Some Selective Agonists and Antagonists of Metabotropic Glutamate Receptors**

| Compound | Clinical status |
|---|---|
| Group I | |
|   Agonists | |
|     S(–)-3,5-Dihydrophenylglycine (DHPG) | |
|     Quisqualate | |
|     LY 367366 | |
|     LY 393675 | |
|     SIB-1893 | |
|   Antagonists | |
|     LY 367385 | |
|     MPEP | |
|     CPCCOEt | |
| Group II | |
|   Agonists | |
|     LY 354740 | In development—anxiety |
|     LY 389795 | |
|     LY 379268 | |
|     DCG-IV | |
|     2R,4R-APDC | |
|   Antagonists | |
|     LY 341495 | |
|     ADBD (LY 307452) | |
|     ADED (LY 310225) | |
| Group III | |
|   Agonists | |
|     L-AP4 | |
|     L-SOP | |
|     S-Homo-AMPA | |
|   Antagonist | |
|     CPPG | |

abused substances (cf. ref. *160*). Detailed reviews of the biochemistry, physiology, and pharmacology of the burgeoning field of mGluRs are available *(160–166)*.

Although rapid progress has been made in the discovery of selective ligands for the eight mGluRs, the pharmacology of mGluRs is not fully developed (*see* Table 6). Although there are some compounds that have varying degrees of selectivity for each of the three groups of mGluRs, selectivity for any one of the eight members of the mGluR family has not yet been achieved. For example, the potent and selective group II agonist LY 354740 has micromolar affinity for two of the group III receptor subtypes. An additional complication arises from differences in efficacy. For example, benzyl-APDC is a selective agonist at mGlu6 receptors but demonstrates antagonist actions at mGlu2, mGlu3, and mGlu5 receptors. Furthermore, although new compounds with putative high selectivities have been reported, such as the stereoselective activation of mGluR4a and mGluR7b by (+)-4-phosphonophenylglycine, the overall selectivity profile and functional activities of these compounds remains to be characterized *(167)*.

A variety of recent studies have examined the potential therapeutic efficacy of group I and group II mGluR ligands in the treatment of drug abuse. In contrast, the availability of selective compounds for the investigation of group III mGluRs is more limited, so less is known about the potential utilities of ligands for these receptors. For example, the group II agonist LY 354740 blocks PCP-induced stereotypy, hyperlocomotion, and disruptions in memory without effects on spontaneous locomotion when given alone *(168)*. Indeed, it appears that group II agonists may preferentially attenuate the motoric effects of PCP in comparison with the locomotor stimulation induced by *d*-amphetamine (cf. ref. *169*). LY 35470 also attenuated the signs of nicotine and morphine withdrawal in dependent animals (cf. refs. *170* and *171*) and can interfere with the development of tolerance to the analgesic effects of morphine *(172)*. The group I antagonists MPEP and SIB 1893, which display selectivity for mGluR5, are effective anticonvulsants with favorable therapeutic indices *(173)*; the role of these compounds in drug abuse has yet to be examined. In addition to a role for mGluR ligands in drug-abuse treatment, there is some evidence for alterations in the sensitivity of groups II and III mGluRs following chronic exposure to drugs of abuse like cocaine that have suggested their involvement in cocaine addiction and withdrawal *(174)*.

## 5. GLUTAMATE RELEASE

Inhibition of glutamate release is another means of functionally dampening excessive glutamate-mediated excitation in pathological states. Agents that block glutamate release—mainly anticonvulsants such as phenytoin, lamotrigine, and zonisamide—are generally believed to do so by interfering with the activity of $Na^+$ or $Ca^{2+}$ channels in nerve terminals *(4,175,176)*, although there is recent evidence for more complex effects on the release machinery *(177,178)*. Interestingly, these drugs may preferentially inhibit glutamate release as opposed to GABA release *(179)*. In addition, agents that interact with adenosine receptors and other presynaptic receptors, including mGluRs, can reduce glutamate release. In the case of adenosine, there is also a preferential effect on excitatory versus inhibitory synapses. Several glutamate release inhibitors are in various phases of development as shown in Table 7.

## 6. NITRIC OXIDE SYNTHASE

Glutamate-gating of NMDA receptors permits $Ca^{2+}$ entry that specifically activates neuronal nitric oxide synthase (nNOS), a $Ca^{2+}$-dependent enzyme that catalyzes the synthesis of the free-radical nitric oxide (NO) from L-arginine *(180)* (Fig. 1). Because NO may act as an intercellular messenger mediating forms of neuronal plasticity, blockade of nNOS could potentially interfere with the long-term consequences of NMDA receptor activation and might thereby have an influence on the behavioral effects of drugs of abuse or on their addictive properties. In fact, the NOS inhibitors *N(G)*-nitro-L-arginine-methyl ester (L-NAME) and 7-nitroindazole have been shown to block some of the behavioral effects of drugs of abuse, such as stereotypy induced by methamphetamine *(181)*, sensitization to the behavioral stimulant effects of cocaine *(182)*, and maintenance of cocaine self-administration in rats *(183)*. L-NAME also blocks sensitization to the convulsant effects of cocaine in mice, an effect that can be prevented by the NOS substrate L-arginine but not by D-arginine *(184)*. Moreover, the specific nNOS inhibitor AR-R 17477 blocked behavioral effects of PCP in models of psychosis without behavioral side effects of its own *(185)*. Nor do these compounds appear to produce PCP-like subjective effects as predicted from PCP discrimination experiments *(186)*. NOS exists in three forms with distinct cellular localizations and functions. Like nNOS, endothelial NOS (eNOS) is a constitutive and $Ca^{2+}$-dependent NOS enzyme that participates in regulation of smooth-muscle tone. A third form of NOS is the $Ca^{2+}$-independent inducible NOS (iNOS) that is activated by cytokines and plays a role in immune functions. In addition to the key roles of eNOS and iNOS in peripheral tissues, both of these NOS forms are also present along with nNOS in the brain. Therefore, indiscriminate inhibition of NOS with nonselective inhibitors such as L-NAME can lead to untoward side effects. It is notable, therefore, that AR-R 17477 was without cardiovascular effects, a major side effect of less selective NOS inhibitors.

Table 7
Some Glutamate-Release Inhibitors

| Compound | Clinical status |
|---|---|
| Phenytoin | Marketed—epilepsy |
| Fosphenytion (phenytoin pro-drug) | Marketed—epilepsy |
| | In development—stroke |
| Lamotrigine | Marketed—epilepsy |
| Zonisamide | Marketed—epilepsy |
| BW-619C89 | Prior development—stroke |
| MS-153 (MS-424) | In development—ischemia |
| Riluzole | Prior development—epilepsy |
| | Marketed—amyotrophic lateral sclerosis |

## 7. NAALADASE

Glutamate carboxypeptidase II (*N*-acetylated α-linked-acidic dipeptidase; EC 3.4.17.21) or NAAL-ADase is a neuropeptidase that catalyzes the cleavage of glutamate from *N*-acetyl-aspartyl-glutamate (NAAG), the most prevalent and widely distributed neuropeptide in the mammalian nervous system *(187)*. Thus, inhibition of NAALADase would decrease glutamate levels and increase levels of NAAG. Theoretically then, NAALADase inhibition would have the potential to negatively modulate glutamatergic neurotransmission by two separate mechanisms. First, NAALADase inhibition would produce decreases in glutamate levels. Second, as NAAG is known to act at mGlu3 receptors *(188)*, NAALADase inhibition could have the additional effect characteristic of mGluR3 agonists of dampening glutamate-mediated neurotransmission.

Until recently, pharmacological tools for selectively inhibiting NAALADase were not available. However, this situation changed with the design and synthesis of 2-(phosphonomethyl) pentanedioic acid (2-PMPA) *(189)*, which is a high-affinity ($K_i$ = 280 p*M*) and selective NAALADase inhibitor *(190)*. In addition to its potential role in the therapy of conditions associated with excessive glutamate, recent reports suggest a role for NAALADase inhibition in modulating the behavioral and toxic effects of drugs of abuse. Thus, 2-PMPA blocks the development of behavioral sensitization resulting from repeated exposure to cocaine in rats without affecting acute stimulant effects of cocaine *(191)*. 2-PMPA also attenuates both the expression and development of cocaine-kindled seizures in mice without producing the motor side effects seen with NMDA receptor antagonists *(192)*.

## 8. OTHER APPROACHES

In the future, it may be possible to modulate brain glutamate systems by treatments that are designed to modify the expression of critical genes involved in glutamate signaling. Indeed, drugs of abuse produce prominent changes in gene transcription that may alter the expression of components of the glutamate-signaling pathways and these changes have been hypothesized to play a role in the development of drug dependence *(193)*. Furthermore, drugs used to treat drug dependence may also operate through similar pathways. For example, antidepressants, widely used to treat drug dependence *(3)*, appear to have the common effect of increasing the levels of growth factors such as brain-derived neurotrophic factor (BDNF), as pointed out by Skolnick *(194)*. Depression is comorbid with drug-dependence syndromes *(195)* and NMDA-receptor antagonists are effective in animal models of depression *(194)*, as they are in models of drug addiction. Therefore, novel approaches to the treatment of drug addiction may be based on the design of compounds that promote BDNF formation as, for example, through the cyclic AMP-dependent response pathway. Increased levels of BDNF could then ultimately result in a dampending of NMDA receptor function through the regulation of NMDA receptor genes (cf. ref. *195*).

## 9. CONCLUSIONS

Glutamate plays a key role in drug dependence and associated behavioral and toxic effects of drugs of abuse. A range of experimental drugs is available to explore further the involvement of glutamate in drug abuse and for the discovery of potential therapeutic candidates. Although these compounds produce a host of common actions and effects, a difference in effects among compounds that appears to derive from the specific molecular targets upon which they act provide a rich diversity for investigatory and development purposes. Both preclinical and clinical findings point to the feasibility of effectively dampening glutamatergic neurotransmission and blocking pathophysiological conditions arising from excessive glutamatergic transmission without major interference with normal neurobehavioral functioning.

## REFERENCES

1. Vanderschuren, L. J. and Kalivas, P. W. (2000) Alterations in dopaminergic and glutamatergic transmission in the induction and expression of behavioral sensitization: a critical review of preclinical studies. *Psychopharmacology* **151,** 99–120.
2. Parsons, C. G., Danysz, W., and Quack, G. (1998) Glutamate in CNS disorders as a target for drug development: An update. *Drug News Perspect.* **11,** 523–569.
3. Witkin, J. M. (1995) Role of *N*-methyl-D-aspartate receptors in behavior and behavioral effects of drugs, in *CNS Neurotransmitters and Neuromodulators, Volume 1: Glutamate* (Stone, T. W., ed.), CRC P, Boca Raton, FL., pp. 323–350.
4. Rogawski, M. A. (1996) Epilepsy, in *Neurotherapeutics: Emerging Strategies* (Pullan, L. and Patel, J., eds.) Humana P, Totowa, NJ, pp. 193–273.
5. Bräuner-Osborne, H., Egebjerg, J., Nielsen, E. Ø., Madsen, U., and Krogsgaard-Larsen, P. (2000). Ligands for glutamate receptors: design and therapeutic prospects. *J. Med. Chem.* **43,** 2609–2645.
6. MacDonald, J. F., Bartlett, M. C., Mody, I., Pahapill, P., Reynolds, J. N., Salter, M. W., et al. (1991) Actions of ketamine, phencyclidine and MK-801 on NMDA receptor currents in cultured mouse hippocampal neurones. *J. Physiol.* **432,** 483–508.
7. Dingledine, R., Borges, K., Bowie, D., and Traynelis, S. F. (1999) The glutamate receptor ion channels. *Pharmacol. Rev.* **51,** 7–61.
8. Rogawski, M. A. (2000) Low affinity channel blocking (uncompetitive) NMDA receptor antagonists as therapeutic agents—toward an understanding of their favorable tolerability. *Amino Acids* **19,** 133–149.
9. Maurice, T., Vignon, J., Kamenka, J. M., and Chicheportiche, R. D. (1991) Differential interaction of phencyclidine-like drugs with the dopamine uptake complex in vivo. *J. Neurochem.* **56,** 553–559.
10. Rothman, R. B. (1994) PCP site 2: a high affinity MK-801-insensitive phencyclidine binding site. *Neurotoxicol. Teratol.* **16,** 343–353.
11. ffrench-Mullen, J. M. H. and Rogawski, M. A. (1989) Interaction of phencyclidine with voltage-dependent potassium channels in cultured rat hippocampal neurons: comparison with block of the NMDA receptor–ionophore complex. *J. Neurosci.* **9,** 4051–4061.
12. Wong, E. H. G., Knight, A. R., and Woodruff, G. N. (1988) [$^3$H]MK-801 labels a site on the *N*-methyl-D-aspartate receptor complex in rat brain membranes. *J. Neurochem.* **50,** 274–281.
13. Witkin, J. M., Gasior, M., Heifets, B., and Tortella, F. C. (1999) Anticonvulsant efficacy of *N*-methyl-D-aspartate antagonists against convulsions induced by cocaine. *J. Pharmacol. Exp. Ther.* **289,** 703–711.
14. Ginski, M. and Witkin, J. M. (1994) Sensitive and rapid behavioral differentiation of *N*-methyl-D-aspartate receptor antagonists. *Psychopharmacology* **114,** 573–582.
15. Tricklebank, M. D., Singh, L., Oles, R. J., Preston, C., and Iversen, S. D. (1989) The behavioural effects of MK-801: a comparison with antagonists acting non-competitively and competitively at the NMDA receptor. *Eur. J. Pharmacol.* **167,** 127–135.
16. Koek, W. and Colpaert, F. C. (1990) Selective blockade of *N*-methyl-D-aspartate (NMDA)-induced convulsions by the NMDA antagonists and putative glycine antagonists: relationships with phencyclidine-like behavioral effects. *J. Pharmacol. Exp. Ther.* **252,** 349–357.
17. Balster R. L. and Willetts, J. (1988) Receptor mediation of the discriminative stimulus properties of phencyclidine and sigma-opioid agonists, in *Psychopharmacology: Transduction Mechanisms of Drug Stimuli,* (Colpaert, F. C. and Balster, R. L., eds.), Springer-Verlag Berlin, pp. 122–135.
18. Grant, K. A., Colombo, G., Grant, J., and Rogawski, M. A. (1996) Dizocilpine-like discriminative stimulus effects of low-affinity uncompetitive NMDA antagonists. *Neuropharmacology* **35,** 1709–1719.

19. Geter-Douglass, B. and Witkin, J. M. (1997) Dizocilpine-like discriminative stimulus effects of competitive NMDA receptor antagonists in mice. *Psychopharmacology* **133**, 43–50.
20. Troupin, A. S., Mendius, J. R., Cheng, F., and Risinger, M. W. (1986) MK-801, in *New Anticonvulsant Drugs* (Meldrum, B. and Porter, R. J., eds.), Libey, London, pp. 191–201.
21. Krystal, J. H., Karper, L. P., Seibyl, J. P., Freeman, G. K., Delaney, R., Bremner, J. D. et al. (1994) Subanesthetic effects of the noncompetitive NMDA antagonist, ketamine, in humans. Psychotomimetic, perceptual, cognitive, and neuroendocrine responses. *Arch. Gen. Psychiatry* **51**, 199–214.
22. Olney, J. W., Labruyere, J., Wang, G., Wozniak, D. F., Price, M. T., and Sesma, M. A. (1991) NMDA antagonist neurotoxicity: mechanism and prevention. *Science* **254**, 1515–1518.
23. Fix, A. S., Horn, J. W., Wightman, K. A., Johnson, C. A., Long, G. G., Storts, R. W., et al. (1993) Neuronal vacuolization and necrosis induced by the noncompetitive *N*-methyl-D-aspartate (NMDA) antagonist MK(+)801 (dizocilpine maleate): a light and electron microscopic evaluation of the rat retrosplenial cortex. *Exp. Neurol.* **123**, 204–215.
24. Palmer, G.C. and Wiszowski, D. (2000) Low affinity use-dependent NMDA receptor antagonists show promise for clinical development. *Amino Acids* **19**, 151–155.
25. Rogawski, M. A. (1993) Therapeutic potential of excitatory amino acid antagonists:channel blockers and 2,3-benzodiazepines. *Trends Pharmacol. Sci.* **14**, 325–331.
26. Parsons, C. G., Quack, G., Bresink, I., Baran, L., Przegalinski, E., Kostowski, W., et al, (1995) Comparison of the potency, kinetics and voltage-dependency of a series of uncompetitive NMDA receptor antagonists in vitro with anticonvulsive and motor impairment activity in vivo. *Neuropharmacology* **34**, 1239–1258.
27. Bubser, M., Zadow, B., Kronthaler, U. O., Felsheim, U., Rückert, G. H., and Schmidt, W. J. (1997) Behavioural pharmacology of the non-competitive NMDA antagonists dextrorphan and ADCI: relations between locomotor stimulation, anticataleptic potential and forebrain dopamine metabolism. *Naunyn-Schmiedeberg's Arch. Pharmacol.* **355**, 767–773.
28. Grant, K. A., Snell, L. D., Rogawski, M. A., Thurkauf, A., and Tabakoff, B. (1992) Comparison of the effects of the uncompetitive *N*-methyl-D-aspartate antagonist (±)-5-aminocarbonyl-10,11-dihydro-5*H*-dibenzo[*a,d*]cyclohepten-5,10-imine (ADCI) with its structural analogs dizocilpine (MK-801) and carbamazepine on ethanol withdrawal seizures. *J. Pharmacol. Exp. Ther.* **269**, 1017–1022.
29. Geter-Douglass, B. and Witkin, J. M. (1999) Behavioral effects and anticonvulsant efficacies of low affinity, uncompetitive NMDA antagonists in mice. *Psychopharmacology* **146**, 280–289.
30. Sanger, D. J., Terry, P., and Katz, J. L. (1992) Memantine has phencyclidine-like but not cocaine-like discriminative stimulus effects in rats. *Behav. Pharmacol.* **3**, 265–268.
31. Danysz, W., Parsons, C. G., Kornhuber, J., Schmidt, W. J., and Quack, G. (1997) Aminoadamantanes as NMDA receptor antagonists and antiparkinsonian agents-preclinical studies. *Neurosci-Biobehav. Rev.* **21**, 455–468.
32. Ditzler, K. (1991) Efficacy and tolerability of memantine in patients with dementia syndrome *Arzneimettelforschung/Drug Res.* **41**, 773–780.
33. Riederer, P., Lange, K. W., Kornhuber, J., and Danielczyk, W. (1991) Pharmacotoxic psychosis after memantine in Parkinson's disease. *Lancet* **338**, 1022–1023.
34. Rabey, J. M., Nissipeanu, P., and Korczyn, A. D. (1992) Efficacy of memantine, an NMDA receptor antagonist, in the treatment of Parkinson's disease. *J. Neural Transm.* **4**, 277–282.
35. Mueller, A. L., Artman, L. D., Balandrin, M. F., Brady E., Chien Y., Delmar, E. G., et al, (1999) NPS 1506, a novel NMDA receptor antagonist and neuroprotectant. Review of preclinical and clinical studies. *Ann. NY Acad. Sci.* **890**, 450–457.
36. Porter, P. H., and Greenamyre, J. T., (1995) Regional variations in the pharmacology of NMDA receptor channel blockers: implications for therapeutic potential. *J. Neurochem.* **64**, 614–623.
37. Bresink, I., Danysz, W., Parsons, C. G., and Mutschler, E. (1995) Different binding affinities of NMDA receptor channel blockers in various brain regions—indication of NMDA receptor heterogeneity. *Neuropharmacology* **34**, 533–540.
38. Baumann, M. H., Rothman, R. B., and Ali, S. F. (2000) Comparative neurobiological effects of ibogaine and MK-801 in rats. *Drug Alcohol Depend.* **59**, 143–151.
39. Rho, J. M., Donevan, S. D., and Rogawski, M. A. (1994) Mechanism of action of the anticonvulsant felbamate: opposing effects on NMDA and GABA$_A$ receptors. *Ann. Neurol.* **35**, 229–234.
40. Popik, P., Layer, R. T., and Skolnick, P. (1995) 100 years of ibogaine:neurochemical and pharmacological actions of a putative anti-addictive drug. *Pharmacol. Rev.* **47**, 235–253.
41. Mash, D. C., Kovera, C. A., Buck, B. E., Norenberg, M. D., Shapshak, P., Hearn, W. L., et al. (1998) Medication development of ibogaine as a pharmacotherapy for drug dependence. *Ann. NY Acad. Sci.* **844**, 274–292.
42. Popik., P., Layer, R. T., Fossom, L.H., Benveniste, M., Geter-Dopuglass, B., Witkin, J. M. et al, (1995) NMDA antagonist properties of the putative anti-addictive drug, ibogaine. *J. Pharmacol. Exp. Ther.* **275**, 753–760.
43. Chen, K., Kokate, T. G., Donevan, S. D., Carroll, F. I., and Rogawski, M. A. (1996) Ibogaine block of the NMDA receptor: in vitro and in vivo studies. *Neuropharmacology* **35**, 423–431.
44. Newman, A. H. (2000) Novel pharmacotherapies for cocaine abuse 1997–2000. *Exp. Opin. Ther. Patents* **10**, 1095–1122.
45. Willetts, J., Balster, R. L., and Leander, J. D. (1990) The behavioral pharmacology of NMDA receptor antagonists. *Trends Pharmacol. Sci.* **11**, 423–428.

46. Kornhuber, J. and Weller, M. (1997) Psychotogenicity and N-methyl-D-aspartate receptor antagonism: implications for neuroprotective pharmacotherapy. *Biol. Psychiatry* **41**, 135–144.
47. Sveinbjornsdottir, S., Sander, J. W. A. S., Upton, D., Thompson, P. J., Patsalos, P. N., Hirt, D., et al. (1993) The excitatory amino acid antagonist D-CPP-ene (SDZ EAA-494) in patients with epilepsy. *Epilepsy Res.* **16**, 165–174.
48. Bullock, R. (1995) Strategies for neuroprotection with glutamate antagonists. Extrapolating from evidence taken from the first stroke and head injury studies. *Ann. NY Acad. Sci.* **765**, 272–278.
49. Carter, A. (1994) Many agents that antagonize the NMDA receptor–channel complex in vivo also cause disturbances of motor coordination. *J. Pharmacol. Exp. Ther.* **269**, 573–580.
50. Löscher, W. and Hönack, D. (1991a) Anticonvulsant and behavioral effects of two novel competitive N-methyl-D-aspartic acid receptor antagonists, CGP 37849 and CGP 39551, in the kindling model of epilepsy. Comparison with MK-801 and carbamazepine. *J. Pharmacol. Exp. Ther.* **256**, 432–440.
51. Löscher, W. and Hönack, D. (1991b) The novel competitive N-methyl-D-aspartate (NMDA) antagonist CGP 37849 preferentially induces phencylidine-like behavioral effects in kindled rats: attenuation by manipulation of dopamine, alpha-1, and serotonin$_{1A}$ receptors. *J. Pharmacol. Exp. Ther.* **257**, 1146–1153.
52. Smith, S. E. and Chapman, A. G. (1993) Acute and chronic anticonvulsant effects of D(–)CPPene in genetically epilepsy-prone rats. *Epilepsy Res.* **15**, 193–199.
53. Willetts, J., Clissold, D. B., Hartman, T. L., Brandsgaard, R. R., Hamilton, G. S., and Ferkany, J. W. (1993) Behavioral pharmacology of NPC 17742, a competitive N-methyl-D-aspartate (NMDA) antagonist. *J. Pharmacol. Exp. Ther.* **265**, 1055–1062.
54. Johnson, J. W. and Ascher, P. (1987) Glycine potentiates the NMDA response in cultured mouse brain neurons. *Nature* **325**, 529–531.
55. Kleckner, N. W. and Dindeldine, R. (1988) Requirement for glycine in activation of NMDA-receptors expressed in *Xenopus oocytes*. *Science* **241**, 444–449.
56. Mothet, J-P., Parent, A. T., Wolosker, H., Brady, R. O., Jr., Linden, D. J., Ferris, C. D., et al. (2000) D-Serine is an endogenous ligand for the glycine site of the N-Mehtyl-D-aspartate receptor. *Proc. Natl.Acad. Sci. USA* **97**, 4926–4931.
57. Viu, E., Zapata, A., Capdevila, J. et al. (2000) Glycine(B) receptor antagonists and partial agonists prevent memory deficits in inhibitory avoidance learning. *Neurobiol. Learn. Mem.* **74**, 146–160.
58. Balster, R. L., Mansbach, R. S., Shelton, K. L., Nicholson, K. L., Grech, D. M., Wiley, J. L., et al. (1995) Behavioural pharmacology of two novel substituted quinoxalinedione glutamate antagonists. *Behav. Pharmacol.* **6**, 577–590.
59. Witkin, J. M., Brave, S., French, D., and Geter-Douglass, B. (1995) Discriminative stimulus effects of R-(+)-3-amino-1-hydroxypyrrolid-2-one, [(+)-HA-966], a partial agonist of the strychnine-insensitive modulatory site of the N-methyl-D-aspartate receptor. *J. Pharmacol. Exp. Ther.* **275**, 1267–1273.
60. Witkin, J. M., Steele T. D., and Sharpe, L. G. (1997) Effects of strychnine-insensitive glycine receptor ligands in rats discriminating either dizocilpine or phencyclidine from saline. *J. Pharmacol. Exp. Ther.* **280**, 46–52.
61. Kotlinska, J. and Bialwa, G. (2000) Memantine and ACPC affect conditioned place preference induced by cocaine in rats. *Pol. J. Pharmacol.* **52**, 179–185.
62. Honer, M., Benke, D., Laube, B., Kuhse, J., Heckendorn, R., Allgeier, H., et al. (1998) Differentiation of glycine antagonist sites of N-methyl-D-aspartate receptor subtypes. *J. Biol. Chem.* **273**, 11,158–11,163.
63. Carling, R. W., Leeson, P. D., Moore, K. W., Smith, J. D., Moyes, C. R., Mawer, I. M., et al. (1993) 3-Nitro-3,4-dihydro-2(1H)-quinolones. Excitatory amino acid antagonists acting at glycine-site NMDA and (RS)-α-amino-3-hydroxy-5-methyl-4-isoxazolepropionic acid receptors. *J. Med. Chem.* **36**, 3397–3408.
64. Kloog, Y., Haring, R., and Sokolovsky, M. (1988) Kinetic characterization of the phencylidine-N-methyl-D-aspartate receptor interaction: evidence for a steric blockade of the channel. *Biochemistry* **27**, 843–848.
65. Ascher, P. and Nowak, L. (1987) Electrophysiological studies of NMDA receptors. *Trends Neurosci.* **10**, 284–287.
66. Salt, T. E. (1989) Modulation of NMDA receptor-mediated responses by glycine and D-serine in the rat thalamus *in vivo*. *Brain Res.* **481**, 403–406.
67. Thomson, A. M., Walker, V. E., and Flynn, D. M. (1989) Glycine enhances NMDA-receptor medicated synaptic potentials in neocortical slices. *Nature* **338**, 422–424.
68. Toth, E. and Lajtha, A. (1986) Antagonism of phencyclidine-induced hyperactivity by glycine in mice. *Neurochem. Res.* **11**, 393–400.
69. Evoniuk, G. E., Hertzman, R. P., and Skolknick, P. (1991) A rapid method for evaluating the behavioral effects of phencyclidine-like dissociative anesthetics in mice. *Psychopharmacology* **105**, 125–128.
70. Contreras, P. (1990) D-Serine antagonized phencyclidine and MK-801-induced stereotyped behavior and ataxia. *Neuropharmacology* **29**, 291–293.
71. Tanii, Y., Nishikawa, T., Hashimoto, A., and Takahaski, K. (1991) Stereoselective inhibition by D- and L-alanine of phencyclidine-induced locomotor stimulation in the rat. *Brain Res.* **563**, 281–284.
72. Tanii, Y., Nishikawa, T., Hashimoto, A., and Takahaski, K. (1994) Stereoselective antagonism by enantiomers of alanine and serine of phencyclidine-induced hyperactivity, stereotypy and ataxia. *J. Pharmacol. Exp. Ther.* **269**, 1040–1048.
73. McBain, C. J., and Mayer, M. L. (1994) N-Methyl-D-aspartic acid receptor structure and function. *Physiol. Rev.* **74**, 723–760.

74. Avenet, P., Léonardon, J., Besnard, F., Graham, D., Depoortere, H., and Scatton, B. (1997) Anagonist properties of eliprodil and other NMDA receptor antagonists at rat NR1A/NR2A and NR1A/NR2B receptors expressed in *Xenopus oocytes*. *Neurosci. Lett.* **223,** 133–136.
75. Brimecombe, J. C., Boeckman, F. A., and Aizenman, E. (1997) Functional consequences of NR2 subunit composition in single recombinant *N*-methyl-D-aspartate receptors. *Proc. Natl. Acad. Sci. USA* **94,** 11109–11024.
76. Fischer, G., Mutel, V., Trube, G., Malherbe, P., Kew, J. N. C., Mohacsi, E., et al. (1997) Ro 25-6981, a highly potent and selective blocker of *N*-methyl-D-aspartate receptors containing the NR2B subunit. Characterization *in vitro*. *J. Pharmacol. Exp. Ther.* **283,** 1285–1292.
77. Tamiz, A. P., Whittemore, E. R., Zhou, Z.-L., Huang, J.-C, Drewe, J. A., Chen, J.-C., et al. (1998) Structure–activity relationships for a series of bis(phenylalkyl) amines: potent subtype-selective inhibitors of *N*-methyl-D-aspartate receptors. *J. Med. Chem.* **41,** 3499–3506.
78. Mott, D., Doherty, J., Zhang, S., Washburn, M., Fendley, M., Lyuboslavsky, P., et al. (1998) Enhancement of proton inhibition: a novel mechanism for NMDA receptor inhibition by phenylethenolamines. *Nature Neurosci.* **1,** 659–667.
79. Scatton B., Avenet, P., Benavides, J., Carter, C., Duverger, D., Oblin, A., et al. (1994) Neuroprotective potential of the polyamine site-directed NMDA receptor antagonists—ifenprodil and eliprodil, in *Direct and Allosteric Control of Glutamate Receptors* (Palfreyman, M. G., Reynolds, I. J., and Skolnick, P., eds.), CRC, Boca Raton Fl, pp. 139–154.
80. Duval, D., Roome, N., Gauffeny, C., Nowicki, J. P., and Scatton, B. (1992) SL 82.0715, an NMDA antagonist acting at the polyamine site, does not induce neurotoxic effects on rat cortical neurons. *Neurosci. Lett.* **137,** 193–197.
81. Deren-Wesolek, A. and Maj, J. (1993) Central effects of SL 82.0715, an antagonist of polyamine site of the NMDA receptor complex. *Pol. J. Pharmacol.* **45,** 467–480.
82. Harty, T. P. and Rogawski, M. A. (1997) Channel block of NMDA receptors by the anticonvulsant ADCI: studies with cloned NR1a/NR2A and NR1a/NR2B subunits. *Soc. Neurosci. Abst.* **23,** 2165.
83. Harty, T. P. and Rogawski, M. A. (2000) Felbamate block of recombinant *N*-methyl-D-aspartate receptors: selectivity for the NR2B subunit. *Epilepsy Res.* **29,** 47–55.
84. Donevan, S. D. and McCabe, R. T. (2000) Conantokin G is an NR2B-selective competitive antagonist of *N*-methyl-D-aspartate receptors. *Mol. Pharmacol.* **58,** 614–623.
85. Gallagher, M. J., Huang, H., and Lynch, D. R. (1998) Modulation of the *N*-methyl-D-aspartate receptor by haloperidol: NR2B-specific interactions. *J. Neurochem.* **70,** 2120–2128.
86. Yamakura, T., Sakimura, K., Mishina, M., and Shimoji, K. (1998) Sensitivity of the *N*-methyl-D-aspartate receptor channel to butyrophenones is dependent on the ε2 subunit. *Neuropharmacology* **37,** 709–717.
87. Mori, H. and Mishina, M. (1995) Structure and function of the NMDA receptor channel. *Neuropharmacology* **34,** 1219–1237.
88. Li, H. and Rogawski, M. A. (1998) GluR5 kainate receptor mediated synaptic transmission in rat basolateral amygdala *in vitro*. *Neuropharmacology* **37,** 1279–1286.
89. Lerma, J. (1998) Kainate receptors: an interplay between excitatory and inhibitory synapses. *FEBS Lett.* **430,** 100–104.
90. Rodriguez-Moreno, A., Lopez-Garcia, J. C., and Lerma, J. (2000) Two populations of kainate receptors with separate signaling mechanisms in hippocampal interneurons. *Proc. Natl. Acad. Sci.* **97,** 1293–1298.
91. Bleakman, D. and Lodge, D. (1998) Neuropharmacology of AMPA and kainate receptors. *Neuropharmacology* **37,** 1187–1204.
92. Rogawski, M. A. and Donevan, S. D. (1999) AMPA receptors in epilepsy and as targets for antiepileptic drugs, in *Jasper's Basic Mechanisms of the Epilepsies:* 3rd ed. Advances in Neurologic Vol. 79 (Delgado-Escueta, A. V. Wilson, W. A., Olsen, R. W., and Porter, R. J., eds.), Lippincott Williams & Wilkins, Philadelphia, pp 947–963.
93. Lees, G. J. (2000) Pharmacology of AMPA/kainate receptor ligands and their therapeutic potential in neurological and psychiatric disorders. *Drugs* **59,** 33–78.
94. Honoré, T. (1991) Inhibitors of kainate and AMPA ionophore receptors, in: *Excitatory Amino Acid Antagonists* (Meldrum, B. S., ed.), Blackwell, Oxford, pp. 180–194.
95. Watjen, F., Nielsen, E. O., Drejer, J., and Jensen, L. H. (1993) Isatin oximes—a novel series of bioavailable non-NMDA antagonists. *Bioorg. Med. Chem. Lett.* **3,** 105–106.
96. Nijholt, I., Blank T., Grafelmann, B., Cepok, S., Kugler, H., and Spiess, J. (1999) NS-257, a novel competitive AMPA receptor antagonist, interacts with kainate and NMDA receptors. *Brain Res.* **821,** 374–382.
97. Witkin, J. M. (1993) Blockade of the locomotor stimulant effects of cocaine and methamphetamine by glutamate antagonists. *Life Sci.* **53,** PL405–PL410.
98. Li, Y., Hu, X. T., Berney, T. G., Vartanian, A. J., Stine, C. D., Wolf, M. E., et al. (1999) Both glutamate receptor antagonists and prefrontal cortex lesions prevent induction of cocaine sensitization and associated neuroadaptations. *Synapse* **34,** 169–180.
99. Cornish, J. L. and Kalivas, P. W. (2000) Glutamate neurotransmission in the nucleus accumbens mediates relapse in cocaine addiction. *J. Neurosci.* **20,** 1–5.
100. Ingwersen, S. H., Öhrström, J. K., Petersen, P. Drustrup, J., Bruno, L., and Nordholm, L. (1994) Human pharmacokinetics of the neuroprotective agent NBQX. *Am. J. Ther.* **1,** 296–303.

101. Ohmori, J., Sakamoto, S., Kubota, H., Shimizu-Sasamata, M., Okada, M., Kawasaki, S., et al. (1994) 6-(1H-Imidazol-1-yl)-7-nitro-2,3(1H,4H)-quinoxalinedione hydrochloride (YM90K) and related compounds:structure–activity relationships for the AMPA-type non-NMDA receptor. *J. Med. Chem.* **37**, 467–475.
102. Umemura, K., Kondo, K., Ikeda, Y, Teraya, Y., Yoshida, H., Homma. M., et al. (1997) Pharmacokinetics and safety of the novel amino-3-hydroxy-5-methylisoxazole-4-propionate receptor antagonist YM90K in healthy men. *J. Clin. Pharmacol.* **37**, 719–727.
103. Ohmori,J., Shimizu-Sasamata, M., Okada, M., and Sakamoto S. (1997) 8(1H-Imidazol-1-yl)-7-nitro-4(5H)-imidazo[1,2-alpha]quinoxalinone and related compounds; synthesis and structure activity relationships for the AMPA-type non-NMDA receptor. *J. Med Chem.* **40**, 2053–2063
104. Ohmori, J., Shimizu-Sasamata, M., Okada, M., and Sakamoto S. (1996) Novel AMPA receptor antagonists: synthesis and structure-activity relationships of 1-hydroxy-7-(1H-imidazol-1-yl)-6-nitro-2,3(1H,4H)-quinoxalinedione and related compounds. *J. Med. Chem.* **39**, 3971–3979.
105. Kohara, A., Okada, M., Tsutsumi, R., Ohno, K., Takahashi, M., Shimizu-Sasamata, M., et al. (1998) In-vitro characterization of YM872, a selective, potent and highly water-soluble alpha-amino-3-hydroxy-5-methylisoxazole-4-propionate receptor antagonist. *J. Pharm. Pharmacol.* **50**, 795–801.
106. Takahashi, M., Ni, J. W., Kawasaki-Yatsugi, S., Toya, T., Yatsugi, S. I., Shimizu-Sasamata, M., et al. (1998) YM872, a novel selective alpha-amino-3-hydroxy-5-methylisoxazole-4-propionic acid receptor antagonist, reduces brain damage after permanent focal cerebral ischemia in cats. *J. Pharmacol. Exp. Ther.* **284**, 467–473.
107. Kawasaki-Yatsugi, S., Ichiki, C., Yatsugi, S., Takahashi, M., Shimizu-Sasamata, M., Yamaguchi, T., et al. (2000) Neuroprotective effects of an AMPA receptor antagonist YM872 in a rat transient middle cerebral artery occlusion model. *Neuropharmacology* **39**, 211–217.
108. Nishiyama, T., Gyermek, L., Lee, C., Kawasaki-Yatsugi, S., and Yamaguchi, T. (1999) The systemically administered competitive AMPA receptor antagonist, YM872, has analgesic effects on thermal or formalin-induced pain in rats. *Adnaesth. Analg.* **89**, 1534–1537.
109. Schoepp, D. D., Lodge D., Bleakman, D., Leander, J. D., Tizzano, J. P., Wright, R. A., et al. (1995) In vitro and in vivo antagonism of AMPA receptor activation by (3S4aR6R,8aR)-6-[2-(1(2)H-tetrazole-5-yl)ethyl]-decahydroisoquinoline-3-carboxylic acid. *Neuropharmacology* **34**, 1159–1168.
110. Bleakman, R., Schoepp, D. D., Ballyk, B., Bufton, H., Sharpe, E. F., Thomas, K., et al. (1996) Pharmacological discrimination of GluR5 and GluR6 kainate receptor subtypes by (3S,4aR,6R,8aR)-6-[2-(1(2)H-tetrazole-5-yl)ethyl]decahydroisdoquinoline-3 carboxylic-acid. *Mol. Pharmacol.* **49**, 581–585.
111. Rogawski, M. A., Kurzman, P., Yamaguchi, S., and Li, H. (2000) Role of AMPA and GluR5 kainate receptors in the development and expression of amygdala kindling in the mouse. *Neuropharmacology,* **40**, 28–35.
112. Carlezon, W. A., Jr., Rasmussen, K., and Nestler, E. J. (1999) AMPA antagonist LY293558 blocks the development, without blocking the expression, of behavioral sensitization to morphine. *Synapse* **31**, 256–262.
113. Kest, B., McLemore, G., Kao, B., and Inturrisi, C. E. (1997) The competitive alpha-amino-3-hydroxy-5-methylisoxazole-4-propionate receptor antagonist LY293558 attenuates and reverses analgesic tolerance to morphine but not to delta or kappa opioids. *J. Pharmacol. Exp. Ther.* **283**, 1249–1255.
114. Rasmussen, K., Kendrick, W. T., Kogan J. H., and Aghajanian, G. K. (1996) A selective AMPA antagonist, LY293558, suppresses morphine withdrawal-induced activation of locus coeruleus neurons and behavioral signs of morphine withdrawal. *Neuropsychopharmacology* **15**, 497–505.
115. Sang, C. N., Hostetter, M. P., Gracely, R. H., Chappell, A. S., Schoepp, D. D., Lee, G., et al. (1998) AMPA/kainate antagonist LY293558 reduces capsaicin-evoked hyperalgesia but not pain in normal skin in humans. *Anesthesiology* **89**, 1060–1067.
116. Mutel, V., Trube, G., Klingelschmidt, A., Messer, J., Bleuel , Z., Humbel U., et al. (1998) Binding characteristics of a potent AMPA receptor antagonist [$^3$H]Ro 48-8587 in rat brain. *J. Neurochem.* **71**, 418–426.
117. Löscher, W., Lehmann, H., Behl, B., Seemann, D., Teschendorf, H. J., Hofmann, H. P., et al. (1999) A new pyrrolyl-quinoxalinedione series of non-NMDA glutamate receptor antagonists: pharmacological characterization and comparison with NBQX and valproate in the kindling model of epilepsy. *Eur. J. Neurosci.* **11**, 250–262.
118. Turski, L., Huth, A., Sheardown, M., McDonald, F., Neuhaus, R., Schneider, H. H., et al. (1998) ZK200775: a phosphonate quinoxalinedione AMPA antagonist for neuroprotection in stroke and trauma. *Proc. Natl. Acad. Sci. USA* **95**, 10,960–10,965.
119. Löscher, W., Rundfeldt, C., and Hönack, D. (1993) Low doses of NMDA receptor antagonists synergistically increase the anticonvulsant effect of the AMPA receptor antagonist NBQX in the kindling model of epilepsy. *Eur. J. Neurosci.* **5**, 1545–1550.
120. Löscher, W. (1998) New visions in the pharmacology of anticonvulsion. *Eur. J. Pharmacol.* **342**, 1–13.
121. Potschka, H., Löscher, W., Wlaz, P., Behl, B., Hofmann, H. P., Treiber, H. J., et al (1998) LU73068, a new non-NMDA and glycine/NMDA receptor antagonist: pharmacological characterization and comparison with NBQX and L-701,324 in the kindling model of epilepsy. *Br. J. Pharmacol.* **125**, 1258–1266.
122. Tarnawa, I., Farkas, S., Berzsenyi, P., Pataki, A., and Andrasi, F. (1989) Electrophysiological studies with a 2,3-benzodiazepine muscle relaxant: GYKI 52466. *Eur. J. Pharmacol.* **167**, 193–199.

123. Bleakman, D., Ballyk, B. A., Schoepp, D. D., Palmer, A. J., Bath, C. P., Sharpe, E. F., et al. (1996) Activity of 2,3-benzodiazepines at native rat and recombinant human glutamate receptors in vitro: stereospecificity and selectivity profiles. *Neuropharmacology* **35,** 1689–1702.
124. Donevan, S. D., Yamaguchi, S., and Rogawski, M. A. (1994) Non-*N*-methyl-D-aspartate receptor antagonism by 3-*N*-substituted 2,3-benzodiazepines: relationship to anticonvulsant activity. *J. Pharmacol. Exp. Ther.* **271,** 25–29.
125. Donevan, S. D. and Rogawski, M. A. (1993) GYKI 52466, a 2,3-benzodiazepine, is a highly selective, noncompetitive antagonist of AMPA/kainate receptor responses. *Neuron* **10,** 51–59.
126. Zorumski, C. F., Yamada, K. A., Price, M. T., and Olney, J. W. (1993) A benzodiazepine recognition site associated with the non-NMDA glutamate receptor. *Neuron* **10:** 61–67.
127. Tarnawa, I. and Vizi, E. S. (1998) 2,3-Benzodiazepine AMPA antagonists. *Restor. Neurol. Neurosurg.* **13,** 41–57.
128. Yamaguchi, S., Donevan, S. D., and Rogawski, M. A. (1993) Anticonvulsant activity of AMPA/kainate antagonists: comparison of GYKI 52466 and NBQX in maximal electroshock and chemoconvulsant models, *Epilepsy Res.* **15,** 179–184.
129. Lodge, D., Bond, A., O'Neill, M. J., Hicks, C. A., and Jones, M. G. (1996) Stereoselective effects of 2,3-benzodiazepines in vivo: electrophysiology and neuroprotection studies. *Neuropharmacology* **35,** 1681–1688.
130. Abraham, G., Solyom, S., Csuzdi, E., Berzsenyi, P., Ling, I., Tarnawa, I., et al. (2000) New non competitive AMPA antagonists. *Bioorg. Med. Chem.* **8,** 2127–2143.
131. Pelletier, J. C., Hesson, D. P., Jones, K. A., and Costa, A. M. (1996) Substituted 1,2-dihydrophthalazines: potent, selective, and noncompetitive inhibitors of the AMPA receptor. *J. Med. Chem.* **39,** 343–346
132. Pei, X.-F., Sturgess, M. A., Valenzuela, C. F., and Maccecchini, M.-L. (1999) Allosteric modulators of the AMPA receptor:novel 6-substituted dihydrophthalazines. *Bioorg. Med. Chem. Lett.* **9,** 539–542.
133. Grasso, S., De Sarro, G., De Sarro, A., Micale, N., Zappalà, M., Puja, G., et al (2000) Synthesis and anticonvulsant activity of novel and potent 6,7-methylenedioxyphthalazin-1(2*H*)-ones. *J. Med. Chem.* **43,** 2851–2859.
134. Chenard, B. L., Menniti, F. S., Pagnozzi, M. J., Shenk, K. d., Ewing, F. E., and Welch, W. M. (2000) Methaqualone derivatives are potent noncompetitive AMPA receptor antagonists. *Bioorg. Med. Chem. Lett.* **10,** 1203–1205.
135. Yamada, K. A. (1998) Modulating excitatory synaptic neurotransmission: potential treatment for neurological disease? *Neurobiol. Dis.* **5,** 67–80.
136. Zivkovic, I., Thompson, D. M., Bertolino, M., Uzunov, D., DiBella, M., Costa, E., et al. (1957) 7-Chloro-3-methyl-3-4-dihydro-2*H*-1,2,4 benzothiadiazine *S,S*-dioxide (IDRA 21): a benzothiadiazine derivative that enhances cognition by attenuating DL-alpha-amino-2,3-dihydro-5-methyl-3-oxo-4-isoxazolepropanoic acid (AMPA) receptor desensitization. *J. Pharmacol. Exp. Ther.* **272,** 300–309.
137. Thompson, D. M., Guidotti, A., DiBella, M., and Costa, E. (1995) 7-Chloro-3-methyl-3,4-dihydro-2*H*-1-2,4-benzothiadiazine *S,S*-dioxide (IDRA 21), a congener of aniracetam, potently abates pharmacologically induced cognitive impairments in patas monkeys. *Proc. Natl. Acad. Sci. USA* **92,** 7667–7671.
138. Hampson, R. E., Rogers, G., Lynch, G., and Deadwyler, S. A. (1998) Facilitative effects of the ampakine CX516 on short-term memory in rats: enhancement of delayed-nonmatch-to-sample performance. *J. Neurosci.* **18,** 2740–2747.
139. Hampson, R. E., Rogers, G., Lynch, G., and Deadwyler, S. A. (1998b) Facilitative effects of the ampakine CX516 on short-term memory in rats: correlations with hippocampal neuronal activity. *J. Neurosci.* **18,** 2748–2763.
140. Ingvar, M., Ambros-Ingerson, J., Davis, M., Granger, R., Kessler, M., Rogers, G. A., et al. (1997) Enhancement by an ampakine of memory encoding in humans. *Exp. Neurol.* **146,** 553–559.
141. Lynch, G., Granger, R., Ambros-Ingerson, J., Davis, C. M., Kessler, M., and Schehr, R. (1997) Evidence that a positive modulator of AMPA-type glutamate receptors improves delayed recall in aged humans. *Exp. Neurol.* **145,** 89–92.
142. Meador-Woodruff, J. H. and Healy, D. J. (2000) Glutamate receptor expression in schizophrenic brain. *Brain Res. Brain. Res. Rev.* **31,** 288–294.
143. Johnson, S. A., Luu, N. T., Herbst, T. A., Knapp, R., Lutz, D., Arai, A., et al. (1999) Synergistic interactions between ampakines and antipsychotic drugs. *J. Pharmacol. Exp. Ther.* **289,** 392–397.
144. Bowie, D. and Smart, T. G. (1993) Thiocyanate ions selectively antagonize AMPA-evoked responses in *Xenopus laevis* oocytes injected with chick brain mRNA. *Neurosci. Lett.* **121,** 68–72.
145. Donevan, S. D. and Rogawski, M. A. (1998) Allosteric regulation of AMPA receptors SCN$^-$ and cyclothiazide at a common modulatory site distinct from that of 2,3-benzodiazepines. *Neuroscience* **87,** 615–629.
146. Chittajallu, R., Braithwaite, S. P., Clarke, V. R., and Henley, J. M. (1999) Kainate receptors: subunits, synaptic localization and function. *Trends Pharmacol. Sci.* **20,** 26–35.
147. Frerking, M. and Nicoll, R. A. (2000) Synaptic kainate receptors. *Curr. Opin. Neurobiol.* **10,** 342–351.
148. Johansen, T. H., Drejer, J., Watjen, F., and Nielsen, E. O. (1993) A novel non-NMDA receptor antagonist shows selective displacement of low-affinity [$^3$H]kainate binding. *Eur. J. Pharmacol.* **246,** 195–204.
149. Verdoorn, T. A., Johansen, T. H., Drejer, J., and Nielsen, E. O. (1994) Selective block of recombinant glur6 receptors by NS-102, a novel non-NMDA, receptor antagonist. *Eur. J. Pharmacol.* **269,** 43–49.
150. Paternain, A. V., Morales, M., and Lerma, J. (1995) Selective antagonism of AMPA receptors unmasks kainate receptor-mediated responses in hippocampal neurons. *Neuron* **14,** 185–189.

151. O'Neill, M. J., Bogaert, L., Hicks, C. A., Bond, A., Ward, M. A., Ebinger. G., et al. (2000) LY377770, a novel iGlu5 Kainate receptor antagonist with neuroprotective effects in global and focal cerebral ischaemia. *Neuropharmacology* **39**, 1575–1588.
152. O'Neill, M. J., Bond, A., Ornstein, P. L., Ward, M. A., Hicks, C. A., Hoo, K., et al. (1998) Decahydroisoquinolines:novel competitive AMPA/Kainate antagonists with neuroprotective effects in global cerebral ischaemia. *Neuropharmacology* **37**, 1211–1222.
153. Vignes, M. and Collingridge, G. L. (1997) The synaptic activation of kainate receptors. *Nature* **388**, 179–182.
154. Clarke, V. R. J., Ballyk, B. A., Hoo, K. H., Mandelzys, A., Pellizzari, A., Bath, C. P., et al (1997) A hippocampal GluR5 kainate receptor regulating inhibitory synaptic transmission. *Nature* **389**, 599–603.
155. Pemberton, K. E., Belcher, S. M., Ripellino, J. A., and Howe, J. R. (1998) High-affinity kainate-type ion channels in rat cerebellar granule cells. *J. Physiol.* **510**, 401–420.
156. Li, P., Wilding, T., Kim, S. J., Calejesan, A. A., Huettner, J. E., and Zhuo, M. (1999) Kainate receptor-mediated sensory synaptic transmission in mammalian spinal cord. *Nature* **397**, 161–164.
157. Simmons R. M. A., Li, D. L., Hoo, K. H., Deverill, M., Omstein, P. L., and Iyengar, S. (1998) Kainate GluR5 receptor subtype mediates the nociceptive response to formalin in the rat. *Neuropharmacology* **37**, 25–36.
158. Sailer, A., Swanson, G. T., Perez-Otano, I, O'Leary, L., Malkmus, S. A., Dyck, R. H., et al. (1999) Generation and analysis of GluR5(Q636R) kainate receptor mutant mice. *J. Neurosci.* **19**, 8757–8764.
159. Li, H. and Rogawski, M. A. (1999) Kainate receptor mediated heterosynaptic facilitation in the amygdala, *Soc. Neurosci. Abst.* **25**, 974.
160. Cartmell, J. and Schoepp, D. D. (2000) Regulation of neurotransmitter release by metabotropic glutamate receptors. *J. Neurochem.* **75**, 889–907.
161. Conn, P. J. and Pin, J.-P. (1997) Pharmacology and functions of metabotropic glutamate receptors. *Annu Rev. Pharmacol. Toxicol.* **37**, 205–237.
162. Anwyl, R. (1999) Metabotropic glutamate receptors: electrophysiological properties and role in placticity. *Brain Res. Rev.* **29**, 83–120.
163. Bordi, F. and Ugolini, A. (1999) Group I metabotropic glutamate receptors: implications for brain diseases. *Prog. Neurobiol.* **59**, 55–79.
164. Schoepp, D. D., Jane, D. E., and Monn, J. A. (1999) Pharmacological agents acting at subtypes of metabotropic glutamate receptors, *Neuropharmacology* **38**, 1431–1476.
165. Monn, J. A. and Schoepp, D. D. (2000) Metabotropic glutamate receptor modulators: recent advances and therapeutic potential. *Annu. Rep. Med. Chem.* **35**, 1–10.
166. Varney, M. A. and Suto, C. M. (2000) Discovery of subtype-selective metabotropic glutamate receptor ligands using functional HTS assays. *Drug Discov. Today* **1(HTS Suppl.)**, 20–26.
167. Gasparini, F., Inderbitzin, W., Francotte, E., Lecis, G., Richert, P., Dragic, Z., et al. (2000) (+)-4-Phosphonophenylglycine (PPG) a new group III selective metabotropic glutamate receptor agonist. *Bioorg. Med. Chem. Lett.* **10**, 1241–1244.
168. Moghaddam, B. and Adams, B. W. (1998) Reversal of phencyclidine effects by a group II metabotropic glutamate receptor agonist in rats. *Science* **281**, 1349–1352.
169. Cartmell, J., Monn, J. A., and Schoepp, D. D. (1999) The metabotropic glutamate 2/3 receptor agonists LY354740 and LY379268 selectively attenuate phencyclidine versus *d*-amphetamine motor behaviors in rats. *J. Pharmacol. Exp. Ther.* **291**, 161–170.
170. Helton, D. R., Tizzano, J. P., Monn, J. A., Schoepp, D. D., and Kallman, M. J. (1997) LY354740: a potent, orally active, highly selective metabotropic glutamate receptor agonist which ameliorates symptoms of nicotine withdrawal. *Neuropharmacology* **36**, 1511–1516.
171. Vandergriff, B. C. and Rasmussen, K. (1999) The selective mGlu2/3 receptor agonist LY354740 attenuates morphine-withdrawal-induced activation of locus coerulus neurons and behavioral signs of morphine withdrawal. *Neuropharmacology* **38**, 217–222.
172. Popik, P., Kozela, E., and Pilc, A. (2000) Selective agonist of group II glutamate metabotropic receptors, LY354740, inhibits tolerance to analgesic effects of morphine in mice. *Br. J. Pharmacol.* **130**, 1425–1431.
173. Chapman, A. G., Nanan, K., Williams, M., and Meldrum, B. S. (2000) Anticonvulsant activity of two metabotropic glutamate Group I antagonists selective for mGlu5 receptor: 2-methyl-6-(phenylethynyl)-pyridine (MPEP), and *(E)*-6-methyl-2-styryl-pyridine (SIB 1893). *Neuropharmacology* **39**, 1567–1574.
174. Neugebauer, V., Zinebi, F., Russell, R., Gallagher, J. P., and Shinnick-Gallagher, P. (2000) Cocaine and kindling alter the sensitivity of group II and III metabotropic glutamate receptors in the central amygdala. *J. Neurophysiol.* **84**, 759–770.
175. Stefani, A., Spadoni, F., and Benardi, G. (1997) Differential inhibition by riluzole, lamotrigine and phenytoin of sodium and calcium currents in cortical neurones:implicaitons for neuroprotective strategies. *Exp. Neurol.* **147**, 115–122.
176. Zhu, W. and Rogawski, M. A. (1999) Zonisamide depresses excitatory synaptic transmission by a presynaptic action. *Epilepsia* **40(Suppl. 7)**, 244.
177. Cunningham, M. O. and Jones, R. S. G. (2000) The anticonvulsant, lamotrigine decreases spontaneous glutamate release but increases spontaneous GABA release in the rat entorhinal cortex in vitro. *Neuropharmacology* **39**, 2139–2146.

178. Cunningham, M. O., Dhillon, A., Wood, S. J., and Jones, R. S. G. (2000) Reciprocal modulation of glutamate and GABA release may underlie the anticonvulsant effect of phenytoin. *Neuroscinece* **95,** 343–351.
179. Waldmeier, P. C., Baumann, P. A., Wicki, P., Feldtrauer, J.-J., Stierlin, C., and Schmutz, M. (1995) Similar potency of carbamazepine, oxcarbazepine, and lamotrigine in inhibiting the release of glutamate and other neurotransmitters. *Neurology* **45,** 1907–1913.
180. Snyder, S. H. (1992) Nitric Oxide, first in a new class of neurotransmitters. *Science* **257,** 494–496.
181. Semba, J., Wantanabe, H., Suhara, T., and Akanuma, N. (2000) Neonatal treatment with L-Name (*NG*-nitro-L-arginine methylester) attenuates stereotyped behavior induced by acute methamphetamine but not development of behavioral sensitization to methamphetamine. *Prog. Neuro-Pyschopharmacol. Biol. Psychiatry* **24,** 1017–1023.
182. Haracz, J. L., MacDonall, J. S., and Sircar, R. (1997) Effects of nitric oxide synthase inhibitors on cocaine sensitization. *Brain Res.* **746,** 183–189.
183. Pulvirenti, L., Balducci, C., and Koob, G. F. (1996) Inhibition of nitric oxide synthesis reduces intravenous cocaine self-administration in the rat. *Neuropharmacology* **35,** 1811–1814.
184. Itzhak, Y. (1995) Cocaine kindling in mice. Responses to *N*-methyl-D,L-aspartate (NMDLA) and L-arginine. *Mol. Neurobiol.* **11,** 217–222.
185. Johansson, C., Deveney, A. M., Reif, D., and Jackson, D. M. (1999) The neuronal selective nitric oxide inhibitor AR-R 17477, blocks some effects of phencyclidine, while having no observable behavioural effects when given alone. *Pharmacol. Toxicol.* **84,** 226–233.
186. Wiley, J. L., Harvey, S. A., and Balster, R. L. (1999) Nitric oxide synthase inhibitors do not substitute in rats trained to discriminate phencyclidine from saline. *Eur. J. Pharmacol.* **367,** 7–11.
187. Neale, J. H., Bzdega, T., and Wroblewska, B. (2000) *N*-Acetylaspartylglutamate: the most abundant peptide neurotransmitter in the mammalian central nervous system. *J. Neurochem.* **75,** 443–452.
188. Wroblewska, B., Wroblewski, J. T., Pshenichkin, S., Surin, A., Sullivan, S. E., and Neale, J. H. (1997) NAAG selectively activates mGluR3 receptors in transfected cells. *J. Neurochem.* **69,** 174–181.
189. Jackson, P. F., Cole, D. C., Slusher, B. S., Stetz, S. L., Ross, L. E., Donazanti, B. A., et al. (1996) Design, synthesis, and biological activity of a potent inhibitor of the neuropeptidase *N*-acetylated α-linked acidic dipeptidase. *J. Med. Chem.* **39,** 619–622.
190. Slusher, B. S, Vornov, J. J., Thomas, A. G., Hurn, P. D., Harukuni, I., Bhardwaj, A., et al. (1999) Selective inhibition of NAALADase, which converts NAAG to glutamate, reduces ischemic brain injury. *Nature Med.* **5,** 1396–1402.
191. Shippenberg, T. S., Rea, W., and Slusher, B. S. (2000) Modulation of behavioral sensitization to cocaine by NAALADase inhibition. *Synapse,* **38,** 161–166.
192. Witkin, J. M., Gasior, M., Zapata, A., Slusher, B. S., and Shippenberg, T. S. (2000) Inhibition of NAALAdase: a novel strategy for preventing sensitization to convulsant stimuli. *FASEB J.* **14,** A1414.
193. Nestler, E. J., Berhow, M. T., and Brodkin, E. S. (1996) Molecular mechanisms of drug addiction: adaptations in signal transduction pathways. *Mol. Psychiatry* **1,** 190–199.
194. Skolnick, P. (1999) Antidepressants for the new millennium. *Eur. J. Pharmacol.* **375,** 31–40.
195. Brandolin, C., Sanna, A., De Bernardi, M. A., Follesa, P., Brooker, G., and Mocchetti, I. (1998) Brain-derived neurotrophic factor and basic fibroblast growth factor downregulate NMDA receptor function in cerebellar granule cells. *J. Neurosci.* **18,** 7953–7961.
195. Regier, D. A., Farmer, M. E., Rae, D. S., Locke, B. Z., Keith , S. K., Judd, L. L., et al. (1990) Comorbidity of mental disorders with alcohol and other drug abuse. *J. Am. Med. Assoc.* **264,** 2511–2518.

# 3
# Glutamate and Neurotoxicity

## B. Joy Snider, MD, PhD and Dennis W. Choi, MD, PhD

## 1. HISTORICAL PERSPECTIVE

The toxic effects of glutamate exposure on neurons were first recognized nearly half a century ago, when Lucas and Newhouse observed that subcutaneous administration of glutamate caused loss of neurons in the inner nuclear layer of the retina in both adult and neonatal mice (1). Olney extended these findings to other regions of brain, including neurons in the roof of the third ventricle, the hypothalamus, and the dentate gyrus (2). Changes evolved rapidly, over minutes in adult mice to several hours in neonates, and were characterized by intracellular edema and pyknotic nuclei, consistent with necrosis. In the next few years the role of glutamate as the major excitatory neurotransmitter in the mammalian central nervous system (CNS) became clear (3–6) and the existence of specific glutamate receptors was demonstrated. Excitotoxicity, the effect of glutamate receptor activation to trigger neuronal cell death, was proposed to play a role in many pathological conditions, in large part based on the observations that injection of glutamate agonists, notably kainate, could result in neuronal death and biochemical abnormalities resembling the pathology seen in disorders such as Huntington's disease (7,8) and epilepsy (9,10). A role for endogenous glutamate release and subsequent glutamate receptor activation in triggering neuronal death under pathological conditions was further suggested by demonstrations that blockade of presynaptic glutamate release could attenuate neuronal injury in oxygen-deprived cultured hippocampal neurons (11) and that a blockade of the N-methyl-D-aspartate (NMDA) subtype of glutamate receptors attenuated neuronal injury in rodent models of global ischemia and hypoglycemic brain damage (12,13). Cell culture models were useful in exploring the ionic changes responsible for glutamate-mediated cell death (see Section 2). More recent observations suggest that receptor-mediated glutamate toxicity may not be limited to neurons, but may also affect oligodendrocytes (14–16). A non-receptor-mediated form of glutamate cytotoxicity due to cystine deprivation and lowering of intracellular glutathione has also been described (17,18), although the levels of sustained exposure required to induce this death are higher than expected in most in vivo situations.

The original description of glutamate-mediated neuronal death in retina and in brain fits with a morphological picture of necrosis, but glutamate receptor activation can also trigger cell death with features of apoptosis. More recent findings suggest that, in some circumstances, glutamate receptor-stimulated $Ca^{2+}$ entry could promote neuronal survival rather than neuronal death and might specifically attenuate some forms of neuronal apoptosis (see Section 2).

## 2. GLUTAMATE NEUROTOXICITY IN VITRO

In vitro systems have made it possible to dissect out some of the ionic alterations and signaling pathways involved in glutamate-induced neuronal death. Three subtypes of ionotropic glutamate receptors have been characterized and the subunits comprising these receptors have been cloned

(*19–21*; see also Chapters 1 and 2). *N*-Methyl-D-aspartate receptors and a subset of α-amino-3-hydroxy-5-methyl-4-isoxazole propionic acid (AMPA) receptors are permeable to $Ca^{2+}$ as well as $Na^+$, whereas the majority of AMPA receptors and kainate receptors are permeable to $Na^+$ but not $Ca^{2+}$. Metabotropic glutamate receptors (mGluRs) are linked to G-proteins rather than ion channels (reviewed in refs.*22* and *23*); activation of these receptors does not appear to mediate excitotoxicity, but it can modulate it in complex ways *(24–27)*.

Activation of ionotropic glutamate receptors causes an initial $Na^+$ influx (with accompanying $Cl^-$ and water influx) and induces cell body swelling during glutamate overexposure (reviewed in ref. *28*; see also ref. *29*). $Ca^{2+}$ influx through NMDA receptors and through the $Ca^{2+}$-permeable subset of AMPA receptors *(30)*, likely augmented by secondary $Ca^{2+}$ influx through voltage-gated $Ca^{2+}$ channels and reverse operation of neuronal $Na^+$–$Ca^{2+}$ exchangers (resulting from membrane depolarization and elevated intracellular $Na^+$) *(31)*, is the predominant factor in the neurodegeneration that occurs over subsequent hours. NMDA-induced neuronal death correlates well with the amount of calcium influx, as measured by the uptake of radiolabeled calcium *(32)*, or with intracellular free levels ($[Ca^{2+}]_i$) measured with low-affinity indicator dyes *(33)*. A similar correlation between $[Ca^{2+}]_i$ and glutamate agonist-induced cell death has been observed in non-neuronal cells transfected with NMDA receptors *(34–35)*.

The precise downstream mechanisms linking intracellular $Ca^{2+}$ overload to cell death are still not entirely clear, but mitochondria probably play an important role. Mitochondrial calcium uptake following glutamate exposure may result in the uncoupling of electron transport from ATP synthesis, with resultant increased production of mitochondrial reactive oxygen species and derangements of energy metabolism *(36–40)*. Indeed, free-radical scavengers attenuate glutamate neurotoxicity in vitro *(41,42)*.

The $Ca^{2+}$ influx triggered by glutamate receptor activation can also directly activate catabolic enzymes: calcium-dependent proteases, phospholipases, and endonucleases. For example, calpain is activated following glutamate receptor activation, and inhibition of calpain attenuates glutamate agonist-induced death in vitro and also reduces neuronal death in transient global ischemia *(43–45)*.

The cellular swelling and calcium overload triggered by the glutamate receptor over activation typically results in neuronal death with features consistent with necrosis, including early cell body swelling and loss of plasma membrane integrity, organelle disruption, and insensitivity to inhibitors of protein synthesis or caspase activity (e.g., refs. *46–49*). Some features of apoptosis, such as positive TUNEL (terminal transferase-mediated dUTP-digoxigenin nick end labeling) staining, internucleosomal DNA fragmentation (DNA laddering), and nuclear and chromatin condensation have been reported in cultured cerebellar granule cells and neocortical neurons exposed to glutamate agonists *(50–52)*. TUNEL staining and DNA ladders have also been observed in neurons dying by excitotoxic necrosis *(49,53)*, and in isolation, they do not support an important role for apoptosis in excitotoxic neuronal death. Under certain circumstances, pharmacological or genetic inhibitors of apoptosis reduce glutamate-mediated neuronal death, but many such studies have used relatively immature neurons. For example, protein and RNA synthesis inhibitors attenuate NMDA-induced neuronal death in cultured retinal ganglion cells and immature neocortical neurons (maintained in culture for 3–5 d) *(54,55)*. The caspase inhibitor Z-VAD.FMK attenuates NMDA-induced death in more mature (15 d in culture) rat neocortical cultures exposed to NMDA in the absence of $Mg^{2+}$ *(56)*, but does not attenuate NMDA-induced neuronal death in murine neocortical cultures *(48)*. Deletion of the *bax* gene attenuates glutamate and kainate-induced death in neocortical neurons grown in culture for 4 d *(57)*. At least some of the glutamate-induced neuronal death in such immature neurons could be secondary to cystine depletion rather than receptor-mediated excitotoxicity, an idea supported by the observation that *bax* gene deletion did not alter the vulnerability of more mature (14 d in culture) murine neocortical neurons to NMDA-mediated excitotoxicity (Gottron and Choi, unpublished observation). Taken together, available observations support a model in which excitotoxic glutamate receptor overactivation favors necrosis, but can also lead to apoptosis under certain circumstances. In particular, factors such as milder insult intensity (see ref. *54*), cell immaturity (associated with fewer glutamate receptors and intrinsically higher propensity to undergo apoptosis) *(58)*, low intracellular $Ca^{2+}$, and low trophic factor availability may favor apoptosis after any insult, including excitotoxic insults. Mixed forms of death may be particularly prominent in vivo, where an initial excitotoxic insult may be followed by loss of trophic factor or surface factor support resulting from damage to inputs or surrounding cells.

$Ca^{2+}$ may not be the only divalent cation important to excitotoxicity, as toxic neuronal uptake of $Zn^{2+}$ released from glutamatergic presynaptic nerve terminals has been suggested to interact importantly with excitotoxicity (for review, see refs. *59–61*). A chelatable pool of zinc is located in synaptic vesicles within glutamatergic nerve terminals throughout the telencephalon *(see ref 59)*. $Zn^{2+}$ directly inhibits the NMDA receptor *(62–64)* and modulates multiple other receptors and channels, the former including GABA, glycine, and purine receptors (reviewed in ref. *65*). It may also induce a delayed upregulation of NMDA receptor function and NMDA receptor-mediated excitotoxicity *(66)*. In addition to its neuromodulatory role, $Zn^{2+}$ can enter postsynaptic neurons in toxic quantities via routes facilitated by AMPA/kainate and NMDA receptor activation *(67)*, and thereby contribute to neuronal death after transient global ischemia *(68,69)* and seizures *(70,71)*.

A central role for $K^+$ efflux in promoting apoptosis has been increasingly suspected, and in neurons, this efflux may be mediated both by the delayed rectifier $I_k$ as well as glutamate receptors *(72–74)*. The blockade of potassium channels reduces neuronal death in focal and global ischemia *(75,76)*, contrary to the conventional expectation that this maneuver would enhance excitotoxicity and thus increase ischemic neuronal death.

Although many studies in cultured neurons, like the early in vivo studies of glutamate toxicity, have focused on neuronal death induced by exogenously added glutamate agonists, cell culture models have also provided insights into the role of endogenously released glutamate in neuronal death after other insults. For example, the blockade of NMDA receptors attenuates oxygen–glucose deprivation-induced death in cultured neocortical neurons *(77,78)* and death induced by exposure to the mitochondrial toxin 3-nitropropionic acid in organotypic corticostriatal slice cultures *(79)*. Oxygen–glucose deprivation-induced neuronal death is associated with enhancement in neuronal $[Ca^{2+}]_i$ and with uptake of radiolabeled $Ca^{2+}$ *(80,81)*.

## 3. GLUTAMATE NEUROTOXICITY IN ANIMAL MODELS OF NEURONAL INJURY

Extracellular glutamate levels are elevated in brain following ischemia *(82)*, seizures *(83)*, and head trauma *(84)*. Although Simon's original observation that NMDA antagonists attenuate hippocampal neuronal death following global ischemia has not been consistently confirmed [*(12)*; but see ref. *85*], AMPA receptor antagonists have reduced hippocampal injury following global ischemia in many studies *(85–88)* and also reduce infarct volume following focal ischemia *(89,90)*. NMDA antagonists, especially if administered prior to the onset of ischemia, reduce infarct size in rodent and feline models of both transient and permanent focal ischemia *(91–93)*. Administration of glutamate antagonists improves neurological outcome in rodent models of traumatic brain injury *(94,95)* and spinal cord injury *(96)*.

Other factors present in the injured nervous system could cause neurons to become vulnerable to glutamate neurotoxicity even when the synaptic release and extracellular concentration of glutamate are not especially elevated [e.g., when neuronal homeostatic mechanisms are compromised by energy depletion *(97)* or mitochondrial dysfunction *(98–100)*]. Glutamate-mediated excitotoxicity could thus contribute, at least in a secondary fashion, to the neuronal loss associated with chronic neurodegenerative diseases such as Huntington's disease *(7,8,101)*, Alzheimer's disease (e.g., see refs. *102* and *103* for reviews), or Parkinson's disease *(104,105)*. In particular, loss of transporter-mediated glutamate uptake has been postulated to promote the excitotoxic death of motor neurons in amyotrophic lateral sclerosis *(106)*.

Glutamate-mediated neuronal death in cultured cells can have mixed features of both necrosis and apoptosis (*see* Section 2); similar observations have been made in models of excitotoxic cell death in vivo. Neurons in adult rat brains typically die a morphologically necrotic death after intrastriatal injection of glutamate receptor agonists *(107,108)*, but these same neurons may exhibit some features of apoptosis, including TUNEL positivity and transient DNA laddering *(109)*; injection of non-NMDA agonists also induces chromatin clumping *(110)*. Selective neuronal death following global ischemia can evolve over 2–3 d *(111,112)* and exhibits many features of apoptosis, although features of necrosis can also be present *(113;* reviewed in ref. *114)*. This death is sensitive to inhibition of caspases or overexpression of the anti-apoptotic gene *bcl-2 (115,116)*. In contrast, neuronal death following focal ischemic insults was thought to evolve rapidly via necrosis, but even in this more fulminant injury, recent experiments have suggested that neuronal death can evolve over several days after the onset of injury *(117,118)* and may exhibit morphological and biochemical features of apoptosis, including sensitivity to antiapoptotic strategies, even when treatment is delayed for up to 6 h after the onset of ischemia *(117–120)*. Apoptosis has also been implicated in neuronal death occurring after brain or spinal cord trauma *(121–124)* or in association with several chronic neurodegenerative disease states, including Alzheimer's disease *(125–127)* and Huntington's disease *(108)*.

## 4. GLUTAMATE NEUROTOXICITY AND MODULATION OF NEURONAL CALCIUM LEVELS: THE CALCIUM SET POINT

As discussed earlier, glutamate-mediated elevations in $[Ca^{2+}]_i$ play a central role in excitotoxic neuronal necrosis. However, developmental neuronal apoptosis has been linked to an opposite change in calcium levels, i.e., a drop below an optimal "set point" *(128)*. This idea has been most extensively studied in sympathetic neurons deprived of nerve growth factor and in cerebellar granule cells switched from high to low potassium media. In cultured sympathetic neurons, $[Ca^{2+}]_i$ at early timepoints correlates with survival: lowering $[Ca^{2+}]_i$ (reduced extracellular $[Ca^{2+}]$, voltage-gated calcium channel blockers, or intracellular $Ca^{2+}$ chelators) enhances apoptosis (128–130), whereas raising $[Ca^{2+}]_i$ (increased extracellular $[Ca^{2+}]$, increased extracellular $K^+$, BayK 8644, or nicotinic receptor agonists) blocks apoptosis. Similarly, lowering extracellular $K^+$ lowers $[Ca^{2+}]_i$ and enhances cerebellar granule apoptosis *(131–133)*, whereas raising $[Ca^{2+}]_i$ by increasing $Ca^{2+}$ release from intracellular stores or increasing $Ca^{2+}$ entry by exposure to elevated $K^+$ or glutamate blocks this apoptosis *(134–136)*. Reduced levels of $[Ca^{2+}]_i$ are found in cultured neocortical neurons undergoing apoptosis following oxygen–glucose deprivation in the presence of glutamate antagonists *(137)*; NMDA antagonists and agonists, respectively, enhance or block neocortical neuronal apoptosis in culture *(138)*. The association of reduced $[Ca^{2+}]_i$ with apoptosis is not limited to neurons. For example, calcium chelators induce apoptosis in astrocytes and lymphoid cells *(139–143)*. Some authors point to a reduction in intracellular calcium stores rather than a reduction in overall $[Ca^{2+}]_i$ as a mediator of apoptosis [e.g., in lymphoid cells undergoing apoptosis after glucocorticoid exposure, intracellular calcium stores were reduced *(144,145)*]. Elevations in $[Ca^{2+}]i$ have been associated with apoptosis as well, particularly with activation of some of the mediators of the programmed cell (e.g., see refs. *(146–149)*. It is possible that deviation of $[Ca^{2+}]_i$ from a calcium set point, either up or down, could trigger apoptosis, with elevations in calcium occurring more universally later in the apoptotic process *(150)*.

If lowering $[Ca^{2+}]_i$ can promote neuronal death via apoptosis, it is possible that glutamate antagonists might be neurotoxic under some conditions. This possibility was supported by the observation of pathological changes, most notably vacuolization in the cingulate gyrus, in rats treated with the NMDA antagonists phenylcyclidine (PCP), MK-801, and ketamine *(151)*, although this may be the result of released circuit overexcitation rather than $Ca^{2+}$ deprivation-induced apoptosis *(152,153)*. More specific support for glutamate antagonist-induced, $Ca^{2+}$ deprivation-induced neuronal apoptosis was provided by observations that NMDA antagonists could induce neuronal apoptosis in culture *(154)* and widespread apoptotic neuronal death in the developing nervous system in vivo *(155)*. These observations not only raise caution against the use of compounds with NMDA antagonist activity in the young brain, but may also, in part, underlie the neuronal death seen in fetal alcohol syndrome *(156)*. Studies of head trauma in infant rats demonstrate the two-edged potential of glutamate antagonists: early excitotoxic neuronal death is reduced, but later apoptotic neuronal death is enhanced when NMDA antagonists are administered *(157)*. As discussed above, apoptosis may not be limited to the developing nervous system and has been described following ischemia or trauma in the adult brain (see above), as well as in neurodegenerative diseases. Recent studies in our laboratory have suggested that ischemic apoptosis and apoptosis associated with proteasome inhibition might be associated with reductions in neuronal $[Ca^{2+}]_i$ and that raising $[Ca^{2+}]_i$ might attenuate these forms of neuronal death *(137, 161)*.

Although there have been as of yet no reports of glutamate antagonists enhancing apoptosis when administered following injury in the adult nervous system, the apoptosis-promoting effects of glutamate antagonists could underlie the disappointing outcome of trials of glutamate antagonists in human disease (e.g., see ref. *158;* reviewed in ref *60*). The timing of the administration of NMDA antagonists may be crucial to the efficacy of this therapeutic approach, with administration in the immediate peri-ischemic period able to reduce intracellular calcium levels and attenuate acute excitotoxicity, whereas later administration might tend to exacerbate the apoptotic component of ischemic cell death. It is plausible that the more complex watersheds in human gyrencephalic brain compared to those of the lissencephalic rodent brain, together with the not uncommonly stuttering onset of human stroke, would favor an increased apoptotic component in human versus rodent ischemic brain injury. These concerns may not be limited to ischemic injury and raise a note of caution regarding the proposed use of NMDA antagonists in diseases such as Huntington's disease *(159,160)* for which proteasome inhibition and resultant reductions in neuronal $[Ca^{2+}]_i$ might occur.

## ACKNOWLEDGMENTS

This work was supported by NIH grant NS 32636 and by the Christopher Reeve Paralysis Foundation (DWC).

## REFERENCES

1. Lucas, D. and Newhouse, J. (1957) The toxic effects of sodium L-glutamate on the inner layers of the retina. *Arch, Ophthalmol.* **58,** 193–201.
2. Olney, J. W. (1969) Brain lesions, obesity, and other disturbances in mice treated with monosodium glutamate. *Science* **164,** 719–721
3. Curtis, D. R., Phillis, J. W., and Watkins, J. (1960) The chemical excitation of spinal neurons by certain acidic amino acids. *J. Physiol. (Lond.)* **150,** 656–682.
4. Crawford, J. M. and Curtis, D. R. (1964) The excitation and depression of mammalian cortical neurons by amino acids. *Br. J. Pharm.* **23,** 323–329.
5. Krnjevic, K. (1974) Chemical nature of synaptic transmission in vertebrates. *Physiol. Res.* **418,** 418–540.
6. Di Chiara, G. and Gessa, G. L. (eds.) (1981), Psychopharmacology Vol. 27 *Glutamate as a Neurotransmitter,* Advances in Biochemical Raven P, New York.
7. Coyle, J. T. and Schwarcz, R. (1976) Lesion of striatal neurones with kainic acid provides a model for Huntington's chorea. *Nature* **263,** 244–246.
8. McGeer, E. G, and McGeer, P. L. (1976) Duplication of biochemical changes of Huntington's chorea by intrastriatal injections of glutamic and kainic acids. *Nature* **263,** 517–519.
9. Nadler, J. V., Perry, B. W., and Cotman, C. W. (1978) Intraventricular kainic acid preferentially destroys hippocampal pyramidal cells. *Nature* **271,** 676–677.
10. Sloviter, R. S. (1983) "Epileptic" brain damage in rats induced by sustained electrical stimulation of the perforant path. I. Acute electrophysiological and light microscopic studies. *Brain Res. Bull.* **10,** 675–697.
11. Rothman, S. (1984) Synaptic release of excitatory amino acid neurotransmitter mediates anoxic neuronal death. *J. Neurosci.* **4,** 1884–1891.

12. Simon, R. P., Swan, J. H., Griffiths, T., and Meldrum, B. S. (1984) Blockade of N-methyl-D-aspartate receptors may protect against ischemic damage in the brain. *Science* **226**, 850–850.
13. Wieloch, T. (1985) Hypoglycemia-induced neuronal damage prevented by an N-methyl-D-aspartate antagonist. *Science* **230**, 681–683.
14. Yoshioka, A., Hardy, M., Younkin, D. P., Grinspan, J. B., Stern, J. L., and Pleasure, D. (1995) Alpha-amino-3-hydroxy-5-methyl-4-isoxazolepropionate (AMPA) receptors mediate excitotoxicity in the oligodendroglial lineage. *J. Neurochem.* **64**, 2442–2448.
15. Matute, C., Sanchez-Gomez, M. V., Martinez-Millan, L., and Miledi, R. (1997) Glutamate receptor-mediated toxicity in optic nerve oligodendrocytes. *Proc. Natl. Acad. Sci. USA* **94**, 8830–8835.
16. McDonald, J. W., Althomsons, S. P., Hyrc, K. L., Choi, D. W., and Goldberg, M. P. (1998) Oligodendrocytes from forebrain are highly vulnerable to AMPA/kainate receptor-mediated excitotoxicity. *Nat. Med.* **4**, 291–297.
17. Oka, A., Belliveau, M. J., Rosenberg, P. A., and Volpe, J. J. (1993) Vulnerability of oligodendroglia to glutamate: pharmacology, mechanisms, and prevention. *J. Neurosci.* **13**, 1441–1453.
18. Chen, C. J., Liao, S. L., and Kuo, J. S. (2000) Gliotoxic action of glutamate on cultured astrocytes. *J. Neurochem.* **75**, 1557–1565.
19. Seeburg, P. H. (1993) The TINS/TiPS Lecture. The molecular biology of mammalian glutamate receptor channels. *Trends Neurosci.* **16**, 359–365.
20. Hollmann, M. and Heinemann, S. (1994) Cloned glutamate receptors. *Annu. Rev. Neurosci.* **17**, 31–108.
21. Seeburg, P. H., Higuchi, M., and Sprengel, R. (1998) RNA editing of brain glutamate receptor channels: mechanism and physiology. *Brain Res. Rev.* **26**, 217–229.
22. Nakanishi, S. and Masu, M. (1994) Molecular diversity and functions of glutamate receptors. *Annu. Rev. Biophys. Biomol. Struct.* **23**, 319–348.
23. Conn, P. J. and Pin, J. P. (1997) Pharmacology and functions of metabotropic glutamate receptors. *Annu. Rev. Pharmacol. Toxicol.* **37**, 205–237.
24. Kerchner, G. A., Kim, A. H., and Choi, D. W., Glutamate-mediated excitotoxicity, *Ionotropic Glutamate Receptors in the CNS*, In (P. Jonas and H. Monyer, (eds.), Springer-Verlag, Berlin, 1999, pp. 443–469.
25. Nicoletti, F., Bruno, V., Catania, M. V., Battaglia, G., Copani, A., Barbagallo, G., et al. (1999) Group-I metabotropic glutamate receptors: hypotheses to explain their dual role in neurotoxicity and neuroprotection. *Neuropharmacology* **38**, 1477–1484.
26. Cartmell, J. and Schoepp, D. D. (2000) Regulation of neurotransmitter release by metabotropic glutamate receptors. *J. Neurochem.* **75**, 889–907.
27. Fagni, L., Chavis, P., Ango, F., and Bockaert, J. (2000) Complex interactions between mGluRs, intracellular $Ca^{2+}$ stores and ion channels in neurons. *Trends Neurosci.* **23**, 80–88.
28. Choi, D. W. (1988) Calcium-mediated neurotoxicity: relationship to specific channel types and role in ischemic damage. *Trends Neurosci.* **11**, 465–469.
29. Olney, J. W., Price, M. T., Samson, L., and Labruyere, J. (1986) The role of specific ions in glutamate neurotoxicity. *Neurosci. Lett.* **65**, 65–71.
30. Turetsky, D. M., Canzoniero, L. M. T., Sensi, S. L., Weiss, J. H., Goldberg, M. P., and Choi, D. W. (1994) Cortical neurones exhibiting kainate-activated $Co^{2+}$ uptake are selectively vulnerable to AMPA/kainate receptor-mediated toxicity. *Neurobiol. Dis.* **1**, 101–110.
31. Yu, S. P. and Choi, D. W. (1997) $Na^+$–$Ca^{2+}$ exchange currents in cortical neurons: concomitant forward and reverse operation and effect of glutamate. *Eur. J. Neurosci.* **9**, 1273–1281.
32. Hartley. D. M., Kurth, M. C., Bjerkness, L., Weiss, J. H., and Choi, D. W. (1993) Glutamate receptor-induced $^{45}Ca^{2+}$ accumulation in cortical cell culture correlates with subsequent neuronal degeneration. *J. Neurosci.* **13**, 1993–2000.
33. Hyrc, K., Handran, S. D., Rothman, S. M., and Goldberg, M. P. (1997) Ionized intracellular calcium concentration predicts excitotoxic neuronal death: observations with low-affinity fluorescent calcium indicators. *J. Neurosci.* **17**, 6669–6677.
34. Grimwood, S., Gilbert, E., Ragan, C. I., and Hutson, P. H. (1996) Modulation of $^{45}Ca^{2+}$ influx into cells stably expressing recombinant human NMDA receptors by ligands acting at distinct recognition sites. *J. Neurochem.* **66**, 2589–2595.
35. Grant, E. R., Bacskai, B. J., Pleasure, D. E., Pritchett, D. B., Gallagher, M. J., Kendrick, S. J., et al. (1997) N-Methyl-D-aspartate receptors expressed in a nonneuronal cell line mediate subunit-specific increases in free intracellular calcium. *J. Biol. Chem.* **272**, 647–656.
36. Dugan, L. L., Sensi, S. L., Canzoniero, L. M., Handran, S. D., Rothman, S. M., Lin, T. S., et al. (1995) Mitochondrial production of reactive oxygen species in cortical neurons following exposure to N-methyl-D-aspartate. *J. Neurosci.* **15**, 6377–6388.
37. Reynolds, I. J. and Hastings, T. G. (1995) Glutamate induces the production of reactive oxygen species in cultured forebrain neurons following NMDA receptor activation. *J. Neurosci.* **15**, 3318–3327.
38. White, R. J. and Reynolds, I. J. (1995) Mitochondria and $Na^+/Ca^{2+}$ exchange buffer glutamate-induced calcium loads in cultured cortical neurons. *J. Neurosci.* **15**, 1318–1328.
39. Schinder, A. F., Olson, E. C., Spitzer, N. C., and Montal, M. (1996) Mitochondrial dysfunction is a primary event in glutamate neurotoxicity. *J. Neurosci.* **16**, 6125–6133.

40. White, R. J. and Reynolds. I. J. (1996) Mitochondrial depolarization in glutamate-stimulated neurons: an early signal specific to excitotoxin exposure. *J. Neurosci.* **16,** 5688–5697.
41. Dykens, J. A., Stern, A., and Trenkner, E. (1987) Mechanism of kainate toxicity to cerebellar neurons in vitro is analogous to reperfusion tissue injury. *J. Neurochem.* **49,** 1222–1228.
42. Monyer, H., Hartley, D. M., and Choi, D. W. (1990) 21-Aminosteroids attenuate excitotoxic neuronal injury in cortical cell cultures. *Neuron* **5,** 121–126.
43. Siman, R., Noszek, J. C., and Kegerise, C. (1989) Calpain I activation is specifically related to excitatory amino acid induction of hippocampal damage. *J. Neurosci.* **9,** 1579–1590.
44. Lee, K. S., Frank, S., Vanderklish, P., Arai, A., and Lynch, G. (1991) Inhibition of proteolysis protects hippocampal neurons from ischemia. *Proc. Natl. Acad. Sci. USA* **88,** 7233–7237.
45. Brorson, J. R., Manzolillo, P. A., and Miller, R. J. (1994) $Ca^{2+}$ entry via AMPA/KA receptors and excitotoxicity in cultured cerebellar Purkinje cells. *J. Neurosci.* **14,** 187–197.
46. Dessi, F., Charriaut-Marlangue, C., Khrestchatisky, M., and Ben-Ari, Y. (1993) Glutamate-induced neuronal death is not a programmed cell death in cerebellar culture. *J. Neurochem.* **60,** 1953–1955.
47. Regan, R. F., Panter, S. S., Witz, A., Tilly, J. L., and Giffard, R. G. (1995) Ultrastructure of excitotoxic neuronal death in murine cortical culture. *Brain Res.* **705,** 188–198.
48. Gottron, F. J., Ying, H. S., and Choi, D. W. (1997) Caspase inhibition selectively reduces the apoptotic component of oxygen-glucose deprivation-induced cortical neuronal cell death. *Mol. Cell. Neurosci.* **9,** 159–169.
49. Gwag, B. J., Koh, J. Y., DeMaro, J. A., Ying, H. S., Jacquin, M., and Choi, D. W. (1997) Slowly triggered excitotoxicity occurs by necrosis in cortical cultures. *Neuroscience* **77,** 393–401.
50. Ankarcrona, M., Dypbukt, J. M., Bonfoco, E., Zhivotovsky, B., Orrenius, S., Lipton, S. A., et al. (1995) Glutamate-induced neuronal death: a succession of necrosis or apoptosis depending on mitochondrial function. *Neuron* **15,** 961–973.
51. Bonfoco, E., Krainc, D., Ankarcrona, M., Nicotera, P., and Lipton, S. A. (1995) Apoptosis and necrosis: two distinct events induced, respectively, by mild and intense insults with *N*-methyl-D-aspartate or nitric oxide/superoxide in cortical cell cultures. *Proc. Natl. Acad. Sci. USA* **92,** 7162–7166.
52. Simonian, N. A., Getz, R. L., Leveque, J. C., Konradi, C., and Coyle, J. T. (1996) Kainate induces apoptosis in neurons. *Neuroscience* **74,** 675–683.
53. Didier, M., Bursztajn, S., Adamec, E., Passani, L., Nixon, R. A., Coyle, J. T., et al. (1996) DNA strand breaks induced by sustained glutamate excitotoxicity in primary neuronal cultures. *J. Neurosci.* **16,** 2238–2250.
54. Dreyer, E. B., Zhang, D., and Lipton, S. A. (1995) Transcriptional or translational inhibition blocks low dose NMDA-mediated cell death. *Neuroreport* **6,** 942–944.
55. Finiels, F., Robert, J. J., Samolyk, M. L., Privat, A., Mallet, J., and Revah, F. (1995) Induction of neuronal apoptosis by excitotoxins associated with long-lasting increase of 12-*O*-tetradecanoylphorbol 13-acetate-responsive element-binding activity. *J. Neurochem.* **65,** 1027–1034.
56. Tenneti, L., D'Emilia, D. M., Troy, C. M., and Lipton, S. A. (1998) Role of caspases in *N*-methyl-D-aspartate-induced apoptosis in cerebrocortical neurons. *J. Neurochem.* **71,** 946–959.
57. Xiang, H., Kinoshita, Y., Knudson, C. M., Korsmeyer, S. J., Schwartzkroin, P. A., and Morrison, R. S. (1998) Bax involvement in p53-mediated neuronal cell death. *J. Neurosci.* **18,** 1363–1373.
58. McDonald, J. W., Behrens, M. I., Chung, C., Bhattacharyya, T., and Choi, D. W. (1997) Susceptibility to apoptosis is enhanced in immature cortical neurons. *Brain Res.* **759,** 228–232.
59. Frederickson, C. J. (1989) Neurobiology of zinc and zinc-containing neurons. *Int. Rev. Neurobiol.* **31,** 145–238.
60. Lee, J. M., Zipfel, G. J., and Choi, D. W. (1999) The changing landscape of ischaemic brain injury mechanisms. *Nature* **399,** A7–A14.
61. Weiss, J. H., Sensi, S. L., and Koh, J. Y. (2000) $Zn^{2+}$: a novel ionic mediator of neural injury in brain disease. *Trends Pharmacol. Sci.* **21,** 395–401.
62. Peters, S., Koh, J.-Y., and Choi, D. W. (1987) Zinc selectively blocks the action of *N*-methyl-D-aspartate on cortical neurons. *Science* **236,** 589–593.
63. Westbrook, G. L. and Mayer, M. L. (1987) Micromolar concentrations of $Zn^{2+}$ antagonize NMDA and GABA responses of hippocampal neurons. *Nature* **328,** 640–643.
64. Christine, C. W. and Choi, D. W. (1988) Zinc alters NMDA receptor-mediated channel events on cortical neurons. *Neurology* **38 (Suppl),** 274–275.
65. Smart, T. G., Xie, X., and Krishek, B. J. (1994) Modulation of inhibitory and excitatory amino acid receptor ion channels by zinc. *Prog. Neurobiol.* **42,** 393–441.
66. Manzerra, P., Behrens, M. I., Canzoniero, L. M., Wang. X. Q., Heidinger, V., Ichinose, T., et al. (2001) Zinc induces a Src family kinase-mediated upregulation of NMDA receptor activity and excitotoxicit. *Proc. Natl. Acad. Sci. USA* **98,** 11051–11061.
67. Choi, D. W. and Koh, J. Y. (1998) Zinc and brain injury. *Annu. Rev. Neurosci.* **21,** 347–375.
68. Tonder, N., Johansen, F. F., Frederickson, C. J., Zimmer, J., and Diemer, N. H. (1990) Possible role of zinc in the selective degeneration of dentate hilar neurons after cerebral ischemia in the adult rat. *Neurosci. Lett.* **109,** 247–252.

69. Koh, J.-Y., Suh, S. W., Gwag, B. J., He, Y. Y., Hsu, C. Y., and Choi, D. W. (1996) The role of zinc in selective neuronal death after transient global cerebral ischemia. *Science* **272,** 1013–1016.
70. Frederickson, C. J., Hernandez, M. D., Goik, S. A., Morton, J. D., and McGinty. J. F. (1988) Loss of zinc staining from hippocampal mossy fibers during kainic acid induced seizures: a histofluorescence study. *Brain Res.* **446,** 383–386.
71. Frederickson. C. J., Hernandez, M. D., and McGinty, J. F. (1989) Translocation of zinc may contribute to seizure-induced death of neurons. *Brain Res.* **480,** 317–321.
72. Yu. S. P., Yeh, C. H., Sensi, S. L., Gwag, B. J., Canzoniero, L. M., Farhangrazi, Z. S., et al. (1997) Mediation of neuronal apoptosis by enhancement of outward potassium current. *Science* **278,** 114–117.
73. Yu, S. P., Farhangrazi, Z. S., Ying, H. S., Yeh, C. H., and Choi, D. W. (1998) Enhancement of outward potassium current may participate in beta-amyloid peptide-induced cortical neuronal death. *Neurobiol. Dis.* **5,** 81–88.
74. Yu, S. P., Yeh, C., Strasser, U., Tian, M., and Choi, D. W. (1999) NMDA receptor-mediated K$^+$ efflux and neuronal apoptosis. *Science* **284,** 336–339.
75. Choi, D. W., Yu, S. P., Wei, L., and Gottron, F. (1998) Potassium channel blockers attenuate neuronal deaths induced by hypoxic insults in cortical culture and transient focal ischemia in the rat. *Soc. Neurosci. Abst.* **24,** 1226.
76. Huang, H., Gao, T. M., Gong, L. W., Zhuang, Z. Y., and Li, X. M. (2001) Potassium channel blocker TEA prevents CA1 hippocampal injury following transient forebrain ischemia in adult rats. *Neurosci. Lett.* **305,** 83–86.
77. Goldberg, M. P., Weiss., J. H., Pham, P. C., and Choi, D. W. (1987) *N*-Methyl-D-aspartate receptors mediate hypoxic neuronal injury in cortical culture. *J. Pharmacol. Exp. Ther.* **243,** 784–791.
78. Kaku, D. A., Goldberg, M. P., and Choi, D. W. (1991) Antagonism of non-NMDA receptors augments the neuroprotective effect of NMDA receptor blockade in cortical cultures subjected to prolonged deprivation of oxygen and glucose. *Brain Res.* **554,** 344–347.
79. Storgaard, J., Kornblit, B. T., Zimmer, J., and Gramsbergen, J. B. (2000) 3-Nitropropionic acid neurotoxicity in organotypic striatal and corticostriatal slice cultures is dependent on glucose and glutamate. *Exp. Neurol.* **164,** 227–235.
80. Goldberg, M. P. and Choi, D. W. (1990) Intracellular free calcium increases in cultured cortical neurons deprived of oxygen and glucose. *Stroke* **21,** III75–III7.
81. Goldberg, M. P. and Choi, D. W. (1993) Combined oxygen and glucose deprivation in cortical cell culture: calcium-dependent and calcium-independent mechanisms of neuronal injury. *J. Neurosci.* **13,** 3510–3524.
82. Benveniste, H., Drejer, J., Schousboe, A., and Diemer, N. H. (1984) Elevation of the extracellular concentrations of glutamate and aspartate in rat hippocampus during transient cerebral ischemia monitored by intracerebral microdialysis. *J. Neurochem.* **43,** 1369–1374.
83. Meldrum, B. S. (1994) The role of glutamate in epilepsy and other CNS disorders. *Neurology* **44,** S14–S23.
84. Katayama, Y., Becker, D. P., Tamura, T., and Hovda, D. A. (1990) Massive increases in extracellular potassium and the indiscriminate release of glutamate following concussive brain injury. *J. Neurosurg.* **73,** 889–900.
85. Buchan, A. M., Lesiuk, H., Barnes, K. A., Li, H., Huang, Z. G., Smith, K. E., et al. (1993) AMPA antagonists: do they hold more promise for clinical stroke trials than NMDA antagonists? *Stroke* **24,** I148–I152.
86. Sheardown, M. J., Nielsen, E. O., Hansen, A. J., Jacobsen, P., and Honore, T. (1990) 2,3-Dihydroxy-6-nitro-7-sulfamoyl-benzo(*F*)quinoxaline: a neuroprotectant for cerebral ischemia. *Science* **247,** 571–574.
87. Diemer, N. H., Jorgensen, M. B., Johansen, F. F., Sheardown, M., and Honore, T. (1992) Protection against ischemic hippocampal CA1 damage in the rat with a new non-NMDA antagonist, NBQX. *Acta Neurol. Scand.* **86,** 45–49.
88. Nellgard, B. and Wieloch, T. (1992) Postischemic blockade of AMPA but not NMDA receptors mitigates neuronal damage in the rat brain following transient severe cerebral ischemia. *J. Cereb, Blood Flow Metab.* **12,** 2–11.
89. Buchan, A. M., Xue, D., Huang, Z. G., Smith, K. H., and Lesiuk, H. (1991) Delayed AMPA receptor blockade reduces cerebral infarction induced by focal ischemia. *Neuroreport* **2,** 473–476.
90. Gill, R., Nordholm, L., and Lodge, D. (1992) The neuroprotective actions of 2,3-dihydroxy-6-nitro-7-sulfamoyl-benzo(*F*)quinoxaline (NBQX) in a rat focal ischaemia model. *Brain Res.* **580,** 35–43.
91. Gotti, B., Duverger, D., Bertin, J., Carter, C., Dupont, R., Frost, J., et al. (1988) Ifenprodil and SL 82.0715 as cerebral anti-ischemic agents. I. Evidence for efficacy in models of focal cerebral ischemia. *J. Pharmacol. Exp. Ther.* **247,** 1211–1221.
92. Ozyurt, E., Graham, D. I., Woodruff, G. N., and McCulloch, J. (1988) Protective effect of the glutamate antagonist, MK-801 in focal cerebral ischemia in the cat. *J. Cereb. Blood Flow Metab.* **8,** 138–143.
93. Park, C. K., Nehls, D. G., Graham, D. I., Teasdale, G. M., and McCulloch, J. (1988) The glutamate antagonist MK-801 reduces focal ischemic brain damage in the rat. *Ann. Neurol.* **24,** 543–551.
94. Hayes. R. L., Jenkins, L. W., Lyeth, B. G., Balster, R. L., Robinson, S. E., Clifton, G. L., et al. (1988) Pretreatment with phencyclidine, an *N*-methyl-D-aspartate antagonist, attenuates long-term behavioral deficits in the rat produced by traumatic brain injury. *J. Neurotrauma* **5,** 259–274.
95. Faden, A. I., Demediuk, P., Panter, S. S., and Vink, R. (1989) The role of excitatory amino acids and NMDA receptors in traumatic brain injury. *Science* **244,** 798–800.
96. Wrathall, J. R., Teng, Y. D., Choiniere, D., and Mundt, D. J. (1992) Evidence that local non-NMDA receptors contribute to functional deficits in contusive spinal cord injury. *Brain Res.* **586,** 140–143.

97. Novelli, A., Reilly, J. A., Lysko. P. G., and Henneberry, R. C. (1988) Glutamate becomes neurotoxic via the N-methyl-D-aspartate receptor when intracellular energy levels are reduced. *Brain Res.* **451,** 205–212.
98. Beal, M. F. (2000) Energetics in the pathogenesis of neurodegenerative diseases. *Trends Neurosci.* **23,** 298–304.
99. Nicholls, D. G., and Ward, M. W. (2000) Mitochondrial membrane potential and neuronal glutamate excitotoxicity: mortality and millivolts. *Trends Neurosci.* **23,** 166–174.
100. Ward, M. W., Rego, A. C., Frenguelli, B. G., and Nicholls, D. G. (2000) Mitochondrial membrane potential and glutamate excitotoxicity in cultured cerebellar granule cells. *J. Neurosci.* **20,** 7208–7219.
101. Sun, Y., Savanenin, A., Reddy, P. H., and Liu, Y. F. (2001) Polyglutamine-expanded huntingtin promotes sensitization of N-methyl-D-aspartate receptors via post-synaptic density 95. *J. Biol. Chem.* **276,** 24,713–24,718.
102. Greenamyre, J. T. (1991) Neuronal bioenergetic defects, excitotoxicity and Alzheimer's disease: "use it and lose it". *Neurobiol. Aging* 12, 334–336; discussion 352–355.
103. Harkany, T., Abraham. I., Konya, C., Nyakas, C., Zarandi, M., Penke, B., et al. (2000) Mechanisms of beta-amyloid neurotoxicity: perspectives of pharmacotherapy.*Rev. Neurosci.* **11,** 329–382.
104. Dickie, B. G., Holmes, C., and Greenfield, S. A. (1996) Neurotoxic and neurotrophic effects of chronic N-methyl-D-aspartate exposure upon mesencephalic dopaminergic neurons in organotypic culture. *Neuroscience* 72, 731–741.
105. Beal, M. F. (1998) Excitotoxicity and nitric oxide in Parkinson's disease pathogenesis. *Ann. Neurol.* **44,** S110–S114.
106. Rothstein, J. D., Martin, L. J., and Kuncl, R. W. (1992) Decreased glutamate transport by the brain and spinal cord in amyotrophic lateral sclerosis. *N. Engl. J. Med.* **326,** 1464–1468.
107. Ferrer, I., Martin, F., Serrano, T., Reiriz, J., Perez-Navarro, E., Alberch, J., et al. (1995) Both apoptosis and necrosis occur following intrastriatal administration of excitotoxins. *Acta Neuropathol.* **90,** 504–510.
108. Portera-Cailliau, C., Hedreen, J. C., Price, D. L., and Koliatsos, V. E. (1995) Evidence for apoptotic cell death in Huntington disease and excitotoxic animal models. *J. Neurosci.* **15,** 3775–3787.
109. Qin, Z. H., Wang, Y., and Chase, T. N. (1996) Stimulation of N-methyl-D-aspartate receptors induces apoptosis in rat brain. *Brain Res.* **725,** 166–176.
110. Portera-Cailliau, C., Price, D. L., and Martin, L. J. (1997) Excitotoxic neuronal death in the immature brain is an apoptosis–necrosis morphological continuum. *J. Comp. Neurol.* **378,** 70–87.
111. Kirino, T. (1982) Delayed neuronal death in the gerbil hippocampus following ischemia. *Brain Res.* **239,** 57–69.
112. Pulsinelli, W. A., Brierley, J. B., and Plum, F. (1982) Temporal profile of neuronal damage in a model of transient forebrain ischemia. *Ann. Neurol.* **11,** 491–498.
113. Zeng, Y. S. and Xu, Z. C. (2000) Co-existence of necrosis and apoptosis in rat hippocampus following transient forebrain ischemia. *Neurosci. Res.* **37,** 113–125.
114. Lipton, P. (1999) Ischemic cell death in brain neurons. *Physiol. Rev.* **79,** 1431–1568.
115. Kitagawa, K., Matsumoto, M., Tsujimoto, Y., Ohtsuki, T., Kuwabara, K., Matsushita, K., et al. (1998) Amelioration of hippocampal neuronal damage after global ischemia by neuronal overexpression of BCL-2 in transgenic mice. *Stroke* 29, 2616–2621.
116. Gillardon, F., Kiprianova, I., Sandkuhler, J., Hossmann, K.-A., and Spranger, M. (1999) Inhibition of caspases prevents cell death of hippocampal CA1 neurons, but not impairment of hippocampal long-term potentiation following global ischemia. *Neuroscience* 93, 1219–1222.
117. Du, C., Hu, R., Csernansky, C. A., Hsu, C. Y., and Choi, D. W. (1996) Very delayed infarction after mild focal cerebral ischemia: a role for apoptosis? *J. Cereb. Blood Flow Metab.* **16,** 195–201.
118. Endres, M., Wang, Z. Q., Namura, S., Waeber, C., and Moskowitz, M. A. (1997) Ischemic brain injury is mediated by the activation of poly(ADP- ribose)polymerase. *J. Cereb. Blood Flow Metab.* **17,** 1143–1151.
119. Fink, K., Zhu, H., Namura, S., Shimizu-Sasamata, M., Endres, M., Ma, J., et al. (1998) Prolonged therapeutic window for ischemic brain damage caused by delayed caspase activation. *J. Cereb. Blood Flow Metab.* **18,** 1071–1076.
120. Snider, B. J., Du, C., Wei, L., and Choi, D. W. (2001) Cycloheximide reduces infarct volume when administered up to six hours after mild focal ischemia in rats. *Brain Res.* **917,** 147–157.
121. Rink, A., Fung, K. M., Trojanowski, J. Q., Lee, V. M., Neugebauer, E., and McIntosh, T. K. (1995) Evidence of apoptotic cell death after experimental traumatic brain injury in the rat. *Am. J. Pathol.* **147,** 1575–1583.
122. Crowe, M. J., Bresnahan, J. C., Shuman, S. L., Masters, J. N., and Beattie, M. S. (1997) Apoptosis and delayed degeneration after spinal cord injury in rats and monkeys. *Nat. Med.* **3,** 73–76.
123. Liu, X. Z., Xu, X. M., Hu, R., Du, C., Zhang, S. X., McDonald, J. W., et al. (1997) Neuronal and glial apoptosis after traumatic spinal cord injury. *J. Neurosci.* **17,** 5395–5406.
124. Yong, C., Arnold, P. M., Zoubine, M. N., Citron, B. A., Watanabe, I., Berman, N. E., et al. (1998) Apoptosis in cellular compartments of rat spinal cord after severe contusion injury. *J. Neurotrauma* 15, 459–472.
125. Lassmann, H., Bancher, C., Breitschopf, H., Wegiel, J., Bobinski, M., Jellinger, K., et al. (1995) Cell death in Alzheimer's disease evaluated by DNA fragmentation in situ. *Acta Neuropathol. (Berl.)* 89, 35–41.
126. Smale, G., Nichols, N. R., Brady, D. R., Finch, C. E., and Horton, W. E., Jr. (1995) Evidence for apoptotic cell death in Alzheimer's disease. *Exp. Neurol.* **133,** 225–230.

127. Anderson, A. J., Su, J. H., and Cotman, C. W. (1996) DNA damage and apoptosis in Alzheimer's disease: colocalization with c-Jun immunoreactivity, relationship to brain area, and effect of postmortem delay. *J. Neurosci.* **16,** 1710–1719.
128. Johnson, E. M., Jr., Koike, T., and Franklin, J. (1992) A "calcium set-point hypothesis" of neuronal dependence on neurotrophic factor. *Exp. Neurol.* **115,** 163–166.
129. Koike, T., Martin, D. P., and Johnson, E. M., Jr. (1989) Role of $Ca^{2+}$ channels in the ability of membrane depolarization to prevent neuronal death induced by trophic-factor deprivation: evidence that levels of internal $Ca^{2+}$ determine nerve growth factor dependence of sympathetic ganglion cells. *Proc. Natl. Acad, Sci. USA* 86, 6421–6425.
130. Koike, T. and Tanaka, S. (1991) Evidence that nerve growth factor dependence of sympathetic neurons for survival *in vitro* may be determined by levels of cytoplasmic free $Ca^{2+}$. *Proc. Natl. Acad. Sci. USA* 88, 3892–3896.
131. Lasher, R. S. and Zagon, I. S. (1972) The effect of potassium on neuronal differentiation in cultures of dissociated newborn rat cerebellum. *Brain Res.* **41,** 482–488.
132. Pearson, H., Graham, M. E., and Burgoyne, R. D. (1992) Relationship between intracellular free calcium concentration and NMDA-induced cerebellar granule cell survival *in vitro. Eur. J. Neurosci.* **4,** 1369–1375.
133. Galli, C., Meucci, O., Scorziello, A., Werge, T. M., Calissano, P., and Schettini, G. (1995) Apoptosis in cerebellar granule cells is blocked by high KC1, forskolin, and IGF-1 through distinct mechanisms of action: the involvement of intracellular calcium and RNA synthesis. *J. Neurosci.* **15,** 1172–1179.
134. Gallo, V., Kingsbury, A., Balázs, R., and Jørgensen, O. S. (1987) The role of depolarization in the survival and differentiation of cerebellar granule cells in culture. *J. Neurosci.* **7,** 2203–2213.
135. Yan, G.-M., Ni, B., Weller, M., Wood, K. A., and Paul, S. M. (1994) Depolarization or glutamate receptor activation blocks apoptotic cell death of cultured cerebellar granule neurons. *Brain Res.* **656,** 43–51.
136. Levick, V., Coffey, H., and D'Mello, S. R. (1995) Opposing effects of thapsigargin on the survival of developing cerebellar granule neurons in culture. *Brain Res.* **676,** 325–335.
137. Babcock, D. J., Gottron, F. J., and Choi, D. W. (1999) Raising intracellular calcium attenuates ischemic apoptosis in vitro. *Soc. Neurosci. Abst.* **25,** 2103.
138. Terro, F., Esclaire, F., Yardin, C., and Hugon, J. (2000) *N*-methyl-D-aspartate receptor blockade enhances neuronal apoptosis induced by serum deprivation. *Neurosci. Lett.* **278,** 149–152.
139. Bansal, N., Houle, A. G., and Melnykovych, G. (1990) Dexamethasone-induced killing of neoplastic cells of lymphoid derivation: lack of early calcium involvement. *J. Cell. Physiol.* **143,** 105–109.
140. Baffy, G., Miyashita, T., Williamson, J. R., and Reed, J. C. (1993) Apoptosis induced by withdrawal of interleukin-3 (IL-3) from an IL-3-dependent hematopoietic cell line is associated with repartitioning of intracellular calcium and is blocked by enforced Bcl-2 oncoprotein production. *J. Biol. Chem.* **268,** 6511–6519.
141. Kluck, R. M., McDougall, C. A., Harmon, B. V., and Haliday, J. W. (1994) Calcium chelators induce apoptosis—evidence that raised intracellular ionised calcium is not essential for apoptosis. *Biochim. Biophys. Acta* **1223,** 247–254.
142. Zhu, W.-H. and Loh, T.-T. (1995) Roles of calcium in the regulation of apoptosis in HL-60 promyelocytic leukemia cells. *Life Sci.* **57,** 2091–2099.
143. Chiesa, R., Angeretti, N., Del Bo, R., Lucca, E., Munna, E., and Forloni, G. (1998) Extracellular calcium deprivation in astrocytes: regulation of mRNA expression and apoptosis. *J. Neurochem.* **70,** 1474–1483.
144. Lam, M., Dubyak, G., and Distelhorst, C. W. (1993) Effect of glucocorticosteroid treatment on intracellular calcium homeostasis in mouse lymphoma cells. *Mol. Endocrinol.* **7,** 686–693.
145. Bian, X., Hughes, F. M., Jr., Huang, Y., Cidlowski, J. A., and Putney, J. W., Jr. (1997) Roles of cytoplasmic $Ca^{2+}$ and intracellular $Ca^{2+}$ stores in induction and suppression of apoptosis in S49 cells. *Am. J. Physiol.* **272,** C1241–C1249.
146. Bennett, M. R. and Huxlin, K. R. (1996) Neuronal cell death in the mammalian nervous system: the calmortin hypothesis. *Gen Pharmacol.* **27,** 407–419.
147. McConkey, D. J. and Orrenius, S. (1997) The role of calcium in the regulation of apoptosis. *Biochem. Biophys. Res. Commun.* **239,** 357–366.
148. Distelhorst, C. W. and Dubyak, G. (1998) Role of calcium in glucocorticosteroid-induced apoptosis of thymocytes and lymphoma cells: resurrection of old theories by new findings. *Blood* **91,** 731–734.
149. Toescu, E. C. (1998) Apoptosis and cell death in neuronal cells: where does $Ca^{2+}$ fit in? *Cell Calcium* **24,** 387–403.
150. Yu, S. P. and Choi, D. W. (2000) Ions, cell volume, and apoptosis. *Proc. Natl. Acad. Sci. USA* **97,** 9360–9362.
151. Olney, J. W., Labruyere, J., and Price, M. T. (1989) Pathological changes induced in cerebrocortical neurons by phencyclidine and related drugs. *Science* **244,** 1360–1362.
152. Onley, J. W., Labruyere, J., Wang, G., Wozniak, D. F., Price, M. T., and Sesma, M. A. (1991) NMDA antagonist neurotoxicity: mechanism and prevention. *Science* **254,** 1515–1518.
153. Sharp, F. R., Butman, M., Koistinaho, J., Aardalen, K., Nakki, R., Massa, S. M., et al. (1994) Phencyclidine induction of the hsp 70 stress gene in injured pyramidal neurons is mediated via multiple receptors and voltage gated calcium channels. *Neuroscience* **62,** 1079–1092.
154. Hwang, J. Y., Kim, Y. H., Ahn, Y. H., Wie, M. B., and Koh, J. Y. (1999) *N*-Methyl-D-aspartate receptor blockade induces neuronal apoptosis in cortical culture. *Exp. Neurol.* **159,** 124–130.

155. Ikonomidou, C., Bosch, F., Miksa, M., Bittigau, P., Vockler, J., Dikranian, K., et al. (1999) Blockade of NMDA receptors and apoptotic neurodegeneration in the developing brain. *Science* **283,** 70–74.
156. Ikonomidou, C., Bittigau, P., Ishimaru, M. J., Wozniak, D. F., Koch, C., Genz, K., et al. (2000) Ethanol-induced apoptotic neurodegeneration and fetal alcohol syndrome. *Science* **287,** 1056–1060.
157. Phol, D., Bittigau, P., Ishimaru, M. J., Stadthaus, D., Hubner, C., Onley, J. W., et al. (1999) *N*-methyl-D-aspartate antagonists and apoptotic cell death triggered by head trauma in developing rat brain. *Proc. Natl. Acad. Sci. USA* **96,** 2508–2573.
158. Davis, S. M., Lees, K. R., Albers, G. W., Diener, H. C., Markabi, S., Karlsson, G. et al. (2000) Selfotel in acute ischemic stroke: possible neurotoxic effects of an NMDA antagonist. *Stroke* **31,** 347–354.
159 Kieburtz, K., Feigin, A., McDermott, M., Como, P., Abwender, D., Zimmerman, C., et al. (1996) A controlled trial of remacemide hydrochloride in Huntington's disease. *Mov. Cement Disord.* **11,** 273–277.
160. Murman, D. L., Giordani, B., Mellow, A. M., Johanns, J. R., Little, R. J. A., Hariharan, M., et al. (1997) Cognitive, behavioral, and motor effects of the NMDA antagonist ketamine in Huntington's disease. *Neurology* **49,** 153–161.
161. Snider, B. J., Tee, L. Y., Canzoniero, L. M., and Choi, D. W. (2002) NMDA antagonists exacerbate neuronal death caused by proteasome inhibition in cultured cortical and striatal neurons. *Eur. J. Neurosci.* **15,** 419–428.

# 4
# Maturational Regulation of Glutamate Receptors and Their Role in Neuroplasticity

## Russell M. Sanchez, PhD and Frances E. Jensen, MD

## 1. INTRODUCTION

Glutamate receptors mediate most excitatory synaptic transmission in the brain. Additionally, they mediate many forms of synaptic plasticity such as those thought to comprise the physiological basis of learning and memory. In the developing brain, glutamate receptor activation is required for appropriate synaptogenesis and activity-driven refinement of functional synaptic networks (1,2). Thus, in early brain development, glutamate receptors additionally mediate highly age-specific forms of neuroplasticity that may not continue into maturity. Notably, activity-driven and maturational changes in the physiological roles glutamate receptors are paralleled by dynamic regulation of their molecular composition and functional properties. In this chapter, we review the glutamate receptor subtypes and discuss the possible relationships between their dynamic regulation during development and their ability to mediate various forms of synaptic plasticity.

## 2. GLUTAMATE RECEPTOR SUBTYPES

Glutamate is an ubiquitous excitatory neurotransmitter in the brain, and there are several subtypes of glutamate receptor (for reviews, see refs 3–5). Glutamate receptors are broadly divided into the ionotropic glutamate receptors, which from glutamate-gated transmembrane ion channels, and the metabotropic glutamate receptors, which, when activated by glutamate, trigger intracellular signaling pathways via receptor-coupled second messengers. The ionotropic glutamate receptors are comprised of three subtypes whose names derive from selective agonist that bind each with highest affinity: the α-amino-3-hydroxy-5-methyl-4-isoxazole propionic acid (AMPA), N-methyl-D-aspartate (NMDA), and kainate (KA) receptors (3). The properties of each and their roles in neuroplasticity will be discussed separately.

### 2.1. NMDA Receptors

NMDA receptors are well known to be critically involved in many forms of activity- driven synaptic plasticity, and these have been extensively reviewed (see, for example, ref. 6). Three general features of NMDA receptors give them a unique role in the activity-dependent regulation of synaptic function. First, NMDA receptors form ion channels that are highly permeable to $Ca^{2+}$ (in addition to $Na^+$ and $K^+$), and the influx of $Ca^{2+}$ through NMDA receptors may trigger $Ca^{2+}$-dependent signaling pathways that regulate synaptic function and synaptogenesis (7,8) Second, NMDA receptors are highly voltage-dependent because their channels are blocked by $Mg^{2+}$ at membrane potentials at or more negative to the resting potential, and $Mg^{2+}$ is extruded from the channels only

From: *Contemporary Clinical Neuroscience: Glutamate and Addiction*
Edited by: Barbara H. Herman et al. © Humana Press Inc., Totowa, NJ

upon depolarization *(9)*. Thus, NMDA receptors require concurrent membrane depolarization (through the activation of non-NMDA ionotropic glutamate receptors) and glutamate binding to conduct appreciable current. This voltage dependence gives NMDA receptors the capacity to activate mechanisms of neuroplasticity in response only to specific patterns of synaptic input. The third key feature of NMDA receptors is that their channel kinetics are much slower than those of non-NMDA ionotropic glutamate receptors, and, therefore, their activation can result in relatively long-lasting membrane depolarization. This prolonged depolarization can further relieve the $Mg^{2+}$ block of NMDA receptor channels, recurrently increasing membrane depolarization and activating high voltage-activated $Ca^{2+}$ channels to additionally increase intracellular $Ca^{2+}$. In addition to their contributions to physiological forms of neuroplasticity, the $Ca^{2+}$ permeability and kinetic properties of NMDA receptors give them a critical role in pathophysiological processes such as ictal seizure discharges and hypoxic/ischemic neuronal injury *(10–12)*.

The precise properties of native NMDA receptors are determined in large part by the particular combination of independently genetically encoded molecular subunits that comprise each receptor *(3)*. NMDA receptors are heteromerically assembled from subunits dubbed NRI, and NR2A, B, C, and D (for reviews, see refs *(13 and 14)*. Evidence from recombinant expression studies indicates that only receptors composed of both NRI and NR2 subunits exhibit the functional properties of native NMDA receptors and, further, that certain properties (such as $Mg^{2+}$ sensitivity or channel kinetics) differ subtly depending on the particular NR2 subunits expressed *(15,16)*. Notably, NRI subunits are expressed throughout the brain, whereas each of the NR2 subunits displays regionally and developmentally specific expression patterns *(17)*. Thus, regional and maturational differences in NMDA receptor properties appear to derive in large part from differences in the particular NR2 subunits expressed.

Differences in the key properties of NMDA receptors that result from different subunit combinations can have profound consequences for neuroplasticity and disease. The properties of NMDA receptors generally are such that their activity is enhanced during early postnatal development, and this is a period in which neuroplasticity is more robust and the brain is highly susceptible to epileptogenesis and excitotoxicity. For example, NMDA receptor-mediated synaptic currents appear generally to be more slowly decaying in the early postnatal brain compared to the adult *(18,19)*. In the forebrain, the ratio of NR2B to NR2A expression is much greater during early brain development compared to adulthood. Recombinant NMDA receptors that contain predominantly NR2B tend to exhibit slower decay times than those containing NR2A *(17,20,21)*, and native NMDA receptors in neurons that express NR2A exhibit more rapid decay kinetics compared to those in neurons that do not express detectable NR2A transcripts *(22)*. Therefore, activation of NMDA receptors in immature forebrain neurons would be expected to induce a much longer-lasting depolarization and possibly increase the capacity for glutamate-mediated neuroplasticity compared to their adult counterparts. Consistent with this notion, the capacity for NMDA receptor-mediated synaptic plasticity has been observed to be enhanced in the immature brain and decreases with maturation *(19,23)*. Additionally, transgenic mice overexpressing NR2B were observed to maintain enhanced NMDA receptor-dependent synaptic plasticity (and learning) into adulthood compared to wild-type mice *(24)*. These data indicate that NMDA receptor-mediated mechanisms of synaptic plasticity can be strongly influenced by relatively subtle differences in the properties of the NMDA receptor channels.

Notably, the NR2D subunit also is expressed at higher levels in subcortical structures in the immature brain compared to the adult brain *(17,25)*. Dimeric receptors composed of NRI and NR2D exhibit decreased channel block by $Mg^{2+}$ and slower decay kinetics compared to receptors composed of other subunit combinations *(25,26)*. Thus, the transient developmental upregulation of NR2D could enhance NMDA receptor-mediated plasticity in the immature brain by decreasing the $Mg^{2+}$-block of NMDA receptor channels at resting membrane potentials and allowing significant membrane depolarization and $Ca^{2+}$ influx in response to any pattern of afferent activation.

More generally, these data taken together indicate that NMDA receptor composition and function is geared toward increased synaptic plasticity in the developing brain, where synaptogenesis and activity-dependent refinement of synaptic networks is ongoing and a high level of plasticity is necessary. This scenario also renders the immature brain more susceptible to NMDA receptor-mediated injury compared to the adult brain *(27)*.

## 2.2. AMPA Receptors

In contrast to NMDA receptors, AMPA receptors mediate fast excitatory neuronal signaling, as they exhibit rapid activation and desensitization, operate linearly near the resting membrane potential, and mostly form ion channels that are virtually impermeable to $Ca^{2+}$ for review, see ref. *(3)*. For these reasons, AMPA receptors historically were viewed as simply transmitting bits of information between neurons, and their role in neuroplasticity was thought to be only in the ability for $Ca^{2+}$-activated mechanisms to adjust the gain of the signals that AMPA receptors transmit. This role for AMPA receptors certainly has been firmly established, as studies over the last several years have revealed numerous posttranslational mechanisms by which AMPA receptor function is regulated by synaptic activity and $Ca^{2+}$-dependent pathways *(28)*.

It is now clear, however, that a subset of AMPA receptors in the brain and spinal cord exhibit relatively high permeability to $Ca^{2+}$ and can directly activate $Ca^{2+}$-dependent mechanisms of neuroplasticity similarly to NMDA receptors. AMPA receptors are thought to be pentamers assembled from any combination of the molecular subunits GluR1, 2,3, and 4 (alternatively, GluRA–D) *(3)*. Notably, each AMPA receptor gene encodes a subunit that will form homomeric channels that are permeable to $Ca^{2+}$ and other divalent cations. However, only the GluR2 mRNA undergoes posttranscriptional editing that results in the replacement of a neutral glutamine (Q) by a charged arginine (R) at a key site within the putative pore-forming region of the AMPA receptor channel to express a $Ca^{2+}$-impermeable channel *(29)*. Recombinant AMPA receptors that lack a GluR2 subunit exhibit significantly greater permeability to $Ca^{2+}$ and other divalent cations compared to those that contain GluR2 *(30–33)*.

In embryonic rat brain, the proportion of Q/R edited to unedited GluR2 increases with age, with virtually 100% of GluR2 being edited in the postnatal brain *(34)* Thus, among native AMPA receptors, only those that lack a GluR2 subunit will exhibit appreciable divalent permeability. In the adult rat brain, the vast majority of AMPA receptors expressed in the forebrain contain GluR2. However, a number of molecular and electrophysiological studies in the last several years indicated that a subset of neurons in the postnatal nervous system express AMPA receptors that may lack GluR2 and exhibit relatively high permeability to $Ca^{2+}$ and that these may have a role similar to that of NMDA receptors in activating $Ca^{2+}$ dependent pathways of neuroplasticity. In spinal cord, for example, strong activation of $Ca^{2+}$-permeable AMPA receptors (with NMDA receptors pharmacologically blocked) was observed to potentiate AMPA receptor-mediated excitatory postsynaptic currents (EPSCs) in a subpopulation of dorsal root ganglion neurons *(35)*. In the hippocampus, $Ca^{2+}$ influx through AMPA receptors was shown to be necessary for the induction of long-term depression (LTD) observed in type II interneurons in area CA3 *(36)*.

Notably,the ratio of expression of GluR2 subunits to that of other AMPA receptor subunits is significantly lower in the immature hippocampus compared to the adult *(37,38)*. and a larger number of principal neurons express divalent-permeable AMPA receptors in the neonatal hippocampus compared to the adult *(38)*. Thus, similar to NMDA receptors, maturational regulation of this key feature of AMPA receptors may confer upon them a specialized role in $Ca^{2+}$-dependent neuroplasticity during early brain development. The expression of these receptors in principal forebrain neurons selectively during early postnatal development suggests their possible role in normal development and age-dependent plasticity, as well as in the enhanced susceptibility of the immature brain to AMPA receptor-mediated epileptogenesis and excitotoxic injury *(38–40)*.

## 2.3. Kainate Receptors

Historically, kainate receptors were viewed as similar to AMPA receptors, largely because of their overlapping pharmacological sensitivities, fast kinetics, and lack of voltage dependence, but they are molecularly and functionally distinct *(41)*. kainate receptors are heteromerically assembled from the molecular subunits GluR5, 6, and 7, and KA1 and KA2. The GluR5 and GluR6 subunits also undergo posttranscriptional editing at the codon for the Q/R site that results in significantly decreased $Ca^{2+}$ permeability *(29)*. Similarly to AMPA receptors, the proportion of Q/R edited subunits increases with maturation *(42)*, and in spinal cord neurons, this has been shown to be correlated with a developmental decrease in $Ca^{2+}$ permeability *(43)*, These observations suggest specialized roles of divalent-permeable kainate receptors in neuroplastic processes during early brain development.

The lack of adequately specific agonists and antagonists historically made it difficult to distinguish kainate receptor-mediated responses from those of AMPA receptors in native brain preparations to examine their potential roles in neuroplasticity. Recently, using an antagonist that specifically blocks GluR5-containing kainate receptors, Collingridge and colleagues were able to determine that a form of hippocampal long-term potentiation (LTP) known to be independent of NMDA receptor activation can be induced by the activation of postsynaptic kainate receptors in CA3 pyramidal neurons *(44)*. Kainate receptor-mediated postsynaptic currents in these neurons in hippocampal slices are extremely small compared to AMPA receptor-mediated or NMDA receptor-mediated EPSCs and require temporal summation following trains of repeated stimulation (with AMPA and NMDA receptors pharmacologically blocked) to be clearly distinguished from noise by conventional voltage-clamp recording methods *(45)*. Thus, at first glance, it would appear unlikely that such small events could result in a rise in intracellular $Ca^{2+}$ that is sufficient to activate signaling pathways that fail to be activated following the much larger events that result from NMDA or AMPA receptor activation in the same cells. However, it is conceivable that kainate receptors could be specifically coupled to second messengers that do not interact with other glutamate receptors or are sequestered from $Ca^{2+}$ entering through AMPA or NMDA receptors. Notably, the antagonist used by Collingridge's group was shown among recombinant homomeric kainate receptors to be highly selective for GluR5-containing receptors, yet *in situ* hybridization studies suggest that CA3 pyramidal neurons do not express mRNA for this subunit in abundance *(46,47)*. It certainly is possible that native kainate receptors respond differently than the homomeric receptors used to determine drug selectivity *(44)*, but, clearly, the precise role for kainate receptors in this form of neuroplasticity remains somewhat controversial.

In addition, similar to NMDA and AMPA receptors, kainate receptors appear to undergo developmental regulation in their expression and function in such manner as to promote neuronal excitability in early brain development. Kidd and Isaac have shown that low-affinity kainate receptors are activated at thalamocortical synapses in the early postnatal period and that their contribution to postsynaptic responses decreases with maturation. Notably, the immature brain is much more sensitive to the epileptogenic effects of kainate compared to the adult *(48,49)*. However, as kainate also is a potent AMPA receptor agonist, it is not yet clear if the maturational state of kainate receptor function is a critical mediator of the developmental changes in kainate sensitivity.

## 3. METABOTROPIC GLUTAMATE RECEPTORS

Whereas the ionotropic glutamate receptors are capable of secondarily directly or indirectly activating mechanisms of synaptic plasticity, the metabotropic receptors are directly coupled to second-messenger pathways that mediate forms of short-term plasticity. There are at least eight cloned metabotropic glutamate receptors (termed mGluR1–mGluR8) *(4)*. These have been classified into three groups based on sequence homology, coupling to second-messenger systems, and pharmacological sensitivities. Group I receptors are coupled to phosphoinositide (PI) hydrolysis that leads to $Ca^{2+}$ mobilization from intracellular stores, whereas groups II and III receptors are negatively coupled to adenylyl cyclase (AC) activity.

Although the consequences of metabotropic glutamate receptor activation vary depending on receptor type, neuronal type, or brain region, some general principles regarding the outcomes of their activation have emerged *(5)*. Postsynaptic group I metabotropic receptor activation, in general, causes and increase in the intrinsic excitability of principal neurons (particularly in hippocampal CA1 and CA3 subfields), mainly via down-modulation of voltage-gated potassium channels *(50)*, groups II and III receptor activation tends to depress excitatory synaptic transmission by inhibiting glutamate release *(51)*.

Evidence suggests that group I postsynaptic metabotropic glutamate receptors are mostly involved in the regulation of synaptic plasticity (for reviews, see refs. 52 and 53). For example, mice lacking mGluR5 show reduced hippocampal CA1 LTP (although CA3 LTP was normal) *(54)*. However, presynaptic metabotropic glutamate receptors also may play roles in synaptic plasticity, as Laezza et al. showed that LTD in CA3 interneurons only resulted from a synergistic effect that required both the activation of presynaptic metabotropic receptors and $Ca^{2+}$ entry through postsynaptic AMPA receptors *(36)*. Thus, either presynaptic or postsynaptic metabotropic glutamate receptors may contribute to glutamate-mediated synaptic plasticity under different conditions.

Interestingly, in the developing brain, as with the ionotropic glutamate receptors, metabotropic glutamate receptor function undergoes maturational regulation in such a manner as to promote neuronal excitability in early postnatal development. Agonist-stimulated PI turnover has been shown to be relatively robust in slices of immature rat brain, increasing from age P1 to P7–P10 before gradually decreasing to adult levels at around P24 *(55)*. This contrasts with the activity of metabotropic receptors negatively coupled to AC, as cyclic AMP accumulation induced by the AC activator forskolin was shown to be inhibited by the nonspecific metabotropic glutamate receptor agonist 1$S$,3$R$(ACPD) in adult but not in neonatal (P1–P15) rat hippocampus *(56,57)*. Notably, metabotropic glutamate receptors negatively coupled to AC are expressed in early postnatal development, but nonspecific metabotropic receptor activation in the neonatal hippocampus increases basal cyclic AMP levels *(58)* and, thus, would be expected to promote neuronal excitability in early postnatal development. Consistent with notion, the nonspecific mGluR agonist 1$S$,3$R$-1-aminocyclopentane-1,3,dicarboxylic acid [(1$S$,3$R$)ACPD], which activates both AC- and PI-coupled metabotropic receptors, was shown to elicit dose- dependent limbic seizures in neonatal (postnatal d 7) rats *(59)*. This proconvulsant effect was similar to that observed for the specific group I mGluR agonist ($R$,$S$)-3,5-dihydroxyphenylglycine (3,5-DHPG) *(60)*, suggesting that it was mediated by AC-coupled metabotropic receptors. Furthermore, specific group II agonists appear to be anticonvulsant, as intraventricular infusion of the group II *agonist* (2$S$,1′$R$,2′$R$,3′$R$)-2-(2,3-dicarboxycyclopropyl)glycine (DCG-IV) decreased the incidence of continuous limbic motor seizures induced by intraventricular kainate *(61)*, and microinjection of ($S$)-4-carboxy-3-hydroxyphenylglycine ]($S$)-4C3HPG], a group I antagonist and group II agonist, into the inferior colliculus inhibited audiogenic seizures in genetically epilepsy-prone rats *(62)*. However, the proconvulsant actions of nonspecific metabotropic glutamate receptor agonists may not be entirely age-selective, as microinjections of (1$S$,3$R$)ACPD in adult rat hippocampus *(63)* also elicited limbic seizures. Compared to the ionotropic glutamate receptors, the developmental pattern of metabotropic glutamate receptor function is not as clearly linked to the developmental patterns of mGluR gene expression and glutamate receptor-mediated pathogenesis.

## 4. SUMMARY

It has been long established that much neuroplasticity in the brain is mediated by the actions of glutamate at glutamate receptors. Additionally, a vast literature exists on the role of glutamate receptors in neurological disease states, and the selective vulnerability of the immature brain. Until recently, experimental evidence suggested that most glutamate-mediated plasticity and pathology relied on the activation of the NMDA subtype of glutamate receptor, mainly due to its high permeability to $Ca^{2+}$. However, it is now clear that all of the glutamate receptors have the capacity to mediate neuroplastic and neuropathological processes and that these processes may be enhanced in the developing brain.

# REFERENCES

1. Fox, K., Schlaggar, B. L., Glazewski, S., and O'Leary, D. D. M. (1996) Glutamate receptor blockade at cortical synapses disrupts development of thalamocortical and columnar organization in somatosensory cortex. *Proc. Natl. Acad. Sci. USA* **93**, 5584–5589.
2. Constantine-Paton, M. (1994) Effects of NMDA receptor antagonists on the developing brain. *Psychopharmacol. Bull.* **30**, 561–565.
3. Dingledine, R., Borges, K., Bowie, D., and Traynelis, S. F. (1999) The glutamate receptor ion channels. *Pharmacol. Rev.* **51(1)**, 7–61.
4. Conn, P. J. and Pin, J.-P. (1997) Pharmacology and functions of metabotropic glutamate receptors. *Annu. Rev. Pharmacol. Toxicol.* **37**, 205–237.
5. Wong, R. K., Bianchi, R., Taylor, G. W., and Merlin, L. R. (1999). Role of metabotropic glutamate receptors in epilepsy. *Adv. Neurol.* **79**, 685–698.
6. Malenka, R. C. and Nicoll, R. A. (1993) NMDA-receptor-dependent synaptic plasticity: multiple forms and mechanisms. *Trends Neurosci.* **16(12)**, 521–527.
7. Aamodt, S. M. and Constantine-Paton, M. (1999) The role of neural activity in synaptic development and its implications for adult brain function. *Adv. Neurol.* **79**, 133–144.
8. Vallano, M. L. (1998) Developmental aspects of NMDA receptor function. *Crit. Rev. Neurobiol.* **12(3)**, 177–204.
9. Ascher, P. and Nowak, L. (1988) The role of divalent cations in the $N$-methyl-D-aspartate responses of mouse central neurons in culture. *J. Physiol.* **399**, 247–266.
10. Chapman, A. G. (1998) Glutamate receptors in epilepsy. *Prog. Brain Res.* **116**, 371–383.
11. Michaelis, E. K. (1998) Molecular biology of glutamate receptors in the central nervous system and their role in excitotoxicity, oxidative stress, and aging. *Prog. Neurobiol.* **54(4)**, 369–415.
12. Meldrum, B. S. (1994) The role of glutamate in epilepsy and other CNS disorders. *Neurology* **44(Suppl. 8)**, S14–S23.
13. Hollman, M. and Heinemann, S. (1994) Cloned glutamate receptors. *Annu. Rev. Neurosci.* **17**, 31–108.
14. Nakanishi, S. (1992) Molecular diversity of glutamate receptors and implications for brain function. *Science* **258**, 597–603.
15. Monyer, H., Sprengel, R., Schoepfer, R., Herb, A., Higuchi, M., Lomeli, H., et al. (1992) Heteromeric NMDA receptors: molecular and functional distinction of subtypes. *Science* **256**, 1217–1221.
16. Ishii, T., Moriyoshi, K., Sugihara, H., Sakurada, K., Kadotani, H., Yokoi, M., et al. (1992) Molecular characterization of the family of the $N$-methyl-D-aspartate receptor subunits. *J. Biol. Chem.* **268**, 2836–2843.
17. Monyer, H., Burnashev, N., Laurie, D. J., Sakmann, B., and Seeburg, P. H. (1994) Developmental and regional expression in the rat brain and functional properties of four NMDA receptors. *Neuron* **12**, 529–540.
18. Hestrin, S. (1992) Developmental regulation of NMDA receptor-mediated synaptic currents at a central synapse. *Nature* **357**, 686–689.
19. Crair, M. C. and Malenka, R. C. (1995) A critical period for long-term potentiation at thalamocortical synapses. *Nature* **375**, 325–328.
20. Zhong, J., Carrozza, D. P., Williams, K., Pritchett, D. B., and Molinoff, P. B. (1995) Expression of mRNAs encoding subunits of the NMDA receptor in developing rat brain. *J. Neurochem.* **64**, 531–539.
21. Burnashev, N., Jonas, P., Helm, P. J., Wisden, W., Monyer, H., Seeburg, P. H., et al. (1992) Calcium-permeable AMPA-kainate receptors in fusiform cerebellar glial cells. *Science* **256**, 1566–1570.
22. Flint, A. C., Maisch, U. S., Weishaupt, J. H., Kriegstein, A. R., and Monyer, H. (1997). NR2A subunit expression shortens NMDA receptor synaptic currents in developing neocortex. *J. Neurosci.* **17(7)**, 2469–2476.
23. Fox, K. (1995) The critical period for long-term in primary sensory cortex. *Neuron* **15**, 485–488.
24. Tang, Y. P., Shimizu, E. Dube, G. R., Rampon, C. Kerchner, G. A., Zhuo, M., et al. (1999) Genetic enhancement of learning and memory in mice. *Nature* **401**, 25–27.
25. Kohr, G., Eckhardt, S., Luddens, H., Monyer, H., and Seeburg, P. H. (1994) NMDA receptor channels: subunit-specific potentiation by reducing agents. *Neuron* **12**, 1031–1040.
26. Wenzel, A., Villa, M., and Benke, D. (1996) Development and regional expression of NMDA receptor subtypes containing the NR2D subunit in rat brain. *J. Neurochem.* **66**, 1240–1248.
27. Ikonomidou, C., Mosinger, J. L., Shahid Salles, K., Labryere, J., and Onley, J. W. (1989) Sensitivity of the developing rat brain to hypobaric/ischemic damage parallels sensitivity to $N$-methylaspartate toxicity. *J. Neurosci.* **9(8)**, 2809–2818.
28. Ziff, E. B. (1999) Recent excitement in the ionotropic glutamate receptor field. *Ann. NY Acad. Sci.* **868**, 465–473.
29. Sommer, B., Kohler, M., Sprengel, R., and Seeburg, P. H. (1991) RNA editing in brain controls a determinant of ion flow in glutamate-gated channels. *Cell* **67**, 11–19.
30. Burnashev, N., Monyer, H., Seeburg, P. H., and Sakmann, B. (1992) Divalent ion permeability of AMPA receptor channels is dominated by the edited form of a single subunit. *Neuron* **8**, 189–198.
31. Jonas, P., Racca, C., Sakmann, B., Seeburg, P. H., and Monyer, H. (1994) Differences in $Ca^{2+}$ permeability of AMPA-type glutamate receptor channels in neocortical neurons caused by differential expression of the GluR-B subunit. *Neuron* **12**, 1281–1289.

32. McBain, C. J. and Dingledine, R. (1993) Heterogeneity of synaptic glutamate receptors on CA3 stratum radiatum interneurones of rat hippocampus. *J. Physiol.* **462**, 373–392.
33. Washburn, M. S., Numberger, M., Zhang, S., and Dingledine, R. (1997) Differential dependence on GluR2 expression of the three characteristic features of AMPA receptors. *J. Neurosci.* **17**, 9393–9406.
34. Pellegrini-Giampietro, D. E., Gorter, J. A., Bennett, M. V. L., and Zukin, R. S. (1997) The GluR2 (GluR-B) hypothesis: Ca(2+)-permeable AMPA receptors in neurological disorders. *Trends Neurosci.* **20**, 464–470.
35. Gu, J. G., Albuquerque, C. J., Lee, C. J., and MacDermott, A. B. (1996) Synaptic strengthening through activation of $Ca^{2+}$-permeable AMPA receptors. *Nature* **381**, 793–796.
36. Laezza, F., Doherty, J. J., and Dingledine, R. (1999) Long-term depression in hippocampal interneurons: joint requirement for pre- and postsynaptic events. *Science* **285**, 1411–1414.
37. Pellegrini-Giampietro, D. E., Bennett, M. V. L., and Zukin, R. S. (1991) Differential expression of three glutamate receptor genes in developing rat brain: an in situ hybridization study. *Proc. Natl. Acad. Sci. USA* **88**, 4157–4161.
38. Sanchez, R. M., Koh, S., Rio, C., Wang, C., Lamperti, E. D., Sharma, D., et al. (2001) Decreased GluR2 expression and enhanced epileptogenesis in immature rat hippocampus following perinatal hypoxia-induced seizures. *J. Neurosci.*, **21**, 8145–8163.
39. McDonald, J. W., Trescher, W. H., and Johnston, M. V. (1992) Susceptibility of brain to AMPA induced excitotoxicity transiently peaks during early postnatal development. *Brain Res.* **583(1–2)**, 54–70.
40. McDonald, J. W., Trescher, W. H., and Johnston, M. V. (1990) The selective ionotropic-type quisqualate receptor agonist AMPA is a potent neurotoxin in immature rat brain. *Brain Res.* **526(1)**, 165–168.
41. Lerma, J., Paternain, A. V., Rodriguez-Moreno, A., and Lopez-Garcia, J. C. (2001) Molecular physiology of kainate receptors. *Physiol. Rev.* **81(3)**, 971–998.
42. Bernard, A., Ferhat, L., Dessi, F., Charton, G., Represa, A., Ben-Ari, Y., et al. (1999) Q/R editing of the rat GluR5 and GluR6 kainate receptors in vivo and in vitro: evidence for independent developmental, pathological and cellular regulation. *Eur. J. Neurosci.* **11**, 604–616.
43. Lee, C. J., Kong, H., Manzini, M. C., Albuquerque, C., Chao, M. V., and MacDermott, A. B. (2001) Kainate receptors expressed by a subpopulation of developing nociceptors rapidly switch from high to low $Ca^{2+}$ permeability. *J. Neurosci.* **21(13)**, 4572–4581.
44. Bortolotto, Z. A., Clarke, V. R., Delany, C. M., Parry, M. C., Smolders, I., Vignes, M., et al. (1999) Kainate receptors are involved in synaptic plasticity. *Nature* **402**, 297–301.
45. Vignes, M. and Collingridge, G. L. (1997) The synaptic activation of kainate receptors. *Nature* **388**, 179–182.
46. Bahn, S., Volk, B., and Wisden, W. (1994) Kainate receptor gene expression in the developing rat brain. *J. Neurosci.* **14**, 5525–5547.
47. Paternain, A. V., Herrera, M. T., Nieto, M. A., and Lerma, J. (2000) GluR5 and GluR6 kainate receptor subunits coexist in hippocampal neurons and coassemble to form functional receptors. *J. Neurosci.* **20(1)**, 196–205.
48. Albala, B. J., Moshe, S. L., and Okada, R. (1984) Kainic-acid-induced seizures: a developmental study. *Brain Res.* **315(1)**, 139–148.
49. Stafstrom, C. E., Thompson, J. L., and Holmes, G. L. (1992) Kainic acid seizures in the developing brain: status epilepticus and spontaneous recurrent seizures. *Brain Res. Dev. Brain Res.* **65**, 227–236.
50. Gerber, U. and Gahwiler, B. H. (1994) Modulation of ionic currents by metabotropic glutamate receptors, in *The Metabotropic Glutamate Receptors* (Conn, P. J. and Patel, J., eds.), Humana, Totowa, NJ, pp. 125–146.
51. Glaum, S. R. and Miller, R. J. (1994) Acute regulation of synaptic transmission by metabotropic glutamate receptors, in *The Metabotropic Glutamate Receptors* (Conn, P. J. and Patel, J. eds.) Humana, Totowa, NJ, pp. 147–172.
52. De Blasi, A. D., Conn, P. J., Pin, J.-P., and Nicoletti, F. (2001). Molecular determinants of metabotropic glutamate receptors signaling. *Trends Pharmacol. Sci.* **22(3)**, 114–120.
53. Pin, J.-P. and Duvoisin, R. (1995) The metabotropic receptors: structure and function. *Neuropharmacology* **34**, 1–26.
54. Lu, Y.-M., Jia, Z., Janus, C., Henderson, J. T., Gerlai, R., Wojtowicz, J. M., et al. (1997) Mice lacking metabotropic glutamate receptor 5 show impaired learning and reduced CA1 long-term potentiation (LTP) but normal CA3 LTP. *J. Neurosci.* **17(13)**, 5196–5205.
55. Nicoletti, E., Aronica, E., Battaglia, G., Bruno, V., Casabona, G., Catania, M. V., et al. (1994) Plasticity of metabotropic glutamate receptors in physiological and pathological conditions, in *The Metabotropic Glutamate Receptors* (Conn, P. J., J Patel, J., eds.), Humana, Totowa, NJ, pp. 243–269.
56. Casabona, G., Genazzani, A. A., Di Stefano, M., Sortino, M. A., and Nicoletti, F. (1992) Developmental changes in the modulation of cyclic AMP formation by the metabotropic glutamate receptor agonist 1S,3R-aminocyclopentane-1,3-dicarboxylic acid in brain slices. *J. Neurochem.* **59**, 1161–1163.
57. Schoepp, D. D. and Johnson, B. G. (1993) Metabotropic glutamate receptor modulation of cAMP accumulation in the neonatal rat hippocampus. *Neuropharmacology* **32**, 1359–1365.
58. Schoepp, D. D., Johnson, B. G., and Monn, J. A. (1996) (1S,3R)-1-aminocyclopentane-1,3-dicarboxylic acid-induced increases in cyclic AMP formation in neonatal rat hippocampus are mediated by a synergistic interaction between phosphoinositide- and inhibitory cyclic AMP-coupled mGluRs. *J. Neurochem.* **66**, 1981–1985.

59. McDonald, J. W., Fix, A. S., Tizzano, J. P., and Schoepp, D. D. (1993) Seizures and brain injury in neonatal rats induced by 1S,3R-ACPD, a metabotropic glutamate receptor agonist. *J. Neurosci.* **13**, 4445–4455.
60. Camon, L., Vives, P., de Vera, N., and Martinez, E. (1998) Seizures and neuronal damage induced in the rat by activation of group I metabotropic glutamate receptors with their selective agonist 3,4-dihydroxyphenylglycine. *J. Neurosci. Res.* **51**, 339–348.
61. Miyamoto, M., Ishida, M., and Shinozaki, H. (1997) Anticonvulsive and neuroprotective actions of a potent agonist (DCG-IV) for group II metabotropic glutamate receptors against intraventricular kainate in the rat. *Neuroscience* **77**, 131–140.
62. Tang, E., Yop, P. K., Chapman, A. G., Jane, D. E., and Meldrum, B. S. (1997) Prolonged anticonvulsant action of glutamate metabotropic receptor agonists in inferior colliculus of genetically epilepsy-prone rats. *Eur. J. Pharmacol.* **327**, 109–115.
63. Sacaan, A. I. and Schoepp, D. D. (1992) Activation of hippocampal metabotropic excitatory amino acid receptors leads to seizures and neuronal damage. *Neurosci. Lett.* **139**, 77–82.

# 5
# Role of the NMDA Receptor in Neuronal Apoptosis and HIV-Associated Dementia

## Marcus Kaul, PhD and Stuart A. Lipton, MD, PhD

## 1. INTRODUCTION

Neuronal injury and apoptosis may account, at least in part, for neurological complications associated with human immunodeficiency virus (HIV)-1 infection ranging from mild cognitive and motor impairment to dementia. The primary cell types infected in the brain are macrophages and microglia. These cells have been found in vivo and in vitro to release neurotoxic factors. Evidence has accumulated that neuronal apoptosis in HIV-related insults occurs predominantly via an indirect pathway comprising a complex cooperation of cytokines, reactive oxygen species and reactive nitrogen species, lipid mediators, and excitotoxins. These molecules lead to excessive stimulation of the *N*-methyl-D-aspartate subtype of glutamate receptor (NMDAR). Of note, chemokine receptors, which, in conjunction with CD4, mediate HIV infection of macrophages/microglia, are present on neurons and astrocytes in addition to macrophages/microglia. Thus, these receptors potentially allow direct interaction between the virus and neurons (Fig. 2). The fact that specific chemokines ameliorate HIV/gp120-induced neuronal apoptosis that is mediated by NMDARs suggests a functional connection between the receptors for chemokines and NMDA. Accordingly, here we review the role of the NMDAR in HIV-1-related and excitotoxic neuronal cell death.

## 2. GLUTAMATE RECEPTORS, EXCITOTOXICITY, AND NEURONAL CELL DEATH

NMDAR belongs to a large and heterogeneous family of membrane proteins, the glutamate receptors. These glutamate receptors recognize the major excitatory neurotransmitter in the central nervous system (CNS), (*S*)-glutamic acid (Glu), and other related excitatory amino acids (EAAs) (*1–3*). To date, four classes of EAA receptors have been identified and many member subunits cloned. These include three "ionotropic" receptor classes [iGluRs, comprised of ligand-gated ion channels termed (*RS*)-2-amino-3-(3-hydroxy-5-methyl-4-isoxazolyl)propionic acid (AMPA), kainic acid (KA), and NMDA receptors] and a G-protein-coupled or "metabotropic" EAA receptor class (mGluRs) (*1,2,4*). Both iGluRs and mGluRs are considered to play important roles in the CNS under normal physiological and pathophysiological conditions. Under physiological conditions, activation of iGluRs in neurons initiates transient depolarization and excitation. AMPARs mediate the fast component of excitatory postsynaptic currents and NMDARs underlie a slower component. Presynaptic release of Glu and consequent depolarization of the postsynaptic neuronal membrane via AMPAR-coupled channels relieve the $Mg^{2+}$ block of the ion channel associated with the NMDAR under resting conditions. This effect allows subsequent controlled $Ca^{2+}$ influx through the NMDAR-coupled ion channel.

From: *Contemporary Clinical Neuroscience: Glutamate and Addiction*
Edited by: Barbara H. Herman et al. © Humana Press Inc., Totowa, NJ

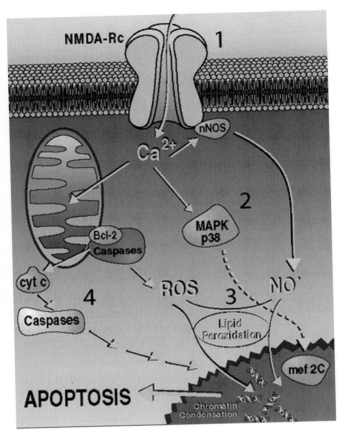

**Fig. 1.** Current model of NMDA receptor-associated neuronal injury. Schematic illustration of NMDAR-related signaling pathways that lead to neuronal apoptosis and may contribute to neurodegenerative disease, including HIV-associated dementia. These pathways can be interrupted to prevent neuronal apoptosis; thus, their study affords the opportunity to develop potential treatments for various neurologic diseases. Drug or molecular therapies are being developed to (1) antagonize NMDA receptors (NMDA-Rc), (2) modulate activation of the p38 mitogen-activated kinase (MAPK)–MEF2C (transcription factor) pathway, (3) prevent toxic reactions of free radicals such as nitric oxide (NO) and reactive oxygen species (ROS) that form peroxynitrite (ONOO−), and (4) inhibit apoptosis-inducing factors, including caspases. Activation of the p38 MAPK–MEF2C pathway appears to occur upstream of the effector caspases (not illustrated here). (Color illustration in insert following p. 142.)

This voltage-dependent modulation of the NMDAR results in activity-driven synaptic modulation *(2,5)*. However, extended and/or excessive NMDAR activation and consequent overexcitation can damage neurons and eventually cause cell death. This process is called excitotoxicity and appears to be favored by sustained elevation of the intracellular $Ca^{2+}$ concentration and/or compromised cellular energy metabolism *(5,6)*.

A role for Glu excitotoxicity in brain disorders was first suggested by the work of Olney following the pioneering work of Lucas and Newhouse in the retina *(6)*. Subsequently, several lines of evidence indicated that excessive stimulation of glutamate receptors contributes to the neuropathological processes in stroke, head and spinal cord injury, Huntington's disease, Parkinson's disease, possibly Alzheimer's disease, amyotrophic lateral sclerosis, multiple sclerosis, glaucoma, and HIV-1 associated dementia *(1,5,7)*. Indeed, excitotoxicity seems to represent a common final pathway in a wide variety of neurodegenerative disorders *(8)*.

NMDAR has attracted particular interest as a major player in excitotoxicity because this receptor, in contrast to most non-NMDARs (AMPA and KA receptors), is highly permeable to $Ca^{2+}$, and excessive $Ca^{2+}$ influx can trigger excitotoxic neuronal injury *(3,9)*. In addition, NMDAR antagonists effectively prevent glutamate neurotoxicity, both in vitro and in vivo in animal studies, as well as in recent phase III clinical trials with the NMDAR open-channel blocker, memantine *(2,5,10)*. However, AMPA and KA receptors can also mediate excitotoxicity and contribute to neuronal damage under certain conditions *(2,5)*. For example, a subpopulation of $Ca^{2+}$- or $Zn^{2+}$-permeable AMPA receptor-coupled channels have been implicated in selective neurodegenerative disorders, such as ischemia, epilepsy, Alzheimer's disease, and amyotrophic lateral sclerosis *(3)*. Also, transgenic mice overexpressing AMPARs display increased damage subsequent to ischemia when compared to control animals *(11)*.

Excessive stimulation of the NMDAR induces several detrimental intracellular signals that contribute to neuronal cell death by apoptosis or necrosis, depending on the intensity of the initial insult *(12)*. Excessive $Ca^{2+}$ influx through NMDAR-coupled ion channels leads to an elevation of the intracellular free-$Ca^{2+}$ concentration to a point that results in $Ca^{2+}$ overload of mitochondria, depolarization of the mitochondrial membrane potential, and a decrease in ATP synthesis. Additionally, excessive intracellular $Ca^{2+}$ stimulates protein kinase cascades and the generation of free radicals, including reactive oxygen species (ROS) and nitric oxide (NO) *(12)*. NO can react with ROS to form cytotoxic peroxynitrite ($OONO^-$) *(12)*, and in alternative redox states, NO can also activate p21ras by S-nitrosylation (transfer of the NO group to a critical cysteine thiol) *(13)*. However, the NO group can also inhibit caspases in cerebrocortical neurons via S-nitrosylation, thereby attenuating apoptosis *(14)*. The scaffolding protein PSD-95 (postsynaptic density-95) links the principal subunit of the NMDAR (NR1) with neuronal nitric oxide synthase (nNOS), a $Ca^{2+}$-activated enzyme, and thus brings nNOS into close proximity to $Ca^{2+}$ via the NMDAR-operated ion channel *(15)* (see Fig. 1).

Importantly, excessive $Ca^{2+}$ influx also activates the stress-related p38 mitogen-activated protein kinase (p38 MAPK)/myocyte enhancer factor 2C (MEF2C transcription factor) pathway and c-Jun N-terminal kinase (JNK) pathways in cerebrocortical or hippocampal neurons. Activation of these pathways has been implicated in neuronal apoptosis *(16,17)*. As stated above, excessive intracellular $Ca^{2+}$ accumulation after NMDAR stimulation leads to depolarization of the mitochondrial membrane potential ($\Delta\Psi_m$) and a drop in the cellular ATP concentration. If the initial excitotoxic insult is fulminant, the cells do not recover their ATP levels and die at this point because of the loss of ionic homeostasis, resulting in acute swelling and lysis (necrosis). If the insult is milder, ATP levels recover, and the cells enter a delayed death pathway requiring energy, known as apoptosis *(12)*.

It has been reported that NMDAR-mediated excitotoxicity leading to neuronal apoptosis also involves activation of the $Ca^{2+}$/calmodulin-regulated protein phosphatase calcineurin, release of cytochrome c from mitochondria, activation of caspase-3, lipid peroxidation, and cytoskeletal breakdown *(12,18,19)*. Inhibition of calcineurin or caspase-3 with FK506 caspase inhibitors, respectively, can attenuate this form of excitotoxicity *(12,19)*. It has been proposed that the adenine nucleotide translocator (ANT) is a part of the mitochondrial permeability transition pore (PTP) and participates in mitochondrial depolarization. Indeed, our group has found that pharmacologic blockade of the ANT with bongkrekic acid prevented collapse of the mitochondrial membrane potential ($\Delta\Psi_m$), as well as subsequent caspase-3 activation and NMDA-induced neuronal apoptosis. However, treatment with bongkrekic acid failed to inhibit the transient drop in ATP concentration (although it hastened the recovery of ATP levels) and did not prevent the liberation of cytochrome c into the cytosol. Thus, initiation of caspase-3 activation and resultant neuronal apoptosis after NMDAR activation require a factor(s) in addition to cytochrome c release *(18)*.

Interestingly, stimulation of specific subtypes of the G-protein-coupled mGluRs interferes with excitotoxic NMDAR-mediated activation of MAPKs and can attenuate subsequent neuronal cell death

*(16)*. Additionally, glial cells, including astrocytes, microglia, and oligodendrocytes, may possess some types of glutamate receptor *(4)*. Both AMPA and KA receptor subtypes as well as mGluRs have been reported on microglia, and functional NMDARs have been reported to exist in some cases on astrocytes and oligodendrocytes (although these latter findings of NMDARs need to be verified). Glial glutamate receptors appear to be involved in interactions between neuronal and glial cells and, hence, may conceivably contribute to synaptic efficacy. Furthermore, under certain pathologic circumstances, such as cerebral hypoxia–ischemia and possibly HIV-1 infection of the brain, astrocytes and oligodendrocytes may undergo glutamate-mediated excitotoxic cell death *(4)*.

## 3. HIV-1 INFECTION AND NMDA RECEPTOR-RELATED NEURONAL APOPTOSIS

HIV-associated dementia eventually develops in approximately half of children and a quarter of adults infected with HIV-1 *(7)*. Neuropathological features that may accompany this cognitive–motor complex include dendritic and synaptic damage, apoptosis and frank loss of neurons, myelin pallor, astrocytosis, and infiltration of macrophages, microglia, and multinucleated giant cells *(7,20,21)*.

Addiction to drugs, such as heroin, cocaine, and methamphetamine, is considered to be a major risk factor for HIV infection *(22)*. Furthermore, it has been proposed that abuse of substances that cause by themselves profound alterations in CNS function could also affect the development of HIV-associated dementia *(23)*. An autopsy study detected evidence of HIV encephalitis, including multinucleated giant cells and HIV p24 antigen, more frequently in HIV-positive drug users than HIV-infected homosexual man *(24)*, and cocaine has been implicated in facilitation of HIV entry into the brain *(25,26)*. However, although an effect of substance abuse on HIV-associated dementia appears likely, direct links between drug addiction and HIV-induced brain injury remain to be elucidated *(23)*.

Macrophages and microglia play a crucial role in HIV-associated dementia because they are the predominant cells productively infected with HIV-1 in AIDS brains *(7)*. Infection of astrocytes has also been rarely observed in pediatric cases (reviewed in ref. 27). It is currently held that HIV-1 infected macrophages migrate into the brain *(28)*, thus allowing viral entry into the CNS, and the presence of macrophages/microglia has been reported to correlate with the severity of HIV-associated dementia *(29)*. Furthermore, we and our colleagues have shown that HIV-1-infected or immune stimulated macrophages/microglia produce neurotoxins *(7,30)*.

The mechanisms that initiate and mediate HIV-associated dementia are not completely understood. HIV-1 apparently enters the CNS soon after peripheral infection, and the virus primarily resides in microglia and macrophages, especially in those located in the perivascular space *(28)*. It is not clear if the migration of infected monocytes and macrophages represents the only pathway for viral entry into the brain. Additionally, infection of monocytoid cells *per se* may not be sufficient to initiate the dementing process *(28)*. In the CNS, HIV-1 is thought to cause immune activation of macrophages/microglia, changes in expression of cytokines, chemokines, and their receptors, and upregulation of endothelial adhesion molecules (reviewed in ref. 28). However, these observations may be the result of the process rather than the inciting event for HIV-1-associated brain pathology. Therefore, it has been proposed that peripheral (non-HIV) infection or other factors may trigger events leading to dementia after HIV-1 infection in the CNS has been established. One such factor could be the increased number of activated monocytes in the circulation that express CD16 and CD69. These activated cells could possibly adhere to the normal endothelium of the brain microvasculature, transmigrate, and then trigger a number of deleterious processes *(28)*.

Infection of cells by HIV-1 can occur after binding of the viral envelope protein gp120 to one of several possible chemokine receptors in conjunction with CD4. Depending on the exact type of gp120, different HIV-1 strains may use CXCR4, CCR3, CCR2, CCR5, or a combination of these chemokine receptors to enter target cells *(31,32)*. Microglia are infected by HIV-1 primarily via CCR5 and CCR3, and possibly via CXCR4 *(33)*. CCR5 and CXCR4, among other chemokine receptors, are also present on neurons and astrocytes, and, in particular, CXCR4 and CCR5 are highly expressed on neurons of macaques and humans *(34)*. In vitro studies strongly suggest that chemokine receptors are directly involved in HIV-associated neuronal damage *(17,35,36)*.

Even in the absence of intact virus, the HIV proteins gp120, gp41, gp160, Tat, Nef, Rev, and Vpr have been reported to initiate neuronal damage both in vitro and in vivo *(37–41)*. In this regard, the viral envelope protein gp120 has been of particular interest, as it is essential for selective binding and signaling of HIV-1 to its target cell and for viral infection *(17,33,35,36,39)*. Additionally, evidence has been provided that gp41, the membrane-spanning region of the viral envelope protein, correlates with the expression of immunologic/type II NOS (iNOS) as well as the degree of HIV-associated dementia *(40)*.

A recurring question has been whether HIV-1 or its component proteins induce neuronal damage predominantly by an indirect route (e.g., via toxins produced by infected or immune-stimulated macrophages and/or astrocytes) or by a direct route (e.g., via binding to neuronal receptors) *(7,17,42)*. Several lines of evidence suggest that HIV-associated neuronal injury involves predominantly an indirect route and resulting excessive activation of NMDARs with consequent excitotoxicity *(30,43,44)*. Analysis of specimens from AIDS patients *(44)* as well as in vivo and in vitro experiments indicate that HIV-1 infection creates excitotoxic conditions, most probably indirectly via induction of soluble factors in macrophage/microglia and/or astrocytes, such as glutamatelike molecules, viral proteins, cytokines, chemokines, and arachidonic acid metabolites *(7,42,45)*.

However, it has also been suggested that HIV-1 or its protein components can directly interact with neurons and also modulate NMDAR function, at least under some conditions *(35,46)*. Picomolar concentrations of soluble HIV/gp120 induce injury, both in vitro and in vivo, and this can lead to apoptosis in primary rodent and human neurons *(37,39)*. Additionally, our group and subsequently several others have shown that gp120 contributes to NMDAR-mediated neurotoxicity *(43)*. Both voltage-gated $Ca^{2+}$ channel blockers and NMDAR antagonists can ameliorate gp120-induced neuronal cell death in vitro, although NMDAR antagonists are more effective in their protection *(43,47)*. Transgenic mice expressing gp120 manifest neuropathological features that are similar in many ways to the findings in brains of AIDS patients, and in these mice, neuronal damage is ameliorated by the NMDAR antagonist memantine *(38,48)*. Memantine acts as an open-channel blocker of the NMDAR-coupled ion channel, but the drug only manifests significant action when the NMDAR is excessively activated, leaving relatively spared normal synaptic transmission. Hence, memantine has proven to be clinically tolerated in a number of human trials and is already available in Europe for other clinical indications. It is also conceivable that other glutamate receptors in addition to NMDARs influence HIV-associated neuronal damage. For example, disparate mGluRs have been found to up- or down-modulate excitotoxic signals triggered by NMDARs *(16)*.

In our hands, the predominant mode of HIV-1- or gp120-induced neurotoxicity of cerebrocortical neurons requires the presence of macrophages/microglia; HIV-1-infected or gp120-stimulated mononuclear phagocytes have been shown to release neurotoxins that lead to excessive stimulation of NMDARs *(17,30)*. These macrophage toxic factors include molecules that directly or indirectly act as NMDAR agonists, such as quinolinic acid, cysteine, and arachidonic acid and its metabolites, such as platelet-activating factor (PAF), a low-molecular-weight amine designated NTox, and perhaps glutamate itself *(7,45,49)*.

Additionally, activated macrophages/microglia and possibly astrocytes produce inflammatory cytokines, including tumor necrosis factor (TNF)-α and interleukin (IL)-1β, arachidonic acid metabolites, and free radicals (ROS and NO) that may contribute to excitotoxic neuronal damage *(7,45)*. TNF-α and IL-1β may amplify neurotoxin production by stimulating adjacent glial cells and by increasing iNOS activity [*(45)*; Fig. 2].

In contrast to these indirect neurotoxic pathways, it has been reported that gp120 can directly interact with neurons when the neurons are exposed to gp120 in the absence of glial cells. Recently,

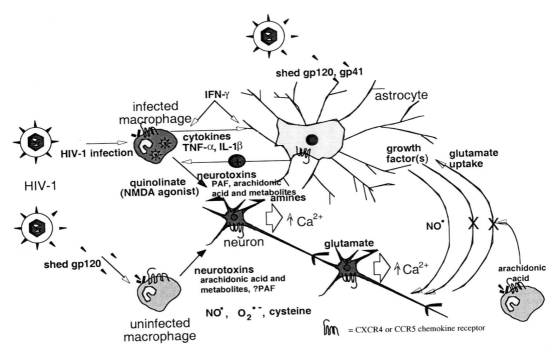

**Fig. 2.** Current model of HIV-associated neuronal injury. Immune activated and HIV-infected brain macrophages/microglia release potentially neurotoxic substances. These substances, emanating from macrophages and also possibly from reactive astrocytes, contribute to neuronal injury and apoptosis as well as to proliferation and activation of astrocytes (astrocytosis). A major mode of entry of HIV-1 into monocytoid cells occurs via the binding of gp120 and, therefore, it is not surprising that gp120 (or a fragment thereof) is capable of activating uninfected macrophages to release similar factors to those secreted in response to frank HIV infection. Macrophages bear CCR5 and possibly CXCR4 chemokine receptors on their surface in addition to CD4, and gp120 binds via these receptors. Some populations of neurons and astrocytes have been reported to also bear CXCR4 and CCR5 receptors on their surface, raising the possibility of direct interaction with gp120. Macrophages and astrocytes have mutual feedback loops (signified by the reciprocal arrows). Cytokines participate in this cellular network in several ways. For example, HIV infection or gp 120 stimulation of macrophages enhances their production of TNF-α and IL-1β (solid arrow). The TNF-α and IL-1β produced by macrophages stimulate astrocytosis. Neuronal injury is primarily mediated by overactivation of NMDARs with resultant excessive intracellular $Ca^{2+}$ levels. This, in turn, leads to overactivation of a variety of potentially harmful enzyme systems, the formation of free radicals, and release of the neurotransmitter—glutamate. Glutamate subsequently overstimulates additional NMDARs on neighboring neurons, resulting in further injury. This final common pathway of neurotoxic action can be blocked in large measure by NMDAR antagonists. For certain neurons, depending on their exact repertoire of ionic channels, this form of damage can also be ameliorated to some degree by calcium channel antagonists or non-NMDAR antagonists. Additionally, agonists of β-chemokine receptors, which are present in the CNS on neurons, astrocytes, and microglia, can confer partial protection against neuronal apoptosis induced by HIV/gp120 or NMDA. IFN-interferon; IL-interleukin; $NO^·$-nitric oxide; $O_2^{·-}$-superoxide anion; TNF-tumor necrosis factor. (Color illustration in insert following p. 142.)

picomolar levels of gp120 was found to act at chemokine receptors on neurons to induce their death *(35)*. Additionally, higher (nanomolar) concentrations of gp120 have been reported to interact with the glycine-binding site of the NMDAR *(50)*, although it is not clear if gp120 exists at this high a level in the HIV-infected brain. Furthermore, gp120 may produce a direct excitotoxic influence via NMDAR-mediated $Ca^{2+}$ oscillations in rat hippocampal neurons *(51)* and may bind to noradrenergic axon terminals in neocortex, where it possibly potentiates NMDA-evoked noradrenaline release *(52)*. Nonetheless, it must be emphasized that many, if not all, of these direct effects on neurons were observed in vitro in the absence of glial cells. However, glial cells are known to modify these death pathways. Thus, we feel that based on work in mixed neuronal/glial systems that simulate the conditions that exist in vivo, the indirect route to neuronal injury is the predominant one.

Along these lines, gp120 has been found to aggravate excitotoxic conditions by impairing astrocyte uptake of glutamate via arachidonic acid that is released from activated macrophages/microglia *(42)*. Metabolites of arachidonic acid, such as prostaglandins, also stimulate a $Ca^{2+}$-dependent release of Glu by astrocytes *(53)*. Moreover, HIV-1 can induce astrocytic expression of the β-chemokine known as macrophage chemotactic protein-1 (MCP-1). This β-chemokine, in turn, attracts additional mononuclear phagocytes and microglia to further enhance the potential for indirect neuronal injury via release of macrophage toxins *(54)*.

Existing evidence suggests that HIV-1 infection and its associated neurological dysfunction involve both chemokine receptors and NMDAR-mediated excitotoxicity. This dual-receptor involvement raises the question of whether G-protein-coupled chemokine receptors and ionotropic glutamate receptors might influence each other's activity. Indeed, the β-chemokine known as "regulated upon activation T-cell expressed and secreted" (RANTES), which binds to the chemokine receptors CCR1, CCR3, and CCR5, can abrogate neurotoxicity induced by gp120 *(17)* or by excessive NMDAR stimulation *(55)*. In turn, excitotoxic stimulation can enhance expression of CCR5 *(56)*. Whether or not these findings reflect a mechanism of feedback or crosstalk of these receptors remains to be elucidated.

## 4. CONCLUSION

Considerable progress has been made in understanding the mechanisms of toxicity associated with overstimulation of NMDARs that lead to pathological neuronal excitation, excessive $Ca^{2+}$ influx, free-radical generation, and apoptosis. Increasing evidence indicates that excitotoxicity represents a common final pathway in many neurological disorders, including HIV-associated dementia. NMDAR antagonists can inhibit both in vitro and in vivo the neurotoxicity of glutamate/NMDA and of HIV/gp120. Additionally, chemokine receptors, essential coreceptors of HIV infection, are present in the CNS on neurons, astrocytes, and microglia, and agonists of β-chemokine receptors can, in part, also confer protection against neuronal apoptosis induced by HIV/gp120 or glutamate/NMDA. These findings suggest the possibility of a functional connection between receptors for chemokines and NMDA. Recently, phase III trials with the NMDAR antagonist memantine have demonstrated benefit in a series of neurodegenerative conditions, including Alzheimer's disease and, what is termed in Europe vascular dementia. Moreover, a large, multi center clinical trial of memantine for HIV-associated dementia is currently being analyzed. In the future, clinical studies may lead to therapeutic applications of chemokines for HIV-associated dementia and other neurodegenerative disorders as well.

## ACKNOWLEDGMENTS

This work was supported in part by NIH grants P01 HD29587 and R01 EY09024 (to SAL) and by fellowships from the DFG and American Foundation for AIDS Research (to MK). We thank several present and former members of the Lipton laboratory who contributed to the work described herein, including the recent work of Dr. Samantha Budd, Dr. Gwenn Garden, Dr. Yun-Beom Choi, and Dr. Michael Yeh, as well as members of the collaborating laboratories of Dr. Jonathan Stamler, Dr. Howard Gendelman, Dr. Pierluigi Nicotera, and Dr. Sten Orrenius.

## REFERENCES

1. Brauner-Osborne, H., Egebjerg, J., Nielsen, E. O., Madsen, U., and Krogsgaard-Larsen, P. (2000) Ligands for glutamate receptors: design and therapeutic prospects. *J. Med. Chem.* **43**, 2609–2645.
2. Bigge, C. F. (1999) Ionotropic glutamate receptors. *Curr. Opin. Chem. Biol.* **3**, 441–447.
3. Weiss, J. H. and Sensi, S. L. (2000) $Ca^{2+}$–$Zn^{2+}$ permeable AMPA or Kainate receptors: possible key factors in selective neurodegeneration. *Trends Neurosci.* **23**, 365–371.

4. Gallo, V. and Ghiani, C. A. (2000) Glutamate receptors in glia: new cells, new inputs and new functions. *Trends Pharmacol. Sci.* **21,** 252–258.
5. Doble, A. (1999) The role of excitotoxicity in neurodegenerative disease: implications for therapy. *Pharmacol. Ther.* **81,** 163–221.
6. Olney, J. W. (1969) Brain lesions, obesity, and other disturbances in mice treated with monosodium glutamate. *Science* **164,** 719–721.
7. Lipton, S. A. and Gendelman, H. E. (1995) Dementia associated with the acquired immunodeficiency syndrome. *N. Engl. J. Med.* **332,** 934–940.
8. Lipton, S. A. and Rosenberg, P. A. (1994) Excitatory amino acids as a final common pathway for neurologic disorders. *N. Engl. J. Med.* **330,** 613–622.
9. Choi, D. W. (1988) Glutamate neurotoxicity and diseases of the nervous system. *Neuron* **1,** 623–634.
10. Choi, D. W., Koh, J. Y., and Peters, S. (1988) Pharmacology of glutamate neurotoxicity in cortical cell culture: attenuation by NMDA antagonists. *J. Neurosci.* **8,** 185–196.
11. Le, D., Das, S., Wang, Y. F., Yoshizawa, T., Sasaki, Y. F., Takasu, M., Names, A., et al. (1997) Enhanced neuronal death from focal ischemia in AMPA-receptor transgenic mice. *Brain. Res. Mol. Brain. Res.* **52,** 235–241.
12. Nicotera, P., Ankarcrona, M., Bonfoco, E., Orrenius, S., and Lipton, S. A. (1997) Neuronal necrosis and apoptosis: two distinct events induced by exposure to glutamate or oxidative stress. *Adv. Neurol.* **72,** 95–101.
13. Yun, H. Y., Gonzalez-Zulueta, M., Dawson, V. L., and Dawson, T. M. (1998) Nitric oxide mediates $N$-methyl-D-aspartate receptor-induced activation of p21ras. *Proc. Natl. Acad. Sci. USA* **95,** 5773–5778.
14. Tenneti, L., DEmilia, D. M., and Lipton, S. A. (1997) Suppression of neuronal apoptosis by S-nitrosylation of caspases. *Neurosci. Lett.* **236,** 139–142.
15. Sattler, R., Xiong, Z., Lu, W. Y., Hafner, M., MacDonald, J. F., and Tymianski, M. (1999) Specific coupling of NMDA receptor activation to nitric oxide neurotoxicity by PSD-95 protein. *Science* **284,** 1845–1848.
16. Mukherjee, P. K., DeCoster, M. A., Campbell, F. Z., Davis, R. J., and Bazan, N. G. (1999) Glutamate receptor signaling interplay modulates stress-sensitive mitogen-activated protein kinases and neuronal cell death. *J. Biol. Chem.* **274,** 6493–6498.
17. Kaul, M. and Lipton, S. A. (1999) Chemokines and activated macrophages in gp120-induced neuronal apoptosis. *Proc. Natl. Acad. Sci. USA* **96,** 8212–8216.
18. Budd, S. L., Tenneti, L., Lishnak, T., and Lipton, S. A. (2000) Mitochondrial and extramitochondrial apoptotic signaling pathways in cereberocortical neurons. *Proc. Natl. Acad. Sci. USA* **97,** 6161–6166.
19. Tenneti, L., D'Emilia, D. M., Troy, C. M., and Lipton, S. A. (1998) Role of caspases in $N$-methyl-D-asoartate-induced apoptosis in cerebrocortical neurons. *J. Neurochem.* **71,** 946–959.
20. Masliah, E., Heaton, R. K., Marcotte, T. D., Ellis, R. J., Wiley, C. A., Mallory, M., et al. (1997) Dendritic injury is a pathological substrate for human immunodeficiency virus-related cognitive disorders. HNRC group. The HIV neurobehavioral research center. *Ann. Neurol.* **42,** 963–972.
21. Petito, C. K. and Roberts, B. (1995) Evidence of apoptotic cell death in HIV encephalitis. *Am. J. Pathol.* **146,** 1121–1130.
22. Goodkin, K., Shapshak, P., Metsch, L. R., McCoy, C. B., Crandall, K. A., Kumar, M., et al. (1998) Cocaine abuse and HIV-1 infection: epidemiology and neuropathogenesis. *J. Neuroimmunol.* **83,** 88–101.
23. Tyor, W. R. and Middaugh, L. D. (1999) Do alcohol and cocaine abuse alter the course of HIV-associated dementia complex? *J. Leukocute Biol.* **65,** 475–481.
24. Bell, J. E., Brettle, R. P., Chiswick, A., and Simmonds, P. (1998) HIV encephalitis, proviral load and dementia in drug users and homosexuals with AIDS. Effect of neocortical involvement. *Brain* **121,** 2043–2052.
25. Zhang, L., Looney, D., Taub, D., Chang, S. L., Way, D., Witte, M. H., et al. (1998) Cocaine opens the blood-brain barrier to HIV-1 invasion. *J.Neurovirol.* **4,** 619–626.
26. Fiala, M., Gan, X. H., Zhang, L., House, S. D., Newton, T., Graves, M. C., et al. (1998) Cocaine enhances monocyte migration across the blood-brain barrier. Cocaine's connection to AIDS dementia and vasculitis? *Adv. Exp. Med. Biol.* **437,** 199–205.
27. Brack-Werner, R. and Bell, J. E. (1999) Replication of HIV-1 in human astrocytes. *NeuroAIDS* **2,** 1–4 (www.sciencemag.org/NAIDS).
28. Gartner, S. (2000) HIV infection and dementia. *Science* **287,** 602–604.
29. Glass, J. D., Fedor, H., Wesselingh, S. L., and McArthur, J. C. (1995) Immunocytochemical quantitation of human immunodeficiency virus in the brain: correlations with dementia. *Ann. Neurol.* **38,** 755–762.
30. Giulian, D., Vaca, K., and Noonan, C. A. (1990) Secretion of neurotoxins by mononuclear phagocytes infected with HIV-1 *Science* **250,** 1593–1596.
31. Bleul, C. C., Farzan, M., Choe, H., Parolin, C., Clark-Lewis, I., Sodroski, J., et al. (1996) The lymphocyte chemoattractant SDF-1 is a ligand for LESTR/fusin and blocks HIV-1 entry. *Nature* **382,** 829–833.
32. Doranz, B. J., Rucker, J., Yi, Y., Smyth, R. J., Samson, M., Peiper, S. C., et al. (1996) A dual-tropic primary HIV-1 isolate that uses fusin and the beta-chemokine receptors CKR5, CKR3, and CKR2b as fusion cofactors. *Cell* **85,** 1149–1158.
33. He, J., Chen, Y., Farzan, M., Choe, H., Ohagen, A., Gartner, S., et al. (1997) CCR3 and CCR5 are co-receptors for HIV-1 infection of microglia. *Nature* **385,** 645–649.

34. Zhang, L., He, T., Talal, A., Wang, G., Frankel, S. S., and Ho, D. D. (1998) In vivo distribution of the human immunodeficiency virus/simian immunodeficiency virus coreceptors: CXCR4, CCR3, and CCR5. *J. Virol.* **72,** 5035–5045.
35. Meucci, O., Fatatis, A., Simen, A. A., Bushell, T. J., Gray, P. W., and Miller, R. J. (1998) Chemokines regulate hippocampal neuronal signaling and gp 120 neurotoxicity. *Proc. Natl. Acad. Sci. USA* **95,** 14,500–14,505.
36. Meucci, O., Fatatis, A., Simen, A. A., and Miller, R. J. (2000) Expression of CX3CR1 chemokine receptors on neurons and their role in neuronal survival. *Proc. Natl. Acad. Sci. USA* **97,** 8075–8080.
37. Brenneman, D. E., Westbrook, G. L., Fitzgerald, S. P., Ennist, D. L., Elkins, K. L., Ruff, M. R., et al. (1988) Neuronal cell killing by the envelope protein of HIV and its prevention by vasoactive intestinal peptide. *Nature* **335,** 639–642.
38. Toggas, S. M., Masliah, E., Rockenstein, E. M., Rall, G. F., Abraham, C. R., and Mucke, L. (1994) Central nervous system damage produced by expression of the HIV-1 coat protein gp120 in transgenic mice. *Nature* **367,** 188–193.
39. Lannuzel, A., Lledo, P. M., Lamghitnia, H. O., Vincent, J. D., and Tardieu, M. (1995) HIV-1 envelope proteins gp120 and gp160 potentiate NMDA [$Ca^{2+}$]$_i$ increase, alter [$Ca^{2+}$]$_i$ homeostasis and induce neurotoxicity in human embryonic neurons. *Eur. J. Neurosci.* **7,** 2285–2293.
40. Adamson, D. C., Wildemann, B., Sasaki, M., Glass, J. D., McArthur, J. C., Christov, V. I., et al. (1996) Immunologic NO synthase: elevation in severe AIDS dementia and induction by HIV-1 gp41. *Science* **274,** 1917–1921.
41. Nath, A., Geiger, J. D., Mattson, M. P., Magnuson, D. S., Jones, M., and Berger, J. R. (1998) Role of viral proteins in HIV-1 neuropathogenesis with emphasis on Tat. *NeuroAids* **1,** 1–3 (www.sciencemag.org/NAIDS).
42. Lipton, S. A. (1997) Neuropathogenesis of acquired immunodeficiency syndrome dementia. *Curr. Opin. Neurol.* **10,** 247–253.
43. Lipton, S. A., Sucher, N. J., Kaiser, P. K., and Dreyer, E. B. (1991) Synergistic effects of HIV coat protein and NMDA receptor-mediated neurotoxicity. *neuron* **7,** 111–118.
44. Sardar, A. M., Hutson, P. H., and Reynolds, G. P. (1991) Deficits of NMDA receptors and glutamate uptake sites in the frontal cortex in AIDS. *Neuroreport* **10,** 3513–3515.
45. Lipton, S. A. (1998) Neuronal injury associated with HIV-1: approaches to treatment. *Annu. Rev. Pharmacol. Toxicol.* **38,** 159–177.
46. Savio, T. and Levi, G. (1993) Neurotoxicity of HIV coat protein gp120, NMDA receptors, and protein kinase C: a study with rat cerebellar granule cell cultures. *J. Neurosci. Res.* **34,** 265–272.
47. Dreyer, E. B., Kaiser, P. K., Offermann, J. T., and Lipton, S. A. (1990) HIV-1 coat protein neurotoxicity prevented by calcium channel antagonists. *Science* **248,** 364–367.
48. Toggas, S. M., Masliah, E., and Mucke, L. (1996) Prevention of HIV-1 gp120-induced neuronal damage in the central nervous system of transgenic mice by the NMDA receptor antagonist memantine. *Brain Res.* **706,** 303–307.
49. Yeh, M. W., Kaul, M., Zheng, J., Nottet, H. L. M., Thylin, M., Gendelman, H. E., et al. (2000) Cytokine-stimulated but not HIV-infected human monocyte-derived macrophages produce neurotoxic levels of the NMDA agonist, L-cysteine. *J. Immunol.* **164,** 4265–4270.
50. Fontana, G., Valenti, L., and Raiteri, M. (1997) Gp120 can revert antagonism at the glycine site of NMDA receptors mediating GABA release from cultured hippocampal neurons. *J. Neurosci. Res.* **49,** 732–738.
51. Lo, T. M., Fallert, C. J., Piser, T. M., and Thayer, S. A. (1992) HIV-1 envelope protein evokes intracellular calcium oscillations in rat hippocampal neurons. *Brain Res.* **594,** 189–196.
52. Pittaluga, A., Pattarini, R., Severi, P., and Raiteri, M. (1996) Human brain N-methyl-D-aspartate receptors regulating noradrenaline release are positively modulated by HIV-1 coat protein gp120. *AIDS* **10,** 463–468.
53. Bezzi, P., Carmignoto, G., Pasti, L., Vesce, S., Rossi, D., Rizzini, B. L., et al. (1998) Prostaglandins stimulate calcium-dependent glutamate release in astrocytes. *Nature* **391,** 281–285.
54. Conant, K., Garzino-Demo, A., Nath, A., McArthur, J. C., Halliday, W., Power, C., et al. (1998) Induction of monocyte chemoattractant protein-1 in HIV-1 tat-stimulated astrocytes and elevation in AIDS dementia. *Proc. Natl. Acad. Sci. USA* **95,** 3117–3121.
55. Brunol, V., Copanil, A., Besong, G., Scoto, G., and Nicoletti, F. (2000) Neuroprotective activity of chemokines against N-methyl-D-aspartate or beta-amyloid-induced toxicity in culture. *Eur. J. Pharmacol.* **399,** 117–121.
56. Galasso, J. M., Harrison, J. K., and Silverstein, F. S. (1998) Excitotoxic brain injury stimulates expression of the chemokine receptor CCR5 in neonatal rats. *Am. J. Pathol.* **153,** 1631–1640.

# II Glutamate
## *Stimulant Drugs of Abuse (Cocaine, Amphetamine, Methamphetamine)*

*Section Editors*

Jerry Frankenheim
Barbara H. Herman

# 6
# Role of Glutamate and Nitric Oxide in the Acquisition and Expression of Cocaine-Induced Conditioned Increases in Locomotor Activity

## Agu Pert, PhD, Robert M. Post, PhD, and Susan R. B. Weiss, PhD

## 1. INTRODUCTION

Some of the behavioral actions of psychomotor stimulants increase in intensity and duration with repetitive administration (behavioral sensitization) *(1,2)*. Behavioral sensitization to psychomotor stimulants appears to be determined by two different but interactive processes or mechanisms. One mechanism is nonassociative in nature (i.e., does not depend on the stimulus context in which the drug is experienced) and is related predominantly to neurobiological adaptations that are induced by repetitive exposure to neuropharmacological agents. These would include alterations in receptor sensitivity and neurotransmitter release capacity, enhanced receptor upregulation, decreases in autoreceptor sensitivity *(1,3,4)*, and cellular and molecular adaptations *(5)*. The primary focus of research directed at identifying the neural adaptations that may underlie psychomotor stimulant-induced behavioral sensitization has been the meso-accumbens dopamine system because it has been shown to be involved in mediating the motoric effects of a variety of drugs *(6)*. The principal findings do indeed support a role for the mesoaccumbens dopamine projections in the development and expression of behavioral sensitization to psychomotor stimulants. First, behavioral sensitization is associated with enhanced in vivo and in vitro release of dopamine in the nucleus accumbens *(3)*. Second, the response to D1 dopamine agonists is augmented in the nucleus accumbens of sensitized rats *(7,8)*. Third, repeated administration of amphetamine directly onto dopamine cell bodies in the ventral tegmental area (VTA) produces behavioral sensitization to a systemic challenge *(9–11)*. Based on such findings, the induction of sensitization has been conceptualized to occur in the VTA, where the psychomotor stimulant acts on dopamine cell bodies to trigger the sequence of cellular events that underlies the development of behavioral sensitization. The enhanced dopamine release and postsynaptic responsiveness to dopamine in the nucleus accumbens and striatum, on the other hand, is thought to mediate the expression of sensitization *(3,12)*.

The second set of factors contributing to behavioral sensitization is associative in nature; that is, they are established through learning processes. Pavlov *(13)* was probably the first to recognize, for example, that drugs could act as unconditioned stimuli. In these early studies, it was found that discrete visual and auditory stimuli associated contiguously with injections of morphine or apomorphine developed, over time, the ability to elicit emesis in dogs when presented alone. Thus, such classical conditioning can confer to neutral stimuli the ability to elicit certain pharmacological actions of a drug. With regard to psychomotor stimulants, the conditioned response comes to resemble, to some degree, the unconditioned motoric effects produced by the drug itself *(14)*.

From: *Contemporary Clinical Neuroscience: Glutamate and Addiction*
Edited by: Barbara H. Herman et al. © Humana Press Inc., Totowa, NJ

An important contributing factor underlying the development of sensitization appears to be the acquisition of progressively increasing conditioned pharmacological effects, which add to the unchanging unconditioned response produced by the drug itself and the enhanced sensitivity caused by neuroadaptive changes in response to repetitive drug exposure. The importance of associative learning processes in the development of psychomotor stimulant-induced sensitization has also been noted by Anagnostaras and Robinson (15). The relative contribution of associative and nonassociative factors to sensitization would depend, of course, on the experimental setting and procedures employed. Not only is the study of drug-associated conditioned effects important for elucidating the mechanisms responsible for behavioral sensitization, but understanding such an associative process may also have relevance for revealing motivational mechanisms operational during addictive behaviors. We have previously proposed (16) that the classical conditioning of drug effects to environmental cues underlies the development of incentive motivation, which is probably the neurobehavioral substrate responsible for craving. Understanding the circuitry that mediates the conditioned drug effects may aid in the development of appropriate pharmacotherapeutic adjuncts for the treatment of drug addiction.

## 2. NEUROBIOLOGY OF COCAINE-INDUCED CONDITIONED INCREASES IN LOCOMOTOR ACTIVITY

We have employed a relatively simple design in some of our studies to evaluate the behavioral and neurobiological variables regulating the acquisition and expression of cocaine-induced conditioned effects. Basically, three groups of rats are employed in this paradigm. In its simplest form, on d 1 the first groups of rats (PAIRED) is injected with cocaine (30–40 mg/kg ip) prior to placement in locomotor activity chambers (scented with peppermint) for 30 min. One hour following return to their home cages, these rats are injected with saline. The second group (UNPAIRED) is treated in a similar fashion, but receives saline prior to placement in the locomotor activity chamber and cocaine in the home cage. The third group (CONTROL) receives saline in both environments. On d 2, all rats are challenged with either saline or 10 mg/kg of cocaine prior to placement in the locomotor activity chamber. In some of our studies, the training sessions were increased to 3, 5, or 7 d when it was necessary to increase the persistence and strength of conditioning.

We have shown significant conditioned effects of cocaine using this design, which is reflected by dramatic increases in locomotor output in the PAIRED group on the test day relative to the other two groups. Evidence from the laboratory (16) indicates that such increases in locomotor activity in the PAIRED group is established through associative learning mechanisms and does not simply reflect the inability to habituate to the environment under the influence of the drug, as has been suggested by some.

One of our main interests has been to define the neurobiological substrates underlying the conditioned effects of cocaine. We have found, for example, that dopamine (DA)-depleting lesions of the nucleus accumbens and amygdala prevented cocaine-induced conditioning after one training session. The amygdala lesions, however, were not effective in preventing conditioning when more extensive training was employed. The dopaminergic components of the amygdala appear to play a more subtle role in the formation of cocaine-conditioned behavior than those in the nucleus accumbens. Dopamine-depleting lesions of the frontal cortex and striatum were not effective in preventing the conditioned locomotor effects of cocaine. Neurotoxin-induced lesions of the raphe and locus ceruleus were equally ineffective. Radio-frequency lesions of the dorsal and ventral hippocampus, as well as the cerebellum, also had little effect on the establishment of cocaine-induced conditioning after one day of training. Such findings strongly suggested that DA function in the nucleus accumbens and, to a lesser degree, in the amygdala, is necessary for the formation of cocaine-conditioned behaviors.

It has been suggested that different neurobiological processes are involved in the acquisition and expression of psychomotor stimulant-induced conditioned increases in motor behavior. With lesions made prior to training, it is not possible to determine whether the deficit seen is related to disruptions in the acquisition process or to the expression of the behavior. Because DA is involved in the stimulatory and appetitive properties of psychomotor stimulants, it is not surprising that the blockade of DA function would lead to decreases in the acquisition of conditioning. For example, neuroleptics coadministered with either amphetamine *(17)* or cocaine *(18,19)* have been found to prevent the development of conditioned locomotor behaviors. More recently, we have found that D1 and D2 DA receptor antagonists are equally effective in preventing the formation of cocaine-induced conditioning *(20)*. Likewise, conditioning in the 1-d design was found only following administration of a combination of D1 and D2 agonists during training, and not when either was administered separately. This would suggest that concurrent D1 and D2 DA receptor occupation is necessary for conditioning to occur.

There are a variety of mechanisms by which DA antagonists could disrupt the acquisition of cocaine conditioning *(20)*. It is most likely, however, that the ability of these drugs to decrease or prevent conditioning to psychomotor stimulants is related to their ability to attenuate the unconditioned effects of the drugs, which are critical in forming the conditioned association. We have suggested that the ability of DA blockers to prevent conditioning is related to their ability to decrease the motivational significance of the unconditioned stimulus (e.g., cocaine). It is well established that the strength of conditioning is directly related to the intensity of the unconditioned stimulus in other conditioning paradigms *(16)*.

Although mixed D1–D2 and selective D1 and D2 antagonists have been found to prevent the establishment of conditioning to cues associated with cocaine, they have been reported to be relatively ineffective in preventing expression once established. Early studies by Beninger and Hahn *(21)* and Beninger and Herz *(18)* found that pimozide did not eliminate the behavioral differential between cocaine-conditioned animals and their controls. We have recently extended these findings by demonstrating that neither D1 nor D2 antagonists are effective in altering the differential in performance between the conditioned and unconditioned rats during the test phase *(20)*.

On the surface, these findings, together with the ability of DA antagonists to block the acquisition of conditioned behaviors, appear to suggest that although intact DA function is critical for the development of conditioning to cocaine-associated cues, it is not necessary for the expression of the conditioned response once established. It is possible that nondopaminergic pathways acquire the ability to elicit such conditioned reactions. The second alternative is that DA is involved in the expression of the conditioned behavior and that the differential in activity seen between the conditioned and unconditioned groups is determined and maintained by increased activity of mesolimbic DA pathways in the former group, despite similar partial blockades of DA receptors in all experimental groups.

Using the 1-d conditioning paradigm described previously, we have evaluated the ability of stimuli associated with cocaine to increase mesolimbic DA functions *(22)*. Microdialysis procedures revealed significant increases in extracellular DA in the nucleus accumbens in the PAIRED group relative to the control groups during test day. Kalivas and Duffy *(23)* also have reported increases in mesolimbic DA elicited by stimuli associated with cocaine. More recent studies in our laboratory have not found such conditioned increases in DA overflow in either the amygdala or striatum, suggesting some regional specificity in the effects of conditioned stimuli on DA function. Lesion studies also appear to support these findings. DA-depleting lesion studies of the nucleus accumbens made immediately after 7 d of conditioning were able to prevent the expression of the conditioned response when rats were tested 7 d postoperatively. Gold et al. *(24)* have reported similar findings.

## 3. ROLE OF GLUTAMATE IN THE CONDITIONED EFFECTS OF COCAINE

Although dopamine appears to be important for the formation of both the associative and nonassociative components of sensitization, it has become apparent that other neurotransmitter systems are involved as well. Of particular importance is glutamate. Karler and colleagues *(25,26)* were the first to report that the development of sensitization to cocaine and amphetamine in mice was prevented by pretreatment with MK-801, a noncompetitive antagonist of the *N*-methyl-D-aspartate (NMDA) type of glutamate receptor. These findings were subsequently confirmed by a number of other investigators in mice and rats using noncompetitive, competitive, and glycine-site antagonists of the NMDA receptor (*see* ref. *27* for review). Interestingly, coadministration of MK-801 with psychomotor stimulants has also been found to prevent cellular adaptations to repetitive drug exposure, such as DA autoreceptor subsensitivity in the VTA *(28)*, D1 receptor supersensitivity in the nucleus accumbens *(28)*, augmented, drug-elicited DA efflux in the nucleus accumbens *(29)*, increases in striatal calmodulin content *(30)*, and increases in VTA tyrosine hydroxylase immunoreactivity *(31)* and tyrosine hydroxylase mRNA levels *(32)*.

Although both noncompetitive and competitive NMDA antagonists have been found to consistently block the development of behavioral sensitization, expression has been somewhat resistant to disruption *(26,33)*. Preferential involvement of NMDA receptors in the developmental phase of sensitization suggests some similarity to long-term potentiation (LTP) in which NMDA receptor activation is required for the induction of LTP *(34)*, whereas its expression involves selective enhancement of the non-NMDA component of the excitatory postsynaptic potential *(35)*.

All of the above findings taken together suggest that glutamate transmission at NMDA receptor sites is a requirement in the cascade of cellular changes leading to sensitization. Kalivas *(12)* has postulated that sensitization to psychomotor stimulants is determined by the initial increase in somatodendritic DA release in the VTA, which stimulates D1 receptors located on cortical afferents in this structure to enhance excitatory amino acid (EAA) release. EAA released from the cortical afferents, in turn, stimulates NMDA receptors on dopaminergic soma and/or dendrites desensitizing D2 somatodendritic D2 autoreceptors. This diminishes the hyperpolarization of dopamine cells that normally accompanies increased somatodendritic dopamine release and permits a further augmentation of dopamine release. These events lead to enhanced NMDA receptor stimulation and presumably increased intracellular concentrations of $Ca^{2+}$ in DA cells, leading to changes in cell function that initiate behavioral sensitization.

This type of postulation represents the nonassociative view in which NMDA antagonists prevent sensitization by interfering with a critical cellular process initiated by repetitive exposure to psychomotor stimulants. Conversely, and/or in addition, NMDA receptor blockade could also prevent the development of sensitization by blocking associative processes (i.e., those involved in learning), especially under experimental circumstances in which such processes may be critical. There is, of course, considerable evidence to suggest that glutamate is important in learning processes. Glutamate, as noted, has been implicated in the induction of LTP, which presumably represents an electrophysiological model of the elementary changes in synaptic plasticity that underlie memory formation *(34–36)*. Competitive and noncompetitive antagonists have been reported to inhibit learning in a variety of paradigms *(37,38)*.

It is likely that glutamate is also involved in the associative process underlying the development of cocaine-induced conditioned increases in locomotor activity that contribute to behavioral sensitization. Sensory information from conditioned stimuli, for example, needs to gain access to meso-accumbens DA neurons, which we, as well as others, have implicated in the acquisition and expression of conditioned psychomotor stimulant effects. Likewise, these neurons need to ultimately activate motor pathways either directly or indirectly. It should be possible to disrupt the acquisition and expression of cocaine-induced conditioned increases in locomotor behavior by altering functional activity at any number of relays or integrative centers of this circuit.

Although neither the nucleus accumbens (DA terminal area) nor VTA (DA perikaryal region) receives input from primary sensory cortical regions, DA activity in this system could be influenced indirectly through other structures such as the amygdala or frontal cortex. The amygdala, for example, receives polysensory information from cortical sensory association areas and project, in turn, to the nucleus accumbens and VTA *(39)*. The prefrontal cortex sends monosynaptic excitatory amino acid inputs to both DA and non-DA cells in the VTA *(40)*. Because corticofugal neurons in general are predominantly glutamatergic in nature, it should be possible to disrupt both the acquisition and expression of cocaine-conditioned behaviors with EAA antagonists. This sort of blockade could prevent the acquisition of conditioning by either degrading sensory input (from the CS) or by interfering with the mechanisms underlying neuronal plasticity associated with conditioning. Expression of conditioned behaviors, once established, might be disrupted by blocking conditioned stimulus input to meso-accumbens DA neurons carried by indirect glutamatergic pathways.

Although glutamate antagonists, as noted above, have been found to prevent the induction of behavioral sensitization, few studies have directly examined the ability of such agents to alter the acquisition or expression of psychomotor stimulant-induced conditioned increases in locomotor activity. Unfortunately, most of the studies that have evaluated the effects of glutamate antagonists on the development of sensitization have employed designs which do not allow definitive conclusions to be drawn regarding the participation of conditioned drug effects *(41–49)*. These are all studies in which drugs were either occasionally administered in the test environment (producing latent inhibition) or did not include appropriate control groups to evaluate the participation of conditioned drug effects.

Stewart and Druhan *(50)*, and Druhan et al. *(51)*, using an appropriate design to reveal conditioned drug effects, reported that pretreatment with MK-801 blocked the development of conditioned activity by both amphetamine and apomorphine. Blockade of conditioned locomotor activity increases by systemic MK-801 have also been found by Wolf and Khansa *(52)* and Segal et al. *(53)*. More recently, Cervo and Samanin *(54)* reported blockade of cocaine-induced conditioning by pretreatment with MK-801 and DNQX (an AMPA [α-amino-3-hydroxy-5-methyl-4-isoxozole propionic acid] receptor antagonist). Morphine-induced conditioned place preference, which also involves the conditioning of drug-effects environmental cues, was likewise prevented by pretreatment with MK-801 *(55)*. The purpose of our present studies described below was to extend the above observations by systematically evaluating the effects of glutamate antagonists, as well as a nitric oxide synthase inhibitor, specifically on the development and expression of cocaine-induced conditioned increases in locomotor activity.

## 4. BLOCKADE OF COCAINE-INDUCED CONDITIONED EFFECTS WITH MK-801

Conditioned effects of cocaine were established and evaluated using the paradigm described above in which PAIRED, UNPAIRED, and CONTROL animals are employed. In the initial study, we used a 1-d conditioning procedure to evaluate the effects of MK-801 on the acquisition of cocaine-conditioned increases in locomotor activity. PAIRED rats were pretreated with either saline or various doses of MK-801 (0.25–1.0 mg/kg ip) 30 min prior to injections of 40 mg/kg of cocaine ip and then placed in activity chambers scented with peppermint for 30 min. One hour following return to their home cages, these rats received two saline injections separated by 30 min. UNPAIRED rats received two saline injections separated by 30 min prior to placement in the activity chamber and either saline or MK-801 injections 1 h after return to their home cages, followed by a second saline injection 30 min later. When tested with 10 mg/kg of cocaine on d 2, PAIRED rats pretreated with saline expressed conditioned increases in locomotor activity, whereas those that had been pretreated with all doses of MK-801 failed to show conditioning (Fig. 1). MK-801 was therefore effective in preventing the acquisition of cocaine-induced conditioning, confirming previous findings.

Repetitive administration of MK-801 itself has been found to produce conditioned increases in locomotor activity *(52)* as well as behavioral sensitization, although in studies in which it is not possible to determine if contextual or noncontextual factors were critical *(43,44,56–58)*. It was of interest, therefore, to evaluate whether more prolonged training, as described above, would still reveal the ability of MK-801 to prevent cocaine-induced conditioning or whether possible conditioned effects associated with repetitive MK-801 would confound such actions. Two groups of PAIRED and UNPAIRED rats were pretreated with either saline or MK-801 (0.25 mg/kg ip) as described above prior to injections of 30

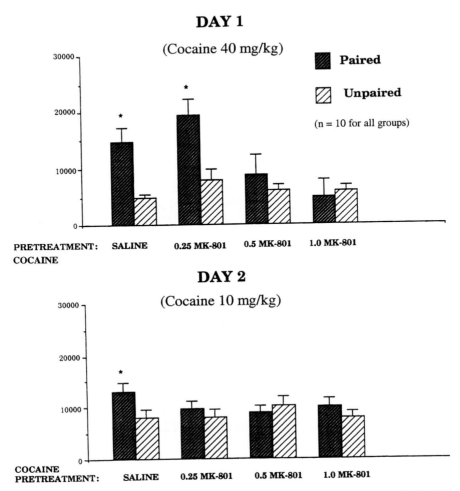

**Fig. 1.** Effects of MK-801 on acquisition of cocaine-induced conditioning following 1 d of training. PAIRED and UNPAIRED rats were pretreated with saline and various doses of MK-801 on d 1. On d 2, all rats were injected with 10 mg/kg of cocaine and tested for 30 min. *$p < 0.05$ for comparisons between PAIRED and UNPAIRED groups with the Scheffe test for post hoc comparisons. All values are means ± SEM.

mg/kg of cocaine for three consecutive days and then tested on d 4 and 11 with saline. Two additional groups were used to assess the conditioned effects of MK-801 alone. One group received MK-801 in the context of the test cage for 3 d, whereas the other group was injected with MK-801 in the home cage. These two groups were also tested with saline in the conditioning chamber on d 4 and 11.

Robust conditioned effects (Fig. 2A,B) were seen in the cocaine PAIRED group pretreated with saline, as well as the cocaine PAIRED group pretreated with MK-801 on the first test day (d 4). Conditioned increases in locomotor activity were also observed in the group injected with MK-801 alone in the test cage (Fig. 2C). When tested on d 11, however, only the saline pretreated PAIRED group still exhibited a conditioned response. The other two PAIRED groups (MK-801 pretreated and MK-801 alone) failed to differ from UNPAIRED controls. It appears that 3 d of conditioning with MK-801 were sufficient to produce conditioned increases in locomotor activity that were able to mask its disruptive effects on the associative processes underlying cocaine-induced conditioned increases in

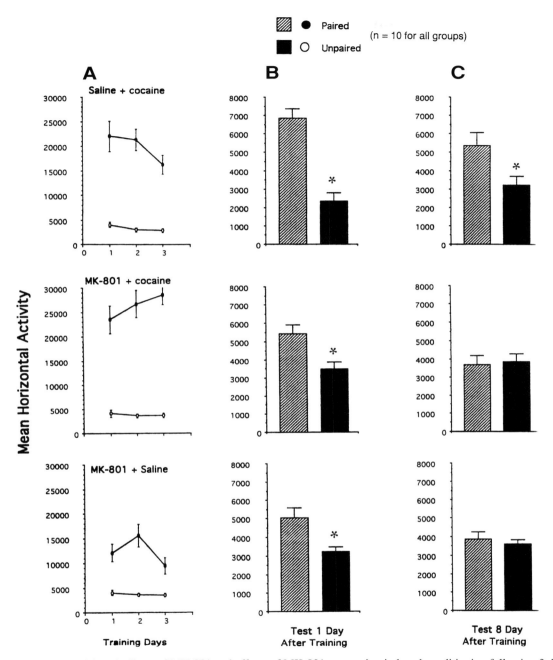

**Fig. 2.** Conditioned effects of MK-801 and effects of MK-801 on cocaine-induced conditioning following 3 d of training. **(A)** Horizontal locomotor activity of PAIRED and UNPAIRED rats during 3 d of training; **(B)** activity of PAIRED and UNPAIRED rats on d 4 following injections of 10 mg/kg cocaine HCL; **(C)** Activity of PAIRED and UNPAIRED rats on d 8 following injections of 10 mg/kg cocaine HCL. *$p < 0.05$ for comparisons between PAIRED and UNPAIRED groups. All values are means ± SEM.

locomotor activity. During the second test day (d 11), the conditioned effects of MK-801 had apparently decayed (Fig. 2B), revealing the initial blockade of cocaine-conditioned effects during the acquisition phase. The conditioned effects of MK-801 may be a confounding variable in studies designed to evaluate its ability to alter the conditioned motoric effects of psychomotor stimulants, especially where the contextual exposure to MK-801 is prolonged.

Kalivas and Alesdatter (45) have reported that MK-801 injections into the ventral tegmental area and amygdala blocked sensitization to cocaine following one injection. Unfortunately, the design used does not permit definitive conclusions to be drawn regarding the effects of NMDA blockade on the contextual (conditioned) or noncontextual components because appropriate control groups were not included. The purpose of the next series of studies was to evaluate the importance of intact NMDA receptor function in the nucleus accumbens, VTA, and amygdala in the acquisition of cocaine-induced conditioned increases in locomotor activity. Rats were implanted with guide cannulae aimed for the basolateral amygdala, ventral tegmental area, or the nucleus accumbens and were divided into four groups. One group (VEHICLE-PAIRED) was injected with artificial cerebrospinal fluid (aCSF) intracerebrally (ic) prior to ip cocaine (30 mg/kg), and the other group (MK-801-PAIRED) was injected ic with 5 nmol of MK-801 immediately prior to ip cocaine (30 mg/kg). Both groups were placed in the conditioning activity chambers for 30 min following cocaine injections. One hour following return to the home cage, these rats were injected ic with aCSF followed by ip saline. The UNPAIRED groups received the same treatments but in opposite contexts. The rats were conditioned for 3 d and then tested following saline injections on d 4. PAIRED rats pretreated with aCSF in the VTA, amygdala, and nucleus accumbens showed significant conditioned increases in locomotor activity on d 4 (Figs. 3–5).

MK-801 injections into the VTA during the training phase were found to prevent the acquisition of cocaine-conditioned increases in locomotor activity (Fig. 5). Similar pretreatment with MK-801 in the amygdala or nucleus accumbens was without effect (Figs. 3 and 4). Thus, it appears that intact glutamate function is necessary in the VTA, but not in the nucleus accmbens or amygdala, for the development of cocaine-induced conditioning. The lack of effect in the amygdala is surprising because this structure has been demonstrated to be important in the formation of stimulus–reward associations (39,59–61). It appears that glutamate input to the VTA, perhaps from the frontal cortex, is necessary for the formation of cocaine-induced conditioning.

Activation of NMDA receptors has been shown to induce nitric oxide (NO) synthesis, which then activates guanylate cyclase and leads to the formation of cyclic GMP in the brain. There is evidence to suggest that NO is involved in synaptic plasticity, including LTP, as well as learning and memory. A variety of studies also have reported that $N^6$-nitro-arginine-methyl ester (L-NAME), a nitric oxide synthase inhibitor, prevents the induction of behavioral sensitization to psychomotor stimulants (48,62), although not always (50,63). Although there is evidence to suggest a role for NO in the acquisition of behavioral sensitization, the designs of the above studies directed at this issue do not allow definitive conclusions to be made regarding whether NO is predominately involved in noncontextual or contextual components.

In order to assess the role of NO in the formation of cocaine-induced conditioning, PAIRED and UNPAIRED rats, as described above, were pretreated with either saline or 100 mg/kg of $N^6$-nitro-L-arginine methyl ester (L-NAME). Twenty-four hours later, the PAIRED animals treated with L-NAME were again injected with 100 mg/kg L-NAME ip followed 1 h later by 30 mg/kg cocaine ip. The UNPAIRED rats treated the day before with L-NAME were injected with saline. Both groups were placed in the locomotor activity chambers for 30 min following injections. One hour following removal from the activity chambers, the rats in these two groups received, respectively, either two injections of saline separated by 1 h or injections of L-NAME and then saline separated by 1 h. The other groups of PAIRED and UNPAIRED rats were treated the same as described above (in the present study), but in place of L-NAME, they received injections of saline. The four groups were treated as described above for three consecutive days (d 2–4). On d 5, all animals were injected with saline and placed in the locomotor activity chambers for 30 min.

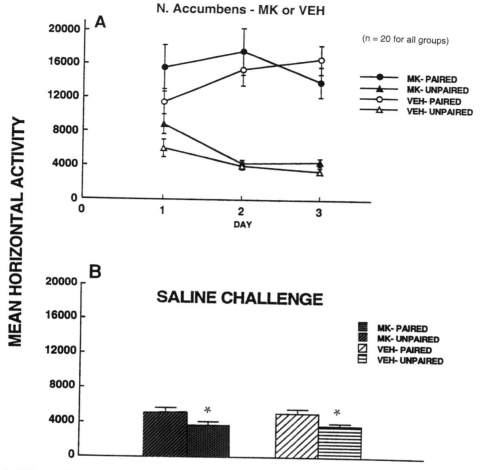

**Fig. 3.** Effects of MK-801 injections into the n. accumbens on the acquisition of cocaine-induced conditioned increases in locomotor activity. **(A)** Locomotor activity of PAIRED and UNPAIRED rats pretreated bilaterally with either 7.5 nmol of MK-801 or 1 µL vehicle during 3 d of conditioning; **(B)** locomotor activity of rats on d 4 following injections of saline. *$p < 0.05$ for comparisons between PAIRED and UNPAIRED rats with the Scheffe test for post hoc comaparisons. All values are means ± SEM.

Treatment with L-NAME had no apparent affect on locomotor behavior during the training phase (Fig. 6A). During the test day (d 5), the PAIRED rats pretreated with L-NAME had total activity output that was significantly lower than the PAIRED rats retreated with saline. Thus, it appears that NO may play some role in the associative processes underlying cocaine-induced conditioning.

Although intact NMDA function appears to be critical for the formation of psychomotor stimulant-induced sensitization and, more specifically, the formation of conditioned drug effects, considerably less is known regarding the role of excitatory amino acid neurotransmission in the expression of the conditioned effects once established. The purpose of the next series of studies was to evaluate the effects of MK-801 and L-NAME on the expression of cocaine-conditioned increases in locomotor activity once established. PAIRED and UNPAIRED rats were trained for three consecutive days as described above The conditioning dose of cocaine was 30 mg/kg. On d 4, the PAIRED and

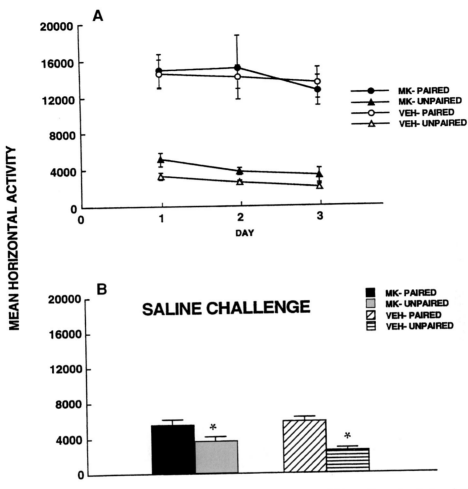

**Fig. 4.** Effects of MK-801 injections into the amygdala on the acquisition of cocaine-induced conditioned increases in locomotor activity. **(A)** Locomotor activity of PAIRED and UNPAIRED rats pretreated bilaterally with either 7.5 nmol of MK-801 or 1 µL vehicle during 3 d of conditioning; **(B)** locomotor activity of rats on d 4 following injections of saline. $*p < 0.05$ for comparisons between PAIRED and UNPAIRED rats with the Scheffe test for post hoc comparisons. All values are means ± SEM.

UNPAIRED rats were pretreated with either saline, 0.1 mg/kg, or 0.25 mg/kg of MK-801 and then injected with 10 mg/kg of cocaine. Following cocaine injections, the animals were placed in the activity chambers for 30 min. Although pretreatment with MK-801 on the test day enhanced cocaine-induced locomotor activity in the UNPAIRED and PAIRED RATS, it failed to eliminate the differential between the two groups following cocaine injection (Fig. 7). Thus, it appears that although MK-801 is effective in preventing the acquisition of cocaine-induced conditioned effects, it is ineffective in preventing expression once established.

In the next series of studies, PAIRED and UNPAIRED rats were trained for 3 d as described above. One hour following home cage injections on d 3, one-fourst of the PAIRED and one-fourth of the UNPAIRED rats were injected with 100 mg/kg of L-NAME ip. On d 4, the same rats were injected again with a similar dose of L-NAME 60 min prior to saline injections. The rats were run for 30 min in

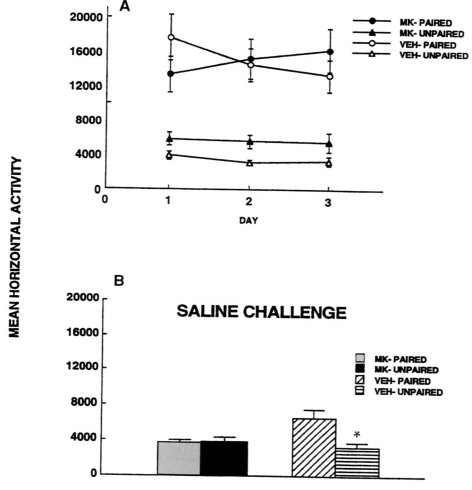

**Fig. 5.** Effects of MK-801 injections into the ventral tegmental area (VTA) on the acquisition of cocaine-induced conditioned increases in locomotor activity. **(A)** Locomotor activity of PAIRED and UNPAIRED rats pretreated bilaterally with either 7.5 nmol of MK-801 or 1 μl vehicle during 3 d of conditioning; **(B)** locomotor activity of rats on d 4 following injections of saline. $*p < 0.05$ for comparisons between PAIRED and UNPAIRED rats with the Scheffe test for post hoc comparisons. All values are means ± SEM.

the activity chambers immediately following saline injections. One-third of the remaining PAIRED and UNPAIRED rats were injected with 20 mg/kg of CPP (a competitive NMDA antagonist) ip on d 4, one-third were injected with 0.25 mg/kg of MK-801, and the remaining one-third received saline injections. Thirty minutes following these injections, the animals were injected with saline and then placed in the activity chambers for 30 min. A control group was also included in this study. Rats in this group received injections of saline in both environments (test cage and home cage) during training and received two saline injections separated by 30 min on the test day.

Rats pretreated with saline, MK-801, CPP, or NAME all showed a significant conditioned response on d 4 (Fig. 8), even though MK-801 enhanced activity in both PAIRED and UNPAIRED, NAME decreased locomotor output in both groups when compared with saline-pretreated animals. Although the studies above failed to reveal the participation of glutamate in the expression of

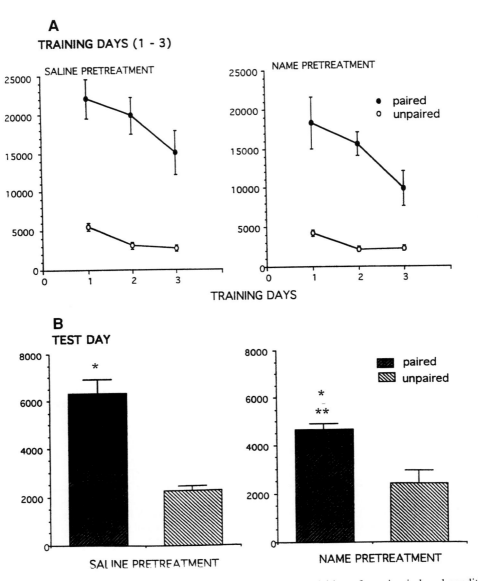

**Fig. 6.** Effects of L-NAME pretreatment during training on the acquisition of cocaine-induced conditioning. **(A)** Horizontal locomotor activity of PAIRED and UNPAIRED rats pretreated with either saline or NAME during 3 d of conditioning with cocaine; **(B)** horizontal locomotor activity of PAIRED and UNPAIRED rats on d 4 following injections of 10 mg/kg of cocaine. *$p < 0.05$ for comparisons between PAIRED and UNPAIRED rats with the Scheffe test for post-hoc comparisons; **$p < 0.05$ for comparison between PAIRED saline-pretreated and PAIRED L-NAME-pretreated rats on the test day. All values are means ± SEM.

cocaine-conditioned behavior, it is possible that the effects of NMDA blockade on the expression of cocaine-conditioned activity might be revealed under more subtle procedures (i.e., when conditioning is not as pronounced). PAIRED and UNPAIRED rats were treated on d 1 as already described. On d 2, one-third of the PAIRED and UNPAIRED rats were injected with saline, one-third with 5 mg/kg of CPP, and the other one-third with 20 mg/kg of CPP. Thirty minutes later, all rats were

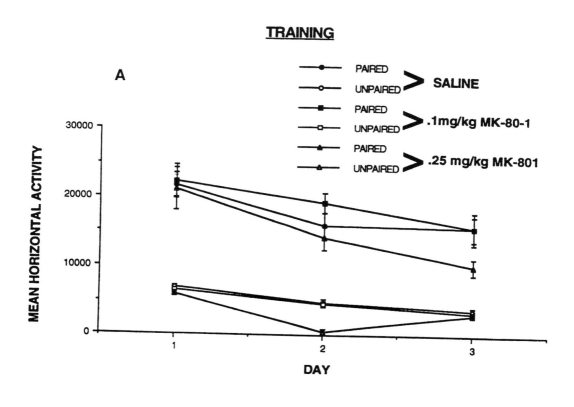
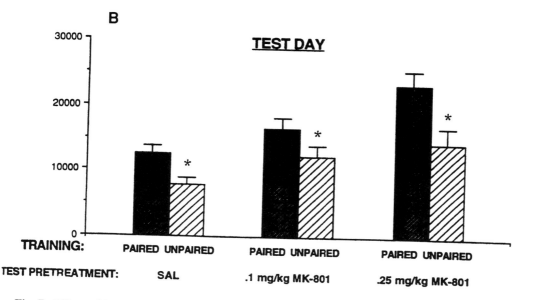

**Fig. 7.** Effects of MK-801 on the expression of cocaine-induced conditioned increases in locomotor activity. (**A**) Locomotor activity of PAIRED and UNPAIRED rats during training that were pretreated with either saline, 0.1 ng/kg or 0.25 ng/kg of MK-801 on d 4 prior to injections of 10 ng/kg of cocaine; (**B**) mean activity scores on the test day of saline and MK-801 pretreated rats. *$p < 0.05$ for comparisons between PAIRED and UNPAIRED rats with the Scheffe test for post hoc comparisons. All values are means ± SEM.

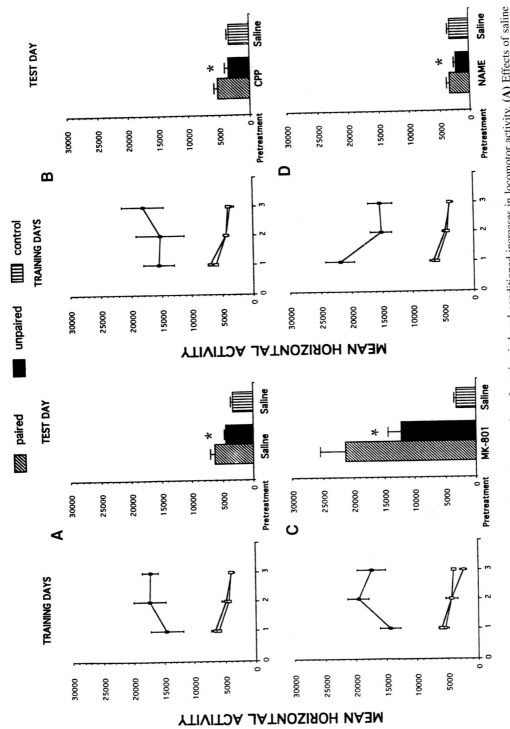

Fig. 8. Effects of CPP, MK-801 and NAME on the expression of cocaine-induced conditioned increases in locomotor activity. (A) Effects of saline pretreatment during the test day on expression following injection of saline; (B) effects of CPP pretreatment (20 mg/kg ip) during the test day on expression; (C) Effects of MK-801 (0.25 ng/kg ip) pretreatment during the test day on expression; (D) effects of NAME pretreatment during the test day on expression. $*p < 0.05$ for comparisons between PAIRED and UNPAIRED rats on the test day with the Scheffe test for post hoc comparisons. All values are means ± SEM.

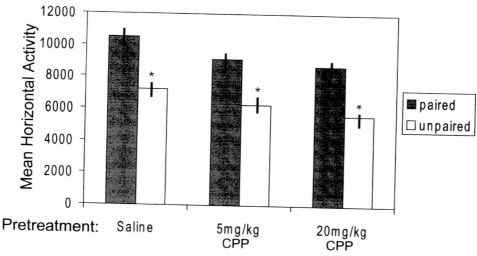

**Fig. 9.** Effects of test-day CPP pretreatment on the expression of cocaine-induced conditioned increases in locomotor activity following one conditioning session. *$p < 0.05$ for comparisons between PAIRED and UNPAIRED rats on the test day with the Scheffe test for post hoc comparisons. All values are means ± SEM.

injected with 10 mg/kg of cocaine and then run for 30 min in the activity chambers. Neither dose of CPP prevented the expression of cocaine-conditioned effects (Fig. 9).

The VTA appears to be the principal brain region involved in the acquisition of cocaine-conditioned increases in locomotor activity as well as context-independent sensitization. The purpose of the next study was to determine whether direct blockade of NMDA receptor function in the VTA would prevent the expression of cocaine-conditioned behavior. Rats were implanted with bilateral cannulae guides aimed for an area 2 mm dorsal to the VTA. Two weeks later, the animals were divided into two groups (PAIRED and UNPAIRED). Animals in each group were treated for 3 d with 30 mg/kg of cocaine, as described above. On d 4, one-half of the PAIRED and UNPAIRED rats were injected bilaterally in the VTA with 5 nmol of MK-801 in 1 μL of aCSF and the other half were injected with the vehicle. Immediately after the intracerebral injections, all rats were injected intraperitoneally with saline and then placed in locomotor chambers for 30 min. Although MK-801 injections into the VTA increased locomotor activity in both the PAIRED and UNPAIRED rats, it failed to eliminate the expression of the conditioned response (Fig. 10).

Although intact NMDA receptor function does not appear to be critical for the expression of sensitization, whether context dependent or context independent, there is evidence that the AMPA subtype of glutamate receptors might be involved *(25,54)*. The ability of DNQX, a relatively selective antagonist of the AMPA receptor, to alter the expression fo cocaine-conditioned increases in locomotor activity was evaluated by the injection of 25 nmol of this antagonist icv prior to the test for conditioning 1 d following three conditioning sessions. DNQX was found to fully prevent the expression of cocaine-induced conditioned effects, even though it had little effect on locomotor activity by itself (Fig. 11). These findings are similar to those reported by Cervo and Samanin *(54)*.

## 5. CONCLUSIONS

Our first series of experiments confirmed findings from previous studies *(52,53)* which found that blockade of glutamate function concurrent with psychomotor stimulant administrations in a specific environmental context prevented the acquisition of conditioning. However, in studies reported here,

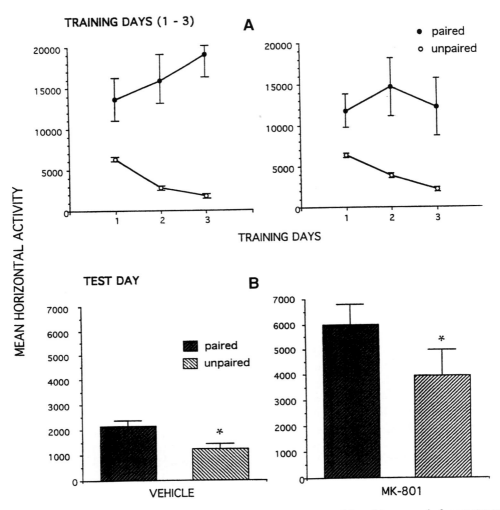

**Fig. 10.** Effects of MK-801 on the expression of cocaine-induced conditioned increases in locomotor activity following direct injections into the VTA during the test day. **(A)** Locomotor activity during 3 d of training for PAIRED and UNPAIRED rats; **(B)** activity of PAIRED and UNPAIRED rats during the test day following pretreatment with bilateral vehicle or MK-801 (7.5 nmol) injections into the VTA. $*p < 0.05$ for comaprisons between the PAIRED and UNPAIRED rats on the test day with the Scheffe test for post hoc comparisons. All values are means ± SEM.

such inhibition of conditioning by MK-801 was only seen following one conditioning session, but not three. Such differences in the ability of MK-801 to prevent cocaine-induced conditioning, depending on the duration of the treatment regimen, appeared to be related to the development of conditioned effects to MK-801, which masked its disrupted effects on cocaine-induced conditioning after three training sessions. Apparently, one conditioning session, sufficient to establish cocaine-conditioned effects, is not enough to produce conditioning to MK-801. The only effect of MK-801 under this training regimen is to inhibit cocaine-induced conditioning. MK-801 also inhibits the establishment of cocaine-conditioned effects when training is extended to 3 d. However, when training is prolonged, the locomotor stimulating effects of MK-801 also become conditioned to the apparatus cues. Thus, when

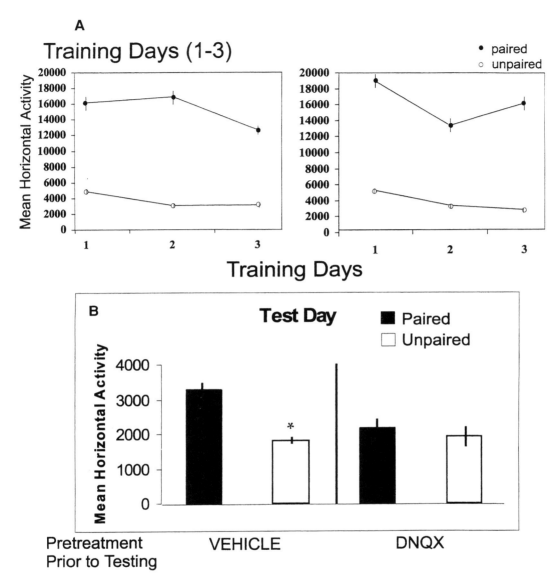

**Fig. 11.** Effects of DNQX on the expression of cocaine-induced conditioned increases in locomotor activity following intracerebroventricular injections during the test day. (**A**) Locomotor activity during 3 d of training for PAIRED and UNPAIRED rats; (**B**) activity of PAIRED and UNPAIRED rats during the test day following pretreatment with injections of vehicle on 25 nmol of DNQX icv. *$p < 0.05$ for comparisons between PAIRED and UNPAIRED rats on the test day with the Scheffe test for post hoc comparisons. All values are means ± SEM.

rats pretreated with MK-801 prior to cocaine are tested 1 d following 3 d of training, they show conditioned MK-801-induced increases in locomotor output, making it appear as though MK-801 was ineffective in preventing cocaine-induced conditioning. The conditioned effects of MK-801 decay rapidly with time, so that when MK-801-pretreated rats are tested 10 d later, the inhibitory effects of this glutamate blocker on the formation of cocaine-induced conditioning are revealed. Indeed, no conditioned effects of MK-801 were found 11 d following training.

As noted above, MK-801 alone has been reported to induce conditioned increases in locomotor behavior *(52)*, as well as behavioral sensitization *(43,44,56–58)*, although in the latter studies, it is not possible to determine whether contextual or noncontextual processes were involved. It is, of course, paradoxical that MK-801 can, at the same time, prevent the establishment of conditioning to psychomotor stimulants (presumably by interfering with glutamatergic processes underlying synaptic plasticity) and induce conditioned effects by itself. MK-801 produces its unconditioned effects by blocking glutamate function; yet, such blockade does not seem to prevent conditioning to this agent, suggesting that nonglutamatergic processes may be involved.

If MK-801 is capable of producing conditioned effects, it would suggest that this glutamate antagonist has motivational or affective properties in order for it to act as an unconditioned stimulus. Indeed, there is evidence from a variety of sources that MK-801 possesses hedonic effects. For example, MK-801 has been found to facilitate intracranial electrical self-stimulation in the medial forebrain bundle as well as the medial frontal cortex *(64)*, produce conditioned place preference *(65–67)*, and support self-administration into both the nucleus accumbens *(68)* and ventral tegmental area *(69)*. As we have suggested for cocaine *(16)*, the conditioned increases in motoric output seen with MK-801 may reflect the energizing function of anticipatory incentive-motivational processes elicited by drug-associated cues (see below).

NMDA receptor activation as well as the release of nitric oxide (a free-radical gas that functions as an intracellular messenger in brain) have been postulated to be necessary for the induction of LTP, which has been suggested to underlie various forms of learning as well as memory. Blockade of NO synthase activity has, for example, been found to prevent the induction of LTP in hippocampal slices *(70–72)* as well as the acquisition and consolidation of inhibitory avoidance learning *(73,74)* and working memory *(75)*. As noted previously, NO synthase inhibitions also have been found to prevent the development of behavioral sensitization to cocaine *(48,62,76)*, although not in all studies *(63,77,78)*. The reasons for the disparity among these findings are not entirely clear, although it has been suggested that some of the differences may be the result of differential participation of contextual (conditioned) sensitization, which may be more susceptible to disruption by NO synthase inhibition *(27,76)*. The above-referenced studies, however, have used designs that do not allow definitive conclusions to be drawn regarding the involvement of contextual or noncontextual factors. Our current series of studies do seem to support the notion that NO is involved, at least to some degree, in the development of cocaine-induced conditioning.

It is generally thought that the VTA is a region critically involved in the initiation of processes underlying psychomotor-induced sensitization. As noted previously, one hypothesis has assumed that sensitization is initiated by increases in somatodendritic DA release in the VTA, which stimulates D1 receptors located on cortical afferents from the MPFC to enhance EAA release, which leads to changes in DA cell function that initiate behavioral sensitization *(12)*. Wolf *(27)* has also emphasized the importance of increased excitatory drive to VTA DA neurons seen following termination of repetitive treatment with psychomotor stimulants, as a critical determinant of sensitization. It has been demonstrated, for example, that the responsiveness of VTA DA neurons to glutamate is enhanced in amphetamine and cocaine-treated rats *(79)*. This effect is apparently related to an increase in the responsiveness of AMPA receptors on VTA DA neurons *(32)*. DA neurons in the VTA also have been found to show enhanced excitatory responses to stimulation of EAAergic afferents from the frontal cortex *(80)*. More recently, Ungless et al. *(81)* have found that a single exposure to cocaine induced LTP of AMPA receptor-mediated currents at excitatory synapses on DA cells in the VTA. Consistent with such findings, Kalivas and Alesdatter *(45)* reported that injections of MK-801 and CPP into the VTA prevented the induction of behavioral sensitization to one injection of cocaine. Regrettably, the design used (e.g., habituation to apparatus cues prior to the cocaine injections and the failure to use UNPAIRED controls) makes it unlikely that cocaine-induced conditioned effects were measured. The possibility that noncontextual processes were responsible for sensitization in the above-cited study is further supported by recent findings demonstrating that excitatory synapses in the VTA are potentiated following a single exposure to cocaine (81). Although glutamate in the VTA appears to be critical for the formation of context-independent sensitization, the present studies also reveal clearly that intact glutamate in this region is also necessary for the formation of cocaine-conditioned effects.

We have previously reported that cocaine-associated cues are able to elicit increases in nucleus accumbens DA overflow when presented along (22). Thus, such conditioned stimuli appear to acquire the ability to activate meso-accumbens DA pathways. Activation of meso-accumbens DA neurons in the VTA is thought to reflect information about the motivationally relevant properties of external stimuli (82–84). An important glutamatergic input to the VTA arises from the prefrontal cortex and forms excitatory synapses on both DA and GABA cells (40). It is possible that information related to cocaine-associated cues is transmitted to the VTA via such excitatory pathways from the frontal cortex. The concurrent activation of glutamate receptors on VTA DA cells by cocaine-induced release of glutamate (possibly via D1 DA receptors) and novel conditioned stimuli may form the basis for the induction of classical conditioning in the present studies. LTP, for example, has been shown to be associative in that activation of one set of synapses can facilitate LTP at an independent set of adjacent active synapses on the same cell if both are activated within a finite temporal interval (36).

Glutamate function in the nucleus accumbens or amygdala does not appear to be critical for the formation of cocaine-induced conditioned behavior, at least with the present design. The inability of MK-801 injections into the amygdala to alter the acquisition of cocaine-conditioned increases in motor activity is somewhat surprising considering that this brain region appears to play a significant role in mediating the effects of stimulus-reward associations on behavior (59,60) and in associating stimuli with specific incentive properties of reward (61). However, such functions may be related to DA mechanisms in this structure and not involve glutamate.

Interestingly, Kalivas and Alesdatter (45) found that MK-801 injections into the amygdala prevented the development of sensitization to cocaine utilizing a 1-d sensitization design. Because it is unlikely that such sensitization was context dependent, glutamate in the amygdala may be necessary for the formation of context-independent sensitization. Conversely, if conditioning did play some role in the sensitization design used by these investigators, our failure to find disruptive effects on conditioning following VTA injections of MK-801 may be the result of the use of a design utilizing prolonged training, which may have overcome subtle deficits induced by MK-801.

Although it is generally agreed that glutamate blockade prevents the development of sensitization, some have reported that the expression of sensitization is not altered by such manipulations (25–27). Other studies have reported that glutamate antagonists do prevent the expression of cocaine-sensitized responses following systemic injections of CPP (85) or focal applications of NMDA antagonists into the striatum and cortex (17,86). The studies reported here clearly indicate that glutamate blockade or inhibition of NO has little effect on the expression of cocaine-induced conditioned increases in locomotor activity. Surprisingly, direct injections of MK-801 into the VTA prior to tests for conditioning were also ineffective in preventing the expression of conditioned effects, suggesting that NMDA in this structure (which are important for the formation of cocaine-induced conditioning) do not mediate the effects of cocaine-associated cues. There are somewhat analogous findings for the induction of LTP. The induction of LTP is thought to require NMDA receptor-mediated processes, whereas expression of LTP involves non-NMDA receptors (35). There is substantial evidence to suggest that the initial increase in sensitivity during LTP involves postsynaptic modifications of AMPA receptor function and localization. It is of interest, therefore, that cocaine recently has been shown to increase the number and function of both of the AMPA receptors in the postsynaptic membrane of DA neurons in the VTA (81). Consistent with such findings, we found that icv DNQX, a relatively selective AMPA receptor antagonist, fully inhibited the expression of cocaine-conditioned increases in locomotor activity. It seems that cocaine-associated cues activate VTA AMPA receptors probably sensitized during the acquisition phase by concurrent glutamate stimulation induced by both cocaine and contextual stimuli.

## 5.1. Relevance to Drug Addiction

Recent theories in drug addition research have emphasized the importance of incentive-motivational mechanisms in the addiction process *(1,87–90)*. The concept of incentive-motivation is derived from the notion that there are objects in the environment to which an organism is attracted. Incentives are thought to "pull" behavior *(91)*. Although the mechanisms for incentive-motivation are innate, the processes necessary for incentive-motivation to influence behavior are acquired through learning. Essentially, organisms need to learn when and how to engage this system. Stimuli that are repeatedly associated with a primary or innate reinforcer, such as food or drugs, are thought to acquire two properties through classical conditioning. The first is secondary reinforcement and the second is incentive-motivation *(92)*. The secondary reinforcing properties of such stimuli when they follow a specific behavior enable them to facilitate and augment future performance of the behavior *(93,94)*. When such stimuli appear prior to a particular behavior, the incentive-motivational properties, developed through classical conditioning, appear to energize and facilitate initiation of the behavior.

Drugs such as cocaine can be thought of as positive incentives similar to food, water, or a sexual partner. The incentive-motivational properties of cocaine are not established, however, until the pharmacological effects are experienced. Such properties are then conferred to stimuli associated with the drug (visual characteristics of the drug, drug-using rituals, paraphernalia, or environment). These drug-associated cues, when encountered subsequently, are able to activate the anticipatory incentive-motivational processes that engage motor programs leading to drug-seeking and drug-taking behaviors.

The energizing function of incentive-motivational stimuli has been known for some time. It has been well established, for example, that stimuli repeatedly associated with positive reinforcers such as food can increase general motor activity or energize behaviors when presented alone *(16,91,95,96)*. Bindra and Palfi *(95)* have suggested that such incentive-motivational activity is characterized by anticipatory excitement often seen during classical appetitive conditioning and appears to be investigatory and goal directed in nature. Bindra *(92)* has also proposed that conditioned incentive-motivational stimuli established by pairing with a positive reinforcer such as food acquire similar appetitive properties and come to energize the appetitive-motivational system as a whole and produce a positive incentive-motivational state. We have also proposed previously *(16)* that the conditioned increases in locomotor activity seen with psychomotor stimulants is not simply a conditioned skeletal response, but reflects the operation of the energizing or the anticipatory function of incentive-motivation. Indeed, observation of rats in response to conditioned cues present in the activity chamber seems to indicate that the increase in locomotor activity does have investigatory and goal-directed properties. Previous studies *(22)* have established that cocaine-associated cues increase the activity to the meso-accumbens system, thereby engaging incentive-motivational processes. The present studies reveal an important role for glutamate in both the acquisition and expression of cocaine-induced increases in locomotor activity. It is proposed that information from conditioned stimuli is transmitted to the VTA from the frontal cortex via glutamatergic pathways to activate the meso-accumbens dopamine system, which is the neural substrate for incentive motivation.

Because most approaches to therapeutics of cocaine-abuse disorders involve after-the-fact interventions, the current series of studies could suggest the importance of treatment aimed at the manifestation of the expression phase of cocaine sensitization. As such, manipulation of the AMPA receptor system might provide a novel target. Studies of the AMPA receptor antagonist topiramate, which is an anticonvulsant with promising effects in the affective disorders, and other AMPA receptor manipulations appear worthy of further research.

## REFERENCES

1. Robinson, T. E. and Becker, J. B. (1986) Enduring changes in brain and behavior produced by chronic amphetamine administration: a review and evaluation of animal models of amphetamine psychosis. *Brain Res. Rev.* **11**, 157–198.

2. Segal, D. S., Geyer, M. A., and Schuckit, M. A. (1981) Stimulant-induced psychosis: an evaluation of animal models, in *Essays in Neurochemistry and Neuropharmacology, Volume 5.* (Youdim, M. B. H., Lovenberg, W., Sharman, D. F., and Lagnado, J. R., eds.), Wiley, London, pp. 95–129.
3. Kalivas, P. W. and Stewart, J. (1991) Dopamine transmission in the initiation and expression of drug- and stress-induced sensitization of motor activity. *Brain Res. Rev.* **16**, 223–244.
4. Pierce, R. C. and Kalivas, P. W. (1997) A circuitry model of the expression of behavioral sensitization to amphetamine-like psychostimulants. *Brain Res. Rev.* **25**, 192–216.
5. Nestler, E. J. (1994) Molecular neurobiology of drug addiction. *Neuropsychopharmacology* **11**, 77–87.
6. Kelly, P. H. and Iversen, S. D. (1975) Selective 6-OHDA-induced destruction of mesolimbic dopamine neurons: Abolition of psychostimulant-induced locomotor activity in rats. *Am. J. Pharmacol.* **40**, 45–56.
7. Henry, D. J. and White, F. J. (1991) Repeated cocaine administration causes a persistent enhancement of D1 dopamine receptor supersensitivity within the rat nucleus accumbens. *J. Pharmacol. Exp. Ther.* **258**, 882–890.
8. Higashi, H., Inanaga, K., Nishi, S., and Uchimura, N. (1989) Enhancement of dopamine actions on rat nucleus accumbens neurones *in vitro* after methamphetamine pretreatment. *J. Physiol. (Lond.)* **408**, 587–603.
9. Kalivas, P. W. and Weber, B. (1988) Amphetamine injection into the ventral mesencephalon sensitizes rats to peripheral amphetamine and cocaine. *J. Pharmacol. Exp. Ther.* **245**, 1095–1102.
10. Vezina, P. (1993) Amphetamine injected into the ventral tegmental area sensitizes the nucleus accumbens dopaminergic response to systemic amphetamine in an *in vivo* microdialysis study in the rat. *Brain Res.* **605**, 332–337.
11. Vezina, P. and Stewart, J. (1990) Amphetamine administered to the ventral tegmental area but not to the nucleus accumbens sensitizes rats to systemic morphine lack of conditioned effects. *Brain Res.* **516**, 99–106.
12. Kalivas, P. W. (1995) Interaction between dopamine and excitatory amino acids in behavioral sensitization to psychostimulants. *Drug Alchohol Depend.* **37**, 95–100.
13. Pavlov, I. P. (1927) *Conditioned Reflexes: An Investigation of the Physiological Activity of the Cerebral Cortex*, Dover, New York.
14. Pert, A., Post, R. M., and Weiss, S. R. B. (1990) Conditioning as a critical determinant of sensitization induced by psychomotor stimulants, in *Neurobiology of Drug Abuse: Learning and Memory* (Erinoff, L., ed.), NIDA Research Monographs, US Government Printing Office, Washington, DC, pp. 208–241.
15. Anagnostaras, S. G. and Robinson, T. E. (1996) Sensitization to the psychomotor stimulant effects of amphetamine modulation by associative learning. *Behav. Neurosci.* **110**, 1397–1414.
16. Pert, A. (1994) Neurobiological mechanisms underlying the acquisition and expression of incentive motivation by cocaine-associated stimuli: relationship to craving, in *Neurobiological Models of Drug Addiction* (Erinoff, L. ed.), NIDA Research Monographs, US Government Printing Office, Washington, DC.
17. Bedingfield, J. B., Calder, L. D., Thai, D. K., and Karler, R. (1997) The role of the striatum in the mouse in behavioral sensitization to amphetamine. *Pharmacol. Biochem. Behav.* **56**, 305–310.
18. Beninger, J. J. and Herz, R. S. (1986) Pimozide blocks establishment but not expression of cocaine-produced environmental specific conditioning. *Life Sci.* **38**, 1424–1431.
19. Weiss, S. R. B., Post, R. M., Pert, A., Woodward, R., and Murman, D. (1989) Context-dependent cocaine sensitization: differential effects of haloperidol on development versus expression. *Pharmacol. Biochem. Behav.* **34**, 655–661.
20. Fontana, D. J., Post, R. M., Weiss, S. R. B., and Pert, A. (1993) The role of D1 and D2 dopamine receptors in the acquisition and expression of cocaine induced conditioned increases in locomotor behavior. *Behav. Pharmacol.* **4**, 375–387.
21. Beninger, R. J. and Hahn, B. C. (1983) Pimozide blocks establishment but not expression of amphetamine-produced environment-specific conditioning. *Science* **220**, 1304–1306.
22. Fontana, D. J., Post, R. M., and Pert, A. (1993) Conditioned increases in mesolimbic dopamine overflow by stimuli associated with cocaine. *Brain Res.* **629**, 31–39.
23. Kalivas, P. W. and Duffy, P. (1990) Effects of acute and daily cocaine treatment on extracellular dopamine in the nucleus accumbens. *Synapse* **5**, 48–58.
24. Gold, L. H., Swerdlow, N. R., and Koob, G. F. (1988) The role of mesolimbic dopamine in conditioned locomotion produced by amphetamine. *Behav. Neurosci.* **102**, 544–552.
25. Karler, R., Calder, L. D., and Turkanis, S. A. (1991) DNQX blockade of amphetamine behavioral sensitization. *Brain Res.* **552**, 295–300.
26. Karler, R., Chaudhry, I. A., Calder, L. D., and Turkanis, S. A. (1990) Amphetamine behavioral sensitization and the excitatory amino acids. *Brain Res.* **537**, 76–82.
27. Wolf, M. E. (1998) The role of excitatory amino acids in behavioral sensitization to psychomotor stimulants. *Prog. Neurobiol.* **54**, 679–720.
28. Wolf, M. E., White, F. J., and Hu, X.-T. (1994) MK-801 prevents alterations in the mesoaccumbens dopamine system associated with behavioral sensitization to amphetamine. *J. Neurosci.* **14**, 1735–1745.
29. Jake-Matthews, C., Jolly, D. C., Queen, A. L., Brose, J., and Vezina, P. (1997) The competitive NMDA receptor antagonist CGS 19755 blocks the development of sensitization of the locomotor and dopamine activating effects of amphetamine. *Soc. Neurosci. Abstr.* **23**, 1092.

30. Gnegy, M. E., Hewlett, G. H. K., and Pimputkar, G. (1996) Haloperidol and MK-801 block increases in striatal calmodulin resulting from repeated amphetamine treatment. *Brain Res.* **734,** 35–42.
31. Berhow, M. T., Horoi, N., and Nestler, E. J. (1996) Regulation of ERK (extracellular signal regulated kinase), part of the neurotrophin signal transduction cascade, in the rat mesolimbic dopamine system by chronic exposure to morphine or cocaine. *J. Neurosci.* **16,** 4707–4715.
32. Zhang, X.-F., Hu, X.-T., White, F. J., and Wolf, M. E. (1997) Increased responsiveness of ventral tegmental area dopamine neurons to glutamate after repeated administration of cocaine or amphetamine is transient and selectively involves AMPA receptors. *J. Pharmacol. Exp. Ther.* **281,** 699–706.
33. Wolf, M. E., Dahlin, S. L., Hu, X.-T., Xue, C.-J., and White, K. (1995) Effects of lesions of prefrontal cortex, amygdala, or fornix on behavioral sensitization to amphetamine: comparison with N-methyl-D-aspartate antagonists. *Neuroscience* **69,** 417–439.
34. Madison, D. V., Malenka, R. C., and Nicoll, R. A. (1991) Mechanisms underlying long-term potentiation of synaptic transmission. *Annu. Rev. Neurosci.* **14,** 379–397.
35. Muller, D. and Lynch, G. (1988) Long-term potentiation differentially affects two components of synaptic responses in hippocampus. *Proc. Natl. Acad. Sci. USA* **59,** 9346–9350.
36. Malenka, R. C. and Nicoll, R. A. (1999) Long-term potentiation—a decade of progress? *Science* **285,** 1870–1874.
37. Venable, N. and Kelly, P. H. (1990) Effects of NMDA receptor antagonists on passive avoidance learning and retrieval in rats and mice. *Psychopharmacology* **100,** 215–221.
38. Ward, L., Mason, S. E., and Abraham, W. C. (1990) Effects of the NMDA antagonists CPP and MK-801 on radial arm maze performance in rats. *Pharmacol. Biochem. Behav.* **35,** 785–790.
39. Ono, T., Nishoyo, N., and Umano, T. (1995) Amygdala role in conditioned associative learning. *Prog. Neurobiol.* **46,** 401–422.
40. Sesack. S. R. and Pickel, V. M. (1992) Prefrontal cortical efferents in the rat synapse on unlabeled neuronal targets of catecholamine terminals in the nucleus accumbens septi and on dopamine neurons in the ventral tegmental area. *J. Comp. Neurol.* **320,** 145–160.
41. De Montis, M. G., Gambarana, C., Ghiglieri, O., and Tagliamonte, A. (1995) Reversal of stable behavioural modifications through NMDA receptor inhibition in rats. *Behav. Pharmacol.* **6,** 562–567.
42. Haracz, J. L., Belanger, S. A., MacDonall, J. S., and Sircar, R. (1995) Antagonists of N-methyl-D-aspartate receptors partially prevent the development of cocaine sensitization. *Life Sci.* **57,** 2347–2357.
43. Ida, I., Asami, T., and Kuribara, H. (1995) Inhibition of cocaine sensitization by MK-801, a noncompetitive N-methyl-D-aspartate (NMDA) receptor antagonist: evaluation by ambulatory activity in mice. *Jpn. J. Pharmacol.* **69,** 83–90.
44. Jeziorski, M., White, F. J., and Wolf, M. E. (1994) MK-801 prevents the development of behavioral sensitization during repeated morphine administration. *Synapse* **16,** 137–147.
45. Kalivas, P. W. and Alesdatter, J. E. (1993) Involvement of N-methyl-D-aspartate receptor stimulation in the ventral tegmental area and amygdala in behavioral sensitization to cocaine. *J. Pharmacol. Exp. Ther.* **267,** 486–495.
46. Morrow, B. A., Taylor, J. R., and Roth, R. H. (1995) R-(+)-HA-966, an antagonist for the glycine/NMDA receptor, prevents locomotor sensitization in repeated cocaine exposures. *Brain Res.* **673,** 165–169.
47. Ohmori, T., Abekawa, T., Muraki, A., and Koyama, T. (1994) Competitive and noncompetitive NMDA antagonists block sensitization to methamphetamine. *Pharmacol. Biochem. Behav.* **48,** 587–591.
48. Pudiak, C. M. and Bozarth, M. A. (1993) L-NAME and MK-801 attenuate sensitization to the locomotor-stimulating effect of cocaine. *Life Sci* **53,** 1517–1524.
49. Tzschentke, T. M. and Schmidt, W. J. (1997) Interactions of MK-801 and GYKI 52466 with morphine and amphetamine in place preference conditioning and behavioural sensitisation. *Behav. Brain Res.* **84,** 99–107.
50. Karler, R., Calder, L. D., Chaudhry, I. A., and Turkanis, S. A. (1989) Blockade of "reverse tolerance" to cocaine and amphetamine by MK-801. *Life Sci.* **45,** 599–606.
50. Stewart, J. and Druhan, J. P. (1993) Development of both conditioning and sensitization of the behavioral activating effects of amphetamine is blocked by the non competitive NMDA receptor antagonist, MK-801. *Psychopharmacology* **110,** 125–132.
51. Druhan, J. P., Jakob, A., and Stewart, J. (1993) The development of behavioral sensitization to apomorphine is blocked by MK-801. *Eur. J. Pharmacol.* **243,** 73–77.
52. Wolf, M. E. and Khansa, M. R. (1991) Repeated administration of MK-801 produces sensitization to its own locomotor stimulant effects but blocks sensitization to amphetamine. *Brain Res.* **562,** 164–168.
53. Segal, D. S., Kuczenski, R., and Florin, S. M. (1995) Does dizocilpine (MK-801) selectively block the enhanced responsiveness to repeated amphetamine administration? *Behav. Neurosci.* **109,** 532–546.
54. Cervo, L. and Samanin, R. (1996) Effects of dopaminergic and glutamatergic receptor antagonists on the establishment and expression of conditioned locomotion to cocaine in rats. *Brain Res.* **731,** 31–38.
54. Kim, H.-S. and Park, W.-K. (1995) Nitric oxide mediation of cocaine-induced dopaminergic behaviors: ambulation-accelerating activity, reverse tolerance and conditioned place preference in mice. *J. Pharmacol. Exp. Ther.* **275,** 551–557.
55. Tzschentke, T. M. and Schmidt, W. J. (1995) N-Methl-D-aspartic acid-receptor antagonists block morphine-induced conditioned place preference in rats. *Neurosci. Lett.* **193,** 37–40.

56. Carey, R. J., Dai, H., Krost, M., and Huston, J. P. (1995) The NMDA receptor and cocaine: evidence that MK-801 can induce behavioral sensitization effects. *Pharmacol. Biochem. Behav.* **51,** 901–908.
57. Duke, M. A., O'Neal, J., and McDougall, S. A. (1997) Ontogeny of dopamine agonist-induced sensitization: role of NMDA receptors. *Psychopharmacology* **129,** 153–160.
58. Wolf, M. E., White, F. J., and Hu, X.-T. (1993) Behavioral sensitization to MK-801 (dizocilpine): neurochemical and electrophysiological correlates in the mesoaccumbens dopamine system. *Behav. Pharmacol.* **4,** 429–442.
59. Cador, M., Robbins, T. W., and Everitt, B. J. (1989) Involvement of the amygdala in stimulus-reward associations: interaction with the ventral striatum. *Neuroscience* **30,** 77–86.
60. Everitt, B. J., Morris, K. A., O'Brien, A., and Robbins, T. W. (1991) The basolateral amygdala–ventral striatal system and conditioned place preference: further evidence of limbic–striatal interactions underlying reward-related processes. *Neuroscience* **42,** 1–18.
61. Gaffan, D. and Harrison, S. (1987) Amygdalectomy and disconnection in visual learning and auditory secondary reinforcement by monkeys. *J. Neurosci.* **7,** 2285–2292.
61. Muller, D., Joly, M., and Lynch, G. (1988) Contributions of quisqualate and NMDA receptors to the induction and expression of LTP. *Science* **242,** 1694–1697.
62. Haracz, J. L., MacDonall, J. S., and Sircar, R. (1997) Effects of nitric oxide synthase inhibitors on cocaine sensitization. *Brain Res.* **746,** 183–189.
63. Abekawa, T., Ohmori, T., and Koyama, T. (1995) Effects of nitric oxide (NO) synthesis inhibition on the development of supersensitivity to stereotypy and locomotion stimulating effects of methamphetamine. *Brain Res.* **679,** 200–204.
64. Corbett, D. (1989) Possible abuse potential of the NMDA antagonist MK-801. *Behav. Brain Res.* **34,** 239–246.
65. Del Pozo, E., Barrios, M., and Baeyens, J. M. (1996) The NMDA receptor antagonist dizocilpine (MK-801) stereoselectively inhibits morphine-induced place preference conditioning in mice. *Psychopharmacology* **125,** 209–213.
66. Hoffman, D. C. (1994) The noncompetitive NMDA antagonist MK-801 fails to block amphetamine-induced place conditioning in rats. *Pharmacol. Biochem. Behav.* **47,** 907–912.
67. Papp, M., Maryl, E., and Maccecchini, M. L. (1996) Differential effects of agents acting at various sites of the NMDA receptor complex in a place preference conditioning model. *Eur. J. Pharmacol.* **317,** 191–196.
68. Carlezon, W. A. and Wise, R. A. (1996) Rewarding actions of phencyclidine and related drugs in nucleus accumbens and frontal cortex. *J. Neurosci.* **16,** 3112–3122.
69. David, V., Durkin, T. P., and Cazala, P. (1998) Rewarding effects elicited by the microinjection of either AMPA or NMDA glutamatergic antagonists into the ventral tegmental area revealed by an intracranial self-administration paradigm in mice. *Eur. J. Neurosci.* **10,** 1394–1405.
70. Bohme, G. A., Bon, C., Stutzmann, J.-M., Doble, A., and Blanchard, J.-C. (1991) Possible involvement of nitric oxide in long-term potentiation. *Eur. J. Pharmacol.* **199,** 379–381.
71. Haley, J. E., Wilcox, G. L., and Chapman, P. F. (1992) The role of nitric oxide in hippocampal long-term potentiation. *Neuron* **8,** 211–216.
72. Schuman, E. M. and Madison, D. V. (1994) Nitric oxide and synaptic function. *Annu. Rev. Neurosci.* **17,** 153–183.
73. Bernabeu, R., Levi de Stein, M., Fin, C., Izquierdo, I., and Medina J. H. (1995) Role of hippocampal NO in the acquisition and consolidation of inhibitory avoidance learning. *NeuroReport* **6,** 1498–1500.
73. Robinson, T. E. and Berridge, K. C. (1993) The neural basis of drug craving: an incentive-sensitization theory of addiction. *Brain Res. Rev.* **18,** 247–292.
74. Fin, C., Da Cunha, C., Bromberg, E., Schmitz, P. K., Bianchin, M., Medina, J. H., et al. (1995) Experiments suggesting a role for nitric oxide in the hippocampus in memory processes. *Neurobiol. Learn. Mem.* **63,** 113–115.
75. Cobb, B. L., Ryan, K. L., Frei, M. R., Guel-Gomez, V., and Mickley G. A. (1995) Chronic administration of L-NAME in drinking water alters working memory in rats. *Brain Res. Bull.* **38,** 203–207.
76. Itzhak, Y. (1997) Modulation of cocaine- and methamphetamine-induced behavioral sensitization by inhibition of brain nitric oxide synthase. *J. Pharmacol. Exp. Ther.* **282,** 521–527.
77. Inoue, H., Arai, I., Shibata, S., and Watanabe, S. (1996) L-Arginine methyl ester attenuates the maintenance and expression of methamphetamine-induced behavioral sensitization and enhancement of striatal dopamine release. *J. Pharmacol. Exp. Ther.* **277,** 1424–1430.
78. Stewart, J., Deschamps, S.-E., and Amir, S. (1994) Inhibition of nitric oxide synthase does not block the development of sensitization to the behavioral activating effects of amphetamine. *Brain Res.* **641,** 141–144.
79. White, F. J., Hu, X.-T., Zhang, X.-F., and Wolf, M. E. (1995) Repeated administration of cocaine or amphetamine alters neuronal responses to glutamate in the mesoaccumbens dopamine system. *J. Pharmacol. Exp. Ther.* **273,** 445–454.
80. Tong, Z.-Y., Overton, P. G., and Clark, D. (1995) Chronic administration of (+)-amphetamine alters the reactivity of midbrain dopaminergic neurons to prefrontal cortex stimulation in the rat. *Brain Res.* **674,** 63–74.
81. Ungless, M. A., Whistler, J. L., Malenka, R. C., and Bonci, A. (2001) Single cocaine exposure *in vivo* induces long-term potentiation in dopamine neurons. *Nature* **411,** 583–587.
82. Overton, P. G. and Clark, D. (1997) Burst firing in midbrain dopaminergic neurons. *Brain Res. Rev.* **25,** 312–334.
83. Schultz, W. (1998) Predictive reward signal of dopamine neurons. *J. Neurophysiol.* **80,** 1–27.

84. White F. J. (1996) Synaptic regulation of mesocorticolimbic dopamine neurons. *Annu. Rev. Neurosci.* **19,** 405–436.
85. Karler, R., Calder, L. D., and Bedingfield, J. B. (1994a) Cocaine behavioral sensitization and the excitatory amino acids. *Psychopharmacology* **115,** 305–310.
86. Karler, R., Bedingfield, J. B., Thai, D. K., and Calder, L. D. (1997) The role of the frontal cortex in the mouse in behavioral sensitization to amphetamine. *Brain Res.* **757,** 228–235.
87. Markou, A., Weiss, F., Gold, L. H., Caine, S. B., Schulteis, G., and Koob, G. F. (1993) Animal models of drug craving. *Psychopharmacology* **112,** 163–182.
88. Marlatt, G. A. (1978) Craving for alcohol, loss of control, and relapse: cognitive–behavioral analysis, in *Alcoholism: New Directins in Behavioral Research and Treatment* (Nathan, P. E., Marlett, G. A., and Loberg, T., eds.), Plenum, New York, pp. 271–314.
89. Stewart, J., deWit, H., and Eikelboom, R. (1984) The role of unconditioned and conditioned drug effects in the self-administration of opiates and stimulants. *Psychol. Rev.* **91,** 251–268.
90. Wise, R. A. and Bozarth, M. A. (1987) A psychomotor stimulant theory of addiction. *Psychol. Rev.* **94,** 469–492.
91. Bolles, R. C. (1967) *Theory of Motivation,* Harper & Row New York.
92. Bindra, D. (1968) Neurophysiological interpretations of the effects of drive and incentive-motivation on general activity and instrumental behavior. *Psychol. Rev.* **75,** 1–22.
93. Davis, W. M. and Smith, S. G. (1976) Role of conditioned reinforcers in the initiation, maintenance, and extinction of drug-seeking behavior. *Pavlovian J. Biol. Sci.* **11,** 222–236.
94. Schuster, C. R. E. and Woods, J. H. (1968) The conditioned reinforcing effects of stimuli associated with morphine. *Int. J. Addict.* **3,** 223–236.
95. Bindra, D. and Palfi, T. (1967) The nature of positive and negative incentive-motivational effects on general activity. *J. Comp. Physiol. Psychol.* **63,** 288–297.
96. Sheffield, F. D. and Campbell, B. A. (1954) The role of experience in the "spontaneous" activity of hungry rats. *J. Comp. Physiol. Psychol.* **47,** 97–100.

# 7
# Interactions of Dopamine, Glutamate, and GABA Systems in Mediating Amphetamine- and Cocaine-Induced Stereotypy and Behavioral Sensitization

### Ralph Karler, PhD, David K. Thai, PhD, and Larry D. Calder, PhD

## 1. INTRODUCTION

The psychomotor stimulants are distinguished by their drug-abuse liability, their motor effects, and behavioral sensitization. How these effects relate to one another is not clear, but in the past, the studies of these stimulants have centered on dopamine and on the nucleus accumbens and the striatum, two areas of the brain that contain dopaminergic terminal fields, which historically have been implicated in the psychomotor effects of these drugs. The ever more detailed explication of neuroanatomy, specifically of the basal ganglia, makes it obvious that, first, the characteristic actions of the psychomotor stimulants must involve transmitters other than dopamine and brain areas other than the accumbens and the striatum (1–6); second, the effects of the stimulants are mediated by the activation of circuits; third, the different pharmacological properties of the stimulants, such as a motor effect and sensitization, involve, in part, different circuits, even though they undoubtedly share some neuroeffector systems. Starting with these working hypotheses, we initiated a study some 12 years ago that was designed to identify specific neuroeffectors and discrete brain areas that mediate a stimulant-induced motor effect and behavioral sensitization. The data presented below represent our observations on the general role of dopamine, glutamate, and GABA, as well as their specific functions in the striatum and in the frontal cortex, in enabling stimulant-induced stereotypy and behavioral sensitization.

From the many previous studies that have addressed the question of the relationship among dopamine, glutamate, and GABA function in the striatum and the accumbens, it has become clear that the neuroanatomy of these structures suggests that they interact but not exactly how (7–11); nevertheless, a variety of techniques have been used to allege that both glutamatergic and GABAergic drugs affect the release of dopamine (12). In general, these reports, which include in vivo dialysis studies, have at least two serious flaws in the interpretation of their data. First, very few studies have attempted to quantitatively correlate changes in recovered transmitters with changes in behavior, and, with a few notable exceptions (see, e.g., ref. 13), this includes the in vivo dialysis studies associated with behavioral sensitization. To simply measure recovered transmitters ignores the crucial difference between statistically significant and biologically significant effects, a point that is underscored by the reports that the quantitative behavioral responses do not necessarily parallel the recovered dopamine (for a review, see ref. 14). Such results mandate a consideration of physiological factors other than dopamine in order to understand stimulant-induced behavioral effects, which leads to the proposition that a stimulant effect involves a circuit of interacting transmitter systems and that a functional change in any one of them could quantitatively modify the behavioral response to these drugs.

From: *Contemporary Clinical Neuroscience: Glutamate and Addiction*
Edited by: Barbara H. Herman et al. © Humana Press Inc., Totowa, NJ

**Table 1**
Qualitative Influence of Systemically Administered Dopamine, Glutamate, and GABA Receptor Antagonists on Amphetamine-Induced Stereotypy in Normal Mice and on the Induction and Expression of Sensitization

| Antagonist treatment | Acute | Sensitization | |
|---|---|---|---|
| | | Induction | Expression |
| Dopamine | | | |
| $D_1$ (SCH-23390) | Block[a] | Block[a] | Block[a] |
| $D_2$ (sulpiride) | Block[a] | Block[a] | Block[a] |
| Ionotropic glutamate | | | |
| NMDA (CPP) | Block[a] | Block[a] | Block[a] |
| Non-NMDA (DNQX) | No effect[a] | Block[a] | Block[b] |
| Metabotropic glutamate | | | |
| MCPG | No effect | No effect | No effect |
| GABA | | | |
| $GABA_A$ (bicuculline) | Block[a] | Block[a] | Block[a] |
| $GABA_B$ (2-hydroxysaclofen) | No effect | No effect | No effect |

[a] True also for cocaine.
[b] Not true for cocaine.

A second defect in many of the studies of interactions among dopamine, glutamate, and GABA derives from an experimental design based on the use of agonists to activate a system. Such studies frequently fail to differentiate between pharmacological effects and physiological function because agonist drugs can activate both innervated and noninnervated receptors; therefore, their effects may not reflect the functional character of a system. For example, because there are glutamate receptors on virtually all brain structures (15), a study of the effects of glutamate agonists, say, on dopamine release may be no more than a pharmacological exercise, with little or no relevance to the physiological response system.

The following studies were designed to circumvent these specific limitations by the use of antagonist rather than agonist drugs to identify those transmitter systems that are necessary for the stimulant drugs to produce a quantitatively defined motor effect—stereotypy. In these studies, the role of the dopamine, glutamate, and GABA systems in stimulant-induced stereotypy and in sensitization in male CF-1 mice was initially investigated by the systemic administration of selective antagonists prior to the systemic administration of the stimulants. Under these conditions, if an antagonist quantitatively altered the motor response or sensitization to the stimulants, it was concluded that the specific receptor system affected by the antagonist was an integral part of the involved circuits. The data from the systemic drug studies represent the general central nervous system (CNS) effects of the antagonists; subsequently, the antagonists were administered intracranially in the striatum and in the frontal cortex of mice to ascertain the role of these two specific structures in the observed systemic effects of these drugs. The data resulting from the studies of the influence of dopamine, glutamate, and GABA antagonists yielded some insights into the complexity of the circuits that are responsible for stimulant-induced stereotypy, as well as for the phenomena that constitute behavioral sensitization.

## 2. SYSTEMIC EFFECTS OF THE ANTAGONISTS

### 2.1. Qualitative Data

The qualitative data presented in Table 1 represent a summary of our quantitative data, which have been presented previously (16–19). In the summarized data, the influence of the antagonists is

# Dopamine, Glutamate, GABA, and Sensitization

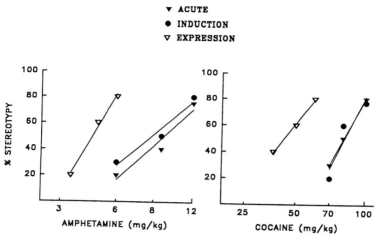

**Fig. 1.** Amphetamine and cocaine dose-response curves for stereotypy in nonsensitized (acute) and in sensitized (expression) animals and for the induction of sensitization.

described in terms of three effects on the responses to amphetamine and cocaine: The "acute" effect refers to the influence of an antagonist on the motor response to the stimulants in nonsensitized animals; "induction" refers to the influence of the antagonists on the ability of the stimulants to produce sensitization; and "expression" represents the effect of the antagonist on the response to the stimulants in previously sensitized animals. These data are based on the use of minimally effective stimulant and antagonist doses in order to maximize the selectivity of the drug effects. To illustrate, the influence of an antagonist on the acute response represents an effect of an antagonist on a stimulant dose that produced about 80% stereotypy in naive animals. The stimulant doses that produced a defined degree of stereotypy were obtained from the dose-response curves shown in Fig. 1. The antagonist doses were also similarly selected from dose-response curves, and an antagonist was considered to block a stimulant effect if it reduced the expected effect from 80% to at least 20% in a dose-dependent manner. Such decreases represent statistically significant effects in relatively small groups of animals; for example, in order to determine if an antagonist can block the acute response and the induction of sensitization, mice were given a single dose of either 12 mg/kg of amphetamine or 100 mg/kg of cocaine because these doses produce stereotypy in about 80% of the naive animals and induce sensitization in about 80% of these animals (Fig. 1). The effect of the antagonist treatment on induction was measured in the same groups of animals challenged 24 h later with either 6 mg/kg of amphetamine or 60mg/kg of cocaine, doses which produce stereotypy in 80% of sensitized animals (Fig. 1). To test a drug effect on expression, animals were sensitized by pretreatment with either 12 mg/kg of amphetamine or 100 mg/kg of cocaine; 24 h later, the mice were given either 6 mg/kg of amphetamine or 60 mg/kg of cocaine in order to evoke a sensitized response. The effect of the prior administration of antagonist drugs on this response was then determined. The experimental design, which included the study of the influence of the antagonists on the three effects of the stimulants, is based on the assumption that these effects do not necessarily involve identical circuits; therefore, the effect of the antagonists may vary among the different conditions. In the data presented, "no effect" of a drug refers to results obtained against both relatively high and low doses of the stimulants; that is, the "no-effect" drugs neither blocked nor enhanced the responses.

The antagonist data listed in Table 1 illustrate that all three of the measured effects induced by the stimulants involve not only the dopaminergic system but also the glutamatergic and GABAergic systems. Although stereotypy and sensitization can be produced by $D_2$ but not $D_1$ agonists, the data demonstrate that either a $D_1$ or the $D_2$ antagonist can block the effects of amphetamine, as has been reported earlier (20); therefore, the systemic data suggest that the $D_1$ system functions in a "permissive" manner to enable $D_2$ effects to manifest themselves. In general, such $D_1/D_2$ interactions have been described in a wide variety of experimental paradigms (21), including sensitization to amphetamine (22,23). Data not shown illustrate that the addition of $D_1$ agonist does not enhance either stereotypy or sensitization caused by amphetamine, which implies that, physiologically, the $D_1$ system is not limiting the $D_2$ effects; it appears that tonic dopamine release provides the necessary $D_1$ activity to enable the activation of the $D_2$ system to express itself maximally (21).

Table 1 delineates the involvement of the ionotropic and metabotropic glutamate systems in the stimulant effects. In general, the subject of the role of the glutamate system in sensitization has been extensively reviewed by Wolf (24), and most of our data are similar to what others have reported. The data in Table 1 indicate that the $N$-methyl-D-aspartate (NMDA) system is involved in all of the effects of the stimulants. That the involvement is not simple was presaged in our initial report that NMDA antagonism affected only induction by amphetamine, not the acute response or the expression of sensitization (16). Later, we used CPP to determine that both the acute response and expression were also blocked by NMDA antagonists (17,19); but to block these effects required much higher doses of the antagonist (see Fig. 4). The original negative data were obtained with the use of dizocilpine (MK-801), relatively high doses of which were deemed unsuitable because of its PCP-like behavioral activity. CPP, on the other hand, is a more selective NMDA antagonist, and relatively high doses of this drug can be used. The dose differential for this drug to block the two amphetamine effects intimates that the mechanism of the induction of sensitization is not tightly coupled to the acute response; yet, a tight coupling is suggested by the dose-response data shown in Fig. 1. The DNQX data in Table 1, however, corroborate the involvement of different circuits for the acute response and induction, because this class of drugs does not affect the acute response, but does block induction, thereby providing a qualitative separation of the two effects. Furthermore, the data in Table 1 indicate that the $D_1$ antagonist SCH-23390 can block all of the effects; but a dose-response analysis of these effects indicate that the median effective dose ($ED_{50}$) to block induction is about 40 times that required to block the acute response or expression (Fig. 3). We have made several other observations that uphold the idea that the two dopaminergic effects—the acute response and induction—represent independent phenomena, because a variety of drugs with ostensibly different mechanisms of action can block induction without affecting the acute response (16,25,26). Furthermore, we have determined that some drugs, such as the muscarinic antagonists, can enhance the acute response to the stimulants without enhancing induction (Karler, unpublished); in addition, we have found that drugs, such as the opioids, when combined with stimulant doses that are alone too low to produce either stereotypy or sensitization, can, nevertheless, cause induction (Karler, unpublished). The totality of the results of these drug studies gives us reason to conclude that the acute response and the induction of sensitization must employ different circuits, despite some common neurotransmitter components.

From the results shown in Table 1, the metabotropic glutamate system does not appear to be required to mediate any of the effects of amphetamine, although there are some data to suggest that this system may be involved (24). The negative data presented, however, are based on the use of limited doses (the highest dose of MCPG used was 25 mg/kg ip), on the use of stereotypy as an end point, and on the use of the mouse. Obviously, these variables are important to consider in defining the role of any given transmitter system in these effects.

The use of GABA antagonists indicated that bicuculline, a $GABA_A$ antagonist, administered systemically in subconvulsant doses can block all three effects of amphetamine (18). In contrast to the effects of bicuculline, the $GABA_B$ antagonist 2-hydroxysaclofen, at least in doses as high as 50 mg/kg ip, was ineffective against any of the amphetamine effects. Although the motor activity of amphetamine and dopamine has been linked to GABA function by others (9,10,27–29), these data demonstrate that the GABA system is also essential to both the induction and expression of sensitization.

The qualitative systemic data presented in Table 1 show that all three of the measured effects of amphetamine require functional $D_1$ and $D_2$ systems, the NMDA glutamate system, and the $GABA_A$ system. The non-NMDA component of the ionotropic glutamate system does not appear to be a factor in the acute response but is necessary for the manifestation of both the induction and expression of sensitization. These data show not only that the three effects of amphetamine investigated require the activation of NMDA and $GABA_A$ systems but that sensitization also involves the introduction of a new participant, the non-NMDA system, in response to amphetamine. Subsequent to these findings, other differences among the systems that constitute the acute effect and sensitization have been noted. A variety of drugs, including calcium-channel blockers, protein-synthesis inhibitors, and nicotinic antagonists, block the induction and expression of sensitization, but these drugs are all ineffective in non-sensitized animals *(25,26,30)*. These data imply that sensitization is not simply an enhanced response derived from an increase in released dopamine, as has been alleged by some investigators based on their results of in vivo dialysis data. The conclusion that sensitization involves the recruitment of new pathways is unavoidable.

The positive findings in Table 1 for amphetamine are also true for cocaine with only one exception—the role of the non-NMDA system in the expression of sensitization. DNQX, which blocks both the induction and expression of amphetamine sensitization, also blocked cocaine induction but failed to block the expression of cocaine sensitization. These results not only illustrate a difference between amphetamine and cocaine, but they also suggest that expression involves different circuits for amphetamine and cocaine. Other differences between amphetamine and cocaine have been reported; for example, nicotinic antagonists can block both the induction and expression of amphetamine sensitization, but these drugs are ineffective against the comparable effects of cocaine *(26)*. Sensitization then is not a single phenomenon, rather it is a general description of a change in drug responsivity, which encompasses a variety of neuroadaptations, a conclusion that is buttressed by earlier observations that exposure to either serotonin or direct-acting dopamine agonists produces a form of sensitization of relatively limited persistence compared to that produced by the psychomotor stimulants *(31–33)*.

## 2.2 Quantitative Data

Figure 1 illustrates the amphetamine and cocaine dose-response relationships for the acute (non-sensitized) response, for the induction of sensitization, and for the expression of sensitization. The "% stereotypy" on the ordinate represents the percentage of mice in a group of 10 that exhibited stereotypy. The dose-response curves for the induction of sensitization were obtained with the same animals that were used to establish the acute effect; these animals were challenged 24 h later with a 6 mg/kg dose of amphetamine or a 60 mg/kg dose of cocaine in order to discover if the acute treatment caused sensitization. To determine the curves for expression, separate groups of animals were given a single dose of either 12 mg/kg of amphetamine or 100 mg/kg of cocaine in order to sensitize them, and the dose-response curves shown for expression were obtained 24 h later. The $ED_{50}$ for the acute response to amphetamine was 8.7 (7.5–10) mg/kg; and for the sensitized animals, 4.9 (4.3–5.5) mg/kg; the corresponding values for cocaine were 80 (69–93) mg/kg and 44 (36–54) mg/kg. These values for the acute and sensitized responses are significantly different and show that sensitization can develop following a single exposure to amphetamine or cocaine, as has been described earlier *(19,33,34)*. The characteristics of the single-dose sensitization were found to be indistinguishable from the conventional repeated daily low-dose treatment procedure *(33)*. That a single dose of the stimulants can produce sensitization is evident from the induction curves compared with the acute dose-response curves, and the data clearly indicate that the induction of sensitization is a dose-related effect and that a single dose of 12 mg/kg of amphetamine or 100 mg/kg of cocaine is sufficient to cause sensitization in the majority of the animals. A single dose of 12 mg/kg of amphetamine apparently induces a maximal degree of sensitization, as implied by the results of attempts to enhance the degree of sensitization by repeated daily exposure to 12 mg/kg of amphetamine. Comparative dose-response data for ampheta-

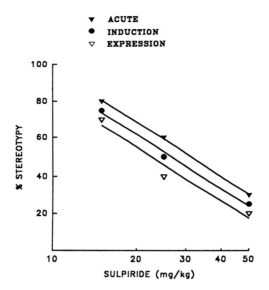

**Fig. 2.** Dose-response curves for systemically administered sulpiride antagonism of amphetamine-induced stereotypy in nonsensitized (acute) and sensitized (expression) mice and of amphetamine-induced sensitization.

mine following one pretreatment and two or three pretreatments were essentially identical, which indicates that a single 12-mg/kg treatment yielded maximal sensitization (Karler, unpublished). Although repeated exposure to a sensitizing dose has no effect on the degree of sensitization, the influence of repeated sensitized responses on other functions has never been evaluated. It should be noted, however, that the single-dose sensitization method avoids a potential problem with the commonly used repeated daily low-dose sensitization procedure; that is, because of individual variability in drug response, some animals are sensitized very early in the daily treatment regimen; yet, because those animals usually continue to receive daily treatment, they are subjected to many episodes of sensitized responses *(30,31)*. The impact of repeated sensitized responses on function is generally not evaluated; nevertheless, such a condition imposes a potentially uncontrolled variable on subsequent functional tests.

Figure 2 represents the dose-response data for the use of sulpiride, the $D_2$ antagonist, in order to block the acute response ($ED_{50}$ = 31 [22–43] mg/kg), as well as the induction ($ED_{50}$ = 26 [17–38] mg/kg) and expression of sensitization ($ED_{50}$ = 24 [15–36] mg/kg). None of these values for sulpiride are significantly different, which indicates that the $D_2$ activity involved in the three effects of amphetamine is the same. These results contrast with those obtained with the $D_1$ antagonist SCH-23390 shown in Fig. 3. The $ED_{50}$ for this drug to block the acute response is 0.055 (0.032–0.072) mg/kg and to block expression 0.046 (0.036–0.061) mg/kg. These two values are not significantly different, suggesting that the $D_1$ activity evoked for comparable effects in the acute response and in the expression of sensitization is identical. This conclusion, however, does not obtain for the induction of sensitization, for the $ED_{50}$ is 2.2 (1.1–4.4) mg/kg, which is almost two orders of magnitude greater and is significantly different from the other two $ED_{50}$ values. These results tell us that the dopaminergic system involved in induction is different from that which accounts for both the acute response and for the expression of sensitization. The relatively high dose of SCH-23390 needed to block induction raises the possibility that the induction of sensitization involves the $D_5$ rather than the $D_1$ system. Although SCH-23390 exhibits equal binding affinity for $D_1$ and $D_5$ receptors, dopamine has been reported to have a 10-fold greater affinity for $D_5$ than $D_1$ receptors *(35)*; this

Table 6
Interaction of Dopamine and GABA Agonists and Antagonists
Administered Intracortically on Amphetamine-Induced Stereotypy
Normal and in Sensitized Mice

| Pretreatment (i.c.) | Amphetamine treatment | |
|---|---|---|
| | Normal response | Sensitized response |
| Dopamine | | |
| + Saline | Block | No effect |
| + Sulpiride | No effect | — |
| + Bicuculline | No effect | — |
| THIP | | |
| + Saline | Block | Block |
| + Sulpiride | Block | — |
| + Bicuculline | No effect | — |

Table 7
Characterization of Bicuculline-Induced Stereotypy Following Intracortical Administration

| Treatment | Acute (% stereotypy) | Induction (% stereotypy) |
|---|---|---|
| Bicuculline (0.005 µg/side) | | |
| + saline | 0 | 0 |
| Amphetamine (6 mg/kg) | 20 | 0 |
| Bicuculline (0.005 µg/side) | | |
| + amphetamine (6 mg/kg) | 100[a] | 0 |
| Bicuculline (0.01 µg/side) | 60 | 0 |
| Bicuculline (0.1 µg/side) | 100 | 0 |
| Bicuculline (0.5 µg/side) | Convulsions | |
| Bicuculline (0.05 µg/side) | | |
| + saline i.c. | 88 | — |
| + CPP (0.1 µg/side) | 25[a] | — |
| Bicuculline (0.05 µg/side) | | |
| + Sulpiride (75 mg/kg ip) | 100 | — |
| + CPP (20 mg/kg ip) | 0[a] | — |

[a] Significantly different from saline control, as determined by a $\chi^2$-test ($p < 0.05$).

The above-described observations demonstrated that bicuculline injected into the frontal cortex enhanced the response to amphetamine administered systemically. The data in Table 7, however, add an additional facet to the role of the GABA system vis-à-vis stereotypy. As shown in Table 7, bicuculline injected into the frontal cortex in relatively high doses produces stereotypy even in the absence of amphetamine. The bicuculline-induced stereotypy either in the presence or the absence of amphetamine does not produce sensitization, providing further evidence of the distinction between the systems that generate the acute response and those that are required for the induction of sensitization. The data shown in Table 7 also indicate that the bicuculline-induced stereotypy is not blocked by sulpiride, suggesting that the stereotypy does not involve dopamine but is blocked by the NMDA antagonist CPP administered either systemically or into the frontal cortex. These results suggest that stereotypy produced by the stimulants is the result of diminishing the physiological GABA inhibitory control over a tonically active NMDA system in the frontal cortex. This view is further supported by the observations in Table 5, which demonstrated that THIP, a $GABA_A$ agonist, injected into the frontal cortex can block the systemic effect of amphetamine. The key role of the $GABA_A$ system in the manifestation of stereotypy is further underscored by the above-described drug studies in the striatum. Stereotypy originating in the striatum appears to result from a GABAergic inhibition of its GABAergic outflow; similarly, in the frontal cortex stereotypy is again the result of the inhibition of a GABAergic system, which physiologically functions to inhibit a tonically active glutamatergic system.

## 5. CONCLUSIONS

First, the purpose of the described study was to define the role of the dopamine, glutamate, and GABA systems in mediating three effects of the psychomotor stimulants: stimulant-induced stereotypy in normal animals, the induction of behavioral sensitization, and the expression of the sensitization. In order to identify the functional involvement of the three neurotransmitter systems, relatively selective antagonists were administered systemically to determine their influence on the three effects of the stimulants. Systemically, all three classes of the antagonists blocked not only the acute response to the stimulants but also the induction and the expression of sensitization; therefore, it appears that all of the measured effects of amphetamine and cocaine require the activation of both $D_1$ and $D_2$ receptors, as well as those of the NMDA glutamate and $GABA_A$ systems. Other classes of receptors, specifically those of the metabotropic glutamate and the $GABA_B$ systems, do not appear to be functional components in any of the measured effects, at least within the limits of the experimental design.

Second, the systemic study of the non-NMDA glutamate system, which is not required for the acute response, demonstrated that it is a functional component of the circuits that constitute both the induction and expression of sensitization. The results with the non-NMDA antagonists illustrate that both sensitization phenomena (induction and expression) can be blocked by these drugs; therefore, sensitization involves systems that are not components of the circuit that produces stereotypy in normal animals. Furthermore, the non-NMDA antagonists, in contrast to the dopamine, the NMDA, and the $GABA_A$ antagonists, cannot completely block the motor response in sensitized animals. The quantitative data suggest that the amphetamine response in sensitized animals consists of two distinct components—the normal response plus the contribution to the total response provided by sensitization. These conclusions suggest that dopaminergically mediated stereotypy in sensitized animals can be produced by the activation of at least two different circuits. The results of the systemic study of the role of the non-NMDA system provide evidence that induction and expression of sensitization depend on partially different circuitries than does the acute response.

Third, the dose-response curves obtained for the above-described qualitative systemic drug effects indicate that the $ED_{50}$ values for the individual antagonists are generally the same for the acute response and for both the induction and the expression of sensitization. The results imply that the function of these systems is common to the three effects investigated. Only the data from the $D_1$ antagonist, SCH-23390, and CPP, the NMDA antagonist, deviate from the general results. In both instances, the $ED_{50}$ values for induction differed from their other two values; in the case of the $D_1$ antagonist, the $ED_{50}$ value to block induction was about two orders of magnitude greater, and for CPP to block induction required about one order of magnitude less drug. Such results support the hypothesis that induction involves a different circuit from those that enable the other two effects. Furthermore, the $D_1$ data on induction provide provocative evidence that this dopamine effect as described above is mediated not by $D_1$ receptors, but by $D_5$ receptors. The quantitative systemic data presented serve to distinguish between the circuit involved in induction and those involved in the acute response and expression; therefore, these data combined with the observed non-NMDA results demonstrate that distinct circuits exist for each of the three effects studied, even though the different circuits appear to share some common neuroeffector systems.

Fourth, the influence of the intrastriatal administration of the dopamine, NMDA, and GABA$_A$ antagonists on amphetamine-induced stereotypy mirrored their systemic effects; that is, these antagonists in the striatum blocked all of the measured effects of amphetamine. Consistent with these observations, the corresponding agonists injected into the striatum all produced stereotypy. The results of the intrastriatal administration of various combinations of dopamine agonists and antagonists indicate that dopamine activates the NMDA glutamate system, which, in turn, activates the GABA$_A$ system. Considering that the major outflow from the striatum is GABAergic, the results of the striatal drug studies suggest that stimulant-induced stereotypy derives from the inhibition of the inhibitory outflow of the striatum.

The results of the cortical studies of the D$_2$ and GABA$_A$ agonists and antagonists demonstrate that stereotypy results from the inhibition of GABA$_A$ activity in the frontal cortex. The data suggest that GABA physiologically functions tonically in the cortex to inhibit stereotypy and that drugs which either enhance stimulant-induced stereotypy or produce stereotypy in the absence of the stimulants do so by diminishing GABA$_A$ activity in the cortex. The data obtained from studies in the striatum also suggest that the stimulant-induced stereotypy is the consequence of the inhibition of the GABAergic outflow of the striatum. The results of the drug studies both in the striatum and in the frontal cortex emphasize the pivotal role that the inhibition of GABAergic activity plays in the manifestation of stereotypy.

The loss of corticodopaminergic inhibition appears to be at least a part of the mechanism of sensitization; that is, the normal stimulant-induced D$_2$ activity in the cortex inhibits systemic amphetamine or cocaine, and this inhibitory activity is mediated by the GABA$_A$ system in the cortex. The disappearance of the GABAergic inhibition in sensitized animals, therefore, results in an enhanced motor-response characteristic of sensitization. Other explanations of sensitization, such as an increase in striatal dopamine receptors or in dopamine release have not stood the test of time. There are, for example, about as many papers that claim that sensitization is associated with an increase in the release of dopamine in the striatum as there are those that claim there is no such increase (*42*, but *see* ref. *43*); therefore, alternative explanations of sensitization must be sought.

Finally, the above presented data is only a partial description of the response systems activated by amphetamine and cocaine that produce stereotypy and sensitization. The description is only partial because other neuroeffector systems besides those investigated are undoubtedly involved, as are other brain structures. Nevertheless, to judge from the limited data available, the complexity of such effects is obvious, but it is the definition of this complexity that may provide additional pharmacological opportunities to interrupt the deleterious effects of amphetamine and cocaine. To illustrate, the data presented indicate that a functional D$_1$ dopamine system is necessary for all of the measured effects of the stimulants, and this raises the possibility that a D$_1$ antagonist might be of value in the treatment of drug abuse. The data, however, also focus on the pivotal role that the GABA$_A$ system plays in the effects of the stimulants; for example, it is the GABA$_A$ system that appears to control the dopaminergically mediated inhibitory outflow from the striatum that results in stereotypy, and it is also the inhibition of the GABA$_A$ system in the frontal cortex that appears to control dopaminergically induced stereotypy. These GABAergic functions take on a special significance in view of the incredible observation that, in the brain, there are probably more than 500 distinct GABA$_A$ receptor subtypes and that specific subtypes appear to be localized in specific brain areas (*44*). Such observations suggest that at least some brain effector systems are controlled by distinct GABA$_A$ receptors. In support of this concept, there already are data that demonstrate that some functions in the brain are dependent on a specific GABA$_A$ receptor subtype (*44*). The recognition of this structural and functional diversity of the GABA$_A$ system raises the possibility that the development of subtype-selective drugs may provide unique clinical uses of GABAergic drugs in dopamine-dependent dysfunctional states, including not only drug abuse but also parkinsonism and schizophrenia. The latter disorders classically have been attributed to either too little or too much dopamine, but our increasing knowledge of the GABA$_A$

system may, in part, shift the focus in these disorders from dopamine to either too much or too little activity of a specific GABA$_A$ system. The development of GABA$_A$ receptor-specific drugs is now in progress *(45)* and these drugs will ultimately provide us with the pharmacological tools necessary to elucidate the role of specific GABA$_A$ receptor systems in CNS function.

## ACKNOWLEDGMENT

This work was supported by a research grant from the National Institute on Drug Abuse, DA00346.

## REFERENCES

1. Nicola, S. M., Surmeier, D. J., and Malenka, R. C. (2000) Dopaminergic modulation of neuronal excitability in the striatum and nucleus accumbens. *Annu. Rev. Neurosci.* **24,** 195–215.
2. Graybiel, A. M., Aosaki, T., Flaherty, A. W., and Kimura, M. (1994) The basal ganglia and adaptive motor control. *Science* **265,** 1826–1831.
3. Amalric, M. and Koob, G. F. (1993) Functionally selective neurochemical afferents and efferents of the mesocorticolimbic and nigrostriatal dopamine system, in *Progress in Brain Research, Volume 99* (Arbuthnott, G. W. and Emson, P. C., eds.), Elsevier Science, Amsterdam, pp. 209–226.
4. Albin, R. L., Makowiec, Z. R., Hollingsworth, L. S., Dure L. S., IV, Penny, J. B., and Young, A. B. (1992) Excitatory amino acid binding sites in the basal ganglia of the rat. A quantitative autoradiographic study. *Neuroscience* **46,** 35–48.
5. Alexander, G. E. and Crutcher, M. D. (1990) Functional architecture of basal ganglia circuits: neural substrates of parallel processing. *Trends Neurosci.* **13,** 266–271.
6. Graybiel, A. M. (1990) Neurotransmitters and neuromodulators in the basal ganglia. *Trends Neurosci.* **13,** 244–254.
7. Moore, R. Y., Bhatnagar, R. K., and Heller, A. (1971) Anatomical and chemical studies of a nigroneostriatal projection in the cat. *Brain Res.* **30,** 119–135.
8. Bouyer, J. J., Park, D. H., Joh, T. H., and Pickel, V. M. (1984) Chemical and structural analysis of the relation between cortical inputs and tyrosine hydroxylase-containing terminals in rat neostriatum. *Brain Res.* **302,** 267–275.
9. Scheel-Krüger, J. (1986) Dopamine–GABA interactions: evidence that GABA transmits, modulates and mediates dopaminergic functions in the basal ganglia and the limbic system. *Acta Neurol. Scand.* **(Suppl. 73),** 1–54.
10. Smith, A. D. and Bolam, J. P. (1990) The neural network of the basal ganglia as revealed by the study of synaptic connections of identified neurons. *Trends Neurosci.* **13,** 259–265.
11. Nieoullon, A. and Kerkorian-Le Goff, L. (1992) Cellular interactions in the striatum involving neuronal systems using "classical" neurotransmitters: possible functional implications. *Movement Disord.* **7,** 311–325.
12. Santiago, M., Machado, A., and Cano, J. (1996) Dopamine release and its regulation in the CNS, in *CNS Neurotransmitters and Neuromodulators: Dopamine* (Stone, T.W., ed.), CRC, Boca Raton, FL, pp. 41–64.
13. Kuczenski, R., Segal, D. S., and Aizenstein, M. L. (1991) Amphetamine, cocaine, and fencamfamine: relationship between locomotor and stereotypy response profiles and caudate and accumbens dopamine dynamics. *J. Neurosci.* **11,** 2703–2712.
14. Segal, D. S. and Kuczenski, R. (1994) Amphetamine and its analogues: psychopharmacology, toxicology and abuse, in *Behavioral Pharmacology of Amphetamine* (Cho, A. K. and Segal, D. S., eds.), Academic, San Diego, Vol. 4, pp. 115–150.
15. Dure, L. S. and Young, A. B. (1995) The distribution of glutamate subtypes in mammalian central nervous system using quantitative in vitro autoradiography, in *CNS Neurotransmitters and Neuromodulators: Glutamate* (Stone, T. W., ed.), CRC Boca Raton, FL, pp. 83–94.
16. Karler, R., Calder, L. D., and Turkanis, S. A. (1991) DNQX blockade of amphetamine behavioral sensitization. *Brain Res.* **552,** 295–300.
17. Karler, R., Calder, L. D., Thai, L. H., and Bedingfield, J. B. (1994) A dopaminergic–glutamatergic basis for the action of amphetamine and cocaine. *Brain Res.* **658,** 8–14.
18. Karler, R., Calder, L. D., Thai, L. H., and Bedingfield, J. B. (1995) The dopaminergic, glutamatergic, GABAergic basis for the action of amphetamine and cocaine. *Brain Res.* **671,** 100–104.
19. Bedingfield, J. B., Calder, L. D., Thai, D. K., and Karler, R. (1997) The role of the striatum in the mouse in behavioral sensitization to amphetamine. *Pharmacol. Biochem. Behav.* **56,** 305–310.
20. Delfs, J. M. and Kelley, A. E. (1990) The role of D$_1$ and D$_2$ dopamine receptors in oral stereotypy reduced by dopaminergic stimulation of the ventrolateral striatum. *Neuroscience,* **39,** 59–67.
21. LaHoste, G. J. and Marshall, J. F. (1996) Dopamine receptor interactions in the brain, in *CNS Neurotransmitters and Neuromodulators: Dopamine* (Stone, T. W., ed.), CRC, Boca Raton, FL, pp. 107–120.
22. Vezina, P. and Stewart, J. (1989) The effect of dopamine receptor blockade on the development of sensitization to the locomotor activating effects of amphetamine and morphine, *Brain Res.* **499,** 108–120.
23. Kuczenski, R. and Segal, D. S. (1998) Sensitization of amphetamine-induced stereotyped behaviors during the acute response. *J. Pharmacol. Exp. Ther.* **288,** 699–709.

24. Wolf, M. E. (1998) The role of excitatory amino acids in behavioral sensitization to psychomotor stimulants. *Prog. Neurobiol.* **54,** 679–720.
25. Karler, R., Finnegan, K. T., and Calder, L. D. (1993) Blockade of behavioral sensitization to cocaine and amphetamine by inhibitors of protein synthesis. *Brain Res.* **603,** 19–24.
26. Karler, R., Calder, L. D., and Bedingfield, J. B. (1996) A novel nicotinic–cholinergic role in behavioral sensitization to amphetamine-induced stereotypy in mice. *Brain Res.* **725,** 192–198.
27. Moroni, F., Corradetti, R., Casamenti, F., Moneti, G., and Pepeu, G. (1981) The release of endogenous GABA and glutamate from the cerebral cortex of the rat. *Naunyn-Schmiedeberg's Arch. Pharmacol.* **316,** 235–239.
28. Groves, P. M. (1983) A theory of the functional organization of the neostriatum and the neostriatal control of voluntary movement. *Brain Res.* **5,** 109–132.
29. Mora, F. and Porras, A. (1993) Effects of amphetamine on the release of excitatory amino acid neurotransmitters in the basal ganglia of the conscious rat. *Can. J. Physiol. Pharmacol.* **71,** 348–351.
30. Karler, R., Turkanis, S. A., Partlow, L. M., and Calder, L. D. (1991) Calcium channel blockers and behavioral sensitization. *Life Sci.* **49,** 165–170.
31. Karler, R., Chaudhry, I. A., Calder, L. D., and Turkanis, S. A. (1990) Amphetamine behavioral sensitization and the excitatory amino acids. *Brain Res.* **537,** 76–82.
32. Karler, R., Calder, L. D., and Turkanis, S. A. (1990) Reverse tolerance to amphetamine evokes reverse tolerance to 5-hydroxytryptophan. *Life Sci.* **46,** 1773–1780.
33. Bedingfield, J. B., Calder, L. D., and Karler, R. (1996) Comparative behavioral sensitization to stereotypy by direct and indirect dopamine agonists in CF-1 mice. *Psychopharmacology* **124,** 219–225.
34. Browne, R. G. and Segal, D. S. (1977) Metabolic and experiential factors in the behavioral response to repeated amphetamine. *Pharmacol. Biochem. Behav.* **6,** 545–552.
35. Sunahara, R., Guan, H.-C., O'Dowd, B., Seeman, P., Laurier, L., Ng, G., et al. (1991) Cloning of the genefora human $D_5$ receptor with higher affinity for dopamine than $D_1$ *Nature* **350,** 614–619.
36. Filip, M., Thomas, M. L., and Cunningham, K. A. (2000) Dopamine $D_5$ receptors in nucleus accumbens contribute to the detection of cocaine in rats. *J. Neurosci.* **20,** 1–4.
37. Dougherty, G. C., Jr. and Ellinwood, E. H., Jr. (1981) Chronic *d*-amphetamine in nucleus accumbens: lack of tolerance and reverse tolerance of locomotor activity. *Life Sci.* **28,** 2295–2298.
38. Hitzemann, R., Wu, J., Hom, D., and Loh, H. (1980) Brain locations controlling the behavioral effects of chronic amphetamine intoxication. *Psychopharmacology* **72,** 93–101.
39. Deutch, A. Y. (1992) The regulation of subcortical dopamine systems by the prefrontal cortex: interactions of central dopamine systems and the pathogenesis of schizophrenia. *J. Neural Transm.* **36,** 61–89.
40. Karler, R., Bedingfield, J. B., Thai, D. K., and Calder, L. D. (1997) The role of the frontal cortex in the mouse in behavioral sensitization to amphetamine. *Brain Res.* **757,** 228–235.
41. Karler, R., Calder, L. D., and Thai, D. K. (1998) The role of dopamine in the mouse frontal cortex: a new hypothesis of behavioral sensitization to amphetamine and cocaine. *Pharmacol. Biochem. Behav.* **61,** 435–443.
42. Kuczenski, R., Segal, D. S., and Todd, P. K. (1997) Behavioral sensitization and extracellular dopamine responses to amphetamine after various treatments. *Psychopharmacology* **134,** 221–229
43. Vanderschuren, L. J. M. J. and Kalivas, P. W. (2000) Alterations in dopaminergic and glutamatergic transmission in the induction and expression of behavioral sensitization: a critical review of preclinical studies. *Psychopharmacology* **151,** 99–120.
44. Sieghart, W. (2000) Unraveling the function of $GABA_A$ receptor subtypes. *Trends Pharmacol. Sci.* **21,** 409–451.
45. Möhler, H., Crestani, F., and Rudolph, U. (2001) GABA(A)-receptor subtypes: a new pharmacology. *Curr. Opin. Pharmacol.* **1,** 22–25.

# 8
# Addiction and Glutamate-Dependent Plasticity

### Marina E. Wolf, PhD

## 1. INTRODUCTION

Addiction may be defined as the gradual evolution from casual or controlled use into a compulsive pattern of drug-seeking and drug-taking behavior. Even after abstinence is achieved, patients remain vulnerable to episodes of craving and relapse triggered by stimuli previously associated with the availability of drug or the act of drug-taking (1). This transition fits the definition of neuroplasticity, defined as the ability of the nervous system to modify its response to a stimulus based on prior experience. However, it is an exceptionally powerful and persistent form of plasticity. Episodic craving persists for years in humans (2), and animal studies show that conditioned responses to cocaine-related stimuli are highly resistant to extinction (3).

Monoamine transporters are the immediate target in brain of psychostimulants like cocaine and amphetamine, and this interaction is becoming increasingly well characterized. However, it remains a mystery how an initial elevation of monoamine levels leads to changes in the nervous system that persist for years. This problem will be the focus of this chapter. We will emphasize studies related to behavioral sensitization, a glutamate-dependent form of drug-induced plasticity that provides a useful model for some aspects of addiction. An overview of sensitization is therefore provided, but it is not comprehensive (for other reviews, see refs. 4 and 5).

## 2. GLUTAMATE AND ADDICTION: CLUES FROM BEHAVIORAL SENSITIZATION

Behavioral sensitization refers to the progressive enhancement of species-specific behavioral responses that occurs during repeated drug administration and persists even after long periods of withdrawal. The relevance of sensitization to addiction has been discussed (6–9). Key points are as follows: (1) sensitization occurs to the reinforcing effects of psychostimulants, not just locomotor effects, (2) sensitization is influenced by the same factors that influence addiction (e.g., stress, conditioning, drug priming), (3) sensitization is accompanied by profound cellular and molecular adaptations in the mesocorticolimbic circuits that are fundamentally involved in motivation and reward and that are implicated in addiction in humans, and (4) like addiction, sensitization is very persistent, e. g., amphetamine sensitization can last up to a year in rats, a species that lives only 1–2 yr (10). Thus, behavioral sensitization provides an animal model for the induction of persistent changes, at both cellular and behavioral levels, in the neural circuitry of motivation and reward as a result of chronic exposure to drugs of abuse.

In retrospect, there are many logical reasons to focus on glutamate in addiction research. An obvious reason is that addiction fits the definition of plasticity, and glutamate is implicated in nearly every

From: *Contemporary Clinical Neuroscience: Glutamate and Addiction*
Edited by: Barbara H. Herman et al. © Humana Press Inc., Totowa, NJ

form of plasticity. A second reason is that glutamate neurons are well placed, anatomically, to govern the output of dopamine (DA) systems. They provide the major source of excitatory drive to DA cell bodies in the midbrain, whereas target cells in DA projection areas, such as the nucleus accumbens (NAc), receive convergent inputs from DA and glutamate nerve terminals (e.g., ref. *11*). Finally, human imaging studies implicate glutamate-rich cortical and limbic brain regions in cocaine-conditioned responses *(12–17)*. As relapse is the major problem in treating cocaine addicts and may be triggered by drug-conditioned cues, it is likely that plasticity in glutamate projections mediating drug-conditioned responses plays a key role in addiction.

Historically, however, the impetus for studying glutamate's role in addiction came from studies of behavioral sensitization. It was first demonstrated in 1989 that the development of behavioral sensitization in rats and mice was prevented if each amphetamine or cocaine injection in a chronic regimen was preceded by systemic injection of the noncompetitive $N$-methyl-D-aspartate (NMDA) receptor antagonist MK-801 *(18)*. Since then, many studies have demonstrated similar effects with different classes of NMDA receptor antagonists and with α-amino-3-hydroxy-5-isoazole propionic acid (AMPA) and metabotropic glutamate receptor antagonists *(4)*. A key observation is that coadministration of glutamate receptor antagonists with psychostimulants also prevents the ability of prior drug exposure to promote drug self-administration, demonstrating that sensitization to drug reinforcing effects is also a glutamate-dependent process (e.g., ref. *19*; reviewed in ref. *4*). Another important observation is that glutamate antagonist treatments that prevent behavioral sensitization also prevent the development of neurochemical and electrophysiological adaptations that normally accompany sensitization (e.g., ref. *20*; reviewed in ref. *4*). This indicates that glutamate receptor stimulation is a necessary step in the cascade of cellular changes leading to sensitization. An encouraging finding, from a therapeutic perspective, is that manipulations of glutamate transmission can reverse behavioral sensitization *(21,22)*. Some glutamatergic drugs, particularly MK-801, may influence sensitization in part through mechanisms related to state-dependent learning *(23–25)*. However, such effects cannot account for the ability of many classes of glutamate receptor antagonists to prevent the development of behavioral sensitization and associated neuroadaptations *(26,27)*.

Microinjection studies indicate that glutamate receptor antagonists are probably acting in the A9/A10 region, which contains DA cell bodies, to prevent the development of sensitization *(28–31)*. We have hypothesized that glutamate receptor antagonists prevent sensitization by attenuating excitatory drive to midbrain DA neurons *(4)*. Supporting this hypothesis, the development of sensitization is associated with a transient increase in excitatory drive to DA neurons *(4)*, whereas it is prevented by lesions of the prefrontal cortex, an important source of glutamate-containing projections to the ventral tegmental area (VTA) *(20,30,32;* but see ref. *33)*. These findings suggest that sensitization may involve drug-induced plasticity at excitatory synapses between glutamate terminals originating in the prefrontal cortex and VTA DA neurons. The simplest version of this model is that drugs of abuse promote long-term potentiation (LTP) at these synapses, increasing excitatory drive to DA neurons and thus influencing transmission in limbic and cortical DA projection areas. Exciting new data support the possible importance of this mechanism (Section 3). However, recent anatomical studies have shown that prefrontal cortex terminals synapse on meso-accumbens GABA neurons rather than meso-accumbens DA neurons *(34)*, suggesting that the route of communication between prefrontal cortex and mesoaccumbens DA neurons may be indirect. Other findings suggest that alterations in GABA transmission in the VTA may contribute to sensitization (*see* ref. *35*).

In the remainder of this chapter we will address three topics: (1) basic mechanisms by which drugs of abuse may "tap into" cellular mechanisms governing plasticity at excitatory synapses, (2) the role of such plasticity in the induction of sensitization in the VTA, and (3) the role of such plasticity in long-term adaptations within the NAc. We are focusing on VTA and NAc because these are critical brain regions for induction and expression of sensitization, respectively; however, both phases of sensitization actually require complex circuitry *(4)*.

## 3. CANDIDATE MECHANISMS FOR STABLE DRUG-INDUCED CHANGES IN THE BRAIN

Based on animal studies, one can identify several categories of adaptations produced by chronic drug administration that are candidates for triggering long-lasting changes in brain. The first is changes in gene expression, leading to altered activity of neurons expressing these genes and, ultimately, to alterations in the activity of neuronal circuits. This subject was the topic of an excellent and very recent review, which focused on two transcription factors strongly implicated in addiction, CREB and ΔFosB *(36)*. These factors mediate both homeostatic and sensitizing adaptations following repeated drug administration. However, their levels return to normal after relatively short withdrawal periods (less than 1 wk for CREB, a month or two for ΔFosB) *(36)*. Although there may be other drug-regulated transcription factors or regulators that are even longer-lived, it seems most appropriate to view these factors as triggers for stable changes that occur through different mechanisms.

One such mechanism is a change in the shape or the number of excitatory synapses. This mechanism contributes to long-lasting changes in synaptic strength as a result of LTP and other forms of learning *(37)*. Recently, changes in dendritic spines have been found after chronic drug administration. Repeated treatment with either amphetamine or cocaine increased dendritic branching, spine density, and the number of branched spines in Golgi-stained medium spiny neurons in the NAc and pyramidal neurons in the prefrontal cortex, effects which persisted at least 1 mo *(38,39)*. Very similar alterations were observed in rats allowed to self-administer cocaine for 1 mo *(40)*. Interestingly, nicotine produced effects similar to those found with cocaine and amphetamine [perhaps more robust; *(41)*], whereas chronic morphine produced effects opposite to those observed after cocaine or amphetamine, i.e., decreases in spine density and dendritic branching *(42)*. The most important point is that these changes in dendritic morphology are identical to changes implicated in other forms of experience-dependent plasticity and are among the most long-lasting reported in response to chronic drug administration; therefore, they are good candidates for mediating its persistence. Drugs of abuse produce other types of morphological change as well. Onn and Grace *(43)* found that withdrawal from repeated amphetamine produced long-lasting (at least 28 d) increases in gap-junction communication in the NAc and prefrontal cortex and that this was associated with increased neuronal synchronization in these brain regions.

Neurotrophic factors may promote synaptic remodeling during learning *(44,45)*. Thus, an intriguing possibility is that drug-induced increases in neurotrophic factor expression are responsible for alterations in dendritic morphology after chronic drug administration *(46)*. For example, three intermittent injections of amphetamine produce a long-lasting (1 mo) increase in basic fibroblast growth factor (bFGF) immunoreactivity in astrocytes of the rat VTA and substantia nigra (SN) that is blocked by coadministration of the glutamate receptor antagonist kynurenic acid (47). A very exciting finding is that induction of bFGF appears necessary for amphetamine sensitization, as intra-VTA administration of a neutralizing antibody to bFGF prior to daily amphetamine injections prevents its development *(48)*. A 2-wk escalating-dose amphetamine regimen, which is more similar to that shown to alter dendritic morphology in NAc and prefrontal cortex (above), also elevates bFGF in these regions (significant effect in NAc, trend in the prefrontal cortex) *(49)*.

The hypothesis of this chapter is that activity-dependent forms of neural plasticity such as LTP or long-term depression (LTD) are the first step in the cascade leading to structural changes that underlie persistent drug-induced modifications in synaptic structure. Drugs of abuse are proposed to alter activity in circuits related to motivation and reward, leading to abnormal induction of LTP or LTD. They may also directly modify the ability of "normal" neuronal activity to elicit appropriate forms of LTP and LTD. Examples of both will be discussed below. This hypothesis is consistent with evidence for many parallels between mechanisms underlying sensitization and activity-dependent plasticty. For example, as neurotrophic factors are one of the many signaling pathways implicated in activity-dependent plasticity (e.g., ref. 45), their proposed involvement in sensitization-related structural changes (see above) is readily incorporated into this hypothesis. The same applies to protein kinase and phosphatase cascades and to transcriptional regulation, both of which are critical for LTP and other types of learning (50–52) as well as for long-term responses to drugs of abuse (36).

A model has been proposed to explain the sequential changes that may lead from LTP and LTD to alteration in the biochemical composition of the postsynaptic membrane and, ultimately, to changes in the structure of dendritic spines (53). Within the first 30 min after induction of LTP, it is proposed that AMPA receptor signaling is enhanced by $Ca^{2+}$-dependent phosphorylation of AMPA receptors, which increases their single-channel conductance, and by insertion of additional AMPA receptors into the postsynaptic membrane. The latter is probably related to $Ca^{2+}$-dependent enhancement of actin-dependent dynamics in the spines, perhaps involving increased spine motility and the formation of synapses with discontinuities within their postsynaptic densities (perforated synapses). A later stage of this process (60 min after LTP and beyond) could result in duplication of spine synapses or the formation of new synapses. An important foundation for this model is the "silent synapse" hypothesis, which postulates that some synapses are silent at normal recording potentials because they contain only NMDA receptors; LTP is proposed to result from insertion of AMPA receptors into the postsynaptic membrane (54). Although this hypothesis was originally developed based on electrophysiological findings, it is supported by recent studies showing that the surface expression of AMPA receptors is a tightly regulated process, with AMPA receptors inserted into synapses during LTP and internalized during LTD (55,56). A more detailed model has recently been proposed based on subunit-specific regulation of AMPA receptor trafficking in hippocampal neurons (57,57a). GluR2/3-containing AMPA receptors appear to undergo continuous recycling, maintaining a constant level of synaptic AMPA receptors under normal conditions. Induction of LTP enables synaptic delivery of GluR1/2 heteromers both to silent synapses and to synapses that already contain AMPA receptors. GluR1 appears to be the "dominant" subunit for this process, with protein–protein interaction domains on its C-terminus preventing GluR1/2 heteromers from being delivered to synapses under normal conditions, but enabling their delivery in response to signaling pathways activated during LTP induction. An intriguing hypothesis is that LTP may also insert into the postsynaptic membrane putative "slot proteins" that serve as binding sites for AMPA receptors. After the insertion of GluR1/2 heteromers during LTP, the resulting enhancement of synaptic strength can be maintained on a longer-term basis by exchange of GluR1/2 heteromers with intracellular GluR2/3 heteromers; the latter are then maintained by constitutive recycling mechanisms (57).

Can drugs of abuse tap into these molecular mechanisms for altering synaptic strength? We have recently obtained evidence that D1 receptors (which are stimulated during administration of cocaine or amphetamine) can influence AMPA receptor subunit phosphorylation and surface expression. These studies were performed in primary cultures prepared from the NAc of postnatal rats. First, using phosphorylation specific antibodies and Western blotting, we found that D1 receptor stimulation increases GluR1 phosphorylation at the protein kinase A (PKA) site, Ser-845 (58,59). Prior studies have shown that D1 DA receptors stimulate GluR1 phosphorylation at the PKA site in striatal neurons (60,61) and that phosphorylation at the PKA site is associated with enhancement of AMPA receptor currents (60,62–64), but ours is the first to demonstrate this effect in the NAc. To study AMPA receptor surface expression in NAc, we labeled surface AMPA receptors by incubating live cells with an antibody recognizing the extracellular portion of GluR1, fixing cultures, and then incubating with fluorescent secondary antibody. D1 receptor stimulation produced a rapid (5 min) increase in punctate surface GluR1 labeling, whereas glutamate decreased surface GluR1 labeling (65). Downregulation of GluR1 by glutamate has also been demonstrated in hippocampal cultures (55). We are in the process of determining if D1 receptor-mediated phosphorylation of GluR1 is linked to D1 receptor regulation of its surface expression.

In summary, our results demonstrate that D1 receptors modulate both the phosphorylation and surface expression of AMPA receptors, the two major mechanisms for modulating the strength of excitatory synapses, and that the latter mechanism is also regulated by glutamate levels. As AMPA receptors are responsible for the majority of excitatory transmission in the NAc, these results suggest that the level of excitatory drive to NAc neurons may be dynamically regulated in response to rapid changes in the activity of both glutamate and DA afferents. This suggests a direct mechanism by which cocaine and amphetamine, through promoting D1 receptor stimulation locally or altering the activity of circuits that provide glutamate input to the NAc, could tap into fundamental mechanisms for LTP and LTD. When D1 receptors are overstimulated during chronic administration of psychostimulants, we hypothesize that adaptive changes in these mechanisms occur that influence the generation of LTP and LTD. This could ultimately lead to persistent changes in the structure and function of excitatory synapses (see above). Supporting this general possibility, a recent study showed that repeated cocaine treatment altered the coupling of D1 receptors to PKA-dependent signaling pathways that modulate AMPA receptor currents *(64)*. Of course, the findings discussed above are most relevant for plasticity occurring at synapses onto neurons containing both D1 and AMPA receptors (e.g., NAc neurons), whereas other mechanisms must account for drug modulation of plasticity in DA neurons (which possess AMPA receptors but lack D1 receptors). The remainder of this chapter will further examine the hypothesis that drugs of abuse produce long-lasting changes in brain function by influencing activity-dependent forms of plasticity such as LTP and LTD and thereby producing changes in synaptic strength in neuronal circuits related to motivation and reward.

## 4. ACTIVITY-DEPENDENT SYNAPTIC PLASTICITY AND THE INDUCTION OF BEHAVIORAL SENSITIZATION IN THE VENTRAL TEGMENTAL AREA

The most direct evidence for the hypothesis presented above comes from studies on the mechanisms underlying the induction of behavioral sensitization. As reviewed above, induction of sensitization is dependent on glutamate transmission in the VTA. Of course, the induction of LTP and LTD share this requirement for glutamate transmission, which was the basis for early hypotheses for involvement of these phenomena in sensitization *(18,66)*.

Many lines of evidence suggest that sensitization involves an increase in glutamate transmission in the VTA *(4)*. This could occur as a result of an increase in glutamate release, an increase in glutamate receptor number, or an increase in glutamate receptor sensitivity. Recent evidence supports a version of the third hypothesis in which sensitization is accompanied by LTP-like changes that increase in the efficiency of glutamate transmission in the VTA.

LTP is expressed as a potentiation of AMPA receptor transmission *(67,68)*. If the early phase of sensitization is associated with LTP at synapses onto midbrain DA neurons, the DA neurons should exhibit increased responsiveness to AMPA. To test this, single-unit recording studies examined the responsiveness of VTA DA neurons to glutamate agonists either 3 or 14 d after discontinuing repeated administration of cocaine or amphetamine *(69,70)*. When glutamate or AMPA was applied directly to DA cell body regions by microiontophoresis, DA neurons recorded from amphetamine- or cocaine-treated rats showed enhanced excitatory responses compared to neurons recorded from saline controls. No difference was observed in responsiveness to NMDA. The enhanced responsiveness to glutamate and AMPA dissipated after 10–14 d of withdrawal, consistent with a role in induction mechanisms.

Based on these electrophysiological results, we predicted that increased AMPA receptor responsiveness should be detectable in microdialysis experiments as an increase in the ability of AMPA to drive meso-accumbens DA cells and, thus, elicit DA release in the NAc. We tested this using rats treated with the same amphetamine regimen used for electrophysiological studies (or repeated saline) and dual-probe microdialysis. We found that intra-VTA administration of a low dose of AMPA produced significantly greater DA efflux in the ipsilateral NAc of amphetamine-treated rats *(71)*. This augmented response to AMPA was transient, because it was present 3 d, but not 10–14 d, after the last injection. It was specific for AMPA, because intra-VTA NMDA administration produced a trend toward increased NAc DA levels that did not differ between groups. Thus, our microdialysis data are in complete agreement with prior electrophysiological studies using the same amphetamine regimen. Both suggest an LTP-like enhancement of AMPA transmission onto VTA DA neurons during the early phase of drug withdrawal.

Of course, an alternative explanation for enhanced responsiveness to AMPA is an increase in AMPA receptor expression. Indeed, there are reports that GluR1 levels in VTA, quantified using Western blots, are increased 16–24 h after discontinuation of repeated cocaine, morphine, ethanol, or stress paradigms (72–74) but not after 3 w withdrawal from cocaine (74). Increased GluR1 was not observed in the substantia nigra after repeated treatment with cocaine or morphine (73; the substantia nigra was not examined in stress experiments), but after repeated ethanol, there was a greater increase in GluR1 in the substantia nigra than in the VTA (72). In contrast, our own quantitative immunoautoradiographic studies (75) found no change in levels of GluR1 immunoreactivity in VTA, substantia nigra, or a transitional area after 16–24 h of withdrawal from repeated amphetamine or cocaine treatment, or after 3 or 14 d of withdrawal from amphetamine; this study used the same amphetamine regimen that resulted in enhanced electrophysiological (70) and neurochemical (71) responsiveness to intra-VTA AMPA at the 3-d withdrawal time and one of the same cocaine regimens used previously (73). Possible reasons for these discrepant results at the protein level have been discussed (75). In contrast, all studies agree that mRNA levels for GluR1 (and GluR2-4) in the VTA are not altered during withdrawal from repeated amphetamine or cocaine (75–77). Taking all data into account, we believe that an overall increase in GluR1 expression in the midbrain is unlikely to account for increased responsiveness of VTA AMPA receptors at short withdrawal times. We propose that more complex mechanisms, related to LTP, are responsible for AMPA receptor plasticity in the VTA. Although mechanisms might involve increased AMPA receptor surface expression, models for this process do not necessary involve increases in total levels of AMPA receptors at a cellular or regional level (Section 3). As a first step toward evaluating this hypothesis, several laboratories have begun to characterize synaptic plasticity in midbrain DA neurons.

Overton et al. (78) recorded from a parasagittal slice preparation containing the substantia nigra and the subthalamic nucleus (STN); the latter sends glutamatergic projections to substantia nigra DA neurons. Tetanic stimulation of the STN region produced long-lasting changes in the amplitude of excitatory postsynaptic currents (EPSPs) in approximately half of the DA neurons evaluated, whereas many of the others showed an immediate but short-lived potentiation of EPSP amplitude (short-term potentiation). No LTP was observed in experiments conducted in the presence of NMDA receptor antagonists. Bonci and Malenka (79) examined the types of plasticity exhibited by both DA and non-DA neurons of the VTA. Many of the non-DA cells contain GABA and project to the forebrain (e.g., ref. 80). Synapses onto DA neurons exhibited paired-pulse depression as well as depression in response to a train of 10 stimuli, whereas non-DA cells displayed facilitation under both conditions. Using a protocol that reliably elicits LTP in hippocampal CA1 pyramidal cells, it was found that DA neurons (but not non-DA neurons) exhibited LTP and that its induction required NMDA but not metabotropic glutamate receptor activation.

Next, two studies showed that DA neurons also exhibit LTD (81,82). Both found that the induction of LTD did not require NMDA receptor or metabotropic glutamate receptor activation. However, it was prevented by the $Ca^{2+}$ chelator BAPTA (81,82) or by voltage-clamping cells at depolarized potentials that prevent the activation of voltage-dependent $Ca^{2+}$ channels (82). Conversely, it was induced by driving $Ca^{2+}$ into the DA neuron with repetitive depolarization; importantly, the LTD induced by this mechanism occluded synaptically driven LTD (81). Together, these results suggest that LTD induction in midbrain DA neurons requires a postsynaptic rise in intracellular $Ca^{2+}$. The L-type $Ca^{2+}$ channel antagonist nifedipine did not block LTD, suggesting a role for high-threshold $Ca^{2+}$ channels

*(82)*. Perhaps the most exciting result from both articles is that D2 receptor activation appears to oppose LTD. Thomas et al. *(82)* applied DA for 15–20 min and blocked LTD. This was mimicked by quinpirole but not SKF 38393, suggesting mediation by D2 receptors. Jones et al. *(81)* found that incubating slices for 15 min with amphetamine blocked LTD and that this was prevented by the D2 receptor antagonist eticlopride. Interestingly, amphetamine had no effect on LTD at hippocampal synapses *(81)* and DA had no effect on LTD in the NAc *(82)*, suggesting specificity for certain circuits. The ability of DA agonists to inhibit LTD in the midbrain is consistent with a role for high-threshold $Ca^{2+}$ channels, because DA inhibits N- and P/Q-type $Ca^{2+}$ currents in midbrain DA neurons *(83)*. The importance of postsynaptic $Ca^{2+}$ for plasticity in DA neurons is consistent with in vivo studies indicating a requirement for VTA $Ca^{2+}$ signaling during induction of sensitization, although these in vivo studies implicated L-type $Ca^{2+}$ channels rather than the high-threshold $Ca^{2+}$ channels implicated in LTD *(84–86)*.

Jones et al. *(81)* proposed that "LTD normally acts to protect VTA dopamine neurons from excessive glutamatergic excitation, but that in the presence of amphetamine this brake is removed, permitting unrestricted excitation of dopamine neurons." Loss of this "braking mechanism" may promote "pathological" strengthening of excitatory synapses on DA neurons via LTP. This situation may be exacerbated by stimulant-induced increases in glutamate levels in the VTA, although exactly how drugs modulate VTA glutamate levels is a matter of some debate. We have shown that amphetamine produces a delayed but persistent increase in extracellular glutamate levels in the VTA *(87–89)* through a mechanism involving glutamate transporters and reactive oxygen species *(90)*. We have argued that this delayed increase in glutamate levels is relevant to sensitization because treatments that prevent amphetamine from eliciting sensitization (pretreatment with MK-801, SCH 23390, or lesions of prefrontal cortex) also prevent amphetamine from eliciting the delayed increase in VTA glutamate levels *(89)*. On the other hand, Kalivas and Duffy *(91,92)* have reported a short-duration increase in VTA glutamate levels in response to cocaine challenge in naive and chronic cocaine-treated rats. They propose that cocaine elevates DA levels in the VTA, leading to activation of D1 receptors on glutamate nerve terminals that promote the release of glutamate. Our data, and other results, are not consistent with this model *(88)*. However, the most important point is that drugs of abuse can elevate VTA glutamate levels and this is likely to be important in triggering subsequent forms of neuroplasticity.

Another independent mechanism has been identified by which DA-releasing psychostimulants could enhance the excitability of DA neurons. Paladini et al. *(93)* found that amphetamine selectively inhibits slow mGluR-mediated inhibitory postsynaptic potentials (IPSPs) in DA neurons recorded from midbrain slices without affecting ionotropic glutamate receptor-mediated EPSCs. Unlike amphetamine's effect on LTD *(81,82)*, this effect does not involve D2 receptors. Rather, the DA released by amphetamine activates postsynaptic $\alpha_1$ adrenergic receptors on DA neurons, leading to desensitization of inositol 1,4,5-triphosphate (InsP3)-mediated $Ca^{2+}$ release from internal stores and, thus, to inhibition of mGluR-mediated IPSPs (which depend on InsP3-induced calcium release). The mechanism of the desensitization is not clear. However, this postsynaptic $\alpha_1$ receptor-mediated effect is functionally relevant because it leads to prolonged responses to repetitive stimulation. The authors speculate that amphetamine may thus increase bursting, which may be relevant to attentional or motivational aspects of behavior. Putting this together with the results of Jones et al. *(81)* and Thomas et al. *(82)*, it appears that DA-releasing psychostimulants may promote excitation of DA neurons through two distinct mechanisms—inhibition of LTD and inhibition of mGluR-mediated IPSPs. Nicotine may also promote LTP in the VTA, albeit by a different mechanism *(94)*.

The critical question is whether drug regimens producing sensitization in a whole animal can influence synaptic plasticity in the VTA. The first study to address this question was published recently *(95)*. The authors prepared midbrain slices from naive mice and mice injected the day before with a single injection of saline or 15 mg/kg cocaine. The relative contribution of AMPA receptors and NMDA receptors to excitatory postsynaptic currents was compared to obtain a measure of potentiated

AMPA receptor transmission. In slices prepared from cocaine-treated mice, the authors found a significantly larger contribution of AMPA receptors (AMPA receptor/NMDA receptor ratio). This could reflect a change in the probability of transmitter release, increased AMPA receptor function, or decreased NMDA receptor function. Arguing against the first possibility, cocaine-treated mice showed no change in their responses to paired pulses. Consistent with an increase in AMPA receptor function, the cocaine-treated mice showed a significant increase in both the amplitude and frequency of miniature AMPA receptor-mediated EPSCs and also showed larger currents in response to exogenously applied AMPA. In contrast, there was no difference between NMDA-induced currents in cocaine- and saline-injected mice. The increase in AMPA/NMDA receptor contributions to postsynaptic currents was transient (present 5 but not 10 d after cocaine injection), like other adaptations at the level of the VTA, and was not seen in hippocampus or on GABA neurons of the VTA. Importantly, the authors confirmed that the single injection of cocaine used in their study produced behavioral sensitization in the mice. Coadministration of MK-801 with cocaine blocked sensitization in the mice, as expected, and also blocked the cocaine-induced change in the AMPA receptor/NMDA receptor ratio. Is the cocaine-induced potentiation equivalent to LTP? This appears to be so, because the cocaine-induced potentiation prevented the subsequent in vitro induction of LTP. Finally, it should be noted that the cocaine-induced potentiation was not accompanied by changes in GluR1 or GluR2 immunoreactivity as determined by Western blotting *(95)*, in agreement with the idea that drugs of abuse modulate AMPA transmission in the VTA through mechanisms more subtle than altered expression of AMPA receptor subunits *(75)*.

In conclusion, it is encouraging that the importance of augmented AMPA receptor transmission in VTA for sensitization has now been demonstrated using widely different experimental approaches, including in vitro electrophysiology *(95)*, in vivo electrophysiology *(69,70)*, and microdialysis *(71)*, and that its dependence on NMDA receptor transmission is consistent with results of behavioral studies showing that intra-VTA administration of NMDA receptor antagonists prevents sensitization *(28,30,31)*.

## 5. DRUGS OF ABUSE INFLUENCE SYNAPTIC PLASTICITY IN THE NUCLEUS ACCUMBENS

The NAc occupies a key position in the neurocircuitry of motivation and reward and is the site of many persistent changes associated with chronic drug exposure *(96)*. Neurons in the NAc consist of medium spiny GABA neurons (90%) and several populations of interneurons (10%). The medium spiny neurons are the output neurons of the NAc and receive convergent DA and glutamate inputs. The interneurons play important roles in information processing within the NAc and may also be regulated by DA and glutamate *(97)*. DA exerts neuromodulatory effects within the NAc, both by directly influencing synaptic transmission and by modulating voltage-dependent conductances *(98)*. There are many controversies about the effects of drugs of abuse on glutamate transmission in the NAc. One debate is whether glutamate transmission, particularly originating in prefrontal cortex, is required for the expression of sensitization. Another is whether psychostimulants increase glutamate levels in the NAc. There are also discrepant findings about the effect of psychostimulants on glutamate receptor subunit expression. These issues are beyond the scope of this chapter (*see* ref. *4*). Rather, this section will focus on the hypothesis that abnormal synaptic plasticity in the NAc, triggered by drug exposure, leads to dysregulation of motivation- and reward-related circuits and thereby contributes to addiction.

Activity-dependent plasticity of the corticostriatal pathway in drug-naive rats has been well characterized. Repetitive activation of corticostriatal glutamatergic fibers produces LTD of excitatory synaptic transmission in the striatum measured using in vitro recording techniques (e.g., ref. *99*). Striatal LTD requires membrane depolarization and action potential discharge of the postsynaptic cell during the conditioning tetanus, coactivation of D1 and D2 receptors, activation of metabotropic glutamate receptors, and release of nitric oxide from striatal interneurons *(99,100)*. Striatal LTP is produced under in vitro conditions that enhance NMDA receptor activation and in vivo after tetanic stimulation of cortical fibers (e.g., refs. *101* and *102*). Interestingly, pulsatile application of DA during a conditioning protocol that normally results in LTD shifts the effect toward potentiation of EPSP amplitude *(103)*. Activity-dependent plasticity has also been demonstrated for excitatory synapses in the NAc. Recordings from NAc slices showed that tetanic stimulation of prefrontal cortical afferents produced both LTP and LTD, although LTP was more frequently observed *(104)*. Tetanization of the fimbria-fornix produces LTP of field potentials in the NAc *(105)*. LTP in NAc neurons is NMDA receptor dependent *(104,106)* and may be modulated by DA (see below). LTD in NAc neurons requires NMDA receptor activation and consequent rises in postsynaptic $Ca^{2+}$ levels, does not require metabotropic glutamate receptor activation, and is not affected by bath application of DA *(82)*. The latter two features distinguish it from LTD in the dorsal striatum (see above).

Striatal and NAc neurons are normally quiescent, and their activation requires synchronous activation of multiple excitatory inputs *(107)*. LTP or LTD in excitatory pathways impinging on these neurons would have profound effects on their output, because these processes would influence the likelihood of synchronized activation. It is therefore exciting that several recent studies have found alterations in corticostriatal plasticity after chronic drug treatment. Studies addressing two different but related questions will be discussed in turn.

One question is whether chronic drug exposure alters the likelihood of LTP or LTD in a manner that outlasts the presence of drug in the brain. Pulvirenti et al. *(108)* compared evoked field responses in the NAc after stimulation of fimbria afferents in rats exposed to either 1 or 5 d of cocaine self-administration and found that the acquisition of cocaine-seeking behavior is associated with enhancement of hippocampal–accumbens transmission. Thomas et al. *(109)* treated mice for 5 d with cocaine, challenged with cocaine after 10–14 d withdrawal to demonstrate behavioral sensitization, and prepared slices of NAc 1 d later. Although no changes in the size of field EPSPs were found, cocaine-treated mice showed a reduction in the amplitude of AMPA receptor-mediated quantal events, specifically at synapses activated by cortical afferents. This was found in the shell but not core. No changes in NMDA receptor-mediated synaptic responses or the probability of transmitter release were observed. Furthermore, the magnitude of LTD that could be evoked in vivo was reduced in cocaine-treated mice, suggesting that the decrease in synaptic strength produced by cocaine shares expression mechanisms with LTD.

These findings suggest that chronic cocaine induces a long-lasting depression of excitatory synaptic transmission in the NAc *(109)*. This is consistent with our findings of decreased responses of NAc neurons to iontophoretic glutamate after 3–14 d withdrawal from amphetamine or cocaine *(69)*. However, the correspondence is not perfect, as the latter effect was not restricted to the NAc shell *(69)* and decreased responses to both AMPA and NMDA were observed in a follow-up study (Hu and White, unpublished findings). Decreased peak amplitudes of AMPA/kainate-induced inward currents have also been observed in acutely dissociated striatal neurons prepared from chronic cocaine-treated rats *(64)*. Repeated cocaine administration decreases glutamate immunolabeling in nerve terminals of the NAc shell *(110)*, an effect that appears more persistent when cocaine is self-administered *(111)*. AMPA receptor subunit expression in the NAc is not altered after short (1–3 d) withdrawals from repeated cocaine or amphetamine *(73,74,112,113)*. However, protein and mRNA levels for the AMPA receptor subunits GluR1 and GluR2 are significantly decreased after 10–14 d withdrawal from repeated amphetamine administration *(112,113)*. In contrast, repeated cocaine treatment and 3 wk of withdrawal produce increased GluR1 levels, decreased GluR3 mRNA levels, and a trend toward increased GluR1 mRNA levels in the NAc *(74,76)*. These studies, along with other results showing alterations in NMDA and metabotropic glutamate receptor expression in the NAc after repeated drug administration *(76,114–116)*, do not lend themselves to the formulation of a simple working model. However, the bulk of evidence seems to suggest that the NAc is more quiescent after long withdrawals from repeated drug administration, perhaps as a result of a combination of enhanced LTD (synaptic level), decreased AMPA receptor expression (cellular level), and other types of changes (e.g., changes in voltage-dependent conductances; *117*). Decreased excitability of the NAc could be related to withdrawal symptoms such as anergia, anhedonia, and depression (see ref. *117*).

A different question is whether the modulatory effects of drugs of abuse, or DA itself, on LTP and LTD are altered after repeated drug exposure. Li and Kauer *(118)* reported that bath application of amphetamine blocked the induction of LTP in NAc slices prepared from naive rats. This effect was reproduced by DA + deprenyl (an MAO-B inhibitor), but not DA alone, consistent with a previous report that DA alone does not modulate NAc LTP *(104)*. However, when slices were prepared from rats previously treated with amphetamine for 6 d, bath application of amphetamine no longer blocked LTP *(118)*. In contrast to these observations for LTP, LTD in the NAc is apparently not subject to acute modulation by DA *(82)*, although LTD itself is promoted by prior exposure to the DA-releasing agent cocaine *(109; above)*. Chronic exposure to ethanol *(119)* or methamphetamine *(120)* has also been reported to alter plasticity in the rat neostriatum. Tetanic stimulation induced LTD in naive rats or saline-treated rats, whereas the same stimulation produced a slowly developing form of LTP was observed in slices prepared 6 d after discontinuation of methamphetamine injections or 15–20 h after ethanol withdrawal. D2 receptor activation depressed the magnitude of LTP in ethanol-withdrawn rats. Drawing upon these and other findings, the authors speculated that the "switch" to LTP might reflect increased NMDA receptor tone coupled with the loss of normal D2 receptor-mediated negative control over LTP induction *(119)*.

It is very exciting that all of these studies support the same hypothesis; that is, that DA receptor activation normally suppresses activity-dependent increases in synaptic strength in the striatal complex, but that this regulatory mechanism is lost after repeated drug exposure. If the normal effect of DA is to keep motivational circuits "in check" despite the high rewarding value of drugs of abuse, loss of this mechanism could contribute in a fundamental way to the loss of control over drug-seeking behavior that characterizes addiction. Indeed, there is good evidence that glutamate transmission in the NAc plays an important role in responses to drug-conditioned cues and in reinstatement of drug self-administration *(121–130)*. However, whereas loss of regulatory DA tone after chronic cocaine would favor potentiation of synaptic transmission, the chronic cocaine-induced enhancement of LTD at some synapses in the NAc would have the opposite effect *(109)*. This underscores the complexity of addiction-related adaptations, which can be viewed as a set of interacting positive and negative feedback loops.

## 6. CONCLUSIONS

LTP and LTD involve a complex cascade of biochemical changes leading to posttranslational modification and altered surface expression of AMPA receptors and, ultimately, to changes in the number and morphology of synapses. Recent work suggests that drugs of abuse may tap into these mechanisms to produce the maladaptive forms of synaptic plasticity that contribute to drug addiction.

## REFERENCES

1. Ehrman, R. N., Robbins, S. J., Childress, A. R., and O'Brien, C. P. (1992) Conditioned responses to cocaine-related stimuli in cocaine abuse patients. *Psychopharmacology* **107,** 523–529.
2. Gawin, F. H. and Kleber, H. D. (1986) Abstinence symptomatology and psychiatric diagnosis in cocaine abusers. Clinical observations. *Arch. Gen. Psychiatry* **43,** 107–113.
3. Weiss, F., Martin-Fardon, R., Ciccocioppo, R., Kerr, T. M., Smith, D. L., and Ben-Shahar, O. (2001) Enduring resistance to extinction of cocaine-seeking behavior by drug-related cues. *Neuropsychopharmacology* **25,** 361–372.
4. Wolf, M. E. (1998) The role of excitatory amino acids in behavioral sensitization to psychomotor stimulants. *Prog. Neurobiol.* **54,** 679–720.

5. Vanderschuren, L. J. M. J. and Kalivas P. W. (2000) Alterations in dopaminergic and glutamatergic transmission in the induction and expression of behavioral sensitization: a critical review of preclinical studies. *Psychopharmacology* **151,** 99–120.
6. Kalivas, P. W. and Stewart, J. (1991) Dopamine transmission in the initiation and expression of drug- and stress-induced sensitization of motor activity. *Brain Res. Rev.* **16,** 223–244.
7. Robinson, T. E. and Berridge, K. C. (1993) The neural basis of drug craving: an incentive-sensitization theory of addiction. *Brain Res. Rev.* **18,** 247–291.
8. Robinson, T. E., Browman, K. E., Crombag, H. S., and Badiani, A. (1998) Modulation of the induction or expression of psychostimulant sensitization by the circumstances surrounding drug administration. *Neurosci. Biobehav. Rev.* **22,** 347–354.
9. Wolf, M. E. (2001) The neuroplasticity of addiction, in *Toward a Theory of Neuroplasticity* (Shaw, C. and McEachern J., eds.), Taylor & Francis, Philadelphia, pp. 359–372.
10. Paulson, P. E., Camp, D. M., and Robinson, T. E. (1991) The time course of transient behavioral depression and persistent behavioral sensitization in relation to regional monoamine concentrations during amphetamine withdrawal in rats. *Psychopharmacology* **103,** 480–492.
11. Sesack, S. R. and Pickel, V. M. (1992) Prefrontal cortical efferents in the rat synapse on unlabeled neuronal targets of catecholamine terminals in the nucleus accumbens septi and on dopamine neurons in the ventral tegmental area. *J. Comp. Neurol.* **320,** 145–160.
12. Grant, S., London, E. D., Newlin, D. B., Villemagne, V. L., Liu, X., Contoreggi, C., et al. (1996) Activation of memory circuits during cue-elicited cocaine craving. *Proc. Natl. Acad. Sci. USA* **93,** 12040–12045.
13. Wang, G.-J., Volkow, N. D., Fowler, J. S., Cervany, P., Hitzemann, R. J., Pappas, N. R., et al. (1999) Regional brain metabolic activation during craving elicited by recall of previous drug experiences. *Life Sci.* **64,** 775–784.
14. Childress, A. R., Mozley, P. D., McElgin, W., Fitzgerald, J., Reivich, M., and O'Brien, C. P. (1999) Limbic activation during cue-induced cocaine craving. *Am. J. Psychiatry* **156,** 11–18.
15. Maas, L. C., Lukas, S. E., Kaufman, M. J., Weiss, R. D., Daniels, S. L., Rogers, V. W., et al. (1998) Functional magnetic resonance imaging of human brain activation during cue-induced cocaine craving. *Am. J. Psychiatry* **155,** 124–126.
16. Garavan, H., Pankiewicz, J., Bloom, A., Cho, J.-K., Sperry, L., Ross, T. J., et al. (2000) Cue-induced cocaine craving: neuroanatomical specificity for drug users and drug stimuli. *Am. J. Psychiatry* **157,** 1789–1798.
17. Wexler, B. E., Gottschalk, C. H., Fulbright, R. K., Prohovnik, I., Lacadie, C. M., Rounsaville, B. J., et al. (2001) Functional magnetic resonance imaging of cocaine craving. *Am. J. Psychiatry* **158,** 86–95.
18. Karler, R., Calder, L. D., Chaudhry, I. A., and Turkanis S. A. (1989) Blockade of "reverse tolerance" to cocaine and amphetamine by MK-801. *Life Sci.* **45,** 599–606.
19. Schenk, S., Valadez, A., McNamara, C., House, D. T., Higley, D., Bankson, M. G., et al. (1993) Development and expression of sensitization to cocaine's reinforcing properties: role of NMDA receptors. *Psychopharmacology* **11,** 332–338.
20. Li, Y., Hu, X.-T., Berney, T. G., Vartanian, A. J., Stine C. D., Wolf, M. E., et al. (1999) Both glutamate receptor antagonists and prefrontal cortex lesions prevent induction of cocaine sensitization and associated neuroadaptations. *Synapse* **34,** 169–180.
21. DeMontis, M. B., Gambarana, C., Ghliglieri, O., and Tagliamonte, A. (1995) Reversal of stable behavioral modifications through NMDA receptor inhibition in rats. *Behav. Pharmacol.* **6,** 562–567.
22. Li, Y., White, F. J., and Wolf, M. E. (2000) Pharmacological reversal of behavioral and cellular indices of cocaine sensitization. *Psychopharmacology* **151,** 175–183.
23. Carlezon, W. A. Jr., Mendrek, A., and Wise, A. (1995) MK-801 disrupts the expression but not the development of bromocriptine sensitization: a state-dependency interpretation. *Synapse* **20,** 1–9.
24. Wise, R. A., Mendrek, A., and Carlezon, W. A., Jr. (1996) MK-801 (dizocilpine): synergist and conditioned stimulus in bromocriptine-induced psychomotor sensitization. *Synapse* **22,** 362–368.
25. Ranaldi, R., Munn, E., Neklesa, T., and Wise, R. A. (2000) Morphine and amphetamine sensitization in rats demonstrated under moderate- and high-dose NMDA receptor blockade with MK-801 (dizocilpine). *Psychopharmacology* **151,** 192–201.
26. Li, Y. and Wolf, M. E. (1999) Can the "state-dependency" hypothesis explain prevention of amphetamine sensitization in rats by NMDA receptor antagonists? *Psychopharmacology* **141,** 351–361.
27. Wolf, M. E. (1999) NMDA receptors and behavioural sensitization: Beyond dizocilpine. *Trends Pharmacol. Sci.* **20,**188–189.
28. Kalivas, P. W. and Alesdatter, J. E. (1993) Involvement of N-methyl-D-aspartate receptor stimulation in the ventral tegmental area and amygdala in behavioral sensitization to cocaine. *J. Pharmacol. Exp. Ther.* **267,** 486–495.
29. Kim, J.-H. and Vezina, P. (1998) Metabotropic glutamate receptors are necessary for sensitization by amphetamine. *Neuro Report* **9,** 403–306.
30. Cador, M., Bjijou, Y., Cailhol, S., and Stinus, L. (1999) Amphetamine-induced behavioral sensitization: Implication of a glutamatergic medial prefrontal cortex-VTA innervation. *Neuroscience* **94,** 705–721.
31. Vezina, P. and Queen, A. L. (2000) Induction of locomotor sensitization by amphetamine requires the activation of NMDA receptors in the rat ventral tegmental area. *Psychopharmacology* **151,** 184–191.

32. Wolf, M. E., Dahlin, S. L., Hu, X.-T., Xue, C.-J., and White, K. (1995) Effects of lesions of prefrontal cortex, amygdala, or fornix on behavioral sensitization to amphetamine: comparison with *N*-methyl-D-aspartate antagonists. *Neuroscience* **69**, 417–439.
33. Tzschentke, T. M. (2001) Pharmacology and behavioral pharmacology of the mesocortical dopamine system. *Prog. Neurobiol.* **63**, 241–320.
34. Carr, D. B. and Sesack, S. R. (2000) Projections from the rat prefrontal cortex to the ventral tegmental area: target specificity in the synaptic associations with mesoaccumbens and mesocortical neurons. *J. Neurosci.* **20**, 3864–3873.
35. Giorgetti, M., Hotsenpiller, G., Froestl, W., and Wolf, M. E. In vivo modulation of ventral tegmental area dopamine and glutamate efflux by local $GABA_B$ receptors is altered after repeated amphetamine treatment. *Neuroscience,* **109**, 585–595.
36. Nestler, E. J. (2001) Molecular basis of long-term plasticity underlying addiction. *Nature Rev.* **2**, 119–128.
37. Anderson, P. and Soleng, A. F. (1998) Long-term potentiation and spatial training are both associated with the generation of new synapses. *Brain Res. Rev.* **26**, 353–359.
38. Robinson, T. E. and Kolb, B. (1997) Persistent structural modifications in nucleus accumbens and prefrontal cortex neurons produced by previous experience with amphetamine. *J. Neurosci.* **17**, 8491–8497.
39. Robinson, T. E. and Kolb, B. (1999) Alterations in the morphology of dendrites and dendritic spines in the nucleus accumbens and prefrontal cortex following repeated treatment with amphetamine or cocaine. *Eur. J. Neurosci.* **11**, 1598–1604.
40. Robinson, T. E., Gorny, G., Mitton, E., and Kolb, B. (2001) Cocaine self-administration alters the morphology of dendrites and dendritic spines in the nucleus accumbens and neocortex. *Synapse* **39**, 257–266.
41. Brown, R. W. and Kolb, B. (2001) Nicotine sensitization increases dendritic length and spine density in the nucleus accumbens and cingulated cortex. *Brain Res.* **899**, 94–100.
42. Robinson, T. E. and Kolb, B. (1999) Morphine alters the structure of neurons in the nucleus accumbens and neocortex of rats. *Synapse* **33**, 160–162.
43. Onn, S.-P. and Grace, A. A. (2000) Amphetamine withdrawal alters bistable states and cellular coupling in rat prefrontal cortex and nucleus accumbens neurons recorded in vivo. *J. Neurosci.* **20**, 2332–2345.
44. Klintsova, A. Y. and Greenough, W. T. (1999) Synaptic plasticity in cortical systems. *Curr. Opin. Neurobiol.* **9**, 203–208.
45. McAllister, A. K., Katz, L. C., and Lo, D. C. (1999) Neurotrophins and synaptic plasticity. *Annu. Rev. Neurosci.* **22**, 295–318.
46. Pierce, R. C. and Bari, A. A. (2001) The role of neurotrophic factors in psychostimulant-induced behavioral and neural plasticity. *Rev. Neurosci.* **12**, 95–110.
47. Flores, C., Rodaros, D., and Stewart, J. (1998) Long-lasting induction of astrocytic basic fibroblast growth factor by repeated injections of amphetamine: Blockade by concurrent treatment with a glutamate antagonist. *J. Neurosci.* **18**, 9547–9555.
48. Flores, C. and Samaha, A. N., and Stewart, J. (2000) Requirement of endogenous basic fibroblast growth factor for sensitization to amphetamine. *J. Neurosci.* **20**, RC55.
49. Flores, C. and Stewart, J. (2000) Changes in astrocytic basic fibroblast growth factor expression during and after prolonged exposure to escalating doses of amphetamine. *Neuroscience* **98**, 287–293.
50. Roberson, E. D., English, J. D., and Sweatt, J. D. (1996) A biochemist's view of long-term potentiation. *Learn. Mem.* **3**, 1–24.
51. Milner, B., Squire, L. R., and Kandel, E. R. (1998) Cognitive neuroscience and the study of memory. *Neuron* **20**, 445–468.
52. Silva, A. J., Kogan, J. H., Frankland, P. W., and Kida, S. (1998) CREB and memory. *Annu. Rev. Neurosci.* **21**, 127–148.
53. Lüscher, C., Nicoll, R. A., Malenka, R. C., and Muller, D. (2000) Synaptic plasticity and dynamic modulation of the postsynaptic membrane. *Nature Neurosci.* **3**, 545–550.
54. Malenka, R. C. and Nicoll, R. A. (1997) Silent synapses speak up. *Neuron* **19**, 473–476.
55. Carroll, R. C., Beattie, E. C., von Zastrow, M., and Malenka, R. C. (2001) Role of AMPA receptor endocytosis in synaptic plasticity. *Nat. Rev. Neurosci.* **2**, 315–324.
56. Sheng, M. and Lee, S. H. (2001) AMPA receptor trafficking and the control of synaptic transmission. *Cell* **105**, 825–828.
57. Shi, S.-H., Hayashi, Y., Esteban, J. A., and Malinow, R. (2001) Subunit-specific rules governing AMPA receptor trafficking to synapses in hippocampal pyramidal neurons. *Cell* **105**, 331–343.
57a. Passafaro, M., Piech, V., and Sheng, M. (2001) Subunit-specific temporal and spatial patterns of AMPA receptor exocytosis in hippocampal neurons. *Nature Neurosci.* **4**, 917–926.
58. Chao, S. Z., Lu, W. X., Lee, H.-K., Huganir, R. L., and Wolf, M. E. (1999) D1 dopamine receptor stimulation increases GluR1 phosphorylation in postnatal nucleus accumbens cultures. *J. Neurochem.,* in press.
59. Chao, S. Z., Lu, W. X., Lee, H.-K., Huganir, R. L., and Wolf, M E. (2002). D1 dopamine receptor stimulation increases GluR1 phosphorylation in postnatal nucleus accumbens cultures. *J. Neurochem.,* in press.
60. Price, C. J., Kim, P., and Raymond, L. A. (1999) D1 dopamine receptor-induced cyclic AMP-dependent protein kinase phosphorylation and potentiation of striatal glutamate receptors. *J. Neurochem.* **73**, 2441–2446.

61. Snyder, G. L., Alien, P. B., Fienberg, A. A., Valle, C. G., Huganir, R. L., Nairn, A. C., et al. (2000) Regulation of phosphorylation of the GluR1 AMPA receptor in the neostriatum by dopamine and psychostimulants in vivo. *J. Neurosci.* **20,** 4480–4488.
62. Yan, Z., Hsieh-Wilson, L., Feng, J., Tomizawa, K., Alien, P. B., Fienberg, A. A., et al. (1999) Protein phosphatase 1 modulation of neostriatal AMPA channels: regulation by DARPP-32 and spinophilin. *Nature Neurosci.* **2,** 13–17.
63. Banke, T. G., Bowie, D., Lee, H.-K., Huganir, R. L., Schousboe, A., and Traynelis, S. F. (2000) Control of GluR1 AMPA receptor function by cAMP-dependent protein kinase. *J. Neurosci.* **20,** 89–102.
64. Bibb, J. A., Chen, J., Taylor, J. R., Svenningsson, P., Nishi, A., Snyder, G. L., et al. (2001) Effects of chronic exposure to cocaine are regulated by the neuronal protein Cdk5. *Nature* **410,** 376–380.
65. Chao, S. Z., Peterson, D. A., and Wolf, M. E. (2000) Dopamine and glutamate receptors regulate surface expression of GluR1 in postnatal nucleus accumbens cultures. *Soc. Neurosci. Abstr.* **26,** 789.
66. Wolf, M. E. and Khansa, M. R. (1991) Repeated administration of MK-801 produces sensitization to its own locomotor stimulant effects but blocks sensitization to amphetamine. *Brain Res.* **562,** 164–168.
67. Muller, D. and Lynch, G. (1988) Long-term potentiation differentially affects two components of synaptic responses in hippocampus. *Proc. Natl, Acad. Sci, USA* **85,** 9346–9350.
68. Kauer, J. A., Malenka, R. C., and Nicoll, R. A. (1988) A persistent postsynaptic modification mediates long-term potentiation in the hippocampus. *Neuron* **1,** 911–917.
69. White, F. J., Hu, X.-T., Zhang, X.-F., and Wolf, M. E. (1995) Repeated administration of cocaine or amphetamine alters neuronal responses to glutamate in the mesoaccumbens dopamine system. *J. Pharmacol. Exp. Ther.* **273,** 445–454.
70. Zhang, X.-F., Hu, X.-T., White, F. J., and Wolf, M. E. (1997) Increased responsiveness of ventral tegmental area dopamine neurons to glutamate after repeated administration of cocaine or amphetamine is transient and selectively involves AMPA receptors. *J. Pharmacol. Exp. Ther.* **281,** 699–706.
71. Giorgetti, M., Hotsenpiller, G., Ward, P., Teppen, T., and Wolf, M. E. (2001) Amphetamine-induced plasticity of AMPA receptors in the ventral tegmental area: effects on extracellular levels of dopamine and glutamate in freely moving rats. *J. Neurosci.* **21,** 6362–6369.
72. Ortiz, J., Fitzgerald, L. W., Charlton, M., Lane, S., Trevisan, L., Guitart, X., et al. (1995) Biochemical actions of chronic ethanol exposure in the mesolimbic dopamine system. *Synapse* **21,** 289–298.
73. Fitzgerald, L. W., Ortiz, J., Hamedani, A. G., and Nestler, E. J. (1996) Drugs of abuse and stress increase the expression of GluR1 and NMDAR1 glutamate receptor subunits in the rat ventral tegmental area: common adaptations among cross-sensitizing agents. *J. Neurosci.* **16,** 274–282.
74. Churchill, L., Swanson, C. J., Urbina. M., and Kalivas, P. W. (1999) Repeated cocaine alters glutamate receptor subunit levels in the nucleus accumbens and ventral tegmental area of rats that develop behavioral sensitization. *J. Neurochem.* **72,** 2397–2403.
75. Lu, W., Monteggia, L. M., and Wolf, M. E. (2001) Repeated administration of amphetamine or cocaine does not alter AMPA receptor subunit expression in the rat midbrain. *Neuropsychopharmacology,* **26,** 1–3.
76. Ghasemzadeh, M. B., Nelson, L. C., Lu, X.-Y., and Kalivas, P. W. (1999) Neuroadaptations in ionotropic and metabotropic glutamate receptor mRNA produced by cocaine treatment. *J. Neurochem.* **72,** 157–165.
77. Bardo M. T., Robinet P. M., Mattingly B. A., and Margulies, J. E. (2001) Effect of 6-hydroxydopamine or repeated amphetamine treatment on mesencephalic mRNA levels for AMPA glutamate receptor subunits in the rat. *Neurosci. Lett.* **302,** 133–136.
78. Overton, P. G., Richards, C. D., Berry, M. S., and Clark D. (1999) Long-term potentiation at excitatory amino acid synapses on midbrain dopamine neurons. *NeuroReport* **10,** 221–226.
79. Bonci, A. and Malenka, R. C. (1999) Properties and plasticity of excitatory synapses on dopaminergic and GABAergic cells in the ventral tegmental area. *J. Neurosci.* **19,** 3723–3730.
80. Carr, D. B. and Sesack, S. R. (2000) GABA-containing neurons in the rat ventral tegmental area project to the prefrontal cortex. *Synapse* **38,** 114–123.
81. Jones, S., Kornblum, J. L., and Kauer, J. A. (2000) Amphetamine blocks long-term synaptic depression in the ventral tegmental area. *J. Neurosci.* **20,** 5575–5580.
82. Thomas, M. J., Malenka, R. C., and Bonci, A. (2000) Modulation of long-term depression by dopamine in the mesolimbic system. *J. Neurosci.* **20,** 5581–5586.
83. Cardozo, D. L. and Bean, B. P. (1995) Voltage-dependent calcium channels in rat midbrain dopamine neurons: modulation by dopamine and GABAB receptors. *J. Neurophysiol.* **74,** 1137–1148.
84. Karler, R., Turkanis, S. A., Partlow, L. M., and Calder, L. D. (1991) Calcium channel blockers in behavioral sensitization. *Life Sci.* **49,** 165–170.
85. Reimer, A. R. and Martin-Iverson, M. T. (1994) Nimodipine and haloperidol attenuate behavioural sensitization to cocaine but only nimodipine blocks the establishment of conditioned locomotion induced by cocaine. *Psychopharmacology* **113,** 404–410.

86. Licata, S. C., Freeman, A. Y., Pierce-Bancroft, A. F., and Pierce, R. C. (2000) Repeated stimulation of L-type calcium channels in the rat ventral tegmental area mimics the initiation of behavioral sensitization to cocaine. *Psychopharmacology* **152,** 110–118.
87. Xue, C.-J., Ng, J. P., Li, Y., and Wolf, M. E. (1996) Acute and repeated systemic amphetamine administration: effects on extracellular glutamate, aspartate, and serine levels in rat ventral tegmental area and nucleus accumbens. *J. Neurochem.* **67,** 352–363.
88. Wolf, M. E. and Xue, C.-J. (1998) Amphetamine and DI receptor agonists produce biphasic effects on glutamate efflux in rat ventral tegmental area: Modification by repeated amphetamine administration. *J. Neurochem.* **70,** 198–209.
89. Wolf, M. E. and Xue, C.-J., (1999) Treatments that prevent amphetamine sensitization also prevent amphetamine-induced glutamate efflux in the rat ventral tegmental area. *J. Neurochem.* **73,** 1529–1538.
90. Wolf, M. E., Xue, C.-J., Li, Y., and Wavak, D. (2000) Amphetamine increases glutamate efflux in the rat ventral tegmental area by a mechanism involving glutamate transporters and reactive oxygen species. *J. Neurochem.* **75,** 1634–1644.
91. Kalivas, P. W. and Duffy, P. (1995) $D_1$ receptors modulate glutamate transmission in the ventral tegmental area *J. Neurosci.* **15,** 5379–5388.
92. Kalivas, P. W. and Duffy, P. (1998) Repeated cocaine administration alters extracellular glutamate in the ventral tegmental area. *J. Neurochem.* **70,** 1497–1502.
93. Paladini, C. A., Fiorillo, C. D., Morikawa, H., and Williams, J. T. (2001) Amphetamine selectively blocks inhibitory glutamate transmission in dopamine neurons. *Nature Neurosci.* **4,** 275–281.
94. Mansvelder, H. D. and McGehee, D. S. (2000) Long-term potentiation of excitatory inputs to brain reward areas by nicotine. *Neuron* **27,** 349–357.
95. Ungless, M. A., Whistler, J. L., Malenka, R. C., and Bonci, A. (2001) Single cocaine exposure *in vivo* induces long-term potentiation in dopamine neurons. *Nature* **411,** 583–587.
96. White, F. J. and Kalivas, P. W. (1998) Neuroadaptations involved in amphetamine and cocaine addiction. *Drug Alcohol Depend.* **51,** 141–153.
97. Meredith, G. E. (1999) The synaptic framework for chemical signaling in nucleus accumbens. *Ann NY Acad. Sci.* **877,** 140–156.
98. Nicola, S. M., Surmeier, D. J., and Malenka, R. C. (2000) Dopaminergic modulation of neuronal excitability in the striatum and nucleus accumbens. *Annu. Rev. Neurosci.* **232,** 185–215.
99. Calabresi, P., Maj, R., Pisani, A., Mercuri, N. B., and Bernardi, G. (1992) Long-term synaptic depression in the striatum: physiological and pharmacological characterization. *J. Neurosci.* **12,** 4224–4233.
100. Calabresi, P., Centonze, D., Gubellini, P., Marfia, G. A., and Bernardi, G. (1999) Glutamate-triggered events inducing corticostriatal long-term depression. *J. Neurosci.* **19,** 6102–6110.
101. Calabresi, P., Pisani, A., Mercuri, N. B., and Bernardi, G. (1992) Long-term potentiation in the striatum is unmasked by removing the voltage-dependent magnesium block of NMDA receptor channels. *Eur. J. Neurosci.* **4,** 929–935.
102. Charpier, S. and Deniau, J. M. (1997) *In vivo* activity-dependent plasticity at cortico-striatal connections: evidence for physiological long-term potentiation. *Proc. Natl. Acad. Sci. USA* **94,** 7036–7040.
103. Wickens, J. R., Begg, A. J., and Arbuthnott, G. W. (1996) Dopamine reverses the depression of rat corticostriatal synapses which normally follows high-frequency stimulation of cortex in vitro. *Neuroscience* **70,** 1–5.
104. Pennartz, C. M. A., Ameerun, R. F., Groenewegen, H. J., and Lopes da Silva, F. H. (1993) Synaptic plasticity in an in vitro slice preparation of the rat nucleus accumbens. *Eur. J. Neurosci.* **5,** 107–117.
105. Boeijinga, P. H., Mulder, A. B., Pennartz, C. M. A., Manshanden, I., and Lopes da Silva, F. H. (1993) Reponses of the nucleus accumbens following fornix/fimbria stimulation in the rat. Identification and long-term potentiation of mono-and polysynaptic pathways. *Neuroscience* **53,** 1049–1058.
106. Kombian, S. B. and Malenka, R. C. (1994) Simultaneous LTP of non-NMDA- and LTD of NMDA-receptor-mediated responses in the nucleus accumbens. *Nature* **368,** 242–246.
107. O'Donnell, P. and Grace, A. A. (1995) Synaptic interactions among excitatory afferents to nucleus accumbens neurons: hippocampal gating of prefrontal cortical input. *J. Neurosci.* **15,** 3622–3639.
108. Pulvirenti, L., Criado, J., Balducci, C., Koob, G. F., and Henriksen, S. J. (1998) Enhanced synaptic efficacy in the nucleus accumbens during the acquisition of cocaine-seeking behavior. *Soc. Neurosci. Abstr.* **42,** 779.
109. Thomas, M. F., Beurrier, C., Bonci, A., and Malenka, R. (2001) Long-term depression in the nucleus accumbens: a neural correlate of behavioral sensitization to cocaine. *Natural Neurosci.* **4,** 1217–1223.
110. Meshul, C. K., Noguchi, K., Emre, N., and Ellison, G. (1998) Cocaine-induced changes in glutamate and GABA immunolabeling within rat habenula and nucleus accumbens. *Synapse* **30,** 211–220.
111. Keys, A. S., Mark, G. P., Emre, N., and Meshul, C. K. (1998) Reduced glutamate immunolabeling in the nucleus accumbens following extended withdrawal from self-administered cocaine. *Synapse* **30,** 393–401.
112. Lu, W., Chen, H., Xue, C.-J., and Wolf, M. E. (1997) Repeated amphetamine administration alters the expression of mRNA for AMPA receptor subunits in rat nucleus accumbens and prefrontal cortex. *Synapse* **26,** 269–280.
113. Lu, W. and Wolf, M. E. (1999) Repeated amphetamine administration alters immunoreactivity for AMPA receptor subunits in rat nucleus accumbens and medial prefrontal cortex. *Synapse* **32,** 119–131.

114. Lu, W., Monteggia, L. M., and Wolf, M. E. (1999) NMDAR1 expression in the rat mesocorticolimbic dopamine system is decreased after withdrawal from repeated amphetamine administration. *Eur. J. Neurosci.* **11,** 3167–3177.
115. Loftis, J. M. and Janowsky, A. (2000) Regulation of NMDA receptor subunits and nitric oxide synthase expression during cocaine withdrawal. *J. Neurochem.* **75,** 2040–2050.
116. Mao, L. and Wang, J. Q. (2001) Differentially altered mGluR1 and mGluR5 mRNA expression in rat caudate nucleus and nucleus accumbens in the development and expression of behavioral sensitization to repeated amphetamine administration. *Synapse* **41,** 230–240.
117. Zhang, X.-F., Hu, X.-T., and White, F. J. (1998) Whole-cell plasticity in cocaine withdrawal: reduced sodium currents in nucleus accumbens neurons. *J. Neurosci.* **18,** 488–498.
118. Li, Y. and Kauer, J. (2000) Effects of amphetamine on long-term potentiation in the nucleus accumbens. *Soc. Neurosci. Abstr.* **26,** 1398.
119. Yamamoto, Y., Nakanishi, H., Takai, N., Shimazoe, T., Watanabe, S., and Kita, H. (1999) Expression of *N*-methyl-D-aspartate receptor-dependent long-term potentiation in the neostriatal neurons in an in vitro slice after ethanol withdrawal of the rat. *Neuroscience* **91,** 59–68.
120. Nishioku, T., Shimazoe, T., Yamamoto, Y., Nakanishi, H., and Watanabe, S. (1999) Expression of long-term term potentiation of the striatum in methamphetamine-sensitized rats. *Neurosci. Lett.* **268,** 81–84.
121. Pulvirenti, L., Maldonado-Lopez, R., and Koob, G. F. (1992) NMDA receptors in the nucleus accumbens modulate intravenous cocaine but not heroin self-administration in the rat. *Brain Res.* **594,** 327–330.
122. Layer, R. T., Uretsky, N. J., and Wallace, L. J. (1993) Effects of the AMPA/kainite receptor antagonist DNQX in the nucleus accumbens on drug-induced conditioned place preference. *Brain Res.* **17,** 267–273.
123. Burns, L. H., Everitt, B. J., Kelley, A. E., and Robbins, T. W. (1994) Glutamate–dopamine interactions in the ventral striatum: role in locomotor activity and responding with conditioned reinforcement. *Psychopharmacology* **115,** 516–528.
124. Kaddis, F. G., Uretsky, N. J., and Wallace, L. J. (1995) DNQX in the nucleus accumbens inhibits cocaine-induced induced conditioned place preference. *Brain Res.* **697,** 76–82.
125. Bell, K. and Kalivas, P. W. (1996) Context-specific cross-sensitization between system cocaine and intra-accumbens AMPA infusion in the rat. *Psychopharmacology* **127,** 377–383.
126. Cornish, J. L., Duffy, P., and Kalivas, P. W. (1999) A role for nucleus accumbens glutamate transmission in the relapse to cocaine-seeking behavior. *Neuroscience* **93,** 1359–1367.
127. Bell, K., Duffy, P., and Kalivas, P. W. (2000) Context-specific enhancement of glutamate transmission by cocaine. *Neuropsychopharmacology* **23,** 335–344.
128. Cornish, J. L. and Kalivas, P. W. (2000) Glutamate transmission in the nucleus accumbens mediates relapse in cocaine addiction. *J. Neurosci.* **20,** RC89.
129. Di Ciano, P. and Everitt, B. J. (2001) Dissociable effects of antagonism of NMDA and AMPA/KA receptors in the nucleus accumbens core and shell on cocaine-seeking behavior. *Neuropsychopharmacology* **25,** 341–360.
130. Hotsenpiller, G., Giorgetti, M., and Wolf, M. E. (2001) Alterations in behavior and glutamate transmission following presentation of stimuli previously associated with cocaine exposure. *Eur. J. Neurosci.* **14,** 1843–1855.

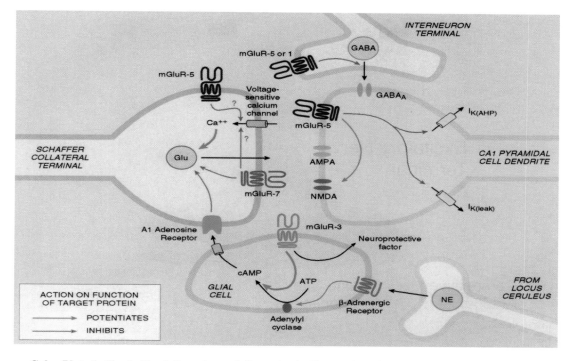

**Color Plate 1, Fig. 1.** (*See* full caption and discussion in Chapter 1, p. 4). Example of physiological roles of mGluRs at the Schaffer collateral synapse in CA1 in the hippocampus.

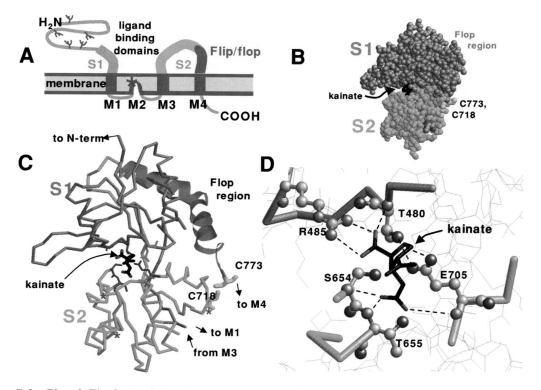

**Color Plate 2, Fig. 3.** (*See* full caption and discussion in Chapter 1, p. 9). Glutamate receptor topology and crystal structure of the agonist-binding pocket.

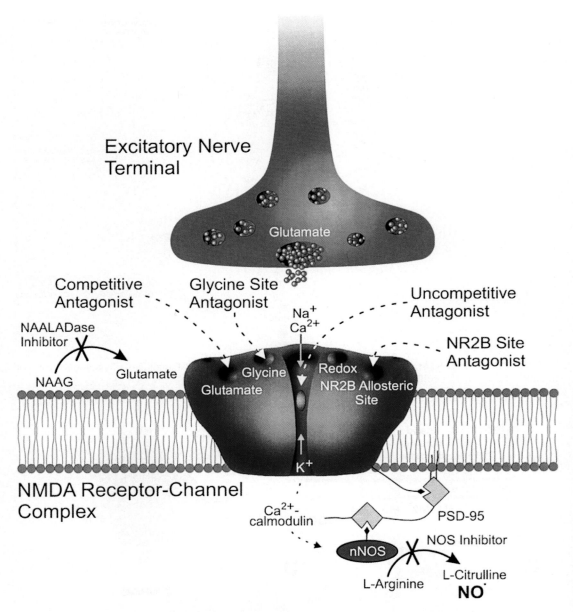

**Color Plate 3, Fig. 1.** (*See* discussion in Chapter 2, p. 25). A representation of the NMDA receptor with associated binding and regulatory domains.

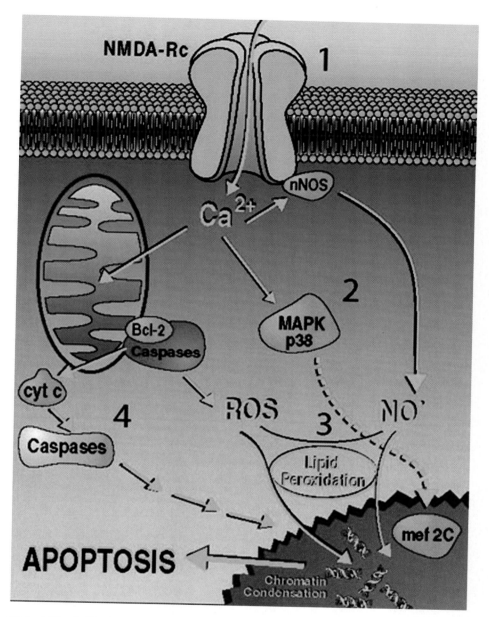

**Color Plate 4, Fig. 1.** (*See* full caption and discussion in Chapter 5, p. 72). Current model of NMDA receptor-associated neuronal injury. Schematic illustration of NMAR-related signaling pathways that lead to neuronal apoptosis and may contribute to neurodegenerative disease, including HIV-associated dementia.

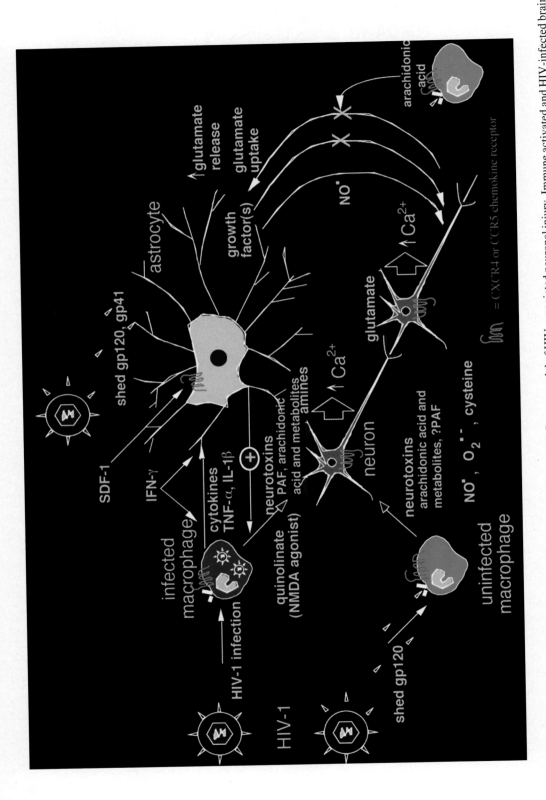

**Color Plate 1, Fig. 2.** (*See* full caption and discussion in Chapter 5, p. 76). Current model of HIV-associated neuronal injury. Immune activated and HIV-infected brain macrophages/microglia release potentially neurotoxic substances.

# 9
# Glutamate and Dopamine Interactions in the Motive Circuit
*Implications for Craving*

### David A. Baker, PhD, Jennifer L. Cornish, PhD, and Peter W. Kalivas, PhD

## 1. INTRODUCTION

Addiction to psychomotor stimulants is marked by a transition in drug consumption from a casual and recreational style of use to a more compulsive and excessive pattern. Acute administration of psychomotor stimulants is associated with numerous effects, including feelings of euphoria and increased energy, which can contribute to repeated recreational consumption. However, chronic administration of psychomotor stimulants results in the emergence of persistent cravings, paranoia, and drug-seeking behaviors that contribute to the development of compulsive drug-taking behavior *(1–4)*. An understanding of the neurobiology of drug addiction will require the identification and characterization of neuroadaptations underlying this transition from casual to chronic drug use.

Efforts to identify the neural basis of drug addiction have focused on the contributions of the meso-accumbens dopamine system, which originates in the ventral tegmental area (VTA) and projects to the nucleus accumbens *(5–8)*. This was due in part to the identification of the molecular site of action for amphetaminelike psychomotor stimulants. Cocaine and methylphenidate bind to monoamine transporters, preventing reuptake of extracellular dopamine (DA), norepinephrine, and serotonin *(9,10)*. Amphetamine, methamphetamine, and MDMA act as false substrates at monoamine transporters promoting the release of cytosolic stores of these neurotransmitters *(11)*. Moreover, a variety of data demonstrate that the acute effects of psychomotor stimulants arise largely from binding to dopamine transporters in the nucleus accumbens. Thus, dopamine receptor antagonists administered into the nucleus accumbens or lesions of meso-accumbens dopamine neurons inhibit the locomotor activating and reinforcing effects of amphetamine and cocaine *(12–14)*. Given the strong evidence that meso-accumbens dopamine is a critical mediator of acute psychostimulant-induced behaviors, it was logical to evaluate the meso-accumbens pathway for long-term neuroadaptations produced by repeated drug administration that may underlie addictive behaviors. Indeed, a number of enduring alterations in presynaplic and postsynaptic dopamine transmission have been characterized *(15,16)*. Many of these neuroadaptations, such as augmented dopamine release and enhanced electrophysiological responsiveness to D1 agonists, are consistent with a sensitization model of addiction and have been presumed to be important neuroadaptations underlying addictive behaviors.

In parallel with the emergence of our understanding, the important role of meso-accumbens dopamine transmission in the acute effects of psychomotor stimulants has been the maturation of our knowledge of the synaptic organization of afferents to spiny cells in the nucleus accumbens. From

From: *Contemporary Clinical Neuroscience: Glutamate and Addiction*
Edited by: Barbara H. Herman et al. © Humana Press Inc., Totowa, NJ

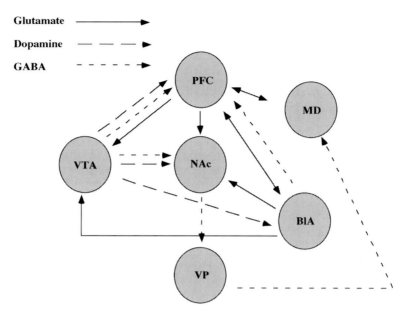

**Fig. 1.** Schematic illustrating a portion of the anatomical connections between regions typically included in the motive circuit: prefrontal cortex (PFC), mediodorsal thalamus, nucleus accumbens (NAc), basolateral amydgala (BLA), ventral pallidum (VP), and ventral tegmental area (VTA).

these anatomical studies, it has become clear that dopamine afferents to spiny cells modulate or gate excitatory input to change the probability of the cells being activated (17). Thus, dopaminergic synapses tend to be located proximal to excitatory synaptic contacts on the dendritic spines (18) and are able to support or diminish excitatory depolarization depending on the quality of the excitatory input (19–21). For example, dopamine transmission supports a strong excitatory volley that maintains a cell in a relatively depolarized resting state (the up state), but also supports a relatively hyperpolarized resting membrane potential (down state) in the absence of substantive excitatory input. The possibility that this intimate anatomical and functional relationship between dopamine and glutamate transmission in the nucleus accumbens might be involved in the behavioral effects of acute and repeated psychostimulant administration has been examined in some detail. Moreover, given that many of the excitatory afferents to the nucleus accumbens (NAc) arise from cortical and allocortical nuclei that also receive dopamine input, it became apparent that changes in dopamine and glutamate transmission elicited by psychostimulants may work in concert via interconnected nuclei termed the motive circuit (*see* Fig. 1; 22). Consistent with this possibility, imaging studies reveal that cocaine or presentation of cocaine-associated environmental stimuli alters brain metabolism in cortical and allocortical brain regions that project to spiny cells in the NAc (3,23–25).

The focus of this chapter is to examine how comodulation of dopamine and glutamate neurotransmission by psychostimulants regulates the activity of the motive circuit and how this may contribute to the long-term behavioral changes associated with addiction, such as craving and drug-seeking. In order to accomplish this goal, short reviews are provided of (1) the anatomy of the circuit, (2) the flow of information through the circuit, and (3) neuroadaptations within the circuit following repeated psychomotor stimulant administration. In particular, this chapter will focus on the contribution of the prefrontal cortex (PFC), basolateral amygdala (BLA), NAc, and ventral tegmental area (VTA).

## GLUTAMATE AND DOPAMINE INTERACTIONS IN THE MOTIVE CIRCUIT

### 2.1. Glutamate and Dopamine Receptors

Extracellular dopamine binds to two major families of receptors; D1-like, which consists of D1 and D5 receptors, and D2-like, which consists of D2, D3, and D4 receptors. The receptor subtypes within each group share molecular and pharmacological properties but can differ in their anatomical distribution. The D1 receptor family is coupled to G proteins and is associated with the activation of adenylyl cyclase. In contrast, the family of D2 receptors are involved in inhibition of adenylyl cyclase activity, inhibition of phosphatidylinositol turnover, increased $K^+$-channel activity, and modulating calcium conductances *(26,27)*.

Glutamate receptors consists of ionotropic and metabotropic receptors. Ionotropic glutamate receptors are classified as $N$-methyl-D-aspartic (NMDA), α-amino-3-hydroxy-5-methyl-4-isoxazole propionic acid (AMPA), and kainate receptors. NMDA receptors are heteromerically formed by NRI and NR2 subunits. In addition to agonist binding, activation of NMDA receptors requires the depolarization of the postsynaptic membrane in order to overcome magnesium blockade of the ion channel; at which point, there is a relatively long lasting increase in $Ca^{2+}$ efflux. AMPA receptors consist of GluR1 GluR2, GluR3, and GluR4 subunits, whereas kainate receptors consist of GluR5–7, KA1, and KA2. Activation of AMPA and kainate receptors opens cation $Na^+$ and $Ca^{2+}$ channels resulting in a rapid rise and decay of synaptic currents. AMPA receptor activation, in particular, is thought to mediate most forms of fast glutamatergic neurotransmission *(28,29)*.

In addition to the ionotropic receptors, eight metabotropic glutamate receptors (mGluR) have been cloned. The corresponding protein products are divided into three families: group I consists of mGluR1 and 5 receptors, group 2 consists of mGluR2 and 3 receptors, and group 3 consists of mGluR4, 6, 7, and 8 receptors. Group 1 receptors are coupled to phosphatidylinositol turnover. Group 2 receptors inhibit forskolin-stimulated formation of cAMP, and group 3 receptors are negatively coupled to adenylate cyclase *(30)*.

### 2.2. Anatomy of the Motive Circuit

The circuitry outlined in Fig. 1 has been termed the motive circuit and is key in translating incoming stimuli into a behavioral response *(31,32)*. A central component of the motive circuit is the meso-accumbens system. Although primarily a dopaminergic projection, as much as 20% of the meso-accumbens pathway contains γ-aminobutyric acid (GABA) instead of dopamine *(33)*. A second critical pathway originating in the VTA is the mesoprefrontal pathway, which sends dopamine projections to the PFC. A portion of these neurons synapse directly onto glutamatergic pyramidal neurons projecting to the nucleus accumbens. Surprisingly, almost 40% of these neurons contain GABAergic *(33)*. In addition to the PFC, the nucleus accumbens receives glutamatergic afferents originating in the hippocampus, mediodorsal thalamus, and basolateral amygdala (34–37). The cells projecting from the nucleus accumbens terminate in the ventral pallidum and ventral mesencephalon and are GABAergic medium spiny neurons.

Ultrastructural anatomical studies indicate that dopamine and glutamate neurotransmission have numerous points of putative interaction within the motive circuit. Both the pyramidal cells in the PFC and medium spiny neurons in the nucleus accumbens are innervated by both transmitters. Within the nucleus accumbens, dopamine and glutamate afferents both synapse onto dendritic spines of the medium spiny neurons. Excitatory afferents synapse onto the head of the spine, whereas dopamine terminals synapse onto the neck of the spine *(38,39)*. A similar orientation has been observed in the PFC, where dopamine synapses on more proximal portions of the pyramidal cell dendrite than glutamatergic afferents from the mediodorsal thalamus *(40)*. Within the VTA, cortical glutamatergic afferents synapse onto both GABAergic and dopaminergic neurons projecting to the nucleus accumbens and PFC *(41)*.

## 2.3. Neuronal Activity in Motive Circuit

Electrophysiological studies have been key in characterizing the contribution that individual nuclei make to the flow of information through the motive circuit. Given the anatomy of the circuit, the following subsections will review the contribution of glutamate and dopamine receptor stimulation in the nucleus accumbens PFC, and BLA to the activity of the medium spiny neurons in the nuecleus accumbens.

### 2.3.1. Nucleus Accumbens

Through a series of experiments over the last decade, Grace and colleagues presented compelling evidence that excitatory afferents to spiny cells from the hippocampus and amygdala serve to gate activity of glutamatergic afferents originating in the PFC *(42)*. The activity of medium spiny neurons exhibit biphasic states characterized by a depolarized state in which the cell is more excitable and a hyperpolarized nonfiring state in which the cell is unlikely to fire *(19,43)*. Hippocampal glutamatergic afferents appear to regulate the transition to the up (relatively depolarized) state. Specifically, stimulation of fimbria fornix produced a long-lasting duration of the up state *(19)*. Similarly, stimulation of BLA resulted in a brief transition to the depolarized state. *(44)*. Conversely, PFC glutamatergic afferents to the accumbens do not appear to alter the frequency of up or down states, but, instead produce action potentials in medium spiny neurons provided the cell is in the up state *(19)*. Stimulation of either NMDA or AMPA receptors in the accumbens produces excitation in medium spiny neuron; however, AMPA but not NMDA receptor antagonists block glutamate-induce excitation of medium spiny neurons *(45)*.

Dopamine neurotransmission in the accumbens also modulates the biphasic states observed in medium spiny neurons. Stimulation of D1 receptor in the NAc is thought to promote a voltage-dependent calcium conductance that prolongs the duration of the up state *(46,47)* However, this occurs only when the neuron is in the up state because the calcium conductance is voltage dependent. In contrast, stimulation of D2-like receptors in the accumbens supports the duration of the down state, as well as decreasing the frequency of transition from the down state to the up state *(42)*. Taken together, this compound effect of dopamine transmission serves to increase the signal-to-noise ratio by supporting either the up or down state. Thus, if there is more depolarizing glutamatergic input, dopamine transmission will increase the duration of depolarization by increasing calcium conductance, and in the relative absence of depolarizing input, dopamine will support an inactive state. This dopaminergic filter on the transit of information is consistent with behavioral studies that have concluded that dopamine serves to "gate" information through the nucleus accumbens *(17)*.

### 2.3.2. Ventral Tegmental Area

The activity of meso-accumbens dopamine neurons is differentially modulated by stimulation of glutamatergic and dopaminergic receptors in the VTA. Meso-accumbens dopaminergic neurons are characterized by long-duration action potentials, irregular- and burst-firing patterns, and slow conduction velocities. Stimulation of NMDA receptors in the VTA regulates burst firing *(48,49)*, whereas stimulation of AMPA receptors appears to regulate the overall firing rate *(50)* Conversely, stimulation of dopamine receptors in the VTA produces an inhibition of activity modulated by D2 autoreceptors. The role of D1 receptor stimulation is unclear because these receptors are located on glutamatergic terminals and GABAergic interneurons. Thus although stimulation of D1 receptors fails to directly alter the activity of meso-accumbens DA neurons *(51,52)* it would be expected to affect overall excitatory and inhibitory tone to the cells.

### 2.3.3. Prefrontal Cortex

Similar to medium spiny neurons in the accumbens, projection (pyramidal) cells in the PFC also exhibit biphasic states *(53,54)*, although the regulation of these states is not well understood. Interestingly, however, several studies have shown that dopamine projections from the VTA modulate this

Similar to humans, both cocaine-paired stimuli and cocaine produce potent reinstatement of drug-seeking seeking behavior. Recently, it was revealed that these two stimuli activate separate nuclei within the motive circuit *(25)*. Specifically, cocaine-associated stimuli resulted in elevated Fos protein expression in a number of regions, including the prefrontal cortex/anterior cingulate, BLA, accumbens, and hippocampus in rats withdrawn from cocaine *(25)*. Conversely, a cocaine-priming injection in these rats produced elevated Fos protein expression in the anterior cingulate/mPFC and VTA. An increase was also obtained in the nucleus accumbens, however, this effect was not significant. These data indicate that the motive circuit is involved in reinstatement produced by both cocaine and cocaine-associated stimuli. In particular, it appears that regardless of the modality of the stimulus, (e.g., pharmacological or environmental), cocaine-seeking behavior is produced by activating the projection from the dorsal PFC to the nucleus accumbens. This is consistent with observations that the expression of behavioral sensitization to cocaine, as well as the increased releasability of glutamate in the nucleus accumbens, is inhibited by selective lesions of the dorsal prefrontal cortex *(119)*. Moreover, cocaine reinstatement of cocaine-seeking behavior via meso-prefrontal modulation of corticofugal projections to the accumbens is indicated by the fact that the blockade of AMPA receptors in the accumbens prevents cocaine-induced reinstatement *(120)*. Consistent with a role of glutamate in modulating the activity of medium spiny neurons to induce craving, stimulation of AMPA receptors produced a selective increase in cocaine-seeking behavior on the cocaine-paired lever *(121)*. Finally, recent evidence from our laboratory indicates that inhibition of the dorsal PFC by microinjecting a combination of $GABA_A$ and $GABA_B$ agonist blocks cocaine-induced reinstatement *(122)*. In contrast, it appears that cocaine-associated stimuli reinstate cocaine-seeking behavior by activating the BLA and the hippocampus, which then regulate the activity of the PFC *(123)*. This idea is consistent with electrophysiological studies revealing that the BLA produces a brief increase in the up state of medium spiny neurons in the accumbens, which would be expected following presentation of a brief but salient stimulus *(44)*. Furthermore, the hippocampus produces a long-lasting increase in the up state of these neurons, consistent with its role in contextual learning *(124)*. Behavioral studies also support this notion. For instance, chronic or reversible lesions of the BLA block cue-induced reinstatement without altering cocaine-induced reinstatement *(123,125)*. Furthermore, cocaine administration in a cocaine-paired, but not in a cocaine-unpaired, environment has been shown to increase extracellular glutamate levels in the accumbens *(126)*.

Reinstatement studies to date support a role for glutamate release in the nucleus accumbens as a critical biological signal to initiate drug-seeking behavior. This poses the interesting possibility that although psychostimulant-induced increases in dopamine transmission critically regulate the acute effects of cocaine, once a behavior is established, it may rely more on glutamate than dopamine transmission. Supporting this notion, the blockade of AMPA, but not dopamine, receptors in the nucleus accumbens inhibits cocaine-induced reinstatement of cocaine-seeking behavior *(120)*. Given the relatively weak or inconsistent effects of dopaminergic drugs *(127)*, the apparent prepotent role of glutamate versus dopamine in the reinstatement of drug-seeking behavior in animal models points to a possible therapeutic intervention in relapse of psychostimulant addicts with agents that diminish glutamate transmission.

## REFERENCES

1. Satel, S. L., Southwick, S. M., and Gawin, F. H. (1991) Clinical features of cocaine-induced paranoia. *Am. J. Psychiatry* **148,** 495–498.
2. Ehrman, R. N., Robbins, S. J., Childress, A. R., and O'Brien, C. P. (1992) Conditioned responses to cocaine-related stimuli in cocaine abuse patients. *Psychopharmacology* **107,** 523–529.
3. Childress, A. R., Mozley, P. D., McElgin, W., Fitzgerald, J., Reivich, M., and O'Brien, C. P. (1999) Limbic activation during cue-induced cocaine craving. *Am. J. Psychiatry* **156,** 11–18.
4. O'Brien, C. P. (1996) Recent developments in the pharmacotherapy of substance abuse. *J Consult. Clin. Psychol.* **64,** 677–686.

5. Robinson, T. E. and Becker, J. B. (1986) Enduring changes in brain and behavior produced by chronic amphetamine administration: a review and evaluation of animal models of amphetamine psychosis. *Brain Res. Rev.* **11,** 157–198.
6. Kalivas, P. W. and Stewart, J. (1991) Dopamine transmission in the initiation and expression of drug- and stress-induced sensitization of motor activity. *Brain Res. Rev.* **16,** 223–244.
7. Koob, G. F. (1988) Drugs of abuse: anatomy, pharmacology and function of reward pathways. *Trends Pharmacol. Sci.* **13,** 177–184.
8. Wise, R. A. and Rompre, P. P. (1989) Brain dopamine and reward. *Annu. Rev. Psychol.* **40,** 191–215.
9. Reid, M. S. and Berger, S. P. (1996) Evidence for sensitization of cocaine-induced nucleus accumbens glutamate release. *Neuroreport* **7,** 1325–1329.
10. Rothman, R., Baumann, M., Dersch, C., Romero, D., Rice, K., Carroll, F., et al. (2001) Amphetamine-type central nervous system stimulants release norepinephrine more potently than they release dopamine and serotonin. *Synapse* **39,** 32–41.
11. Kuczenski, R., 1983, Biochemical actions of amphetamine and other stimulants, in *Stimulants: Neurochemical, Behavioral and Clinical Perspective.* (Creese, I., ed.), Raven, New York, pp. 31–61.
12. Kelly, P. H. and Iversen, S. D. (1976) Selective 6-OHDA-induced destruction of mesolimbic dopamine neurons: abolition of psychostimulant-induced locomotor activity in rats. *Eur. Pharmacol.* **40,** 45–56.
13. Roberts, D. C. S. and Koob, G. F. (1982) Disruption of cocaine self-administration following 6-hydroxydopamine lesions of the ventral tegmental area in rats. *Pharmacol. Biochem. Behav.* **17,** 901–904.
14. Baker, D. A., Fuchs, R. A., Specio, S. E., Khroyan, T. V., and Neisewander, J. L. (1998) Effects of intraaccumbens administration of SCH-23390 on cocaine induced locomotion and conditioned place preference. *Synapse* **30,** 181–193.
15. Nestler, E. J. and Aghajanian, G. K. (1997) Molecular and cellular basis of addiction. *Science* **278,** 58–63.
16. White, F. J. and Kalivas, P. W. (1998) Neuroadaptations involved in amphetamine and cocaine addiction. *Drug Alcohol Depend.* **51,** 141–154.
17. LeMoal, M. and Simon, H. (1991) Mesocorticolimbic dopaminergic network: functional and regulatory roles. *Physiolog. Rev.* **71,** 155–234.
18. Sesack, S. R. and Pickel, V. M. (1990) In the medial nucleus accumbens, hippocampal and catecholaminergic terminals converge on spiny neurons and are in apposition to each other. *Brain Res.* **527,** 266–272.
19. O'Donnell, P. and Grace, A. A. (1995) Synaptic interactions among excitatory afferents to nucleus accumbens neurons: hipppocampal gating of prefrontal cortical input. *J. Neurosci.* **15,** 3622–3639.
20. Levine, M. S., Li, Z., Cepeda, C., Cromwell, H. C., and Altemus, K. L. (1996) Neuromodulatory actions of dopamine on synaptically-evoked neostriatal responses in slices. *Synapse* **24,** 65–78.
21. Kiyatkin, E. and Rebec, G. (1999) Striatal neuronal activity and responsiveness to dopamine and glutamate after selective blockade of D1 and D2 Dopamine receptors in freely moving rats. *J. Neurosci.* **19,** 3594–3609.
22. Pierce, R. C. and Kalivas, P. W. (1997) A circuitry model of the expression of behavioral sensitization to amphetamine-like psychostimulants, *Brain Res. Rev.* **25,** 192–216.
23. Grant, S., London, E. D., Newlin, D. B., Villemagne, V. L., Liu, X., Contoreggi, C., et al. (1996) Activation of memory circuits during cue-elicited cocaine craving, *Proc. Natl. Acad. Sci. (USA)* **93,** 12,040–12,045.
24. Breiter, H. C., Gollub, R. L., Weisskoff. R. M., Kennedy, D. N., Makris, N., Berke, J. D., et al. (1997) Acute effects of cocaine on human brain activity and emotion. *Neuron* **19,** 591–611.
25. Neisewander, J. L., Baker, D. A., Fuchs, R. A., Tran-Nguyen, L. T. L., Palmer, A., and Marshall, J. F. (2000) Fos protein expression and cocaine seeking behavior in rats after exposure to a cocaine self-administration environment. *Neuroscience* **20,** 798–805.
26. Jackson, D. M. and Westlind-Danielsson, A. (1994) Dopamine receptors: molecular biology, biochemistry and behavioural aspects. *Pharmacol. Ther.* **64,** 291–370.
27. Vallone, D., Picetti, R., and Borrelli, E. (2000) Structure and function of dopamine receptors. *Neurosci. Biobehav. Rev.* **24,** 125–132.
28. Nakanishi, S., Nakajima, Y., Masu, M., Ueda, Y., Nakahara, K., Watanabe, D., et al. (1998) Glutamate receptors: brain function and signal tansduction. *Brain Res. Rev.* **26,** 230–235.
29. Bigge, C. F. (1999) Ionotropic glutamate receptors. *Curr. Opin. Chem Biol.* **3,** 441–447.
30. Cartmell, J. and Schoepp, D. D. (2000) Regulation of neurotransmitter release by metabotropic glutamate receptors. *J. Neurochem.* **75,** 889–907.
31. Kalivas, P. W., Churchill, L., and Klitenick, M. A., (1993) The circuitry mediating the translation of motivational stimuli into adaptive motor responses, in *Limbic Motor Circuits and Neuropsychiatry* Kalivas, P. W. and Barnes, C. D., eds.), CRC Boca Raton, FL, pp. 237–287.
32. Mogenson, G. J., Brudzynski, S. M., Wu, M., Yang, C. R., and Yim. C. C. Y., (1993) From motivation to action: a review of dopaminergic regulation of limbic-nucleus accumbens-pedunculopontine nucleus circuitries involved in limbic-motor integration, in *Limbic Motor Circuits and Neuropsychiatry* Kalivas, P. W. and Barnes, C. D., eds), CRC P, Boca Raton FL, pp. 193–236.
33. Carr, D. B. and Sesack, S. R. (2000) GABA-containing neurons in the rat ventral tegmental area project to the prefrontal cortex. *Synapse* **38,** 114–123.

34. Sesack, S. R. and Bunney, B. S. (1989) Pharmacological characterization of the receptor mediating electrophysiological responses to dopamine in the rat medial prefrontal cortex: a microiontophoretic study. *J. Pharmacol. Exp. Ther.* **248**, 1323–1333.
35. Kelley, A. E., Domesick, V. B., and Nauta, W. J. H. (1982) The amygdalostriatal projection in the rat-an anatomical study by anterograde and retrograde tracing methods. *Neuroscience* **7**, 615–630.
36. Kelley, A. E. and Domesick, V. B. (1982) The distribution of the projection from the hippocampal formation to the nucleus accumbens in the rat. An anterograde and retrograde horseradish peroxidase study. *Neuroscience* **7**, 2321–2335.
37. Kelley, A. E., Winniock. M., and Stinus, L. (1986) Amphetamine, apomorphine and investigatory behavior in the rat: analysis of the structure and pattern of responses. *Psychopharmacology* **88**, 66–74.
38. Smiley, J. F. and Goldman-Rakic, P. S. (1993) Heterogeneous targets of dopamine synapses in monkey prefrontal cortex demonstrated by serial section electron microscopy: a laminar analysis using the silver-enhanced diaminobenzidine sulfide (SEDS) immunolabeling technique. *Cereb. Cortex* **3**, 223–238.
39. Carr, D. B. and Sesack, S. R. (1996) Hippocampal afferents to the rat prefrontal cortex: synaptic targets and relation to dopamine terminals. *J. Comp. Neurol.* **369**, 1–15.
40. Yang. C., Seamans, J., and Gorelova, N. (1999) Developing a neuronal model for the pathophysiology of schizophrenia based on the nature of electrophysiological actions of dopamine in the prefrontal cortex. *Neuropsychopharmacology* **21**, 161–194.
41. Carr, D. and Sesack, S. (2000) Projections from the rat prefrontal cortex to the ventral tegmental area: target specificity in the synaptic associations with mesoaccumebens and mesocortical neurons. *J. Neurosci.* **20**, 3864–3873.
42. O'Donnell, P., Greene, J., Pabello, N., Lewis, B. L., and Grace, A. A. (1999) Modulation of cell firing in the nucelus accumbens. *Ann. NY Acad. Sci.* **877**, 157–175.
43. Yim, C. Y. and Mogenson, G. J. (1988) Neuromodulatory action of dopamine in the nucleus accumbens: an in vivo intracellular study. *Neuroscience* **26**, 403–411.
44. Grace. A. A. (2000) Gating of information flow within the limbic system and the pathophysiology of schizophrenia. *Brain Res. Brain Res. Rev.* **31**, 330–341.
45. Hu, X. T. and White, F. J. (1996) Glutamate receptor regulation of rat nucleus accumbens neurons in vivo. *Synapse* **23**, 208–218.
46. Cepeda, C., Colwell, C. S., Itri, J. N., Chandler, S. H., and Levine, M. S. (1998) Dopaminergic modulation of NMDA-induced whole cell currents in neostriatal neurons in slices: contribution of calcium conductances. *J. Neurophysiol.* **79**, 82–94.
47. Hernandez-Lopez, S., Bargas, J., Surmeier, D. J., Reyes, A., and Galarraga, E. (1997) D1 receptor activation enhances evoked discharge in neostriatal medium spiny neurons by modulating an L-type $Ca^{2+}$ conductance. *J. Neurosci.* **17**, 3334–3342.
48. Chergui, K., Charlety, P. J., Akaoka, H., Saunier, C. F., Brunet, J.-L., Buda, M., et al. (1993) Tonic activation of NMDA receptors causes spontaneous burst discharge of rat midbrain dopamine neurons in vivo. *Eur. J. Neurosci.* **5**, 137–144.
49. Tong. Z.-Y., Overton, P. G., and Clark, D. (1996) Stimulation of the prefrontal cortex in the rat induces patterns of activity in midbrain dopaminergic neurons which resemble natural burst events. *Synapse* **22**, 195–208.
50. White. F. J. (1996) Synaptic regulation of mesocorticolimbic dopamine neurons. *Annu. Rev. Neurosci.* **19**, 405–436.
51. White, F. J., and Wang, R. Y. (1984) Pharmacological characterization of dopamine autoreceptors in the rat ventral tegmental area: microiontophoretic studies. *J. Pharmacol. Exp. Ther.* **231**, 275–280.
52. Wachtel, S. R., Hu, S.-T., Galloway, M. P., and White. F. J. (1989) D1 dopamine receptor stimulation enables the postsynaptic, but not autoreceptor effects of D2 dopamine agonists in nigrostriatal and mesoaccumbens dopamine systems. *Synapse* **4**, 327–346.
53. Bazhenov, M., Timofeev, I., Steriade, M., and Sejnowski, T. J. (1998) Cellular and network models for intrathalamic augmenting responses during 10-Hz stimulation. *J. Neurophysiol.* **79**, 2730–2748.
54. Timofeev, I., Grenier, F., and Steriade, M. (1998) Spike-wave complexes and fast components of cortically generated seizures. IV. Paroxysmal fast runs in cortical and thalamic neurons. *J. Neurophysiol.* **80**, 1495–1513.
55. Sesack, S. R., Deutch. A. Y., Roth, R. H., and Bunney, B. S. (1989) Topographical organization of the efferent projections of the medial prefrontal cortex in rat: an anterograde tract-tracing study with *Phaseolus vulgais* leucoagglutinin. *J. Comp. Neurol.* **290**, 213–242.
56. Peterson, S. L., Olsta, S. A., and Matthews, R. T. (1990) Cocaine enhances medial prefrontal cortex neuron response to ventral tegmental area activation. *Brain Res. Bull.* **24**, 267–273.
57. Pirot, S., Godbout, R., Mantz, J., Tassin, J. P., Glowinski, J., and Thierry, A. M. (1992) Inhibitory effects of ventral tegmental area stimulation on the activity of prefrontal cortical neurons: evidence for the involvement of both dopaminergic and GABAergic components. *Neuroscience* **49**, 857–865.
58. Carr, D. B., O'Donnell, P., Card, J. P., and Sesack, S. R. (1999) Dopamine terminals in the rat prefrontal cortex synapse on pyramidal cells that project to the nucleus accumbens. *J. Neurosci.* **19**, 11,049–11,060.
59. Lavin, A. and Grace, A. A. (2001) Stimulation of D1-type dopamine receptors enhances excitability in prefrontal cortical pyramidal neurons in a state-dependent manner. *Neuroscience* **104**, 335–346.
60. Groenewegen, H. J., Berendse, H. W., Wolters, J. G., and Lohman. A. H. (1990) The anatomical relationship of the prefrontal cortex with the striatopallidal system, the thalamus and the amygdala: evidence for a parallel organization. *Prog. Brain Res.* **85**, 95–116.

61. McDonald, A. J., Mascagni, F., and Guo, L. (1996) Projections of the medial and lateral prefrontal cortices to the amygdala: a *Phaseolus vulgaris* leucoagglutinin study in the rat. *Neuroscience* **71**, 55–76.
62. Rosenkranz, J. A. and Grace, A. A. (1999) Modulation of basolateral amygdala neuronal firing and afferent drive by dopamine receptor activation in vivo. *J. Neurosci.* **19**, 11,027–11,039.
63. Yamamoto, B. K. and Davy, S. (1992) Dopaminergic modulation of glutamate release in striatum as measured by microdialysis. *J. Neurochem* **58**, 1736–1742.
64. Kalivas, P. W. and Duffy, T. (1998) Repeated cocaine administration alters extracellular glutamate levels in the ventral tegmental area. *J. Neurochem.* **70**, 1497–1502.
65. Carter, A. J. and Muller, R. E. (1991) Pramipexole, a dopamine D2 autoreceptor agonist, decreases the extracellular concentration of dopamine in vivo. *Eur. J. Pharmacol.* **200**, 65–72.
66. You, Z. B., Herrera-Marschitz, M., Nylander, I., Goiny, M., O'Connor, W. T., Ungerstedt, U., et al. (1994) The striatonigral dynorphin pathway of the rat studied with in vivo microdialysis—II. Effects of dopamine D1 and D2 receptor agonists. *Neuroscience* **63**, 427–434.
67. Roth, R. H., Tam, S.-Y., Ida, Y., Yang, J.-X., and Deutch, A. Y. (1988) Stress and the mesocorticolimbic dopamine systems. *Ann. NY Acad. Sci.* **537**, 138–147.
68. Pennartz, C. M., Delleman-ven der Weel, M. J., Kitai, S. T., and Lopes da Silva, F. H. (1992) Presynaptic dopamine D1 receptors attenuate excitatory and inhibitory limbic inputs to the shell region of the rat nucleus accumbens studied in vitro. *J. Neurophysiol.* **67**, 1325–1334.
69. Nicola, S. M. and Kombian, S. B. (1996) Psychostimulants depress excitatory synaptic transmission in the nucleus accumbens via presynaptic D1-like dopamine receptors. *J. Neurosci.* **16**, 1591–1604.
70. Harvey, J. and Lacey, M. G. (1997) A postsynaptic interaction between dopamine D1 and NMDA receptors promotes presynaptic inhibition in the rat nucleus accumbens via adenosine release. *J. Neurosci.* **17**, 5271–5280.
71. Taber, M. T. and Fibiger, H. C. (1997) Activation of the mesocortical dopamine system by feeding: lack of a selective response to stress. *Neuroscience* **77**, 295–298.
72. Yim, C. Y. and Mogenson, G. J. (1980) Effect of picrotoxin and nipecotic acid on inhibitory response of dopaminergic neurons in the ventral tegmental area to stimulation of the nucleus accumbens. *Brain Res.* **199**, 466–472.
73. Hu, G., Duffy, P., Swanson, C., Behnam Ghasemzadeh, M., and Kalivas. P. W. (1999) The regulation of dopamine transmission by metabotropic glutamate receptors. *J. Pharmacol.* **289**, 412–416.
74. Xi, Z., Shen, H., Carson, D., Baker, D., and Kalivas, P. (2001) Inhibition of glutamate transmission by group II metabotropic glutamate receptors (mGluR2/3): enduring neuroadaptations by cocaine. *Soc. Neurosci. Abstr.* **27**, #997.3.
75. Kalivas, P. W. (1993) Neurotransmitter regulation of dopamine neurons in the ventral tegmental area. *Brain Res. Rev.* **18**, 75–113.
76. Chen, N. N. and Pan, W. H. (2000) Regulatory effects of D2 receptors in the ventral tegmental area on the mesocorticolimbic dopaminergic pathway. *J. Neurochem.* **74**, 2576–2582.
77. Cameron, D. L. and Williams, J. T. (1993) Dopamine D1 receptors facilitate transmitter release. *Nature* **366**, 344–347.
78. Kalivas, P. W. and Duffy, P. (1995) D1 receptors modulate glutamate transmission in the ventral tegmental area. *J. Neurosci.* **15**, 5379–5388.
79. Rosales, M. G., Martinez-Fong, D., Morales, R., Nunez, A., Flores, G., Gongora-Alfaro, J. L., et al. (1997) Reciprocal interaction between glutamate and dopamine in the pars reticulata of the rat substantia nigra: a microdialysis study. *Neuroscience* **80**, 803–810.
80. Karreman, M. and Moghaddam, B. (1996) The prefrontal cortex regulates the basal release of dopamine in the limbic striatum: an effect mediated by ventral tegmental area. *J. Neurochem.* **66**, 589–598.
81. Kretschmer, B. D. (1999) Modulation of the mesolimbic dopamine system by glutamate: role of NMDA receptors. *J. Neurochem.* **73**, 839–848.
82. Takahata, R. and Moghaddam, B. (1998) Glutamatergic regulation of basal and stimulus-activated dopamine release in the prefrontal cortex. *J. Neurochem.* **71**, 1443–1449.
83. Enrico, P., Bouma, M., de Vries, J. B., and Westerink, B. H. C. (1998) The role of afferents to the ventral tegmental area in the handling stress-induced increase in the release of dopamine in the medial prefrontal cortex: a dual-probe microdialysis study in the rat brain. *Brain Res.* **779**, 205–213.
84. Svensson, P. and Hurd, Y. L. (1998) Specific reductions of striatal prodynorphin and $D_1$ dopamine receptor messenger RNAs during cocaine abstinence. *Mol. Brain Res.* **56**, 162–168.
85. Taber, M. T., Das, S., and Fibiger, H. C. (1995) Cortical regulation of subcortical dopamine release: mediation via the ventral tegmental area. *J. Neurochem.* **65**, 1407–1410.
86. Takahata, R. and Moghaddam, B. (2000) Target-specific glutamatergic regulation of dopamine neurons in the ventral tegmental area. *J. Neurochem.* **75**, 1775–1778.
87. Kalivas, P. W., Duffy, P., and Barrow, J. (1989) Regulation of the mesocorticolimbic dopamine system by glutamic acid receptor subtypes. *J. Pharmacol. Exp. Ther.* **251**, 378–387.
88. Wedzony, K., Klimek, V., and Golembiowska, K. (1993) MK-801 elevates the extracellular concentration of dopamine in the rat prefrontal cortex and increases the density of striatal dopamine D1 receptors. *Brain Res.* **622**, 325–329.

89. Nishijima, K., Kashiwa, A., and Nishikawa, T. (1994) Preferential stimulation of extracellular release of dopamine in rat frontal cortex to striatum following competitive inhibition of the N-methyl-D-aspartate receptor. *J. Neurochem.* **63,** 375–378.
90. Verma, A. and Moghaddam, B. (1996) NMDA receptor antagonists impair prefrontal cortex function as assessed via spatial delayed alteration performance in rats: modulation by dopamine. *J. Neurosci.* **16,** 373–379.
91. Santiago, M., Machado, A., and Cano, J. (1993) Regulation of the prefrontal cortical dopamine release by $GABA_A$ and $GABA_B$ receptor agonists and antagonists. *Brain Res.* **630,** 28–31.
92. Abekawa, T., Ohmori, K., and Koyama, T. (2000) D1 dopamine receptor activation reduces extracellular glutamate and GABA concentrations in the medial prefrontal cortex. *Brain Res* **867,** 250–254.
93. Wolf, M. E. (1998) The role of excitatory amino acids in behavioral sensitization to psychomotor stimulants. *Prog. Neurobiol.* **54,** 679–720.
94. Vanderschuren, L. and Kalivas, P. (2000) Alterations in dopaminergic and glutamatergic signaling in the induction and expression of behavioral sensitization. *Psychopharmacology* **15,** 99–120
95. Mello, N. K. and Negus, S. S. (1996) Preclinical evaluation of pharmacotherapies for treatment of cocaine and opioid abuse using drug self-administration procedures. *Neuropsychopharmacology* **14,** 375–424.
96. White, F. J. and Wolf, M. E. (1991) Psychomotor stimulants, in *The Biological Basis of Drug Tolerance and Dependence* (Pratt J., ed.), Academic, London, pp. 153–197.
97. White, F. J. and Wang, R. Y. (1984) Electrophysiological evidence for A10 dopamine autoreceptor sensitivity following chronic d-amphetamine treatment. *Brain Res.* **309,** 283–292.
98. Gao, W. Y., Lee, T. H., King, G. R., and Ellinwood, E. H. (1998) Alterations in baseline activity and quinpirole sensitivity in putative dopamine neurons in the substantia nigra and ventral tegmental area after withdrawal from cocaine pretreatment. *Neuropsychopharmacology* **18,** 222–232.
99. Nestler, E. J., Terwilliger, R. Z., Walker, J. R., Sevarino, K. A., and Duman, R. S. (1990) Chronic cocaine treatment decreases levels of the G protein subunits $G_{i\alpha}$ and $G_{o\alpha}$ in discrete regions of rat brain. *J. Neurochem.* **55,** 1079–1082.
100. Striplin, C. and Kalivas, P. W. (1993) Robustness of G protein changes in cocaine sensitization shown with immunoblotting. *Synapse* **14,** 10–15.
101. Sorg, B. A., Chen, S.-Y., and Kalivas, P. W. (1993) Time course of tyrosine hydroxylase expression following behavioral sensitization to cocaine. *J. Pharmacol. Exp. Ther.* **266,** 424–430.
102. Lu, W. X. and Wolf, M. E. (1997) Expression of dopamine transporter and vesicular monoamine transporter 2 mRNAs in rat midbrain after repeated amphetamine administration. *Mol. Brain Res.* **49,** 137–148.
103. Shilling, P. D., Kelsoe, J. R., and Segal, D. S. (1997) Dopamine transporter mRNA is upregulated in the substantia nigra and the ventral tegmental area of amphetamine-sensitized rats. *Neurosci. Lett.* **236,** 131–134.
104. Kalivas, P. W. and Duffy, P. (1993) Time course of extracellular dopamine and behavioral sensitization to cocaine. I. Dopamine axon terminals. *J. Neurosci.* **13,** 266–275.
105. White, F. J., Hu, X.-T., Zhang, X.-F., and Wolf, M. E. (1995) Repeated admnistration of cocaine or amphetamine alters neuronal responses to glutamate in the mesoaccumbens dopamine system. *J. Pharmacol. Exp. Ther.* **273,** 445–454.
106. Zhang, X.-F., Hu, X.-T., White, F. J., and Wolf, M. E. (1997) Increased responsiveness of ventral tegmental area dopamine neurons to glutamate after repeated administration of cocaine or amphetamine is transient and selectively involves AMPA receptors. *J. Pharmacol. Exp. Ther.* **281,** 699–706.
107. Fitzgerald, L. W., Ortiz, J., Hamedani, A. G., and Nestler, E. J. (1996) Drugs of abuse and stress increase the expression of GluR1 and NMDAR1 glutamate receptor subunits in the rat ventral tegmental area: common adaptations among cross-sensitizing agents. *J. Neurosci.* **16,** 274–282.
108. Churchill, L., Swanson, C. J., Urbina, M., and Kalivas, P. W. (1999) Repeated cocaine alters glutamate receptor subunits levels in the nucleus accumbens and ventral tegmental area of rats that develop behavioral sensitization. *J. Neurochem.* **72,** 2397–2403.
109. Tong, Z.-Y., Overton, P. G., and Clark, D. (1995) Chronic administration of (+)-amphetamine alters the reactivity of midbrain dopaminergic neurons to prefrontal cortex stimulation in the rat. *Brain Res.* **674,** 63–74.
110. Hooks, M. S., Duffy, P., Striplin, C., and Kalivas, P. W. (1994) Behavioral and neurochemical sensitization following cocaine self-administration. *Psychopharmacology* **115,** 265–272.
111. Sorg, B. A., Davidson, D. L., Kalivas, P. W., and Prasad, B. M. (1997) Repeated daily cocaine alters subsequent cocaine-induced increase of extracellular dopamine in the medial prefrontal cortex. *J. Pharmacol. Exp. Ther.* **281,** 54–61.
112. Robinson, T. E. and Becker, J. B. (1982) Behavioral sensitization is accompanied by an enhancement in amphetamine-stimulated dopamine release from striatal tissue in vitro. *Eur. J. Pharmacol.* **85,** 253–254.
113. Pierce, R. C., Duffy, P., and Kalivas, P. W. (1995) Sensitization to cocaine and dopamine autoreceptor subsensitivity in the nucleus accumbens. *Synapse* **20,** 33–36.
114. Pierce, R. C. and Kalivas, P. W. (1997) Repeated cocaine modifies the mechanism by which ampehtamine releases dopamine. *J. Neurosci.* **17,** 3254–3261.
115. Iwata, S.-I., Hewlett, G. H. K., Ferrell, S. T., Kantor, L., and Gnegy, M. E. (1997) Enhanced dopamine release and phosphorylation of synapsin I and neuromodulin in striatal synaptosomes after repeated amphetamine. *J. Pharmacol. Exp. Ther.* **283,** 1445–1452.

116. Kantor, L. and Gnegy, M. (1998) $Ca^{2+}$, $K^+$ and calmodulim kinase II affect amphetamine-mediated dopamine release in sensitized rats. *FASEB* **12,** A159.
117. Goldman-Rakic, P. S. (1995) Cellular basis of working memory. *Neuron* **14,** 477–485.
118. Jentsch, K. and Taylor, J. (1999) Impulsivity resulting form frontostriatal dysfunction in drug abuse: implications for the control of behavior by reward-related stimuli. *Psychopharmacology* **146,** 373–390.
119. Pierce, R. C., Reeder, D. C., Hicks, J., Morgan, Z. R., and Kalivas, P. W. (1998) Ibotenic acid lesions of the dorsal prefrontal cortex disrupt the expression of behavioral sensitization to cocaine. *Neuroscience* **82,** 1103–1114.
120. Cornish, J. L. and Kalivas, P. W. (2000) Glutamate Transmission in the nucleus accumbens mediates relapse in cocaine addiction. *J. Neurosci. (Online)* **20,** RC89.
121. Cornish, J. L., Duffy, P., and Kalivas, P. W. (1999) a role of necleus accumbens glutamate transmission in the relapse to cocaine-seeking behavior. *Neuroscience* **93,** 1359–1368.
122. McFarland, K. and Kalivas, P. (2001) Circuitry mediating cocaine-induced reinstatement of drug-seeking behavior. *J. Neurosci.* **21,** 8655–8663.
123. Meil, W. M. and See, R. E. (1997) Lesions of the basolateral amygdala abolish the ability of drug associated cues to reinstate responding during withdrawal from self-administered cocaine. *Behav. Brain Res.* **87,** 139–148.
124. Fanselow, M. S. (2000) Contextual fear, gestalt memories, and the hippocampus. *Behav. Brain Res.* **110,** 73–81.
125. Grimm, J. and See, R. (2000) Dissociation of primary and secondary reward-relevant limbic nuclei in an animal model of relapse. *Neuropsychopharmacology* **22,** 473–479.
126. Bell, K., Duffy, P., and Kalivas, P. W. (2000) Context-specific enhancement of glutamate transmission by cocaine. *Neuropsychopharmacology* **23,** 335–344.
127. Rothman, R. B. and Glowa, J. R. (1997) A review of the effects of dopaminergic agents on humans, animals, and drug-seeking behavior, and Its implications for medication development. *Mol. Neurobiol.* **11,** 1–3.

# 10
## Glutamate Cascade from Metabotropic Glutamate Receptors to Gene Expression in Striatal Neurons
*Implications for Psychostimulant Dependence and Medication*

### John Q. Wang, MD, PhD, Limin Mao, MD, and Yuen-Sum Lau, PhD

## 1. INTRODUCTION

Glutamate is a major excitatory neurotransmitter in the central nervous system (CNS), which regulates a variety of neuronal activities through interaction with glutamate receptors. Glutamate receptors are divided into two major families: ionotropic (ligand-gated ion channels) and metabotropic (G-protein coupled) receptors based on their biochemical, pharmacological, and molecular profiles (1). Compared to thoroughly investigated roles of the ionotropic receptors, functional studies on metabotropic glutamate receptors (mGluRs) are just emerging in recent years. At present, eight subtypes of mGluRs (mGluR1–8) have been cloned from rat brain tissues. Like ionotropic receptors, mGluR subtypes are heterogeneous in their distribution, pharmacology, and connections with intracellular effectors. According to their sequence homology, pharmacology, and intracellular responses to activation of the mGluRs expressed in the *Xenopus* oocyte system, the eight subtypes are currently classified into three functional groups (2). Activation of group I mGluRs (mGluR1/5) increases phosphoinositide (PI) hydrolysis via stimulation of phospholipase C, resulting in the subsequent $Ca^{2+}$ release from internal stores and protein kinase C activation. Activation of group II (mGluR2/3) and group III (mGluR4/6/7/8) mGluRs inhibits the adenylyl cyclase and cAMP formation in an adenylate cyclase-dependent fashion. The linkages to diverse intracellular effectors, such as cAMP and $Ca^{2+}$, allow mGluRs to be vigorously involved in the slower effects in intracellular and intranuclear compartments, such as DNA transcription (gene expression), as opposed to rapid signal transmission in synapses mediated by ionotropic glutamate receptors.

For decades, the striatum has been the focus of experimental animal studies on brain mechanisms underlying behavioral properties of drugs of abuse. It was found that an increase in the release of dopamine from mesolimbic and mesostriatal pathways initiates changes in locomotor behaviors following administration of the psychostimulants cocaine and amphetamine. Recent molecular studies reveal that these dopamine stimulants induce gene expression in striatal medium-sized spiny neurons, which project to the substantia nigra and pallidal area. Among inducible genes, the immediate early genes (IEG), c-*fos* and *zif/268*, and the neuropeptide genes, preprodynorphin (PPD), substance P (SP), and preproenkephalin (PPE), have been investigated most extensively. Alterations in these gene expression are thought to participate in adaptive changes in cellular physiology (neuroplasticity) related to behavioral responsiveness to subsequent drug exposures.

Although dopamine transmission has been documented as a prime mediator of inducible gene expression in the striatum, increasing evidence suggests an equally important role of glutamate system

From: *Contemporary Clinical Neuroscience: Glutamate and Addiction*
Edited by: Barbara H. Herman et al. © Humana Press Inc., Totowa, NJ

in drug actions *(3–5)*. The striatum is innervated with abundant glutamatergic projections from widespread areas of the forebrain. MGluRs are densely expressed in the striatum both presynaptically and postsynaptically. Available data from functional studies show that pharmacological and genetic manipulations of mGluR activity alter local transmitter release and behaviors in normal and drug-treated animals *(6)*. In particular, activation of mGluRs seems to be essential for amphetamine-stimulated striatal gene expression *(3)*. This indicates that the contribution of mGluRs to the development of neuroplasticity is important for addictive action of psychostimulants. Thus, as a powerful modulator of psychostimulant effects, mGluRs are considered to be promising targets for the development of novel therapeutic agents for neurologic disorders derived from abused drugs.

This chapter reviews, first, the role of mGluRs in the regulation of behavioral and particularly genomic responses in striatal neurons to dopamine stimulation, following a summary of anatomical organization of striatal mGluRs. The possible presynaptic and postsynaptic mechanisms that process mGluR modulatory effects are then discussed in detail. Finally, a possible potential of mGluRs as targets for the development of therapeutic drugs for addiction treatment concludes this chapter.

## 2. ORGANIZATION OF mGluR GLUTAMATERGIC TRANSMISSION IN THE STRIATUM

The striatum represents one of the subcortical areas that is innervated with the highest density of glutamatergic projections, using glutamate as a releasing transmitter *(90)*. From widespread areas of cerebral cortex and thalamus, glutamatergic afferents project to the dorsal striatum (caudoputamen). The areas that send glutamatergic projections to the ventral striatum (nucleus accumbens) include the amygdala, thalamus, ventral subiculum, hippocampus, and prefrontal cortex. Extrinsic glutamatergic terminals make asymmetrical (excitatory) synaptic contacts with intrinsic striatal neurons, including medium-sized spiny projection neurons (striatonigral and striatopallidal neurons) *(7–10)* and aspiny interneurons (GABAergic or cholinergic neurons) *(11,12)*. Because the projection neurons are also major synaptic targets of dopaminergic terminals from the midbrain *(13,14)*, functional activity of a single striatal projection neuron can be modulated postsynaptically by both glutamate and dopamine inputs. Further evidence supporting this is that glutamatergic and dopaminergic terminals are found to directly converge on the dendritic spines of the same striatal output neurons *(9,15)*. In sharp contrast to popular synapses between extrinsic terminals and intrinsic neurons, axoaxonic synapses between the two major incoming terminals are hardly visible throughout the entire striatum *(9,13,15–17)*. This restricts the possibility of presynaptic interactions between glutamate and dopamine terminals in a classic synaptic fashion. However, effective interactions between the two terminals can still take place through nontraditional synaptic contacts *(9,17)*. Through diffusion of the transmitter away from the synapse where it is released, dopamine terminal function can be modified presynaptically by interaction of diffused glutamate with extrasynaptic heteroreceptors, and vice versa.

Parallel with ample glutamatergic afferents, mGluRs are densely found in the striatum. Extensive morphological studies have been conducted in recent years to define *(1)* presynaptic versus postsynaptic localization of mGluRs, (2) localization of mGluRs on specific phenotype of striatal neurons or incoming terminals, (3) colocalization of mGluRs with other receptors of interest, and (4) subtype-specific localization. Although the investigation of these issues is far from complete, final elucidation of them is prerequisite to evaluate which subtype of mGluRs on which phenotype of intrinsic neurons or extrinsic terminals is involved in which specific functional activity. Quantitative receptor autoradiography reveals high levels of the mGluR-binding site in the striatal region *(18)*. Because a lesion of corticostriatal projections has a little effect on mGluR-binding quantity, 90% of mGluRs are thought to locate on postsynaptic striatal neurons *(19)*. Studies with *in situ* hybridization *(20–23)* and immunocytochemistry *(24–26)* show the presence of mGluR1–5 and mGluR7 in rat striatal neurons with high levels of mGluR3/5/7 and low to moderate levels of mGluR1/2/4 *(23,27,28)*. mGluR1/3/5/7 are colocalized with the vast majority (60–70%) of either striatonigral neurons or striatopallidal neurons *(29–32)*. Interestingly, there seemingly exists an mGluR1/5 segregation between two major projection neurons: MGluR1 is primarily present in striatonigral neurons, whereas mGluR5 is present in striatopallidal neurons *(29,31)*. mGluR2 is only expressed in a small population of large polygonal neurons, likely cholinergic interneurons *(23)*. With lesion or double-label immunohistochemistry with presynaptic terminal markers, mGluR2/3/4/7, but not mGluR1/5, subtypes are present on incoming glutamatergic and/or dopaminergic terminals *(28,32–34)*. Although it is not yet conclusive, group I mGluRs are considered predominantly postsynaptic, whereas group II/III mGluRs are primarily presynaptic in the striatum.

## 3. REGULATION OF SPONTANEOUS AND DOPAMINE-DEPENDENT MOTOR ACTIVITIES BY mGluRs

Studies on physiological roles of striatal mGluRs started with behavioral investigation. Unlike the ionotropic glutamate receptor agonists that cause controversial effects on motor activity when injected into the striatum, the mGluR agonists in early studies seem to consistently induce stimulation of spontaneous activity. Sucaan et al. *(35,36)* first reported in rats that acute injection of a nonsubtype-selective mGluR agonist, ACPD, into the unilateral dorsal striatum at a high dose range (500–2000 nmol) caused rotation contralateral to the injection side. Similar findings were generally repeated afterward in other laboratories *(37–40)*. Because these behavioral changes are blocked by a nonsubtype-selective mGluR antagonist, MCPG, but not by the ionotropic glutamate receptor antagonists, selective activation of mGluRs is believed to mediate the ACPD effect. Increased locomotion is also observed after bilateral ACPD injection into the nucleus accumbens *(41,42)*.

The recently developed subtype-specific agents with confirmed selectivity and effectiveness in vivo *(43)* have greatly facilitated pharmacological studies to evaluate the subtype-specific role of mGluRs. Several studies using these agents indicate that behavioral stimulation by ACPD is mediated via selective activation of group I, but not group II/III mGluRs. Infusion of a selective group I agonist, 3,5-dihydroxyphenylglycine (DHPG), into either the dorsal or ventral striatum at moderate doses (20–80 nmol) induces hyperlocomotion and characteristic stereotypical behaviors *(44,45)*. Such behavioral responses are sensitive to a selective group I antagonist, *N*-pheny1-7-(hydroxyimino)cyclopropa *[b]* chromen-1a-carboxamide (PHCCC), but not to a selective group II/III antagonist, (*RS*)-α-methylserine-*O*-phosphate monophenyl ester (MSOPPE) *(44,45)* or to the antagonists for NMDA and kainate/AMPA receptors *(46)*. Inhibition of intracellular $Ca^{2+}$ release is also effective in blocking behavior induced by DHPG *(45)*. In contrast to stimulative effects of DHPG, intrastriatal infusion of a selective group II agonist, (2*S*,2′*R*,3′*R*)-2-(2′,3′-dicarboxycyclopropy1) glycine (DCG-IV), or a selective group III agonist, L- 2-amino-4-phosphonobutyrate (L-AP4), caused no change in motor activity or sedation *(44,47,48)*.

Although dopamine receptors (both $D_1$ and $D_2$) have been thought to be directly responsible for behavioral stimulation induced by dopamine stimulants, several studies in recent years implicate mGluRs in the modulation of acute cocaine-and amphetamine-stimulated motor behaviors in a subtype-specific fashion. Activation or blockade of group I mGluRs seems to have a minimal effect on acute amphetamine-stimulated behaviors (Zhou and Wang, unpublished observations). Thus, even though the group I agonist itself causes motor stimulation, dopamine stimulants and the group I agonist are believed to induce acute motor stimulation via different mechanisms. The group II agonists show different effects on amphetamine action in the two reports. Schoepp's group did not observe significant effects of subcutaneous injection of the systemically active group II agonists, LY354740 and LY379268, on amphetamine-evoked motor activity *(49)*. In contrast, we *(47)* found a significant attenuation of amphetamine-stimulated behavior following injection of the group II agonist DCG-IV directly into the dorsal striatum, similar to the result seen after MCPG injection into the nucleus accumbens *(50)*. The differences in results among the above studies are likely due to different experimental conditions and agonists employed. Especially, systemically administered LY compounds have a widespread influence on all centrally located group II receptors, and, as a result, the final outcome of motor activity may not reflect as a change caused by selective activation of group II receptors in the striatum. Intracaudate injection of a group III agonist (L-AP4) also blocked hyperlocomotion induced by cocaine, amphetamine, or apomorphine (48). The behavioral activity induced by cocaine was much more sensitive to L-AP4 than that induced by amphetamine or apomorphine. At 100 nmol, L-AP4 completely blocked cocaine effect, whereas amphetamine- and apomorphine-stimulated behaviors were blocked only by 28% and 31%, respectively. The different susceptibility of these stimulants to L-AP4 blockade may be related to their different terminal characteristics resulting in the enhancement of dopamine release. Cocaine increases extracellular dopamine levels by blocking the dopamine transporter (51). Therefore, when dopamine release is inhibited by L-AP4 (52,53), the capacity of cocaine to increase extracellular dopamine is diminished. Unlike cocaine, amphetamine increases extracellular dopamine concentrations primarily through a direct stimulation of dopamine release from terminals in a $Ca^{2+}$-independent fashion (54). Thus, L-AP4 that affects cAMP pathway may have limited effects on the $Ca^{2+}$-independent dopamine release induced by amphetamine (53,55).

In addition to the regulation of acute motor stimulation, mGluRs might be involved in the regulation of behavioral sensitization to repeated stimulant exposure. Based on the linkages to multiple second-messenger systems and prominent roles in modifying long-lasting neuroadaptive activities, mGluRs are particularly possible to regulate neuronal sensitization. Indeed, one report found that MCPG injected into the ventral tegmental area blocks the induction of amphetamine sensitization (56). Most recently, an increase in mGluR1 and a decrease in mGluR5 mRNA levels are found in the striatum of rats that developed behavioral sensitization to repeated administration of amphetamine (57). The altered gene expression may substrate changes in receptor functions important for initiation and/or expression of sensitization. Apparently, the study on the roles of mGluR in processing chronic drugs effects (sensitization, dependence, rewarding effects, self-administration, addiction, etc.) is still at its infant stage and remains as a promising avenue for future elucidation of receptor mechanisms underlying drug abuse.

## 4. REGULATION OF CONSTITUTIVE STRIATAL GENE EXPRESSION BY mGluRs

A particularly interesting role that mGluRs may play in regulating the overall neuronal function is the modulation of intracellular metabotropic activity, such as gene expression. Bridged by multiple second-messenger systems, mGluR stimulation may modulate transcription rate of target DNAs, a biochemical process called stimulus-transcription coupling. The genomic responses to mGluR stimulation can serve as an integral component of the molecular/cellular mechanisms underlying neuronal plasticity. Thus, investigation of mGluR regulation of gene expression may provide valuable insight into the formation of striatal neuroplasticity related to addictive properties of drugs.

Immediate early genes (IEGs) have been prime targets during the last decade in exploring receptor-mediated gene expression because they are readily inducible in response to various physiological and pharmacological stimuli (58). The first experiment that investigated mGluRs regulation of striatal IEG expression was carried out in primary cultures of rat striatal neurons (59). In that study, enhancement of mGluR activity by ACPD perfusion elevated basal levels of IEG c-*fos* mRNA, indicating a positive linkage between mGluRs and constitutive c-*fos* gene expression. A recent study performed in vivo established an excellent similarity (60). Striatal mGluRs stimulation by local ACPD injection elevates c-*fos* as well as another IEG *zif/268* mRNA expression in the rat striatum, which is sensitive to MCPG, but not to the antagonists for NMDA (CPP) or dopamine D1 receptors (SCH-23390). This confirms an existence of positive stimulus-transcription coupling between mGluRs and IEG expression in striatal neurons in vivo.

One of the most noticeable IEG roles in mature neurons is to exercise as a third messenger in the stimulus-transcription cascade to initiate the so-called late-response gene expression *(61)*. In this scenario, IEGs are rapidly induced via their own stimulus-transcription coupling mechanisms. Induced IEGs, in turn, function as powerful transcription factors to regulate expression of many other genes, which usually last longer and more directly contribute to long-lasting alterations in cellular physiology.

Indeed, acute injection of ACPD into the dorsal striatum dose-dependently elevates striatonigral PPD/SP and striatopallidal PPE mRNA levels in the rat striatum *(62)*. The increase in opioid peptide mRNA expression is blocked by MCPG. Thus, striatal opioid peptide expression, like the two IEG inductions, is also positively linked to mGluR activation. Moreover, compared to rapid and transient induction of c-*fos* and *zif/268* expression, which usually peaks 1 h and recovers 3 h after mGluR stimulation *(60)*, neuropeptide induction shows a more delayed and prolonged dynamic pattern, as it is evident at 2 or 3 h, and lasts more than 10 h, after ACPD injection *(62)*. The temporal pattern of IEG and opioid induction appears to indicate that the early induced IEG transcription factors trigger, although may not maintain, the subsequent opioid induction. In support of this, intracaudate injection of c-*fos* antisense oligonucleotides, which knocks down Fos protein quantity, reduces dynorphinlike immunoreactivity induced by dopamine stimulation *(63)*.

The regulation of neuropeptide mRNA expression by mGluRs is also subgroup-specific. For example, DHPG is more potent than ACPD in inducing PPD, SP, and PPE mRNA expression in the striatum *(64)*. Moreover, the group I antagonist PHCCC, which alone has no effect on the expression of three mRNA levels, attenuates DHPG-stimulated PPD, PPE, and, to a lesser extent, SP expression. Thus, a positive glutamatergic tone seems to exist on the group I mGluRs, which facilitates neuropeptide gene expression, although it is not tonically active. In contrast to group I receptors, group II/III receptors may be negatively linked to gene expression. However, this notion remains to be proven experimentally.

Presynaptic and postsynaptic mechanisms underlying DHPG stimulation of gene expression are not well understood with limited data so far obtained from a rather *complex* in vivo animal model. It is first assumed that exogenous DHPG directly stimulates the group I mGluRs on the projection neurons because this group of mGluRs seems to be deliberately arranged as postsynaptic receptors in striatal region *(65)*. In support of this, ACPD and DHPG increase PPD/PPE mRNA expression in dissociated striatal neurons cultured from rat embryos or neonatal rat pups *(66)*. Alternatively, DHPG could presynaptically alter transmitter release, which, in turn, stimulates striatal gene expression. Infusion of ACPD, usually at high concentrations (millimolar range), increases extracellular levels of dopamine *(67–70)*. The increase in dopamine release seems due to activation of group I, but not group II/III, mGluRs because DHPG increases, whereas DCG-IV and L-AP4 decrease, extracellular levels of striatal dopamine *(52,53,71)*. However, dopamine $D_1$ receptor blockade does not affect DHPG-stimulated behaviors *(45)* or ACPD-stimulated IEG expression *(60)*. Thus, dopamine release, if there is any, is not a critical component for orchestrating DHPG-induced gene expression.

In addition to dopamine, altered glutamate release could be another presynaptic element contributing to DHPG-induced gene expression. Infusion of ACPD at high concentrations or the group I agonists facilitates striatal glutamate release in conscious or anesthetized rats *(72–74)*. The group II/III agonists, on the other hand, suppress glutamate release *(75,76)*. It is possible that DHPG increases glutamate release that, in turn, interacts with postsynaptic group I mGluRs to alter gene expression. Additionally, the released glutamate can activate ionotropic glutamate receptors to regulate gene expression. However, in the recent attempts to evaluate the relative importance of ionotropic receptors in this scenario, we found that the NMDA and kainate/AMPA antagonists are ineffective to alter DHPG stimulation of motor activity *(46)* and gene expression *(60)*. This argues against the participation of ionotropic glutamate receptors in the DHPG effects.

Intracellularly, activation of group I mGluRs increases the PI hydrolysis, which gives rise to diacylglycerol and inositol 1,4,5-trisphosphate, as illustrated in Fig. 1. The former activates protein kinase C

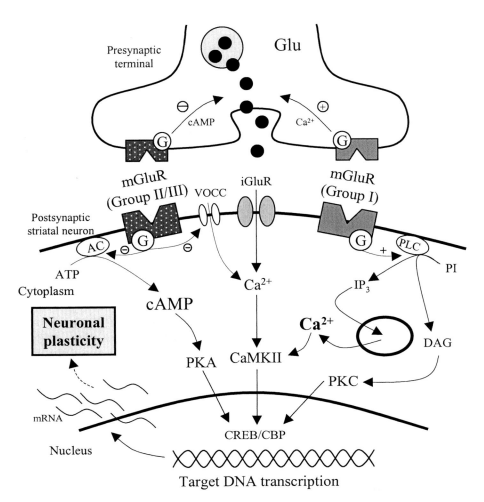

Fig. 1. Schematic illustration of the roles of mGluRs subtypes in the regulation of gene expression related to neuronal plasticity. Postsynaptically, group I receptors enhance intracellular $Ca^{2+}$ levels via the release of $Ca^{2+}$ from the intracellular store, as opposed to $Ca^{2+}$ influx through voltage-operated $Ca^{2+}$ channels (VOCC) or ionotropic glutamate receptors (iGluR) that are $Ca^{2+}$ permeable—in particular, NMDA receptors. Signal $Ca^{2+}$ facilitates gene expression through activation of $Ca^{2+}$/calmodulin-dependent protein kinase II (CaMKII). Alternatively, group I receptors would facilitate gene expression via the diacylglycerol (DAG)/protein kinase C (PKC) pathway. Group II/III receptors, in contrast, suppress gene expression by inhibiting the cAMP/protein kinase A (PKA) pathway, the well-known pathway to induce gene expression in response to $D_1$ dopamine receptor stimulation. Presynaptically, group I receptors augment glutamate (Glu) release, whereas group II/III receptors inhibit glutamate release, through which mGluRs can indirectly regulate gene expression in postsynaptic striatal neurons. AC, adenylyl cyclase; CBP, CREB-binding protein; CREB, cAMP response element-binding protein; G, G-protein; $IP_3$, inositol 1,4,5-triphosphate; PI, phosphoinositide; PLC, phospholipase C.

(PKC), whereas the latter release $Ca^{2+}$ into the cytoplasm from intracellular $Ca^{2+}$ stores. The elevated $Ca^{2+}$ activates $Ca^{2+}$/calmodulin-dependent protein kinase II (CaMKII), which, in turn, phosphorylates the transcription factor, cAMP response element-binding protein (CREB). The phosphorylated CREB after dimerizing with CREB-binding protein (CBP) eventually triggers target DNA transcription by interacting with the specific site in the promoter region of DNAs in a sequence-specific manner.

Hence, $Ca^{2+}$–CaMKII–CREB–CBP forms an effective cascade to transmit the $Ca^{2+}$ signal into the gene expression *(77,78)*. Activation of group I mGluRs could also stimulate gene expression through the diacylglycerol-dependent PKC pathway. MGluR stimulation in cultured striatal neurons causes a translocation of PKC from a cytoplasmic (soluble) form, prevalent under resting conditions, to a membrane-bound form, a mode of activated PKC *(79)*, and concomitant IEG induction, which is sensitive to PKC inhibitors *(59)*.

## 5. REGULATION OF DOPAMINE-DEPENDENT STRIATAL GENE EXPRESSION BY mGluRs

Cocaine and amphetamine stimulate striatal gene expression via a well-known $D_1$/cAMP/PKA pathway. This pathway may crosstalk with the group I mGluR-associated $Ca^{2+}$/CaMKII pathway to produce effective stimulation on gene expression. Coexpression of group I mGluRs and $D_1$ dopamine receptors on the same striatal neurons *(65)* provides anatomical basis for the crosstalks. One study in rats found that the pharmacological blockade of mGluRs with MCPG attenuates acute amphetamine-stimulated PPD and PPE mRNA expression *(80)*, indicating an mGluR dependency of dopamine-stimulated gene expression. The group I mGluRs may be responsible for this dependency because the group I agonist mimics the stimulative effects of amphetamine on PPD/PPE expression (see above). In mutant mice deficient in mGluR1 subtype, acute amphetamine produces significantly less PPD, but not SP, mRNA induction as compared to that in the wild-type mice *(81)*. Thus, mGluR1 regulation of dopamine-dependent gene expression is specific to PPD. This specific regulation seems to closely echo the anatomical observation showing segregative expression of mGluR1 in PPD-containing striatonigral neurons *(29)*. It is currently unclear whether mGluR5 that is primarily expressed by PPE-containing neurons can preferentially regulate PPE expression, a goal of our future study with mGluR5 knockout mice. In addition, the notable finding that group I mGluRs participate in gene induction, but not in motor stimulation (see above), in response to acute administration of amphetamine implies a preferential involvement of this group of mGluRs in the regulation of stimulated intracellular activity (gene induction). Thus, group I mGluRs can be considered as prime receptors important for gene expression-related neuroplasticity.

Little is known, at present, about the roles of group II/III mGluRs in the regulation of dopamine-dependent gene expression. Through the inhibition of adenylyl cyclase, group II/III receptors are expected to confine excitatory responses of the cAMP/PKA pathway to $D_1$ stimulation and thus limit dopamine-stimulated gene expression. However, no attempt has been made to detect this issue, even though this issue is obvious important for evaluation of potential clinical use of group II/III agents for the treatment of drug addiction.

Overall, studies concerning the mGluRs regulation of constitutive and dopamine-stimulated gene expression in striatal neurons are limited so far. Among numerous genes that could be inducible by stimulants, only two sets of genes (i.e., IEGs and opioid peptides) have been investigated in relation to mGluRs' activity. As to the significance of these gene activities in mediating long-term drug effects, induced IEGs are probably involved in the facilitation of late-response genes, such as the opioid genes, as described above. Opioid peptide induction may be more directly involved in resetting the responsiveness of striatal neurons to subsequent stimulant administration. It has been suggested that induced dynorphin suppresses, whereas induced enkephalin facilitates, repeated drug actions. This is supported by data from extensive behavioral studies showing that the dynorphin (κ) receptor agonists and enkephalin (δ) receptor antagonists attenuate, whereas the κ antagonists and δ agonists augment, chronic behavioral activities (sensitization, reinforcement, rewarding effects, discriminative stimulus effects, etc.) induced by cocaine, amphetamine, or morphine *(82–84)*. Overexpression of CREB and thus dynorphin in the nucleus accumbens decreases the rewarding behavior of cocaine *(85)*. These data support the notion that mGluR1-sensitive dynorphin gene induction contributes to inhibition, whereas mGluR5-sensitive enkephalin gene induction, contributes to facilitation, of chronic drug actions (Fig. 2). Besides opioid

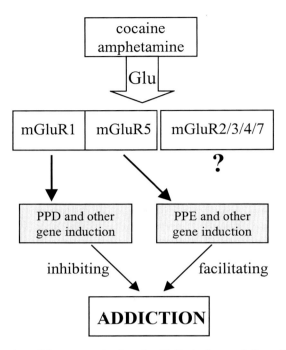

**Fig. 2.** Schematic illustration of the proposed roles of mGluRs in the regulation of PPD and PPE gene expression in striatal neurons induced by psychostimulants, cocaine, and amphetamine. MGluR1 and mGluR5 distinctively contribute to gene induction responsible for inhibition and facilitation of drug action, respectively. The importance of other mGluR subtypes in the regulation of inducible gene expression remains to be determined experimentally.

genes, more studies are needed to explore the regulation of other potential gene expression in the forebrain by mGluRs, especially by group II/III receptors, following acute as well as chronic drug administration. Taken together, inducible genes can be tentatively grouped into two sets of genes, i.e., genes responsible for inhibition and facilitation of chronic drug actions (Fig. 2). Elucidation of these gene activities and their relationship to mGluRs in response to drug administration will provide invaluable information for therapeutic agent development by targeting at mGluRs.

## 6. MGluRs AS THERAPEUTIC TARGETS

mGluRs have been proposed to be novel targets for the development of therapeutic drugs for a variety of neurologic disorders, such as epilepsy, ischemia, and several types of neurodegenerative diseases (86,87). With regard to the neuropsychiatric disorders stemming from substance abuse, mGluRs can also be worthwhile targets. Because current research has clearly indicated that mGluRs play an integral role in regulating striatal neuron functions and, especially, gene expression in the presence of various abused substances (alcohol, morphine, nicotine, and psychostimulants), development of agents that interact with mGluRs could provide novel therapeutic tools for drug-related illnesses. For example, mGluR5 participates in the mediation of drug-induced gene expression that tends to facilitate chronic drug action. The development of agents that selectively block mGluR5 and thus facilitatory gene induction can then have a potential to abate chronic effects of drugs. On the other hand, mGluR1 mediates the induction of a different set of genes that tend to inhibit chronic drug action. Therefore, agents that increase mGluR1 activity can achieve a same effect as mGluR5 blockers in preventing drug abuse. Similarly, agents that can modify group II/III receptor activity can also be considered when the roles of these subtypes in the regulation of gene induction are investigated thoroughly. In addition to their roles in gene expression, mGluRs are also known to be powerful presynaptic modulators for drug-stimulated dopamine and glutamate release. The modulation is subgroup-specific, as the group I agonists facilitate, whereas the group II/III agonists suppress, the stimulated release of dopamine and glutamate. Therefore, agents that act presynaptically to modify mGluR subgroup activity can be designed to control the harmful excessive release of dopamine and/or glutamate. Ideally, agents that inhibit presynaptic group I and activate group II/III mGluRs (group I antagonist with group II/III agonist activity) can be developed to maximally limit stimulated dopamine/glutamate release. There are clear advantages to target mGluRs rather than ionotropic glutamate receptors for the sake of medication. MGluRs, as the modulatory receptors in nature, have little impact on fast synaptic signal transmission. This is probably due to that mGluRs localize at the periphery of the postsynaptic membrane as opposed to the "core" localization of ionotropic receptors in the synapse (88,89). Thus, mGluR agents should have minimal impact on normal synaptic function, which assures them that no general depression or cognitive side effects usually associated with the therapeutic use of the ionotropic glutamate receptor antagonists, especially in chronic therapy, will occur. Additionally, mGluR-targeted drugs should have no major peripheral adverse effects because mGluRs are not present in target organs of the autonomic nervous system. Along with rapid progresses in improving brain-blood barrier penetration, subtype specificity, and other in vivo properties, mGluR agents would be expected to offer significant usefulness in clinical treatment of the disorders relevant to drugs of abuse in the future.

In summary, as an important element of glutamatergic transmission enriched in striatal structure, mGluR activity is linked to many aspects of drug actions. Initial studies reveal that mGluRs effectively modify the typical motor behaviors and transmitter release induced by cocaine or amphetamine. Recent experiments demonstrate an mGluR-sensitive stimulus-transcription coupling in the regulation of IEG and neuropeptide gene expression under normal and drug-treated conditions. The regulation of drug actions by mGluRs is subtype-specific. Usually, a dual modulation of motor activity, transmitter release, and gene expression exists in response to drug administration, depending on the mGluR subtype involved. MGluR involvement in drug actions suggests a novel avenue to develop effective therapeutic agents for drug-abuse treatment by targeting at mGluRs. Because nuclear gene expression lies in heart of signal transduction, the development of small molecules that affect gene expression via influencing activities of mGluRs and associative signal transduction pathways represents a novel gene therapy approach to a variety of neurologic disorders induced by abused drugs.

## ACKNOWLEDGMENTS

The studies of the author mentioned in this chapter were supported by grants (DA10355 and MH61469) from the NIH and a UMRB grant from the University of Missouri.

## REFERENCES

1. Monaghan, D., Bridges, R., and Cotman, C. (1989) The excitatory amino acid receptors: their classes, pharmacology, and distinct properties in the function of the central nervous system. *Annu. Rev. Pharmacol. Toxicol.* **29**, 365–402.
2. Conn, P. J. and Pin, J. P. (1997) Pharmacology and functions of metabotropic glutamate receptors. *Annu. Rev. Pharmacol. Toxicol.* **37**, 205–237.
3. Wang, J. Q. and Mao, L. (1999) Pharmacological regulation of striatal gene expression by metabotropic glutamate receptors. *Acta Pharmacol. Sin.* **20**, 577–584.
4. Wang, J. Q. and McGinty, J. F. (1996) Glutamatergic and cholinergic regulation of immediate early gene and neuropeptide gene expression in the striatum, in *Pharmacological Regulation of Gene Expression in the CNS* (Merchant, K. M., ed.), CRC, Boca Raton, FL, pp. 81–113.
5. Wang, J. Q. and McGinty, J. F. (1999) Glutamate/dopamine interactions mediate the effects of psychostimulant drugs. *Addict. Biol.* **4**, 141–150.
6. Vezina, P. and Kim, J. H. (1999) Metabotropic glutamate receptors and the generation of locomotor activity: Interaction with midbrain dopamine. *Neurosci. Biobehav. Rev.* **23**, 577–589.

7. Dube, L., Smith, A. D., and Bolam, J. P. (1988) Identification of synaptic terminals of thalamic or cortical origin in contact with distinct medium size spiny neurons in the rat neostriatum. *J. Comp. Neurol.* **267,** 455–471.
8. Frotscher M., Rinner U., Hassler R., and Wagner A. (1981) Termination of cortical afferents on identified neurons in the caudate nucleus of the cat: a combined Golgi-electron microscope degeneration study. *Exp. Brain Res.* **41,** 329–337.
9. Smith, Y., Bennett, B. D., Bolam, J. P., Parent, A., and Sadikot, A. F. (1994) Synaptic relationships between dopaminergic afferents and cortical or thalamic input in the sensorimotor territory of the striatum in monkey. *J. Comp. Neurol.* **344,** 1–19.
10. Somogyi, P., Bolam, J. P., and Smith, A. D. (1981) Monosynaptic cortical input and local axon collaterals of identified striatonigral neurons. A light and electron microscopic study using the Golgi–peroxidase transport–denegeration procedure. *J. Comp. Neurol.* **195,** 567–584.
11. Lapper, S. R. and Bolam, J. P. (1992) Input from the frontal cortex and the parafascicular nucleus to cholinergic interneurons in the dorsal striatum of the rat. *Neuroscience* **51,** 533–542.
12. Meredith, G. E. and Wouterlood, F. G. (1990) Hippocampal and midline thalamic fibres and terminals in relation to the choline acetyltransferase-immunoreactive neurons in nucleus accumbens of the rat: a light and electron microscopic study. *J. Comp. Neurol.* **296,** 204–216.
13. Freund, T. F., Powell, J. F., and Smith, A. D. (1984) Tyrosine hydroxylase–immunoreactive boutons in synaptic contact with identified striatonigral neurons, with particular reference to dendritic spines. *Neuroscience* **13,** 1189–1197.
14. Pickel, V. M., Chan, J., and Sesack, S. R. (1992) Cellular basis for interactions between catecholaminergic afferents and neurons containing Leu-enkephalin-like immunorecractivity in rat caudate-putamen nuclei. *J. Neurosci. Res.* **31,** 212.
15. Bouyer, J. J., Park, D. H., Joh, T. H., and Pickel, V. M. (1984) Chemical and structural analysis of the relation between cortical inputs and tyrosine hydroxylase-containing terminals in rat neostriatum. *Brain Res.* **302,** 267–274.
16. Sesack, S. R. and Pickel, V. M. (1990) In the rat medial nucleus accumbens, hippocampal and catecholaminergic terminals converge on spiny neurons and are in apposition to each other. *Brain Res.* **527,** 266–272.
17. Totterdell, S. and Smith, A. D. (1989) Convergence of hippocampal and dopaminergic input onto identified neurons in the nucleus accumbens of the rat. *J. Chem. Neuroanat.* **2,** 285.
18. Albin, R. L., Makowiec, R. L., Hollingsworth, Z. R., Dure, L. S., IV, Penney, J. B., and Young, A. B. (1992) Excitatory amino acid binding sites in the basal ganglia of the rat: a quantitative autoradiographic study. *Neuroscience* **46,** 35–48.
19. Wullner, U., Testa, C. M., Catania, M. V., Young, A. B., and Penney, J. B., Jr. (1994) Glutamate receptors in striatum and substantia nigra: effects of medial forebrain bundle lesions. *Brain Res.* **645,** 98–102.
20. Ohishi, H., Shigemoto, R., Nakanishi, S., and Mizuno, N. (1993) Distribution of the messenger RNA for a metabotropic glutamate receptor, mGluR2, in the central nervous system of the rat. *Neuroscience* **53,** 1009–1018.
21. Shigemoto, R., Nakanishi, S., and Mizuno, N. (1992) Distribution of the mRNA for a metabotropic glutamate receptor (mGluR1) in the central nervous system: an in situ hybridization study in adult and developing rat. *J. Comp. Neurol.* **322,** 121–135.
22. Tanabe, Y., Nomura, A., Masu, M., Shigemoto, R., Mizuno, N., and Nakanishi, S. (1993) Signal transduction, pharmacologic properties, and expression patterns of two rat metabotropic glutamate receptors, mGluR3 and mGluR4. *J. Neurosci.* **13,** 1372–1378.
23. Testa, C. M., Standaert, D. G., Young, A. B., and Penney, J. B., Jr. (1994) Metabotropic glutamate receptor mRNA expression in the basal ganglia of the rat. *J. Neurosci.* **14,** 3005–3018.
24. Fotuhi, M., Sharp, A. H., Glatt, C. E., Hwang, P. M., von Krosigk, M., Snyder, S. H., et al. (1993) Differential localization of phosphoinositide-linked metabotropic glutamate receptor (mGluR1) and the inositol 1,4,5-trisphosphate receptor in rat brain. *J. Neurosci.* **13,** 2001–2012.
25. Martin, L. J., Blackstone, C. D., Huganir, R. L., and Price, D. L. (1992) Cellular localization of a metabotropic glutamate receptor in rat brain. *Neuron* **9,** 259–270.
26. Shigemoto, R., Nomura, S., Ohishi, H., Sugihara, H., Nakanishi, S., and Mizuno, N. (1993) Immunohistochemical localization of a metabotropic glutamate receptor, mGluR5, in the rat brain. *Neurosci. Lett.* **163,** 53–57.
27. Bradley, S. R., Standaert, D. G., Levey, A. I., and Conn, P. J. (1999) Distribution of group III mGluRs in rat basal ganglia with subtype-specific antibodies. *Ann. NY Acad. Sci.* **868,** 531–534.
28. Kosinski, C. M., Bradley, S. R., Conn. P. J., and Levey, A. I. (1999) Localization of metabotropic glutamate receptor 7 mRNA and mGluR7a protein in the rat basal ganglia. *J. Comp. Neurol.* **415,** 266–284.
29. Kerner, J. A., Standaert, D. G., Penny, J. B., Jr., Young, A. B., and Landwehrmeyer, G. B. (1997) Expression of group one metabotropic glutamate receptor subunit mRNAs in neurochemically identified neurons in the rat neostriatum, neocortex, and hippocampus. *Mol. Brain Res.* **48,** 259–269.
30. Tallaksen-Greene, S. J., Kaatz, K. W., Romano, C., and Albin, R. L. (1998) Localization of mGluR1a-like immunoreactivity and mGluR5-like immunoreactivity in identified populations of striatal neurons. *Brain Res.* **780,** 210–217.
31. Testa, C. M., Standaert, D. G., Landwehrmeyer, G. B., Penney, J. B., and Young, A. B. (1995) Differential expression of mGluR5 metabotropic glutamate receptor mRNA by rat striatal neurons. *J. Comp. Neurol.* **354,** 241–252.
32. Testa, C. M., Friberg, I. K., Weiss, S. W., and Standaert, D. G. (1998) Immunohistochemical localization of metabotropic glutamate receptors mGluR1a and mGluR2/3 in the rat basal ganaliz. *J. Comp. Neurol.* **390,** 5–19.

33. Bradley, S. R., Standaert, D. G., Rhodes, K. J., Rees, H. D., Testa, C. M., Levey, A. I. et al. (1999) Immunohistochemical localization of subtype 4a metabotropic glutamate receptors in the rat and mouse basal ganglia. *J. Comp. Neurol.* **407,** 33–46.
34. Petralia, R. S., Wang, Y. X., Niedzielski, A. S., and Wenthold, R. J. (1996) The metabotropic glutamate receptors, mGluR2 and mGluR3, show unique postsynaptic, presynaptic and glial localization. *Neuroscience* **71,** 949–976.
35. Sacaan, A. I., Monn, J. A., and Schoepp, D. D. (1991) Intrastriatal injection of a selective metabotropic excitatory amino acid receptor agonist induces contralateral turning in the rat. *J. Pharmacol. Exp. Ther.* **259,** 1366–1370.
36. Sacaan, A. I., Bymaster, F. P., and Schoepp, D. D. (1992) Metabotropic glutamate receptor activation produces extrapyramidal motor system activation that is mediated by striatal dopamine. *J. Neurochem.* **59,** 245–251.
37. Feeley Kearney, J. A. and Albin, R. L. (1995) Adenosine A2 receptor-mediated modulation of contralateral rotation induced by metabotropic glutamate receptor activation. *Eur. J. Pharmacol.* **287,** 115–120.
38. Feeley Kearney, J. A., Frey, K. A., and Albin, R. L. (1997) Metabotropic glutamate agonist- induced rotation: a pharmacological, Fos immunohistochemical, and [$^{14}$C]-2-deoxyglucose antoradiographic study. *J. Neurosci.* **17,** 4415–4425.
39. Kaatz, K. W. and Albin, R. L. (1995) Intrastriatal and intrasubthalamic stimulation of metabotropic glutamate receptors: a behavioral and Fos immunohistochemical study. *Neuroscience* **66,** 55–65.
40. Smith, I. D. and Beninger, R J. (1996) Contralateral turning caused by metabotropic glutamate receptor stimulation in the dorsal striatum is reversed by MCPG, TTX, and cis- Fluenthixol. *Behav. Neurosci.* **110,** 282–286.
41. Attarian, S. and Amalric, M. (1997) Microinjection of the metabotropic glutamate receptor agonist 1S, 3R-1-aminocyclopentane-1,3-dicarboxylic acid into the nucleus accumbens induces dopamine-dependent locomotor activation in the rat. *Eur. J. Neurosci.* **9,** 809–816.
42. Kim, J. H. and Vezina, P. (1997) Activation of metabotropic glutamate receptors in the rat nucleus accumbens increases locomotor activity in a dopamine-dependent manner. *J. Pharmacol. Exp. Ther.* **283,** 962–968.
43. Schoepp, D. D., Jane, D. E., and Monn, J. A. (1999) Pharmacological agents acting at subtypes of metabotropic glutamate receptors. *Neuropharmacology* 38, 1431–1476.
44. Swanson, C. J. and Kalivas, P. W. (2000) Regulation of locomotor activity by metabotropic glutamate receptors in the nucleus accumbens and ventral tegmental area. *J. Pharmacol. Exp. Ther.* **292,** 406–414.
45. Wang, J. Q. and Mao, L. (2000) Sustained behavioral stimulation following selective activation of group I metabotropic glutamate receptors in the rat striatum. *Pharmacol. Biochem. Behav.* **65,** 439–447.
46. Mao, L. and Wang, J. Q. (2000) Motor stimulation following bilateral injection of the group I metabotropic glutamate receptor agonist into the dorsal striatum of rats: evidence against dependence on ionotropic glutamate receptors. *Psychopharmacology* **148,** 367–373.
47. Mao, L. and Wang, J. Q. (1999) Protection against acute amphetamine-induced behavior by microinjection of a group II metabotropic glutamate receptor agonist into the dorsal striatum of rats. *Neurosci. Lett.* **270,** 103–106.
48. Mao, L. and Wang, J. Q. (2000) Distinct inhibition of acute cocaine-stimulated motor activity following microinjection of a group III metabotropic glutamate receptor agonist into the dorsal striatum of rats. *Pharmacol. Biochem. Behav.* **66,** 1–9.
49. Cartmell, J., Monn, J. A., and Schoepp, D. D. (1999) The metabotropic glutamate 2/3 receptor agonist LY354740 and LY379268 selectively attenuate phencyclidine versus d-amphetamine motor behaviors in rats. *J. Pharmacol. Exp. Thre.* **291,** 161–170.
50. Kim, J. H. and Vezina, P. (1998) Metabotropic glutamate receptors in the rat nucleus accumbens contribute to amphetamine-induced locomotion. *J. Pharmacol. Exp. Ther.* **284,** 317–322.
51. Taylor, D. and Ho, B. T. (1978) Comparison of inhibition of monoamine uptake by cocaine, methylphenidate and amphetamine. *Res. Commun. Chem. Pathol. Pharmacol.* **21,** 67–75.
52. Hu, G., Duffy, P., Swanson, C., Ghasemzadeh, M. B., and Kalivas, P. W. (1999) The regulation of dopamine transmission by metabotropic glutamate receptors. *J. Pharmacol. Exp. Ther.* **289,** 412–416.
53. Mao, L., Lau, Y. S., and Wang, J. Q. (2000) Activation of group III metabotropic glutamate receptors inhibits basal and amphetamine-stimulated dopamine release in rat dorsal striatum: an *in vivo* microdialysis study. *Eur. J. Pharmacol.* **404,** 289–297.
54. Fisher, J. F. and Cho, A. K. (1979) Chemical release of dopamine from striatal homogenates: evidence of an exchange diffusion model. *J. Pharmacol. Exp. Ther.* **208,** 203–209.
55. Trombley, P. Q. and Westbrook, G. L. (1992) L-AP4 inhibits calcium currents and synaptic transmission via a G-protein-coupled glutamate receptor. *J. Neurosci.* **12,** 2043–2050.
56. Kim, J. H. and Vezina, P. (1998) Metabotropic glutamate receptors are necessary for sensitization by amphetamine. *NeuroReport* **9,** 403–406.
57. Mao, L. and Wang, J. Q. (2001) Differentially altered mGluR1 and mGluR5 mRNA expression in rat caudate nucleus and nucleus accumbens in the development and expression of behavioral sensitization to repeated administration of amphetamine. *Synapse* **41,** 230–240.
58. Morgan, J. I. and Curran, T. (1989) Stimulus-transcription coupling in neurons: role of cellular immediate early genes. *TINS* **12,** 459–462.

59. Vaccarino, F. M., Hayward, M. D., Nestler, E. J., Duman, R. S., and Tallman, J. F. (1992) Differential induction of immediate early genes by excitatory amino acid receptor types in primary cultures of cortical and striatal neurons. *Mol. Brain Res.* **12,** 233–241.
60. Wang, J. Q. (1998) Regulation of immediate early gene *c-fos* and *zif/268* messenger RNA expression in rat striatum by metabotropic glutamate receptor. *Mol. Brain Res.* **57,** 46–54.
61. Morgan, J. I. and Curran, T. (1995) Immediate-early genes: ten years on. *TINS* **18,** 66–67.
62. Wang, J. Q. and McGinty, J. F. (1998) Metabotropic glutamate receptor agonist increases neuropeptide mRNA expression in rat striatum. *Mol. Brain Res.* **54,** 262–270.
63. Cole, R. L., Konradi, C., Douglass, J., and Hyman S. E. (1995) Neuronal adaptation to amphetamine and dopamine: molecular mechanisms of prodynorphin gene regulation in rat striatum. *Neuron* **14,** 813–823.
64. Mao, L. and Wang, J. Q. (2001) Selective activation of group I metabotropic glutamate receptors upregulates preprodynorphin, substance P and preproenkephalin mRNA expression in rat dorsal striatum. *Synapse* **39,** 82–94.
65. Takumi, Y., Matsubara, A., Rinvik, E., and Ottersen, O. P. (1999) The arrangement of glutamate receptors in excitatory synapses. *Ann. NY Acad. Sci.* **868,** 474–482.
66. Mao, L. and Wang, J. Q. (2001) Upregulation of preprodynorphin and preproenkephalin mRNA expression by selective activation of group I metabotropic glutamate receptors in characterized primary cultures of rat striatal neurons. *Mol. Brain Res.* **86,** 125–137.
67. Arai, I., Shimazoe, T., Shibata, S., Inoue, H., Yoshimatsu, A., and Watanabe, S. (1996) Enhancement of dopamine release from the striatum through metabotropic glutamate receptor activation in methamphetamine sensitized rats. *Brain Res.* **729,** 277–280.
68. Ohno, M. and Watanabe, S. (1995) Persistent increase in dopamine release following activation of metabotropic glutamate receptors in the rat nucleus accumbens. *Neurosci. Lett.* **200,** 113–116.
69. Taber, M. T. and Fibiger, H. C. (1995) Electrical stimulation of the prefrontal cortex increases dopamine release in the nucleus accumbens of the rat: modulation by metabotropic glutamate receptors. *J. Neurosci.* **15,** 3896–3904.
70. Verma, A. and Moghaddam, B. (1998) Regulation of striatal dopamine release by metabotropic glutamate receptors. *Synapse* **28,** 220–226.
71. Bruton, R. K., Ge, J., and Barnes, N. M. (1999) Group I mGlu receptor modulation of dopamine release in the rat striatum. *Eur. J. Pharmacol.* **369,** 175–181.
72. Liu, J. and Moghaddan, B. (1995) Regulation of glutamate efflux by excitatory amino acid receptors: evidence for tonic inhibitory and phasic excitatory regulation. *J. Pharmacol. Exp. Ther.* **274,** 1209–1215.
73. Moroni, F., Cozzi, A., Lombardi, G., Sourtcheva, S., Leonardi, P., Carfi, M., et al. (1998) Presynaptic mGluR1 type receptors potentiate transmitter output in the rat cortex. *Eur. J. Pharmacol.* **347,** 189–195.
74. Samuel, D., Pisano, P., Forni, C., Nieoullon, A., and Goff, L. K. (1996) Involvement of the glutamatergic metabotropic receptors in the regulation of glutamate uptake and extracellular excitatory amino acid levels in the striatum of chloral hydrate-anesthetized rats. *Brain Res.* **739,** 156–162.
75. Battaglia, G., Monn, J. A., and Schoepp, D. D. (1997) In vivo inhibition of veratridine- evoked release of striatal excitatory amino acids by the group II metabotropic glutamate receptor agonist LY354740 in rats. *Neurosci. Lett.* **229,** 161–164.
76. Cozzi, A., Attucci, S., Peruginelli, F., Maura, M., Luneia, R., Pellicciari, R., et al. (1997) Type 2 metabotropic glutamate (mGlu) receptors tonically inhibit transmitter release in rat caudate nucleus: in vivo studies with (2S,1′S,2′S,3′R)-2-(2′-carboxy-3′-phenylcyclopropyl)glycine, a new potent and selective antagonist. *Eur. J. Neurosci.* **9,** 1350–1355.
77. Bading, H., Ginty, D. D., and Greenberg, M. E. (1993) Regulation of gene expression in hippocampal neurons by distinct calcium signaling pathways. *Science* **260,** 181–186.
78 Greenberg, M. E., Thospson, M. A., and Sheng, M. (1992) Calcium regulation of immediate early gene transcription. *J. Physiol.* **86,** 99–108.
79. Vaccarino, F. M., Liljequist, S., and Tallman, J. F. (1991) Modulation of protein kinase C translocation by excitatory and inhibitory amino acids in primary cultures of neurons. *J. Neurochem.* **57,** 391–396.
80. Wang, J. Q. and McGinty, J. F. (1996) Intrastriatal injection of the metabotropic glutamate receptor antagonist MCPG attenuates acute amphetamine-stimulated neuropeptide mRNA expression in rat striatum. *Neurosci. Lett.* **218,** 13–16.
81. Mao, L., Conquet, F., and Wang, J. Q. (2001) Augmented motor activity and reduced striatal preprodynorphin mRNA induction in response to acute amphetamine administration in mGluR1 knockout mice. *Neuroscience,* **106,** 303–312.
82. Kuzmin, A. V., Gerrits, M. A., and Van Ree, J. M. (1998) Kappa-opioid receptor blockade with nor-binaltorphimine modulates cocaine self-administration in drug-naive rats. *Eur. J. Pharmacol.* **358,** 197–202.
83. Shipperberg, T. S. and Heidbreder, C. (1995) The delta-opioid receptor antagonist naltrindole prevents sensitization to the conditioned rewarding effects of cocaine. *Eur. J. Pharmacol.* **280,** 55–61.
84. Shippenberg, T. S., LeFevours, A., and Herdbreder, C. (1996) Kappa-opioid receptor agonists prevent sensitization to the conditioned rewarding effects of cocaine. *J. Pharmacol. Exp. Ther.* **276,** 545–554.
85. Carlezon, W. A., Jr., Thome, J., Olson, V. G., Lane-Ladd, S. B., Brodkin, E. S., Hiroi, N., et al. (1998) Regulation of cocaine reward by CREB. *Science* **282,** 2272–2275.

86. Knopfel, T., Kuhn, R., and Allgeier, H. (1995) Metabotropic glutamate receptors: novel targets for drug development. *J. Med. Chem.* **38,** 1417–1426.
87. Nicholetti, F., Bruno, V., Copani, A., Casabona, G., and Knopfel, T. (1996) Metabotropic glutamate receptors: a new target for the therapy of neurodegenerative disorders? *TINS* **19,** 267–271.
88. Lujan, R., Nusser, Z., Roberts, J. D., Shigemoto, R., and Somogyi, P. (1996) Perisynaptic localization of metabotropic glutamate receptors mGluR1 and mGluR5 on dendrites and dendritic spines in the rat hippocampus. *Eur. J. Neurosci.* **8,** 1488–1500.
89. Nusser, Z., Mulvihill, E., Streit, P., and Somogyi, P. (1994) Subsynaptic segregation of metabotropic and ionotropic glutamate receptors as revealed by immunogold localization. *Neuroscience* **61,** 421–427.
90. McGeorge, A. J. and Fanll, R. L. M. (1988) The organization of the projection from the cerebral cortex to the stratum in the rat. *Neuroscience* **29,** 503–537.

# 11
# Glutamate Neurotransmission in the Course of Cocaine Addiction

## Luigi Pulvirenti, MD

## 1. NEURAL SUBSTRATES OF COCAINE ABUSE

A critical issue for the understanding of drug dependence is what neurochemical changes occur during the various phases of the course of addiction. Psychostimulant dependence has been a major focus of investigation, especially following the cocaine epidemic of the 1980s. Despite early misconceptions that considered cocaine as a recreational drug devoid of abuse and addictive potential, the recent cocaine epidemic has revealed the harmful consequences of cocaine use and its high abuse liability. Therefore, much experimental attention has been devoted over the past two decades to the search for the neural substrates responsible for cocaine abuse.

With the use of experimental models of intravenous drug self-administration, it has become clear that cocaine and other psychostimulant drugs such as amphetamine are readily self-administered by various species of laboratory animals *(1)*. The acute reinforcing properties of psychostimulant drugs appear to depend on dopamine neurotransmission within areas of the limbic forebrain *(2–4)*. In particular, the nucleus accumbens of the ventral striatum seems a critical structure for the maintenance of cocaine self-administration. The nucleus accumbens is the main projection area of the dopamine neurons originating within the mesencephalic ventral tegmental area (A10) and is critically involved in motivated behavior *(5,6)*. Neurochemical studies have shown that dopamine outflow in the nucleus accumbens is increased after administration of cocaine *(7)*, and an increase in nucleus accumbens dopamine concentration was shown in rats self-administering cocaine by microdialysis studies *(8–10)*. Furthermore, behavioral evidence suggests that lesions of the dopamine terminals within the nucleus accumbens extinguished cocaine self-administration *(11)*, and competitive antagonism at the dopamine receptor site within the same region reduced the reinforcing properties of cocaine *(12)*.

Understanding the neurochemical determinants underlying the acute reinforcing properties of cocaine still leaves several unanswered questions regarding the intimate mechanisms leading to the development of the full cocaine dependence syndrome. The natural history of drug dependence consists of different phases *(13)*. Although the positive reinforcing effects of cocaine as well as of other abused drugs are likely to represent critical factors in the initiation of drug self-administration *(14)*, during this phase important neurochemical events are thought to occur within critical brain sites to contribute to the increased motivation to seek drugs that characterizes compulsive use *(13)*.

Indeed, with more prolonged drug exposure, the period of acquisition of drug intake in rats self-administering cocaine intravenously is followed by a period of maintenance of a relatively low and stable level of intake or by escalation to higher levels of intake *(15)*. This seems to depend on the duration of daily access to drug self-administration *(15)*. This transition from moderate to excessive drug intake

From: *Contemporary Clinical Neuroscience: Glutamate and Addiction*
Edited by: Barbara H. Herman et al. © Humana Press Inc., Totowa, NJ

is particularly significant because it is considered the first step leading to the loss of control, which represents the cardinal feature of the addictive process.

In addition to the issue of uncontrollable use, relapse into drug-seeking behavior following a period of abstinence has recently received much dedicated attention because it is believed to represent one of the major factors leading to the perpetuation of the addictive cycle. In this respect, extinction procedures provide a measure of the motivational properties of drugs as reflected by the persistence of drug-seeking behavior in the absence of the drug. However, extinction procedures also provide a powerful means of assessing the incentive-motivational properties of drug-paired stimuli or noncontingent drug administration in reinstating responding *(16,17)*.

The long-lasting adaptive changes occurring in response to drug exposure are therefore thought to play a key role in the addictive cycle and they may represent the basis for clinically relevant phenomena, including drug craving and conditioned reinforcement. The intimate neurobiological mechanisms underlying such forms of neuroadaptation within critical neural sites remain obscure. However, there is now such evidence suggesting that neurochemical effectors other than dopamine may play a critical role, and a growing body of evidence accumulated over the last few years has indicated that excitatory amino acids may be relevant in this respect.

## 2. GLUTAMATE NEUROTRANSMISSION AND THE LIMBIC SYSTEM

Excitatory amino acids represent the main excitatory neurotransmitters within the mammalian central nervous system and exert their effects through the interaction with ionotropic receptors of the *N*-methyl-D-aspartate (NMDA) and non-NMDA type and metabotropic receptors. In addition, glutamate also stimulates the synthesis of the diffusible messenger nitric oxide (NO). Through activation of NMDA receptors, glutamate promotes the influx of calcium into the postsynaptic neuron. In turn, binding to calmodulin stimulates the activity of nitric oxide synthase *(18)*. Nitric oxide is a gaseous neurotransmitter that acts as a retrograde messenger and affects the release of various neurotransmitters through increases in cyclic GMP *(18)*. Therefore, NO acts as an important intracellular effector of excitatory amino acid neurotransmission through NMDA receptors.

The nucleus accumbens is functionally linked to a number of structures within the limbic system and is part of a ventrotegmental–accumbens–pallidal circuit that seems to be critically involved in drug-seeking behavior *(19)*. In addition to the dopamine-containing projections originating from the ventral tegmental area, the nucleus accumbens also receives primary neuronal afferents from allocortical areas such as the amygdaloid complex and the hippocampal formation as well as from the frontal and prelimbic cortexes *(20)*. These fibers use excitatory amino acids as neurotransmitters *(21)*. Ultrastructural studies revealed that, within the nucleus accumbens, these fibers form both presynaptic contact with tyrosine hydroxylase-containing terminals of the axons of A10 neurons and postsynaptic contacts onto medium spiny output neurons of the nucleus accumbens. The latter also receive convergent input from the A10 region *(22)*. In addition, electrophysiological studies suggest that nucleus accumbens neurons are mostly quiescent under normal conditions, and allocortical afferents are thought to play a major role in driving their activity *(23,24)*. Glutamate inputs from the hippocampus appear to be responsible for the induction of a depolarized state of nucleus accumbens neurons from which an action potential is thought to be generated when these neurons are excited by convergent glutamate inputs from the prefrontal cortex *(25)*. Thus, there is substantial anatomical and electrophysiological evidence to suggest that excitatory amino acid neurotransmission may modulate the integrated function of the nucleus accumbens, possibly through an interaction with dopamine.

Excitatory amino acid-containing fibers also interconnect other areas of the mesolimbic system. The prefrontal cortex, for example, in addition to sending glutamate fibers to the nucleus accumbens, also provides dense innervation of the ventral tegmental area *(26)*. In contrast, the ventral subiculum of the hippocampus sends a vast contingent of fibers to the nucleus accumbens, but not to the ventral tegmental area *(20,22)*. Finally, the excitatory efferent fibers of the amygdala to the nucleus accumbens originate mainly within the basolateral nucleus, whereas the central nucleus projects to the ventrotegmental area *(27,28)*.

In vitro studies have provided much evidence regarding the modulation of the dopamine system by excitatory amino acids. Electrophysiological evidence indicates that glutamate induces a current-dependent increase of firing of quiescent nucleus accumbens neurons *(24)*. Activation of non-NMDA receptors appear particularly important in this respect and activation of NMDA receptors seem to come into play after neurons have been primed with activation of other receptors *(24)*. This observation has prompted speculation that the stimulation of NMDA receptor function in the nucleus accumbens might gain particular significance during periods of behaviorally relevant neural inputs, probably of limbic origin. Activation of metabotropic receptors within the nucleus accumbens, in contrast, does not appear to modify the activity of nucleus accumbens neurons, but it reduced the excitation induced by glutamate acting at the level of ionotropic receptors *(24)*. Anatomical distribution of metabotropic receptors and the pharmacological profile of ACPD *(29,30)* suggests that group I/group II metabotropic receptors might be more relevant for these effects playing a potential role of feedback mechanism *(24)*.

The dopamine-releasing activity of excitatory amino acids has been confirmed in freely moving animals with microdialysis, and it has been shown that both NMDA and non-NMDA receptors may play a role *(31,32)*. However, a reduction of extracellular dopamine in the nucleus accumbens following local application of glutamate agonists has also been reported *(33)*. The hippocampus appears to play a major role in the modulation of dopamine function within the ventral striatum. For example, chemical and electrical stimulation of the ventral subiculum of the hippocampus increased dopamine levels in the nucleus accumbens, an effect that appears to be mediated by glutamate receptors within the ventral striatum *(34)* or, indirectly, via a subiculum–accumbens–ventral tegmental area indirect pathway *(35,36)*. These observations are also in agreement with the findings that activation of both NMDA and non-NMDA receptors within the ventral tegmental area appears to stimulate the firing rate of midbrain dopamine neurons *(37,38)*. Much evidence from various research lines suggests that this phenomenon may be relevant for psychostimulant sensitization (for a review, see ref. *39*).

Finally, the increase of extracellular dopamine content in the nucleus accumbens produced by systemic administration of cocaine has been shown to be reduced by local infusion of both the NMDA receptors antagonist AP-5 and the non-NMDA receptor antagonist 6-cyano-7-nitroquinoxaline-2,3-dione (CNQX) *(40)*. However, the interaction between dopamine and excitatory amino acids appear to be complex because, in untreated animals, local infusion of NMDA antagonists has been shown to increase dopamine release *(41)*. It is also noteworthy that increasing endogenous extracellular glutamate concentrations through local microinfusion of the selective glutamate reuptake inhibitor transpyrrolidine-2,4-dicarboxylic acid (PDC) produced an increase of dopamine release in the rat striatum *(42)* and an increase in locomotor activity has been reported after microinfusion of PDC within the core of the nucleus accumbens *(43)*. Investigation using in vivo microdialysis has also shown that activation of dopamine receptors with amphetamine or the direct dopamine agonists SKF 38393 and quinpirole increased the amount of extracellular glutamate within the nucleus accumbens *(44)* and cocaine and amphetamine appear to preferentially stimulate glutamate release in the nucleus accumbens *(45)*. However, the functional relevance of the increase of extracellular glutamate concentration after administration of psychostimulant drugs awaits further investigation.

## 3. GLUTAMATE NEUROTRANSMISSION AND COCAINE ABUSE

From a functional standpoint, excitatory amino acid neurotransmission within the limbic system appears to play an important role in the modulation of various behavioral consequences of exposure to psychostimulant drugs. Earlier studies had proposed a behavioral stimulatory role for excitatory amino acids in the nucleus accumbens on the basis of the increase in locomotor activity produced by local infusion of glutamate agonists, an effect accompanied by an increase in dopamine turnover

*(46,47)*. Also, the locomotor hyperactivity produced by intranucleus accumbens infusion of glutamate agonists was blocked by a dopamine receptor antagonist *(46)*, suggesting that pharmacological activation of nucleus accumbens excitatory amino acid receptors produced psychomotor stimulation, probably by facilitating dopamine neurotransmission at the same site, in accordance with neurochemical findings *(31,32)*.

To investigate more closely the physiological significance of these observations, the reinforcing and locomotor-activating effects of psychostimulant drugs were tested after temporary reduction of nucleus accumbens excitatory amino acid neurotransmission obtained through local infusion of glutamate receptor antagonists. Microinfusion of both NMDA and non-NMDA receptor antagonists within the nucleus accumbens reduced the expression of the acute psychomotor-activating properties of cocaine and amphetamine administered systemically or amphetamine and dopamine infused into the nucleus accumbens (48–50) as well as novelty-induced locomotion in rats *(51,52)*. These findings suggest that intact glutamate neurotransmission within the nucleus accumbens may be essential for the full expression of the acute locomotor-activating properties of psychostimulant drugs *(53)*. Similarly, earlier studies pointed toward a tonic "permissive" role of excitatory amino acid in the expression of the acute reinforcing properties of cocaine. In fact, microinfusion of AP-5 within the core of the nucleus accumbens reduced the interreinforcement interval of cocaine self-administered intravenously by rats *(54)*.

Anatomically, two subregions of the nucleus accumbens, the core and the shell, have recently been characterized on the basis of histochemical differences and separate neural connections. Investigation of the dopamine system within the core and the shell at the neurochemical and behavioral levels has revealed important differences (for a review, see ref. 55). The modulation of the nucleus accumbens function within the core and shell also appears to be differentially modulated by glutamate receptors. Local blockade of NMDA receptors within the nucleus accumbens core significantly reduced cocaine-induced locomotor activity while leaving spontaneous locomotion unaffected. In contrast, the blockade of NMDA receptors within the nucleus accumbens shell did not significantly modify cocaine-induced locomotor activity but *increased* spontaneous locomotion *(56)*. This finding is also in accordance with previous observations indicating that intranucleus accumbens infusion of the glutamate receptor agonists NMDA and AMPA or the glutamate receptor antagonists CNQX and AP-5 all *increased* spontaneous locomotion. By comparison, the microinfusion of both antagonists *decreased* amphetamine-induced locomotor activity *(57)*, although no differentiation was made in that study regarding core/shell infusions. These findings are in accordance with the observation that activity of nucleus accumbens neurons is associated with a high level of cortical arousal *(58)* and, therefore, it is likely that glutamate receptors may play a more substantial role in the regulation of nucleus accumbens neurons during periods of behaviorally relevant neural inputs of limbic origin.

The regulation of the integrated function of the nucleus accumbens by glutamate neurotransmission is therefore likely to depend on (1) the behavioral state of the organism and (2) the specific substructure where glutamate acts within the nucleus accumbens. The relative strength of each of these components is likely to determine the behavioral outcome achieved by increased or decreased excitatory amino acid neurotransmission within this structure.

Within the context of cocaine addiction, much evidence has been accumulated that antagonism of glutamate receptors affects various measures of cocaine self-administration. Support for the role of nucleus accumbens NMDA receptors in reinforcement comes from studies showing that the noncompetitive NMDA receptor antagonists dizocilpine (MK-801), phencyclidine, and the competitive NMDA receptor antagonist 3-(2-carboxypiperazin-4-yl)propyl-1-phosphonic acid (CPP) support intracranial self-administration within the frontal cortex and the shell of the nucleus accumbens of the rat *(59)*. The fact that microinfusion of the NMDA receptor antagonist AP-5 within the core of the nucleus accumbens appeared to reduce the interreinforcement interval of cocaine self-administration in a fixed ratio schedule *(54)* suggests the possibility that a dissociation between the core and the shell of the nucleus accumbens may exist with regard to the modulation of reinforcement, as it appears to exist with regard to the modulation of spontaneous and cocaine-induced locomotor activity *(56)*.

Systemic pretreatment with dizocilpine within the dose range of 0.1–0.15 mg/kg, reduced responding for cocaine self-administration in a fixed ratio schedule *(60)*, increased the breaking point of progressive ratio responding for cocaine *(61)*, and failed to reinstate cocaine-seeking behavior after extinction *(62)*. In contrast, within the dose range of 0.2–0.3 mg/kg, dizocilpine decreased responding for cocaine maintained both on a fixed ratio and on a progressive ratio schedule and reinstated cocaine-seeking behavior following extinction *(60–62)* Therefore it has been suggested that dizocilpine's ability to enhance the reinforcing effects of cocaine may lie within a narrow dose range on an inverted-U function and suggests that discrepancies in the outcome of dizocilpine's effects in the literature may result from using different doses in differently sensitive paradigms *(61)*. Dizocilpine has also been reported to produce variable effects on dopamine release *(60,63,64)*.

In a series of electrophysiological studies, French and co-workers *(65–69)* have shown that dizocilpine and phencycline are potent activators of ventral tegmental area A 10 neurons and this effect is not shared and is even *blocked* by competitive NMDA antagonists. More recent results have also shown that centrally acting competitive NMDA receptor antagonists, including CGS 19755, significantly modified the firing pattern of dopamine neurons, reducing the incidence and intensity of burst firing *(70)*. These observatons suggest that competitive and non-competitive NMDA receptor antagonists may produce different effects on mesolimbic dopamine neurotransmission.

Altogether the use of dizocilpine as well as phencyclidine has provided valuable initial relevant information that have prompted much research on the role of excitatory amino acids on drug addiction and sensitization. However, the steepness of the dose-response curve produced by dizocilpine is associated with profound and bizarre behavioral effects, including motor impairment and ataxia, which have been reported at the dose of 0.25 mg/kg *(63,71)*. In addition, the peculiar stimulating effects of dizocilpine on the firing of mesolimbic dopamine neurons are not shared with and are even reversed by other NMDA antagonists. These observations, together with the lack of potential for clinical application of dizocilpine, have prompted call for caution *(63)* in the interpretation of the effects produced by dizocilpine in the context of the investigation for the *physiological* role of NMDA neurotransmission.

Recently, a comparative investigation of the effects of systemically administered site-specific NMDA receptor antagonists has examined the effects produced by competitive and noncompetitive NMDA antagonists *(72)* in rats self-administering cocaine. The noncompetitive NMDA antagonist memantine was found to reduce responding for cocaine on a fixed-ratio schedule and produce a sizable (although reportedly nonsignificant) reduction of the breaking point in a progressive-ratio schedule. In addition, the competitive NMDA receptor antagonist CGP 39551 and the NMDA/glycine recognition-site antagonist L-701,324 did not modify responding for cocaine self-administration on a fixed-ratio schedule *(72)*.

These results are similar to those obtained with the noncompetitive NMDA antagonist dextromethorphan *(73)*. Dextromethorphan is a widely used antitussive agent that has been shown to act within the central nervous system as a noncompetitive antagonist at the NMDA receptor complex. In rats trained to self-administer cocaine, dextromethorphan reduced the maintenance of cocaine self-administration and significantly reduced the maximum number of responses performed by rats to obtain a dose of cocaine in a progressive-ratio schedule ("breaking point") The effects produced by dextromethorphan are similar to those produced by memantine, another noncompetitive NMDA antagonist *(72)* and suggest that these drugs may effectively modulate cocaine self-administration by suppressing the motivational strength to obtain the drug.

Comparatively less experimental evidence is available on the role of non-NMDA receptors in animals self-administering cocaine. The non-NMDA receptor antagonist DNQX reduced lever pressing for cocaine self-administration, but the effects appeared to be the result of general suppression of operant behavior because responding for food reinforcement was also reduced by DNQX *(60)* More recently, the role of glutamate, in the cocaine-dependence cycle has been investigated across multiple behavioral measures using a number of different behavioral paradigms of cocaine self-administration in order to extend previous findings to the various phases of the course of cocaine addiction.

## 4. GLUTAMATE-DEPENDENT SYNAPTIC PLASTICITY AND COCAINE ADDICTION

An issue that deserves dedicated attention is the possibility that the relative contribution of excitatory amino acid neurotransmitters may become particularly significant during specific periods of exposure to cocaine. Excitatory amino acids are considered critical mediators of neural plasticity within the central nervous system, and this makes them ideal candidates as participating agents in the development of adaptive changes, which may represent integral parts of the addictive process

Glutamate dependence is a feature of long-term potentiation (LTP), a form of synaptic enhancement associated with repeated use of specific synaptic connections *(74)*. Originally described within the hippocampus, more recent studies have shown LTP in other brain areas, including the nucleus accumbens. Tetanic stimulation of allocortical afferent fibers making monosynaptic connections with nucleus accumbens cells produced LTP in a slice preparation and in the intact animal effects that seemed to be mediated by activation of ionotropic glutamate receptors *(75–76)*. Excitatory amino acid-dependent synaptic changes in the nucleus accumbens is of enormous potential importance because it reveals that forms of synaptic enhancement may occur within this structure, and these changes may be relevant for the development of the addictive process.

In experiments where behavioral techniques were used in combination with electrophysiological analysis, evoked field responses measured in the nucleus accumens shell after stimulation of fimbria afferents were examined in rats exposed to the very first days of self-administration of cocaine. Significantly increased paired-pulse facilitaion of a specific component of the potential (a long-latency component termed P25) as well as a marked potentiation of the same potential after tetanic stimulation of the fimbria were observed in rats self-administering cocaine, but not in yoked controls. These effects were prevented by systemic administration of the NMDA antagonist CGS 19755 and the non-NMDA antagonist NBQX *(77)*. Hence, the modifications of nucleus accumbens synaptic efficacy produced by repeated stimulation of the neural firing of fimbric origin may be part of the neural plastic changes representing the early critical events leading to the development of cocaine addiction. The fact that these changes depend on excitatory amino acid neurotransmission suggests the hypothesis that pharmacological manipulation of glutamate receptors may effectively modify at least part of the neuroadaptive phenomena that occur during the course of drug addiction.

As discussed earlier, the natural history of drug addiction consists of several phases that include the acquisition and the maintenance of stable drug intake, the transition from moderate to excessive drug intake, and extinction and relapse. The protracted withdrawal state, in particular, is of importance becasue, during this period, phenomena such as drug-craving, cue-precipitated drug-seeking behavior, and conditioned reinforcement may occur and these lead into relapse of drug abuse both in animals and in humans *(77–79)*.

One of the major precipitating factors leading into relapse of drug use is exposure to enteroceptive and environmental cues previously associated with the abused drug. Reinstatement of drug-seeking behavior induced by a priming systemic administration of the abused drug or by environmental cues have been well characterized in rodents. Although the intimate neurobiological determinants of this phenomenon have not been fully explored, nucleus accumbens dopamine seems to facilitate operant responding elicited by a conditioned stimulus *(80)*. Interestingly, excitotoxic lesions of the amygala disrupt conditioned responding for food and sexual reinforcement *(81)* and, moreover, intranucleus accumbens infusion of AP-5 reduced the facilitatory effects of amphetamine coinfused within the nucleus accumbens on conditioned reinforcement *(50)*. Taken together, these results suggest that neural messages from the amygdala reaching the nucleus accumbens, and probably other areas of the limbic system, are capable of reinstating responding for the primary reinforcer in the presence of a conditioned stimulus. This is probably through the activation of a glutamate mechanism. Considering that activation of nucleus accumbens dopamine neurotransmission elicits is similar effects *(82)*, it is possible that excitatory amino acid afferents to the nucleus accumbens may primarily drive neural activity of nucleus accumbens neurons, allowing allocortical messages to find access to the motivational/motor effectors of the ventral striatum. The various forms of excitatory amino acid-dependent neural plasticity shown to occur within the nucleus accumbens may, therefore, participate in the concert of cellular events whose behavioral outcome may ultimately be represented by drug-seeking behavior or drug-craving. Indeed recent observations suggest a specific role for glutamate-containing pathways in relapse into cocaine self-administration. Theta-burst electrical stimulation of hippocampal glutamatergic fibers produced reinstatement of operant responses associated with cocaine in rats previously trained to self-administer the drug *(83)*. Interestingly, in the same study, stimulation of the medial forebrain bundle, a pathway critical for reinforcement, did not elicit reinstatement of cocaine-seeking behavior. This suggests that separate neural systems may subserve drug-induced positive reinforcement and incentive properties. Glutamate-containing pathways originating from the hippocampus appear to be more specifically involved in the incentive-motivational aspects of drug addiction, thought to be important for relapse. Interestingly, however, glutamate blockade within the basolateral amygdala failed to affect relapse of cocaine-seeking behavior *(84)*, thus suggesting a specificity for allocortico-limbic glutamate pathways in reinstatement of cocaine-seeking behavior and, possibly, craving. Further studies on the role played by specific excitatory amino acids receptor subtypes on selected aspects of these phases of the cocaine-addiction cycle will permit a closer analysis of the cellular mechanisms through which excitatory amino acids modulate specific aspects of cocaine-seeking behavior when the drug is no longer available.

Through activation of NMDA receptors, glutamate promotes the influx of calcium into the postsynaptic neuron, which, binding to calmodulin, stimulates the activity of nitric oxide synthase *(18)*. Nitric oxide (NO) is a gaseous neurotransmitter that acts as a retrograde messenger and affects the release of various neurotransmitters via increases in cyclic GMP *(18)*. Therefore, NO acts as an important intracellular effector of excitatory amino acid neurotransmission through NMDA receptors. Electrophysiological and behavioral evidence suggests that nitric oxide plays a significant role in various forms of synaptic plasticity and in learning and memory *(85)*. More recently, evidence from different experimental approaches indicates that NO may play a role in the behavioral effects of psychostimulant drugs. Neurochemical studies indicate that endogenous NO facilitates the efflux of dopamine within the striatum *(86)* and perfusion of the nucleus accumbens with NMDA through a microdialysis probe produced a NO-dependent increase of dopamine release *(87)*. In addition, methamphetamine-induced dopamine release in the caudate/putamen of the rat can be reduced by concurrent administration of the NO synthase inhibitor L-NAME *(88)*. Further evidence for a functional interaction between endogenous glutamate neurotransmission, NO, and dopamine within the striatal complex comes from the observation that the glutamate reuptake inhibitor PDC potentiates endogenous NO-facilitated dopamine efflux in the rat striatum *(89)*. Electrophysiological evidence suggests that blockade of NO synthesis with L-NAME reduced NMDA-induced burst firing of rat midbrain dopamine neurons *(90)*. Behavioral studies, in addition, suggest that functional integrity of NO signaling is essential for the full expression of cocaine-induced behavior, including locomotion and conditioned place preference, psychostimulant sensitization, and cocaine kindling *(91–94)*. Within the context of cocaine addiction, earlier studies have shown that administration of L-NAME reduced responding for cocaine in both a fixed-ratio and a progressive-ratio schedule *(95)*. In addition, L-NAME appears to reduce the increase in responding for cocaine during the extinction phase *(96)*. These effects of blockade of NO synthase on cocaine self-administration suggest the possibility that nitric oxide may represent an important component of the cocaine-abuse cycle.

In conclusion, the recent development of pharmacological probes allowing the selective exploration of specific aspects of glutamate neurotransmission will allow a better characterization of still unexplored aspects of excitatory amino acid neurotransmission in the context of addiction. This includes the role played by metabotropic receptors, intracellular effects of NMDA activation such as NO, and the availability of drugs reducing glutamate release. Importantly, these studies will be critical for the characterization of novel potential therapeutic perspective, strategically tailored to affect specific aspects of synaptic plasticity associated with the different phases of the cocaine-abuse cycle.

## ACKNOWLEDGMENTS

I wish to thank Diane Braca for editorial assistance. This work was partially supported by NIH grant DA 10072. This is publication 14414-NP of The Scripps Research Institute.

## REFERENCES

1. Pickens, R. and Thompson, T. (1968) Cocaine-reinforced behavior in rats: effects of reinforcement magnitude and fixed-ratio size. *J. Pharmacol. Exp. Ther.* **161,** 122–129.
2. Yokel, R. A. and Wise, R. A. (1975) Increased lever pressing for amphetamine after pimozide in rats: implications for a dopamine theory of reward. *Science* **187,** 547–549.
3. Ettenberg, A., Pettit, H. O., Bloom, F. E., and Koob, G. F. (1982) Heroin and cocaine intravenous self-administration in rats: mediation by separate neural systems. *Psychopharmacology* **78,** 204–209.
4. DeWit, H. and Wise, R. A. (1977) Blockade of cocaine reinforcement in rats with the dopamine receptor blocker pimozide, but not with the noradrenergic blockers phentolamine and phenoxybenzamine. *Can. J. Psychol.* **31,** 195–203.
5. Le Moal, M. and Simon, H. (1991) Mesocorticolimbic dopaminergic network: functional and regulatory roles. *Physiol Rev.* **71,** 155–234.
6. Robbins, T. W. and Everitt, B. J. (1992) Functions of dopamine in the dorsal and ventral striatum. *Semin. Neurosci.* **4,** 119–127.
7. Di Chiara, G. and Imperato, A. (1988) Drugs abused by humans preferentially increase synaptic dopamine concentrations in the mesolimbic system of freely moving rats. *Proc. Natl. Acad. Sci. USA.* **85,** 5274–5278.
8. Di Ciano, P., Blaha, C. D., and Phillips, A. G. (1996) Changes in dopamine oxidation currents in the nucleus accumbens during unlimited-access self-administration of d-amphetamine by rats. *Behav. Pharmacol.* **7,** 714–729.
9. Hurd, Y. L., Weiss, F., Koob, G. F., Anden, N. E., and Ungerstedt, U. (1989) Cocaine reinforcement and extracellular dopamine overflow in rat nucleus accumbens: an in vivo microdialysis study. *Brain Res.* **498,** 199–203.
10. Pontieri, F. E., Tanda, G., and Di Chiara, G. (1995) Intravenous cocaine, morphine, and amphetamine preferentially increase extracellular dopamine in the "shell" as compared with the "core" of the rat nucleus accumbens, *Proc. Natl. Acad. Sci. USA* **92,** 12,304–12,308.
11. Pettit, H. O., Ettenberg, A., Bloom, F. E., and Koob, G. F. (1984) Destruction of dopamine in the nucleus accumbens selectively attenuates cocaine but not heroin self-administration in rats. *Psychopharmacology* **84,** 167–173.
12. Maldonado, R., Robledo, P., Chover, A. J., and Koob, G. F. (1993) D-1 dopamine in the nucleus accumbens accumbens modulate cocaine self-administration in the rat. *Pharmacol. Biochem. Behav.* **45,** 239–242.
13. American Psychiatric Association (1994) *Diagnostic and Statistical Manual of Mental Disorders,* American Psychiatric Association Washington, DC.
14. Wise, R. A. (1988) The neurobiology of craving: implications for the understanding and treatment of addiction. *J. Abnorm. Psychol.* **97,** 118–132.
15. Ahmed, S. H. and Koob, G. F. (1998) Transition from moderate too excessive drug intake: a change in hedonic set point. *Science* **282,** 298–300.
16. Gerber, G. J. and Stretch, R. (1975) Drug-induced reinstatement of extinguished self-administration behavior in monkey. *Pharmacol. Biochem. Behav.* **3,** 1055–1061.
17. Stewart, J. and DeWit, H. (1987) Reinstatement of drug-taking behavior as a method of assessing incentive motivational properties of drugs, in *Methods of Assessing the Reinforcing Properties of Abused Drugs* (Bozarth, M. A., ed.), Springer-Verlag, New York, pp. 211–227.
18. Garthwaite, J. (1991) Glutamate, nitric oxide and cell-cell signalling in the nervous system. *Trends Neurosci.* **14,** 60–67.
19. Koob, G. F. (1992) Drugs of abuse: anatomy, pharmacology and function of reward pathways. *Trends Pharmacol. Sci.* **13,** 177–184.
20. Kelley, A. E. and Domesick, V. B. (1982) The distribution of the projection from the hippocampal formation to the nucleus accumbens in the rat: an anterograde and retrograde horseradish peroxidase study. *Neuroscience* **7,** 2321–2335.
21. Kelley, A. E., Domesick, V. B., and Nauta, W. J. H. (1982) The amygdalostriatal projection in the rat: an anatomical study by anterograde and retrograde tracing methods. *Neuroscience* **7,** 615–630.

22. Sesack, S. R. and Pickel, V. M. (1990) In the rat medial nucleus accumbens hippocampal and catecholaminergic terminals converge on spiny neurons and are in apposition to each other. *Brain Res.* **527,** 266–279.
23. Pennartz, C. M. A., Boejinga, P. H., Kitai, S. T., and Lopes da Silva, F. H. (1991) Contribution of NMDA receptors to postsynaptic potentials and paired-pulse facilitation in identified neurons of the rat nucleus accumbens in vivo. *Exp. Brain Res.* **86,** 190–198.
24. Hu, X. T. and White, F. J. (1996) Glutamate receptor regulation of rat nucleus accumbens neurons in vivo. *Synapse* **23,** 208–218.
25. O'Donnell, P. and Grace, A. A. (1995) Synaptic interactions among excitatory afferents to nucleus accumbens neurons: hippocampal gating of prefrontal cortical input. *J Neurosci.* **15,** 3622–3639.
26. Sesack, S. R. and Pickel, V. M. (1992) Prefrontal cortical efferents in the rat synapse on unlabeled neuronal targets of catecholamine terminals in the nucleus accumbens septi and on dopamine neurons in the ventral tegmental area. *J. Comp. Neurol.* **330,** 145–160.
27. Wallace, D. M., Magnuson, D. J., and Gray, T. S. (1992) Organization of amygdaloid projection to brainstem dopaminergic, noradrenergic and adrenergic cell groups in the rat brain. *Brain Res Bull.* **28,** 447–454.
28. Shinonaga, Y., Takada, M., and Mizuno, N. (1994) Topographic organization of collateral projections from the basolateral amygdaloid nucleus to both the prefrontal cortex and nucleus accumbens in the rat. *Neuroscience* **58,** 389–397.
29. Schoepp, D. D., Johnson, B. G., Sacaan, A. I., True, R. A., and Monn, J. A. (1992) In vitro and in vivo pharmacology of 1$S$,3$R$- and 1$R$,3$S$-ACPD: evidence for the role of metabotropic glutamate receptors in striatal motor function. *Mol. Neuropharmacol.* **2,** 33–37.
30. Testa, C. M., Standaert, D. G., Young, A. B., and Penney, J. B. (1994) Metabotropic glutamate receptor mRNA expression in the basal ganglia of the rat. *J Neurosci.* **14,** 3005–3018.
31. Imperato, A., Honore, T., and Jensen, L. H. (1990) Dopamine release in the nucleus caudatus and nucleus accumbens is under glutamatergic control through non-NMDA receptors: a study in freely-moving rats. *Brain Res.* **530,** 223–228.
32. Youngren, K. D., Daly, D. A., and Moghaddam, B. (1993) Distinct actions of endogenous excitatory amino acids on the outflow of dopamine in the nucleus accumbens. *J. Pharmacol. Exp. Ther.* **264,** 289–293.
33. Taber, M. T. and Fibiger, H. C. (1996) Glutamate receptor agonists decrease extracellular dopamine in the rat nucleus accumbens in vivo, *Synapse* **24,** 165–172.
34. Taepavarapruk, P. Floresco, S. B., and Phillips, A. G. (2000) Hyperlocomotion and increased dopamine efflux in the nucleus accumbens evoked by electrical stimulation of the ventral subiculum: role of ionotropic glutamate and dopamine D1 receptors. *Psychopharmacology* **151,** 242–251.
35. Legault, M. and Wise, R. A. (1999) Injections of $N$-methyl-D-aspartate into the ventral hippocampus increase extracellular dopamine in the ventral tegmental area and nucleus accumbens. *Synapse* **31,** 241–249.
36. Floresco, S. B., Todd, C. L., and Grace, A. A. (2001) Glutamatergic afferents from the hippocampus to the nucleus accumbens regulate activity of ventral tegmental area dopamine neurons. *J. Neurosci.* **21,** 4915–4922.
37. Chergui, K., Charlety, P. J., Akaoka, K., Saunier, C. F., Brunet, J. L., Buda, M., et al. (1993) Tonic activation of NMDA receptors causes spontaneous burst discharge of rat midbrain dopamine neurons in vivo. *Eur. J. Neurosci.* **5,** 137–144.
38. Christoffsen, C. L. and Meltzer, L. T. (1995) Evidence for the $N$-methyl-D-aspartate and AMPA subtype of the glutamate receptors on substantia nigra dopamine neurons: Possible preferential role for $N$-methyl-D-aspartate receptors. *Neuroscience* **67,** 373–381.
39. White, F. J. and Kalivas, P. W. (1998) Neuroadaptations involved in amphetamine and cocaine addiction. *Drug Alcohol Depend.* **51,** 141–153.
40. Pap, A. and Bradbery, C. W. (1995) Excitatory amino acid antagonists attenuate the effects of cocaine on extracellular dopamine in the nucleus accumbens. *J. Pharmacol. Exp. Ther.* **274,** 127–133.
41. Taber, M. T. and Fibiger, H. C. (1995) Electrical stimulation of the prefrontal cortex increases dopamine release in the nucleus accumbens of the rat: modulation by metabotropic glutamate receptors. *J. Neurosci.* **15,** 3896–3904.
42. Segovia, G., Del Arco, A., and Mora, F. (1997) Endogenous glutamate increases extracellular concentrations of dopamine, GABA and taurine through NMDA and AMPA/kainate receptors in the striatum of freely moving rats: a microdyalisis study. *J. Neurochem.* **69,** 1476–1483.
43. Kim, J.-H. and Vezina, P. (1999) Blockade of glutamate reuptake in the rat nucleus accumbens increases locomotor activity. *Brain Res.* **819,** 165–169.
44. Dalia, A., Uretsky, N. J., and Wallace, L. J. (1998) Dopaminergic agonists administered into the nucleus accumbens: effects on extracellular glutamate and on locomotor activity. *Brain Res.* **788,** 111–117.
45. Reid, M. S., Ksu, K., and Berger, S. P. (1997) Cocaine and amphetamine preferentially stimulate glutamate release in the limbic system: studies on the involvement of dopamine. *Synapse* **2,** 95–105.
46. Donzanti, B. A. and Uretsky, N. J. (1983) Effects of excitatory amino acids on locomotor activity after bilateral microinjection into the rat nucleus accumbens: possible dependence on dopaminergic mechanisms. *Neuropharmacology* **22,** 971–981.
47. Hamilton, M. H., De Belleroche, J. S., Gardiner, I. M., and Herber, L. J. (1986) Stimulatory effect of $N$-methyl asparate on locomotor activity and transmitter release from rat nucleus accumbens. *Pharmacol. Biochem. Behav.* **25,** 943–948.

48. Pulvirenti, L., Swerdlow, N. R., and Koob, G. F. (1989) Microinjection of a glutamate antagonist into the nucleus accumbens reduces psychostimulant locomotion in rats. *Neurosci. Lett.* **103**, 197–245.
49. Pulvirenti, L., Swerdlow, N. R., and Koob, G. F. (1991) Nucleus accumbens NMDA antagonist decreases locomotor activity produced by cocaine, heroin, or accumbens dopamine, but not caffeine. *Pharmacol. Biochem, Behav.* **40**, 1841–1845.
50. Kelley, A. E. and Throne, L. C. (1992) NMDA receptors mediate the behavioral effects of amphetamine infused into the nucleus accumbens. *Brain Res. Bull.* **39**, 247–254.
51. Mogenson, G. J. (1987) Limbic-motor integration, *in Psychobiology and Physiological Psychology* (Sprague, J. and Epstein, A. N., eds.), Academic P, New York, vol. 12, pp. 117–170.
52. Hooks, M. S. and Kalivas, P. W. (1994) Involvement of dopamine and excitatory amino acid transmission in novelty-induced motor activity. *J. Pharmacol. Exp. Ther.* **269**, 976–988.
53. Freed, W. (1994) Glutamatergic mechanisms mediating stimulant and antipsychotic drug effects. *Neurosci. Biobehav. Rev.* **18**, 111–120.
54. Pulvirenti, L., Maldonado, R., and Koob, G. F. (1992) NMDA receptors in the nucleus accumbens modulate intravenous cocaine, but not heroin self-administration, in the rat. *Brain Res.* **594**, 327–330.
55. Heimer, L., Alheid, G. F., de Olmos, J. S., Groenewegen, H. J., Haber, S. N., Harlan, R. E., et al. (1997) The accumbens: beyond the core-shell dichotomy. *J. Neuropsychiatry Clin. Neurosci.* **9**, 354–381.
56. Pulvirenti, L., Berrier, R., Kriefeldt, M., and Koob, G. F. (1994) Modulation of locomotor activity by NMDA receptors in the nucleus accumbens core and shell regions of the rat. *Brain Res.* **664**, 231–236.
57. Burns, L. H., Everitt, B. J., Kelley, A. E., and Robbins, T. W. (1994) Glutamate–dopamine interactions in the ventral striatum: role of locomotor activity and responding with conditioned reinforcement. *Psychopharmacology* **115**, 516–528.
58. Callaway, C. W. and Henriksen, S. J. (1992) Neuronal firing in the nucleus accumbens is associated with the level of cortical arousal. *Neuroscience* **51**, 547–553.
59. Carlezon, W. A., Jr. and Wise, R. A. (1996) Rewarding actions of phencyclidine and related drugs in nucleus accumbens shell and frontal cortex. *J. Neurosci.* **16**, 3112–3122.
60. Pierce, R. C., Meil, W. M., and Kalivas, P. W. (1997) The NMDA antagonist dizocilpine enhances cocaine reinforcement without influencing mesoaccumbens dopamine transmission. *Psychopharmacology* **133**, 188–195.
61. Ranaldi, R., French, E., and Roberts, D. C. S. (1996) Systemic pretreatment with MK-801 (dizocilpine) increases breaking point for self-administration of cocaine on a progressive ratio schedule in rats. *Psychopharmacology* **128**, 83–88.
62. De Vries, T. J., Schoffelmeer, A. N., Binnekade, R., Mulder, A. H., and Vanderschuren, L. J. (1998) MK-801 reinstates drug-seeking behaviour in cocaine-trained rats. *NeuroReport* **9**, 637–640.
63. Wolf, M. E. (1999) NMDA receptors and behavioral sensitization: beyond dizocilpine. *Trends Pharmacol. Sci.* **20**, 188–189.
64. Miller, D. W. and Abercrombie, E. D. (1996) Effects of MK-801 on spontaneous and amphetamine-stimulated dopamine release in striatum measured with in vivo microdialysis in awake rats. *Brain Res. Bull.* **40**, 57–62.
65. French, E. D. (1986) Effects of phencyclidine on ventral tegmental A10 neurons in the rat. *Neuropharmacology* **25**, 241–248.
66. French, E. D. and Ceci, A. (1990) Non-competitive NMDA antagonists are potent activators of ventral tegmental A10 dopamine neurons. *Neurosci. Lett.* **19**, 159–162.
67. French, E. D. (1992) Competitive NMDA antagonists attenuate phencyclidine-induced excitation of A10 dopamine neurons. *Eur. J. Pharmacol.* **217**, 1–7.
68. French, E. D., Mura, A., and Wang, T. (1993) MK-801, phencyclidine (PCP) and PCP-like drugs increase burst firing of rat A10 dopamine neurons: comparison to competitive NMDA antagonists. *Synapse* **13**, 108–116.
69. French, E. D. (1994) Phencyclidine and the midbrain dopamine system: electrophysiology and behavior. *Neurotoxicol. Teratol.* **16**, 355–362.
70. Connelly, S. T. and Shepard, F. D. (1997) Competitive NMDA receptor antagonists differentially affect dopamine cell firing pattern. *Synapse* **25**, 234–242.
71. Wolf, M. E., White, F. J., and Hu, T.-X. (1994) MK-801 prevents alterations in the mesoaccumbens dopamine system associated with behavioral sensitization to amphetamine. *J. Neurosci.* **14**, 1735–1745.
72. Hyttiaa, P., Backstrom, P., and Liljequist, S. (1998) Site-specific NMDA receptor antagonists produce differential effects on cocaine self-administration in rats. *Eur. J. Pharmacol.* **378**, 9–16.
73. Pulvirenti, L., Balducci, C., and Koob, G. F. (1997) Dextromethorphan reduces cocaine self-administration in the rat. *Eur. J. Pharmacol.* **321**, 279–283.
74. Bliss, T. V. P. and Collingridge, G. L. (1993) A synaptic model of memory: long-term potentiation in the hippocampus. *Nature* **361**, 31–39.
75. Pennartz, C. M. A., Ameerun, R. F., Groenewegen, H. J., and Lopes da Silva, F. H. (1993) Synaptic plasticity in an in vitro slice preparation of the rat nucleus accumbens. *Eur. J. Neurosci.* **5**, 107–117.
76. Kombian, S. B. and Malenka, R. C. (1994) Simultaneous LTP of non-NMDA- and LTD of NMDA-receptor-mediated responses in the nucleus accumbens. *Nature* **368**, 242–246.
77. Pulvirenti, L., Criado, J., Balducci, C., Koob, G. F., and Henriksen, S. J. (1998) Enhanced synaptic efficacy in the nucleus accumbens during the acquisition of cocaine self-administration. *Soc. Neurosci. Abst.* **24**, 779.

78. Ehrman, R. N., Robbins, S. J., Childress, A. R., and O'Brien, C. P. (1992) Conditioned response to cocaine related stimuli in cocaine abuse patients. *Psychopharmacology* **107,** 523–529.
79. Worley, C. A., Vadadez, A., and Schenk, S. (1994) Reinstatement of extinguished cocaine-taking behavior by cocaine and caffeine. *Pharmacol. Biochem. Behav.* **48,** 217–221.
80. Taylor, J. R. and Robbins, T. W. (1984) Enhanced behavioral control by conditioned reinforcers following microinjection of d-amphetamine into the nucleus accumbens. *Psychopharmacology* **84,** 405–412.
81. Cador, M., Robbins, T. W., and Everitt, B. J. (1989) Involvement of the amygdala in stimulus–reward associations: interaction with the ventral striatum, *Neuroscience* **30,** 77–86.
82. Everitt, B. J., Cador, M., and Robbins, T. W. (1989) Interactions between the amygdala and ventral striatum in stimulus–reward associations: studies using a second-order schedule of sexual reinforcement. *Neuroscience* **30,** 63–75.
83. Vorel, S. R., Liu, X., Hayes, R. J., Spector, J. A., and Gardner, E. L. (2001) Relapse to cocaine-seeking after hippocampal theta burst stimulation. *Science* **292,** 1175–1177.
84. See, R. E., Kruzich, P. J., and Grimm, J. W. (2001) Dopamine, but not glutamate, receptor blockade in the basolateral amygdala attenuates conditioned reward in a rat model of relapse to cocaine-seeking behavior. *Psychopharmacology* **101,** 301–310.
85. Huang, E. P. (1997) Synaptic plasticity: a role for nitric oxide in LTP. *Curr. Biol.* **7,** R141–R143.
86. West, A. R. and Galloway, M. P. (1997) Endogenous nitric oxide facilitates striatal dopamine and glutamate efflux in vivo: role of ionotropic glutamate receptor-dependent mechanisms. *Neuropharmacology* **36,** 1571–1581.
87. Ohno, M. and Watanabe, S. (1995) Persistent increase in dopamine release following activation of metabotropic glutamate receptors in rat nucleus accumbens. *Neurosci. Lett.* **200,** 113–116.
88. Bowyer, J. F., Clausing, P., Gough, B., Slikker, W., and Holson, R. R. (1995) Nitric oxide regulation of methamphetamine-induced dopamine release in caudate/putamen. *Brain Res.* **699,** 62–70.
89. West, A. R. and Galloway, M. P. (1997b) Inhibition of glutamate reuptake potentiates endogenous nitric oxide-facilitated dopamine efflux in the rat striatum: an in vivo microdialysis study. *Neurosci. Lett.* **230,** 21–24.
90. Cox, B. A. and Johnson, S. W. (1998) Nitric oxide facilitates *N*-methyl-D-aspartate-induced burst firing in dopamine neurons from rat midbrain slices. *Neurosci. Lett.* **255,** 131–134.
91. Kim, H. and Park, W. (1995) Nitric oxide mediation of cocaine-induced dopaminergic behaviors: ambulation-accelerating activity, reverse tolerance and conditioned place preference in mice. *J. Pharmacol. Exp. Ther.* **275,** 551–557.
92. Itzhak, Y. and Ali, S. F. (1996) The neuronal nitric oxide synthase inhibitor, 7-nitroindazole, protects against methamphetamine-induced neurotoxicity in vivo. *J. Neurochem.* **67,** 1770–1773.
93. Itzhak, Y. (1996) Attenuation of cocaine kindling by 7-nitroindazole, an inhibitor of brain nitric oxide synthase. *Neuropharmacology* **35,** 1065–1073.
94. Haracz, J. L., MacDonall, J. S., and Sicar, R. (1997) Effects of nitric oxide synthase inhibitors on cocaine sensitization. *Brain Res.* **746,** 183–189.
95. Pulvirenti, L., Balducci, C., and Koob, G. F. (1996) Inhibition of nitric oxide synthesis reduces cocaine self-administration in the rat. *Neuropharmacology* **35,** 1811–1814.
96. Orsini, C., Izzo, E., Koob, G. F., and Pulvirenti, L. (2001) Inhibition of nitric oxide synthase reduces extinction and relapse of cocaine-seeking behavior in the rat. Submitted.

# 12
# Glutamate and the Self-Administration of Psychomotor-Stimulant Drugs

## Paul Vezina, PhD and Nobuyoshi Suto, MA

## 1. INTRODUCTION

Psychomotor-stimulant drugs such as the amphetamines and cocaine are self-administered by humans and laboratory animals and produce locomotor activation. Repeated exposure to these drugs produces long-term enhancements in their ability to elicit these locomotor responses so that subsequent re-exposure to the drug, weeks to months later, produces greater behavioral activation than seen initially. Most importantly, previous exposure to such sensitizing regimens of amphetamine injections has also been reported to produce long-lasting enhancements in animals' predisposition to self-administer the drug. The long-term neurobiological changes associated with these enhancements may also figure importantly in the reinstatement of drug taking in individuals that have been drug-free for some time.

Considerable evidence links meso-accumbens dopamine (DA) neurons to the locomotion produced and the self-administration supported by psychomotor stimulants like amphetamine. It is generally agreed that this drug produces effects in both the cell body and terminal regions of these neurons. In the ventral tegmental area (VTA, site of the cell bodies of these neurons), amphetamine appears to initiate the neuronal events underlying behavioral sensitization. In the nucleus accumbens (NAc, their major subcortical projection), it initiates the neuronal events underlying its acute and, at least in part, the expression of its sensitized behavioral effects. Results obtained in this laboratory and reviewed in this chapter suggest a direct relation between the sensitization of meso-accumbens DA neuron reactivity and the excessive pursuit and self-administration of drugs observed in sensitized animals.

Interactions between excitatory amino acid (EAA) and DA pathways in the basal ganglia have been known for some time to contribute importantly to the generation of motor behaviors. In particular, the role played by ionotropic glutamate receptors (iGluRs) in such interactions and in the production of locomotion has received considerable attention particularly in brain areas such as the VTA, where EAA afferants are known to modulate the activity of DA neurons, and the NAc, where descending EAA projections and ascending DA mesencephalic projections come in close apposition to each other and coinnervate intrinsic neurons projecting to motor-output regions. Given such an anatomical arrangement, it is not surprising that considerable evidence now indicates a critical contribution by the EAA glutamate both to the induction and the expression of sensitization by amphetamine. In particular, the role played by iGluRs has received considerable attention.

Recently, the growing importance of the metabotropic glutamate receptor (mGluR) in the generation of motor behaviors and various forms of plasticity has begun to emerge. The known coupling of the mGluR to second-messenger systems and its demonstrated role in the long-term modulation of synaptic transmission make it an attractive candidate for more than simply the generation of

From: *Contemporary Clinical Neuroscience: Glutamate and Addiction*
Edited by: Barbara H. Herman et al. © Humana Press Inc., Totowa, NJ

locomotion involving EAA–DA interactions. Importantly, these characteristics make it an ideal contributor to the induction and expression of locomotor plasticity involving these neurotransmitters. Indeed, available evidence obtained with nonselective ligands already indicates an important role for the mGluR in amphetamine sensitization.

## 2. AMPHETAMINE, LOCOMOTION, MESO-ACCUMBENS DA, AND SENSITIZATION

### 2.1. Amphetamine-Induced Locomotion

It is now generally agreed that ascending meso-accumbens DA neurons are critical for the locomotor stimulation produced by psychomotor stimulants like amphetamine. This drug increases extracellular levels of DA primarily by causing reverse transport of DA and preventing its uptake via the DA transporter *(1,2)*. Infusions of amphetamine into the NAc increase locomotor activity *(3,4)*. This effect is prevented by injections of DA receptor antagonists into or 6-OHDA lesions of the DA nerve terminals in this site *(3,5–7)*.

### 2.2. Locomotor Sensitization by Amphetamine: DA and Induction

Repeated exposure to amphetamine produces long-term enhancements in its ability to elicit locomotor activity so that subsequent re-exposure to the drug, weeks to months later, produces greater behavioral activation than seen initially. It is an action of amphetamine in the VTA that is responsible for the induction of this locomotor sensitization. For example, sensitization of locomotion is produced by repeated infusions of the drug into the VTA, but not when it is infused repeatedly into a number of DA neuron terminal fields, including the NAc *(4,8–12)*. Because of these findings, efforts to determine the neurobiological basis of locomotor sensitization by psychomotor-stimulant drugs have concentrated on the VTA for its induction and the NAc for its expression. Psychomotor stimulants increase the extracellular content of DA in the VTA *(13)* and this event, resulting in the activation of local D1 DA receptors, is critical for the induction of sensitization by this drug (Fig. 1A; *see* refs *14, 17* and *18*). In addition, a number of short-term changes in meso-accumbens DA neurotransmission (observed 1 h to 3 d after the last drug injection) have been reported in this site, although their contribution to the induction or the expression of sensitization remains unknown. These include changes in basal extracellular DA levels *(13)*, tyrosine hydroxylase protein content *(19,20)*, levels of the G-protein subunits $G_{o\alpha}$ and $G_{i\alpha}$ *(21)*, sensitivity of D2 DA autoreceptors, and the number and firing rate of A10 DA neurons *(22,23)*.

### 2.3. Locomotor Sensitization by Amphetamine: DA and Expression

The change in meso-accumbens DA neurotransmission most consistently associated with the expression of locomotor sensitization to psychomotor-stimulant drugs, on the other hand, is not seen 3–4 d after the last drug injection but rather appears to increase with time. Enhanced drug-induced increases in levels of extracellular NAc DA have been demonstrated 1 wk to 3 mo following the last drug injection, indicating that this change may be associated with the persistence of behavioral sensitization to this drug *(23–29;* cf. ref. *30)*. The increase in extracellular content of DA in the NAc produced by systemic amphetamine is also enhanced by previous exposure to the drug in the VTA *(18,31)*. Functional supersensitivity of D1 but not D2, DA receptors in the NAc has also been shown to occur following previous exposure to amphetamine, administered systemically *(32)* or in the VTA *(33)*. Unlike the increases in extracellular content of DA described above, this change, postsynaptic to DA neuron terminals in the NAc, is seen 1 d to 1 mo after the last drug injection and diminishes after longer withdrawal periods.

Given that behavioral sensitization is observed at all times following pre-exposure, it would appear that different substrates may underlie enhanced responding at different withdrawal times. Indeed,

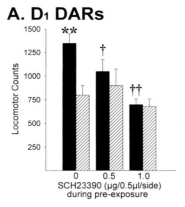

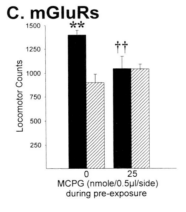

**Fig. 1.** Activation of D1 DA, *N*-methyl-D-aspartate (NMDA), and metabotropic glutamate (mGlu) receptors in the VTA is necessary for the induction of locomotor sensitization by amphetamine. **(A)** The effect of administering the D1 DA receptor antagonist, SCH-23390, into the VTA during pre-exposure on the induction of locomotor sensitization by amphetamine; **(B)** the effect of administering the NMDA receptor antagonist, AP-5; **(C)** the effect of administering the mGluR antagonist *(RS)*-MCPG. Rats were administered the D1 DA and NMDA receptor antagonists into the VTA prior to each of the systemic amphetamine injections (1.0 mg/kg ip) during pre-exposure (A and B). In the mGluR experiment (C), the receptor antagonist was administered in cocktail with amphetamine (2.5 µg/0.5 µl/side) into the VTA during pre-exposure. Some time (up to 2 wk following the last pre-exposure injection, all rats were challenged with a systemic injection of amphetamine (0.5–1.0 mg/kg ip). No receptor antagonists were administered on this test. Data are shown as group mean (+SEM) 2-h-session total locomotor counts. Bars: amphetamine (filled) and saline (hatched) pre-exposed. **$p<0.01$, ***$p<0.001$: significantly different from saline pre-exposed rats at the specified dose of the receptor antagonist. †$p<0.05$, ††$p<0.01$, †††$p<0.001$: significantly different from rats previously exposed to amphetamine alone. *n* group=4–10. (Adapted from refs. *14* [A], *15* [B], and *16* [C].)

when viewed in the context of the basal ganglia and the interconnections between its various nuclei, it is clear that meso-accumbens DA neurotransmission, whether associated with the acute or repeated effects of psychomotor stimulants, necessarily impacts the generation of various behaviors by means of interactions with other neuronal systems. As described below, those using glutamate as a neurotransmitter have received particular attention.

## 3. GLUTAMATE–DA INTERACTIONS: LOCOMOTION AND PSYCHOMOTOR-STIMULANT SENSITIZATION

### 3.1. Generation of Locomotor Activity: Glutamate–DA Interactions in VTA and NAc

Interactions between glutamate and DA and their impact on the generation of locomotor behaviors have been highlighted in both the VTA and the NAc. The NAc, in addition to being the major subcortical terminal field of DA perikarya located in the VTA, also receives extensive glutamatergic projections directly from the prefrontal cortex and limbic structures such as the hippocampal formation and amygdala *(34,35)*. Ultrastructural studies of the NAc indicate that some of the terminals of the descending EAA projections from cortex and those of ascending DA mesencephalic projections not only come in close apposition to each other but form synaptic contacts with the same intrinsic NAc neurons as well *(36,37)*. Considering that these latter neurons project to motor-output regions *(38)*, such an anatomical arrangement provides the basis for possible behaviorally relevant interactions between DA and glutamate at the level of nerve terminals in the NAc. The nature of these interactions and the manner in which they influence behavior appear complex and are not competely understood. Considerable evidence has nonetheless accumulated indicating that DA and glutamate, acting via iGluRs and mGluRs, do interact in the NAc and that these interactions contribute importanlty to the generation of locomotor behaviors (for reviews, see, refs. *39–41)*.

The VTA, site of the DA perikarya projecting to the NAc, also receives EAA inputs from a number of sources, including the prefrontal cortex, subthalamic nucleus, laterodorsal tegmentum, and habenula *(37,42–48)*. EAAs in the VTA have been shown to regulate, by acting at iGluRs on DA cell bodies and dendrites *(49)*, the firing of meso-accumbens DA neurons and, consequently, the DA released by their terminals in the NAc *(50–54)*. Not surprisingly, infusion into the VTA of agonists selective for iGluRs has also been shown to produce DA-dependent increases in locomotor activity *(55–57)*. Activation of mGluRs (groups I and II) in the VTA has also been reported to increase locomotor activity *(58)*, but the mechanisms underlying these effects remain unclear. (1S,3R)-ACPD, a broad-spectrum agonist of mGluRs, has been reported to current-dependently increase the firing of DA neurons when microiontophoretically applied to the VTA in anesthetized rats *(59)*. However, when infused into the VTA of freely moving rats, this agonist failed to affect extracellular levels of DA in the NAc *(54)*.

### 3.2. Locomotor Sensitization by Psychomotor Stimulants: Glutamate–DA Interactions and Induction

In addition to contributing to the generation of acute locomotor effects, activation of glutamate receptors appears to play an important role in the induction and the expression of locomotor sensitization by psychomotor-stimulant drugs (for review and references, *see* ref. *60*). There have, for example, been many reports that preceding injections of amphetamine or cocaine with systemic injections of NMDA or non-NMDA, iGluR antagonists blocks the induction of locomotor sensitization normally produced by these psychomotor stimulants (e.g.,refs. *32* and *61–64)*. Consistent with the induction of sensitization by amphetamine in the VTA, blocking NMDA receptors selectively in this site has been shown to prevent the induction of locomotor sensitization by systemic cocaine *(65)*, systemic amphetamine *(15;* see Fig. 1B), or intra-VTA amphetamine *(66)*. Because the activation of D1 DA receptors in this site is necessary for the induction of sensitization by amphetamine *(14,17,18)*, this activation, by increasing glutamate release from afferent terminals in the VTA *(67;* cf. ref. *68)*, is likely also to be an important regulator of glutamate's contribution to sensitization. Thus, glutamate may act at NMDA receptors, and perhaps AMPA *(64)* receptors, expressed by DA perikarya in the VTA *(69)* to initiate long-term intracellular changes in these neurons. The consequences of such changes could be transported to the terminals of these neurons in the NAc, where amphetamine is known to produce enhanced responding in sensitized animals *(11,12)*. The transient increase in the VTA both in glutamate receptor subunit expression *(70)* and in the reactivity of DA neurons to glutamate *(59,71)* may well signal the momentary but necessary recruitment of NMDA receptors to the induction of sensitization. Consistent with this view, lesions of the medial prefrontal cortex and the resulting removal of this site's excitatory amino acid projections to the VTA block the induction of stimulant sensitization *(63,72)*. Alternatively, such lesions have also been reported to prevent the *expression* of locomotor sensitization *(73;* cf. ref.*74)*, suggesting that changes in the afferent regulation of DA neurons in the VTA may be altered in sensitized animals. It has been reported, for example, that animals previously exposed to cocaine show long-term enhancements in D1 DA receptor mediated increases in extracellular glutamate content in the VTA *(75;* cf. ref. 76) and, interestingly, long-term decrements in D1 DA receptor mediated increases in extracellular GABA levels in this site *(77)*.

Ventral tegmental area mGluRs have also been shown to contribute to the induction locomotor sensitization by psychomotor stimulants. When bilaterally co-injected with amphetamine into the VTA, the broad-spectrum mGluR antagonist *(RS)*-MCPG was found to block the induction of locomotor sensitization normally produced by amphetamine in this site (see Fig. 1C; ref. *16)*. This receptor antagonist may have blocked the induction of sensitization by preventing D1 DA receptor-dependent increases in extracellular glutamate levels from phasically activating mGluRs expressed by DA perikarya *(49,78)*. As with iGluRs, such activation of mGluRs could initiate long-term intracellular changes in these neurons. Because mGluRs are also known to modulate iGluR function in different brain regions *(79–81)*, including the substantia nigra, and the activation of iGluRs is necessary for the induction of stimulant sensitization, it is also possible that *(RS)*-MCPG in the VTA produced this effect by preventing the recruitment of iGluRs to the sensitization process. Finally, as shown for D1 DA and $A_1$ adenosine receptors in the regulation of GABA release in the VTA *(77)*, mGluRs co-expressed with D1 DA receptors on glutamate afferent terminals may contribute to the regulation of extracellular levels of glutamate in this site. *(RS)*-MCPG may have blocked amphetamine sensitization by perturbing this contribution and preventing changes in the afferent regulation of DA neurons by glutamate.

The results of a number of recent cell recording experiments using selective ligands suggest that specific mGluR subtypes may play a role in some of these potential effects. Rapid activation of the group I mGluR1 subtype has been shown, for example, to produce hyperpolarization of DA neurons in the VTA that turns to depolarization following prolonged activation [reflecting possible desensitization *(82)*]. Interestingly, amphetamine selectively blocks the mGluR1-mediated hyperpolarization *(83)* that would be expected to promote more somatodendritic DA release by this drug. Also, selective inhibition of the mGluR1-mediated hyperpolarization by adenosine was also observed following cocaine exposure, an effect that would be expected to result in more effective burst firing mediated by glutamate afferents *(84)*. Activation of other mGluR subtypes has been shown to inhibit glutamate release in the VTA [group II *(85,86)*] and substantia nigra [group III *(87)*] and the former effect is enhanced soon after morphine pre-exposure *(88)*.

Taken together, the above findings clearly indicate that glutamate and DA can interact in a number of ways in the VTA to influence the induction and expression of sensitization by psychomotor stimulants.

### 3.3. Locomotor Sensitization by Psychomotor Stimulants: Glutamate–DA Interactions and Expression

A number of studies have shown that glutamate neurotransmission in the NAc is also altered by repeated exposure to psychomotor stimulants. For example, previous exposure to such drugs leads to enhanced locomotor responding to NAc infusions of AMPA 1 d to 3 wk later *(89,90)*. Enhanced cocaine-induced NAc glutamate overflow in stimulant pre-exposed animals has also been reported 10 d to 3 wk later *(90,91)*. In animals tested with amphetamine 2 d following pre-exposure, only increased aspartate overflow *(92)* or no effect was observed *(76,93)*. Interestingly, these effects (enhanced NAc glutamate overflow at later, but not early, withdrawal times and enhanced responding to NAc AMPA at 1 d to 3 wk withdrawal) parallel the time-course observed with DA overflow and supersensitivity of D1 DA receptors in the NAc and may reflect interactions in this site between glutamate and DA in the expression of sensitization by psychomotor stimulants. Glutamate and DA do appear to interact to influence each other's synaptic release in the striatum and NAc *(94–101)*. In addition, despite the development of D1 DA receptor supersensitivity in the NAc, local infusions of D1 DA receptor agonists fail to produce sensitized locomotor responding in amphetamine pre-exposed rats *(102,103) unless* glutamate reuptake is simultaneously inhibited in this site *(104)*. Thus, while reflecting complex effects, these data collectively indicate that glutamate and DA can interact in a number of ways, both presynaptically and postsynaptically, in the NAc to influence the expression of sensitization by psychomotor stimulants (for discussion, see refs. *39* and *60*). NAc mGluRs may play an important role in such effects. Rats exposed 2 wk earlier to systemic amphetamine have been reported to show enhanced lomotor activity in response to *(RS)*-MCPG, but not 1*S*,3*R*-ACPD, in the NAc *(105)*. Although these findings do not, by themselves, preclude the need for an intact dopaminergic afferentation of the NAc, they are consistent with the production of long-term changes in glutamatergic neurotransmission in the NAc of animals previously exposed to psychomotor stimulants and the regulation of these changes by mGluRs. In these animals, the blockade of NAc mGluRs may thus increase locomotor output by disinhibiting the release of glutamate from glutamate afferent terminals (for discussion, see ref. *41*). Consistent with this possibility, similar results have been obtained in this laboratory with the selective group II mGluR antagonist LY341495. Activation of group II mGluRs is known to inhibit glutamate release in the NAc *(106)*.

## 4. SENSITIZATION AND THE SELF-ADMINISTRATION OF PSYCHOMOTOR STIMULANT DRUGS

### 4.1. Facilitation of Psychomotor-Stimulant Self-Administration

It is now well established that animals previously exposed to sensitizing injection regimens of morphine, amphetamine, cocaine, or treatments with environmental stressors are subsequently more sensitive to the rewarding or incentive-motivational effects of these drugs as measured by the conditioned place-preference procedure *(107–109)*. In addition, these animals are more susceptible to self-administer amphetamine and cocaine *(110–118)*. The long-term neurobiological changes associated with these effects have also been suggested to play an important role in the reinstatement of drug-taking in individuals that have been drug free for some time *(109,119–124)*. By examining those neuronal events known to influence locomotor sensitization to such drugs, it may be possible to gain an understanding of those neuronal systems underlying predisposition to self-administer these very drugs and liability for reinstatement of this behavior even in drug-free individuals.

The neurotransmitter system most studied in this context has been that comprising the mesoaccumbens DA neurons. As noted earlier, there is general agreement that this system mediates the locomotor activity produced and the self-administration supported by psychomotor-stimulant drugs as well as by more natural incentives *(119,125,126)*. For example, rats will self-administer amphetamine into the NAc *(127,128)*, whereas infusions of DA receptor antagonists, 6-OHDA, or, kainic acid into this site have been found to disrupt iv self-administration of amphetamine and cocaine *(129–133)*. These manipulations are also known to block the locomotor effects of psychomotor stimulant drugs *(18,134–136)*. It is possible, therefore, that the sensitization observed following drug exposure is due to an exaggeration of processes that are normally set in motion when an animal initially encounters biologically significant stimuli and interacts with them. These processes may serve to facilitate or promote subsequent encounters with these stimuli. If activity in meso-accumbens DA neurons underlies the incentive valence of stimulant and opiate drugs and these neurons become sensitized when repeatedly exposed to such drugs, it would be expected that the subsequent administration of one of these drugs would produce an enhanced incentive to pursue the drug (see refs. *119, 123,* and *137*).

Although phasic decreases in extracellular DA levels in the NAc may play a role in triggering successive responses for drug in well-trained animals *(138,139)*, these levels have also been shown to be tonically increased in these animals during the self-administration period *(140,141)*. Other studies have also shown that extinguished heroin or cocain self-administration behavior can be reinstated or primed by a noncontingent iv injection of the administered drug, intra-VTA morphine, or intra-NAc amphetamine, suggesting that it is synaptic DA, and not its absence, that promotes self-administration behavior *(120–122, 142–146)*.

The evidence that previous exposure to psychomotor stimulants enhances the self-administration of these drugs has, until recently, been limited to that obtained in studies assessing how readily animals will acquire such behaviors under relatively simple schedules of reinforcement with low drug doses for intravenous injections (e.g., refs. *110–116*). This approach obviated the need to train animals to self-administer the drug before testing. It was reasoned that exposure to the drug during such training could compromise existing differences in drug pre-exposure history between animals in different groups. Indeed, self-administration of cocaine has been shown to produce locomotor sensitization *(147,148)*. *One interesting problem with this approach, however, has been its failure to generalize to higher drug doses.* For example, when high intravenous drug doses are made available, drug and saline pre-exposed animals do not to differ in their acquisition of self-administration behaviors *(117,118)*. It would seem, therefore, that even though previous exposure to a drug may decrease the threshold dose able to initiate and maintain responding, all animals irrespective of drug pre-exposure history will readily self-administer a large dose of the drug when given the opportunity to do so on a simple schedule of reinforcement. These findings suggest that sensitization is linked not so much to the act of drug-taking itself but rather to enhanced incentive to engage in this behavior. Given the earlier report of Mendrek et al. *(117)*, we reasoned that because subthreshold drug doses can elicit responding in animals previously exposed to the drug, these animals would be expected to work more than drug-naive animals to obtain a high dose of the drug. This could be observed if the demand characteristics required to obtain the drug were increased progressively, as with a progressive ratio (PR) schedule of reinforcement *(118,137,149)*. According to this view, sensitized individuals would be expected to more actively seek drug, to maintain drug-taking behaviors more readily, and to do so in a greater variety of situations than nonsensitized individuals. Clearly, identification of the neuronal events leading to sensitization would be of extreme importance for the understanding of those events leading to such behaviors, ultimately characteristic of drug-craving and abuse.

The results of more recent studies support this view and have begun to shed some light on the neuronal mechanisms underlying the excessive pursuit and self-administration of psychomotor-stimulant drugs. For example, when rats previously exposed to systemic amphetamine were placed on a PR schedule and made to work more for each successive infusion of a high dose (200 μg/kg/infusion) of the drug, they emitted a far greater number of lever presses (often threefold higher) than saline pre-exposed animals *(117,118)*. In one of these studies, the difference between groups was maintained not only during six consecutive days of PR self-administration, but it was still apparent close to 3 wk later *(118)*. In addition, these long-term consequences of prior exposure to amphetamine on the subsequent self-administration of the drug have been found to be accompanied by enhanced levels of extracellular DA in the NAc *(118,149)*. Because these latter results were obtained after animals received a systemic priming injection of amphetamine and exhibited lever pressing for saline (*see* refs. *123* and *124*), they clearly were not due to the presence of higher brain levels of self-administered drug. Rather, they appear to reflect a role for this neurotransmitter in the expression of long-term enhancements in animals' predisposition or incentive to pursue the drug. It was recently reported that cocaine-experienced animals actually exhibited less DA overflow in the NAc relative to controls following an ip injection of cocaine on a reinstatement test *(150)*. It must be noted, however, that in this latter experiment, controls had previously received the drug only as yoked injections and had never been exposed to the levers prior to the reinstatement test, making it difficult to interpret the behavioral and biochemical results obtained and understand how they are related to each other. The above data obtained with the PR schedule, on the other hand, were obtained in groups with equal access to the drug lever but differing in their prior exposure to a sensitizing-drug regimen.

As shown in Fig. 2, the above findings have been extended in the laboratory to include an assessment of the effect of previous exposure to amphetamine when it is infused into the VTA. Rats previously exposed to amphetamine in this site, but not in the NAc (not shown here) or in sites surrounding the VTA, emitted significantly more presses and obtained more amphetamine throughout PR testing *(149)*. In addition, consistent with reports of sensitized locomotor and NAc DA responding to systemic cocaine following exposure to amphetamine administered systemically or into the VTA *(9)*, rats previously exposed to VTA amphetamine were also found to consistently work more and to obtain more iv infusions of cocaine *(151)*.

## 4.2. Facilitation of Psychomotor Stimulant Self-Administration: Glutamate–DA Interactions and Induction

The above findings indicate that, in a manner similar to the sensitization of amphetamine's locomotor and NAc DA activating effects, this drug acts in the VTA, but not in the NAc, to produce changes that subsequently promote not only its pursuit and self-administration but that of other psychomotor stimulants as well.

Given the critical role of D1 DA receptors in the VTA in the induction of sensitization of the locomotor and NAc DA effects of amphetamine (*see* Section 2.2.), the role played by these receptors in the facilitation of cocaine self-administration by amphetamine in the VTA was investigated in the laboratory. It was found that, consistent with the above view, blocking D1 DA receptors in the VTA prevented the induction of sensitization by amphetamine in this site as assessed by the subsequent self-administration of cocaine on a PR schedule of reinforcement (Fig. 3). Whereas rats previously exposed to VTA amphetamine worked more and obtained more cocaine than VTA saline pre-exposed animals, rats previously exposed to VTA amphetamine+SCH-23390 failed to show this enhanced self-administration of cocaine. Previous exposure to the D1 DA receptor antagonist alone in the VTA was without effect on the subsequent self-administration of cocaine *(151)*. In a similar experiment, Pierre and Vezina *(152)* showed that preceding each systemic amphetamine pre-exposure injection with a sc injection of SCH-23390 prevented the facilitation of acquisition of self-administration of a low dose of the drug.

These findings were then extended to the case of NMDARs and mGluRs in the VTA. Again, as outlined earlier, activation of both of these receptors in the VTA has been shown to be necessary for the induction of locomotor sensitization by amphetamine in this site. In a series of experiments conducted in the laboratory, it was found that facilitation of cocaine self-administration by amphetamine was prevented by infusing NMDAR and mGluR antagonists with amphetamine into the VTA during pre-exposure. As in the above experiments, whereas rats previously exposed to VTA amphetamine worked more and obtained more cocaine than VTA saline pre-exposed animals, rats previously exposed to VTA amphetamine with CPP (the NMDAR antagonist; Fig. 4) or MCPG (the mGluR antagonist; Fig. 5) failed to show this enhanced self-administration of cocaine. In the case of CPP, this effect was dose dependent. Previous exposure to either glutamate receptor antagonist alone in the VTA was without effect on the subsequent self-administration of cocaine *(153,154)*. The findings obtained with CPP are also consistent with others reported earlier showing that a systemically administered NMDAR antagonist could prevent the facilitation by systemic cocaine of acquisition of self-administration of a low dose of this drug *(155)*.

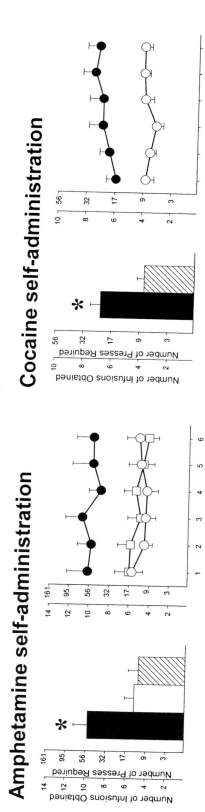

Fig. 2. VTA amphetamine pre-exposure facilitates the self-administration of amphetamine and cocaine. Different groups of rats were administered three infusions of VTA amphetamine (2.5 μg/0.5 μl/side; solid bars), VTA saline (hatched bars), or amphetamine in sites adjacent to the VTA (open bar). Infusions were made once every third day. Starting 10 d later, rats were trained to self-administer amphetamine (200 μg/kg/infusion, iv) or cocaine (300 μg/kg/infusion, iv) and then tested on a progressive-ratio schedule of reinforcement for 6 d. Bar graphs were derived from means of the values obtained for each subject on each of the six PR test days. The group mean number of infusions obtained on each of the test days is shown to the right of the bar graphs (VTA amphetamine, filled circles; VTA saline, open circles; amphetamine outside VTA, open squares). *$p<0.05$: significantly different from VTA saline pre-exposed animals. $n=7$–$11$/group. (Adapted from refs. 149 [amphetamine] and 151 [cocaine].) For detailed description of procedures, see also ref. 118.

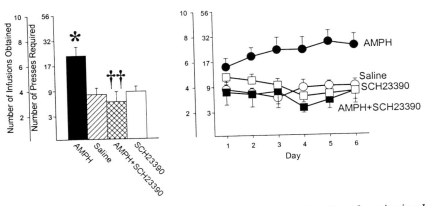

**Fig. 3.** Previous exposure to VTA amphetamine facilitates the self-administration of cocaine in a D1 DA receptor-dependent manner. Different groups of rats were administered three infusions into the VTA of amphetamine (AMPH; 2.5 μg/0.5 μl/side), saline, amphetamine + the D1 DA receptor antagonist SCH-23390 (0.25 μg/0.5 μl/side) or SCH-23390 alone. Infusions were made once every third day. Starting 10 d later, rats were trained to self-administer cocaine (300 μg/kg/infusion, iv) and then tested on a PR schedule of reinforcement for 6 d. Bar graphs were derived from means of the values obtained for each subject on each of the six PR test days. The group mean number of infusions obtained on each of the test days is shown to the right of the bar graphs. SCH-23390 completely prevented the facilitation of cocaine self-administration by VTA amphetamine. *$p<0.05$: significantly different from VTA saline pre-exposed animals; ††$p<0.01$: significantly different from rats previously exposed to VTA amphetamine alone. $n=7–10$/group. (Adapted from ref. *151*.)

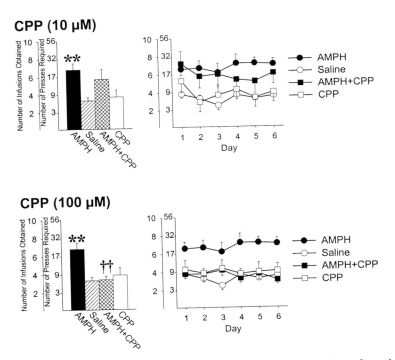

**Fig. 4.** Previous exposure to VTA amphetamine facilitates the self-administration of cocaine in an NMDA receptor-dependent manner. Different groups of rats were administered three infusions into the VTA of amphetamine (AMPH), saline, amphetamine + the NMDA receptor antagonist CPP or CPP alone. Remaining procedures and illustration of data are as described in Fig. 3. CPP dose-dependently prevented the facilitation of cocaine self-administration by VTA amphetamine. **$p<0.01$: significantly different from VTA saline pre-exposed animals; ††$p<0.01$: significantly different from rats previously exposed to VTA amphetamine alone. $n=7–12$/group (Adapted from refs. *153* and *154*.)

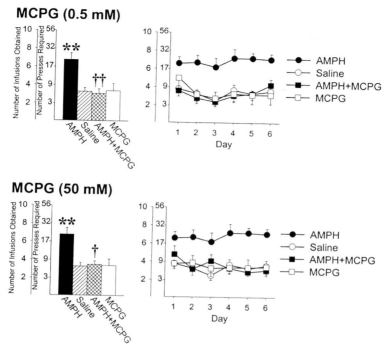

**Fig. 5.** Previous exposure to VTA amphetamine facilitates the self-administration of cocaine in an mGluR-dependent manner. Different groups of rats were administered three infusions into the VTA of amphetamine (AMPH), saline, amphetamine + the mGluR antagonist MCPG or MCPG alone. Remaining procedures and illustration of data are as described in Fig. 3. MCPG, at both concentrations tested, completely prevented the facilitation of cocaine self-administration by VTA amphetamine. **$p<0.01$: significantly different from VTA saline pre-exposed animals, †$p<0.05$, ††$p<0.01$: significantly different from rats previously exposed to VTA amphetamine alone. $n=8–12$/group. (Adapted from ref. *154*.)

Taken together, the above findings support the view that facilitation of psychomotor-stimulant self-administration represents an instance of amphetamine sensitization that is induced and expressed via the same neuronal mechanisms leading to and underlying enhanced locomotor and DA responding to the drug. These neuronal mechanisms, initiated and thus recruited by antecedent pharmacological as well as (it is important to remember) nonpharmacological events, may underlie individuals' predisposition to initiate (in the drug-free individual) or to resume (in the drug-experienced individual) the pursuit and self-administration of drugs of abuse.

The above-reviewed findings highlight consistent parallels between the induction of sensitization of the locomotor and NAc DA activating effects of amphetamine and the induction of facilitation of its ability (as well as that of other psychomotor stimulants) to support self-administration. Thus, in addition to initiating those changes leading to locomotor and NAc DA sensitization, amphetamine appears able by its actions in the VTA to also induce enduring changes that subsequently promote its self-administration as well as that of other psychomotor-stimulant drugs. Importantly, manipulations known to prevent the induction of sensitized locomotor and, in some cases, NAc DA responding *to* amphetamine (blocking D1 DA, NMDA, and mGluR receptors prior to each amphetamine pre-exposure injection) also prevent the development of enhanced responding *for* a psychomotor stimulant in a drug self-administration paradigm. Although much remains to be clarified regarding the manner in which glutamate and DA interact in the VTA (see above), it is clear that each of these neurotransmitters, by virtue of its interactions with the other, plays a critical role in initiating those changes that lead ultimately to the generation of sensitized behavioral output following the repeated administration of psychomotor stimulant drugs.

Finally, in addition to playing a critical role in the development of these effects in the VTA, glutamate–DA interactions in the NAc may also be important for the subsequent expression of excessive seeking and self-administration of these drugs. It was recently reported that the activation of AMPA receptors in the NAc mediates relapse in animals trained to self-administer cocaine *(156,157)*. In a manner consistent with the enhanced locomotor responding to NAc infusions of AMPA observed in rats pre-exposed to cocaine *(90)*, animals previously exposed to amphetamine have been found to exhibit enhanced lever pressing for amphetamine when primed with NAc AMPA *(158)*.

## ACKNOWLEDGMENT

This work was supported by USPHS grants DA-9397 and DA-9860 (PV)

## REFERENCES

1. Sulzer, D., Chen, T.-K., Lau, Y. Y., Kristensen, H., Rayport, S., and Ewing, A. (1995) Amphetamine redistributes dopamine from synaptic vesicles to the cytosol and promotes reverse transport. *J. Neurosci.* **15,** 4102–4108.
2. Sulzer, D. and Rayport, S. (1990) Amphetamine and other psychostimulants reduce pH gradients in midbrain dopaminergic neurons and chromaffin granules: a mechanism of action. *Neuron* **5,** 797–808.
3. Pijnenberg, A. J. J., Honig, W. M. M., van der Heyden, J. A. M., and van Rossum, J. M. (1976) Effects of chemical stimulation of the mesolimbic dopamine system upon locomotor activity. *Eur. J. Pharmacol.* **35,** 45–58.
4. Vezina, P. and Stewart, J. (1990) Amphetamine administered to the ventral tegmental area but not to the nucleus accumbens sensitizes rats to systemic morphine: lack of conditioned effects. *Brain Res.* **516,** 99–106.
5. Roberts, D. C. S., Zis, A. P., and Fibiger, H. C. (1975) Ascending catecholaminergic pathways and amphetamine induced locomotion: Importance of dopamine and apparent non-involvement of norepinephrine. *Brain Res.* **93,** 441–445.
6. Kelly, P. H. and Iversen, S. D. (1976) Selective 6-OHDA-induced destruction of mesolimbic dopamine neurons: abolition of psychostimulant induced locomotor activity in rats. *Eur. J. Pharamacol.* **40,** 45–56.
7. Joyce, E. M. and Koob, G. F. (1981) Amphetamine-, scopolamine- and caffeine-induced locomotor activity following 6-hydroxydopamine lesions of the mesolimbic dopamine system. *Psychopharmacology* **73,** 311–313.
8. Dougherty, G. G., Jr. and Ellinwood, E. H., Jr. (1981) Chronic *d*-amphetamine in nucleus accumbens: lack of tolerance or reverse tolerance of locomotor activity. *Life Sci.* **28,** 2295–2298.
9. Kalivas, P. W. and Weber, B. (1988) Amphetamine injected into the ventral mesencephalon sensitizes rats to peripheral amphetamine and cocaine. *J. Pharmacol. Exp. Ther.* **245,** 1095–1102.
10. Hooks, M. S., Jones, G. H., Liem, B. J., and Justice, J. B., Jr. (1992) Sensitization and individual differences to intraperitoneal amphetamine, cocaine or caffeine following repeated intracranial amphetamine infusions. *Pharmacol. Biochem. Behav.* **43,** 815–823.
11. Perugini, M. and Vezina, P. (1994) Amphetamine administered to the ventral tegmental area sensitizes rats to the locomotor effects of nucleus accumbens amphetamine. *J. Pharmacol. Exp. Ther.* **270,** 690–696.
12. Cador, M., Bjijou, Y., and Stinus, L. (1995) Evidence of a complete independence of the neurobiological substrates of the induction and expression of behavioral sensitization to amphetamine. *Neuroscience* **65,** 385–395.
13. Kalivas, P. W. and Duffy, P. (1993) Time course of extracellular dopamine and behavioral sensitization to cocaine. II. Dopamine perikarya. *J. Neurosci.* **13,** 276–284.
14. Stewart, J. and Vezina, P. (1989) Microinjections of SCH-23390 into the ventral tegmental area and substantia nigra pars reticulata attenuate the development of sensitization to the locomotor effects of systemic amphetamine. *Brain Res.* **495,** 401–406.
15. Vezina, P. and Queen, A. L. (2000) Induction of locomotor sensitization by amphetamine requires the activation of NMDA receptors in the rat ventral tegmental area. *Psychopharmacology* **151,** 184–191.
16. Kim, J. H. and Vezina, P. (1998) Metabotropic glutamate receptors are necessary for sensitization by amphetamine. *NeuroReport* **9,** 403–406.
17. Bjijou, Y., Stinus, L., Le Moal, M., and Cador, M. (1996) Evidence for selective involvement of dopamine $D_1$ receptors in the ventral tegmental area in the behavioral sensitization induced by intraventral tegmental area injections of *d*-amphetamine. *J. Pharmacol. Exp. Ther.* **277,** 1177–1187.
18. Vezina, P. (1996) $D_1$ dopamine receptor activation is necessary for the induction of sensitization by amphetamine in the ventral tegmental area. *J. Neurosci.* **16,** 2411–2420.

19. Beitner-Johnson, D. and Nestler, E. J. (1991) Morphine and cocaine exert common chronic actions on tyrosine hydroxylase in dopaminergic brain reward regions. *J. Neurochem.* **57,** 344–347.
20. Sorg, B. A., Shiouh-yi, C., and Kalivas, P. W. (1993) Time course of tyrosine hydroxylase expression following behavioral sensitization to cocaine. *J. Pharmacol. Exp. Ther.* **266,** 424–430.
21. Nestler, E. J., Terwilliger, R. Z., Walker, J. R., Sevarino, K. A., and Duman, R. S. (1990) Chronic cocaine treatment decreases levels of the G protein subunits $G_{i\alpha}$ and $G_\alpha$ in discrete regions of rat brain. *J. Neurochem.* **55,** 1079–1082.
22. Henry, D. J., Greene, M. A., and White, F. J. (1989) Electrophysiological effects of cocaine in the mesoaccumbens dopamine system: repeated administration. *J. Pharmacol. Exp. Ther.* **251,** 833–839.
23. Wolf, M. E., White, F. J., Nassar, R., Brooderson, R. J., and Khansa, M. R. (1993) Differential development of autoreceptor subsensitivity and enhanced dopamine release during amphetamine sensitization. *J. Pharmacol. Exp. Ther.* **264,** 249–255.
24. Robinson, T. E. (1988) Stimulant drugs and stress: factors influencing individual differences in the susceptibility to sensitization, in *Sensitization in the Nervous System* (Kalivas, P. W. and Branes, C. D., eds.), Telford, Caldwell, NJ, pp. 145–173.
25. Hamamura, T., Akiyama, K., Akimoto, K., Kashihara, K., Okumura, K., Ujike, H., et al. (1991) Co-administration of either a selective D1 or D2 dopamine antagonist with methamphetamine prevents methamphetamine-induced behavioral sensitization and neurochemical change, studied by in vivo intracerebral dialysis. *Brain Res.* **546,** 40–46.
26. Robinson, T. E. (1991) The neurobiology of amphetamine psychosis: evidence from studies with an animal model, in *Biological Basis of Schizophrenia* (Nakazawa, T., ed.), Japan Scientific Societies, Tokyo.
27. Segal, D. S. and Kuczenski, R. (1992) In vivo microdialysis reveals a diminished amphetamine-induced DA response corresponding to behavioral sensitization produced by repeated amphetamine pretreatment. *Brain Res.* **571,** 330–337.
28. Kalivas, P. W. and Duffy, P. (1993) Time course of extracellular dopamine and behavioral sensitization to cocaine. I. Dopamine axon terminals. *J. Neurosci.* **13,** 266–275.
29. Paulson, P. E. and Robinson, T. E. (1995) Amphetamine-induced time-dependent sensitization of dopamine neurotransmission in the dorsal and ventral striatum: a microdialysis study in behaving rats. *Synapse* **19,** 56–65.
30. Kuczenski, R., Segal, D., and Todd, P. K. (1997) Behavioral sensitization and extracellular dopamine responses to amphetamine after various treatments. *Psychopharmacology* **134,** 221–229.
31. Vezina, P. (1993) Amphetamine injected into the ventral tegmental area sensitizes the nucleus accumbens dopaminergic response to systemic amphetamine: an in vivo microdialysis study in the rat. *Brain Res.* **605,** 332–337.
32. Wolf, M. E., White, F. J., and Hu, X.-T. (1994) MK-801 prevents alterations in the mesoaccumbens dopamine system associated with behavioral sensitization to amphetamine. *J. Neurosci.* **14,** 1735–1745.
33. Hu, X.-T., Koeltzow, T. E., Cooper, D. C., Robertson, G. S., White, F. J., and Vezina, P (2002). Repeated ventral tegmental area amphetamine administration alters $D_1$ dopamine receptor signaling in the nucleus accumbens. *Synapse,* in press.
34. Christie, M. J., Summers, R. J., Stephenson, J. A., Cook, C. J., and Beart, P. M. (1987) Excitatory amino acid projections to the nucleus accumbens septi in the rat: a retrograde transport study utilizing D[3H]aspartate and [3H]GABA. *Neuroscience* **22,** 425–439.
35. Meredith, G. E., Pennartz, C. M., and Groenewegen, H. J. (1993) The cellular framework for chemical signalling in the nucleus accumbens. [review] *Prog. Brain Res.* **99,** 3–24.
36. Sesack, S. R. and Pickel, V. M. (1990) In the rat medial nucleus accumbens, hippocampal and catecholaminergic terminals converge on spiny neurons and are in apppositon to each other. *Brain Res.* **527,** 266–279.
37. Sesack, S. R. and Pickel, V. M. (1992) Prefrontal cortical efferents in ther rat synapse on unlabeled neuronal targets of catecholamine terminals in the nucleus accumbens septi and on dopamine neurons in the ventral tegmental area. *J. Comp. Neurol.* **320,** 145–160.
38. Mogenson, G. J., Brudzynski, S. M., Wu, M., Yang, C. R., and Yim, C. C. Y. (1993) From motivation to action: a review of dopaminergic regulation of limbic – nucleus accumbens – ventral pallidum – pedunculopontine nucleus circuitries involved in limbic–motor integration, in *Limbic Motor Circuits and Neuropsychiatry* (Kalivas. P. W. and Barnes, C. D., eds), CRC, Boca Raton, FL, pp. 193–236.
39. Freed, W. J. (1994) Glutamatergic mechanisms mediating stimulant and antipsychotic drug effects. *Neurosci. Biobehav. Rev.* **18,** 111–120.
40. Schmidt, W. J. and Kretschmer, B. D. (1997) Behavioural pharmacology of glutamate receptors in the basal ganglia. *Neurosci. Biobehav. Rev.* **21,** 381–392.
41. Vezina, P. and Kim, J.-H. (1999) Metabotropic glutamate receptors and the generation of locomotor activity: interactions with midbrain dopamine. *Neurosci. Biobehav. Rev.* **23,** 577–589.
42. Phillipson, O. (1979) Afferent projections to the ventral tegmental area of Tsai and interfascicular nucleus: a horseradish peroxidase study in the rat. *J. Comp. Neurol.* **187,** 117–144.
43. Christie, M. J., Bridge, S., James, L. B., and Beart, P. M. (1985) Excitotoxin lesions suggest an aspartatergic projection from rat medial prefrontal cortex to ventral tegmental area. *Brain Res.* **333,** 169–172.
44. Sesack, S. R., Deutch, A. Y., Roth, R. H., and Bunney, B. S. (1989) Topographical organization of the efferent projections of the medial prefrontal cortex in the rat: an anterograde tract-tracing study with *Phaseolus vulgaris* leucoagglutinin. *J. Comp. Neurol.* **290,** 213–242.

45. Cornwall, J., Cooper, J. D., and Phillipson, O. T. (1990) Afferent and efferent connections of the laterodorsal tegmental nucleus in the rat. *Brain Res. Bull.* **25,** 271–284.
46. Kalivas, P. W. (1993) Neurotransmitter regulation of dopamine neurons in the ventral tegmental area. *Brain Res. Rev.* **18,** 75–113.
47. Oakman, S. A., Faris, P. L., Kerr, P. E., Cozzari, C., and Harman, B. K. (1995) Distribution of pontomesencephalic cholinergic neurons projecting to substantia nigra differs significantly from those projecting to ventral tegmental area. *J. Neurosci.* **15,** 5859–5869.
48. Smith, Y., Charara, A., and Parent, A. (1996) Glutamatergic inputs from the pedunculopontine nucleus of midbrain dopaminergic neurons in primates: *phaseolus vulgaris*–leucoagglutinun anterode labeling combined with postembedding glutamate and GABA immunohistochemistry.*J.Comp. Neurol.* **364,** 254–266.
49. Mercuri, N. B., Stratta, F., Calabresi, P., and Bernardi, G. (1992) Electrophysiological evidence for the presence of ionotropic and metabotropic excitatory amino acid receptors on dopaminergic neurons of the rat mesencephalon. *Funct. Neurol.* **7,** 231–234.
50. Gariano, R. F. and Groves, P. M. (1988) Burst firing induced in mid-brain dopamine neurons by stimulation of the medial prefrontal and anterior cingulate cortices. *Brain Res.***462,** 194–198.
51. Suaud-Chagny, M. F., Chergui, K., Chouvet, G., and Gonon, J. (1992) Relationship between dopamine release in the rat nucleus accumbens and the discharge activity of dopaminergic neurons during local in vivo application of amino acids in the ventral tegmental area. *Neuroscience,* **49,** 63–72.
52. Murase, S., Grenhoff, J., Chouvet, G., Gonon, F. G., and Svensson, T. H. (1993) Prefrontal cortex regulates burst firing and transmitter release in rat mesolimbic dopamine neurons studied in vivo. *Neurosci. Lett.* **157,** 53–56.
53. Karreman, M. and Moghaddam, B. (1996) The prefrontal cortex regulates the basal release of dopamine in the limbic striatum: an effect mediated by ventral tegmental area. *J. Neurochem.* **66,** 589–598.
54. Karreman, M., Westerink, B. H. C., and Moghaddam, B. (1996) Excitatory amino acid receptors in the ventral tegmental area regulate dopamine release in the ventral striatum. *J. Neurochem.* **67,** 601–607.
55. Pycock, C. J. and Dawbarn, D. (1980) Acute motor effects of N-methyl-D-aspartic acid and kainic acid applied focally to mesencephalic dopamine cell body regions in the rat. *Neurosci. Lett.* **18,** 85–90.
56. Kalivas, P. W., Duffy, P., and Barrow, J. (1989) Regulation of the mesocoticolimbic dopamiine system by glutamic acid receptor subtypes. *J. Pharmacol. Exp. Ther.* **251,** 378–387.
57. Schenk, S. and Partridge, B. (1997) Sensitization and tolerance in psychostimulant self-administration. *Pharmacol. Biochem. Behav.* **57,** 543–550.
58. Swanson, C. J. and Kalivas, P. W. (2000) Regulation of locomotor activity by metabotropic glutamate receptors in the nucleus accumbens and ventral tegmental area. *J. Pharmacol. Exp. Ther.* **292,** 406–414.
59. Zhang, X. F., Hu, X. T., White, F. J., and Wolf, M. E. (1997) Increased responsiveness of ventral tegmental area dopamine neurons to glutamate after repeated administration of cocaine or amphetamine is transient and selectively involves AMPA receptors. *J. Pharmacol. Exp. Ther.* **281,** 699–706.
60. Wolf, M. E. (1998) The role of excitatory amino acids in behavioral sensitization to psychomotor stimulants. *Prog. Neurobiol.* **54,** 679–720.
61. Karler, R., Calder, L. D., and Turkanis, S. A. (1991) DNQX blockade of amphetamine behavioral sensitization. *Brain Res.* **552,** 295–300.
62. Stewart, J. and Druhan, J. P. (1993) Development of both conditioning and sensitization of the behavioral activating effects of amphetamine is blocked by the noncompetitive NMDA receptor antagonist, MK-801. *Psychopharmacology* **110,** 125–132.
63. Wolf, M. E., Drahlin, S. L., Hu, X. T., Xue, C. J., and White, K. (1995) Effects of lesions of prefrontal cortex, amygdala, or fornix on behavioral sensitization to amphetamine: Comparison with N-methyl-D-aspartate antagonists. *Neuroscience* **69,** 417–439.
64. Li, Y., Vartanian, A. J., White, F. J., Xue, C. J., and Wolf, M. E. (1997) Effects of the AMPA receptor antagonist NBQX on the development and expression of behavioral sensitization to cocaine and amphetamine. *Psychopharmacology* **134,** 266–276.
65. Kalivas, P. W. and Alesdatter, J. E. (1993) Involvement of NMDA receptor stimulation in the VTA and amygdala in behavioral sensitization to cocaine. *J. Pharmacol. Exp. Ther.* **267,** 486–495.
66. Cador, M., Bjijou, Y., Cailhol, S., and Stinus, L. (1999) d-Amphetamine-induced behavioral sensitization: implications of a glutamatergic medial prefrontal cortex-ventral tegmental area innervation. *Neuroscience* **94,** 705–721.
67. Kalivas, P. W. and Duffy, P. (1995) $D_1$ receptors modulate glutamate transmission in the ventral tegmental area. *J. Neurosci.* **15,** 5379–5388.
68. Wolf, M. E. and Xue, C.-J. (1998) Amphetamine and D1 dopamine receptor agonists produce biphasic effects on glutamate efflux in rat ventral tegmental area: modification by repeated amphetamine administration. *J. Neurochem.* **70,** 198–209.
69. Wang, T. and French, E. D. (1993) Electrophysiological evidence for the existence of NMDA and non-NMDA receptors on rat ventral tegmental dopamine neurons. *Synapse* **13,** 270–277.

70. Fitzgerald, L. W., Ortiz, J., Hamedani, A. G., and Nestler, E. J. (1996) Drugs of abuse and stress increase the expression of GluR1 and NMDAR1 glutamate receptor subunits in the rat ventral tegmental area: common adaptations among cross-sensitizing agents. *J. Neurosci.* **16**, 274–282.
71. White F. J., Hu, X.-T., Henry, D. J., and Zhang, X.-F. (1995) Repeated administration of cocaine or amphetamine alters neuronal responses to glutamate in the mesoaccumbens dopamine system. *J. Pharmacol. Exp. Ther.* **273**, 445–454.
72. Li, Y., Hu, X.-T., Berney, T. G., Vartanian, A. J., Stine, C. D., Wolf, M. E., et al. (1999) Both glutamate receptor antagonists and prefrontal cortex lesions prevent induction of cocaine sensitization and associated neuroadaptations. *Synapse* **34**, 169–180.
73. Pierce, R. C., Reeder, D. C., Hicks, J., Morgan, Z. R., and Kalivas, P. W. (1998) Ibotenic acid lesions of the dorsal prefrontal cortex disrupt the expression of behavioral sensitization to cocaine. *Neuroscience* **82**, 1103–1114.
74. Li, Y. and Wolf, M. E. (1997) Ibotenic acid lesions of prefrontal cortex do not prevent expression of behavioral sensitization to amphetamine. *Behav. Brain Res.* **84**, 285–289.
75. Kalivas, P. W. and Duffy, P. (1998) Repeated cocaine administration alters D-1 dopamine receptor regulation of extracellular glutamate levels in the ventral tegmental area. *J. Neurochem.* **70**, 1497–1502.
76. Xue, C. J., Ng, J. P., Li, Y., and Wolf, M. E. (1996) Acute and repeated systemic amphetamine administration: effects on extracellular glutamate, aspartate, and serine levels in rat ventral tegmental area and nucleus accumbens. *J. Neurochem.* **67**, 352–363.
77. Bonci, A. and Williams, J. T. (1996) A common mechanism mediates long-term changes in synaptic transmission after chronic cocaine and morphine. *Neuron* **16**, 631–639.
78. Martin, L. J., Blackstone, C. D., Huganir, R. L, and Price, L. (1992) Cellular localization of a metabotropic glutamate receptor in rat brain. *Neuron* **9**, 259–270.
79. Aniksztejn, L., Otani, S., and Ben-Ari, Y. (1992) Quisqualate metabotropic receptors modulate NMDA currents and facilitate induction of long-term potentiation through protein kinase C. *Eur. J. Neurosci.* **4**, 500–505.
80. Glaum, S. R. and Miller, R. J. (1993) Activation of metabotropic glutamate receptors produces reciprocal regulation of ionotropic glutamate and GABA responses in the nucleus of the tractus solitarius of the rat. *J. Neurosci.* **13**, 1636–1641.
81. Colwell, C. S. and Levine, M. S. (1994) Metabotropic glutamate receptors modulate NMDA receptor function in neostriatal neurons. *Neuroscience* **61**, 497–507.
82. Fiorillo, C. D. and Williams, J. T. (1998) Glutamate mediates an inhibitory postsynaptic potential in dopamine neurons. *Nature* **394**, 78–82.
83. Paladini, C. A., Fiorillo, C. D., Morikawa, H., and Williams, J. T. (2001) Amphetamine selectively blocks inhibitory glutamate transmission in dopamine neurons. *Nature Neurosci.* **4**, 275–281.
84. Fiorillo, C. D. and Williams, J. T. (2000) Selective inhibition by adenosine of mGluR IPSPs in dopamine neurons after cocaine treatment. *J. Neurophysiol.* **83**, 1307–1314.
85. Bonci, A., Grillner, P., Siniscalchi, A., Mercuri, N. B., and Bernardi, G. (1997) Glutamate metabotropic receptor agonists depress excitatory and inhibitory transmission on rat mesencephalic principal neurons. *Eur. J. Neurosci.* **9**, 2359–2369.
86. Wigmore, M. A. and Lacey, M. G. (1998) Metabotropic glutamate receptors depress glutamate-mediated synaptic input to rat midbrain dopamine neurons in vitro. *Br. J. Pharmacol.* **123**, 667–674.
87. Wittmann, M., Marino, M. J., Bradley, S. R., and Conn, P. J. (2001) Activation of group III mGluRs inhibits GABAergic and glutamatergic transmission in the substantia nigra pars reticulata. *J. Neurophysiol.* **85**, 1960–1968.
88. Manzoni, O. J. and Williams, J. T. (1999) Presynaptic regulation of glutamate release in the ventral tegmental area during morphine withdrawal. *J. Neurosci.* **19**, 6629–6636.
89. Bell, K. and Kalivas, P. W. (1996) Context-specific cross-sensitization between systemic cocaine and intra-accumbens AMPA infusion in the rat. *Psychopharmacology* **127**, 377–383.
90. Pierce, R. C., Bell, K., Duffy, P., and Kalivas, P. W. (1996) Repeated cocaine augments excitatory amino acid transmission in the nucleus accumbens only in rats having developed behavioral sensitization. *J. Neurosci.* **16**, 1550–1560.
91. Reid, M. S. and Berger, S. P. (1996) Evidence for sensitization of cocaine-induced nucleus accumbens glutamate release. *NeurOreport* **7**, 1325–1329.
92. Robinson, S. E., Kunko, P. M., Smith, J. A., Wallace, M. J., Mo, Q., and Maher, J. R. (1997) Extracellular aspartate concentration increases in nucleus accumbens after cocaine sensitization. *Eur. J. Pharmacol.* **319**, 31–36.
93. Pierce, R. C., Duffy, P., and Kalivas, P. W. (1996) Changes in excitatory amino acid transmission in the nucleus accumbens associated with behavioral sensitization to cocaine during early withdrawal. *Neurosci.* **1**, article #100 08.
94. Youngren, K. D., Daly, D. A., and Moghaddam, B. (1993) Distinct actions of endogenous excitatory amino acids on the outflow of dopamine in the nucleus accumbens. *J. Pharmacol. Exp. Ther.* **264**, 289–293.
95. Smith, J. A., Mo, Q., Guo, H., Kunko, P. M., and Robinson, S. E. (1995) Cocaine increases extraneuronal levels of aspartate and glutamate in the nucleus accumbens. *Brain Res.* **683**, 264–269.
96. Taber, M. T., Baker, G. B., and Fibiger, H. C. (1996) Glutamate receptor agonists decrease extracellular dopamine in the rat nucleus accumbens in vivo. *Synapse* **24**, 165–172.
97. Kalivas, P. W. and Duffy, P. (1997) Dopamine regulation of extracellular glutamate in the nucleus accumbens. *Brain Res.* **761**, 173–177.

98. Reid, M. S., Hsu, K., Jr., and Berger, S. P. (1997) Cocaine and amphetamine preferentially stimulate glutamate release in the limbic system: studies on the involvement of dopamine. *Synapse* **27**, 95–105.
99. Segovia, G., Del Acro, A., and Mora, F. (1997) Endogenous glutamate increases extracellular concentrations of dopamine, GABA, and taurine through NMDA and AMPA/kainate receptors in striatum of the freely moving rat: a microdialysis study. *J. Neurochem.* **69**, 1476–1483.
100. West, A. R. and Galloway, M. P. (1997) Inhibition of glutamate reuptake potentiates endogenous nitric oxide-facilitated dopamine efflux in the rat striatum: an in vivo microdialysis study. *Neurosci. Lett.* **230**, 21–24.
101. Dalia, A., Uretsky, N. J., and Wallace, L. J. (1998) Dopaminergic agonists administered into the nucleus accumbens: effects on extracellular glutamate and on locomotor activity. *Brain Res.* **788**, 111–117.
102. Perugini, M. and Vezina, P. (1994) Lack of sensitization to the locomotor effects of direct $D_1$ dopamine receptor agonists in rats having been pre-exposed to amphetamine. *Soc. Neurosci. Abst.* **20**, 825.
103. Pierce, R. C. and Kalivas, P. W. (1995) Amphetamine produces sensitized increses in locomotion and extracellular dopamine preferentially in the nucleus accumbens shell of rats administered repeated cocaine.*J. Pharmacol. Exp. Ther.* **275**, 1019–1029.
104. Kim, J. H., Perugini, M., Austin, J. D., and Vezina, P. Previous exposure to amphetamine enhances the subsequent locomotor response to a $D_1$ dopamine receptor agonist when glutamate reuptake is inhibited. *J. Neurosci.* **21**, 1–6.
105. Kim, J. H. and Vezina, P. (1998) The metabotropic glutamate receptor antagonist *(RS)*-MCPC produces hyperlocomotion in amphetamine pre-exposed rats. *Neuropharmacology* **37**, 189–197.
106. Manzoni, O. J., Michel, J.-M., and Bockaert, J. (1997) Metabotropic glutamate receptors in the rat nucleus accumbens. *Eur. J. Neurosci.* **9**, 1514–1523.
107. Gaiardi, M., Bartoletti, M., Bacchi, A., Gubellini, C., Costa, M., and Babbini, M. (1991) Role of repeated exposure to morphine in determining its affective properties: place and taste conditioning studies in rats. *Psychopharmacology* **103**, 183–186.
108. Lett, R. T. (1989) Repeated exposures intensify rather that diminish the rewarding effects of amphetamine, morphine, and cocaine. *Psychophramacology* **98**, 357–362.
109. Shippenberg, T. S. and Heidbreder, C. H. (1995) Sensitization to the conditioned rewarding effects of cocaine: pharmacological and temporal characteristics. *J. Pharmacol. Exp. Ther.* **273**, 808–815.
110. Horger B. A., Shelton, K., and Schenk, S. (1990) Pre-exposure sensitizes rats to the rewarding effects of cocaine. *Pharmacol. Biochem. Behav.* **37**, 707–711.
111. Horger, B. A., Giles, M. K., and Schenk, S. (1992) Preexposure to amphetamine and nicotine predisposes rats to self-administer a low dose of cocaine. *Psycopharamacology* **107**, 271–276.
112. Piazza, P. V., Deminiere, J., Le Moal, M., and Simon, H. (1989) Factors that predict individual vulnerability to amphetamine self-administration. *Science* **245**, 1511–1513.
113. Piazza, P. V., Maccari, S. Deminière, J.-M., Le Moal, M., Mormède, P., and Simon, H. (1991) Corticosterone levels determine individual vulnerability to amphetamine self-administration. *Proc. Natl. Acad. Sci. USA* **88**, 2088–2092.
114. Pierre, P. J. and Vezina, P. (1997) Predisposition to self-administer amphetamine: the contribution of response to novelty and prior exposure to the drug. *Psychopharmacology* **129**, 277–284.
115. Valadez, A., and Schenk, S. (1994) Persistence of the ability of amphetamine pre-exposure to facilitate acquistion of cocaine self-administration. *Pharmacol. Biochem. Behav.* **47**, 203–205.
116. Woolverton, W. L., Cervo, L., and Johanson, C. E. (1984) Effects of repeated mehtamphetamine administration on methamphetamine self-administration in rhesus monkeys. *Pharmacol. Biochem. Behav.* **21**, 737–741.
117. Mendrek, A., Blaha, C., and Phillips, A. G. (1998) Pre-exposure to amphetamine sensitizes rats to its rewarding properties as measured by a progressive ratio schedule. *Psychopharmacology* **135**, 416–422.
118. Lorrrain, D. S., Arnold, G. M., and Vezina, P. (2000) Previous exposure to amphetamine increases incentive to obtain the drug: Long-lasting effects revealed by the progressive ratio schedule. *Behav. Brain Res.* **107**, 9–19.
119. Robinson T. E. and Berridge, K. C. (1993) The neural basis of drug craving: an incentive-sensitization theory of addiction. *Brain Res. Rev.* **18**, 247–291.
120. Shaham, Y. and Stewart, J. (1995) Stress reinstates heroin-seeking in drug-free animals: an effect mimicking heroin, not withdrawal. *Psychopharmacology* **119**, 334–341.
121. Shaham, Y., Rajabi, H., and Stewart, J. (1996) Relapse to heroin-seeking under opioid maintenance: the effects of opioid withdrawal, heroin priming and stress. *J. Neurosci.* **16**, 1957–1963.
122. Shaham, Y. and Stewart, J. (1996) Effects of opioid and dopamine receptor antagonists on relapse induced by stress and re-exposure to heroin in rats. *Psychopharmacology* **125**, 385–391.
123. Stewart, J., deWit, H., and Eikelboom, R. (1984) The role of unconditioned and conditioned drug effects in the self-administration of opiates and stimulants. *Psychol. Rev.* **91**, 251–268.
124. Stewart, J. and deWit, H. (1987) Reinstatement of drug-taking behavior as a method of assessing incentive motivational properties of drugs, in *Methods of Assessing the Reinforcing Properties of Abused Drugs* (Bozarth, M. A., ed.), Springer-Verlag, New York, pp. 211–227.
125. Koob, G. F. and Bloom, F. E. (1988) Cellular and molecular mechanisms of drug dependence. *Science* **242**, 715–723.

126. Wise, R. A., and Bozarth, M. A. (1987) A psychomotor stimulant theory of addiction. *Psychol. Rev.* **94,** 469–492.
127. Bozarth, M. and Wise, R. A. (1981) Heroin reward is dependent on a dopaminergic sustrate. *Life Sci.* **29,** 1881–1886.
128. Hoebel, B. G., Monaco, A. P., Hernandez, L., Aulisi, E. F., Stanley, B. G., and Lenard, L. (1983) Self-injection of amphetamine directly into the brain. *Psychopharmacology* **81,** 158–163.
129. Pettit, H. O., Ettenberg, A., Bloom, F. E., and Koob, G. F. (1984) Destruction of dopamine in the nucleus accumbens selectively attenuates cocaine but not heroin self-administration in rats. *Psychopharmacology* **84,** 167–173.
130. Roberts, D. C. S., Koob, G. F., Klonoff, P. and Fibiger, H. C. (1980) Extinction and recovery of cocaine self-administration following 6-OHDA lesions of the nucleus accumbens. *Pharmacol. Biochem. Behav.* **12,** 781–787.
131. Zito, K. A., Vickers, G., and Roberts, D. C. S. (1985) Disruption of cocaine and heroin self-administration following kainic acid lesions of the nucleus accumbens. *Pharmacol. Biochem. Behav.* **23,** 1029–1036.
132. Woolverton, W. L. and Virus, R. M. (1989) The effects of D1 and D2 dopamine antagonists on behavior maintained by cocaine or food. *Pharmacol. Biochem. Behav.* **32,** 691–697.
133. Caine, S. B. and Koob, G. F. (1994) Effects of dopamine D1 and D2 antagonists on cocaine self-administration under different schedules of reinforcement in the rat. *J. Pharmacol. Exp. Ther.* **270,** 209–218.
134. Clarke, P. B., Jakubovic, A., and Fibiger, H. C. (1988) Anatomical analysis of the involvement of mesolimbocortical dopamine in the locomotor stimulant actions of *d*-amphetamine and apomorphine. *Psychopharmacology* **96,** 511–520.
135. McCreary, A. C. and Marsden, C. A. (1993) Dopamine D1 receptor antagonism by SCH-23390 prevents expression of conditioned sensitization following repeated administration of cocaine. *Neuropharmacology* **32,** 387–391.
136. Meyer, M. E., Cottrell, G. A., Van Hartesveldt, C., and Potter, T. J. (1993) Effects of dopamine D1 antagonists SCH-23390 and SKF-83566 on locomotor activities in rats. *Pharmacol. Biochem. Behav.* **44,** 429–432.
137. Vezina, P., Pierre, P. J., and Lorrain, D. S. (1999) The effect of previous exposure to amphetamine on drug-induced locomotion and self-administration of a low dose of the drug. *Psychopharmacology* **147,** 125–134.
138. Dakis, C. A., and Gold, M. S. (1985) New concepts in cocaine addiction: the dopamine depletion hypothesis. *Neurosci. Behav. Rev.* **9,** 469–77.
139. Hurd, Y. L., Weiss, F., Koob, G., and Ungerstedt, U. (1990) The influence of cocaine self-administration on in vivo dopamine and acetylcholine neurotransmission in rat caudateputamen. *Neurosci. Lett.* **109,** 227–233.
140. Pettit, H. O. and Justice, J. B., Jr. (1989) Dopamine in the nucleus accumbens during cocaine self-administration as studied as by in vivo microdialysis. *Pharmacol. Biochem. Behav.* **34,** 899–904.
141. Wise, R. A., Newton, P., Leeb, K., Burnette, B., Pocock, D., and Justice, J. B., Jr. (1995) Fluctuations in nucleus accumbens dopamine concentration during intravenous cocaine self-administration in rats. *Psychopharmacology* **120,** 10–20.
142. De Wit, H. and Stewart, J. (1983) Drug reinstatement of heroin-reinforced responding in the rat. *Psychopharmacology* **79,** 29–31.
143. Erb. S., Shaham, Y., and Stewart, J. (1996) Stress reinstates cocaine-seeking behavior after prolonged extinction and a drug-free period. *Psychopharmacolog* **128,** 408–412.
144. Stewart, J. (1984) Reinstatement of heroin and cocaine self-administration behavior in the rat by intracerebral application of morphine in the ventral tegmental area. *Pharmacol. Biochem. Behav.* **20,** 917–923.
145. Stewart, J. and Vezina, P. (1988) A comparison of the effects of intra-accumbens injections of amphetamine and morphine on reinstatement of heroin intravenous self-administration behavior. *Brain Res.* **457,** 287–294.
146. Stewart, J. and Wise, R. A. (1992) Reinstatement of heroin self-administration habits: morphine prompts and naltrexone discourages renewed responding after extinction. *Psychopharmacology* **108,** 79–84.
147. Hooks, M. S., Duffy, P., Striplin, C., and Kalivas, P. W. (1994) Behavioral and neurochemical sensitization following cocaine self-administration. *Psychopharmacology* **115,** 265–272.
148. Phillips, A. G. and Di Ciano, P. (1996) Behavioral sensitization is induced by intravenous self-administration of cocaine by rats. *Psychopharmacology* **124,** 279–281.
149. Vezina, P., Lorrain, D. S., Arnold, G. M., Austin, J. D., and Suto, N. (2002) Sensitization of midbrain dopamine neuron reactivity promotes the pursuit of amphetamine. *J. Neurosci.*, in press.
150. Neisewander, J. L., O'Dell, L. E., Tran-Nguyen, T. L., Castaneda, E., and Fuchs, R. A. (1996) Dopamine overflow in the nucleus accumbens during extinction and reinstatement of cocaine self-administration behavior. *Neuropsychopharmacology* **15,** 506–14.
151. Suto, N., Austin, J. D., Kramer, M. K., Tanabe, L. M., and Vezina, P. (2000) Previous exposure to amphetamine in the VTA leads to excessive cocaine self-administration in a D1 DA receptor dependent manner. *Soc. Neurosci. Abst.* **26,** 793.
152. Pierre, P. J., and Vezina, P. (1998) D1 dopamine receptor blockade prevents the facilitation of amphetamine self-administration induced by prior exposure to the drug. *Psychopharmacology* **138,** 159–166.
153. Suto, N, Austin, J. D., Tanabe, L. M., and Vezina, P. (2001) Previous exposure to VTA amphetamine promotes the self-administration of cocaine: $D_1$ dopamine and NMDA glutamate receptor dependence. *Drug and Alcohol Dependence* **63,** 631.
154. Suto, N., Austin, J. D., Tanabe, L. M., Svoboda, R. A., and Vezina, P. (2001) VTA amphetamine pre-exposure facilitates cocaine self-administration: glutamate receptor dependence. *Soc. Neurosci. Abst.,* **27,** 979.3.

155. Schenk, S., Valadez, A., McNamara, C., House, D. T., Higley, D., Bankson, M. S., et al. (1993) Development and expression of sensitization to cocaine's reinforcing properties: role of NMDA receptors. *Psychopharmacology* **111,** 332–338.
156. Cornish, J. L., Duffy, P., and Kalivas, P. W. (1999) A role for nucleus accumbens glutamate transmission in the relapse to cocaine-seeking behavior. *Neuroscience* **93,** 1359–1367.
157. Cornish, J. L., and Kalivas, P. W. (2000) Glutamate transmission in the nucleus accumbens mediates relapse in cocaine addiction. *J. Neurosci.* **20,** 1–5.
158. Vezina, P., Suto, N., Austin, J. D., Tanabe, L. M., and Creekmore, E. (2001) Previous exposure to amphetamine enhances cocaine self-administration as well as its reinstatement by nucleus accumbens AMPA. *Soc. Neurosci. Abst.* **27,** 979.17.

# 13
# Roles of Glutamate, Nitric Oxide, Oxidative Stress, and Apoptosis in the Neurotoxicity of Methamphetamine

## Jean Lud Cadet, MD

## 1. INTRODUCTION

Oxygen and nitric oxide (NO) are essential elements for normal life. Indeed, the reduction of molecular oxygen represents one of the most important generators of energy for aerobic organisms. This occurs through the four-election reduction of dioxygen to yield water. Nevertheless, these substances can also participate in deleterious reactions that negatively impact lipid, protein, and nucleic acid. Thus, normal physiological function depends on a balance between these potentially toxic substances and the scavenging systems that aerobic organisms have developed to counteract their deleterious effects. Both exogenous and endogenous causes can tilt that balance. In the present chapter, I will elaborate on the thesis that the neurodegenerative effects of methamphetamine are due to reactive oxygen species (ROS) overproduction in monoaminergic systems in the brain. I will also discuss the possible role of glutamate and of NO in the cascade that leads to methamphetamine (METH)-induced neurotoxicity. Moreover, this chapter will review briefly recent data that provide conclusive evidence that METH can also cause cell death in various regions of the brain.

## 2. OXIDATIVE MECHANISMS IN THE BRAIN

Free radicals (FR) are compounds that have at least one unpaired electron. They are very reactive and can cause damage to nucleic acids, lipids, and proteins *(1)*. Damaging oxyradical species can be produced through both endogenous and exogenous sources. Exogenous sources include xenobiotics, radiation, and chemical toxins, whereas endogenous ones include mitochondrial respiration, cytochrome P-450 reactions, phagocytic oxidative bursts, and peroximal leakage.

Aerobic organisms take up oxygen, which is used by the mitochondria in a process that makes water. Molecular oxygen is itself a free radical because it has two unpaired electrons. Because of spin restriction, molecular oxygen takes up one electron at a time in a process that results in the formation of superoxide radicals ($O_2^{\cdot-}$ hydrogen peroxide ($H_2O_2$), and hydroxyl radicals (•OH).

Superoxide is also generated by the actions of enzymes such as amino acid oxidase, cytochrome oxidases, monoamine oxidases, xanthine oxidase, and aldehyde oxidase. Auto-o-xidation of catecholamines, leukoflavins, tetrahydropterins, hydroquinones, and ferrodoxins also produce superoxides *(2,3)*. Superoxide is also formed during reperfusion subsequent to ischemia. Culcasic et al. *(4)* reported that stimulation of glutamate receptors can also cause the production of superoxide radicals. This might be the main cause of the cell death due to hypoxic insults to the brain. Interestingly, stimulation of glutamate receptors and the enzymatic reaction of nitric oxide synthase (NOS) can also cause the production of superoxide radicals. Superoxide radicals are converted to $H_2O_2$ by superoxide dismutase (SOD). $H_2O_2$ is also made during reactions catalyzed by the actions of amino acid oxidase

From: *Contemporary Clinical Neuroscience: Glutamate and Addiction*
Edited by: Barbara H. Herman et al. © Humana Press Inc., Totowa, NJ

and monoamine oxidase. Moreover, it is produced during the auto-o-xidation of ascorbate and catecholamines *(5)*. $H_2O_2$ is not a free radical because it does not contain unpaired electrons. Its reaction with transition metals leads to the formation of the highly toxic hydroxyl radical *(6)*. Hydroxyl radicals can damage sugars, amino acids, phospholipids, and nucleic acids. It can start the process of membrane lipid peroxidation that can lead to membrane damage. The accumulation of ions and water secondary to membrane dysfunction can lead to activation of proteolytic enzymes, swelling, and the activation of the cell death machinery.

Nitric oxide is a ubiquitous free radical *(7,8)*. NO is made through the action of NOS. NO can exist in several redox states, including the nitrosonium ($NO^+$) ions *(9)*. NO can form peroxynitrite ($ONOO^-$) by interacting with $O_2^{\cdot-}$; $ONOO^-$ is neurotoxic because of its interaction with thiol groups *(10)*. NO has been implicated in the toxic effects of glutamate in several in vitro models *(11–14;* see also below).

Oxidative damage to DNA includes damage to bases (thymine glycol) and to the phosphodiester backbone of DNA. Such attacks of the DNA structure can lead to loss of genetic information. Because differentiated neurons do not go through the full cell cycle, oxidative stress can result in accumulation of damaged DNA over time. These changes in DNA structures might play key roles in the aging process.

## 3. SCAVENGING SYSTEMS IN THE BRAIN

In the brain, the antioxidant defense system consists of the enzymes SOD, glutathione peroxidase (GSH-Px), and catalase (CAT) *(15)*. Nonenzymatic dietary antioxidants include ascorbic acid (vitamin C), α-tocopherol (vitamin E), and β-carotene. Other antioxidants are glutathione (GSH) and sulfydryl-containing, proteins. These antioxidants interact in parallel or sequential pathways to maintain oxidative homeostasis.

The first line of defense against $O_2^{\cdot-}$ is SOD, which catalyzes its reaction to $H_2O_2$ *(2,3)*. $H_2O_2$ is eliminated by CAT and glutathione peroxidase. CAT is a heme-containing enzyme that converts $H_2O_2$ to water and molecular oxygen. Because CAT has been found in peroxisomes, lysosomes, and mitochondria, it might provide inadequate protection against $H_2O_2$ produced in other cellular compartments. GSH-Px, a seleno-metaloenzyme, is another line of defense against $H_2O_2$ in these locations. GSH-Px prevents further propagation of radical chain reactions that can cause lipid peroxidation, membrane destabilization, and impairment of membrane function. It is to be noted that the levels of CAT and GSH-Px in the brain are relatively low in contrast to the oxidative burden created by the high rate of oxidative metabolism in this organ. GSH is a tripeptide thiol (L-γ-glutamyl-L-cysteinyl-glycine), which is an essential antioxidant. Depletion of GSH causes cellular demise because of dysfunction of mitochondria and of other subcellular organelles. SOD is found throughout the brain *(15)*.

Vitamin E is present in vegetable oils and is particularly rich in wheat germ. It is absorbed with fats in the gut and carried in the plasma by lipoproteins. Vitamin E is lipid soluble and is found in plasma membrane. Reactions that involve vitamin E lead to the formation of the tocopheroxyl radical by removal of the phenolic hydrogen. Vitamin C can reduce the radical to the original tocopherol. The tocopheroxyl radical can also be reduced by GSH-dependent reactions. Vitamin C *(16)* is a water-soluble antioxidant, which is present in high concentrations in fruits and vegetables. The brain has a very high concentration of vitamin C. It is highly concentrated in vesicular compartments of monoaminergic neurotransmitters. Ascorbate can also react with ROS to produce dehydroscorbate, which can then be reduced by GSH-dependent enzymatic reactions.

## 4. THE POTENTIAL ROLE OF OXYGEN-BASED RADICALS IN METH NEUROTOXICITY

### *4.1. Toxic Effects of Methamphetamine*

Methamphetamine is an illicit drug that has a marked increase in its abuse throughout the world *(17–19)*. Its acute administration can cause neuropsychiatric and neurological complications, including paranoia, psychosis, coma, and death *(20)*. Abrupt cessation of use can cause withdrawal symptoms that can progress to a suicidal depressive state *(21)*. The acute effects of the drug might be caused by increases in the levels of synaptic dopamine (DA) *(22)*, whereas the long-term effects are probably due to persistent perturbations in monoaminergic systems *(23)*. For example, administration of toxic doses of METH can cause marked depletion of dopaminergic and serotonergic markers in rodents *(24–29)*. These include neostriatal DA levels *(26,30–32)*, striatal tyrosine hydroxylase activity *(33)*, and DA uptake sites *(31,32,34,35)*. Nonhuman primates *(36,37)* and human METH users *(38,39)* also show marked toxic effects of the drug in their brains.

### 4.2. METH-Induced Toxicity and Oxygen-Based Radicals

Although the cellular and molecular events involved in METH-induced neurotoxicity remain to be elucidated, a number of investigators *(32,40–43)* have hinted to a role for oxygen-based free radicals in the actions of this drug. For example, administration of antioxidants, such as ascorbic acid or vitamin E, caused attenuation of METH-induced neurotoxicity *(41,44)*, whereas inhibition of SOD by diethyldithiocarbamate increased its neurotoxicity *(41)*. In the following, we provide further evidence that both superoxide radicals and NO are indeed involved in the neurodegenerative effects of this amphetamine analog. In order to test the role of superoxide radicals in the neurotoxic effects of METH Cadet et al. *(30)* and Hirata et al. *(45)* made use of transgenic (Tg) mice that express the human CuZn-SOD gene *(46)*. These mice have been shown to have much higher CuZn-SOD activity than wild-type animals from similar backgrounds *(46)*. Moreover, homozygous SOD-Tg mice have a mean increase of 5.7-fold, whereas heterozygous SOD-Tg mice have a mean increase of 2.5-fold in comparison to wild-type mice. In these mice, the toxic effects of METH were significantly attenuated in gene dosage fashion, with the homozygous mice showing greater protection *(45)*.

These results suggest that there is production of the superoxide radicals in the striata of mice treated with METH. This increase in superoxide radicals is probably due to release of DA in the brain after METH administration *(47,48)* and subsequent DA oxidation within DA terminals. Catecholamines have been shown to generate free radicals in the brain *(49)*. Specifically, the metabolic breakdown of DA, serotonin, and norepinephrine by monoamine oxidase results in the production of $H_2O_2$ *(50)*. In addition, catecholamines can autoxidize to form quinones *(51)*. Quinone by products can damage proteins via nucleophilic attacks on their side chains *(52)*. Further redoxcycling of dopaquinone formed during DA catabolism would enhance the concentration of oxygen-based radicals within DA terminals; this could cause their demise through membrane destabilization or through changes in calcium homeostasis. The proposition that quinones might be involved in toxicity is supported by a recent article that documented their production in DA terminals after administration of the METH *(53)*. In addition to superoxide radicals, the formation of hydroxyl radicals has also been implicated in neurotoxic effects *(42)*. Thus, it appears that METH administration might be associated with a toxic cascade that involves the production of superoxides, hydrogen peroxide, and hydroxyl radicals *(54)*. Further evidence for this view was provided by recent data that the antioxidant melatonin can also protect against METH neurotoxicity *(43,55)*.

### 4.3. Role of Glutamate and Nitric Oxide in METH Neurotoxicity

Cellular damage can also occur through glutamate toxicity *(56,57)*. The cytotoxic effects of glutamate occur through receptor-mediated excitotoxicity and receptor-independent events. Excitotoxicity appears to depend on the production of NO *(12,58)*. This occurs through conversion of L-arginine to L-citrulline in a reaction catalyzed by NOS; this reaction can be blocked by L-arginine analogs such as $N^{\omega}$-nitro-L-arginine (NOArg) *(7,8)*. It has been suggested that the oxygen-based pathways and the L-arginine–NO pathways might act alone or might collaborate in some instances to cause degenerative changes in a number of pathological states *(59)*.

A number of investigators have studied the role of glutamate in METH neurotoxicity *(60–63;* and see Chapter 14). Stephans and Yamamoto *(61)* first published a paper demonstrating that the glutamate antagonist MK801 can reduce the toxic effects of METH in mice. That group subsequently extended these findings to other antagonists *(62).* Weihmuller et al. *(64)* then supported this thesis by demonstrating that MK801 had its protective effects by attenuating METH-induced DA overflow. Other studies using lesions of intrinsic striatal cells that contain glutamate receptors also provided further support for the role of glutamate in this model *(63).* Nevertheless, recent observations that these compounds might attenuate METH-induced temperature elevation had cast some doubt onthe robustness of the glutamate effects *(65–67).* It is to be pointed out that some drugs that decrease body temperature can actually exacerbate METH toxicity *(68),* thus indicating that there is not a one-to-one correspondence between protection and attenuated temperature response.

Because glutamate appears to cause some of its neurotoxic effects via the production of NO *(69),* several groups of investigators have sought to determine if NO was also involved in METH neurotoxicity *(55,70–72;* and Chapter 15). Using a cell culture system, *(70)* showed that blockade of NO formation by NOS inhibitors provided significant protection against METH-induced cell death. Subsequently, Itzak et al. *(73),* using NOS knockout mice, also found substantial protection against the toxic effects of METH on dopaminergic markers. These results indicate that blocking of NO formation can result in the attenuation of the toxic effects of the drug. Thus, application of METH might have caused sustained release of NO that mesencephalic cells could not tolerate. The role of NO in the toxic effects of METH was further supported by the recent observations that METH can cause overexpression of neuronal NOS in the mouse brain *(74).* The present data are also consistent with those of others that have shown that NO is toxic to a number of cell types in vitro *(11,69).* Results from this laboratory *(70)* and those of others *(40,75)* indicate that METH can affect both DA and non-DA cells in vitro. The effects of METH on non-DA cells are consistent with the recent observations that METH can kill intrinsic nonmonoaminergic cell bodies in the cortex and striatum of mice *(76–78).* Taken together, these observations suggest that the amphetamines might affect more than just the monoaminergic systems in the brain; this would be consistent with the demonstrated role for NO and $O_2^{\cdot-}$, both of which could affect more than just monoaminergic systems in vivo. These issues are being investigated in our laboratory.

### 4.4. Methamphetamine-Induced Toxicity Involves Activation of Poly (ADP-ribose) Polymerase

The production of ROS is known to cause DNA damage *(79).* Associated with this damage is the activation of the enzyme poly(ADP-ribose) polymerase (PARP) with subsequent depletion of cellular energy stores and cell death *(79).* NO is thought to cause damage in a similar way *(80).* Sheng et al. *(70)* were able to show the involvement of PARP in METH toxicity because benzamide, nicotinamide, 3-aminobenzamide, and theophylline, which all inhibit the activity of the enzyme, also block METH-induced cell death. These studies show that the production of NO and superoxide radicals is implicated in METH-induced toxicity and that the cascade to cell death also includes PARP activation.

## 5. APOPTOSIS AND METH NEUROTOXICITY
### 5.1. Role of bcl-2 Gene Family

The evidence summarized above had suggested that METH could cause neuronal death. In order to evaluate this further, we used a cell culture system that consisted of an immortalized neural cell line from the rat mesencephalon *(81).* Exposure of these cells to METH caused the loss of viability in a dose-dependent fashion as measured by a fluorescent assay. However, this approach did not fully clarify the mode of cell death caused by METH.

Cell death can occur either via a necrotic or an apoptotic process *(82,83).* Apoptotic cell death has been shown to be attenuated or inhibited by the overexpression of *bcl-2 (84,85).* In order to better identify the mechanism that was involved in METH-induced loss of cellular viability, we used a flow cytometric approach to evaluate the presence of DNA strand breaks *(81)*. METH was found to cause DNA breaks in a fashion similar to what has been reported during apoptosis. *bcl-2* overexpression also attenuated these changes.

Additional confirmation that METH can cause apoptosislike changes was provided by DNA gel electrophoresis, which shows a ladder-type pattern that is also consistent with the apoptotic nature of the process involved in METH-induced cellular damage. These ladder patterns were not observed in *bcl-2* overexpressing cells. We confirmed the apoptotic nature of METH-induced cellular damage by staining cells with acridine orange in order to assess morphological changes. METH was found to cause chromatin condensation and nuclear fragmentation in a very high percentage of the cells. All of these changes were abrogated in *bcl-2* overexpressing cells.

Therefore, using cytometry, fluorescence microscopy, and DNA electrophoresis, we have been able to provide concordant evidence that METH kills cells via a process that involves internucleosomal DNA breaks and chromatin condensation, all of which are consistent with an apoptotic process. The present results indicate, in addition, that the use of METH can cause activation of caspase-activated DNase (CAD) that has been shown to be involved in fragmenting DNA after treatment with apoptosis-causing agents *(86,87)*. Nevertheless, the possibility that METH might have blocked the synthesis of inhibitors of CAD such as ICAD *(87)* needs also to be taken into consideration.

## 5.2. Involvement of p53

*p53* is a tumor-suppressor gene whose activation has been associated with apoptosis *(88–92)*. Mutations in the *p53* gene have been identified in both inherited and sporadic forms of cancer *(93–96)*. Moreover, the wild-type p53 protein has been shown to be involved in both apoptosis and cell-cycle arrest after toxic insults *(89,91,94,96)*. Exposure of cells to γ-irradiation or etoposide causes DNA damage that is associated with accumulation of p53 protein *(97–99)*. Depending on the extent of DNA damage, p53 activation can cause apoptosis instead of cell-cycle arrest *(89,91,92,96,100)*. Although the involvement of p53 in cell death and toxicity has been assessed mostly through in vitro experiments, we reasoned that this process might be a more general phenomenon associated with the cell-death cascade, neurotoxicity, or neurogeneration in vivo. We thus postulated that if the p53 protein is an important determinant of METH-induced neurotoxicity, animals lacking the gene for the p53 protein should show protection against the toxic effects of the drug. Our results support this prediction and demonstrate that the *p53*-knockout phenotype, in fact, does attenuate the long-term neurotoxic effects of METH on striatal dopaminergic terminals and on midbrain DA cell bodies *(28)*.

These results were the first demonstration that the long-term deleterious effects of a drug acting on dopaminergic systems could be attenuated by the absence of the p53 protein in mice engineered not to produce that protein. We also showed that METH could cause accumulation of p53. It is thus tempting to speculate that the loss of DA terminals induced by the administration of METH may be associated with a process that resembles apoptosis. This idea is consistent with the idea of synaptic apoptosis recently promulgated by Mattson et al. *(101)*.

The manner by which METH causes increased accumulation of p53 is not clear. However, the possibility that this might be in response to the acute oxidative stress caused by the drug should be considered *(102)*. For example, METH is known to cause marked increases in the release of DA within the striatum *(47,63)*. The metabolic breakdown of DA leads to the generation of oxygen-based free radicals such as the superoxide anion ($O_2^{\cdot-}$) and hydrogen peroxide ($H_2O_2$) with subsequent formation of hydroxyl radicals *(23,49,50,54)*. Hydroxyl radicals could then cause DNA lesions, which are known to be associated with the accumulation of p53 in vitro *(97–99)*. This reasoning suggests that other processes that are thought to involve free-radical formation might be attenuated in *p53–/–* mice or could lead to p53 accumulation in vivo. This is indeed the case, because postischemic neuronal damage is attenuated in *p53–/–* mice *(103)*, and excitotoxic damage, which is thought to involve free-radical production *(104)*, is also associated with p53 protein accumulation *(105)*. This reasoning is supported further by the recent observation that excitotoxic damage by kainic acid is prevented in *p53–/–* mice *(106)*. It is therefore important to note that the use of glutamate antagonists has been shown to prevent the toxic effects of METH on dopaminergic systems *(61,62)*. It has also been reported that the administration of toxic doses of METH is associated with increases in the extracellular levels of glutamate in the striatum *(107)*. Thus, either the persistent oxidative stress caused by the increase in DA metabolism, the possible excitotoxic damage caused by increased glutamate level, or both could cause the increase in p53-like immunoreactivity after METH administration. The increase in p53 in intrinsic striatal cells might also be of relevance to our recent observation that METH can cause apoptosis of intrinsic striatal cells *(76)*.

It is also of interest to relate the observed protection against METH-induced toxicity in the p53-knockout mice to recent reports that these mice show decreases in Bax protein, but no changes in bcl-2 in their brains *(108)*. Because Bax is a known inhibitor of the antioxidant effects of *bcl-2* and a promotor of apoptosis *(109)*, it is possible that the protective effects observed in the knockout mice might be the result of the relative greater abundance of *bcl-2* in these animals *(108)*. This reasoning is consistent with our recent demonstration that *bcl-2* can protect against the apoptotic effects of METH in vitro *(81)*.

## ACKNOWLEDGMENT

This work is supported by the NIH/NIDA Intramural Research Program.

## REFERENCES

1. Stadtman, E. R. (1993). Oxidation of free amino acid residues in proteins by radiolysis and by metal catalyzed reactions. *Annu. Rev. Biochem.* **62,** 797–821.
2. Fridovich, I. (1983). Superoxide radical: an endogenous toxicant. *Annu. Rev. Pharmacol.* **23,** 239–257.
3. Fridovich, I. (1986). Biological effects of the superoxide radical. *Arch. Biochem. Biophys.* **247,** 1–11.
4. Culcasi, M., Lafon-Cazal, M., Pietri, S., and Bockaert, J. (1994). Glutamate receptors induce a burst of superoxide via activation of nitric oxide synthase in arginine-depleted neurons. *J. Biol. Chem.* **269,** 12,589–12,593.
5. Chance, B., Sies, H., and Boveris, H. (1979). Hydroperoxide metabolism in mammalian organs. *Physiol. Rev.* **59,** 527–605.
6. Mellow-Filho, A. C. and Meneghini, R. (1984). *In vivo* formation of single strand breaks in DNA by hydrogen peroxide is mediated by the Haber–Weiss reaction. *Biochem. Biophys. Acta* **781,** 56–63.
7. Moncada, S., Palmer, R. M. J., and Higgs, E. A. (1991). Nitric oxide physiology, pathophysiology and pharmacology. *Pharmacol. Rev.* **43,** 109–142.
8. Nathan, C. (1992). Nitric oxide as a secretory product of mammalian cells. *FASEB J.* **6,** 3051–3064.
9. Lipton, S., Choi, Y.-B., Pan, Z-H., Lei, S. Z., Chen, H. S., Sucker, N. J., et al. (1993). A redox-based mechanism for the neuroprotective and neurodestructive effects of nitric oxide and related nitroso-compounds. *Nature* **364,** 626–632.
10. Radi, R., Beckamn, J. S., Bush, K. M., and Freeman, B. A. (1991). Peroxynitrite oxidation of sulfhydryls. The cytotoxic potential of superoxide and nitric oxide. *J. Biol. Chem.* **266,** 4244–4250.
11. Dawson, V. L., Dawson, T. M., Bartley, D. A., Uhl, G. R., and Snyder, S. M. (1993). Mechanisms of nitric oxide-mediated neurotoxicity in primary brain cultures. *J. Neurosci.* **13,** 2651–2661.
12. Gu, H. M., Zeng, J. X., Zhao, X. N., and Zhang, Z. X. (1999). The role of NO and B-50 in neurotoxicity of excitatory amino acids. *J. Basic Clin. Physiol. Pharmacol.* **10,** 327–336.
13. Lafon-Cazal, M., Culcasi, M., Gaven, F., Pietri, S., and Bockaert, J. (1993a). Nitric oxide, superoxide, and peroxynitrite: putative mediators of NMDA-induced cell death in cerebellar granule cells. *Neuropharmacology* **32,** 1259–1266.
14. Lafon-Cazal, M., Pietri, S., Culcasi, M., and Bockaert, J., (1993b). NMDA-dependent superoxide production and neurotoxicity. *Nature* **364,** 535–537.
15. Mavelli I., Rigo, A., Federico, R., Ciriolo, M. R., and Rotilio, G. (1982). Superoxide dismutase, glutathione peroxidase and catalase in developing rat brain. *Biochem. J.* **204,** 535–540.
16. Frei, B., Englan, L., and Ames, B. N. (1989). Ascorbate is an outstanding antioxidant in human blood plasma. *Proc. Natl. Acad. Sci. USA* **86,** 6377–6381.
17. Miller, M. A. (1991) Trends and patterns of methamphetamine smoking in Hawaii. *NIDA Res. Monogr.* **115,** 72–83.
18. Greberman, S. B. and Wada, K. (1994) Social and legal factors related to drug abuse in the United States and Japan. *Public Health Rep.* **109,** 731–737.
19. Shaw, K. P. (1999) Human methamphetamine-related fatalities in Taiwan during 1991–1996. *J. Forensic Sci.* **44,** 27–31.

20. Lan, K. C., Lin, Y. F., Yu, F. C., Lin, C. S., and Chu, P. (1998) Clinical manifestations and prognostic features of acute methamphetamine intoxication. *J. Formos Med. Assoc.* **97**, 528–533.
21. Murray, J. B., (1998) Pscophsiological aspects of amphetamine–methamphetamine abuse. *J. Psychol.* **132**, 227–237.
22. Stephans, S. E. and Yamamoto, B. Y. (1995). Effect of repeated methamphetamine administration on dopamine and glutamate efflux in rat prefrontal cortex. *Brain Res.* **700**, 99–106.
23. Cadet, J. L. and Brannock, C. (1998) Free radicals and the pathobiology of brain dopamine systems. *Neurochem. Int.* **32**, 117–131.
24. Ricaurte, G. A., Guillery, R. W., Seiden, L. S., Schuster, C. R., and Moore, R. Y. (1982) Dopamine nerve terminal degeneration produced by high doses of methylamphetamine in the rat brain. *Brain Res.* **235**, 93–103.
25. Matsuda, L. A. Schmidt, C. J., Gibb, J. W., and Hanson, G. R. (1988) Effects of methamphetamine on central monominergic systems in normal and ascorbic acid-deficient guinea pigs. *Biochem. Pharmacol.* **37**, 3477–3484.
26. O'Callaghan, J. P. and Miller, D. E. (1994). Neurotoxicity profiles of substituted amphetamines in the C57BL/6J mouse. *J. Pharmacol. Exp. Ther.* **270**, 741–751.
27. Hirata, H. and Cadet, J. L. (1997). Methamphetamine-induced serotonin neurotoxicity is attenuated is attenuated in p53-knockout mice. *Brain Res.* **768**, 345–348.
28. Hirata, H. and Cadet, J. L. (1997). p53 knockout mice are protected against the long-term effects of methamphetamine on dopaminergic terminals and cell bodies. *J. Neurochem.* **69**, 780–790.
29. Fukumura, M., Cappon, G. D., Pu, C., Broening, H. W., and Vorhees, C. V. (1998) A single dose model of methamphetamine-induced neurotoxicity in rats: effects on neostriatal monoaminesand glial fibrillary acidic protein. *Brain Res.* **806**, 1–7.
30. Cadet, J. L., Sheng, P., Ali, S., Rothman, R., Carlson, E., and Epstein, C. (1994). Attenuation of methamphetamine-induced neurotoxicity in copper/zinc superoxide dismutase transgenic mice. *J. Neurochem.* **62**, 380–383.
31. Ricaurte, G. A., Schuster, C. R., and Seiden, L. S. (1980). Long-term effects of repeated methylamphetamine administration on dopamine and serotonin neurons in the rat brain: regional study. *Brain Res.* **193**, 153–163.
32. Wagner, G. C., Ricaurte, G. A., Seiden, L. S., Schuster, C. R., Miller, C. R., Miller, R. J., and Westley, J. (1980). Long-lasting depletions of striatal dopamine and loss of dopamine uptake sites following repeated administration of methamphetamine. *Brain Res.* **181**, 151–160.
33. Hotchkiss, A. J., and Gibb, J. W. (1980). Long-term effects of multiple doses of methamphetamine on tryptophan hydrosylase and tyrosine hydroxylase activity in rat brain. *J. Pharmacol. Exp. Ther.* **214**, 257–262.
34. Nakayama, M., Loyama, T., and Yamashita, I. (1993). Long-lasting decreases in dopamine uptake sites following repeated administration of methamphetamine in the rat striatum. *Brain Res.* **601**, 209–212.
35. Steranka, L. R. and Sanders-Bush, E. (1980). Long-term effects of continuous exposure to amphetamine in brain dopamine concentration and synaptosomal uptake in mice. *Eur. J. Pharmacol.* **65**, 439–443.
36. Preston, K. L., Wagner, G. C., Schuster, C. R., and Seiden, L. S. (1985). Long-term effects of repeated methylamphetamine administration on monoamine neurons in the rhesus monkey brain. *Brain Res.* **338**, 243–248.
37. Villemagne, V., Yuan, J., Wong, D. F., Dannals, R. F., Hatzidimitriou, G., Mathews, W. B., et al. (1998). Brain dopamine neurotoxicity in baboons treated with doses of methamphetamine comparable to those recreationally abused by humans: evidence from [$^{11}$C]WIN-35, 428 positron emission tomography studies and direct *in vitro* determinations. *J. Neurosci.* **18**, 419–427.
38. Wilson, J. M., Kalasinsky, K. S., Levey, A. I., Bergeron, C., Reiber, G., Anthony, R. M., et al. (1996) Striatal dopamine nerve terminal markers in human, chronic methamphetamine users. *Nature Med.* **2**, 699–703.
39. McCann, U. D., Wong, D. F., Yokoi, F., Villemagne, V., Dannals, R. F., and Ricaurte, G. A. (1998) Reduced striatal dopamine transporters density in abstinent methamphetamine and methcathinone users: evidence from positron emission tomography studies with [$^{11}$C]WIN-35, 428, *J. Neurosci.* **18**, 8417–8422.
40. Cubells, J. F., Rayport, S., Rajndron, G., and Sulzer, D. (1994). Methamphetamine neurotoxicity involves vacuolation of endocytic organelles and dopamine-dependent intracellular oxidative stress. *J. Neurosci.* **14**, 2260–2271.
41. DeVito, M. J., and Wagner, G. C. (1989). Methamphetamine-induced neuronal damage: a possible role for free radicals. *Neuropharmacology* **28**, 1145–1150.
42. Giovanni, A., Liang, L. P., Hastings, T. G., and Zigmond, M. J. (1995). Estimating hydroxyl radical content in rat brain using systemic and intraventricular salicylate: impact of methamphetamine. *J. Neurochem.* **64**, 1819–1825.
43. Hirata, H., Asanuma, M., and Cadet, J. L. (1998). Melatonin attenuates methamphetamine-induced toxic effects on dopamine and serotonin terminals in mouse brain. *Synapse* **30**, 150–155.
44. Wagner, G. C., Lucot, J. B., Schuster, C. R., and Seiden, L. S., (1983) Alpha-methyltyrosine attenuates and reserpine increases methamphetamine-induced neuronal changes. *Brain Res.* **270**, 285–288.
45. Hirata, H., Ladenheim, B., Carlson, E., Epstein, C., and Cadet, J. L. (1996). Autoradiographic evidence for methamphetamine-induced striatal dopaminergic loss in mouse brain: attenuation in CuZn-superoxide dismutase transgenic mice. *Brain Res.* **714**, 95–103.
46. Epstein, C. J., Avraham, K. B., Lovett, M., Smith, S., Elroy-Stein, O., Rotman, G., et al. (1987). Transgenic mice with increased CuZn-superoxide dismutase activity: animal model of dosage effects in Down syndrome. *Proc. Natl. Acad. Sci. USA* **84**, 8044–8048.

47. Baldwin, H. A., Colado, M. I., Murry, T. K., De Souza, R. J., and Green, A. R. (1993). Striatal dopamine release *in vivo* following neurotoxic doses of methamphetamine and effect of the neuroprotective drugs, chlormethiazole and dizocilpine. *Br. J. Pharmacol.* **108,** 590–596.
48. Marshall, J. F., O'Dell, S. J., and Weihmuller, F. B. (1993). Dopamine–glutamate interactions in methamphetamine-induced neurotoxicity. *J. Neural Transm.* **91,** 241–254.
49. Cadet, J. L. (1988). A unifying hypothesis of movement and madness: involvement of free radicals in disorders of the isodendritic core. *Med. Hypotheses* **27,** 87–94.
50. Cohen, G. and Heikkila, R. E. (1974). The generation of hydrogen peroxide, superoxide radical and hydroxyl radical by hydroxydopamine, dialuric acid, and related cytotoxic agents. *J. Biol. Chem.* **249,** 2447–2452.
51. Graham, D. G. (1978). Oxidative pathways for catecholamines in the genesis of neuromelanin and cytotoxic quinones. *Mol. Pharmacol.* **14,** 633–643.
52. Graham, D. G., Tiffany, S. M., Bell, W. R., and Gutknecht, W. F. (1978). Auto-oxidation versus covalent binding of quinones as the mechanism of toxicity of dopamine, 6-OH–dopamine, and related compounds towards C300 neuroblastoma cells *in vitro. Mol. Pharmacol.* **14,** 644–653.
53. LaVoie, M. J. and Hastings, T. G. (1999). Dopamine quinone formation and protein modification associated with the striatal neurotoxicity of methamphetamine: evidence against a role for extracellular dopamine. *J. Neurosci.* **19,** 1484–1491.
54. Jayanthi, S., Ladenheim, B., and Cadet, J. L. (1998). Methamphetamine-induced changes in antioxidant enzymes and lipid peroxidation in copper/zinc-superoxide dismutase transgenic mice. *Ann NY Acad. Sci.* **844,** 92–102.
55. Itzhak, Y. and Ali, S. F. (1996) The neuronal nitric oxide synthase inhibitor, 7-nitoindazole, protects against methamphetamine-induced neurtoxicity *in vivo. J. Neurochem.* **67,** 1770–1773.
56. Choi, D. W., Maulucci-Gedde, M., and Kriegstein, A. R. (1987). Glutamate neurotoxicity in cortical cell culture. *J. Neurosci.* **7,** 357–368.
57. Choi, D. W., Koh, J. Y., and Peter, S. (1988). Pharmacology of glutamate neurotoxicity in cortical cell culture: attenuation by NMDA antagonists. *J. Neurosci.* **8,** 185–196.
58. Dawson, V. L., Dawson, T. M., London, E. D., Bredt, D. S., and Snyder, S. H. (1991). Nitric oxide mediates glutamate neurotoxicity in primary cortical cultures. *Proc. Natl. Acad. Sci. USA* **88,** 6368–6371.
59. Beckman, J. S. (1991). The double-edged role of nitric oxide in brain and superoxide-mediated injury. *J. Dev. Physiol.* **15,** 53–59.
60. Stephans, S. E. and Yamamoto, B. Y. (1994). Methamphetamine-induced neurotoxicity: roles for glutamate and dopamine efflux. *Synapse* **17,** 203–209.
61. Sonsalla, P. K., Nicklas, W. J., and Heikkila, R. E. (1989). Role for excitatory amino acids in methamphetamine-induced dopaminergic toxicity. *Science* **243,** 398–400.
62. Sonsalla, P. K., Riordan, D. E., and Heikkila, R. E. (1991). Competitive and noncompetitive antagonists at *N*-methyl-D-aspartate receptors protect against methamphetamine-induced dopaminergic damage in mice. *J. Pharmacol. Exp. Ther.* **256,** 506–512.
63. O'Dell, S. J., Weihmuller, F. B., McPherson, R. J., and Marshall, J. F. (1994) Excitotoxic striatal lesions protect against subsequent methamphetamine-induced dopamine depletions. *J. Pharmacol. Exp. Ther.* **269,** 1319–1325.
64. Weihmuller, F. B., O'Dell, S. J., and Marshall, J. F. (1992). MK-801 protection against methamphetamine-induced striatal dopamine terminal injury is associated with attenuated dopamine overflow. *Synapse* **11,** 155–163.
65. Ali, S. F., Newport, G. D., Holson, R. R., Slikker, W., Jr., and Bowyer, J. F. (1994). Low environmental temperatures or pharmacologic agents that produce hypothermia decrease methamphetamine neurotoxicity in mice. *Brain Res.* **658,** 33–38.
66. Bowyer, J. F., Tank, A. W., Newport, G. D., Slikker, W., Jr., Ali, S. F., and Holson, R. R. (1992) The influence of environmental temperature on the transient effects of methamphetamine on dopamine levels and dopamine release in striatum. *J. Pharmacol Exp. Ther.* **260,** 817–824.
67. Bowyer, J. F., Davied, D. L., Schumued, L., Broening, H. W., Newport, G. D., Slikker, W., Jr., et al. (1994) Further studies of the role of hyperthermia in methamphetamine neurotoxicity. *J. Pharmacol. Exp. Ther.* **268,** 1571–1580.
68. Wagner, G. C., Carelli, R. M., and Jarvis, M. F. (1985). Pretreatment with ascorbic acid attenuates the neurotoxic effects of methamphetamine in rats. *Res. Commun. Chem. Pathol. Pharmacol.* **47,** 221–228.
69. Dawson, V. L. and Dawson, T. M. (1996). Nitric oxide neurotoxicity. *J. Chem. Neuroanat.* **10,** 179–190.
70. Sheng, P., Cerruti, C., Ali, S., and Cadet, J. L. (1996) Nitric oxide is a mediator of methamphetamine (METH)-induced neurotoxicity: *in vitro* evidence from primary cultures of mesencephalic cells. *Ann. NY Acad. Sci.* **801,** 174–186.
71. Ali, S. F. and Itzhak, Y. (1998). Effects of 7-nitroindazole, an NOS inhibitor on methamphetamine-induced dopaminergic and serotonergic neurotoxicity in mice. *Ann. NY Acad. Sci.* **844,** 122–130.
72. Itzhak, Y., Martin, J. L., Black, M. D., and Ali, S. F. (1998). Effect of melatonin on methamphetamine-and 1-methyl-4-phenyl-1,2,3,6-tetrahydropyridine-induced dopaminergic neurotoxicity and methamphetamine-induced behavioral sensitization. *Neuropharmacology* **37,** 781–791.
73. Itzhak, Y., Gandia, C., Huang, P. L., and Ali, S. F. (1998). Resistance of neuronal nitric oxide synthase-deficient mice to methamphetamine-induced dopaminergic neurotoxicity. *J. Pharmacol. Exp. Ther.* **284,** 1040–1047.

74. Deng, X. and Cadet, J. L. (1999) Methamphetamine administration causes overexpression of nNOS in the mouse striatum. *Brain Res.* **851,** 254–257.
75. Bennett, B. A., Hyde, C. E., Pecore, J. R., and Coldfelter, J. E. (1993). Differing neurotoxic potencies of methamphetamine, mazindol, and cocaine in mesencephalic cultures. *J. Neurochem.* **60,** 1444–1452.
76. Deng, X., Ladenheim, B., Tsao, L., and Cadet, J. L. (1999). Null mutation of c-fos causes exacerbation of methamphetamine-induced neurotoxicity. *J. Neurosci.* **19,** 10,107–10,115.
77. Eisch, A. J., Schmued, L. C., and Marshall, J. F. (1998). Characterizing cortical neuron injury with Fluoro–Jade labeling after a neurotoxic regimen of methamphetamine. *Synapse* **30,** 329–333.
78. Pu, C., Broening, H. W., and Vorhees, C. (1996). Effect of methamphetamine on glutamate-positive neurons in the adult and developing rat somatosensory cortex. *Synapse* **23,** 328–334.
79. Schraufstatter, I. U., Hinshaw, D. B., Hyslop, P. A., Spragg, R. G., and Conchrane, C. G. (1986). Oxidant injury of cells: DNA strand-breaks activate polyadenosine diphosphate–ribose polymerase and lead to depletion of nicotinamide adenine dinucleotide. *J. Clin. Invest.* **77,** 1312–1320.
80. Zhang, J., Dawson, V. L., Dawson, T. M., and Snyder, S. H. (1994). Nitric oxide activation of poly (ADP-ribose) synthetase in neurotoxicity. *Science* **263,** 687–689.
81. Cadet, J. L., Ordonez, S. V., and Ordenez, J. V. (1997) Methamphetamine induces apoptosis in immortalized neural cells: protection by the proto-ocogene, bcl-2. *Synapse* **25,** 176–184.
82. Bonfoco, E., Kraine, D., Anfarcona, M., Nicotera, P., and Lipton, S. A. (1995) Apoptosis and necrosis: two distinct events induced, respectively, by mild and intense insults with N-methyl-D-asparate or nitric oxide/superoxide in cortical cell cultures. *Proc. Natl. Acad. Sci. USA.* **32,** 7162–7166.
83. Caron-Leslie, L. A. M., Evans, R. B., and Cidlowski, J. A. (1994). Bcl-2 inhibits glucocorticoid apoptosis but only partially blocks calcium ionophore or cycloheximide-regulated apoptosis in S49 cells. *FASEB* **8,** 639–645.
84. Hockenbery, D. M., Oitvai, Z. N., Yin, X. M., Milliman, C. L., and Korsmeyer, S. J. (1993) Bcl-2 functions in an antioxidant pathway to prevent apoptosis. *Cell* **75,** 241–251.
85. Kane, D. J., Safarian, T. A., Anton, R., Hahn, H., Gralla, E. B., Valentine, J. S., et al. (1993) bcl-2 Inhibition of neural death: decreased generation of reactive oxygen species. *Science* **262,** 1274–1277.
86. Uegaki, K., Otomo, T., Sakahira, H., Shimizu, M., Yumoto, N., Kyogoku, Y., et al. (2000). Structure of the CAD domain of caspase-activated DNase and interaction with the CAD domain of its inhibitor. *J. Mol. Biol.* **297,** 1121–1128.
87. Sakahira, H., Iwamatsu, A., and Nagata, S. (2000). Specific chaperone-like activity of inhibitor of caspase-activated DNase for caspase-activated DNase. *J. Biol. Chem.* **275,** 8091–8096.
88. Clarke, A. R., Puride, C. A., Harrison, D. J., Morris, R. G., Bird, C. C., Hooper, M. L., et al. (1993). Thymocyte apoptosis induced by p53-dependent and independent pathway. *Nature* **362,** 849–852.
89. Lowe, S. W., Ruley, H. E., Jacks, T., and Housman, O. E. (1993) p53-dependent apoptosis modulates the cytotoxicity of anticancer agents. *Cell* **74,** 1957–1967.
90. Hermeking, H. and Eick, D. (1994) Mediation of c-Myc-induced apoptosis by p53. *Science* **265,** 2091–2093.
91. Morgenbesser, S. D., Williams, B. O., Jacks, T., and Depinho, R. A. (1994) p53-dependent apoptosis produced by Rb-deficiency in the developing mouse lens. *Nature* **371,** 72–74.
92. Wagner, A. J., Kokontis, J. M., and Hay, N. (1994). Myc-mediated apoptosis requires wild-type p53 in a manner independent of cell cycle arrest and the ability of p53 to induce p21 waf1/cip1. *Genes Dev.* **8,** 2817–2830.
93. Hollstein, M., Sidransky, D., Vogelstein, B., and Harris, C. C. (1991) p53 mutations in human cancers. *Science* **253,** 49–53.
94. Levine, A. J., Momand, J., and Finlay, C. A. (1991) The p53 tumor suppressor gene. *Nature* **351,** 453–456.
95. Donehower, L. A. and Bradley, A. (1993) The tumor suppressor p53. *Biochem. Biophys. Acta* **1155,** 181–205.
96. Lowe, S. W., Schmitt, E. M., Smith, S. W., Osborne, B. A., and Jacks, T. (1993) p53 is required for radiation-induced apoptosis in mouse thymocytes. *Nature* **362,** 847–849.
97. Kastan, M. B., Onyewere, O., Sidransky, D., Vogelstein, B., and Craig, R. W. (1991) Participation of p53 protein in the cellular response to DNA damage. *Cancer Res.* **51,** 6304–6311.
98. Fritsche, M., Haessler, C., and Brandner, G. (1993) Induction of nuclear accumulation of the tumor-suppressor protein p53 by DNA-damaging agents. *Oncogene* **8,** 307–318.
99. Zhan, Q., Carrier, F., and Fornace, A. J. (1993) Induction of cellular p53 activity by DNA-damaging agents and growth arrest. *Mol. Cell Biol.* **13,** 4242–4250.
100. Yonish-Rouach, E., Resnitzky, D., Lotem, J., Sachs, L., Kimchi, A., and Oren, M. (1991) Wild-type p53 induced apoptosis of myeloid leukaemic cells that is inhibited by interleukin-6. *Nature* **352,** 345–347.
101. Mattson, M. P., Keller, J. N., and Begley, J. G. (1998) Evidence of synaptic apoptosis. *Exp. Neurol.* **153,** 35–48.
102. Wood, A. K. and Youle, R. J. (1995). The role of free radicals, and p53 in neuron apoptosis *in vivo*. *J. Neurosci.* **15,** 5851–5859.
103. Crumrine, R. C., Thomas, A. L., and Morgan, P. F. (1994) Attenuation of p53 expression protects against focal ischemic damage in transgenic mice. *J. Cereb. Blood Flow Metab.* **14,** 887–891.
104. Coyle, J. T., and Puttfarcken, P. (1993) Oxidative stress, glutamate, and neurodegenerative disorders. *Science* **262,** 689–695.

105. Sakhi, S., Bruce, A., Sun, N., Tocco, G., Baudry, M., and Schreiber, S. S. (1994) p53 induction is associated with neuronal damage in the central nervous system. *Proc. Natl. Acad. Sci. USA* **91,** 7525–7529.
106. Morrison, R. S. Wenzel, H. J., Kinoshita, Y., Robbins, C. A., Donehower, L. A., and Schwartzkroin, P. A. (1996) Loss of p53 tumore suppressor gene protects neurons from kainate-induced cell death. *J. Neurosci.* **16,** 1337–1345.
107. Nash, J. F. and Yamamoto, B. K. (1992) Methamphetamine neurotoxicity and striatal glutamate release: comparison to 3,4-methylenedioxymethamphetamine. *Brain Res.* **581,** 237–243.
108. Miyashita, T., Krajewski, S., Krajewski, M., Wang, H. G., Lin, H. K., Liebermann, D. A., et al. (1994) Tumor suppressor p53 is a regulator of bcl-2 and bax gene expression *in vitro* and *in vivo*. *Onogene* **9,** 1799–1805.
109. Oltvai, Z. N., Milliman, C. L., and Korsmeyer, S. J. (1993) Bcl-2 heterodimerizes *in vivo* with a conserved homolog, Bax, that accelerates programmed cell death. *Cell* **74,** 609–619.

# 14
# Methamphetamine Toxicity
*Roles for Glutamate, Oxidative Processes, and Metabolic Stress*

## Kristan B. Burrows, PhD and Bryan K. Yamamoto, PhD

## 1. INTRODUCTION

Methamphetamine (MA) is a sympathomimetic amine with potent effects on the peripheral and central nervous systems, resulting in psychomotor activation, mood elevation, anorexia, increased mental alertness, enhanced physical endurance, and hyperthermia. The mood-elevating and positive-reinforcing effects most likely contribute to the high abuse liability of this drug. Indeed, MA abuse has increased across the United States at an alarming rate since the late 1980s. MA-related emergencies have increased sixfold in the past decade and 4–5 million people in the United States now report using MA at some time in their lives *(1)*, highlighting the urgency for research on the pharmacology and toxicity of this drug.

Preclinical studies have revealed that single or repeated administration of a high dose of MA is neurotoxic to both rodents and nonhuman primates. High doses of MA result in a long-lasting depletion of dopamine (DA) content and a decrease in the appearance of other markers associated with DA neurotransmission in the striatum (Table 1). In contrast, DA terminals outside the extrapyramidal motor system are relatively unaffected. In recent years, similar changes in the striatal DA system have been found in human MA abusers *(11,12)*. In contrast to this selective destruction, MA administration is also associated with widespread decreases in serotonin (5-HT) terminal markers in areas including the cortex, striatum, hippocampus, amygdala, hypothalamus, thalamus, and brainstem *(8,13)*. Because these biochemical effects have been reported to endure for months *(14,15)*, these changes are well accepted as evidence of neurotoxicity (for review, *see* ref. *16*). Owing to the similarity between the relatively selective destruction of the striatal DA system in Parkinson's disease and following MA administration, a majority of the research on the underlying mechanisms of MA toxicity has focused on the ability of MA to damage DA terminals. Although damage to 5-HT terminals has been thoroughly characterized, less is known about factors mediating the toxicity to the 5-HT system after MA. Therefore, a major focus of this chapter will be on mechanisms of damage to DA neurons. The differences between this damage and damage to 5-HT neurons are addressed in the last section of the chapter.

Because amphetamines produce a massive release of DA, DA itself has been implicated in mediating the long-term effects of MA neurotoxicity. There is considerable evidence that DA can produce neurotoxicity *(17–19)*. Furthermore, inhibition of dopaminergic transmission through the inhibition of tyrosine hydroxylase *(20)*, the blockade of transporter-mediated DA release with uptake blockers *(6,20,21)*, and antagonism of DA receptors *(22–24)* all attenuate the long-term DA depletions produced by MA. However, high extracellular DA alone does not account for the toxicity of substituted amphetamines *(25,26)*. For example, although the local perfusion of MA into the striatum produces a

From: *Contemporary Clinical Neuroscience: Glutamate and Addiction*
Edited by: Barbara H. Herman et al. © Humana Press Inc., Totowa, NJ

**Table 1**
**Evidence for Damage to DA and 5-HT Nerve Terminals Following MA Administration**

- Loss of DA uptake sites *(2)*
- Flurescent swollen tyrosine hydroxylase-positive axons *(3)*
- Fink–Heimer silver staining *(4,5)*
- Decrease in tyrosine hydroxylase-immunoreactive fibers *(6,7)*
- Depletion of DA and 5-HT tissue concentrations *(4,8–10)*

marked and sustained increase in DA release, intrastriatal MA perfusion does not produce long-term depletions of striatal DA or 5-HT tissue content *(25)*. Consequently, additional factors likely mediate MA-induced damage to brain monoaminergic systems.

Glutamate and other excitatory amino acids have been linked to a number of neurodegenerative disorders, including Huntington's disease, brain hypoxia/ischemia, and epilepsy *(27,28)*. Glutamate also appears to mediate the toxicity produced by MA. Sonsalla et al. *(29)* were the first to implicate excitatory amino acids by demonstrating that an *N*-methyl-D-aspartate (NMDA) receptor antagonist, MK-801, blocks the decreases in tyrosine hydroxylase activity and DA tissue content after MA. Their original findings have since been extended by others using both noncompetitive and competitive NMDA receptor antagonists *(30–32)*. Our laboratory was the first to demonstrate that MA itself, or *d*-amphetamine administered to iprindole-treated rats, increases the extracellular concentration of striatal glutamate measured in vivo *(33,34)*. These results have been confirmed subsequently by others *(35–37)*. We have recently examined the acute and long-term effects of systemic administration of MA compared to the local intrastriatal perfusion of MA. Although both routes of administration acutely increase DA release to a similar degree, only the systemic administration of MA increases extracellular concentrations of glutamate and produces lasting depletions in striatal DA content *(25)*. These results support the hypothesis that glutamate release is obligatory in the neurotoxic cascade that follows MA administration, but the mechanisms that appear to culminate in excitotoxicity and damage the nigrostriatal DA system are still unclear.

## 2. BRAIN CIRCUITRY AND MECHANISMS OF GLUTAMATE RELEASE

Although evidence indicates that MA does not directly increase the release of glutamate in the striatum *(25)*, several studies suggest that activation of the corticostriatal pathway following MA administration may be responsible for increased striatal extracellular glutamate concentrations. The increases in striatal extracellular glutamate that are typically observed after MA are tetrodotoxin (TTX) sensitive (Fig. 1), suggesting that MA-induced changes in glutamate are impulse mediated. Moreover, unilateral ablation of motor and premotor cortexes decreases striatal glutamate acitivity by eliminating a majority of corticostriatal efferents *(38)* and protects against MA-induced damage to DA terminals (Fig. 2). In addition, MA treatment increases extracellular concentrations of glutamate and decreases glutamate immunolabeling of nerve terminals in both the motor cortex and striatum, suggesting that a release of neuronal glutamate occurs in both these regions *(36,39,40)*.

The presence or absence of increases in glutamate release within specific cortical subregions may be predictive of dopaminergic damage in their respective terminal fields. For example, the medial prefrontal cortex and nucleus accumbens are DA-rich areas resistant to the toxic effects of MA *(41)*. Accordingly, MA does not alter extracellular glutamate concentrations within the medial prefrontal cortex or within its primary target, the nucleus accumbens *(9,42)*. In contrast, we have found that a neurotoxic regimen of MA produces a gradual but marked and significant increase in extracellular glutamate concentrations in the somatosensory (parietal) cortex of the rat (Fig. 3). A delayed rise in extracellular glutamate concentrations also occurs in the lateral striatum *(33,42)*, the major terminal field of

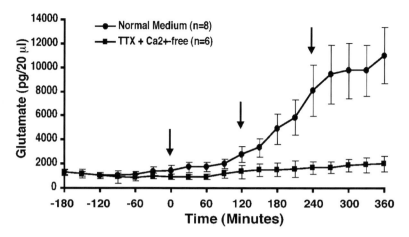

**Fig. 1.** Intrastriatal perfusion with TTX$^+$ and Ca$^{2+}$-free medium blocks the increase in extracellular glutamate levels following repeated administration of MA (arrows indicate injection of 7.5 mg/kg MA at times 0, 120, and 240 min). Bar indicates time of perfusion.

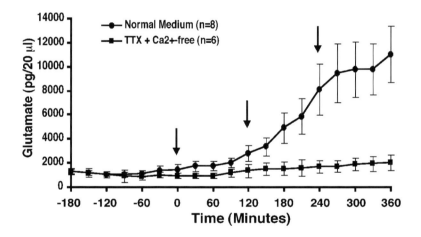

**Fig. 2.** Unilateral cortical ablation prevents the loss of striatal DA content 1 wk following MA administration (10 mg/kg × 4 doses over 8 h). Removal of cortical inputs to the striatum did not alter DA tissue content. *$p<0.05$ verses other groups.

these cortical regions *(43)*. MA also produces silver staining and reactive gliosis in these striatal and cortical regions, suggestive of a correlation between glutamate release and lasting neuronal damage *(6,44,45)*. In addition, MA alters binding to NMDA receptors specifically within the striatum and somatosensory cortex *(46)* and degenerates cell bodies in this cortical region *(47,48)*. Together, these data indicate that MA increases glutamatergic activity, specifically within the corticostriatal pathway, that, in turn, may produce damage to dopaminergic striatal nerve terminals and nonmonoaminergic cortical cell bodies.

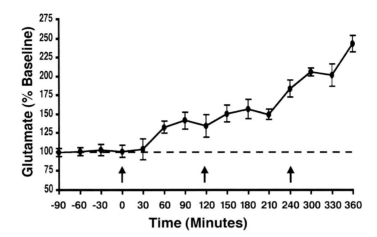

**Fig. 3.** Administration of MA (arrows indicate injection of 7.5 mg/kg MA at times 0, 120, and 240 min) increased extracellular glutamate levels in the partietal cortex. This rise temporally correlates with the increase in striatal glutamate (*see* Fig. 1).

Activation of the outflow pathways of the basal ganglia and the subsequent stimulation of thalamocortical and corticostriatal projections are indirect mechanisms through which MA may increase striatal glutamate release. Along these lines, O'Dell et al. *(49)* demonstrated that excitotoxic lesions of striatal output neurons prevents MA toxicity, indicative of the necessity for an intact extrapyramidal outflow loop to mediate MA-induced damage to striatal DA terminals. This is consistent with the observation that MA-induced glutamate release is dependent on DA receptor activation because D2 receptor antagonism with haloperidol attenuates the increase in extracellular glutamate concentrations following systemic MA *(10)*.

The substantia nigra pars reticulata and globus pallidus are major targets of the efferent projections from the striatum *(43)* that send convergent inputs to the ventral thalamus *(50,51)*. The ventral thalamus then diverges to cortical areas *(51,52)*, which, in turn, project back to the striatum *(53,54)*. The convergence of striatal efferent outflow onto the ventral thalamus makes this a unique area to target for the examination of changes in striatal activity. Figure 4A illustrates that corticostriatal glutamate release in modulated by a series of nigrothalamic, thalamocortical, and corticostriatal pathways. In fact, these pathways appear to mediate amphetamine-induced ascorbate release from corticostriatal terminals *(55)*. We have found that MA administration (10 mg/kg, ip, every 2 h over a 6-h period), significantly decreases the extracellular concentrations of γ-aminobutyric acid (GABA) in ventral thalamus as measured by in vivo microdialysis *(40)*. One interpretation is that MA disinhibits the ventral thalamus through decreases in GABA efflux from nigrothalamic neurons. The decreased GABA efflux may be the result of an increase in DA transmission produced by MA in the basal ganglia. A hypothesized scenario is that the MA-induced increase in nigral DA transmission, via D1 receptor activation in the substantia nigra pars reticulata, stimulates GABA release *(56–58)* and GABA-mediated inhibition of these nigral neurons *(59)*. Consequently, the activation of $GABA_A$ receptors in the substantia nigra *(60)* will inhibit GABAergic nigrothalamic transmission and ultimately increase corticostriatal glutamate release (Fig. 4B). Collectively, these data suggest that MA indirectly increases the excitatory drive to the cortex via activation of the striatal output neurons, leading to stimulation of corticostriatal glutamate activity and toxicity to dopamine terminals in the striatum.

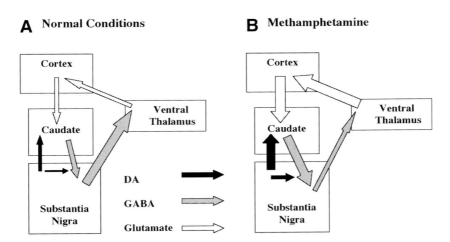

**Fig. 4.** Circuitry diagram demonstrating the hypothesized effects of striatal activation on glutamate release from corticostriatal afferents. **(A)** Under normal conditions, tonic activity in the substantia nigra regulates both DA and glutamate release in the striatum; **(B)** following MA, increased DA release in the striatum inhibits GABAergic outflow from the substantia nigra and leads to disinhibition of thalamocortical afferents and subsequent activation of the corticostriatal pathway.

## 3. MECHANISMS OF METHAMPHETAMINE TOXICITY

As discussed above, increases in the extracellular concentrations of both DA and glutamate within the striatum appear to contribute to MA-induced damage to dopaminergic nerve terminals. Many of the manipulations used to identify the role of DA in mediating MA-induced damage also modify MA-induced changes in the extracellular concentrations of glutamate. These findings suggest that the release of both DA and glutamate are obligatory in the MA toxicity cascade *(9,10,33,34)*. For example, D2 receptor blockade decreases glutamate release in the striatum without altering DA overflow, whereas administration of a DA uptake inhibitor decreases the ability of MA to release DA without affecting the striatal increase in glutamate efflux *(10)*. Although these treatments differentially affect the release of DA and glutamate, both are neuroprotective and demonstrate the importance of these neurotransmitters as comediators of MA toxicity.

There are several ways that the actions of DA and glutamate may synergize to mediate the toxicity of MA. High-dose MA treatment has been found to induce the endogenous formation of oxidizing compounds in brain regions susceptible to toxicity *(61,62)*, implicating oxidative stress as an underlying cause of terminal damage. In support of this finding, DA exacerbates glutamate-induced cell death in vitro via an oxidative mechanism *(63)*. Dopaminergic lesions of the nigrostriatal pathway in vivo decreases the excitotoxic effect of intrastriatal infusion of excitatory amino acids *(64,65)*, further implicating interactions between DA and glutamate in MA toxicity. In addition, efflux of both glutamate and DA can lead to the formation of reactive oxygen species and a shift in mitochondrial membrane potential to compromise mitochondrial function and produce metabolic stress and subsequent cell death *(66–69)*.

Overall, there is substantial support for the hypothesis that increased DA and glutamate efflux leads to excitotoxic, oxidative, and metabolic stress and that substrates that attenuate the consequences of such stressors (glutamate receptor antagonists, antioxidants, free-radical scavengers, or substrates for the electron-transport chain) are neuroprotective. Evidence for the ability of DA and glutamate to induce excitotoxic, oxidative, and metabolic stress, as well as evidence for their involvement in MA toxicity, are discussed below (Fig. 5).

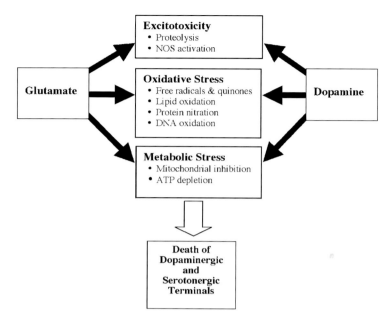

**Fig. 5.** Glutamate and dopamine contribute to MA toxicity by influencing several factors, including excitotoxic stress, oxidative stres, and metabolic stress.

## 3.1. Excitotoxicity

Increased extracellular glutamate concentrations and overstimulation of ionotropic glutamate receptors leads to a cascade of events that culminate in excitotoxic cell death (for review, see refs, 70 and 71). Initially, stimulation of α-amino-3-hydroxy-5-methyl-4-isoazole propionic acid (AMPA) receptors increases intracellular $Na^+$, resulting in depolarization and removal of the voltage-gated $Mg^+$ block from the NMDA receptor channel (72). Further glutamate stimulation at the NMDA receptor increases intracellular $Ca^{2+}$ and a subsequent sequestration of $Ca^{2+}$ within the mitochondira via activation of a $Ca^{2+}$-ATPase. Because the oxidation of pyruvate drives both $Ca^{2+}$ sequestration and ATP synthesis, an increase in intracellular $Ca^{2+}$ can shift the balance between these two processes and interrupt ATP synthesis. This eventually leads to the depletion of energy stores, collapse of the mitochondrial membrane potential, and a consequent rise in intracellular $Ca^{2+}$ levels as $Ca^{2+}$ is released from mitochondrial stores.

The NMDA receptor activation and elevated levels of intracellular $Ca^{2+}$ that result from increased extracellular glutamate concentrations can activate a number of enzymes, including calpain, endonucleases, phospholipase $A_2$, xanthine oxidase, nitric oxide synthase, and arachidonate (73,74). Each of these enzymes can elicit a sequence of destructive events that lead to the formation of intracellular reactive oxygen species and eventual cell death (27,70). In addition, the free-radical species that are generated further enhance glutamate release, inhibit glutamate reuptake (75–77), and thus promote a feed-forward cycle to augment glutamate-mediated damage.

A consequence of an increase in intracellular $Ca^{2+}$ is the activation of a $Ca^{2+}$-dependent protease, calpain. Calpain activation is mediated by excitatory amino acid release and results in the proteolysis of axonal spectrin, a major component of the cytoskeleton (78,79). Activation of calpain is a primary mechanism that contributes to several types of neurodegenerative condition, including glutamate-induced neurotoxicity associated with traumatic brain injury, ischemia, and hyperthermia (80–83). Glutamate-mediated activation of calpain also catalyzes the conversion of xanthine dehydrogenase to xanthine oxidase. Xanthine oxidase, in turn, promotes the catabolism of xanthine and hypoxanthine to uric acid, yielding oxygen free radicals in the process *(84)*. We have recently shown that MA treatment increases the concentration of uric acid in the striatum, providing evidence that glutamate-mediated excitotoxic stress accompanies MA administration *(85)*.

In summary, excitotoxic mechanisms may underlie, in part, the damage to dopaminergic nerve terminals following high-dose MA administration. MA-induced neurotoxicity specifically involves the activation of several glutamate-mediated enzymes, including calpain, xanthine oxidase, and nitric oxide synthase (*see* Section 3.2). Activation of these enzymes, and other glutamate- and $Ca^{2+}$- mediated systems, could result in the formation of reactive oxygen species. Together with free radicals that may be formed as a result of increased DA release, these neurotoxic oxygen species can actively participate in cell death.

## 3.2. Oxidative Stress

There is indirect and direct evidence that MA produces oxidative stress. Oxidative stress is defined as the cytotoxic consequences of reactive oxygen species (e.g., $\cdot O_2^-$, $\cdot OH$) generated as byproducts of oxidative metabolism. Evidence that indirectly supports the contention that MA leads to oxidative stress is that MA administration results in the production of hydroxyl radicals ($\cdot OH$) in the striatum *(85–87)*. Conversely, antioxidants (e.g.,ascorbic acid) and the spin-trap agent, phenyl-*t*-butylnitrone, prevent the striatal toxicity produced by MA *(85,88,89)*. Overexpression of the human Cu/Zn-superoxide dismutase gene, which degrades $\cdot O_2^-$, also confers protection against the DA-depleting effects of MA *(90)*. In addition, because it is thought that the immediate early gene c-*fos* plays a protective role in the brain by activating a variety of antioxidant enzyme systems *(91,92)* or by increasing the levels of trophic factors in the brain (for review, *see* ref. *93*), the induction of c-*fos* following MA administration *(94–96)* and the exacerbation of toxicity in c-*fos* knockout mice *(97)* further support for the role of oxidative processes in MA-induced damage.

More direct evidence of free-radical-mediated damage by MA would indicate the presence of oxidized proteins (protein nitration), lipids (lipid peroxidation), and DNA (nucleotide oxidation) *(98)*. In fact, all three types of cellular damage occur after MA administration. DA-dependent intracellular oxidataion following exposure to MA produces degeneration of neurite outgrowth in DA neuron cultures *(99)* and induces apoptosis in intrinsic nondopaminergic neurons in the striatum and frontal cortex of mice in vivo, as determined by TUNEL staining for DNA fragmentation *(97)*. Additionally, MA treatment increases lipid peroxidation in the striatum as evidenced by an increase in malonyldialdehyde production *(85,100)*. Conversely, inhibition of lipid peroxidation attenuates the toxicity produced by MA *(101)*. Furthermore, MA treatment results in protein nitration as evidenced by the formation of 3-nitrotyrosine from peroxnitrite production *(102,103)*.

The mechanistic underpinnings of MA-induced oxidative stress may involve dopamine and glutamate. The increase in cytosolic and extracellular DA produced by MA may induce cytotoxicity via the generation of free-radical species and quinones. DA is enzymatically metabolized to form $H_2O_2$ that is then nonenzymatically catalyzed by iron to form $\cdot OH$ *(29)*. In addition, DA autoxidation produces cytotoxic quinones, which attack thiol-containing proteins and result in the formation of 5-cysteinyl adducts of DA *(104)*. Consistent with these in vitro findings, intrastriatal injection of high concentrations of DA results in neurotoxicity and in the in vivo formation of protein-bound cysteinyl adducts of DA, both of which are prevented by the coadministration of antioxidants *(105)*. Similar to the effects of MA administration in vivo, free radicals and DA quinones rapidly decrease DA transporter function and inactivate tyrosine hydroxylase in vitro *(106–108)*. Therefore, the massive increase in the extracellular concentrations of DA, such as that produced by MA, could result in the production of hydroxyl free radicals, oxidative stress, and eventual damage to DA terminals.

A compromise in endogenous antioxidant mechanisms (e.g., glutathione) by MA may also contribute to oxidative stress. MA decreases glutathione peroxidase activity *(109)*. Although total glutathione content in the striatum is reduced in the long run after MA *(110)*, we have shown that both reduced glutathione and oxidized glutathione are acutely increased in the striatum following a neurotoxic regimen of MA *(111)*, it is possible that MA-induced oxidative stress results in the rapid recruitment of the endogenous glutathione antioxidant system followed by a lasting decrease associated with neurotoxicity and long-term dopaminergic damage.

Glutamate and glutamate receptor activation also can cause neuronal death through these oxidative mechanisms *(112)*. Several lines of evidence indicate that glutamate exposure and subsequent nitric oxide production lead to a depletion of endogenous antioxidant and energy stores and an accumulation of intracellular peroxides leading to oxidative stress and cell death—a phenomenon known as oxidative glutamate toxicity *(113)*. Glutamate-mediated activation of NMDA receptors, neuronal NOS, and the production of excess nitric oxide *(114)* can produce neurotoxicity *(115,116)*. Nitric oxide reacts with $\cdot O_2^-$ to form the oxidant, peroxynitrite (ONOO$^-$) *(112,117)*. Peroxynitrite and its decomposition product nitrite may contribute to toxicity via oxidation of DA and protein modification *(118)*. Conversely, inhibition of nitric oxide synthesis by administration of the neuronal NOS inhibitor 7-nitroindazole, in vivo, protects against DA damage caused by MPTP administration *(119,120)* and attenuates excitotoxicity following intrastriatal administration of NMDA *(116)*. Moreover, inhibition of neuronal NOS also protects against MA-induced toxicity both in vitro and in vivo *(121–124)* presumably resulting from the attenuation of hydroxyl radical formation and the consequent decrease in formation of 8-hydroxy-2-deoxyguanosine as well as 3-nitrotyrosine *(116,120)*.

In general, a substantial amount of evidence supports the hypothesis that MA administration leads to the endogenous formation of reactive oxygen species through both dopaminergic and glutamatergic mechanisms and that these reactive compounds mediate toxicity to dopaminergic nerve terminals. However, intimately related to the glutamate-dependent dependent production of oxidative stress and its role in MA toxicity are the effects of glutamate on cellular bioenergetics and the production of metabolic stress.

### 3.3. Metabolic Stress

Mitochondrial dysfunction, metabolic stress, and disruption of bioenergetic systems that result from high concentrations of extracellular glutamate also contribute to MA-induced neurotoxicity. Alterations in brain energy utilization by low doses of amphetamine and related analogs were reported originally in the 1970s. The results of these early experiments show that low doses of amphetamine and MA rapidly increase metabolism in the cerebral cortex or the whole brain as measured by lactate formation and changes in high-energy substrates such as ATP and phosphocreatine *(125)*. More recent studies have demonstrated that amphetamine and MA increase local cerebral glucose utilization in multiple brain regions within 45 min of drug administration *(126,127)*. In contrast, high-dose treatment with MA decreases cerebral glucose metabolism for weeks to months following drug administration, suggesting that initial increases in energy utilization are followed by lasting impairments in metabolism *(128)*.

Methamphetamine and amphetamine alter energy utilization in a brain-region-specific manner, in that acute increases in glucose utilization appear to be greatest in those brain regions most susceptible to the toxic effects of MA. Our laboratory has demonstrated that MA increases the extracellular concentrations of lactate in the striatum but not in the prefrontal cortex, the latter area being relatively resistant to the long-term DA-depleting effects of MA *(129)*. MA also rapidly and transiently decreases complex IV (cytochrome-*c* oxidase) activity and ATP concentrations in the striatum but not the hippocampus, a region resistant to the DA-depleting effects of MA *(130,131)*. Because brain-region-dependent changes in metabolism appear to be correlated with depletions of DA, the selective effect of MA-induced energy consumption and subsequent energy depletion may be related to MA-induced glutamate release, oxidative stress, and the long-term depletions of DA.

Stimulant-induced increases in the extracellular concentrations of monoamines may contribute to mitochondrial inhibition. Elevated extracellular DA may compromise mitochondrial function via autoxidation to form quinones and/or the enzymatic degradation of DA to form $H_2O_2$ and the generation of hydroxyl radicals *(132,133)*. This hypothesis is especially interesting given the finding that decreased cytochrome-*c* oxidase activity is restricted to DA-rich brain regions (striatum, nucleus accumbens, and substantia nigra) *(130)*. Reactive oxygen species and DA-derived quinones are known to directly inhibit mitochondrial enzymes associated with energy production *(66,134,135)*. Although DA-mediated inhibition of energy production has not been demonstrated to occur in vivo, in vitro incubation of rat brain mitochondria with DA or DA-derived quinones decreases state 3 (ATP-synthesis coupled) and increases state 4 respiration *(67)*. These studies indicate that reactive DA byproducts may increase proton leakage across the mitochondrial membrane and inhibit the production of energy stores.

Several additional mechanisms could underlie the compromise in metabolic function that follows MA administration. Psychostimulants may increase neuronal energy utilization through the sustained sodium-dependent reversal of monoamine transporters, hyperlocomotion, and the production of hyperthermia *(136–138)*. The majority of ATP in the neuropil is devoted to the maintenance of ion (e.g., $Na^+$) gradients and the restoration of the membrane potential following depolarization *(139–142)*. Therefore, sustained activation of the ATP-dependent $Na^+/K^+$-ATPase following prolonged neurotransmitter release may lead indirectly to the depletion of substrates for the electron-transport chain. Such a decrease in available precursors may slow or halt the production of ATP through a decline in complex IV activity.

Depletion of striatal ATP stores could significantly contribute to elevated glutamate levels and further potentiate damage following MA administration (for review, *see* ref. 27). For example, a loss of $Na^+/K^+$ ATPase activity could lead to depolarization and release of neuronal glutamate from vesicular stores. In addition, energy failure could contribute to excess extracellular glutamate levels by disrupting or reversing the ATP-dependent glutamate transporter. The conversion of glutamate to glutamine in glia is also ATP dependent. Thus, depletion of energy stores could increase intraglial concentrations of glutamate. Increased intracellular glutamate concentrations could disrupt the concentration-dependent uptake of glutamate into glia, resulting in the accumulation of extracellular glutamate. Thus, in addition to activation of the corticostriatal pathway, MA administration could lead indirectly to elevated extracellular glutamate concentrations by disrupting bioenergetic systems and depleting energy (ATP) stores.

As discussed previously, increased extracellular glutamate concentrations after MA and subsequent NMDA receptor activation may lead to metabolic inhibition via classic excitotoxic mechanisms. Direct inhibition of mitochondrial function induces NMDA receptor-mediated excitotoxic damage that has similarities with damage resulting from MA administration. Almeida et al. *(143)* reported that neurons exposed to glutamate in vitro had decreased glutathione and ATP content, increased lactate dehydrogenase activity, decreased mitochondrial enzyme activity (succinate cytochrome-*c* reductase and cytochrome-*c* oxidase), and decreased oxygen consumption. Interestingly, increases in the extracellular concentrations of lactate, decreases in ATP content, and inhibition of cytochrome-*c* oxidase have all been found to occur in vivo following MA administration *(129–131)*. Similarly, local striatal perfusion of mitochondrial inhibitors acutely increases the extracellular concentration of DA and glutamate, depletes ATP, and produces an accumulation of lactate *(25,144–146)*. The long-term effects of malonate infusions include damage to striatal DA and, to a lesser extent, 5-HT terminals and a potentiated depletion of DA produced by both systemic and central administration of MA *(25,147)*. Furthermore, removal of excitatory corticostriatal afferents or administration of glutamate receptor antagonists attenuates striatal damage induced by either MA or the metabolic inhibitors malonate and 3-nitropropionic acid *(148,149)*. In addition, MA toxicity appears to be dependent on an increase in the release of nitric oxide, via glutamate activation of NMDA receptors, and the subsequent activation of the NOS pathway *(150–152)*. The subsequent production of nitric oxide can lead to the formation of reactive oxygen species (peroxynitrite) and to mitochondrial dysfunction by directly inhibiting complex IV of the electron-transport chain, cytochrome-*c* oxidase *(151,153)*. Thus, metabolic stress appears to be an important mediator in the excitotoxicity following direct inhibition of mitochondrial enzymes by malonate or 3-nitroproprionic, or indirectly following MA administration.

Several studies have demonstrated that manipulation of energy availability, via metabolic inhibition or support of bioenergetic systems, can alter the lasting effects of MA administration. Chan et al. *(131)* reported that inhibition of metabolism by pretreatment with 2-deoxyglucose exacerbates both MA-induced ATP loss and long-term reduction of striatal DA content (but see ref. *155*). Similarly, the local inhibition of complex II via intrastriatal perfusion with malonate synergizes with the local administration of MA to enhance DA toxicity compared to the perfusion of either drug alone *(25)*. Conversely, pretreatment with nicotinamide attenuates both the acute decrease in striatal ATP and the lasting DA depletions following amphetamine administration *(155)*. In addition, the local intrastriatal perfusion of substrates for the electron-transport chain (ubiquinone or nicotinamide) for several hours following MA administration attenuates the long-term loss of DA content *(129)*. Taken together, these data indicate that metabolic deficits and a depletion of energy stores is critical to the loss of monoamine nerve terminals following amphetamine and that the restoration or supplementation of energy production can attenuate the toxicity to MA.

## 4. GLUTAMATE MEDIATION OF DOPAMINE AND 5-HT TOXICITY

Although MA has been found to damage DA terminals in the striatum and 5-HT terminals in multiple brain regions, factors that mediate damage to these monoamine systems may differ on a fundamental level. A growing body of evidence suggests that DA-containing nerve terminals are inherently more vulnerable to damage following metabolic inhibition compared with 5-HT containing terminals. Additionally, increased extracellular glutamate may have a more direct effect in mediating toxicity to DA systems following MA administration.

Glutamate overflow and subsequent activation of the NOS pathway may differentially mediate DA and 5-HT toxicity. Glutamate overflow is not correlated with the depletion of 5-HT content in different brain regions after MA. In fact, 3,4-methylenedioxy-methamphetamine (MDMA), a more selective 5-HT toxin structurally similar to MA, damages striatal 5-HT terminals but does not result in glutamate overflow in this region *(33)* Abekawa et al. *(149)* report that administration of the NOS inhibitor L-NAME protects against MA-induced DA loss in the striatum, but it does not attenuate 5-HT toxicity in the striatum, nucleus accumbens, and medial frontal cortex of the same animals *(150)*. However, pretreatment with a different NOS inhibitor, $N^{\omega}$-nitro-L-arginine (L-NOARG), partially protects against long-term 5-HT depletion induced by MDMA in frontal cortex and parietal cortex, but not in other brain regions *(152)*. The interaction between glutamate and lasting depletion of 5-HT may, therefore, be brain region dependent.

The differential role of glutamate in mediating DA versus 5-HT toxicity also is evidenced by the inherent vulnerabilities of these systems to metabolic stress. In cultured mesencephalic neurons and synaptosomal preparations, inhibitors of oxidative phosphorylation decrease DA uptake to a greater degree compared to uptake of GABA, 5-HT, and norepinephrine *(156)*. Inherent differences in the effects of mitochondrial inhinition on neurotransmitter release in vivo may predict lasting toxicity to these systems. MA decreases cytochrome-*c* oxidase activity in DA-rich areas, but not in regions where MA toxicity manifests as a loss of 5-HT *(130)*, implicating DA release in mediating metabolic stress following MA. Furthermore, the local perfusion of the succinate dehydrogenase inhibitor malonate increases DA overflow more than 100-fold, whereas 5-HT release increases merely 5-fold *(157)*. In addition to differentially affecting the release of monoamines, intrastriatal infusions of malonate preferentially damage DA systems compared to GABA- or 5-HT-containing nerve terminals *(25,157,158)*. Coperfusion of MA and malonate synergize to produce even greater depletions of DA without affecting 5-HT tissue levels *(25,157)*, suggestive of the correlation between the degree of transmitter release and the differential toxic profiles of mitochondrial inhibitors on monoamine systems.

The possible mediation of serotonergic damage by extracellular glutamate is less studied and remains unclear. However, there is some evidence that indicates an NMDA receptor mediation of 5-HT loss. Pretreatment with the NMDA receptor antagonist MK-801 blocks both 5-HT and DA loss after MA, and 5-HT depletion following MDMA *(159)*. However, the protective effects of MK-801 may be related to the attenuation of stimulant-induced hyperthermia and, thus, may not be selectively mediated by the glutamate pathway *(160–162)*. Further studies are necessary to clarify the mechanism by which glutamate receptor antagonists convey neuroprotection.

The majority of available data are consistent with the conclusion that dopaminergic neurons are inherently more sensitive than 5-HT neurons to damage mediated by metabolic stress. In addition, vulnerability to mitochondrial inhibition may underlie DA-specific neurodegenerative disorders such as Parkinson's disease *(163)*. Although the etiology of the vulnerability of DA versus 5-HT neurons to excitotoxic, metabolic, and oxidative insults is not known, the ability of DA to autoxidize, combined with the enzymatic oxidation of DA to form $H_2O_2$, may lead to elevated concentrations of intracellular reactive oxygen species that render DA neurons more vulnerable to metabolic inhibition or excitotoxic events.

## 5. CONCLUSIONS

Substantial evidence supports the hypothesis that an increase in extracellular glutamate following MA administration is an obligatory step in the cascade of events culminating in striatal DA terminal loss. Several different mechanisms may contribute to this rise in extracellular glutamate, including a circuit-mediated increase in corticostriatal activity, a decrease in glutamate uptake into glia, and an increase in vesicular release following disruption of the membrane potential via a loss of $Na^+/K^+$-ATPase activity. Increased glutamate overflow likely contributes to the toxicity of amphetamines by initiating an excitotoxic response. Together, MA-mediated DA release and NMDA receptor activation can lead to the formation of intracellular reactive oxygen species and inhibition of metabolic function. Both oxidative and metabolic stress have been implicated in mediating the damage to DA terminals following MA administration, and substrates that attenuate the consequences of such stressors (antioxidants, free-radical scavengers, or substrates for the electron-transport chain) are neuroprotective. Additional evidence points to an inherent vulnerability of DA terminals to metabolic stress when compared with 5-HT systems, suggesting that factors which mediate the neurotoxic effect of MA on DA and 5-HT terminals may be substantially different.

Although the toxicity of MA was first recognized almost 30 yr ago *(164,165)*, the mechanisms culminating in DA loss are still under investigation. Recent evidence of DA terminal dysfunction in human MA abusers *(11,12,166)* indicates that MA abuse may have lasting consequences. It is not known if MA abuse is a risk factor in Parkinson's disease. Nevertheless, the possibility exists that MA-induced damage to the nigrostriatal DA system could result in an earlier onset of symptoms in individuals predisposed to develop Parkinson's disease. Further clinical and preclinical studies are obviously necessary to elucidated the risks, consequences, and treatment of stimulant-induced damage to the nigrostriatal DA system. A more basic understanding of factors that influence changes in the dopaminergic and serotonergic systems following MA exposure will hopefully lead to novel therapies designed to reverse or attenuate the excitotoxic, metabolic, and oxidative effects of this abused drug.

## ACKNOWLEDGMENTS

This work was supported in part by grants DA07606, DA05984, and DAMD 17-99-1-9479.

## REFERENCES

1. Services U.S.D.H.H. (1997) *Proceedings of the National Consensus Meeting on the Use, Abuse and Sequelae of Abuse of Methamphetamine with Implications for Prevention, Treatment, and Research,* Substance Abuse and Mental Health Services Administration, Center for Substance Abuse Treatment, Washington, DC.

2. Wagner, G. C., Ricaurte, G. A., Seiden, L. S., Schuster, C. R., Miller, R. J., and Westly, J. (1980) Long-lasting depletion of striatal dopamine uptake sites following repeated administration of methamphetamine. *Brain Res.* **171,** 151–160.
3. Ellison, G., Eison, M. S., Huberman, H. S., and Daniel, F. (1978) Long-term changes in dopaminergic innervation of caudate nucleus after continuous amphetamine administration. *Science* **201,** 276–278.
4. Ricaurte, G. A., Guillery, R. W., Seiden, L. S., and Schuster, C. R. (1982) Dopamine nerve terminal degeneration produced by high doses of methamphetamine in the rat brain. *Brain Res.* **235,** 93–103.
5. Ricaurte, G. A., Seiden, L. S. and Schuster, C. R. (1984) Further evidence that amphetamines produce long-lasting dopamine neurochemical deficits by destroying dopamine nerve fibers. *Brain Res.* **303,** 359–364.
6. Pu, C., Fisher, J. E., Cappon, G. D., and Vorhees, C. V. (1994) The effects of amfonelic acid, a dopamine uptake inhibitor, on methamphetamine-induced dopaminergic terminal degeneration and astrocytic response in rat striatum. *Brain Res.* **649,** 217–224.
7. Ryan, L. J., Martone, M. E., Linder, J. C., and Groves, P. M. (1988) Continuous amphetamine administration induces tyrosine hydroxylase immunoreactive patches in the adult neostriatum. *Brain Res. Bull.* **21,** 133–137.
8. Ricaurte, G. A., Schuster, C. R., and Seiden, L. S. (1980) Long-term effects of repeated methylamphetamine administration on dopamine and serotonin neurons in rat brain. *Brain Res.* **193,** 153–163.
9. Stephans, S. and Yamamoto, B. (1996) Methamphetamine pretreatment and the vulnerability of the striatum to methamphetamine neurotoxicity. *Neuroscience* **72(3),** 593–600.
10. Stephans, S. E. and Yamamoto, B. K. (1994) Methamphetamine-induced neurotoxicity: roles for glutamate and dopamine efflux. *Synapse* **17,** 203–209.
11. McCann, U. D., Wong, D. F., Yokoi, F., Villemagne, V., Dannals, R. F., and Ricaurte, G. A. (1998) Reduced striatal dopamine transporter density in abstinent methamphetamine and methcathinone users: evidence from positron emission tomography studies with [11C]WIN-35,428. *J. Neurosci.* **18(20),** 8417–8422.
12. Wilson, J. M., Kalasinsky, K. S., Levey, A. I., Bergeron, C., Reiber, G., Anthony, R. M., et al. (1996) Striatal dopamine nerve terminal markers in human, chronic methamphetamine users. *Nature Med.* **2(6),** 699–703.
13. Seiden, L. S., Commins, D. L., Vosmer, G. L., Axt, K. J., and Marek, G. J. (1998) Neurotoxicity in dopamine and 5-HT terminal fields: a regional analysis in nigrostriatal and mesolimbic projections. *ANYAS* **537,** 161–172.
14. Bittner, S. E., Wagner, G. C., Aigner, T. G., and Seiden, L. S. (1981) Effects of a high dose treatment of methamphetamine on caudate dopamine and anorexia in rats. *Pharmacol. Biochem. Behav.* **14,** 481–486.
15. Seiden, L. S., Fishman, M. W., and Schuster, C. R. (1975/76) Long-term methamphetamine induced changes in brain catecholamines in tolerant rhesus monkeys. *Drug Alcohol Depend.* **1,** 215–219.
16. Seiden, L. S. and Ricaurte, G. A. (1987) Neurotoxicity of methamphetamine and related drugs, in *Psychopharmacology: The Third Generation of Progress* (Meltzer, H. Y., ed.), Raven, New York, pp. 359–366.
17. Filloux, F. and Townsend, J. J. (1993) Pre- and post-synaptic neurotoxic effects of dopamine demonstrated by intrastriatal injection. *Exp. Neurol.* **119,** 79–88.
18. Michel, P. and Hefti, F. (1990) Toxicity of 6-hydroxydopamine and dopamine for dopaminergic neurons in culture. *J. Neurosci. Res.* **26(4),** 428–435.
19. Rosenberg, P. A. (1988) Catecholamine toxicity in cerebral cortex in dissociated cell culture. *J. Neurosci.* **8,** 2887–2894.
20. Schmidt, C. J. and Gibb, J. W. (1985) Role of the dopamine uptake carrier in the neurochemical response to methamphetamine: effects of amfonelic acid. *Eur. J. Pharmacol.* **109,** 73–80.
21. Marek, G. J., Vosmer, G., and Seiden, L. S. (1990) Dopamine uptake inhibitors block long-term neurotoxic effects of methamphetamine upon dopaminergic neurons. *Brain Res.* **513,** 274–279.
22. Buening, M. and Gibb, J. W. (1974) Influence of methamphetamine and neuroleptic drugs on tyrosine hydroxylase activity. *Eur. J. Pharmacol.* **26,** 30–34.
23. Hotchkiss, A. J. and Gibb, J. W. (1980) Long-term effects of multiple doses of methamphetamine on tryptophan hydroxylase and tyrosine hydroxylase activity in rat brain. *J. Pharmacol. Exp. Ther.* **214(2),** 257–262.
24. Sonsalla, P. K., Gibb, J. W., and Hanson, G. R. (1986) Roles of D1 and D2 dopamine receptor subtypes in mediating the methamphetamine-induced changes in monoamine systems. *J. Pharmacol. Exp. Ther.* **238,** 932–937.
25. Burrows, K. B., Nixdorf, W. L., and Yamamoto, B. K. (2000) Central administration of methamphetamine synergizes with metabolic inhibition to deplete striatal monoamines. *J. Pharmacol. Exp. Ther.* **292(3),** 853–860.
26. LaVoie, M. and Hastings, T. (1999) Dopamine quinone formation and protein modification associated with the striatal neurotoxicity of methamphetamine: evidence against a role for extracellular dopamine. *J. Neurosci.* **19(4),** 1484–1491.
27. Lipton, S. A. and Rosenberg, P. A. (1994) Excitatory amino acids as a final common pathway for neurologic disorders. *N. Engl. J. Med.* **330(9),** 613–622.
28. Olney, J. W. (1990) Excitotoxic amino acids and neuropsychiatric disorders. *Annu. Rev. Pharmacol. Toxicol.* **30,** 47–71.
29. Sonsalla, P. K., Nicklas, W. J., and Heikkila, R. E. (1989) Role for excitatory amino acids in methamphetamine-induced nigrostriatal dopaminergic toxicity. *Science* **243,** 398–400.

30. Baldwin, H. A., Colado, M. I., Murray, T. K., De Souza, R. J., and Green, A. R. (1993) Striatal dopamine release in vivo following neurotoxic doeses of methamphetamine and effect of the neuroprotective drugs chloromethiazole and dizocilpine. *Br. J. Pharmacol.* **108,** 590–596.
31. Fuller, R. W., Hemrick-Luecke, S. K., and Ornstein, P. L. (1992) Protection against amphetamine-induced neurotoxicity toward striatal dopamine neurons in rodents by LY274614, an excitatory amino acid antagonist. *Neuropharmacology* **31(10),** 1027–1032.
32. Weihmuller, F. B., O'Dell, S. J., and Marshall, J. F. (1992) MK-801 protection against methamphetamine-induced striatal dopamine terminal injury is associated with attenuated dopamine overflow. *Synapse* **11,** 155–163.
33. Nash, J. F. and Yamamoto, B. K. (1992) Methamphetamine neurotoxicity and striatal glutamate release: comparison to 3,4-methylenedioxymethamphetamine. *Brain Res.* **581,** 237–243.
34. Nash, J. F. and Yamamoto, B. K. (1993) Effect of *d*-amphetamine on the extracellular concentrations of glutamate and dopamine in iprindole-treated rats. *Brain Res.* **627,** 1–8.
35. Abekawa, T., Ohmori, T., and Koyama, T. (1994) Effect of NO synthase inhibition on behavioral changes induced by a single administration of methamphetamine. *Brain Res.* **666,** 147–150.
36. Bowyer, J. F., Gough, B., Slikker, W., Jr., Lipe, G. W., Newport, G. D., and Holson, R. R. (1993) Effects of a cold environment or age on methamphetamine-induced dopamine release in the caudate putamen of female rats. *Pharmacol. Biochem. Behav.* **44,** 87–98.
37. Mora, F. and Porras, A. (1993) Effects of amphetamine in the release of excitatory amino acid neurotransmitters in the basal ganglia of the conscious rat. *Can. J. Pharmacol.* **71,** 348–351.
38. Hassler, R., Haug, P., Nitsch, C., Kim, S. J., and Paik, K. (1982) Effect of motor and premotor cortex ablation on concentrations of amino acids, monoamines, and acetylcholine, and on the ultrastructure in rat striatum: a confirmation of glutamate as the specific cortico-striatal transmitter. *J. Neurochem.* **38,** 1087–1098.
39. Burrows, K. B. and Meshul, C. K. (1997) Methamphetamine alters presynaptic glutamate immunoreactivity in the caudate nucleus and motor cortex. *Synapse* **27,** 133–144.
40. Yamamoto, B. K., Gudelsky, G. A., and Stephans, S. E. (1998) Amphetamine neurotoxicity: roles for dopamine, glutamate, and oxidative stress, in *Neurochemical Markers of Degenerative Diseases & Drug Addiction* (Qureshi, G. A., Parvez, H., Caudy, P., and Parvez, S., eds.), VSP, Utrecht, Vol. **7,** pp. 223–244.
41. Broening, H. W., Pu, C., and Vorhees, C. (1997) Methamphetamine selectively damages dopaminergic innervation to the nucleus accumbens core while sparing the shell. *Synapse* **2,** 153–160.
42. Abekawa, T., Ohmori, T., and Koyama, T. (1994) Effects of repeated administration of a high dose of methamphetamine on dopamine and glutamate release in rat striatum and nucleus accumbens. *Brain Res.* **643,** 276–281.
43. Alexander, G. E. and Crutcher, M. D. (1990) Functional architecture of basal ganglia circuits: neural substrates of parallel processing. *Trends Neurosci.* **13(7),** 266–271.
44. Pu, C., Broening, H. W., and Vorhees, C. V. (1996) Effect of methamphetamine on glutamate-positive neurons in the adult and developing rat somatosensory cortex. *Synapse* **23,** 328–334.
45. Herbert, M. A. and O'Callaghan, J. P. (2000) Protein phosphorylation cascades associated with methamphetamine-induced glial activation. *ANYAS* **914,** 238–262.
46. Eisch, A. J., O'Dell, S. J., and Marshall, J. F. (1996) Striatal and cortical NMDA receptors are altered by a neurotoxic regimen of methamphetamine. *Synapse* **22,** 217–225.
47. Eisch, A. J. and Marshall, J. F. (1998) Methamphetamine neurotoxicity: dissociation of striatal dopamine terminal damage from parietal cortical cell body injury. *Synapse* **30,** 433–445.
48. O'Dell, S. J. and Marshall, J. F. (2000) Repeated administration of methamphetamine damages cells in the somatosensory cortex: overlap with cytochrome oxidase-rich barrels. *Synapse* **37,** 32–37.
49. O'Dell, S. J., Weihmuller, F. B., McPherson, R. J., and Marshall, J. F. (1994) Excitotoxic striatal lesions protect against subsequent methamphetamine-induced dopamine depletions. *J. Pharmacol. Exp. Ther.* **269(3),** 1319–1325.
50. Deniau, J. M. and Chevalier, G. (1992) The lamellar organization of the rat substantia nigra pars reticulata: distribution of projection neurons. *Neuroscience* **46,** 361–377.
51. Donoghue, J. P. and Parham, C. (1983) Afferent conditions of the lateral agranular field of the rat motor cortex. *J. Comp. Neurol.* **217,** 390–404.
52. Cicirata, F., Anagaut, P., Ciopni, M., Serapide, M. F., and Papale, A. (1986) Functional organization of the thalamic projections to the motor cortex. An anatomical and electrophysical study in the rat. *Neuroscience* **19,** 81–99.
53. Albin, R. L., Young, A. B., and Penney, J. B. (1989) The functional anatomy of basal ganglia disorders. *Trends Neurosci.* **12,** 366–375.
54. Graybiel, A. M. (1990) Neurotransmitters and neuromodulators in the basal ganglia. *Trends Neurosci.* **13(7),** 244–254.
55. Basse-Tomusk, A. and Rebec, G. V. (1990) Corticostriatal and thalamic regulation of amphetamine-induced ascorbate release in the neostriatum. *Pharmacol. Biochem. Behav.* **35,** 55–60.
56. Matuszewich, L. and Yamamoto, B. K. (1999) Modulation of GABA release by dopamine in the substantia nigra. *Synapse* **32,** 29–36.

57. Rosales, M. G., Martinez-Fong, D., Morales, R., Nunez, A., Flores, G., Gongora, A., et al. (1997) Reciprocal interaction between glutamate and dopamine in the pars reticulata of the rat substantia nigra: a microdialysis study. *Neuroscience* **80**, 803–810.
58. Timmerman, W. and Westerink, B. H. C. (1995) Extracellular γ-aminobutyric acid in the substantia nigra reticulata measured by microdialysis in awake rats: effects of various stimulants. *Neurosci. Lett.* **197**, 21–24.
59. Radnikow, G. and Misgeld, U. (1998) Dopamine D1 receptors facilitate $GABA_A$ synaptic currents in the rat substantia nigra pars reticulata. *J. Neurosci.* **18**, 2009–2016.
60. Nicholson, L. F., Faull, R. L., Waldvogel, H. J., and Dragunow, M. (1992) The regional, cellular and subcellular localization of $GABA_A$/benzodiazepine receptors in the substantia nigra of the rat. *Neuroscience* **50**, 355–370.
61. Commins, D. L., Axt, K. J., Vosmer, G., and Seiden, L. S. (1987) 5,6-Dihydroxytryptamine, a serotonergic neurotoxin, is formed endogenously in the rat brain. *Brain Res.* **403(1)**, 7–14.
62. Seiden, L. S. and Vosmer, G. L. (1984) Formation of 6-hydroxydopamine in caudate nucleus of the rat brain after a single large dose of methylamphetamine. *Pharmacol. Biochem. Behav.* **21**, 29–31.
63. Hoyt, K., Reynolds, I. and Hastings, T. (1997) Mechanisms of dopamine-induced cell death in cultured rat forebrain neurons: interactions with and differences from glutamate-induced cell death. *Exp. Neurol.* **143**, 269–281.
64. Chapman, A. G., Durmuller, N., Lees, G. J., and Meldrum, B. S. (1989) Excitotoxicity of NMDA and kainic acid is modulated by nigrostriatal dopaminergic fibers. *Neurosci. Lett.* **107**, 256–260.
65. Filloux, F. and Wamsley, J. K. (1991) Dopaminergic modulation of excitotoxicity in rat striatum: evidence from nigrostriatal lesions. *Synapse* **8**, 281–288.
66. Ben-Schachar, D., Zuk, R., and Glinka, Y. (1995) Dopamine neurotoxicity: inhibition of mitochondrial respiration. *J. Neurochem.* **64**, 718–723.
67. Berman, S. B. and Hastings, T. G. (1999) Dopamine oxidation alters mitochondrial respiration and induces permeability transition in brain mitochondria: implications for Parkinson's disease. *J. Neurochem.* **73(3)**, 1127–1137.
68. Dugan, L. L., Sensi, S. L., Conzoniero, L., Handran, S. D., Rothman, S. M., Lin, T. S., et al. (1995) Mitochondrial production of reactive oxygen species in cortical neurons following exposure to N-methyl-D-aspartate. *J. Neurosci.* **15**, 6377–6388.
69. Reynolds, I. J. and Hastings, T. G. (1995) Glutamate induces the production of reactive oxygen species in cultured forebrain neurons following NMDA receptor activation. *J. Neurosci.* **15(5)**, 3318–3327.
70. Fonnum, F. (1998) Excitotoxicity in the brain. *Arch. Toxicol.* **20(Suppl.)**, 387–395.
71. Nicholls, D. G. and Budd, S. L. (1998) Neuronal excitotoxicity: the role of mitochondria. *BioFactors* **8**, 287–299.
72. Choi, D. W. (1988) Glutamate neurotoxicity and diseases of the nervous system. *Neuron* **1**, 623–634.
73. Dumius, A., Sebben, M., Haynes, L., Pin, J. P., and Bockaert, J. (1988) NMDA receptors activate the arachidonic acid cascade system in striatal neurons. *Nature* **336**, 68–70.
74. Lasarewicz, J. W., Wroblewski, J. T., Palmer, M. E., and Costa, E. (1988) Activation of N-methyl-D-aspartate-sensitive glutamate receptors stimulates arachidonic acid release in primary cultures of cerebellar granule cells. *Neuropharmacology* **27**, 765–769.
75. Pellegrini-Giampietro, D. E., Cherici, G., Alesiani, M., Carla, V., and Moroni, F. (1990) Excitatory amino acid release and free radical formation may cooperate in the genesis of ischemia-induced neuronal damage. *J. Neurosci.* **10**, 1035–1041.
76. Volterra, A., Trotti, D., Tromba, C., Floridi, S., and Racagni, G. (1994) Glutamate uptake inhibition by oxygen free radicals in cortical astrocytes. *J. Neurosci.* **14**, 2924–2932.
77. Williams, J. H., Errington, M. L., Lynch, M. A., and Bliss, T. V. P. (1989) Arachidonic acid induces long-term activity-dependent enhancement of synaptic transmission in the hippocampus. *Nature* **341**, 739–742.
78. Bi, X., Change, V., Siman, R., Tocco, G., and Baudry, M. (1996) Regional distribution and time-course of calpain activation following kainate-induced seizure activity in adult rat brain. *Brain Res.* **726**, 98–108.
79. Siman, R. and Noszek, J. C. (1988) Excitatory amino acids activate calpain I and induce structural protein breakdown in vivo. *Neuron* **1**, 279–287.
80. Buki, A., Siman, R., Trojanowski, J. Q., and Povlishock, J. T. (1999) The role of calpain-mediated spectrin proteolysis in traumatically induced axonal injury. *J. Neuropathol. Exp. Neurol.* **58**, 365–375.
81. Minger, S. L., Gegddes, J. W., Holtz, M. L., Craddock, S. D., Whireheart, S. W., Siman, R., et al. (1998) Glutamate receptor antagonists inhibit calpain-mediated cytoskeletal proteolysis in focal cerebral ischemia. *Brain Res.* **810**, 181–189.
82. Morimoto, T., Ginsberg, M. D., Dietrich, W. D., and Zhao, W. (1997) Hyperthermia enhances spectrin breakdown in transient focal cerebral ischemia. *Brain Res.* **746(1–2)**, 43–51.
83. Pike, B. R., Zhao, X., Newcomb, J. K., Posmantur, R. M., Wang, K. K., and Hayes, R. L. (1998) Regional calpain and caspase-3 proteolysis of alpha-spectrin after traumatic brain injury. *NeuroReport* **9**, 2437–2442.
84. Dykens, J. A., Stern, A., and Trenkner, E. (1987) Mechanism of kainate toxicity to cerebellar neurons in vitro is analogous to reperfusion tissue injury. *J. Neurochem.* **49**, 1222–1228.
85. Yamamoto, B. K. and Zhu, W. (1998) The effects of methamphetamine on the production of free radicals and oxidative stress. *J. Pharmacol. Exp. Ther.* **287(1)**, 107–114.

86. Fleckenstein, A. E., Wilkins, D. G., Gibb, J. W., and Hanson, G. R. (1997) Interaction between hyperthermia and oxygen radical formation in the 5-hydroxytryptaminergic response to a single methamphetamine administration. *J. Pharmacol. Exp. Ther.* **283,** 281–285.
87. Giovanni, A., Liang, L. P., Hastings, T. G., and Zigmond, M. J. (1995) Estimating hydroxyl radical content in rat brain using systemic and intraventricular salicylate: impact of methamphetamine. *J. Neurochem.* **64,** 1819–1825.
88. Cappon, G. D., Broening, H. W., Pu, C., Morford, L. and Vorhees, C. V. (1996) Alpha-phenyl-*N-tert*-butyl nitrone attenuates methamphetamine-induced depletion of striatal dopamine without altering hyperthermia. *Synapse* **24(2),** 173–181.
89. De Vito, M. J. and Wagner, G. C. (1989) Methamphetamine-induced neuronal damage: a possible role for free radicals. *Neuropharmacology* **28(10),** 1145–1150.
90. Cadet, J. L., Sheng, P., Ali, S. F., Rothman, R., Carlson, E., and Epstein, C. (1994) Attenuation of methamphetamine-induced neurotoxicity in copper/zinc superoxide dismutase transgenic mice. *J. Neurochem.* **62(1),** 380–383.
91. Li, Y. and Jaiswal, A. K. (1992) Regulation of human NAD(P)H: quinone oxidoreductase gene. Role of AP1 binding site contained within human antioxidant response element. *J. Biol. Chem.* **267,** 15,097–15,104.
92. Pinkus, R., Weiner, L. M., and Daniel, V. (1995) Role of quinone-mediated generation of hydroxyl radicals in the induction of glutathione S-transferase gene expression. *Biochemistry* **34,** 81–88.
93. Herdegen, T. and Leah, J. D. (1998) Inducible and constitutive transcription factors in the mammalian nervous system: control of gene expression by Jun, Fos and Krox, and CREB/ATF proteins. *Brain Res. Rev.* **28,** 370–490.
94. Hirata, H., Asanuma, M., and Cadet, J. L. (1998) Superoxide radicals are mediators of the effects of methamphetamine on Zif268 (Egr-1, NGFI-A) in the brain: evidence from using CuZn superoxide dismutase transgenic mice. *Mol. Brain Res.* **58,** 209–216.
95. Merchant, K. M., Hanson, G. R., and Dorsa, D. M. (1994) Induction of neurotensin and c-fos mRNA in distinct subregions of rat neostriatum after acute methamphetamine: comparison with acute haloperidol effects. *J. Pharmacol. Exp. Ther.* **269(2),** 806–812.
96. Sheng, P., Cerruti, C., Ali, S., and Cadet, J. L. (1996) Nitric oxide is a mediator of methamphetamine (METH)-induced neurotoxicity. In vitro evidence from primary cultures of mesencephalic cells. *ANYAS* **801,** 174–186.
97. Deng, X., Ladenheim, B., Tsao, L., and Cadet, J. (1999) Null mutation of c-fos causes exacerbation of methamphetamine-induced neurotoxicity. *J. Neurosci.* **19(22),** 10,107–10,115.
98. Halliwell, B. (1992) Reactive oxygen species and the central nervous system. *J. Neurochem.* **59(5),** 1609–1623.
99. Cubells, J. F., Rayport, S., Rajendran, G., and Sulzer, D. (1994) Methamphetamine neurotoxicity involves vacuolation of endocytic organelles and dopamine-dependent intracellular oxidative stress. *J. Neurosci.* **14(4),** 2260–2271.
100. Acikgoz, O., Gonenc, S., Kayatekin, B. M., Uysal, N., Pekcetin, C., Semin, I., et al. (1998) Methamphetamine causes lipid peroxidation and an increase in superoxide dismutase activity in the rat striatum. *Brain Res.* **813,** 200–202.
101. Tsao, L. I., Ladenheim, B., Andrews, A. M., Chiueh, C. C., Cadet, J. L., and Su, T. P. (1998) Delta opioid peptide [D-Ala$^2$, D-leu$^5$] enkephalin blocks the long-term loss of dopamine transporters induced by multiple administrations of methamphetamine: involvement of opioid receptors and reactive oxygen species. *J. Pharmacol. Exp. Ther.* **287(1),** 322–330.
102. Imam, S. Z. and Ali, S. F. (2000) Selenium, an antioxidant, attenuates methamphetamine-induced dopaminergic toxicity and peroxynitrite generation. *Brain Res.* **855(1),** 186–191.
103. Imam, S. Z., Crow, J. P., Newport, G. D., Islam, F., Slikker, W. J., and Ali, S. F. (1999) Methamphetamine generates peroxynitrite and produces dopaminergic neurotoxicity in mice: protective effects of peroxynitrite decomposition catalyst. *Brain Res.* **837(1–2),** 15–21.
104. Fornstedt, B., Brun, A., Rosengren, E., and Carlsson, A. (1989) The apparent autoxidation rate of catechols in dopamine-rich regions of human brains increases with the degree of depigmentation of substantia nigra. *J. Neural Transm.* **1,** 279–295.
105. Hastings, T., Lewis, D., and Zigmond, M. (1996) Role of oxidation in the neurotoxic effects of intrastriatal dopamine injections. *Proc. Natl. Acad. Sci. USA* **93,** 1956–1961.
106. Berman, S., Zigmond, M., and Hastings, T. (1996) Modification of dopamine transporter function: effect of reactive oxygen species and dopamine. *J. Neurochem.* **67,** 593–600.
107. Fleckenstein, A. E., Metzger, R. R., Beyeler, M. L., Gibb, J. W., and Hanson, G. R. (1997) Oxygen radicals diminish dopamine transporter function in rat striatum. *Eur. J. Pharmacol.* **334,** 111–114.
108. Kuhn, D. M., Arthur, R. E., Thomas, D. M., and Elferink, L. A. (1999) Tyrosine hydroxylase is inactivated by catechol-quinones and converted to a redox-cycling quinoprotein: relevance to Parkinson's disease. *J. Neurochem.* **73,** 1309–1317.
109. Jayanthi, S., Ladenheim, B., and Cadet, J. L. (1998) Methamphetamine-induced changes in antioxidant enzymes and lipid peroxidation in copper/zinc-superoxide dismutase transgenic mice. *ANYAS* **844,** 92–102.
110. Moszczynska, A., Turenne, S., and Kish, S. J. (1998) Rat striatal levels of the antioxidant glutathione are decreased following binge administration of methamphetamine. *Neurosci. Lett.* **255(1),** 49–52.
111. Harold, C., Wallace, T., Friedman, R., Gudelsky, G., and Yamamoto, B. K. (2000) Methamphetamine selectively alters brain antioxidants. *Eur. J. Pharmacology* **400,** 99–102.
112. Lafon-Cazal, M., Pietri, S., Culcasi, M., and Bockaert, J. (1993) NMDA-dependent superoxide production and neurotoxicity. *Nature* **364,** 535–537.

113. Murphy, T. H., Miyamoto, M., Sastre, A., Schnaar, R. L., and Coyle, J. T. (1989) Glutamate toxicity in a neuronal cell line involves inhibition of cystine transport leading to oxidative stress. *Neuron* **2(6),** 1547–1558.
114. Garthwaite, J., Charles, S. L., and Chess-Williams, R. (1988) Endothelium-derived relaxing factor release on activation of NMDA receptors suggests a role as intracellular messenger in the brain. *Nature* **336,** 385–387.
115. Dawson, V. L. and Dawson, T. M. (1996) Nitric oxide neurotoxicity. *J. Chem. Neuroanat.* **10,** 179–190.
116. Schulz, J. B., Matthews, R. T., Jenkins, B. G., Ferrante, R. J., Siwek, D., Henshaw, D. R., et al. (1995) Blockade of neuronal nitric oxide synthase protects against excitotoxicity in vivo. *J. Neurosci.* **15,** 8419–8429.
117. Radi, R., Beckman, J. S., Bush, K. M., and Freeman, B. A. (1991) Peroxynitrite oxidation of sulfhydryls. The cytotoxic potential of superoxide and nitric oxide. *J. Biol. Chem.* **266(7),** 4244–4255.
118. LaVoie, M. J. and Hastings, T. G. (1999) Peroxynitrite- and nitrite-induced oxidation of dopamine: implications for nitric oxide in dopaminergic cell loss. *J. Neurochem.* **73(6),** 2546–2554.
119. Przedborski, S., Jackson-Lewis, V., Yokoyama, R., Shibata, T., Dawson, V. L. and Dawson, T. M. (1996) Role of neuronal nitric oxide in 1-methyl-4-phenyl-1,2,3,6-tetrahydropyridine (MPTP)-induced dopaminergic neurotoxicity. *Proc. Natl. Acad. Sci. USA* **93(10),** 4565–4571.
120. Schulz, J. B., Matthews, R. T., Muqit, M. M., Browne, S. E., and Beal, M. F. (1995) Inhibition of neuronal nitric oxide synthase by 7-nitroindazole protects against MPTP-induced neurotoxicity in mice. *J. Neurochem.* **64(2),** 936–939.
121. Di Monte, D. A., Royland, J. E., Jakowec, M. W., and Langston, J. W. (1996) Role of nitric oxide in methamphetamine neurotoxicity: protection by 7-nitroindazole, an inhibitor of neuronal nitric oxide synthase. *J. Neurochem.* **67(6),** 2443–2450.
122. Itzhak, Y. and Ali, S. F. (1996) The neuronal nitric oxide synthase inhibitor, 7-nitroindazole, protects against methamphetamine-induced neurotoxicity in vivo. *J. Neurochem.* **67,** 1770–1773.
123. Itzhak, Y., Gandia, C., Huang, P. L., and Ali, S. F. (1998) Resistance of neuronal nitric oxide synthase-deficient mice to methamphetamine-induced dopaminergic neurotoxicity. *J. Pharmacol. Exp. Ther.* **284,** 1040–1047.
124. Sheng, P., Ladenheim, B., Moran, T. H., Wang, X. B., and Cadet, J. L. (1996) Methamphetamine-induced neurotoxicity is associated with increased striatal AP-1 DNA-binding activity in mice. *Mol. Brain Res.* **42,** 171–174.
125. Sylvia, A., LaManna, J., Rosenthal, M., and Jobsis, F. (1977) Metabolite studies of methamphetamine effects based upon mitochondrial respiratory state in rat brain. *J. Pharmacol. Exp. Ther.* **201,** 117–125.
126. Pontieri, F. E., Crane, A. M., Seiden, L. S., Kleven, M. S., and Porrino, L. J. (1990) Metabolic mapping of the effects of intravenous methamphetamine administration in freely moving rats. *Psychopharmacology* **102(2),** 175–182.
127. Porrino, L. J., Lucignani, G., Dow-Edwards, D., and Sokoloff, L. (1984) Correlation of dose-dependent effects of acute amphetamine administration on behavior and local cerebral metabolism in rats. *Brain Res.* **307(1–2),** 311–320.
128. Huang, Y. H., Tsai, S. J., Su, T. W., and Sim, C. B. (1999) Effects of repeated high-dose methamphetamine on local cerebral glucose utilization in rats. *Neuropsychopharmacology* **21(3),** 427–434.
129. Stephans, S. E., Whittingham, T. S., Douglas, A. J., Lust, W. D., and Yamamoto, B. K. (1998) Substrates of energy metabolism attenuate methamphetamine-induced neurotoxicity in striatum. *J. Neurochem.* **71,** 613–621.
130. Burrows, K. B., Gudelsky, G., and Yamamoto, B. K. (2000) Rapid and transient inhibition of mitochondrial function following methamphetamine or MDMA administration. *Eur. J. Pharmacol.* **398(1),** 11–18.
131. Chan, P., Di Monte, D. A., Luo, J. J., DeLanney, L. E., Irwin, I., and Langston, J. W. (1994) Rapid ATP loss caused by methamphetamine in the mouse striatum: relationship between energy impairment and dopaminergic neurotoxicity. *J. Neurochem.* **62,** 2484–2487.
132. Graham, D. G., Tiffany, S. M., Bell, W. B., and Gutknecht, W. F. (1978) Autoxidation versus covalent binding of quinones as the mechanism of toxicity of dopamine, 6-hydroxy dopamine, and related compounds toward C1300 neuroblastoma cells in vitro. *Mol. Pharmacol.* **14,** 644–653.
133. McLaughlin, B. A., Nelson, D., Erecinska, M., and Chesselet, M. F. (1998) Toxicity of dopamine to striatal neurons in vitro and potentiation of cell death by a mitochondrial inhibitor. *J. Neurochem.* **70,** 2406–2415.
134. Yagi, T. and Hatefi, Y. (1987) Thiols in oxidative phosphorylation: thiols in the FO of ATP synthase essential for ATPase activity. *Arch. Biochem. Biophys.* **254(1),** 102–109.
135. Zhang, Y., Marcillat, O., Giulivi, C., Ernster, L., and Davies, K. (1990) The oxidative inactivation of mitochondrial electron transport chain components and ATPase. *J. Biol. Chem.* **265(27),** 16,330–16,336.
136. Fischer, J. F., and Cho, A. K. (1979) Chemical release of dopamine from striatal homogenates: evidence for an exchange diffusion model. *J. Pharmacol. Exp. Ther.* **208(2),** 203–209.
137. Huether, G., Zhou, D., and Ruther, E. (1997) Causes and consequences of the loss of serotonergic presynapses elicited by the consumption of 3,4-methylenedioxymethamphetamine (MDMA, "ecstasy") and its congeners. *J. Neural Transm.* **104,** 771–794.
138. Raiteri, M., Cerrito, F., Cervoni, A. M., and Levi, G. (1979) Dopamine can be released by two mechanisms differentially affected by the dopamine transport inhibitor nomifensine. *J. Pharmacol. Exp. Ther.* **208(2),** 195–202.
139. Erecinska, M. and Silver, I. (1989) ATP and brain function. *J. Cereb. Blood Flow Metab.* **9,** 2–19.
140. Hevner, R., Duff, R., and Wong-Riley, M. (1992) Coordination of ATP production and consumption in brain: parallel regulation of cytochrome oxidase and $Na^+$, $K^+$-ATPase. *Neurosci. Lett.* **138,** 188–192.

141. Siesjo, B. K. (1978) *Brain Energy Metabolism,* Wiley, New York.
142. Wong-Riley, M. (1989) Cytochrome oxidase: an endogenous metabolic marker for neuronal activity. *Trends Neurosci.* **12(3),** 94–101.
143. Almeida, A., Heales, S. J. R., Bolanos, J. P., and Medina, J. M. (1998) Glutamate neurotoxicity is associated with nitric oxide-mediated mitochondrial dysfunction and glutathione depletion. *Brain Res.* **790,** 209–216.
144. Beal, M., Brouillet, E., Jenkins, B., Henshaw, R., Rosen, B., and Hyman, B. (1993) Age-dependent striatal excitotoxic lesions produced by the endogenous mitochondrial inhibitor malonate. *J. Neurochem.* **61,** 1147–1150.
145. Beal, M. F., Hyman, B. T., and Koroshetz, W. (1993) Do defects in mitochondrial energy metabolism underlie the pathology of neurodegenerative diseases? *Trends Neurosci.* **16(4),** 125–130.
146. Messam, C., Greene, J., Greenamyre, J., and Robinson, M. (1995) Intrastriatal injections of the succinate dehydrogenase inhibitor, malonate, cause a rise in extracellular amino acids that is blocked by MK-801. *Brain Res.* **684(2),** 221–224.
147. Albers, D., Zeevalk, G., and Sonsalla, P. (1996) Damage to dopaminergic nerve terminals in mice by combined treatment on intrastriatal malonate with systemic methamphetamine or MPTP. *Brain Res.* **718(1–2),** 217–220.
148. Beal, M. F., Brouillet, E., Jenkins, B. G., Ferrante, R. J., Kowall, N. W., Miller, J. M., et al. (1993) Neurochemical and histologic characterization of striatal excitotoxic lesions produced by the mitochondrial toxin 3-nitropropionic acid. *J. Neurosci.* **13(10),** 4181–4192.
149. Ludolph, A. C., Seeling, M., Ludolph, A. G., Sabri, M. I., and Spencer, P. S. (1992) ATP deficits and neuronal degeneration induced by 3-nitropropionic acid. *ANYAS* **648,** 3000–3002.
150. Abekawa, T., Ohmori, T., and Koyama, T. (1996) Effects of nitric oxide synthesis inhibition on methamphetamine-induced dopaminergic and serotonergic neurotoxicity in the rat brain. *J. Neural Transm.* **103(6),** 671–680.
151. Lizasoain, I., Moro, M. A., Knowles, R. G., Darley-Usmar, V., and Moncada, S. (1996) Nitric oxide and peroxynitrite exert distinct effects on mitochondrial respiration which are differentially blocked by glutathione or glucose. *Biochem. J.* **314(3),** 877–880.
152. Zheng, Y. and Laverty, R. (1998) Role of brain nitric oxide in (+/–) 3,4-methylenedioxymethamphetamine (MDMA)-induced neurotoxicity in rats. *Brain Res.* **795,** 257–263.
153. Cleeter, M. W., Cleeter, J. M., Darley-Usmar, V. M., Moncada, S., and Schepira, A. H. (1994) Reversible inhibition of cytochrome c oxidase, the terminal enzyme of the mitochondrial respiratory chain, by nitric oxide. Implications for neurodegenerative diseases. *FEBS Lett.* **345(1),** 50–54.
154. Callahan, B. T., Yuan, J., Stover, G., Hatzidimitriou, G., and Ricaurte, G. A. (1998) Effects of 2-deoxy-D-glucose on methamphetamine-induced dopamine and serotonin neurotoxicity. *J. Neurochem.* **70(1),** 190–197.
155. Wan, F. J., Lin, H. C., Kang, B. H., Tseng, C. J., and Tung, C. S. (1999) D-Amphetamine-induced depletion of energy and dopamine in the rat striatum is attenuated by nicotinamide pretreatment. *Brain Res. Bull.* **50(3),** 167–171.
156. Marey-Semper, I., Gelman, M., and Levi-Strauss, M. (1993) The high sensitivity to rotenone of striatal dopamine uptake suggests the existence of a constitutive metabolic deficiency in dopaminergic neurons from the substantia nigra. *Eur. J. Neurosci.* **5(8),** 1029–1034.
157. Nixdorf, W. L., Burrows, K. B., Gudelsky, G. A., and Yamamoto, B. K. (2001) Differential enhancement of serotonin and dopamine depletions by inhibition of energy metabolism: comparisons between methamphetamine and 3,4-methylenedioxymethamphetamine. *J. Neurochem.* **77,** 647–654.
158. Zeevalk, G., Manzino, L., Hoppe, J., and Sonsalla, P. (1997) In vivo vulnerability of dopamine neurons to inhibition of energy metabolism. *Eur. J. Pharmacol.* **320,** 111–119.
159. Finnegan, K. T. and Taraska, T. (1996) Effects of glutamate antagonists on methamphetamine and 3,4-methylenedioxymethamphetamine-induced striatal dopamine release in vivo. *J. Neurochem.* **66,** 1949–1958.
160. Albers, D. S. and Sonsalla, P. K. (1995) Methamphetamine-induced hyperthermia and dopaminergic neurotoxicity in mice: pharmacological profile of protective and nonprotective agents. *J. Pharmacol. Exp. Ther.* **275,**(3) 1104–1114.
161. Farfel, G. M., and Seiden, L. S. (1995) Role of hypothermia in the mechanism of protection against serotonergic toxicity. II. Experiments with methamphetamine, *p*-chloroamphetamine, fenfluramine, dizocilpine and dextromethorphan. *J. Pharmacol. Exp. Ther.* **272(2),** 868–875.
162. Sonsalla, P., Albers, D., and Zeevalk, G. (1998) Role of glutamate in neurodegeneration of dopamine neurons in several animal models of parkinsonism. *Amino Acids* **14,** 69–74.
163. DiMauro, S. (1993) Mitochondrial involvement in Parkinson's disease: the controversy continues. *Neurology* **43,** 2170.
164. Fibiger, H. C. and McGeer, E. G. (1971) Effect of acute and chronic methamphetamine treatment on tyrosine hydroxylase activity in brain and adrenal medulla. *Eur. J. Pharmacol.* **16,** 176–180.
165. Koda, L. Y. and Gibb, J. W. (1973) Adrenal and striatal tyrosine hydroxylase activity after methamphetamine. *J. Pharmacol. Exp. Ther.* **185,** 42.
166. McCann, U., Szabo, Z., Scheffel, U., Dannals, R., and Ricaurte, G. (1998) Positron emission tomographic evidence of toxic effect of MDMA ("Ecstasy") on brain serotonin neurons in human beings. *Lancet* **352,** 1433–1437.

# 15
# Nitric Oxide-Dependent Processes in the Action of Psychostimulants

## Yossef Itzhak, PhD, Julio L. Martin, PhD and Syed F. Ali, PhD

## 1. INTRODUCTION

Nitric oxide (NO) is considered a retrograde messenger involved in synaptic plasticity and neurotoxicity *(1–3)*. In the brain, NO is generated by the neuronal, inducible, and endothelial isoforms of nitric oxide synthase (nNOS, iNOS, and eNOS, respectively). nNOS and eNOS are calcium dependent, whereas iNOS, present primarily in microglia and macrophages, is calcium independent. There is a particular interest on the role of nNOS in central nervous system (CNS) pathology because the stimulation of nNOS is associated primarily with the activation of the *N*-methyl-D-aspartate (NMDA) type of glutamate receptors. The increase of calcium influx caused by NMDA receptor activation leads to binding of calcium to calmodulin, which then stimulates nNOS *(1–3)*. L-arginine is the substrate of NOS; the conversion of L-arginine to L-citrulline causes the release of NO. Several studies have indicated that NO modulates the release of various neurotransmitters such as dopamine, gultamate, and norepinephrine *(4–6)*. Thus, there is a reason to consider NO as a major neuromodulator of synaptic transmission in brain.

Psychostimulants such as cocaine and methamphetamine (METH) cause an increase in synaptic dopamine (DA) level in the caudate nucleus (dorsal striatum) and nucleus accumbens (NAc; ventral striatum). Cocaine binds to the dopamine transporter and blocks the reuptake of DA. Amphetamines not only inhibit the reuptake of DA but also cause further release of DA via reversal of the DA transporter.

Psychostimulant-induced augmentation in extracellular DA level with in the NAc, after repeated administration of the drug, is one leading hypothesis that pertains to psychostimulant-induced behavioral sensitization *(7–9)*. However, other studies have shown persistent behavioral sensitization that coincides with diminished DA response in the NAc *(10)*. In recent years, increasing evidence suggests the involvement of excitatory amino acids (e.g., glutamate and glutamate receptor subtypes) in the induction and expression of sensitization to psychostimulants *(11,12)*. For example, the induction of behavioral sensitization to cocaine and amphetamine appears to involve activation of NMDA and non-NMDA glutamate receptors, as well as DA receptors *(13–15;* and chapter7).

Because evidence has linked the activation of NMDA receptors to the stimulation of nNOS, we hypothesized that NO is involved in the effects of psychostimulants. To test this hypothesis, we investigated the effect of nNOS inhibitors on (1) the psychomotor-stimulating effect of cocaine and METH, (2) the rewarding effect of cocaine, and (3) the dopaminergic neurotoxicity caused by a high dose of METH. In addition, we investigated the effects of cocaine and METH on mice deficient of the nNOS gene—nNOS knockout mice. Both the pharmacological and genetic manipulation of nNOS support the hypothesis that nNOS plays a major role in the psychomotor-stimulating and rewarding effects of psychostimulants as well as in METH-induced dopaminergic neurotoxicity.

From: *Contemporary Clinical Neuroscience: Glutamate and Addiction*
Edited by: Barbara H. Herman et al. © Humana Press Inc., Totowa, NJ

## 2. NO IS REQUIRED FOR THE INDUCTION AND EXPRESSION OF BEHAVIORAL SENSITIZATION TO PSYCHOSTIMULANTS

### 2.1. Sensitization to Cocaine and Methamphetamine

Postsynaptic DA receptors in the dorsal striatum and NAc are thought to be involved in psychostimulant-induced hyperactivity (16). The D1-class receptors may have a more prominent role than the D2-like DA receptors in the action of psychostimulants (17). However, repeated exposure to psychostimulants that cause behavioral sensitization is more complex and apparently involves multiple neurotransmitter systems. The development of behavioral sensitization to amphetamines has been linked to an amphetamine-induced psychosis in humans (7,11,18,19) and the development of drug craving (11,20). Behavioral sensitization that develops in animal models may persist for a long period, suggesting that drug-induced neuroadaptation, cellular changes, and neural plasticity produced by chronic drug use are long-lasting. An important aspect in the development of behavioral sensitization to psychostimulants is the emergence of a context-dependent locomotion or conditioning. Pairing a specific environment with the injection stimulus of the psychostimulant results in augmentation in locomotor activity when the animals are paired again in the same environment but with a saline injection. This paradigm is thought to be relevant to drug-seeking behavior and relapse in association with a specific environment (21,22).

### 2.2. Effect of 7-Nitroindazole on Cocaine- and Methamphetamine-Induced Sensitization

To investigate the role of NO in the development of sensitization to cocaine and METH, we first examined the effect of the nNOS inhibitor 7-nitroindazole (7-NI) on the psychomotor-stimulating effect of these drugs. Swiss Webster mice were injected with (1) vehicle/saline (control), (2) vehicle/cocaine (15 mg/kg) once a day, (3) 7-NI (25 mg/kg)/cocaine, and (4) 7-NI/saline for 5 consecutive days. Injections were paired with the locomotor-activity cages (test cage) on the first day and the fifth day, whereas the other injections were delivered in the home cage. This regimen allowed the development of conditioned locomotion as determined by the administration of saline injection in the test cage (23). Animals treated with vehicle/cocaine developed marked locomotor sensitization to a challenge cocaine (15 mg/kg) (Fig. 1) and cross-sensitization to challenge METH (0.5 mg/kg) injection given after a 10-d drug-free period. This treatment also produced context-dependent hyperlocomotion as evident by the sensitized response to a challenge saline injection given on d 8 in the test cage (23). The pretreatment with 7-NI 30 min before cocaine administration (for 5 d) completely blocked the induction of sensitization to cocaine (Fig. 1), the cross-sensitization to METH, and the conditioned locomotion (23). In addition, the coadministration of 7-NI with cocaine on d 5 blocked the expression of the sensitized response to cocaine. 7-NI when given alone, either acutely or repeatedly for 5 d, had no significant effect on locomotor activity (23). Similarly, mice treated with METH (1 mg/kg) for 5 d developed sensitization to challenge METH (0.5 mg/kg) (Fig. 1), cross-sensitization to challenge cocaine (15 mg/kg), and context-dependent hyperlocomotion. The pretreatment with 7-NI attenuated the sensitized response to METH (Fig. 1) and the cross-sensitization to cocaine as revealed after a 10-d drug-free period. However, 25 mg/kg 7-NI did not block the conditioned hyperlocomotion caused by pairing METH injections with the test cage (23).

The findings that 7-NI blocked (1) the sensitized response to cocaine and METH on d 5 and (2) the sensitized response to a challenge injection of the drugs on d 15 (Fig. 1) suggest that the nNOS inhibitor blocked the expression and induction of sensitization to psychostimulants. However, 7-NI had a differential effect on the conditioned locomotion caused by cocaine and METH; the nNOS inhibitor attenuated only the conditioned hyperlocomotion caused after cocaine administration.

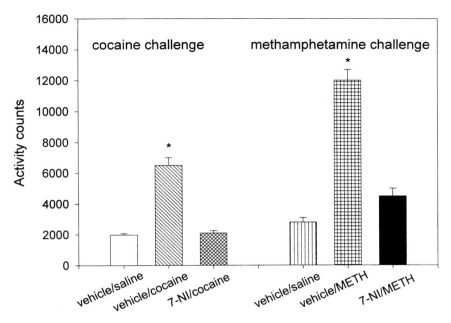

**Fig. 1.** Effect of 7-NI on cocaine- and METH-induced locomotor sensitization. Swiss Webster mice were treated with either vehicle/saline (control), vehicle/cocaine (15 mg/kg), 7-NI (25 mg/kg)/cocaine, vehicle/METH (1.0 mg/kg), or 7-NI/METH for 5 d. After a 10-d drug-free period, animals were challenged with either cocaine (15 mg/kg) or METH (0.5 mg/kg). Locomotor-activity counts were measured by infrared cell beam interruption (30 min for cocaine and 60 min for METH). The sensitized response to cocaine and METH was attenuated in the groups pretreated with 7-NI, suggesting a blockade of the induction of sensitization. *$p<0.05$ compared to vehicle/saline group. (Results adapted from ref. *23*.)

## 2.3. Effect of Cocaine on nNOS (–/–) Mice

To further investigate the role of NO in the induction and expression of behavioral sensitization to cocaine, we studied the effect of repeated cocaine administration to nNOS(–/–) mice *(24)*. nNOS(–/–) and wild-type mice were administered cocaine (15 mg/kg) for 5 d. Injections were paired with the locomotor activity cages on d 1 and 5. Male nNOS(–/–) mice were as sensitive to the acute effect of cocaine on d 1 as the wild-type mice. However, they did not develop a sensitized response to cocaine either on d 5 or d 15 after a 10-d drug-free period (Fig.2). Also, the male nNOS(–/–) mice did not develop conditioned hyperlocomotion *(24)*. Female nNOS(–/–) mice did not show hyperlocomotion after the administration of 15 mg/kg cocaine on d 1, nor did they develop a sensitized response to 15 mg/kg cocaine given on d 5 or 15, after a 10 d drug-free period (Fig.2). The female nNOS(–/–) mice were responsive only to the high dose of 30 mg/kg cocaine *(24)*. In contrast to the nNOS(–/–) mice, heterozygote nNOS(+/–) mice (male and female) developed a sensitized response to cocaine (15 mg/kg) as for wild-type mice *(24)*.

Investigation of [$^3$H] cocaine disposition in the striatum and frontal cortex of the nNOS(–/–), nNOS(+/–), and wild-type mice revealed neither gender nor strain differences in the drug disposition *(24)*. Also, striatal DA levels and its metabolites [determined by high-performance liquid chromatography (HPLC)] and the DA transporter-binding sites (determined by [$^3$H] mazindol binding) in the nNOS(–/–), nNOS(+/–), and wild-type mice were very similar, suggesting no major differences between nNOS(–/–) and wild-type mice in the development of the striatal dopaminergic system *(24)*.

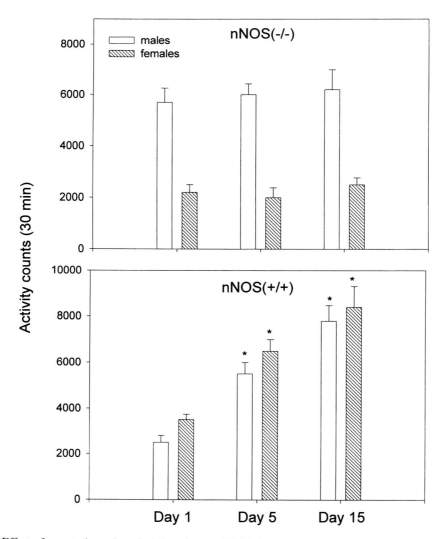

**Fig. 2.** Effect of repeated cocaine administration to nNOS(–/–) and wild-type nNOS(+/+) mice. Animals were administered cocaine (15 mg/kg) for 5 consecutive days. After a 10-d drug-free period, they were challenged with cocaine (15 mg/kg). Locomotor activity was measured on d 1,5, and 15. In the nNOS(–/–) mice, cocaine-induced locomotor activity was on d 5 and 15 was not significantly different from d 1, suggesting a lack of sensitization to cocaine. The female nNOS(–/–) were less responsive to cocaine compared to the male nNOS(–/–) mice. In the nNOS(+/+) mice (male and female), cocaine-induced locomotor activity on d 5 and 15 was significantly higher compared to d 1 ($p < 0.05$), suggesting the development of sensitization to cocaine. (Data adapted from ref. 24.

The only difference found between the nNOS(–/–)and wild-type mice was the absence of [$^3$H]L-nitroarginine-binding sites in the striatum of nNOS(–/–) mice (24).This finding confirmed the absence of nNOS in the knockout mice compared to the wild-type mice. Thus, the resistance of the nNOS (–/–) mice to cocaine-induced behavioral sensitization is primarily the result of the deletion of the nNOS gene rather than alterations in basal dopaminergic activity. Taken together, it appears that the pharmacological blockade of nNOS by 7-NI and the genetic manipulation of nNOS resulted in a similar outcome with respect to cocaine sensitization.

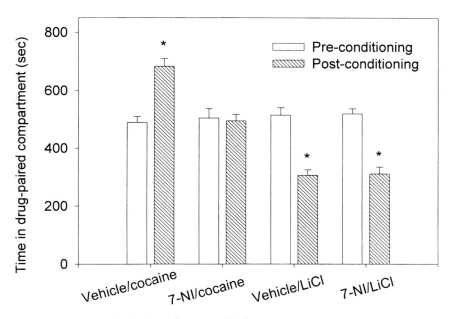

**Fig. 3.** Effect of 7-NI on cocaine-induced CPP and LiCl-induced CPA. Swiss Webster mice were administered vehicle/cocaine (20 mg/kg), 7-NI (25 mg/kg)/cocaine, vehicle/LiCl (150 mg/kg), or 7-NI (25 mg/kg)/LiCl every other day for 4 d; vehicle/saline injections were given in alternate days for 4 d. On the test day, vehicle/cocaine-treated mice developed significant preference for the drug-paired compartment (*$p<0.01$) compared to the time spent in the same compartment before the CPP. The vehicle/LiCl-treated mice developed significant aversion from the drug-paired compartment compared to the time spent in the same compartment before the CPA. 7-NI pretreatment completely blocked cocaine-induced CPP, but it had no effect on LiCl-induced CPA. (Date adapted from ref. *25* and *27.*)

## 3. NO IS INVOLVED IN THE REWARDING EFFECTS OF COCAINE, ALCOHOL, AND NICOTINE, BUT NOT IN THE AVERSIVE EFFECT OF LiCl

### 3.1. Effect of 7-Nitroindazole on the Acquisition of Reward and Aversion

Repeated administration of a drug with rewarding properties in a specific environment (e.g., black or white compartment of a cage) causes conditioned place preference (CPP). A drug with aversive properties such as LiCl elicits conditioned place aversion (CPA); animals spend less time in the drug-paired compartment compared to the time they spent in the same compartment before drug administration. The rewarding effect of cocaine, alcohol, and nicotine was investigated in the CPP paradigm *(25–27)*. In the first experiment, Swiss Webster mice were treated with cocaine (20 mg/kg) in one compartment of the cage and with saline in the other compartment. Drug and saline injections were given in alternate days for a total of 8 d; there were four drug sessions and four saline sessions. On d 9, animals received a saline injection and had the choice of moving freely between the two compartments of the cage. Cocaine-experienced mice spent significantly more time in the drug-paired compartment compared to the time they spent in the same compartment before the CPP (increase of 192 ± 12 s; Fig. 3). Animals pretreated with the nNOS inhibitor 7-NI (25 mg/kg) 30 min before cocaine did not show any preference to the drug-paired compartment, suggesting that inhibition of nNOS blocked the rewarding effect of cocaine (Fig.3). Animals treated with 7-NI/saline did not show preference for the drug-paired compartment, suggesting that 7-NI alone has no rewarding or aversive properties *(25)*. In a second experiment, the effect of 7-NI on alcohol-induced CPP was investigated *(26)*. DBA mice were treated with either vehicle/ethanol(2.5 g/kg) or 7-NI (25 mg/kg)/ethanol and confined to one compartment of the cage for 10 min. After 6 h, animals received vehicle/saline injections and were confined to the other compartment. There was a total of four drug and four saline sessions. On the test day, DBA mice showed a clear preference for the ethanol-paired compartment, suggesting the development of CPP. The pretreatment with 7-NI completely blocked ethanol-induced CPP (26). In the third experiment, the effect of 7-NI on the rewarding effect of nicotine and the aversive effect of LiCl was investigated (27). Swiss Webster mice were treated with (1) vehicle/nicotine (0.5 mg/kg),(2) 7-NI (25 mg/kg)/nicotine, (3) vehicle/LiCl (150 mg/kg), and (4) 7-NI (25 mg/kg)/LiCl. Drug and vehicle/saline injections were given on alternate days for a total of 8 d. Nicotine produced a significant CPP in Swiss Webster mice, whereas LiCl produced marked aversion from the drug-paired compartment (Fig.3 and ref. 27). The pretreatment with 7-NI completely blocked nicotine-induced CPP, but it had no significant effect on LiCl-induced CPA (Fig. 3 and ref. 27). Taken together these findings indicate that the blockade of nNOS by 7-NI attenuates the rewarding effects of cocaine, ethanol, and nicotine. The observation that 7-NI had no effect on LiCl-induced aversion suggests that NO has rather a specific role on reward-processing mechanisms. One possibility is that inhibition of nNOS modulates dopaminergic transmission associated with the rewarding effects of cocaine, ethanol, and nicotine. The acquisition of CPP and CPA requires the application of learning and memory. The finding that 7-NI had differential effect on the acquisition of CPP and CPA suggests that the blockade of drug-induced CPP by 7-NI is not due to impairment of learning and memory.

### *3.2. Absence of Cocaine-Induced CPP in nNOS(–/–)Mice*

The role of nNOS in the rewarding effect of cocaine was further investigated in nNOS(–/–) mice (25). Knockout mice (male and female) and wild-type mice were treated with cocaine (20 mg/kg) in one compartment of the cage and with saline in the other compartment. Following four cocaine sessions and four saline sessions, animals were tested for the development of CPP. The wild-type mice developed marked CPP, as was evident by a significant increase in the time spent in the cocaine-paired compartment on the test day. However,the nNOS (–/–) mice did not develop CPP; they spent the same amount of time in the cocaine-paired compartment as they did before the injections of cocaine were initiated (25). Thus, both the pharmacological blockade of nNOS by 7-NI and the deletion of the nNOS gene resulted in a similar outcome with relevance to cocaine-induced reward.

## 4. ROLE OF nNOS IN METHAMPHETAMINE-INDUCED DOPAMINERGIC NEUROTOXICITY

### *4.1. Amphetamines-Induced Neurotoxicity and Free Radicals*

In addition to the development of addiction to the rewarding effects of amphetamines, data from animal and human studies suggest that they produce dopaminergic and serotoninergic neurotoxicity. Various substituted amphetamines have a differential effect on dopaminergic and serotoninergic transmission. METH causes toxic effects on both DA and 5-HT neurons in animals (28–30). Anatomic evidence of METH-induced neurotoxicity stems from studies showing degeneration of nerve fibers (31–32) and loss of tyrosine hydroxylase (33)and 5-HT-immunoreactive axons and axon terminals (34). Also, depletion in dopaminergic markers in postmortem chronic METH users had been reported (35). Recent PET studies in METH users showed significant decrease in striatal DA transporters(36).

The mechanism(s) underlying the neurotoxicity of substituted amphetamines is unclear. Increases in the level of the neurotransmitters DA (37–38), serotonin (39,40), and glutamate (41–43) have been implicated in METH-induced neurotoxicity. Recently, increasing evidence suggests the involvement of free radicals and, specifically, reactive oxygen species in the neurotoxic effects of METH (44–47). The hypothesis that DA depletion may be associated with oxidative stress has been extended to the pathophysiology of Parkinson's disease (48).

Peroxynitrite (ONOO−), formed from the interaction between NO and superoxide radicals, is considered as a major neurotoxin *(49)*. NO has a major role in excitotoxic neuronal injury. Overstimulation of nNOS leads to activation of the nuclear enzyme poly(ADP-ribose) synthase that leads to energy depletion and cell death *(50)*. The blockade of nNOS by 7-NI reduced NMDA receptor-mediated formation of hydroxy radicals and neurotoxicity *(51)*; cortical cultures from nNOS knockout (KO) mice were protected from NMDA receptor-mediated neurotoxicity *(52)*. The role of NO/peroxynitrite in MPTP-induced dopaminergic neurotoxicity is evident from the findings that the nNOS inhibitor 7-NI blocked MPTP-induced neurotoxicity in mice *(53)* and primates *(54)*, and nNOS KO mice were protected from MPTP neurotoxicity *(55)*.

We hypothesized that a similar mechanism may underlie the protective effect of nNOS inhibition against METH-, glutamatergic-, and MPTP-induced neurotoxicity (e.g., diminished production of superoxide radicals and peroxynitrite). This hypothesis is supported by findings suggesting the involvement of NO/peroxynitrite in amphetamines neurotoxicity: (1) NOS inhibitors blocked METH-induced neurotoxicity in cultures of fetal rat mesencephalon *(56)*; (2) repeated administration of amphetamine to rats increased the production of NO in the striatum and cortex *(57,58)*; (3) METH-induced neurotoxicity in mice was associated with a marked increase in striatal peroxynitrite *(59)* and 3-nitrotyrosine *(60)*, which is a marker form peroxynitrite formation; (4) inactivation of tyrosine hydroxylase caused by peroxynitrite-induced nitration *(61,62)* and increase in the vulnerability of dopaminergic neurons to cell loss after exposure to peroxynitrite *(63)* are believed to be associated with METH neurotoxicity; (5) the nNOS inhibitor 7-NI blocked METH neurotoxicity in vivo *(64,65)*.

## 4.2. Methamphetamine-Induced Hyperthermia

The role of temperature in amphetamine-induced neurotoxicity has been extensively investigated. Several studies have demonstrated that attenuation of METH-induced hyperthermia protected against METH neurotoxicity *(66–69)*. Other evidence, however, suggests mechanism(s) of protection that is independent of a change in body temperature *(70)*. Recently, it became debatable whether 7-NI protection against METH neurotoxicity is dependent on attenuation of METH-induced hyperthermia. In our initial study, we found that a single injection of 7-NI (25 mg/kg) had no effect on METH-induced hyperthermia *(64)*. DiMonte et al. *(65)* also reported that 7-NI afforded protection in a temperature-independent manner. However, others reported that 7-NI (50 mg/kg) produced hypothermia, and it was suggested that the protective effect of 7-NI is due to attenuation of METH hyperthermia *(71)*.

## 4.3. Effect of Various nNOS Inhibitors on Methamphetamine-Induced Hyperthermia and Neurotoxicity

Recently, we compared the effect of multiple injections of a low dose (25 mg/kg) and a high dose (50 mg/kg) of 7-NI on METH-induced hyperthermia *(72)*. Results are summarized in Table 1. Whereas three injections of the low dose of 7-NI had no effect on normal body temperature, the high dose produced hypothermia (Table 1). The low dose of 7-NI did not affect METH-induced hyperthermia after the first and second injection, but it attenuated the hyperthermia caused by the third METH injection (Table 1). This finding suggests that the effect of the third injection may be due to accumulation of 7-NI. This assumption is supported by the finding that the high dose of 7-NI (50 mg/kg) blunted the hyperthermia caused after the first injection of METH and, subsequently (the third injection of the high dose), caused hypothermia (Table 1). Thus, there is a marked difference in the response to 25 and 50 mg/kg 7-NI, and the major difference between our first published study *(64)* and the report by Callahan and Ricaurte *(71)* is the dosage of 7-NI. Whereas the total amount of 7-NI we used in the original study was 75 mg/kg (25 mg/kg × 3), the total amount used in the Callahan and Ricaurte study was 200 mg/kg (50 mg/kg × 4). Clearly, the low dose of 7-NI had a minimal effect on METH-induced hyperthermia (only after the third injection), whereas the high dose had an immediate hypothermic effect (Table 1). However, one may argue that even attenuation of the hyperthermia caused by the last injection of METH may be regarded as a thermoregulator effect that contributed to neuroprotection. To address this issue, we investigated the effect of two other nNOS inhibitors, 3-Br-7-nitroindazole (3-Br-7-NI) and S-methylthiocitrulline (SMTC) on METH-induced hyperthermia and depletion of dopaminergic markers *(72)*.

3-Br-7-NI is considered a selective nNOS inhibitor which is more potent than 7-NI *(73)*. First, we investigated the effect of a single injection of 20 mg/kg 3-Br-7-NI on striatal nNOS activity. Thirty minutes following the administration of the nNOS inhibitor, mice were sacrificed and the striatum was prepared for nNOS enzyme assays. Using a concentration of 3 µ$M$ L-[$^3$H]arginine (equal to the $K_m$ value), we found 63 ± 5% inhibition of enzyme activity compared to control value (28.3±3.5 pmol/mg protein/min). Thus, we chose the dose of 20 mg/kg for the subsequent experiments. Three injections of 3-Br-7-NI (20 mg/kg) had no effect on normal temperature (Table 1). The pretreatment with 3-Br-7-NI (20 mg/kg) 30 min before each METH injection (5 mg/kg) had no effect on the persistent hyperthermia caused by METH (Table 1). However, 3-Br-7-NI provided protection against the depletion of dopaminergic markers in the striatum (Table 1). These findings suggest that inhibition of nNOS provided protection against METH neurotoxicity with no thermoregulator effect.

S-Methylthiocitrulline is a potent nNOS inhibitor with 17-fold selectivity for nNOS compared to eNOS *(74)*. A dose of 10 mg/kg SMTC provided protection against MPTP neurotoxicity *(75)*. Three consecutive injections of SMTC (10 mg/kg) affected neither normal body temperature nor METH-induced hyperthermia (Table 1). However, SMTC afforded protection against depletion of dopaminergic markers in the striatum (Table 1). These findings further support the hypothesis that inhibition of nNOS does not blunt METH hyperthermia, but it affords protection against METH neurotoxicity.

### 4.4. Susceptibility of nNOS(–/–) Mice to Methamphetamine-Induced Dopaminergic Neurotoxicity and Hyperthermia

To investigate whether nNOS(–/–) mice are resistant to METH-induced dopaminergic neurotoxicity, the effect of multiple injections of METH (5 mg/kg × 3) on nNOS (–/–), nNOS(+/–), and nNOS(+/+) mice was investigated *(76)*. Whereas nNOS(–/–) mice were completely protected against depletion of dopaminergic markers (Fig. 4), a limited depletion (25–35%) was observed in nNOS(+/–) mice and a large depletion (up to 68%) was observed in the nNOS(+/+) mice (Fig. 4).

Previously, we have shown that the first injection of METH did not produce hyperthermia in the nNOS(–/–) mice, but it caused the same increase in temperature of the nNOS(+/–) and nNOS(+/+) mice *(76)*. Yet, the nNOS(+/–) mice were relatively protected against neurotoxicity compared with the wild-type mice *(76)*. It is possible that complete deficiency in nNOS brings about more resistance to hyperthermia and neurotoxicity, whereas pharmacological manipulation of nNOS and "partial ablation" of nNOS [e.g., nNOS(+/–)] provide protection against METH neurotoxicity without any effect on METH-induced hyperthermia. Recently, we investigated the effect of the first, second, and third injections of METH (5 mg/kg each) on the body temperature of the nNOS(–/–) and wild-type mice. The results are summarized in Fig. 5. Although the first injection of METH to nNOS(–/–) mice had no significantly effect on temperature, the second and the third injections raised the temperature significantly to a similar extent as in the wild-type mice (Fig. 5). Thus, it appears that the nNOS(–/–) mice are not completely resistant to METH-induced hyperthermia; rather, there is a progressive increase in their sensitivity to hyperthermia. Accordingly, the resistance of the nNOS(–/–) mice to METH-induced dopaminergic neurotoxicity is not entirely due to the absence of METH-induced hyperthermia but rather to the reduced level of NO and probably to the diminished production of peroxynitrite.

### 5. CONCLUSIONS

The involvement of nNOS in the psychomotor-stimulating effects of cocaine and methamphetamine is supported by the findings that (1) the nNOS inhibitor 7-NI blocked the induction and expression of locomotor sensitization to cocaine and methamphetamine and (2) of the resistance of the

**Table 1**
**Effect of nNOS Inhibitors on METH-Induced Hyperthermia and Depletion of Dopaminergic Markers**

| | First injection | | | | Second injection | | | | Third injection | | | |
|---|---|---|---|---|---|---|---|---|---|---|---|---|
| Time (min) | 0 | 30 | 60 | 120 | 0 | 30 | 60 | 120 | 0 | 30 | 60 | 120 |
| Treatment | Temperature(°C) | | | | | | | | | | | |
| **7-NI (25 mg/kg)** | | | | | | | | | | | | |
| Saline | 36.2±0.1 | 36.3±0.2 | 36.1±0.2 | 36.4±0.2 | | 36.3±0.2 | 36.4±0.1 | 36.3±0.1 | | 36.5±0.2 | 36.3±0.2 | 36.2±0.1 |
| METH | 36.3±0.2 | 38.3±0.3* | 38.5±0.2* | 37.3±0.3 | | 38.8±0.2* | 38.3±0.3* | 37.2±0.2 | | 37.1±0.2 | 36.4±0.2 | 36.5±0.3 |
| **7-NI (50 mg/kg)** | | | | | | | | | | | | |
| Saline | 36.4±0.1 | 35.8±0.1 | 35.3±0.2* | 35.1±0.1* | | 34.8±0.2* | 34.4±0.1* | 34.2±0.2* | | 34.5±0.2* | 34.1±0.2* | 34.6±0.3 |
| METH | 36.3±0.2 | 37.2±0.2 | 36.5±0.1 | 36.0±0.2 | | 36.1±0.2 | 36.6±0.3 | 36.2±0.2 | | 35.3±0.2* | 34.6±0.2 | 35.7±0.3 |
| **3-Br-7-NI (20 mg/kg)** | | | | | | | | | | | | |
| Saline | 36.4±0.1 | 36.3±0.2 | 36.2±0.2 | 36.3±0.2 | | 36.5±0.2 | 36.2±0.1 | 36.3±0.2 | | 36.2±0.2 | 36.5±0.1 | 36.3±0.2 |
| METH | 36.2±0.2 | 38.7±0.2* | 38.5±0.3* | 36.9±0.2 | | 38.4±0.3* | 39.2±0.3* | 37.2±0.1 | | 38.8±0.3* | 38.3±0.2* | 36.8±0.2 |
| **SMTC (10 mg/kg)** | | | | | | | | | | | | |
| Saline | 36.1±0.2 | 36.4±0.3 | 36.2±0.2 | 36.5±0.1 | | 36.2±0.3 | 36.3±0.2 | 36.5±0.2 | | 36.2±0.1 | 36.4±0.3 | 36.3±0.2 |
| METH | 36.3±0.2 | 38.7±0.2* | 39.3±0.2* | 37.3±0.2 | | 38.7±0.2* | 38.4±0.1* | 37.5±0.3 | | 38.5±0.1* | 38.7±0.2* | 36.9±0.2 |
| Vehicle/METH | 36.1±0.2 | 38.5±0.3* | 38.3±0.2* | 37.5±0.2 | | 39.1±0.1* | 38.5±0.3* | 37.0±0.3 | | 38.8±0.2* | 38.9±0.2* | 37.2±0.2 |

Striatal dopaminergic markers

| | Dopamine | DOPAC | HVA | DAT |
|---|---|---|---|---|
| | | (ng/100 mg tissue) | | (fmol/mg protein) |
| Control | 610±41 | 102±7 | 89±5 | 1605±92 |
| Vehicle/METH | 213±11* | 41±4* | 33±2* | 642±53* |
| 3-Br-7-NI/METH | 549±33 | 103±5 | 83±7 | 1524±104 |
| SMTC/METH | 530±59 | 88±9 | 81±6 | 1476±122 |

*Note:* Saline or METH were administered 30 min after the administration of the nNOS inhibitor. Rectal temperature was monitored for 2h after METH injection. The interval between each pair of injections was 3h. Administration of the low dose of 7-NI (25 mg/kg) did not blunt METH-induced hyperthermia after the first and second injections, but it attenuated the hyperthermia caused after the third injection of METH. The high dose of 7-NI (50 mg/kg) caused a decrease in temperature and also abolished METH-induced hyperthermia. Administration of 3-Br-7-NI or SMTC affected neither normal temperature nor METH-induced hyperthermia. (*$p<0.05$ compared to temperature measured at time 0). The striatal dopaminergic markers, dopamine, DOPAC, and HVA were measured by HPLC and the dopamine transporter (DAT)-binding sites by [$^3$H]mazindol binding. METH caused a significant depletion in the dopaminergic markers (*$p < 0.05$ compared to control values), whereas the nNOS inhibitors 3-Br-7-NI and SMTC afford a full protection.

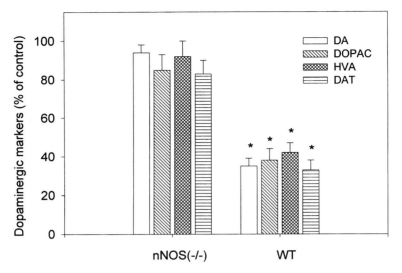

**Fig. 4.** Effect of METH on depletion of dopaminergic markers in nNOS(–/–) and wild-type (WT) mice. Animals received 5 mg/kg METH (ip; q3h × 3) and sacrificed after 72 h. Striatal content of dopamine (DA) and its metabolites 3,4–dihydroxyphenylacetic acid (DOPAC) and homovanilic (HVA) were quantified by HPLC/electrochemical detection. The density of the dopamine transporter (DAT) binding sites was determine by [$^3$H]mazindol binding in the presence of desipramine. Results are presented as the percentage of control (saline) values for each group of mice. Whereas the level of dopaminergic markers in the nNOS(–/–) mice were not significantly affected by METH treatment, a depletion of 60–65% was observed in the WT mice. *$p < 0.02$ compared to control values. (Data adapted from ref. 76.)

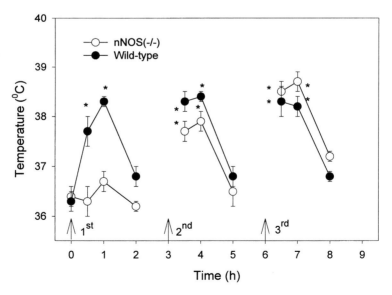

**Fig. 5.** Effect of METH on rectal temperature of nNOS(–/–) and wild-type mice. Animals received 5 mg/kg METH (ip; q3h × 3). Rectal temperature was measured before METH administration (time 0) and then after 30, 60, and 120 min following each injection of METH. The first injection of METH had no significant effect on rectal temperature of the nNOS (–/–) mice, but it caused significant hyperthermia in the wild-type mice (*$p < 0.05$ compared to time 0). However, the second and third injections of METH caused significant hyperthermia in both the nNOS(–/–) and wild-type mice (*$p < 0.05$ compared to time 0). Arrows represent time of METH (5 mg/kg) injections.

nNOS(–/–) mice to cocaine-induced locomotor sensitization. The role of nNOS in the rewarding effect of cocaine and other abused substances (e.g., alcohol and nicotine) is suggested by the following observations. (1) The nNOS inhibitor 7-NI blocked cocaine-, ethanol-, and nicotine-induced CPP. (2) 7-NI had no effect on LiCI-induced conditioned placed aversion. (3) nNOS(–/–) mice were resistant to cocaine-induced conditioned place preference. The involvement of nNOS in methamphetamine-induced dopaminergic neurotoxicity is supported by the findings that (1) several nNOS inhibitors such as 7-NI, 3-Br-7-NI, and SMTC attenuated the depletion of dopaminergic markers with no significant effect on METH-induced hyperthermia and (2) nNOS(–/–) were resistant to METH-induced dopaminergic neurotoxicity. In all the behavioral paradigms tested and in the neurochemical studies, there was a good correlation between the pharmacological blockade of nNOS and the deletion of the nNOS gene. Therefore, nNOS has an important role in mediating both the rewarding and neurotoxic effects of psychostimulants. Further studies are necessary to determine whether selective nNOS inhibitors may be useful therapeutics for the management of addiction to psychostimulants.

## ACKNOWLEDGEMENTS

This work was supported by UPHS award DA08584 from the National Institute on Drug Abuse.

## REFERENCE

1. Garthwaite, J. (1991) Glutamate, nitric oxide and cell-cell signaling in the nervous system. *Trends Neurosci.* **14,** 60–67.
2. Snyder, S. H. (1992) Nitric oxide: first in a new class of neurotransmitter? *Science* **257,** 494–496.
3. Yun, H.-Y., Dawson, V. L. and Dawson, T. M. (1997) Nitric oxide in health and disease of the nervous system. *Mol. Psychiatry* **2,** 300–310.
4. Strasser, A., McCarron, R. M., Ishii, H., Stanimirovic, D. and Spatz, M. (1994) L-Arginine induces dopamine release from the striatum *in vivo*. *NeuroReport* **5,** 2298–2300.
5. Montague, P. R., Gancayco, C. D., Winn, M. J., Marchase, R. B., and Fridlander, M. J. (1994) Role of NO production in NMDA receptor-mediated neurotransmitter release in cerebral cortex. *Science* **263,** 973–977.
6. Lonart, G., Cassels, K. L. and Johnson, K. M. (1993) Nitric oxide induces calcium-dependent [$^3$H]dopamine release from striatal slices. *J. Neurosci. Res.* **35,** 192–198.
7. Robinson, T. E. and Becker, J. B. (1986) Enduring changes in brain and behavior produced by chronic amphetamine administration: a review and evaluation of animal models of amphetamine psychosis. *Brain Res. Rev.* **11,** 157–198.
8. Kalivas, P. W. and Stewart, J. (1991) Dopamine transmission in the initiation and expression of drug-and stress-induced sensitization of motor activity. *Brain Res. Rev.* **16,** 223–244.
9. White, F. J. and Wolf, M. E. (1991) Psychomotor stimulants, in *The Biological Bases of Drug Tolerance and Dependence*. (Pratt, J., ed.), Academic, London, pp. 153–197.
10. Segal, D. S. and Kuczenski, R. (1992) *In vivo* microdialysis reveals a diminished amphetamine-induced DA response corresponding to behavioral sensitization produced by repeated amphetamine pretreatment. *Brain Res.* **57,** 330–337.
11. Pierce, R. C. and Kalivas, P. W. (1997) A circuitry model of the expression of behavioral sensitization to amphetamine-like pyschostimulants. *Brain Res. Rev.* **25,** 192–216.
12. Wolf, M. A. (1998) The role of excitatory amino acids in behavioral sensitization to psychomotor stimulants. *Prog. Neurobiol.* **54,** 679–720.
13. Karler, R., Calder, L. D., Chaudhry, I. A., and Turkanis, S. A. (1989) Blockade of "reverse tolerance" to cocaine and amphetamine by MK-801. *Life Sci.* **45,** 599–606.
14. Karler, R., Chaudhry, I. A., Calder, L. D., and Turkanis, S. A. (1990) Amphetamine behavioral sensitization and excitatory amino acids. *Brain Res.* **537,** 76–82.
15. Karler, R., Calder, L. D., and Bedingfield, J. B. (1994) Cocaine behavioral sensitization and excitatory amino acids. *Psychopharmacology* **115,** 305–310.
16. Baker, D. A., Khroyan, T. V., O'Dell, L. E., Fuchs, R. A., and Neisewander, J. L. (1996) Differential effects of intra-accumbens sulpiride on cocaine-induced locomotion and conditioned place preference. *J. Pharmacol. Exp. Ther.* **279,** 392–401.
17. White, F. J. and Kalivas, P. W. (1998) Neuroadaptations involved in amphetamine and cocaine addiction. *Drug Alcohol Depend.* **51,** 141–153.
18. Segal, D. S. and Kuczenski, R. (1997) An escalating dose "binge" model of amphetamine psychosis: behavioral and neurochemical characteristics. *J. Neurosci.* **7,** 2551–2566.
19. Segal, D. S. and Kuczenski, R. (1997) Repeated binge exposure to amphetamine and methamphetamine: behavioral and neurochemical characterization. *J. Pharmacol. Exp. Ther.* **282,** 561–573.

20. Robinson, T. E. and Berridge, K. C. (1993) The neural basis of drug craving: an incentive-sensitization theory of addiction. *Brain Res. Rev.* **18,** 247–291.
21. Stewart, J. and Vezina, P. (1988) Conditioning and behavioral sensitization, in *Sensitization in the Nervous System* (Kalivas, P. W. and Barnes, C. D., eds.), Telford, Caldwell, NJ, pp. 207–224.
22. O'Brien, C. P., Childerss, A. R., McLellan, T., and Ehrmann, R. (1990) Integrating systemic cue exposure with standard treatment in recovering drug-dependent patient. *Addict. Behav.* **15,** 355–365.
23. Itzhak, Y. (1997) Modulation of cocaine- and methamphetamine-induced behavioral sensitization by inhibition of brain nitric oxide synthase. *J. Pharmacol. Exp. Ther.* **282,** 521–527.
24. Itzhak, Y., Martin, J. L., Black,. M. D, Ali, S. F., and Huang P. L. (1998) Resistance of neuronal nitric oxide synthase-deficient mice to cocaine-induced locomotor sensitization. *Psychopharmacology* **140,** 378–386.
25. Itzhak, Y., Martin, J. L., Black, M. D., and Huang, P. L. (1998) The role of neuronal nitric oxide synthase in cocaine-induced conditioned place preference. *NeuroReport* **9,** 2485–2488.
26. Itzhak, Y. and Martin, J. L. (2000) Blockade of alcohol-induced locomotor sensitization and conditioned place preference in DBA mice by 7-nitroindazole. *Brain Res.* **858,** 402–407.
27. Martin, J. L. and Itzhak, Y. (2000) 7-Nitroindazole blocks nicotine-induced conditioned place preference but not LiCl-induced conditioned place aversion. *NeuroReport* **11,** 947–949.
28. Hotchkiss, A. and Gibb, J. W. (1980) Long-term effects of multiple doses of methamphetamine on tryptophan hydroxylase and tyrosine hydroxylase activity in rat brain. J. Pharmacol. Exp. Ther. **214,** 257–263.
29. Ricaurte, G. A., Schuster, C. R., and Seiden, L. S. (1980) Long-term effects of repeated methamphetamine administration on dopaminergic and serotoninergic neurons in rat brain: a regional study. *Brain Res.* **193,** 153–160.
30. Gibb, J. W., Hanson, G. R., and Johnson, M. (1994) Neurochemical mechanisms of toxicity, in *Amphetamine and Its Analogs, Psychopharmacology. Toxicology and Abuse* (Cho, A.K. and Segal, D.S., eds.). Academic, P San Diego, pp. 269–289.
31. Ricaurte, G. A., Guillery, R. W., Seiden, L. S., Schuster, C. R., and Moore, R. Y. (1982) Dopamine nerve terminal degeneration produced by high dose of methamphetamine in rat brain. *Brain Res.* **235,** 93–103.
32. Commins, D. L., Axt, K. J., Vosmer, G., and Seiden, L. S. (1987) 5,6-Dihydroxytryptamine a serotonin neurotoxin is formed endogenously in the rat brain. *Brain Res.* **403,** 7–14.
33. Hess, A., Desiderio, C., and McAuliffe, W. G. (1989) Acute neuropathological changes in caudate nucleus produced by MPTP and methamphetamine: immunohistochemical studies. *J. Neurocytol.* **19,** 338–342.
34. Axt, K. J. and Molliver, M. E. (1991) Immunocytochemical evidence for methamphetamine-induced serotoninergic axon loss in the rat brain. *Synapse* **9,** 302–313.
35. Wilson, J. M., Kalasinsky, K. S., Levey, A. I., Bergeron, C., Reiber, G., Anthony, R. M., et al. (1996) Striatal dopamine nerve terminal markers in human, chronic methamphetamine users. *Nature Med.* **2,** 699–703.
36. McCann, U. D., Wong, D. F., Yokoi, F., Villemagne, V., Dannals, R. F., and Ricaurte, G. A. (1998) Reduced striatal dopamine transporter density in abstinent methamphetamine and methcathinon users: evidence from positron emission tomography studies with [$^{11}$C]WIN-35, 428. *J. Neurosci.* **18,** 8417–8422.
37. Gibb, J. W. and Kogan, F. J. (1979) Influence of dopamine synthesis on methamphetamine-induced changes in striatal and adrenal tyrosine hydroxylase activity. *Naunyn-Schmieddebergs Arch. Pharmacol.* **310,** 185–187.
38. Seiden, L. S. and Vosmer, G. (1984) Formation of 6-hydroxydopamine in caudate nucleus of the rat brain after a single large dose of methamphetamine. *Pharmacol. Biochem. Behav.* **21,** 29–31.
39. Commins, D. L., Vosmer, G., Virus, R. M., Woolverton, W. L., Schuster, C. R., and Seiden, L. S. (1987) Biochemical and histological evidence that methylenedioxymeth-amphetamine (MDMA) is toxic to neurons in rat brain. *J. Pharmacol. Exp. Ther.* **241,** 338–348.
40. Berger, U. V., Reinhard, G., and Molliver, M. E. (1992) The neurotoxic effects of *p*-chloroamphetamine in rat brain are blocked by prior depletion of serotonin. *Brain Res.* **578,** 177–185.
41. Sonsalla, P. K., Nicklas, W. J., and Heikkila, R. E. (1989) Role of excitatory amino acids in methamphetamine induced nigrostriatal dopaminergic toxicity. *Science* **243,** 398–400.
42. Nash, J. F. and Yamamoto, B. K. (1992) Methamphetamine neurotoxicity and striatal glutamate release: comparison to 3, 4-methylenedioxymethamphetamine. *Brain Res.* **581,** 237–243.
43. Marshall, J. F., O'Dell, S. J., and Weihmuller, F. B. (1993) Dopamine–glutamate interactions in methamphetamine-induced neurotoxicity. *J. Neural Transm.* **91,** 241–254.
44. Cadet, J. L. and Brannock C. (1998) Free radicals and pathobiology of brain dopamine systems. *Neurochem. Int.* **32,** 117–131.
45. Fleckenstein. A. E., Metzger, R. R., Willkins, D. G., Gibb, J. W., and Hanson, G. R. (1997) Rapid and reversible effects of methamphetamine on dopamine transporters. *J. Pharmacol. Exp. Ther.* **282,** 834–888.
46. Hanson, G. R., Gibb, J. W., Metzger, R. R., Kokoshka, J. M., and Fleckenstein, A. E. (1998) Methamphetamine-induced rapid and reversible reduction in the activities of tryptophan hydroxylase and dopamine transporters: oxidative consequence? *Ann. NY Acad. Sci.* **30,** 103–107.
47. Yamamoto, B. K. and Zhu, W. (1998) The effect of methamphetamine on the production of free radicals and oxidative stress. *J. Pharmacol. Exp. Ther.* **287,** 107–114.

48. Jenner, P. (1998) Oxidative mechanisms in nigral cell death in Parkinson's disease. *Movement Disord.* **13,** 24–34.
49. Beckman, J. S., Beckman, T. W., Chen, J., Marshall, P. M., and Freeman, B. A. (1990) Apparent hydroxy radical production by peroxynitrite: implications for endothelial injury from nitric oxide and superoxide. *Proc. Natl. Acad. Sci. USA* **87,** 1621–1624.
50. Zhang, J., Dawson, V. L., Dawson, T. M., and Snyder, S. H. (1994) Nitric oxide activation of poly(ADP-ribose) synthetase in neurotoxicity. *Science* **263,** 687–689.
51. Schulz, J. B., Matthews, R. T., Jeniks, B. G., Ferrante, R. J., Siwek, D., Henshaw, D. R., et al. (1995) Blockade of neuronal nitric oxide synthase protects against excitotoxicity *in vivo. J. Neurosci.* **15,** 8419–8429.
52. Dawson, V. L., Kizushi, V. M., Huang, P. L., Snyder, S. H., and Dawson, T. M. (1996) Resistance to neurotoxicity in cortical cultures from neuronal nitric oxide synthase- deficient mice. *J. Neurosci.* **16,** 2479–2487.
53. Schulz, J. B., Matthews, R. T., Muqit, M. M. K., Browne, S. E., and Beal, M. F. (1995) Inhibition of neuronal nitric oxide synthase protects against MPTP-induced neurotoxicity in mice. *J. Neurochem.* **64,** 936–939.
54. Hantraye, P., Brouillet, E., Ferrante, R., Palfi, S., Dolan, R., Matthews, R. T. et al. (1996) Inhibition of neuronal nitric oxide synthase prevents MPTP-induced parkinsonism in baboons. *Nature Med.* **2,** 1017–1021.
55. Przedborski, S., Jackson-Lewis, V., Yokoyama, R., Shibata, T., Dawson, V. L., and Dawson, T. M. (1996) Role of neuronal nitric oxide in 1-methyl-4-phenyl-1,2,3,6-tetrahydropyridine (MPTP)-induced dopaminergic neurotoxicity. *Proc. Natl. Acad. Sci. USA* **93,** 4565–4571.
56. Sheng, P., Cerruti, C., Ali, S. F., and Cadet, J. L. (1996) Nitric oxide is a mediator of methamphetamine (METH)-induced neurotoxicity, In vitro evidence from primary cultures of mesencephalic cells. *Ann. NY Acad. Sci.* **801,** 174–186.
57. Lin H.-C., Kang, B.-H., Wong, C.-S., Mao, S.-P., and Wan, F.-J. (1999) Systemic administration of D-amphetamine induced a delayed production of nitric oxide in the striatum of rats. *Neurosci. Lett.* **276,** 141–144.
58. Bashkatova, V., Kraus, M., Prast, H., Vanin, A., Rayevsky, K., and Philippu, A. (1999) Influence of NOS inhibitors on changes in ACH release and NO level in brain elicited by amphetamine neurotoxicity. *NeuroReport* **10,** 3155–3158.
59. Imam, S. Z., Crow, J. P., Newport, G. D., Islam, F., Slikker, W., and Ali, S. F. (1999) Methamphetamine generates peroxynitrite and produces dopaminergic neurotoxicity in mice: protective effect of peroxynitrite decomposition catalyst. *Brain Res.* **837,** 15–21.
60. Imam, S. Z. and Ali, S. F. (2000) Selenium, an antioxidant, attenuates methamphetamine-induced dopaminergic toxicity and peroxynitrite generation. *Brain Res.* **855,** 186–191.
61. Ara, J., Przedborski, S., Naini, A. B., Jackson-Lewis, V., Trifiletti, R. R., Horwitz, J., et al. (1998) Inactivation of tyrosine hydroxylase by nitration following exposure to peroxynitrite and 1-methyl-1,2,3,6-tetrahydropyridine (MPTP). *Proc. Natl. Acad. Sci. USA* **95,** 7659–7663.
62. Kuhn, D. M., Aretha, C. W., and Geddes, T. J. (1999) Peroxynitrite inactivation of tyrosine hydroxylase: mediation by sulfhydryl oxidation, not tyrosine nitration. *J. Neurosci.* **19,** 10,289–10,294.
63. La Voie, M. J. and Hastings, T. G. (1999) Peroxynitrite- and nitrite-induced oxidation of dopamine: implications for nitric oxide in dopaminergic cell loss. *J. Neurochem.* **73,** 2546–2554.
64. Itzhak, Y. and Ali, S. F. (1996) The neuronal nitric oxide synthase inhibitor, 7-nitroindazole, protects against methamphetamine-induced neurotoxicity *in vivo. J. Neurochem.* **67,** 1770–1773.
65. Di Monte, D. A., Royland, J. E., Jakowec, M. W., and Langston, J. W. (1996) Role of nitric oxide in methamphetamine neurotoxicity: protection by 7-nitroindazole, an inhibitor of neuronal nitric oxide synthase. *J. Neurochem.* **67,** 2443–2450.
66. Bowyer, J. F., Tank, A. W., Newport, G. D., Slikker, W., Jr., Ali, S. F., and Holson, R. R. (1992) The influence of environmental temperature on the transient effects of methamphetamine on dopamine levels and dopamine release in rat striatum. *J. Pharmacol. Exp. Ther.* **260,** 817–824.
67. Ali, S. F., Newport, G. D., Holson, R. R., Slikker, W., Jr., and Bowyer, J. F. (1994) Low environmental temperature or pharmacologic agents that produce hypothermia decrease methamphetamine neurotoxicity in mice. *Brain Res.* **658,** 33–38.
68. Miller, D. B. and O'Callaghan, J. P. (1994) Environmental-, drug- and stress-induced alterations in body temperature affect the neurotoxicity of substituted amphetamines in the C57BL/6J mouse. *J. Pharmacol. Exp. Ther.* **270,** 752–760.
69. Malberg, J. E. and Seiden, L. S. (1998) Small changes in ambient temperature causes large changes in 3,4-methylenedioxymethamphetamine (MDMA)-induced serotonin neurotoxicity and core body temperature in the rat. *J. Neurosci.* **18,** 5086–5094.
70. Malberg, J. E., Sabol, K. E., and Seiden, L. S. (1996) Co-administration of MDMA with drugs that protect against MDMA neurotoxicity produces different effects on body temperature in the rat. *J. Pharmacol. Exp. Ther.* **278,** 258–267.
71. Callahan. B. T. and Ricaurte, G. A. (1998) Effect of 7-nitroindazole on body temperature and methamphetamine-induced dopamine toxicity. *NeuroReport* **9,** 2691–2695.
72. Itzhak, Y., Martin, J. L., and Ali, S. F. (2000) nNOS inhibitors attenuate methamphetamine-induced dopaminergic neurotoxicity but not hyperthermia in mice. *NeuroReport* **11,** 2943–2946.
73. Bland-Ward, P. A., Pitcher, A., Wallas, P., Gaffen, Z., Babbedge, R. C., and Moore, P. K. (1994) Isoform selectivity of indazole-based nitric oxide synthase inhibitors. *Br. J. Pharmacol.* **112,** 351.

74. Furfine, E. S., Harmon, M. F., Path, J. E., Knowles, R. G., Salter, M., Kiff, R. J., et al. (1994) Potent and selective inhibition of human nitric oxide synthase. *J. Biol. Chem.* **269,** 26,677–26,683.
75. Matthews, R. T., Yang, L. Y., and Beal, M. F. (1997) *S*-Methylthiocitrulline, a neuronal nitric oxide synthase inhibitor, protects against malonate and MPTP neurotoxicity. *Exp. Neurol.* **143,** 282–286.
76. Itzhak, Y., Gandia, C., Huang, P. L., and Ali, S. F. (1998) Resistance of neuronal nitric oxide synthase-deficient mice to methamphetamine-induced dopaminergic neurotoxicity. *J. Pharmacol. Exp. Ther.* **284,** 1040–1047.

# 16
# Effects of Novel NMDA/Glycine-Site Antagonists on the Blockade of Cocaine-Induced Behavioral Toxicity in Mice

## Rae R. Matsumoto, PhD and Buddy Pouw, MD

## 1. INTRODUCTION

Abuse of cocaine can produce many adverse effects, including convulsions and lethality, especially in overdose situations. Previous efforts to develop pharmacotherapies for cocaine overdose have been limited in success, and there are, currently no effective treatments for this medical emergency. Much of the immediate toxicity associated with a cocaine overdose results from overstimulation of the central and peripheral nervous systems. Resultant symptoms such as tachycardia, hypertension, hyperthermia, seizures, respiratory depression, and cardiovascular collapse, if severe enough, can lead to death. All of the aforementioned symptoms of a serious cocaine overdose involve processes that are regulated by glutamatergic systems *(1–4)*, which appear overactivated in an overdose situation. Therefore, even if cocaine does not appear to directly bind to glutamate receptors, antagonism of these receptors is a viable drug development strategy for cocaine overdose by preventing and reversing end-stage events such as convulsions, respiratory distress, and cardiovascular collapse that immediately precede death (Fig. 1).

Many studies have now implicated glutamate receptors in the behavioral toxicity of cocaine and have evaluated the potential utility of glutamate antagonists, particularly *N*-methyl-D-aspartate (NMDA) receptor antagonists for the treatment of overdose. Because many excellent reviews are available on this and related topics *(5–7)*, this chapter focuses on the utility of recently characterized NMDA/glycine-site antagonists for the treatment of cocaine overdose. For comparison, the effects of targeting other sites on the NMDA receptor complex and other glutamatergic receptors are also summarized.

## 2. COCAINE AND NMDA RECEPTORS

### 2.1. Overview of Glutamate Receptors and Importance of NMDA Receptors

The classification of glutamate receptor subtypes and the molecular subunits that can assemble together to form them are summarized in Fig. 2. As described in earlier chapters in this volume, glutamate receptors can be classified as ionotropic or metabotropic receptors. Ionotropic receptors are ligand-gated ion channels that are comprised of five subunits. The ionotropic receptors for glutamate are further subdivided into NMDA and non-NMDA receptors, with additional subtypes of non-NMDA ionotropic glutamate receptors being designated as α-amino-3-hydroxy-5-methylisoxazole-4-proprionic acid (AMPA) or kainate receptors. In their native forms, NMDA receptors are thought

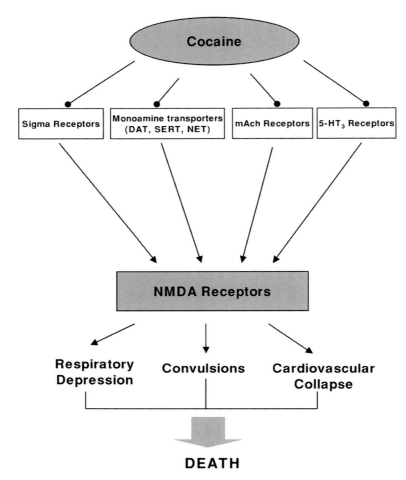

**Fig. 1.** Hypothesized involvement of NMDA receptors in a cocaine overdose. Cocaine binds to its known target sites (sigma receptors, monoamine transporters, muscarinic cholinergic receptors, and 5-hydroxytryptamine$_3$ [5-HT$_3$] receptors), which triggers a series of physiological changes eventually leading to the overstimulation of NMDA receptors. The overstimulation of NMDA receptors is responsible for the onset of many symptoms of a drug overdose, including convulsions, respiratory depression, and cardiovascular collapse, which if severe enough can cause death.

to contain at least one NR1 subunit, with the rest of the receptor assembled from a combination of NR2 and NR3 subunits. AMPA receptors, on the other hand, are thought to assemble from combinations of four subunits, designated GluR1–4. In contrast, kainate receptors are formed from subunit assemblies that may include GluR5–7 and KA1,2.

In contrast to ionotropic receptors, metabotropic receptors are seven transmembrane spanning proteins that are coupled to G proteins. In the case of metabotropic receptors for glutamate, inositol phosphate or adenylyl cyclase serve as intracellular second messengers. At least eight subtypes of metabotropic receptors for glutamate have been cloned (Fig. 2). Although the functional significance of each of these subtypes is still being delineated *(8)* based on similarities in sequence homology and coupling to second-messenger systems, the subtypes have been divided into three groups, designated groups I–III.

# Novel NMDA/Glycine-Site Antagonists and Cocaine-Induced Behavior

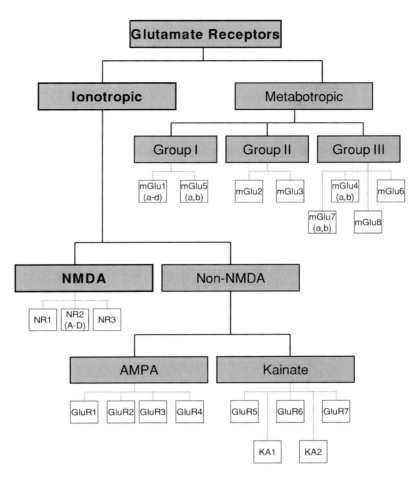

**Fig. 2.** Classification of glutamate receptor subtypes. Different classes of glutamate receptor are shown along with the molecular subunits and subtypes that comprise each of them.

Of the various subtypes of glutamate receptors, the NMDA receptor appears to have a predominant role during a cocaine overdose, with other glutamate receptor subtypes having a modulatory influence. These data are summarized in detail in the sections that follow.

## 2.2. Overview of Drugs Targeting Different Sites on the NMDA Receptor Complex

The NMDA receptor contains several antagonist binding and modulatory sites that can serve as targets for pharmacotherapies: glutamate-binding sites, glycine coagonist sites, sites within the ionophore, and allosteric modulatory sites. Antagonists that target each of these sites have been tested in animal models of cocaine overdose (Table 1). The receptor is depicted schematically in Fig. 3 and representative antagonists for each of these target sites are listed.

Antagonists that interact with each of the pharmacological binding sites on the NMDA receptor complex are associated with slightly different subjective effects and side-effect profiles. The most widely cited examples are the clinically intolerable side effects produced by many NMDA receptor channel blockers, such as MK-801 *(9,10)*. Competitive antagonists can also produce

Table 1
Summary of Chemical Names of Compounds

**NMDA Receptor**
*Glutamate site*
| | |
|---|---|
| D-AP7 | D(−)-2-Amino-7-phosphonoheptanoic acid |
| CGS 19755 | *Cis*-4-phosphonomethyl-2-piperidine carboxylic acid |
| CPP | 3-(2-Carboxypiperazin-4-yl)-propyl-l-phosphonic acid |
| LY 235959 | 3$S$,4a$R$,6$S$,8a$R$,-6-Phosphonomethyl-decahydroisoquinoline-3-carboxylic acid |
| LY 233536 | 6-(1(2)$H$-Tetrazol-5-yl)methyldecahydroisoquinoline-3-carboxylic acid |
| LY 274614 | 6-Phosphonomethyl-decahydroisoquinoline-3-carboxylic acid |
| NPC 12626 | 2-Amino-4,5-(1,2-cyclohexyl)-7-phosphonoheptanoic acid |
| NPC 17742 | 2$R$,4$R$,5$S$-2-Amino-4, 5-(1,2-cyclohenyl)-7-phophonoheptanoic acid |

*Glycine site*
| | |
|---|---|
| ACEA-0762 | 5-Aza-7-chloro-4-hydroxy-3-(m-phenoxyphenyl)quinoline-2(1$H$)-one |
| ACEA-1021 | 5-Nitro-6,7-dichloro-1,4-dihydro-2,3-quinoxalinedione |
| ACEA-1031 | 5-Nitro-6,7-dibromo-1,4-dihydro-2,3-quinoxalinedione |
| ACEA-1328 | 5-Nitro-6,7-dimethyl-1,4-dihydro-2,3-quinoxalinedione |
| ACPC | 1-Amino-1-cyclopropane carboxylic acid |
| 7-CKA | 7-Chlorokynurenic acid |
| DCQX | 6,7-Dichloroquinoxaline-2,3-dione |
| $R$(+)-HA-966 | $R$(+)-3-Amino-1-hydroxypyrrolidin-2-one |

*Channel blockers*
| | |
|---|---|
| ADCI | 5-Aminocarbonyl-10,11-dihydro-5h-dibenzo[$a,d$]cyclohepten-5,10-imine |
| Dextrorphan | (+)-3-Hydroxy-$N$-methyl morphinan |
| Ketamine | 2-(2-Chlorophenyl)-2-(methylamino)-cyclohexanone |
| Memantine | 3,5-Dimethyladamantan-1-amine |
| MK-801 | (5$R$,10$S$)-(+)-5-methyl-10,11-dihydro-5$H$-dibenzo[$a,d$]cyclohepten-5,10-imine |
| PCP (phencyclidine) | 1-(1-Phenylcyclohexyl)piperidine |
| SKF 10,047 | $N$-Allyl-normetazocine |
| TCP | 1-[1-(2-Thienyl)cyclohexyl]piperidine |

*Allosteric modulators*
| | |
|---|---|
| CP-101,606 | (1$S$,2$S$)-1-(4-Hydroxyphenyl)-2-(4-Hydroxy-4-phenylpiperidino)-1-propanol |
| Eliprodil | α-(4-Chlorophenyl)-4-[(4-fluorophenyl) methyl]-1-piperidine ethanol |
| Haloperidol | 4-[4-(4-Chlorophenyl)-4-hydroxy-1-piperidinyl]-1-(4-fluorophenyl)-1-butanone |
| Ifenprodil | 2-(4-Benzylpiperidino)-1-(4-hydroxyphenyl)-1-propanol |

**AMPA/Kainate Receptor**
| | |
|---|---|
| GYKI 52466 | 1-(4-Aminophenyl)-4-methyl-7,8-methylenedioxy-5$H$-2,3-benzodiazepine |
| NBQX | 1,2,3,4-Tetrahydro-6-nitro-2,3-dioxo-benzo[$f$]quinoxaline-7-sulfonamide |

**Metabotropic Receptor**
| | |
|---|---|
| AIDA | 1-Aminoindan-1,5-dicarboxylic acid |
| L(+)-AP-3 | 2-Amino-3-phosphonopropionic acid |
| L-CGG-I | (2$S$,1′$S$,)-2′-(Carboxycyclopropyl)-glycine |
| ($S$)-4C3HPG | ($S$)-4-Carboxy-3-hydroxyphenylglycine |
| $S$(+)-MCPA | ($S$)-α-Methyl-3-carboxyphenylalanine |
| MSOP | α-Methylserine-$O$-phosphate |

**Other**
| | |
|---|---|
| ACEA-1011 | 5-Chloro-7-trifluoromethyl-1,2,3,4-tetrahydroquinoxaline-2,3,-dione |
| Riluzole | 2-Amino-6-(trifluoromethoxy)-benzothiazole |

clinically unacceptable side effects, although they are generally less severe than those produced by channel blockers *(11,12)*. In contrast, antagonists that interact with other binding sites on the NMDA receptor complex, such as the glycine and allosteric modulatory sites, tend to be associated with the fewest problematic side effects *(12–14)*. One reason for this may be that many channel blockers act like a switch to turn the receptor "off" when the ionophore is blocked. Given the extensive

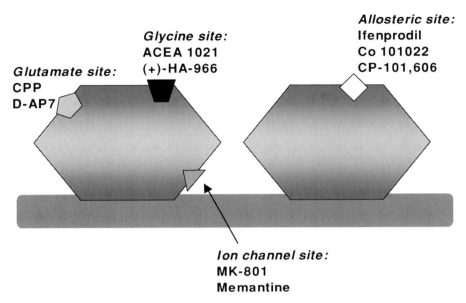

**Fig. 3.** Pharmacological model of NMDA receptor. There are several ligand-binding and modulatory sites on the receptor complex that can serve as targets for antagonists. Competitive antagonists interact with the glutamate-binding site. Antagonists may also bind to the glycine coagonist site. In addition, there are binding sites within the ionophore that, when occupied, elicits antagonist actions. Antagonists may also act through allosteric modulatory sites on the receptor complex.

distribution of glutamate in the nervous system and its involvement in many vital functions, in terms of achieving a therapeutic effect, reducing activity at the receptor may be more desirable than shutting off its function. Allosteric modulators, therefore, tend to produce relatively few side effects, although they are sometimes less efficacious than other classes of NMDA receptor antagonists. In contrast to switching off the NMDA receptor by blocking the ionophore or subtly modulating its activity, antagonists that interact with the glutamate and glycine coagonist sites offer better opportunities for precise regulation of the level of activity at the receptor complex. Because glutamate is the major neurotransmitter for NMDA receptors, interfering with its access to the receptor complex exerts definite antagonist actions without totally shutting down activity at the receptor, except perhaps at high antagonist doses. Antagonists that target the glycine coagonist site are thought to offer additional mechanistic advantages over those blocking the glutamate-binding site. Because the neurotransmitter actions of glutamate are rapidly detectable (within tens of milliseconds) and it can exert its actions in many parts of the nervous system, one potential complication of a system of this sort is that any dysfunction that occurs has the potential to escalate and spiral out of control very quickly. Because in practice this does not occur as often as might be expected, some sort of homeostatic mechanism must exist to help maintain activity at the receptor complex within an acceptable range. We hypothesize that the glycine coagonist site evolved for this function, to serve as a gain control to increase or decrease activity at the receptor complex within this acceptable range. As such, antagonism of NMDA/glycine sites would not only reduce activity at the receptor but also engage homeostatic mechanisms to help self-regulate activity at the receptor to maintain and re-establish function within an acceptable range. This feature may contribute to the improved therapeutic margin and protective actions of NMDA/glycine-site antagonists over many other classes of NMDA receptor antagonists.

## 2.3. Drug Development Potential of NMDA/Glycine-Site Antagonists

Five major structural classes of NMDA/glycine-site antagonists have been identified: indoles, tetrahydroquinolines, benzoazepines, quinoxalinediones, and pyridazinoquinolines *(15)*. The structural class most rigorously studied in the context of cocaine is the quinoxalinediones. Confirming data using drugs from other structural classes are also available and are described in the sections that follow.

A major advantage of NMDA/glycine-site antagonists is that they lack the serious side effects that have compromised the clinical potential of many antagonists that target other binding sites on the NMDA receptor complex. NMDA/glycine-site antagonists are bioavailable and well tolerated at therapeutic doses in many animal models of neurological disorders, including pain (16–22) epilepsy *(23)*, parkinsonism *(24,25)*, stroke *(26,27)*, and traumatic head injury *(28,29)*.

In addition, NMDA/glycine-site antagonists do not appear to produce phencyclidinelike effects that have been problematic for many earlier NMDA receptor antagonists *(10,30–32)*. NMDA/glycine-site antagonists do not kill neurons in the cingulate and retrosplenial cortex after repeated administration *(33,34)*. They do not produce psychotomimetic effects in humans *(35)*. They do not produce discriminative stimulus effects *(30,36–39)* or other actions associated with rewarding properties in animals *(40,41)*. Together, these features suggest a highly improved side-effect profile for NMDA/glycine-site antagonists over other classes of NMDA receptor antagonists that have previously undergone clinical trials.

As with all NMDA receptor antagonists, a major potential limitation of NMDA/glycine-site antagonists is the possibility that they will impair memory and cognitive function with prolonged use. Because of the well-documented cognitive deficits caused by previously tested classes of NMDA receptor antagonists, most studies related to the effects of NMDA/glycine-site compounds on learning and memory have focused on agonists that appear to enhance cognitive function *(42–45)*. However, a recent study reported that NMDA/glycine-site antagonists and partial agonists can also prevent memory deficits under certain conditions *(46)*, suggesting that further studies are needed to fully characterize the effects of agonists versus antagonists at NMDA/glycine sites on cognitive function. Nevertheless, given the importance of NMDA receptors in learning and memory, a potential adverse effect of NMDA/glycine-site antagonists on cognitive function is a complication that may limit their therapeutic use to acute or short-term administration *(14)*. Therefore, this chapter focuses on the potential utility of NMDA/glycine-site antagonists, as compared to other glutamate receptor antagonists, for the treatment of acute cocaine overdose.

## 3. COCAINE-INDUCED CONVULSIONS

The incidence of convulsions in individuals who abuse cocaine is significant. Cocaine-induced convulsions can result from exposure to large doses of cocaine, typically in an overdose situation. In addition, convulsions can manifest as a result of kindling, in which increased sensitivity to cocaine is produced upon repeated exposure to the drug, eventually leading to seizures upon exposure to doses of cocaine that had previously been subconvulsive. These cocaine-kindled seizures have been associated with concomitant increases in NMDA receptor binding in the brain *(47,48)*. In addition, NMDA receptor antagonists have been reported to significantly attenuate kindled seizures resulting from subchronic exposure to cocaine *(47,49,50)*.

*N*-methyl-D-aspartate receptor antagonists have also been reported to attenuate convulsions resulting from acute exposure to cocaine. The results from the acute studies are summarized in the subsections that follow. Most of the acute studies employ a similar methodology involving pretreatment of animals with an antagonist, which is then challenged with a convulsive dose of cocaine. The percentage of animals exhibiting cocaine-induced convulsions were recorded and reported, unless otherwise specified.

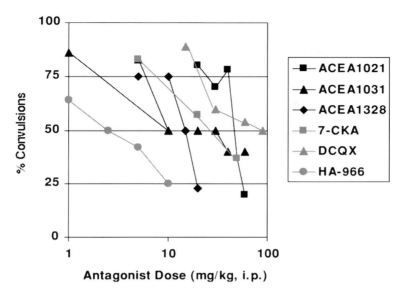

**Fig. 4.** Effect of pretreatment of NMDA/glycine-site antagonists on cocaine-induced convulsions. Pretreatment of mice with NMDA/glycine-site antagonists (ACEA1021, ACEA1031, ACEA1328, 7-chlorokynurenic acid, DCQX, (+)-HA-966) reduces the percentage of animals that exhibit cocaine-induced convulsions. Reductions to 50% or lower represent statistically significant changes. Includes data previously reported *(51,52)*.

## 3.1. NMDA/Glycine-Site Antagonists and Cocaine-Induced Convulsions

### 3.1.1. Effects of NMDA/Glycine-Site Antagonists

Many of the NMDA/glycine-site antagonists that have been tested against cocaine-induced convulsions have come from the quinoxalinedione structural class. The following quinoxalinediones, which act as NMDA/glycine-site antagonists, all significantly attenuate cocaine-induced convulsions in mice: ACEA-1021, ACEA-1031, ACEA-1328, and DCQX *(51,52)*. In addition, the structurally related NMDA/glycine-site antagonists ACEA-0762 and 7-chlorokynurenic acid and the structurally unrelated compounds $R(+)$-HA-966 and ACPC (1-amino-1-cyclopropane carboxylic acid) also significantly attenuate cocaine-induced convulsions in rodents *(51–54* and unpublished data). The protective effects of a select group of NMDA/glycine-site antagonists against cocaine-induced convulsions in mice are illustrated in Fig. 4.

### 3.1.2. Reversal by D-Cycloserine

Two additional lines of evidence suggest that the protective effects of the quinoxalinediones involve antagonism of NMDA/glycine sites. First, the glycine-site partial agonist D-cycloserine can pharmacologically antagonize the anticonvulsive effects produced against cocaine by ACEA-1021, ACEA-1031, and ACEA-1328 *(51,52)*. D-Cycloserine was used to confirm pharmacological antagonism in these experiments because its actions are specific to the glycine site on NMDA receptors (as opposed to glycine that can also act at the classical, strychnine-sensitive glycine receptor) and it is bioavailable after systemic administration *(55)*. Second, 5-nitroquinoxaline-2,3-dione, a compound that has very low affinity for NMDA/glycine sites and is lacking substituents at the 6,7-positions [but is otherwise identical to ACEA-1021, ACEA-1031, and ACEA-1328 *(56)*], fails to significantly attenuate cocaine-induced convulsions under comparable conditions *(51)*. This negative control confirms that simply having a quinoxalinedione structure is not sufficient to attenuate the convulsive effects of cocaine in mice. In order to provide protection, the compounds must antagonize activity at the NMDA receptor complex.

## 3.2. Comparison to Other NMDA Receptor Antagonists

Antagonists that target other sites on the NMDA receptor complex, including the glutamate-binding site, ionophore, and allosteric modulatory sites have also been tested and shown to be capable of attenuating cocaine-induced convulsions. These studies are summarized in the following subsections.

### 3.2.1. Effects of Competitive Antagonists

Competitive antagonists block the access of glutamate to its binding sites on the NMDA receptor complex. The following competitive antagonists have been reported to significantly attenuate cocaine-induced convulsions in mice: CGS 19755, CPP, D-AP7, LY 233536, LY 235959, LY 274614, NPC 12626, and NPC 17742 *(51,53,54,57,58)*.

### 3.2.2. Effects of Channel Blockers

The most widely studied NMDA receptor antagonist is MK-801, a channel blocker that is also referred to as dizocilpine. Although MK-801 has been reported to significantly attenuate cocaine-induced convulsions in numerous studies, it is effective only over a very limited dose range *(51,53,54,57–59)*. In addition, any potential benefits provided by MK-801 are compromised by its phencyclidinelike side effects and the production of seizures when administered alone at higher doses *(10,51,54)*. Other classical channel blockers that attenuate cocaine-induced convulsions include TCP, SKF 10,047, phencyclidine, ketamine, and dextrorphan *(54,57,60)*. However, like MK-801, the therapeutic potential of these compounds are compromised by problematic side effects that may include psychotomimetic effects and abuse potential *(61–63)*.

The clinically available antitussive agent dextromethorphan metabolizes to the NMDA receptor channel blocker dextrorphan. However, dextromethorphan does not appear to be a promising therapeutic candidate for the treatment of cocaine overdose. At doses that attenuate convulsions in other animal models of seizure disorders, dextromethorphan (up to 75 mg/kg, ip) fails to significantly attenuate cocaine-induced convulsions in mice *(53* and unpublished data). Furthermore, when it was tested against nontoxic doses of cocaine, dextromethorphan provided protection at some dose combinations and exacerbated the actions of cocaine at other dose combinations *(64)*. Because dextromethorphan possesses multiple mechanisms of action, some of which are not well characterized, it is believed that some of them may counteract the benefits that are provided through the NMDA blocking actions of the metabolite dextrorphan.

Memantine and ADCI (5-aminocarbonyl-10, 11-dihydro-5h-dibenzo[*a,d*]cyclohepten-5,10-imine) are also NMDA receptor antagonists that act, at least in part, by blocking the ionophore. Both of these compounds significantly attenuate cocaine-induced convulsions *(51,54,65)*, but they have a much more favorable side-effect profile than classical channel blockers such as MK-801 *(9,66)*. Memantine has been used therapeutically in Europe for many years *(67)*. In contrast to MK-801, whose adverse side effects are thought to be related to its high affinity, slow kinetics, and/or channel-trap mechanism of block, memantine interacts with moderate affinity, faster kinetics, and partial trapping actions *(68)*. Therefore, of the various channel blockers, memantine has the most favorable clinical profile, with the low-affinity channel blocker ADCI also having potential.

### 3.2.3. Effects of Allosteric Modulators

Compounds that act as allosteric modulators at NMDA receptors, including ifenprodil, eliprodil, Co 101022, CP-101,606, haloperidol, and nylidrin all significantly attenuate cocaine-induced convulsions *(51,54)*. Furthermore, Co 101022 and haloperidol have been reported to completely eliminate cocaine-induced convulsions when administered at higher doses *(51)*. Although these compounds are derived from diverse chemical classes and all have interactions with receptors and binding sites in addition to

NMDA, the only feature they share in common is their ability to interact with NMDA receptors comprised of NR1/NR2B subunits *(13,69,70;* R. Woodward, CoCensys, Inc., personal communication). Therefore, it appears that allosteric modulatory sites on the NMDA receptor complex are also targets for preventing convulsions resulting from a cocaine overdose.

### 3.3. Comparison to AMPA/Kainate Receptor Antagonists

The ability of NMDA receptor antagonists to attenuate the convulsive effects of cocaine suggests excessive stimulation of glutamate receptors in an overdose. It is therefore plausible that antagonism of other glutamate receptors may also have a beneficial effect in an overdose situation. As mentioned above, glutamate acts through two different ionotropic receptors, namely NMDA and non-NMDA. The non-NMDA receptors are further divided into two subtypes, AMPA and kainate. Although AMPA and kainate receptors can be distinguished from one another based on their amino acid sequences and differential sensitivities to certain drugs, antagonists for AMPA versus kainate receptors are currently unavailable. This section summarizes data showing that mixed AMPA/kainate receptor antagonists have no effect on cocaine-induced convulsions on their own, although they can enhance the protective actions of NMDA/glycine-site antagonists.

#### 3.3.1. Effects of AMPA/Kainate Receptor Antagonists Alone

1,2,3,4-Tetrahydro-6-nitro-2,3-dioxo-benzo [*f*]quinoxaline-7-sulfonamide (NBQX) and 1-(4-aminopheny1)-4-methyl-7,8-methylenedioxy-5H-2,3-benzodiazepine (GYKI 52466) are AMPA/kainate antagonists with high affinity and selectivity for iontropic non-NMDA receptors, as compared to other glutamatergic receptors *(71–73)*. At doses that provide protective effects in other animal models of seizure *(74,75)*, GYKI 52466 is unable to significantly attenuate cocaine-induced convulsions in mice *(76)*. NBQX produces equivocal results, providing protection in some studies *(54)* and none at all in others *(51,52)*. These largely negative findings with non-NMDA receptor antagonists contrast with the ability of NMDA receptor antagonists to consistently mitigate the behavioral toxicity of cocaine. They are, however, consistent with reports that other non-NMDA glutamatergic antagonists, such as glutamate release inhibitors and metabotropic antagonists, fail to inhibit cocaine-induced convulsions (see below). Together, the data suggest that during a cocaine overdose, a generalized overactivation of all glutamatergic receptors does not occur.

#### 3.3.2. Interaction Between AMPA/Kainate and NMDA Receptor Antagonists

Although, alone, non-NMDA receptor antagonists fail to attenuate cocaine-induced convulsions, they are capable of enhancing the protective actions of NMDA receptor antagonists. When the AMPA/kainate receptor antagonist NBQX is combined with the NMDA/glycine-site antagonist ACEA-1021, its presence improves the protective median effective dose ($ED_{50}$) for ACEA-1021 *(76)*. In addition, the mixed NMDA/non-NMDA receptor antagonist ACEA-1011 also provides significant protection from the convulsive effects of cocaine at doses much lower than would be predicted based on its potency at NMDA receptors alone *(76)*. Therefore, the data suggest that AMPA/kainate receptor antagonists have a modulatory influence on the ability of NMDA receptor antagonists to attenuate cocaine-induced behaviors, an interaction between NMDA and non-NMDA receptors that is similar to that reported in other systems *(77,78)*.

### 3.4. Comparison to Metabotropic Receptor Compounds

Metabotropic glutamate receptors can be divided into three groups: group I (mGluR1$_{a-d}$ and mGluR5$_{a,b}$), group II (mGluR2 and mGluR3), and group III (mGluR4$_{a,b}$, mGluR6, mGluR7$_{a,b}$, and mGluR8). Using conventional animal models of epilepsy, anticonvulsant effects have been reported with group I antagonists, group II agonists, and group III antagonists *(79–82)*. Thus far, compounds specific to any of these groups have not been reported to attenuate the acute convulsive effects of cocaine, although it is possible that nonselective metabotropic compounds may possess some anticocaine activity. These data are summarized in the subsections that follow.

### 3.4.1. Effects of Systemic Administration

Systemic administration of the metabotropic antagonist L(+)-AP-3 [L(+)-2-amino-3-phosphonoproprionic acid] fails to reduce the convulsive effects of cocaine in mice. However, because of uncertainties about the bioavailability of L(+)-AP-3 and other metabotropic glutamate antagonists via this route of administration *(83),* they are generally administered via an intracerebral or intracerebroventricular route when targeting the brain.

### 3.4.2. Effects of Central Administration

Intracerebroventricular administration of group I antagonists such as AIDA (1-aminoindan-1,5-dicarboxylic acid) and L(+)-AP-3 fail to alter the convulsive effects of cocaine. Group III antagonists, including $S$(+)-MCPA ($\alpha$-methyl-3-carboxyphenylalanine) and MSOP ($\alpha$- methylserine-$O$-phosphate) are also ineffective against cocaine-induced convulsions. The group II agonist L-CGG-I [(2$S$,1′$S$,2′$S$)-2-(carboxycyclopropyl)-glycine] likewise fails to prevent the convulsive effects of cocaine. However, intracerebroventricular administration of the mixed group II agonist/group I antagonist *(S)*-4C3HPG (4-carboxy-3-hydroxyphenylglycine) is capable of significantly attenuating cocaine-induced convulsions.

Although this initial group of data yielded primarily negative results, additional studies involving readily bioavailable subtype-specific and nonspecific compounds are still needed to fully evaluate the role of metabotropic glutamate receptors in cocaine overdose. Moreover, because metabotropic glutamate receptors are generally thought to modulate synaptic transmission *(84),* they may, similar to AMPA/kainate receptors, lack efficacy on their own against cocaine-induced convulsions. but they may modulate the actions of NMDA receptor antagonists. Studies to examine interactions such as these have yet to be conducted.

### *3.5. Comparison to Glutamate Release Inhibitors*

The glutamate release inhibitor riluzole is unable to attenuate cocaine-induced convulsions *(51).* It has also been reported to have no significant effect on seizure latency, and at higher doses, riluzole worsens the severity of the convulsions *(51).* Therefore, riluzole does not appear to have much promise as a pharmacotherapy for treating convulsions resulting from an acute cocaine overdose.

### *3.6. Summary of Glutamate Antagonists on Cocaine-Induced Convulsions*

$N$-methyl-D-aspartate receptor antagonists attenuate cocaine-induced convulsions in mice regardless of the specific site(s) on the receptor complex with which they interact. In contrast with the consistent ability of NMDA receptor antagonists to attenuate the convulsive effects of cocaine, antagonists at other glutamate receptors, namely, AMPA/kainate and metabotropic, fail to elicit significant protection against cocaine-induced convulsions. The glutamate release inhibitor riluzole is also ineffective against cocaine-induced convulsions. Together, the data suggest that there is a somewhat specific overactivation of NMDA receptors, rather than a global overstimulation of gutamate *per se,* in association with a cocaine overdose. Therefore, among the glutamatergic antagonists, NMDA receptor antagonists are expected to be the most promising pharmacotherapeutic option for treating the convulsive effects of cocaine. In addition, the ability of the AMPA/kainate antagonist NBQX to enhance the protective effects of NMDA receptor antagonists suggests that non-NMDA receptors have a modulatory effect. Therefore, polytherapy involving non-NMDA receptor antagonists may further enhance the protective actions afforded through NMDA receptor antagonists.

## 4. COCAINE-INDUCED LETHALITY

Ideally, pharmacotherapies for the treatment of cocaine overdose should protect against the ultimate toxic endpoint, death. Since death from a cocaine overdose can occur suddenly or with short latency, a

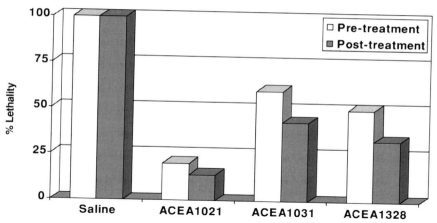

**Fig. 5.** Effect of NMDA/glycine-site antagonists on cocaine-induced lethality. Pretreatment (white bars) of mice with NMDA/glycine-site antagonists (ACEA1021, ACEA1031, and ACEA1328) significantly reduces the percentage of animals that die following a normally lethal dose of cocaine (saline bars). A comparably significant reduction in deaths from a cocaine overdose is observed even when mice are posttreated (dark bars) with NMDA/glycine-site antagonists. Antagonist treatments resulting in reductions of 50% or greater are statistically significant. Includes data previously reported *(51,52)*.

critical feature of an effective pharmacotherapy is that it be fast-acting. The following sections summarize data demonstrating the ability of NMDA/glycine site antagonists to attenuate cocaine-induced lethality in mice, with a select group also showing efficacy when administered as a posttreatment.

## 4.1. NMDA/Glycine-Site Antagonists and Cocaine-Induced Lethality

### 4.1.1. Effects of Pretreatment

Pretreatment of mice with the following NMDA/glycine-site antagonists have been reported to significantly attenuate cocaine-induced lethality: ACEA-1021, ACEA-1031, ACEA-1328, DCQX, and $R(+)$-HA-966 *(51)*. In contrast, structurally related quinoxalinediones that lack antagonist actions at NMDA receptors fail to attenuate the lethal effects of cocaine *(51)*, again emphasizing the importance of NMDA receptor blocking ability.

### 4.1.2. Effects of Posttreatment

Posttreatment with ACEA-1021, ACEA-1031, or ACEA-1328 in mice already injected with a normally lethal dose of cocaine prevents death in a significant proportion of animals *(51)* (Fig. 5). In these studies, NMDA/glycine-site antagonists are able to rescue a significant proportion of mice from death even when the posttreatments are delayed until the mice are already convulsing and expected to live for only another 2–4 min. To further evaluate the potential utility of these NMDA/glycine-site antagonists in overdose situations, in a separate study mice were also first sensitized to cocaine by treating them with a moderate, initially subconvulsive dose of cocaine for 7 d. Two hours after the last of seven exposures to cocaine, the mice received an additional normally lethal overdose of cocaine and were subsequently treated with ACEA-1021 after the onset of convulsions. Despite the severity of the convulsions and the beginning collapse of many organ systems, a significant proportion of mice were rescued when posttreated with the NMDA/glycine-site antagonist ACEA-1021 *(52)*. Therefore, ACEA-1021 appears capable of reversing the lethal effects of cocaine in a murine species and it is possible that a similar efficacy of this medication may be found in treating cocaine overdoses in humans. Additional studies in more advanced nonhuman species on the phylogenetic scale are needed to further evaluate this notion, prior to preliminary tests in humans.

## 4.2. Comparison to Other NMDA Receptor Antagonists

### 4.2.1. Effects of Competitive Antagonists

Pretreatment of mice with the following competitive NMDA receptor antagonists have also been reported to significantly attenuate cocaine-induced lethality: CPP and D-AP7 *(51)*. Despite their efficacy, these compounds have not been tested as posttreatments because earlier complications of competitive antagonists in clinical trials diminish the likelihood that these drugs will become clinically available.

### 4.2.2. Effects of Channel Blockers

At doses that attenuate the convulsive effects of cocaine pretreatment of mice with NMDA receptor channel blockers fail to prevent cocaine-induced lethality. In addition to having no effect on the overall proportion of animals surviving an overdose, MK-801 and memantine also do not alter the latency to death *(51)*. Although MK-801 produces serious side effects that could conceivably compromise its ability to prevent death from a cocaine overdose, the inability of the well-tolerated channel blocker memantine to provide protection demonstrates that improving the side-effect profile of this class of drugs still does not improve efficacy.

### 4.2.3. Effects of Allosteric Modulators

Allosteric modulators are somewhat effective in preventing cocaine-induced lethality. At least one dose of CP-101,606, haloperidol, and ifenprodil have been reported to significantly attenuate cocaine-induced lethality when they are administered as a pretreatment *(51)*. However, Co 101022 and nylidrin fail to prevent death in significant proportions of animals, even at doses that protect against cocaine-induced convulsions *(51)*. The tendency of allosteric modulators to produce variable effects against the lethal end point as compared to their profound attenuation of the convulsive effects of cocaine may be attributed in part of the distribution of NR2B subunits in the body. The high concentration of NR2B subunits in areas of the brain such as the hippocampus and cortex *(85)* are consistent with the powerful anticonvulsant actions of this class of drugs. In contrast, the lethal effects of cocaine involve neural mechanisms in addition to NR2B-related ones (e.g., involvement of medullary respiratory and cardiovascular regions) and organ systems in addition to the brain, which may account for the weaker effects of the allosteric modulators against the lethal effects of cocaine. Differences in the efficacy of allosteric modulators in preventing cocaine-induced lethality also likely stem from differences in their ability to interact with receptors and binding sites other than NMDA *(86–88;* R. Woodward, CoCensys, Inc. personal communication). Interactions with sites such as $Na^+$ channels, dopamine D2 receptors, and sigma receptors are among those that could positively or negatively influence the efficacy of allosteric modulators, against the lethal effects of cocaine *(7,50,89–91)* Therefore, for the allosteric modulators, the protective effects that they afford through NMDA receptor antagonism must be balanced with the nonspecific influences of the compounds through other binding sites to fully understand their effectiveness, or lack thereof, in protecting against the lethal effects of cocaine.

## 4.3. Comparison to AMPA/Kainate Receptor Antagonists

Pretreatment of mice with AMPA/kainate receptor antagonists such as NBQX or GYKI 52466 fail to alter the lethal effects of cocaine *(76)*. Together with the lack of effect of these compounds on their own against cocaine-induced convulsions, the data suggest that, in terms of the glutamatergic component in a cocaine overdose, the effect is somewhat specific to NMDA receptors.

## 4.4. Comparison to Glutamate Release Inhibitors

The glutamate release inhibitor riluzole also fails to reduce the proportion of animals exhibiting cocaine-induced lethality *(51)*. In addition, riluzole has no effect on the latency to death *(51)*.

## 4.5 Summary of Cocaine-Induced Lethality

N-methyl-D-aspartate glycine-site and competitive antagonists are capable of reducing the lethal effects of cocaine. Moreover, posttreatments with NMDA/glycine-site antagonists retain the ability to prevent the lethal effects of cocaine, suggesting that they have the clinical potential to treat life-threatening overdoses. In contrast, antagonists that target other sites on the NMDA receptor complex, namely, the ionophore or allosteric modulatory sites, are less effective at preventing cocaine-induced lethality. Furthermore, targeting non-NMDA receptors provides no protective actions against cocaine-induced lethality. Although metabotropic glutamate receptor compounds have not been tested against-induced lethality, their poor bioavailable and lackluster actions against the less toxic convulsive end point suggest that the existing compounds would not be promising candidates in severe cocaine overdoses. Therefore, when considered together with the pattern observed against cocaine-induced convulsions, pharmacotherapies that target NMDA receptors, particularly the glycine site, appear most promising for the treatment of acute cocaine overdoses.

## 5. OVERALL SUMMARY AND CLINICAL IMPLICATIONS

Of the various subtypes of glutamate receptor, the NMDA receptor appears the most promising target for the development of drugs to treat acute cocaine overdose. NMDA receptor antagonists significantly attenuate cocaine-induced convulsions and lethality in preclinical studies in rodents. In contrast, AMPA/kainate and metabotropic glutamate receptor antagonists are limited in effectiveness on their own against cocaine overdose, but may provide modulatory influences that could be beneficial as part of a polytherapy. An overactivation of NMDA receptors under overdose conditions would be consistent with the appearance of convulsions, respiratory distress, and cardiovascular collapse as major symptoms of a cocaine overdose. All of these systems are regulated by NMDA receptors and are consistent with the established role of these receptors in epilepsy, the maintenance of respiratory rhythms, and cardiovascular function (1–4).

Although it is logical that NMDA receptor antagonists would attenuate the symptoms of a cocaine overdose, it is important to note that antagonists that specifically target the glycine coagonist site appear to have the most favorable clinical potential at this time. Although all classes of NMDA receptor antagonists (competitive, glycine, channel blockers, allosteric modulators) attenuate cocaine-induced convulsions, only the NMDA/glycine-site and competitive antagonists robustly prevent the lethal effects of cocaine. When comparing these latter two classes of compounds, the NMDA/glycine-site antagonists have a more favorable side-effect profile, suggesting that they would have a better therapeutic index than competitive antagonists.

Although studies that have been conducted to date have not shown direct interactions between cocaine and NMDA receptors, such interactions are generally demonstrated using receptor-binding studies. It is, therefore, important to note that radioligands are not available for selectively labeling each of the pharmacological binding sites on the NMDA receptor complex. Furthermore, radioligands that are selective for different subunits that form the receptor complex, especially those that recognize different isoforms, are currently not available. It is, therefore, possible that cocaine can directly interact with specific components of the NMDA receptor complex, but such interactions have not been detected thus far.

Lacking evidence for direct interactions between cocaine and the NMDA receptor complex, the NMDA receptor antagonists appear to prevent the behavioral toxicity of cocaine by intervening at end-stage events (e.g., attenuating convulsions, respiratory depression, cardiovascular collapse) rather than early events (e.g., interfering with cocaine's access to receptors). Although future efforts to target initiating events could be successful, most earlier efforts to prevent the behavioral toxicity of cocaine by interfering with the access of the drug to known binding sites have been largely disappointing. In contrast, the preclinical data involving NMDA receptor antagonists are encouraging. For example, some of the NMDA/glycine-site antagonists rescued the animals from death even when they were administered *after* a cocaine overdose, when the animals only had 2–4 min left to survive. The data therefore indicate that targeting end-stage events via NMDA receptors may be a viable alternative to complement traditional drug development efforts that have focused on early initiating events. Such a strategy directed at terminal stages in an overdose offers several practical advantages as well. For example, pharmacotherapeutic interventions could be delayed until the very late stages of a drug overdose. Because most drug overdoses share common end-stage events, this strategy may be effective against not only cocaine but many other drugs of abuse. In addition, treatment options targeting end-stage events would not be as dependent on knowledge of the intoxicating substance, which can be difficult to ascertain when multiple drugs have been ingested and/or the victim is unconscious. Therefore, pharmacotherapies that target late-stage events in an overdose offer several practical advantages over treatments that are limited to counteracting initiating mechanisms.

In conclusion, the data suggest that NMDA/glycine-site antagonists have the potential to treat overdose to cocaine, and perhaps other drugs of abuse, in humans. Drugs that target other sites on the NMDA receptor, although producing significant effects in preclinical tests, were either less efficacious than NMDA/glycine-site antagonists and/or have a less favorable side-effect profile. Compounds that target other glutamate receptor subtypes such as AMPA/kainate may have potential utility as part of a polytherapy to enhance the therapeutic actions of NMDA receptor antagonists.

## ACKNOWLEDGMENTS

Portions of the work described herein were supported by the Oklahoma Center for the Advancement of Science and Technology. Lynn Doan is gratefully acknowledged for her expert technical assistance during studies involving the metabotropic antagonists.

## REFERENCES

1. Crambes, A., Monassier, L., Chapleau, D., Roegel, J.-C., Feldman, J., and Bousquet, P. (1996) GABAergic and gluataminergic modulation of centrally evoked arrhythmias in rats. *Hypertension* **27,** 148–154.
2. Olney, J. W., Collins, R. C., and Sloviter, R. S. (1986) Excitotoxic mechanisms of epileptic brain damage. *Adv. Neurol.* **44,** 857–877.
3. West, M. and Huang, W. (1994) Spinal cord excitatory amino acids and cardiovascular autonomic responses. *Am. J. Physiol.* **267,** H865–H873.
4. Yao, S. T., Finkelstein, D. I., and Lawrence, A. J. (1999) Nitrergic stimulation of the locus coeruleus modulates blood pressure and heart rate in the anaesthetized rat. *Neuroscience* **91,** 621–629.
5. Rockhold, R. (1998) Glutamatergic involvement in psychomotor stimulant action. *Prog. Drug Res.* **50,** 155–192.
6. Witkin, J. M. (1994) Pharmacotherapy of cocaine abuse: preclinical development. *Neurosci. Biobehav. Rev.* **18,** 121–142.
7. Witkin, J. M. and Katz, J. L. (1992) Preclinical assessment of cocaine toxicity: mechanisms and pharmacotherapy. *NIDA Res. Monogr.* **123,** 44–69.
8. Schoepp, D. D., Jane, D. E., and Monn, J. A. (1999) Pharmacological agents acting at subtypes of metabotropic glutamate receptors. *Neuropharmacology* **38,** 1431–1476.
9. Kornhuber, J. and Weller, M. (1997) Psychotogenicity and N-methyl-D-aspartate receptor antagonism: implications for neuroprotective pharmacotherapy. *Biol. Psychiatry* **41,** 135–144.
10. Vanderschuren, L. J. M. J., Schoffelmeer, A. N. M., Mulder, A. H., and DeVries, T. J. (1998) Dizocilpine (MK-801): use or abuse? *Trends Pharmacol. Sci.* **19,** 79–81.
11. Bullock, R. (1995) Strategies for neuroprotection with glutamate antagonists. Extrapolating from evidence taken from the first stroke and head injury studies. *Ann. NY Acad. Sci.* **765,** 272–278.
12. Koek, W. and Colpaert, F. C. (1990) Selective blockade of N-methyl-d-aspartate (NMDA)-induced convulsions by NMDA antagonists and putative glycine antagonists: relationship with phencyclidine-like behavioral effects. *J. Pharmacol. Exp. Ther.* **252,** 349–357.
13. Chazot, P. L. (2000) CP-101606 Pfizer Inc. *Curr. Opin. Invest. Drugs* **1,** 370–374.
14. Dannhardt, G. and Kohl, B. K. (1998) The glycine site on the NMDA receptor: structure–activity relationships and possible therapeutic application. *Curr. Med. Chem.* **5,** 253–263.
15. Tranquillini, M. E. and Reggiani, A. (1999) Glycine-site antagonists and stroke. *Expert Opin. Invest. Drugs* **8,** 1837–1848.

16. Christensen, D., Idanpaan-Heikkila, J. J., Guilbaud, G., and Kayser, V. (1998) The antinociceptive effect of combined systemic administration of morphine and the glycine/NMDA receptor antagonist, (+)-HA966 in a rat model of peripheral neuropathy. *Br. J. Pharmacol.* **125,** 1641–1650.
17. Dickenson, A. H. and Aydar, E. (1991) Antagonism at the glycine site on the NMDA receptor reduces spinal nociception in the rat. *Neurosci. Lett.* **121,** 263–266.
18. Lutfy, K., Shen, K.-Z., Woodward, R. M., and Weber, E. (1996) Inhibition of morphine tolerance by NMDA receptor antagonists in the formalin test. *Brain Res.* **731,** 171–181.
19. Lutfy, K., Doan, P., Nguyen, M., and Weber, E. (1998) ACEA-1328, an NMDA receptor antagonist, increases the potency of morphine and U50,488H in the tail flick test in mice. *Pharmacol. Res.* **38,** 453–460.
20. Lutfy, K. and Weber, E. (1996) Attenuation of nociceptive responses by ACEA-1021, a competitive NMDA receptor/glycine site antagonist, in the mice. *Brain Res.* **743,** 17–23.
21. Millan, M. J. and Seguin, L. (1993) (+)-HA-966, a partial agonist at the glycine site coupled to NMDA receptors, blocks formalin-induced pain in mice. *Eur. J. Pharmacol.* **238,** 445–447.
22. Quartaroli, M., Carignani, C., Dal Forno, G., Mugnaini, M., Ugolini, A., Arban, R., et al. (1999) Potent antihyperalgesic activity without tolerance produced by glycine site antagonist of $N$-methyl-D-aspartate receptor GV196771A. *J. Pharmacol. Exp. Ther.* **290,** 158–169.
23. Woodward, R. M., Huettner, J. E., Tran, M., Guastella, J., Keana, J. F. W., and Weber, E. (1995) Pharmacology of 5-chloro-7-trifluoromethyl-1,4-dihydro-2,3-quinoxalinedione: a novel systemically active ionotropic glutamate receptor antagonist. *J. Pharmacol. Exp. Ther.* **275,** 1209–1218.
24. Kanthasamy, A. G., Kanthasamy, A., Matsumoto, R. R., Vu, T. Q., and Truong, D. D. (1997) Neuroprotective effects of the strychnine-insensitive glycine site NMDA antagonist *(R)*-HA-966 in an experimental model of Parkinson's disease. *Brain Res.* **759,** 1–8.
25. Konieczny, J., Ossowska, K., Schulze, G., Coper, H., and Wolfarth, S. (1999) L-701,324, a selective antagonist at the glycine site of the NMDA receptor, counteracts haloperidol- induced muscle rigidity in rats. *Psychopharmacology* **143,** 235–243.
26. Takano, K., Tatlisumak, T., Formato, J. E., Carano, R. A. D., Bergmann, A. G., Pullan, L. M., et al. (1997) Glycine site antagonist attenuates infarct size in experimental focal ischemia: postmortem and diffusion mapping studies. *Stroke* **28,** 1255–1263.
27. Warner, D. S., Martin, H., Ludwig, P., McAllister, A., Keana, J. F. W., and Weber, E. (1995) In vivo models of cerebral ischemia: effects of parenterally administered NMDA receptor glycine site antagonists. *J. Cereb. Blood Flow Metab.* **15,** 188–196.
28. Di, X. and Bullock, R. (1996) Effect of the novel high-affinity glycine-site $N$-methyl-D-aspartate antagonist ACEA-1021 on $_{125}$I–MK-801 binding after subdural hematoma in the rat: an in vivo autoradiographic study. *J. Neurosurg.* **85,** 655–661.
29. Tsuchida, E. and Bullock, R. (1995) The effect of the glycine site-specific $N$-methyl-D-aspartate antagonist ACEA 1021 on ischemic brain damage caused by acute subdural hematoma in the rat. *J. Neurotrauma* **12,** 279–288.
30. Geter-Douglass, B. and Witkin, J. M. (1997) Dizocilpine-like discriminative stimulus effects of competitive NMDA receptor antagonists in mice. *Psychopharmacology* **133,** 43–50.
31. Kantak, K. M., Edwards, M. A., and O'Connor, T. P. (1998) Modulation of the discriminative stimulus and rate-altering effects of cocaine by competitive and noncompetitive $N$-methyl-D-aspartate antagonists. *Pharmacol. Biochem. Behav.* **59,** 159–169.
32. Sveinbjornsdottir, S., Sander, J. W. A. S., Upton, D., Thompson, P. J., Patsalos, P. N., Hirt, D., et al. (1993) The excitatory amino acid antagonist D-CPP-ene (SDZ EAA-494) in patients with epilepsy. *Epilepsy Res.* **16,** 165–174.
33. Auer, R. N. (1997) Structural neurotoxicologic investigation of the glycine antagonist 5- nitro-6,7-quinoxalinedione (ACEA-1021). *Neurotoxicology.* **18,** 53–62.
34. Hawkinson, J. E., Huber, K. R., Sahota, P. S., Hsu, H. H., Weber, E., and Whitehouse, M. J. (1997) The $N$-methyl-D-aspartate (NMDA) receptor glycine site antagonist ACEA- 1021 does not produce pathological changes in rat brain. *Brain Res.* **744,** 227–234.
35. Albers, G. W., Clark, W. M., Atkinson, R. P., Madden, K., Data, J. L., and Whitehouse, M. J. (1999) Dose escalation study of the NMDA glycine-site antagonist Licostinel in acute ischemic stroke. *Stroke* **30,** 508–513.
36. Balster, R. L., Mansbach, R. S., Shelton, K. L., Nicholson, K. L., Grech, D. M., Wiley, J. L., et al. (1995) Behavioral pharmacology of two novel substituted quinoxalinedione glutamate antagonists. *Behav. Pharmacol.* **6,** 577–589.
37. Singh, L., Menzies, R., and Tricklebank, M. D. (1990) The discriminative stimulus properties of (+)-HA-966, an antagonist at the glycine/$N$-methyl-D-aspartate receptor. *Eur. J. Pharmacol.* **186,** 129–132.
38. Wiley, J. L., Li, H., and Balster, R. L. (1997) Discriminative stimulus effects of site-selective $N$-methyl-D-aspartate antagonists in NPC 17742-trained rats and squirrel monkeys. *Psychopharmacology* **132,** 382–388.
39. Witkin, J. M., Steele, T. D., and Sharpe, L. G. (1997) Effects of strychnine-insensitive glycine receptor ligands in rats discriminating dizocilpine or phencyclidine from saline. *J. Pharmacol. Exp. Ther.* **280,** 46–52.
40. Kotlinska, J. and Biala, G. (2000) Memantine and ACPC affect conditioned place preference induced by cocaine in rats. *Pol. J. Pharmacol.* **52,** 179–185.

41. Papp, M., Moryl, E., and Maccecchini, M. L. (1996) Differential effects of agents acting at various sites of the NMDA receptor complex in a place preference conditioning model. *Eur. J. Pharmacol.* **317,** 191–196.
42. File, S., Fluck, E., and Fernandes, C. (1999) Beneficial effects of glycine (Bioglycin) on memory and attention in young and middle aged adults. *J. Clin. Psychopharmacol.* **19,** 506–512.
43. Meyer, R. C., Knox, J. Purwin, D. A., Spangler, E. L., and Ingram, D. K. (1998) Combined stimulation of the glycine and polyamine sties of the NMDA receptor attenuates NMDA blockage-induced learning deficits of rats in a 14-unit T-maze. *Psychopharmacology* **135,** 290–295.
44. Schwartz, B. L., Hashtroudi, S., Herting, R. L., Schwartz, P., and Deutsch, S. I. (1996) D- Cycloserine enhances implicit memory in Alzheimer patients. *Neurology* **46,** 420–424.
45. Tsai, G., Falk, W. E., Gunther, J., and Coyle, J. T. (1999) Improved cognition in Alzheimer's disease with short-term D-cycloserine treatment. *Am. J. Psychiatry* **156,** 467–469.
46. Viu, E., Zapata, A., Capdevila, J., Skolnick, P., and Trullas, R. (2000) Glycine(B) receptor antagonists and partial agonists prevent memory deficits in inhibitory avoidance learning. *Neurobiol. Learn. Mem.* **74,** 146–160.
47. Itzhak, Y. and Stein, R. (1992) Sensitization to the toxic effects of cocaine in mice is associated with the regulation of *N*-methyl-D-aspartate receptors in the cortex. *J. Pharmacol. Exp. Ther.* **262,** 464–470.
48. Itzhak, Y. and Martin, J. L. (2000) Cocaine-induced kindling is associated with elevated NMDA receptor binding in discrete mouse brain regions. *Neuropharmacology* **39,** 32–39.
49. Karler, R., Calder, L. D., Chaudhry, I. A., and Turkanis, S. A. (1989) Blockade of "reverse tolerance" to cocaine and amphetamine by MK-801. *Life Sci.* **45,** 599–606.
50. Shimosato, K., Marley, R. J., and Saito, T. (1995) Differential effects of NMDA receptor and dopamine receptor antagonists on cocaine toxicities. *Pharmacol. Biochem. Behav.* **51,** 781–788.
51. Brackett, R. L., Pouw, B., Blyden, J. F., Nour, M., and Matsumoto, R. R. (2000) Prevention of cocaine-induced convulsions and lethality in mice: effectiveness of targeting different sites in the NMDA receptor complex. *Neuropharmacology* **39,** 407–418.
52. Matsumoto, R. R., Brackett, R. L., and Kanthasamy, A. G. (1997) Novel NMDA/glycine site antagonists attenuate cocaine-induced behavioral toxicity. *Eur. J. Pharmacol.* **338,** 233–242.
53. Witkin, J. M. and Tortella, F. C. (1991) Modulators of *N*-methyl-D-aspartate protect against diazepam or phenobarbital-resistant cocaine convulsions. *Life Sci.* **48,** PL51–PL56.
54. Witkin, J. M., Gasior, M., Heifets, B., and Tortella, F. C. (1999) Anticonvulsant efficacy of *N*-methyl-D-aspartate antagonists against convulsions induced by cocaine. *J. Pharmacol. Exp. Ther.* **289,** 703–711.
55. Wood, P. L. (1995) The co-agonist concept: is the NMDA-associated glycine receptor saturated in vivo? *Life Sci.* **57,** 301–310.
56. Keana, J. F. W., Kher, S. M., Cai, S. X., Dinsmore, C. M., Glenn, A. G., Guastella, J., et al. (1995) Synthesis and structure-activity relationships of substituted 1,4-dihydroquinoxaline-2,3- diones: antagonists of *N*-methyl-D-aspartate (NMDA) receptor glycine sites and nonglutamate receptors. *J. Med. Chem.* **38,** 4367–4379.
57. Rockhold, R. W., Oden, G., Ho, I. K., Andrew, M., and Farley, J. M. (1991) Glutamate receptor antagonists block cocaine-induced convulsions and death. *Brain Res. Bull.* **27,** 721–723.
58. Ushijima, I., Kobayashi, T., Suetsugi, M., Watanabe, K., and Yamada, M. (1998) Cocaine: evidence for NMDA-, beta-carboline- and dopaminergic-mediated seizures in mice. *Brain Res.* **797,** 347–350.
59. Derlet, R. W. and Albertston, T. E. (1990) Anticonvulsant modification of cocaine-induced toxicity in the rat. *Neuropharmacology* **29,** 255–259.
60. Ritz, M. C. and George, F. R. (1997) Cocaine-induced convulsions: pharmacological antagonism at serotonergic, muscarinic and sigma receptors. *Psychopharmacology.* **129,** 299–310.
61. Jentsch, J. D. and Roth, R. H. (1999) The neuropsychopharmacology of phencyclidine: from NMDA receptor hypofunction to the dopamine hypothesis of schizophrenia. *Neuropsychopharmacology.* **20,** 201–225.
62. Nicholson, K. L., Hayes, B. A., and Blaster, R. L. (1999) Evaluation of the reinforcing properties of phencyclidine-like discriminative stimulus effects of dextromethorphan and dextrorphan in rats and rhesus monkeys. *Psychopharmacology* **146,** 49–59.
63. Weiner, A. L., Vieira, L., McKay, C. A., and Bayer, M. J. (2000) Ketamine abusers presenting to the emergency department: a case series. *J. Emerg. Med.* **18,** 447–451.
64. Jhoo, W. K., Shin, E. J., Lee, Y. H., Cheon, M. A., Oh, K. W., Kang, S. Y., et al. (2000) Dual effects of dextromethorphan on cocaine-induced conditioned place preference in mice. *Neurosci. Lett.* **288,** 76–80.
65. Seidleck, B. K., Thurkauf, A., and Witkin, J. M. (1994) Evaluation of ADCI against convulsant and locomotor stimulant effects of cocaine: comparison with the structural analogs dizocilpine and carbamazepine. *Pharmacol. Biochem. Behav.* **47,** 839–844.
66. Rogawski, M. A., Yamaguchi, S., Jones, S. M., Rice, K. C., Thurkauf, A., and Monn, J. A. (1991) Anticonvulsant activity of the low-affinity uncompetitive *N*-methyl-D-aspartate antagonist (±)-5-aminocarbonyl-10,11-dihydro-5H-dibenzo[*a,d*] cyclohepten-5,10-imine (ADCI): comparison with the structural analogs dizocilpine (MK-801) and carbamazepine. *J. Pharmacol. Exp. Ther.* **259,** 30–37.

67. Parsons, C. G., Danysz, W., and Quack, G. (1999) Memantine is a clinically well tolerated *N*-methyl-D-aspartate (NMDA) receptor antagonist—a review of preclinical data. *Neuropharmacology* **38**, 735–767.
68. Blanpied, T. A., Boeckman, F. A., Aizenman, R., and Johnsom, J. W. (1997) Trapping channel block of NMDA-activated responses by amantadine and memantine. *J. Neurophsyiol.* **77**, 309–323.
69. Avenet, P., Leonardon, J., Besnard, F., Graham, D., Depoortere, H., and Scatton, B. (1997) Antagonist properties of eliprodil and other NMDA receptor antagonists at rat NR1A/NR2A and NR1A/NR2B receptors expressed in *Xenopus* oocytes. *Neurosci. Lett.* **223**, 133–136.
70. Whittemore, E. R., Ilyin, V. I., Konkoy, C. S., and Woodward, R. M. (1997) Subtype-selective antagonism of NMDA receptors by nylidrin. *Eur. J. Pharmacol.* **337**, 197–208.
71. Sheardown, M. J., Nielsen, E. O., Hansen, A. J., Jacobsen, P., and Honore, T. (1990) 2,3-Dihydroxy-6-nitro-7-sulfamoyl-benzo(*F*)quinoxaline: a neuroprotectant for cerebral ischemia. *Science* **247**, 571–574.
72. Rogawski, M. A. (1993) Therapeutic potential of excitatory amino acid antagonists: channel blockers and 2,3-benzodiazepine. *Trends Pharmacol. Sci.* **14**, 325–331.
73. Zormuski, C. F., Yamada, K. A., Price, M. T., and Olney, J. W. (1993) A benzodiazepine recognition site associated with the non-NMDA glutamate receptor. *Neuron* **10**, 61–67.
74. Bagetta, G., Iannone, M., Palma, E., Nistico, G., and Dolly, J. O. (1996) *N*-methyl-D-aspartate and non-*N*-methyl-D-aspartate receptors mediate seizures and CAl hippocampal damage induced by dendrotoxin-K in rats. *Neuroscience* **71**, 613–624.
75. Smith, S. E., Durmuller, M., and Meldrum, B. S. (1991) The non-*N*-methyl-D-aspartate receptor antagonists, GYKI 52644 and NBQX are anticonvulsant in two animal models of reflex epilepsy. *Eur. J. Pharmacol.* **201**, 179–182.
76. Pouw, B. and Matsumoto, R. R. (1999) Effects of AMPA/kainate glutamate receptor antagonists on cocaine-induced convulsions and lethality in mice. *Eur. J. Pharmacol.* **386**, 181–186.
77. Lippert, K., Welsch, M., and Krieglstein, J. (1994) Over-additive protective effect of dizocilpine and NBQX against neuronal damage. *Eur. J. Pharmacol.* **253**, 207–213.
78. Zarnowski, T., Kleinrok, Z., Turski, W. A., and Czuczawar, S. J. (1993) 2,3-Dihydroxy-6-nitro-7-sulamoylbenzo(*f*)quinoxaline enhances the protective ability of common anti-epileptic drugs against maximal electroshock-induced seizures in mice. *Neuropharmacology* **32**, 895–900.
79. Ghauri, M., Chapman, A. G., and Meldrum, B. S. (1996) Convulsant and anticonvulsant actions of agonists and antagonists of group III mGluRs. *Neuro Report* **7**, 1469–1474.
80. Klodzinska, A., Chojnacka-Wojcik, E., and Pilc, A. (1999) Selective group II glutamate metabotropic receptor agonist LY354740 attenuates pentetrazole- and picrotoxin-induced seizures. *Pol. J. Pharmacol.* **51**, 543–545.
81. Moldrich, R. X., Talebi, A., Beart, P. M., Chapman, A. G., and Meldrum, B. S. (2001) The mGlu(2/3) agonist 2*R*,4*R*-4-aminopyrrolidine-2,4-dicarboxylate is anti- and proconvulsant in DBA/2 mice. *Neurosci. Lett.* **299**, 125–129.
82. Tang, E., Yip, P. K., Chapman, A. G., Jane, D. E., and Meldrum, B. S. (1997) Prolonged anticonvulsant action of glutamate metabotropic receptor agonists in inferior colliculus of genetically epilepsy-prone rats. *Eur. J. Pharmacol.* **327**, 109–115.
83. Knopfel, T., Kuhn, R., and Allgeier, H. (1995) Metabotropic glutamate receptors: novel targets for drug development. *J. Med. Chem.* **38**, 1417–1426.
84. Cartmell, J. and Schoepp, D. D. (2000) Regulation of neurotransmitter release by metabotropic glutamate receptors. *J. Neurochem.* **75**, 889–907.
85. Charton, J. P., Herkert, M., Becker, C. M., and Schroder, H. (1999) Cellular and subcellular localization of the 2B subunit of the NMDA receptor in the adult telencephalon. *Brain Res.* **816**, 609–617.
86. Chenard, B. L., Bordner, J., Butler, T. W., Chambers, L. K., Collins, M. A., De Costa, D. L., et al. (1995) (1*S*,2*S*)-1-(4-Hydroxyphenyl)-2-(4-hydroxy-4-piperidino)-1-propanol: a potent new neuroprotectant which blocks *N*-methyl-D-aspartate responses. *J. Med. Chem.* **38**, 3138–3145.
87. Hashimoto, K. and London, E. D. (1993) Further characterization of [$^3$H]ifenprodil binding to $\sigma$ receptors in rat brain. *Eur. J. Pharmacol.* **236**, 159–163.
88. Westlind-Danielsson, A., Ericsson, G., Sandell, L., Elinder, F., and Arhem, P. (1992) High concentrations of the neuroleptic remoxipride block voltage-activated Na$^+$ channels in central and peripheral nerve membranes. *Eur. J. Pharmacol.* **224**, 57–62.
89. Matsumoto, R. R., McCracken, K. A., Friedman, M. J., Pouw, P., de Costa, B. R., and Bowen, W. D. (2001) Conformationally-restricted analogs of BD1008 and an antisense oligodeoxynucleotide targeting $\sigma_1$ receptors produce anti-cocaine effects in mice. *Eur. J. Pharmacol.* **419**, 163–174.
90. Matsumoto, R. R., McCracken, K. A., Pouw, B., Miller, J., Bowen, W. D., Williams, W., et al. (2001) N-Alkyl substituted analogs of the $\sigma$ receptor ligand BD1008 and traditional $\sigma$ receptor ligands affect cocaine-induced convulsions and lethality in mice. *Eur. J. Pharmacol.* **411**, 261–273.
91. McCracken. K. A., Miller, J., Bowen, W. D., Zhang, Y., and Matsumoto, R. R. (2000) Brain sigma$_1$ receptors are involved in the behavioral effects of cocaine. *NIDA Res. Monogr.* **180**, 291.

# 17
# Clinical Studies Using NMDA Receptor Antagonists in Cocaine and Opioid Dependence

### Adam Bisaga, MD and Marian W. Fischman, PhD

## 1. INTRODUCTIONAL

In the last 10 yr, a substantial amount of preclinical research has been dedicated to studying the role of excitatory amino acids, in particular, glutamate, in laboratory models of alcohol and drug dependence (1–4). As a result, we now have a wealth of information suggesting that glutamatergic neurotransmission is a major component of the neuroadaptive changes that mediate the development, maintenance, and expression of patterns of animal behavior that are believed to be a model of human alcohol and drug dependence. The data are sufficient to warrant further research that would apply this preclinical knowledge to the clinical field and advance the development of potential pharmacotherapeutics (cf. ref. 5). In this chapter, background information will be presented on the stages of clinical medication development and will review available human data relevant for the medication development for opioid and cocaine dependence.

## 2. PHASES OF CLINICAL MEDICATION DEVELOPMENT

Opioid and cocaine dependences are complex behavioral disorders with associated psychiatric and psychosocial sequels. Experimental paradigms that use human volunteers and laboratory animals have been developed to model clinical disorders for the purpose of medication development. Owing to the complexity of the clinical syndromes, only selected aspects of each disorder can be modeled (e.g., physical withdrawal, cue-induced craving, or acute intoxication). A useful method of classifying available models for the purpose of medication development is to relate them to characteristics of the disorder as they emerge during treatment.

The two major stages of treatment where pharmacotherapeutics can be used are (1) induction of initial abstinence and (2) maintenance of abstinence. Initial abstinence can be accomplished either abruptly (forced abstinence usually in the context of detoxification) or through a gradual decrease of use (extinction of competitive drug-seeking). Early abstinence following the successful detoxification is associated with the emergence of affective and physical disturbances, intensive drug-craving, and heightened susceptibility to stress and drug-related environmental stimuli. These abnormal responses, which frequently contribute to relapse after a period of abstinence, can be individually modeled in the laboratory.

Research using laboratory animals has added greatly to our understanding of the effects of opioids and cocaine on the brain and behavior and to the development of potential medication. Traditionally, results of preclinical studies and uncontrolled treatment data were used to justify the initiation of clinical treatment trials. Although animal models provide an important screening tool in the search for

From: *Contemporary Clinical Neuroscience: Glutamate and Addiction*
Edited by: Barbara H. Herman et al. © Humana Press Inc., Totowa, NJ

potentially useful pharmacotherapies, the vast majority of compounds for cocaine dependence have been ineffective when subsequently tested in controlled clinical trials *(6–8)*. Clearly, given the need for effective drug-abuse pharmacotherapy, a less costly, more efficient, and scientifically sound adjunct to traditional clinical procedures is necessary. These new medication development paradigms could be employed to further expand on the results obtained from laboratory animal models before progressing to the large-scale efficacy trials. Our group has proposed using human laboratory paradigms and early phase II studies as such an intermediary step in the medication development process *(9,10)*.

Testing potential medications in a human laboratory setting provides a rigorous environment where a great deal of information relevant to medication development can be obtained. In this setting, experiments can be designed to approximate conditions that occur in naturalistic settings outside the laboratory while controlling many extraneous influences unrelated to experimental manipulations. Laboratory models (1) permit concurrent collection of multiple measures, (2) use standardized procedures for testing volunteers under identical conditions, (3) greatly limit the impact of ongoing environmental events, and (4) permit testing the same participant under both experimental and control (usually the placebo) conditions *(9)*. As a result, the impact of individual differences between participants is minimized, pharmacological effects can be measured more objectively, and fewer research participants are required. An added benefit of these studies is their ability to address the safety of potential treatments by assessing the effects of medication in combination with a given drug of abuse. Several laboratory paradigms have been developed for the purpose of testing promising medications in human volunteers. These include challenge studies, abstinence induction studies, self-administration and choice studies, cue-exposure studies, drug-discrimination studies, and brain-imaging studies with a pharmacological challenge (e.g., refs. *9* and *11–13*). Most frequently, laboratory studies use volunteers who are experienced drug users but are not seeking treatment or do not express the desire to stop using at the time of the study *(9)*.

Compounds that are promising in laboratory studies can then be tested in clinical treatment trials where a small number of subjects are utilized. These trials can be used as a screening model to establish preliminary efficacy of the test medication. Clearly, large controlled clinical trials are necessary for the definitive analysis of safety and efficacy, but they are too expensive, time-consuming, and difficult to be useful as an initial method for screening novel medications. In the past, uncontrolled studies were frequently used in the screening process, but the likelihood of finding false-positive results was so high that the value of such studies was very limited *(14)*.

The improved early phase II clinical trials are placebo-controlled and include a single-blind placebo lead-in phase to remove the most noncompliant patients and reduce attrition in the randomized trial. Manual-guided psychosocial intervention is used to promote compliance. Clinically meaningful dichotomous measures of improvement (such as "relapse"), in addition to continuous measures such as percentage of alcohol-free days, are used to assess outcome. These trials have adequate length to evaluate the durability of treatment responses and use 95% confidence limits of differences between the medication and placebo in order to obtain a measure of effect size *(10)*.

## 3. LABORATORY STUDIES

### 3.1. Models of Physiological Dependence: Development of Medication to Assist in Detoxification

Physiological dependence, and the emergence of a withdrawal syndrome upon cessation of drug use, is a common feature of several substance-dependence syndromes (in particular, opioid dependence). Medications are frequently used to alleviate the signs and symptoms of withdrawal, a treatment known as medical detoxification. Most patients who enter treatment for opioid dependence need to undergo detoxification, but its efficacy is not predictive of the long-term course of opioid dependence. Without further treatment, however, the majority of patients would relapse in the first few months after detoxification. Sufficient relief of withdrawal symptoms is necessary in order to retain patients until detoxification is complete, and a good experience during this phase of treatment may foster patients' commitment to long-term treatment.

There are three human laboratory models of opioid dependence. Traditionally, the effect of novel treatments has been studied in opioid-dependent individuals who are maintained on an opioid while residing on an inpatient research unit. After at least 1 wk of stabilization, opioid administration is abruptly withdrawn and an experimental treatment begins. In most paradigms, the severity of the withdrawal syndrome is then measured for at least 10 d (15). A modification of this experimental approach is the substitution test, where volunteers are maintained on a stable dose of an opioid, which is then substituted with the study compound for at least 24 h, after which an opioid is reintroduced (16). The third model is the precipitation procedure, during which an antagonist is given to opioid-dependent patients and the withdrawal syndrome is acutely precipitated (17,18).

$N$-Methyl-D-aspartate (NMDA) receptor antagonists, such as dextromethorphan (DXM) and memantine, have been stuided in laboratory models of opioid dependence. In the first of these studies, Isbel and Fraser (19) studied the effects of dextromethorphan (DXM) (d-methyl Dromoran) after abrupt morphine withdrawal and in the morphine-substitution test. DXM had no effect on the severity of withdrawal when given in two single doses at the 28th and 32nd h of abstinence, after abrupt cessation of morphine maintenance in opioid-dependent volunteers. An attempt to substitute morphine used to maintain opioid dependence with 75–100 mg DXM given every 6 h was also made, but the time of emergence and the intensity of withdrawal were similar to the control condition. Unfortunately, many details and statistical analyses that would permit close examination of the presented data were not included in this early study.

The effects of DXM were later studied by Rosen et al. (20) in the model of naloxone-precipitated withdrawal. Opiate-dependent volunteers were stabilized on methadone (25 mg/dy) and received a series of naloxone injections. DXM was given in single doses of 60, 120, and 240 mg in random order just before the precipitation of withdrawal. In this study, there was an interindividual variability in response to DXM. Volunteers with higher withdrawal scores in the placebo condition appeared to have lower withdrawal scores after DXM pretreatment; however, there was no overall effect of DXM on any of the withdrawal measures. At the conclusion of the study, two of the three volunteers who were detoxified from methadone using 60 mg DXM every 4 h had experienced considerable withdrawal. One of the limitations of this study is that the dose of DXM used may be too low to obtain a sufficient blockade of NMDA receptors (21,22).

Jasinski recently studied the effects of DXM on the severity of naloxone-precipitated withdrawal (23). The model used was acute physical dependence, for which opioids were given to nondependent individuals for 1 d only and withdrawal was precipitated with a large dose of naloxone given on the second day. In this study, 60 mg morphine was given orally every 6 h on the first day and before naloxone injection on the second day. The order of treatment conditions was random and counterbalanced with a washout between experiments of only 1 d. In this model, 30, 60, or 120 mg of DXM did not alter the severity of precipitated withdrawal. It has to be noted, however, that in this study, DXM was given not only before the precipitation of withdrawal but also before each dose of morphine; therefore, the effects on the development, as well as the expression, of opioid dependence were tested.

Another NMDA receptor antagonist that has been studied in the model of precipitated withdrawal is memantine. Bisaga et al. (24) studied eight opioid-dependent inpatient volunteers, who were maintained throughout the study on 120 mg of oral morphine per day. After a period of stabilization, naloxone challenges were administered 2–3 d apart, with memantine (60 mg) or placebo given 6 h before each challenge. In this study, a multiple baseline design was used. Memantine was administered before only one challenge (6-h challenge) and placebo was given during all baseline challenges and two post-memantine challenges (54 and 128 h postmemantine). Participants were randomly assigned to receive one, two, three, or four baseline challenges. The severity and duration of withdrawal measured was reduced at 6 and 54 h, but not 128 h after memantine administration as compared to the last baseline challenge. The effect of memantine was most evident using observer-rated measures of physical signs of withdrawal. Memantine was safe in combination with morphine and produced positive subjective effects. The subjective effects of memantine were short-lasting, as opposed to the inhibitory effects of memantine on the physical signs of opioid withdrawal that lasted at least 3 d. The difference in results from the previous studies may be due to the different agent studied, or possibly to the different design used, in that only a single dose of memantine was given and the severity of precipitated withdrawal was assessed repeatedly during the 5 d after memantine administration.

Two other available medications that modulate glutamatergic neurotransmission have been studied using single doses in the model of naloxone-precipitated withdrawal. Rosen et al. *(25)* evaluated the effects of lamotrigine, an anticonvulsant that attenuates glutamate release, and *d*-cycloserine, an antibiotic that has an affinity for the glycine-binding site of the NMDA receptor. The effects of lamotrigine (250 and 500 mg) pretreatment did not differ from palcebo. Cycloserine given at doses that have competitive NMDA-receptor (NMDAR) antagonism properties (375 and 750 mg) also did not alter the severity of withdrawal as compared to placebo.

In summary, results of most studies using human laboratory models of opioid dependence and withdrawal did not confirm the inhibitory effects on NMDAR antagonists. One study in which the positive effects were found, differed in the methodology but may be more relevant to the study of the nature of opioid dependence, which is a long-lasting phenomenon *(24)*.

### 3.2. Models of Drug Self-Administration

Drug self-administration and choice paradigms are designed to model drug-taking, one of the main components of addictive behavior. Several factors that are known to influence drug-taking in clinical situations can be controlled in the laboratory setting. These include environmental cues, subject characteristics (like the state of withdrawal or satiation), the availability of the drug, the dose and frequency of administration, and the availability of alternatives to drug-taking. Some of the models, primarily models of opioid dependence, have predictive validity in that medications effective in the treatment of opioid dependence are able to inhibit heroin self-administration in the laboratory setting *(12,26)*. The laboratory model of cocaine self-administration has not been validated because no effective medications to treat cocaine dependence are available.

Collins et al. *(27)* studied the self-administration of smoked cocaine in cocaine-dependent individuals maintained on 20 mg memantine or placebo in a double-blind crossover design. Participants had an opportunity to choose a dose of cocaine or a menetary alternative. When given memantine, participants chose cocaine with the same frequency as when given palcebo. Cocaine's subjective effects, however, including "high," "potency," and "quality," were significantly increased under memantine maintenance. This finding contrasts animal studies that show dose-dependent attenuating effects of memantine *(28)* or dextromethorphan *(29)* on cocaine self-administration, but is consistent with animal studies that used high-affinity channel blockers and showed synergism with cocaine's effects *(30)*. Therefore, the possibility of phylogenetic differences in the central representation of glutamate receptors needs to be considered *(31)*.

## 4. CLINICAL TRIALS

### 4.1. NMDA Receptor Antagonists in the Treatment of Drug Withdrawal

#### 4.1.1. Opiate Detoxification

Laboratory animal studies consistently show that NMDAR antagonists—in, particular channel blockers—attenuate the severity of opioid withdrawal in dependent animals *(1,3)*. The results of human laboratory studies are less decisive. Several preliminary trials have evaluated the utility of clinically available agents for opioid detoxification.

The first published clinical study was conducted in Turkey by Koyuncuoglu and Saydam *(32)*. Forty-eight patients seeking opioid detoxification were randomly assigned to dextromethorphan (360 mg/d) or chlorpromazine (96 mg/d), which was used as an active control. Both group received diazepam as an adjunct treatment. There was a major group difference in treatment retention. At the end of the 8-d trial, 68% of patients treated with DXM were still in treatment, whereas none treated with chlorpromazine completed the study. In fact, in the chlorpromazine group, 71% of patients left in the first 24 h. Patients who received DXM had significantly lower observer-rated abstinence scores across several days of treatment as compared to the control group. Conclusions from this study are limited by the fact that there was no placebo comparison group and by the use of chlorpromazine, a potentially aversive agent, as an active control, even though chlorpromazine is frequently used for detoxifications in settings where methadone is not available. A high dropout rate and related statistical confounds further limit conclusions from this study.

Subsequently, the same research group conducted two other open-label, outpatient, 8-d-long heroin detoxification studies. One study tested DXM in combination with chlorpromazine and diazepam (56 patients) *(33)*. In the other study, DXM was used in combination with tizanidine (a glutamate release inhibitor), chlorpromazine, and diazepam (47 subjects) *(34)*. Medication target doses were identical in both studies: 15 mg DXM every hour, 25–50 mg chlorpromazine every 6 h, and 10 mg diazepam every 6 h. Treatment completion rates were 68% and 96%, respectively. The remaining patients relapsed to heroin use in the first 3 d of detoxification, when abstinence symptoms were most severe. Overall, the completion rates reported in these three studies are equal to or higher than completion rates reported in other studies *(35,36)*; however, the extent of DXM's effect is confounded by the use of other medications that attenuate the withdrawal symptoms.

Another open-label pilot study evaluated the effects of DXM in six heroin-dependent patients seeking inpatient detoxification *(37)*. DXM was given in five daily doses of 75 mg each for up to 6 d. No other medications were used except for occasional, as needed, doses of ibuprofen, acetaminophen, and hydroxizine. Four patients completed the study with no withdrawal signs and symptoms (including heroin-craving) present beyond the fourth day of treatment. DXM was well tolerated with minimal side effects. Two other patients requested a change to methadone on the first day of DXM treatment, but none of them completed methadone-assisted detoxification.

Memantine has also been used for opioid detoxification in five inpatients who were opioid dependent, as confirmed by a naloxone challenge *(38)*. Memantine was given in single daily doses of 30 or 60 mg for 2–3 d. All patients completed detoxification, confirmed by a negative naloxone challenge, but they experienced mild–moderate withdrawal symptoms for 1–2 d and requested treatment with additional adjunct medication (clonidine, clonazepam, zolpidem, prochloperazine).

Another compound that has NMDAR antagonist properties and has been used in opioid detoxification is ibogaine, a naturally occuring alkaloid *(39)*. Ibogaine is a compound that has not been approved for clinical use in most countries, but ibogaine treatment has been carried out for more than 30 yr in informal settings *(40)*. Published data on the effects of ibogaine are mostly uncontrolled, retrospective case series *(41–44)*.

In an open-label study, Alper et al. *(41)* reported that 25 out of 33 opioid-dependent patients who received a single dose of ibogaine had complete resolution of withdrawal signs and symptoms, including drug-craving, within 24 h of receiving ibogaine and this improvement was maintained over 3 d of observation. In a recent prospective study, 27 opiate and cocaine-dependent individuals treated with ibogaine reported statistically significant improvement in depressive symptoms and craving for heroin and cocaine as compared to pretreatment baseline *(43)*. No control group was used however, so it is not known if the symptom improvement after ibogaine treatment would differ from the improvement that is a part of a natural course of the withdrawal syndrome. Additionally, the withdrawal severity and many other details of the study were not reported. Treatment outcomes reported after brief treatment with ibogaine (usually only single doses) are indeed impressive, although the safety of this treatment is not known, as several fatalities were reported *(41)*.

In summary, results of treatment trials conducted so far are generally positive and encouraging although none of these studies was adequately controlled and, therefore, it is unclear if these results will be confirmed in more definitive trials.

### 4.1.2. Treatment of Cocaine Withdrawal

The nature of cocaine withdrawal, and its contribution to continuous use, has been debated in recent years. It is unlikely that a brief treatment, as in the case of detoxification, may have a significant effect on the long-term course of the illness and few studies have recently assessed the effects of medication on acute cocaine withdrawal symptoms and short-term abstinence. Several such studies were conducted in the 1980s and one of the compounds tested, amantadine, was later found to have NMDAR antagonist properties (45). In initial studies, amantadine appeared to reduce early symptoms of cocaine withdrawal (46,47), but this effect declined after the first 2 wk, and several other studies reported negative results (48,49) (see ref. 50 for review). In the most recent controlled study, amantadine has been found to be more effective than placebo in reducing cocaine use. This effect was present only in patients with more severe cocaine-withdrawal symptoms at baseline, which suggests that amantadine may reduce early symptoms of cocaine withdrawal (51).

## 4.2. NMDA Receptor Antagonists in Promoting and Extending Abstinence

It is now widely believed that a long-term behavioral and pharmacological treatment is necessary for the effective treatment of opioid and cocaine dependence and to prevent relapse after a short-term detoxification procedure. Currently, there are two separate approaches to the long-term treatment of drug dependence: (1) gradual cessation of drug use (e.g., methadone treatment for opioid dependence) and (2) maintenance of abstinence (e.g., naltrexone treatment for opioid dependence). It is unclear if the same pharmacological approach may be used to induce and later to maintain the abstinence. Interestingly, it appears that NMDAR antagonists, secondary to their inhibitory effects of drug dependence/withdrawal, as well as their inhibitory effects on drug-induced reinforcement, may be effective in both contexts.

Abstinence-oriented studies using NMDAR antagonists are only beginning to be conducted. Preliminary data from one such study have recently been presented (52). Memantine (30 mg/d) or placebo has been given to 27 heroin-dependent patients who completed detoxification. Patients were treated and monitored in an inpatient drug-treatment program for 21 d. Memantine-treated patients had lower scores on self-report measures of heroin-craving and state anxiety than patients treated with placebo. All patients who received memantine completed the study with opiate-negative urines, whereas 40% of patients who received placebo discontinued the study and were later found to have relapsed to heroin use.

No recent data are available for cocaine treatment. In the past, amantadine has been used in studies to assess its effectiveness in the treatment of cocaine abuse. Although some studies showed beneficial effects of amantadine on cocaine-craving and use (53–55), the majority of studies reported negative findings (56–60).

## 5. SUMMARY AND DIRECTIONS FOR THE FUTURE

Opiate and cocaine addictions are complex disorders with a variety of processes that contribute to the maintenance of pathological behavior and relapse in abstinent individuals. Current animal paradigms are simplistic models of particular aspects of the disease and may be insufficient to predict the clinical utility and justify initiation of efficacy trials. We argue for the development of human laboratory paradigms and improvement of the methodology of small $N$ clinical trials, both of which may be useful in medication development for substance-use disorders. This strategy may be preferable to the traditional strategy of initiation of expensive efficacy trials based on preclinical research for cocaine (but not for opioid dependence). The vast majority of medications for cocaine dependence have been ineffective when tested in controlled clinical trials. Particular aspects of clinical syndromes can be modeled using human research volunteers and the more systematic medication discovery can be applied.

Preclinical research brings overwhelming evidence documenting the role of glutamatergic neurotransmission in the neurobiology of opioid and cocaine dependence and suggests that glutamatergic agents, in particular NMDAR antagonists, may be a useful addition to the treatment of these disorders. To date, the progress in applying results of preclinical studies to clinical medication development has been slow but noteworthy. The results from clinical studies of opioid dependence, both detoxification and relapse prevention, are encouraging (32–34,37,52). Efficacy trials of NMDAR antagonists for opioid dependence, including currently available dextromethorphan and memantine, are justified based on these findings.

Preliminary data in the treatment of cocaine dependence are very limited, and, at this point, it is too early to assess the potential usefulness of NMDAR antagonists for cocaine dependence. It is possible that NMDAR antagonists may be more useful in maintenance of abstinence than in induction of maintenance, but more preliminary studies are needed.

## ACKNOWLEDGMENT

Dr. Marian W. Fischman had passed shortly after this chapter was completed. She was an esteemed colleague and mentor, a pioneer in the field of cocaine research, and a wonderful human being. She will be greatly missed.

## REFERENCES

1. Herman, B. H., Vocci, F., and Bridge, P. (1995) The effects of NMDA receptor antagonists and nitric oxide synthase inhibitors on opioid tolerance and withdrawal. Medication development issues for opiate addiction. *Neuropsychopharmacology* **13,** 269–293.
2. Trujillo, K. A. (1995) Effects of noncompetitive *N*-methyl-D-aspartate receptor antagonists on opiate tolerance and physical dependence. *Neuropsychopharmacology* **13,** 301–307.
3. Bisaga, A. and Popik, P. (2000) In search of a new pharmacological treatment for drug and alcohol addiction: *N*-methyl-D-aspartate (NMDA) antagonists. *Drug Alcohol Depend.* **59,** 1–15.
4. Bisaga, A., et al. (2000) Therapeutic potential of NMDA receptor antagonists in the treatment of alcohol and substance use disorders. *Expert Opin. Investig. Drugs* **9,** 2233–2248.
5. Herman, B. H., et al. (2001) *Glutamate and Addiction* (Herman, B. H., ed.), Human, Totowa, NJ.
6. Bergman, J., et al. (2000) Agonist efficacy, drug dependence, and medications development: preclinical evaluation of opioid, dopaminergic, and GABAA-ergic ligands. *Psychopharmacology (Ber.)* **153,** 67–84.
7. McCance-Katz, E. F., Kosten, T. A., and Kosten, T. R. (2001) Going from the bedside back to the bench with ecopipam: a new strategy for cocaine pharmacotherapy development. *Psychopharmacology (Ber.)* **155,** 327–329.
8. McCance-Katz, E. F. (1997) Medication development for the treatment of cocaine dependence: issues in clinical efficacy trials. *NIDA Res. Monogr.* **175,** 36–72.
9. Fischman, M. W. and Foltin, R. W. (1998) Cocaine self-administration research: implication for rational pharmacotherapy, in *Cocaine Abuse: Behavior, Pharmacology, and Clinical Applications* (Higgins, S. T. and Katz, J. L., eds.), Academic, San Diego, CA, pp. 181–207.
10. Nunes, E. V. (1997) Methodologic recommendations for cocaine abuse clinical trials: a clinician-researcher's perspective. *NIDA Res Monogr* **175,** 73–95.
11. Bigelow, G. E. and Walsh, S. L. (1998) Evaluation of potential pharmacotherapies: response to cocaine challenge in the human laboratory, in *Cocaine Abuse: Behavior, Pharmacology, and Clinical Applications* (Higgins, S. T. and Katz, J. L., eds.) Academic, San Diego, CA, pp. 209–238.
12. Comer, S. D., Collins, E. D., and Fischman, M. W. (2001) Buprenorphine sublingual tablets: effects on IV heroin self-administration by humans. *Psychopharmacology (Ber.)* **154,** 28–37.
13. Ernst, M. and London, E. D. (1997) Brain imaging studies of drug abuse: therapeutic implications. *Semin. Neurosci.* **9,** 120–130.
14. Kosten, T. A. (1992) Pharmacotherapies, in *Clinician's Guide to Cocaine Addiction: Theory, Research, and Treatment* (Kosten, T.A. and Kleber, H.D., eds.), Guilford, New York.
15. Kolb, L. and Himmelsbach, C. K. (1938) A critical review of the withdrawal treatments with method of evaluating abstinence syndromes. *Am. J. Psychiatry* **94,** 759–799.

16. Fraser, H. F. and Isbell, H. (1960) Human pharmacology and addction liabilities of pentazocine and levophenacylmorphan. *Bull. Narcot.* **12**, 15–23.
17. Jasinski, D. R., Martin, W. R., and Haertzen, C. A. (1967) The human pharmacology and abuse potential of *N*-allylnoroxymorphone (naloxone). *J. Pharmacol. Exp. Ther.* **157**, 420–426.
18. Rosen, M. I., McMahon, T. J., Hameedi, F. A., Pearsall, H. R., Woods, S. W., Kreek, M. J., and Kosten, T. R. (1996) Effect of clonidine pretreatment on naloxone-precipitated opiate withdrawal. *J. Pharmacol. Exp. Ther.* **276**, 1128–1135.
19. Isbell, H. and Fraser, H. F. (1953) Actions and addiction liabilities of dromoran derivatives in man. *J. Pharmacol. Exp. Ther.* **106**, 524–530.
20. Rosen, M. I., McMahon, T. J., Woods, S. W., Pearsall, H. R., and Kosten, T. R. (1996) A pilot study of dextromethorphan in naloxone-precipitated opiate withdrawal. *Eur. J. Pharmacol.* **307**, 251–257.
21. Steinberg, G. K., Bell, T. E., and Yenari, M. A. (1996) Dose escalation safety and tolerance study of the *N*-methyl-D-aspartate antagonist dextromethorphan in neurosurgery patients. *J. Neurosurg.* **84**, 860–866.
22. Cornish, J. W., Herman, B. H., Ehrman, R. N., Robbins, S. J., Childress, A. R., Bead, V., Esmonde, C. A., Martz, K., Poole, S., Caruso, F. S., Vocci, F., and O'Brien, P. A randomized, double-bline, placebo-controlled safety study of high-dose dextromethorphan in methadone-maintained male patients. *Drug Alcohol Depend.*, in press.
23. Jasinski, D. R. (2000) Abuse potential of morphine/dextromethorphan combinations. *J. Pain Symptom Manage.* **19**, (Suppl) 26-S30.
24. Bisaga, A., Comer, S. D., Ward, A. S., Popik, P., Kleber, H. D., and Fischman, M. W. (2001) The NMDA antagonist memantine attenuates the expression of opioid physical dependence in humans. *Psychopharmacology (Berl.)* **157(1)**, 1–10.
25. Rosen, M. I., Persall, H. R., and Kosten, T. R. (1998) The effect of lamotrigine on naloxone-precipitated opiate withdrawal. *Drug Alcohol Depend.* **52**, 173–176.
26. Mello, N. K., Mendelson, J. H., Kuehnle, J. C., and Sellers, M. S. (1981) Operant analysis of human heroin self-administration and the effects of naltrexone. *J. Pharmacol. Exp. Ther.* **216**, 45–54.
27. Collins, E. D., Ward, A. S., McDowell, D. M., Foltin, R. W., and Fischman, M. W. (1998) The effects of memantine on the subjective, reinforcing and cardiovascular effects of cocaine in humans. *Behav. pharmacol.* **9(7):** 587–598.
28. Hyytia, P., Backstrom, P., and Liljequist, S. (1999) Site-specific NMDA receptor antagonists produce differential effects on cocaine self-administration in rats. *Eur. J. Pharmacol.* **378**, 9–16.
29. Pulvirenti, L., Balducci, C., and Koob, G. F. (1997) Dextromethorphan reduces intravenous cocaine self-administration in the rat. *Eur. J. Pharmacol.* **321**, 279–283.
30. Kantak, K. M., Edwards, M. A., and O'Connor, T. P. (1998) Modulation of the discriminative stimulus and rate-altering effects of cocaine by competitive and noncompetitive *N*-methyl-D-aspartate antagonists. *Pharmacol. Biochem. Behav.* **59**, 159–169.
31. Herman, B. H. and O'Brien, C. P. (1997) Clinical medication development for opiate addiction: focus on nonopioids and opioid antagonists for the amelioration of opiate withdrawal symptoms and relapse prevention. *Semin. Neurosci.* **9**, 158–172.
32. Koyuncuoglu, H. and Saydam, B. (1990) The treatment of heroin addicts with dextromethorphan. A double-blind comparison of dextromethorphan with chloropromazine. *Int. J. Clin. Pharmacol. Ther.* **28**, 147–152.
33. Koyuncuoglu, H. (1991) The treatment with dextromethorphan of heroin addicts, in *Drug Addiction and AIDS* (Loimer, N., Schmid, R., and Springer, A., eds.), Springer-Verlog, Vienna, pp. 320–329.
34. Koyuncuoglu, H. (1995) The combination of tizanidine markedly improves the treatment with dextromethorphan of heroin addicted outpatients. *Int. J. Clin. Pharmacol. Ther.* **33**, 13–19.
35. O'Connor, P. G., Carroll, K. M., Shi, J. M., Schottenfeld, R. S., Kosten, T. R., and Rounsaville, B. J. (1997) Three methods of opioid detoxification in a primary care setting. A randomized trial. *Ann. Intern. Med.* **127**, 526–530.
36. Washton, A. M. and Resnick, R. B. (1980) Clonidine for opiate detoxification: outpatient clinical trials. *Am. J. Psychiatry* **137**, 1121–1122.
37. Bisaga, A., Gianelli, P., and Popik, P. (1997) Opiate withdrawal with dextromethorphan. *Am. J. Psychiatry* **154**, 584.
38. Bisaga, A., Comer, S. D., Akerele, E. O., Kleber, H. D., and Fischman, M. W. (1999) The clinical use of memantine in opioid detoxification. *NIDA Res. Monogr.* **180**, 227.
39. Popik, P., Layer, R. T., and Skolnick, P. (1995) 100 years of ibogaine: neurochemical and pharmacological actions of a putative anti-addictive drug. *Pharmacol. Rev.* **47**, 235–253.
40. De Rienzo, P., Beal, D., and Staff, P. M. (1997) The Ibogaine Story: *Report on the Staten Island Project.* Autonomedia, Brooklyn, NY.
41. Alper, K. R., Lostof, H. S., Frenken, G. M., Luciano, D. J., and Bastiaans, J. (1999) Treatment of acute opioid withdrawal with ibogaine. *Am. J. Addict.* **8**, 234–242.
42. Kovera, C. A., Kovera, M. B., Singleton, E. G., Ervin, F. R., Williams, I. C., and Mash, D. C. (1998) Decreased drug craving during inpatient detoxification with ibogaine. *NIDA Res. Monogr.* **179**, 294.
43. Mash, D. C., et al. (2000) Ibogaine: complex pharmacokinetics, concerns for safety, and preliminary efficacy measures. *Ann. NY Acad. Sci.* **914**, 394–401.
44. Sheppard, S. G. (1994) A preliminary investigation of ibogaine: case reports and recommendations for further study. *J. Subst. Abuse Treat.* **11**, 379–385.

45. Kornhuber, J., Weller, M., Schoppmeyer, K., and Riederer, P. (1994) Amantadine and memantine are NMDA receptor antagonists with neuroprotective properties. *J. Neural Transm.* **43(Suppl.),** 91–104.
46. Giannini, A. J., Folts, D. J., Feather, J. N., and Sullivan, B. S. (1989) Bromocriptine and amantadine in cocaine detoxification. *Psychiatry Res.* **29,** 11–16.
47. Tennant, F. S., Jr. and Sagherian, A. A. (1987) Double-blind comparison of amantadine and bromocriptine for ambulatory withdrawal from cocaine dependence. *Arch. Intern. Med.* **147,** 109–112.
48. Gawin, F. H., Morgan, C., Kosten, T. R., and Kleber, H. D.(1989) Double-blind evaluation of the effect of acute amantadine on cocaine craving. *Psychopharmacology (Berl.)* **97,** 402–403.
49. Morgan, C., Kosten, T., Gawin, F., and Kleber, H. (1988) A pilot trial of amantadine for ambulatory withdrawal for cocaine dependence. *NIDA Res. Monogr.* **81,** 81–85.
50. Thompson, D. F. (1992) Amantadine in the treatment of cocaine withdrawal. *Ann. Pharmacother.* **26,** 933–934.
51. Kampman, K. M., Volpicelli, J. R., Alterman, A. I., Cornish, J., and O'Brien, C. P. (2000) Amantadine in the treatment of cocaine-dependent patients with severe withdrawal symptoms. *Am. J. Psychiatry* **157,** 2052–2054.
52. Bespalov, A., Zvartau, E. E., Krupitsky, E. M., Mosolov, D. V., and Burakov, A. M. (2001) A pilot study of memantine for the treatment of heroin dependence. *Drug Alcohol Depend* **63,** 14 (abstract).
53. Alterman, A. I., Droba, M., Antelo, R. E., Cornish, J. W., Sweeney, K. K., Parikh, G. A., and O'Brien, C. P. (1992) Amantadine may facilitate detoxification of cocaine addicts. *Drug Alcohol Depend.* **31,** 19–29.
54. Ziedonis, D. M. and Kosten, T. R. (1991) Pharmacotherapy improves treatment outcome in depressed cocaine addicts. *J. Psychoactive Drugs* **23,** 417–425.
55. Shoptaw, S., Kintaudi, K., Charuvastra, V. C., Rawson, R. A., and Ling, W. (1998) Amantadine hydrochloride is effective treatment for cocaine dependence. *NIDA Res. Monogr.* **179,** 55.
56. Kolar, A. F., Brown, B. S., Weddington, W. W., Haertzen, C. C., Michaelson, B. S., and Jaffe, J. H. (1992) Treatment of cocaine dependence in methadone maintenance clients: a pilot study comparing the efficacy of desipramine and amantadine. *Int. J. Addict.* **27,** 849–868.
57. Kosten, T. R., Morgan, C. M., Falcione, J., and Schottenfeld, R. S. (1992) Pharmacotherapy for cocaine-abusing methadone-maintained patients using amantadine or desipramine. *Arch. Gen. Psychiatry* **49,** 894–898.
58. Weddington, W. W., Brown, B. S., Haertzen, C. A., Hess, J. M., Kolar, A. F., and Mahaffey, J. R. (1989) Comparison of amantadine and desipramine combined with psychotherapy for treatment of cocaine dependence. *NIDA Res. Monogr.* **95,** 483–484.
59. Handelsman, L., Limpitlaw, L., Williams, D., Schmeidler, J., Paris, P., and Stimmel, B. (1995) Amantadine does not reduce cocaine use or craving in cocaine-dependent methadone maintenance patients. *Drug Alcohol Depend.* **39,** 173–180.
60. Kampman, K., Volpicelli, J. R., Alterman, A., Cornish, J., Weinreib, R., Epperson, L., Sparkman, T., and O'Brien, C. P. (1996) Amantadine in the early treatment of cocaine dependence: a double- blind, placebo-controlled trial. *Drug Alcohol Depend.* **41,** 25–33.

# 18
## The Role of mGluR5 in the Effects of Cocaine
## Implications for Medication Development

### Mark P. Epping-Jordan, PhD

Investigations over the past several years have yielded insights into the contributions of metabotropic glutamate receptors (mGluRs) to the effects of psychostimulants. Recently, an important role was demonstrated for the mGluR subtype 5 in the locomotor stimulant and reinforcing effects of cocaine *(1)*. The data obtained in that report will be summarized here, followed by a brief discussion of the implications of this evidence for the development of medications to treat cocaine dependence.

mGluRs can be divided into three groups based on amino acid sequence homology, receptor pharmacology, and intracellular signal transduction mechanisms *(2,3)*. Group I consists of mGluR1 and mGluR5, which stimulate phospholipase C and phosphoinositide hydrolysis, while group II, consisting of mGluR2 and mGluR3, and group III, consisting of mGluR4, 6, 7, and 8, are negatively coupled to adenylyl cyclase. All of the mGluR subtypes, except for mGluR6 and mGluR8, are expressed in widespread regions of the brain, with several subtypes being expressed at high levels in brain regions associated with the behavioral effects of drugs of abuse *(4–10)*.

Direct activation or blockade of mGluRs mediate both baseline and psychomotor stimulant-induced locomotor activity *(11,12)*. Bilateral injections into the nucleus accumbens (NAc) of the non-selective mGluR agonist 1-amino-1,3-cyclopentane-*trans*-1,3-dicarboxylic acid (ACPD) increased locomotor activity in the rat *(11,12)*, and this effect was blocked by co-injection of mGluR antagonist (RS)-α-methyl-4-carboxyphenylglycine (MCPG) *(11)*. In addition, the locomotor stimulant effects of both the indirect dopamine (DA) agonist amphetamine as well as the direct DA receptor agonist apomorphine were completely blocked by co-injection of MCPG into the NAc *(13)*. Furthermore, bilateral injections of ACPD into the ventral tegmental area, the source of DA projections into the NAc, increased locomotor activity in rats, through activation of both Group I and Group II mGluRs *(12)*. These data support a role for mGluRs in the control of baseline and drug-induced locomotor activation; however, the contributions of specific mGluR subtypes remain unknown.

The expression pattern of mGluR5 suggests a possible role in the effects of drugs of abuse. mGluR5 is highly expressed in both the intrinsic and efferent projection neurons in the NAc and striatum *(9)*, brain regions associated with the behavioral effects of psychostimulants. Approximately 65% of striatonigral projection neurons and 50–60% of striatopallidal projection neurons, as well as a majority of striatal interneurons express mGluR5 *(10)*. Over 80% of the NAc core and shell neurons projecting to the ventral pallidum and about 50% of the NAc shell neurons projecting to the ventral tegmental area, the primary efferent accumbal pathways, express mGluR5 mRNA *(14)*. The expression of mGluR5 in significant proportions of accumbal and striatal projection neurons and interneurons suggest that the activity of mGluR5 influences the information that is processed within these structures and that is transmitted to their output pathways.

From: *Contemporary Clinical Neuroscience: Glutamate and Addiction*
Edited by: Barbara H. Herman et al. © Humana Press Inc., Totowa, NJ

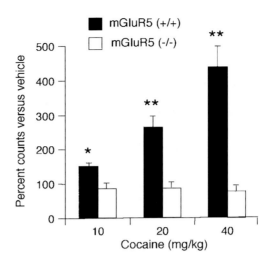

**Fig. 1.** Locomotor response to cocaine administration in mGluR5 (+/+) ($n = 14$) and (−/−) ($n = 16$) mice. Horizontal activity was measured during a 45-min period and was calculated as a percentage of baseline locomotor activity (vehicle treatment during the same period). Values represent mean activity counts ± SEM.*, $P < 0.05$ versus saline vehicle; **, $p < 0.01$ versus vehicle. (Dunnett's test after two-way repeated measures analysis of variance) [Reprinted with permission from Chiamulera et al. (1).]

While the expression and distribution of mGluR5 have been reasonably well-defined, the specific role played by mGluR5 in the effects of cocaine has been examined only recently. Increases in mGluR5 mRNA in the NAc shell and dorsolateral striatum (15) and decreases in mGluR5 protein levels in the medial NAc were observed 3 wk after cessation of repeated cocaine administration (16). These changes in expression suggest an involvement of mGluR5 in the neuroadaptations resulting from repeated cocaine administration; however, the functional consequences of altered mGluR5 expression remain unknown. In order to determine more specifically the role of mGluR5 to the effects of cocaine, the locomotor stimulant and rewarding effects of cocaine were examined in mGluR5 null mutant (−/−) mice and their wild-type (+/+) siblings. In addition, the effect of the selective mGluR5 antagonist 2-methyl-6-(phenylethynyl)-pyridine (MPEP) (17) on cocaine self-administration in C57BL/6J mice was examined [1].

Baseline and cocaine-induced locomotor activity were examined in mGluR5 (−/−) and (+/+) mice. On each test day, all mice received an injection of saline vehicle and were immediately placed into a locomotor activity chamber for 15 min. Mice were then removed from the apparatus and received an injection of either saline or cocaine (10, 20, or 40 mg/kg, ip) and were again placed into the activity chamber for an additional 45 min. Each subject was tested with vehicle alone and all three cocaine doses on different test days with at least 1 wk between test sessions.

Baseline locomotor activity did not differ significantly between (+/+) and (−/−) mice [(+/+), 2393–479; (−/−), 2710–642, mean horizontal activity counts/45-min session − SEM]. Cocaine induced a significant, dose-dependent increase in horizontal activity in (+/+) mice, however, cocaine had no effect on locomotor activity in mGluR5 (−/−) mice at any dose tested (Fig. 1). Because mice received repeated injections of cocaine, it is possible that some of the increase in locomotor activity observed in the (+/+) mice can be attributed to behavioral sensitization (18) Nevertheless, locomotor activity was not altered by cocaine administration in mGluR5 (−/−) mice, under the conditions tested. These results suggest that mGluR5 does not contribute to baseline locomotor activity; however, they indicate that mGluR5 is required for the expression of cocaine-induced locomotor activation (1).

Examination of the effect of mutation of the mGluR5 gene on the reinforcing effects of cocaine in mice may help to establish the role of mGluR5 in human cocaine addiction. Therefore, intravenous cocaine self-administration was investigated in mGluR5 (–/–) and (+/+) mice. Because the self-administration task is complex and requires learning, memory, and motor coordination, which could be disrupted by the mGluR5 mutation, mice were first trained in a discriminated, two-lever, food-reinforced operant task. This training allowed for the assessment of differences in learning rate and/or the ability to perform the operant task separate from the assessment of the reinforcing effects of cocaine. Acquisition of the food-reinforced operant task did not differ between (+/+) and (–/–) mice (Fig. 2A), indicating that the reinforcing properties of food were unchanged in the (–/–) mice.

After the food training, mice were surgically implanted with a catheter in the right jugular vein and, following recovery, were allowed to self-administer cocaine. mGluR5 (+/+) mice acquired stable self-administration across a wide range of cocaine doses; however, (–/–) mice did not acquired cocaine self-administration at any of the doses tested (Fig. 2B). Despite retraining to criterion on the food task between each cocaine dose, lever responding for cocaine in (–/–) mice extinguished within three to five sessions at all doses. Taken together, data from the food training and cocaine self-administration studies suggest that while the reinforcing properties of food remained unchanged, the reinforcing effects of cocaine were absent in mGluR5 (–/–) mice *(1)*.

It is possible that the absence of the reinforcing effect of cocaine in (–/–) mice was due to developmental alterations resulting from the constitutive mutation of the mGluR5 gene rather than to the lack of mGlu5 receptors in the adult mice. Therefore, the effects of the selective mGluR5 antagonist MPEP on cocaine self-administration and operant responding for food in C57BL/6J mice were examined. MPEP dose-dependently reduced cocaine self-administration in C57BL/6J mice (Fig. 2C); however, the most effective MPEP dose (30mg/kg, iv) had no effect on the rate of food-reinforced lever responding in the same mice (Fig. 2D) *(1)*. These data suggest that the effects of MPEP were specific to the reinforcing effects of cocaine and were not due to nonspecific disruption of motivation, to the ability to perform the operant task, or to general malaise induced by MPEP.

Cocaine is known to block the re-uptake of monoamine neurotransmitters, and the locomotor stimulant and reinforcing effects of cocaine have been linked to its ability to increase extracellular levels of DA in the NAc *(19,20)*. Mutation of the mGluR5 gene may have disrupted the ability of cocaine to increase DA levels thereby eliminating cocaine-induced increases in locomotor activity and cocaine self-administration in mGluR5 (–/–) mice. Therefore, extracellular levels of DA in the NAc (Fig. 3A) were examined in awake, freely moving mGluR5 (+/+) and (–/–) mice after injection of either cocaine or saline vehicle. Basal levels of extracellular NAc DA did not differ significantly between mGluR5 (–/–) and (+/+) mice. Cocaine induced significant increases in extracellular DA that did not differ between (–/–) and (+/+) mice (Fig. 3B), suggesting that mutation of the mGluR5 gene did not alter baseline or cocaine-induced increases in NAc DA levels *(1)*.

Because DA neurotransmission appears to be critical to the behavioral effects of cocaine *(19,20)*, the effects of mutation of the mGluR5 gene on the expression and brain distribution of DA receptors and the DA transporter (DAT) were examined. The expression and distribution of DA D1-like and D2-like receptors and the DAT and the expression of D1 and D2 DA receptor mRNA did not differ between mGluR5 (+/+) and (–/–) mice *(1)*. These data suggest that mutation of the mGluR5 gene did not alter the expression or distribution of DA receptors or of the DAT.

The data summarized above indicate an important role for mGluR5 in the locomotor and reinforcing effects of cocaine. However, the specific mechanisms of the mGluR5 contribution to the effects of cocaine are not known. Emerging evidence has identified several possible mechanisms through which mGluR5 may influence pathways involved in the effects of cocaine. This evidence may provide important insights into the development of medications to treat cocaine dependence.

As mentioned above, mGluR5 is highly expressed in several brain regions that contribute to the behavioral effects of cocaine, including the NAc, striatum, ventral tegmental area, and cortex *(9)*.

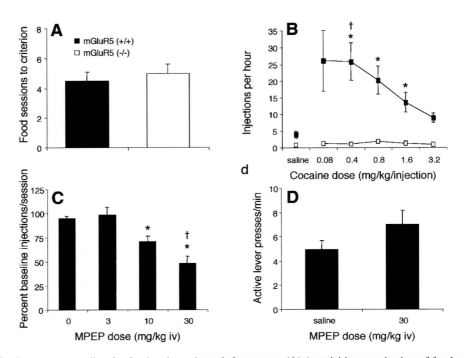

**Fig. 2.** Operant responding for food and cocaine reinforcement. **(A)** Acquisition to criterion of food-reinforced lever pressing in mGluR5 (+/+) ($n = 5$) and (–/–) ($n = 6$) mice. Mice earned 50 food reinforcers (whole milk with sucrose, 60 g/L) under each schedule of reinforcement starting with a fixed-ratio 1 time-out 11-s schedule of reinforcement (FRITO11s) through FRITO15s, FRITO20s, and finally FR2TO20s, identical to the schedule of reinforcement during cocaine self-administration. Mean number of days to criterion (50 reinforcers in 1 h under FR2TO20s) did not differ between genotypes. Values represent mean ± SEM (Student's $t$ test). **(B)** Mice implanted with jugular catheters were allowed to self-administer cocaine to stability. Mice were given access to various doses of cocaine (0.0, 0.08, 0.4, 0.8, 1.6, and 3.2 mg/kg/injection) during single daily 1 h sessions in a Latin square design, and the number of injections at each dose was determined twice during at least two separate sessions. mGluR5 (–/–) mice failed to acquire self-administration, so they were retrained to lever press for food to criterion in between each cocaine dose so that they had a nonzero response rate during the first one or two sessions of access to each dose. Only mice that completed testing at all doses, including saline, were included in the analyses. Values represent mean number of injections per session ± SEM. *, $p < 0.05$ versus saline group with same genotype; †, $p < 0.05$ (+/+) at 0.4 mg/kg/injection dose versus (+/+) at 3.2 mg/kg/injection dose (Student's $t$ tests with Bonferroni correction after two-way repeated measures analysis of variance). **(C)** Effects of MPEP on cocaine self-administration in C57Bl/6J mice ($n = 5$). Values represent percentage of baseline number of injections per 1 h session. MPEP dose-dependently decreased cocaine self-administration, *, $p < 0.05$ compared to saline; †, $p < 0.05$ compared to 3 mg/kg MPEP dose (means comparisons after appropriate one way ANOVA). **(D)** Effects of MPEP (30 mg/kg, iv) on food-reinforced operant responding in C57Bl/6J mice. MPEP had no effect on the number of lever presses/min in C57Bl/6J mice ($n = 5$) responding for food under a FR2TO20s schedule of reinforcement. [Reprinted with permission from Chiamulera et al. (1).]

Furthermore, the distribution of mGluR5 suggests potential roles in learning, memory, motivation, and motor control, processes involved in the complex behaviors that contribute to the development of cocaine dependence. The locomotor and reinforcing effects of cocaine are regulated by both DA and glutamate transmission in the NAc (11,21) and glutamate and DA may act in synergy to mediate the effects of cocaine (1).

Recent evidence has demonstrated interactions between dopamine and metabotropic glutamate receptor intracellular signaling pathways. These interactions may have important implications for the

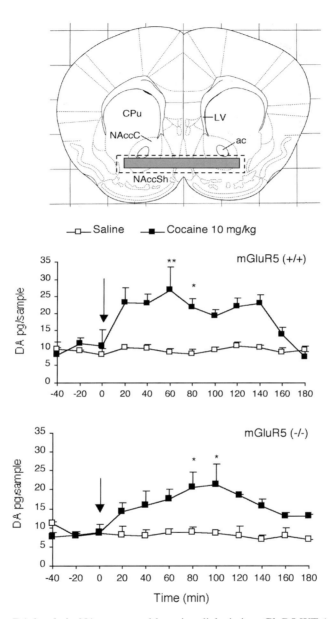

**Fig. 3.** Extracellular DA levels in NAc measured by microdialysis in mGluR5 WT ($n = 3$) and KO ($n = 3$) mice. **(A)** Location of the dialysis probe. Coordinates were calculated to position the probe at the level of the NAc (A: + 1.3 mm, V: –4.6 mm from bregma). The solid and dashed boxes indicate the minimum and maximum extent of the probe placement. Location of the probe was verified in each mouse at the end of the experiment. **(B)** DA levels analysis: 10 μL dialysate samples were collected every 20 min from mGluR5 WT (saline, $n = 3$; cocaine, $n = 3$) and KO mice (saline, $n = 3$; cocaine $n = 3$). Values represent mean pg/sample DA ± SEM. *, $p < 0.05$; **, $p < 0.01$ versus saline at the same timepoint. (Student's $t$ test with Bonferroni correction after two-way repeated measures ANOVA for each genotype). Extracellular DA levels were not significantly different between mGluR5 WT and KO groups (analysis of variance for repeated measures on data of mGluR5 WT and KO groups treated with cocaine: not significant). Abbreviations: ac, anterior commissure; CPu, caudate-putamen; LV, lateral ventricle; NAccC, nucleus accumbens core; NAccSh, nucleus accumbens shell. [Reprinted with permission from Chiamulera et al.*(1)*.]

role of mGluR5 in cocaine dependence and in the development of mGluR agents as therapeutics for the treatment of cocaine addiction.

Incubation of striatal synaptosomes with the Group 1 mGluR agonist (S)-3,5-dihydroxyphenylglycine (DHPG) transiently, but significantly decreased DA uptake through the DAT *(22)*, an effect blocked by the selective mGluR5 antagonist MPEP. This effect appeared to be mediated by mGluR5 activation of protein kinase C and calcium calmodulin-dependent protein kinase II *(22)*. Although the majority of anatomical evidence suggests that mGluR5 is expressed postsynaptically, there is some evidence for localization of mGluR5 in presynaptic axon terminals *(9)*. These results suggest that mGluR5 may interact directly with the DAT to influence DA transmission in the striatum; however, it will be important to evaluate this effect in vivo.

Conflicting evidence exists regarding the interactions between Group 1 mGluR activation and D1-like DA receptor-induced formation of cyclic adenosine monophosphate (cAMP). Evidence from striatal slice preparations showed that D1-like DA receptor agonist-induced cAMP formation was inhibited by the non-selective mGluR agonist ACPD *(23)*, whereas, in cultured striatal neurons, the Group 1 agonist DHPG potentiates cAMP formation induced by the D1-like agonist SKF 38393 *(24)*. Finally, recent data indicate that stimulation of Group I mGluRs with DHPG activates casein kinase 1 (Ck1) and cyclin-dependent kinase 5 (Cdk5), two kinases involved in dopamine signaling in the neostriatum *(25)*. In addition, DHPG treatment in acutely dissociated neurons was shown to increase the activation of voltage-gated $Ca^+$ channels *(25)*. Taken together these results suggest an interaction between Group I mGluRs and DA receptors, although the specific nature of this interaction remains unclear.

The evidence reviewed in this chapter suggests an important and specific role for the Group I metabotropic glutamate receptor subtype mGluR5 in the effects of cocaine. Mice lacking the mGluR5 gene did not self-administer cocaine and showed no alteration in locomotor activity following cocaine treatment. The selective mGluR5 antagonist MPEP significantly reduced cocaine self-administration, but had no effect on operant responding for food, suggesting that antagonism of mGluR5 does not affect the reinforcing effects of natural reinforcers. This evidence suggests that antagonists of mGluR5 may be developed as effective pharmacological treatments for cocaine addiction. Evidence that mGluR5 plays a modulatory role in both glutamate *(3)* and DA receptor signaling *(22–25)* suggests that antagonists of mGluR5 may not induce the aversive side effects associated with direct DA or ionotropic glutamate receptor antagonists. Nevertheless, a significant amount of work remains to be done in order to understand the specific contributions of mGluR5 and the other mGluR subtypes to the effects of cocaine and other drugs of abuse and to the development of addiction.

In conclusion, there are several lines of evidence suggesting that mGluR5 contributes to the effects of cocaine in rodents. mGluR5 is expressed in several brain regions known to contribute to the behavioral effects of cocaine *(9, 10, 14)*. Cocaine did not alter locomotor activity in mice bearing a null mutation of the mGlu5 receptor gene *(1)*. mGluR5 (–/–) mice did not self-administer cocaine, an effect that was not attributable to a learning deficit or to an alteration of brain reward systems *(1)*. Pharmacological blockade of mGluR5 in mice selectively decreases cocaine self-administration *(1)*. Subsequent investigation will lead to a greater understanding of the specific mechanisms through which mGluR5 contributes to the effects of cocaine.

## ACKNOWLEDGMENTS

The figures contained in this manuscript were reprinted from Chiamluera, et al. (2001) *Nature Neuroscience* **4(9)**, 873–874 with permission from the Nature Publishing Group. The author wishes to acknowledge the outstanding contributions of Christian Chiamulera, Alessandro Zocchi, Clara Marcon, Cécilia Cottiny, Stefano Tacconi, Mauro Corsi, Francesco Orzi, and Francois Conquet to the work described here. The author also wishes to thank Egbert Welker, Jean-Pierre Hornung, and GlaxoSmithKline R & D for their support. During the preparation of this manuscript the author was supported, in part, by grant number GEO-204 from the NLS.

## REFERENCES

1. Chiamulera, C., Epping-Jordan, M. P., Zocchi, A., Marcon, C., Cottiny, C., Tacconi, S., Corsi, M., Orzi, F., and Conquet, F. (2001) Reinforcing and locomotor stimulant effects of cocaine are absent in mGluR5 null mutant mice. *Nature Neurosci.* **4,** 873–874.
2. Nakanishi., S. (1992) Molecular diversity of glutamate receptors and implications for brain function. *Science* **258,** 597–603.
3. Conn, J. P. and Pin, J.-P. (1997) Pharmacology and functions of metabotropic glutamate receptors. *Annu. Rev. Pharmacol. Toxicol.* **37,** 205–237.
4. Albin R. L., Makowiec R. L., Hollingsworth Z., Dure L. S., Penney J. B., and Young, A. B. (1992) Excitatory amino acid binding sites in the basal ganglia of the rat: a quantititative autoradiographic study. *Neuroscience* **46,** 35–48.
5. Shigemoto R., Nakanishi S., and Mizuno N. (1992) Distribution of the mRNA for a metabotropic glutamate receptor (mGluR1) in the central nervous system: an in situ hybridization study in adult and developing rat. *J. Comp. Neurol.* **322,** 121–135.
6. Ohishi H., Shigemoto R., Nakanishi S., and Mizuno N. (1993) Distribution of the mRNA for a metabotropic glutamate receptor, mGluR2, in the central nervous system of the rat. *Neuroscience* **53,** 1009–1018.
7. Ohishi H., Shigemoto R., Nakanishi S., and Mizuno N. (1993) Distribution of the mRNA for a metabotropic glutamate receptor (mGluR3) in the rat brain: an in situ hybridization study. *J. Comp. Neurol.* **335,** 252–266.
8. Testa, C. M., Standaert, D. G., Young, A. B., and Penney, J. B. (1994) Metabotropic glutamate receptor mRNA expression in the basal ganglia of the rat. *J. Neurosci.* **14,** 3005–3018.
9. Romano, C., Sesma, M. A., McDonald, C. T., O Malley, K. L., Van den Pol, A. N., and Olney, J. W. (1995) Distribution of metabotropic glutamate receptor mGluR5 immunoreactivity in rat brain. *J. Comp. Neurol.* **355,** 455–469.
10. Tallaksen-Greene, S. J., Kaatz, K. W., Romano, C., and Albin, R. L. (1998) Localization of mGluR1a-like immunoreactivity and mGluR5-like immunoreactivity in identified populations of striatal neurons. *Brain Res.* **780,** 210–217.
11. Vezina, P., and Kim J. H., (1999) Metabotropic glutamate receptors and the generation of locomotor activity: interactions with brain dopamine. *Neurosci. Biobehav. Rev.* **23,** 577–589.
12. Swanson, C. J., and Kalivas, P. W. (2000) Regulation of locomotor activity by metabotropic glutamate receptors in the nucleus accumbens and ventral tegmental area. *J. Pharmacol. Exp. Ther.* **292,** 406–414.
13. Kim, J. H., and Vezina, P. (1998) Metabotropic glutamate receptors in the rat nucleus accumbens contribute to amphetamine-induced locomotion. *J. Pharmacol. Exp. Ther.* **284,** 317–322.
14. Lu, X-Y., Ghasemzadeh, M. B., and Kalivas, P. W. (1999) Expression of glutamate receptor subunit/subtype messenger RNAs for NMDAR1, GluR1, Glur2 and mGluR5 by accumbal projection neurons. *Molecular Brain Res.* **63,** 287–296.
15. Ghasemzadeh, M. B., Nelson, L. C., Lu, X. Y., and Kalivas, P. W. (1999) Neuroadaptations in ionotropic and metabotropic glutamate receptor mRNA produced by cocaine treatment. *J. Neurochem.* **72,** 157–165.
16. Swanson, C. J., Baker, D. A., Carson, D., Worley, P. F., and Kalivas, P. W. (2001) Repeated cocaine attenuates Group I metabotropic glutamate receptor-mediated glutamate release and behavioral activation: A potential role for Homer. *J. Neurosci.* **21,** 9043–9052.
17. Gasparini, F., Lingenhohl, K., Stoehr, N., et al. (1999) 2-Methyl-6-(phenyletynyl)-pyridine (MPEP), a potent, selective, and systemically active mGlu5 receptor antagonist. *Neuropharmacology* **38,** 1493–1503.
18. Robinson, T. E. (1984) Behavioral sensitization: characterization of enduring changes in rotational behavioral produced by intermittent injections of amphetamine in male and female rats. *Psychopharmacology* **84,** 466–475.
19. Koob, G. F., Sanna, P. P., and Bloom, F. E. (1998) Neuroscience of addiction. *Neuron* **21,** 467–476.
20. Zocchi, A., Orsini, C., Cabib, S., and Puglisi-Allegra, S. (1998) Parallel strain-dependent effect of amphetamine on locomotor activity and dopamine release in the nucleus accumbens: an in vivo study in mice. *Neuroscience* **82,** 521–528.
21. White, F. J., and Kalivas, P. W. (1998) Neuroadaptations involved in amphetamine and cocaine addiction. *Drug Alcohol Depend.* **51,** 141–153.
22. Page, G., Peeters, M., Najimi, M., Maloteaux J-M., and Hermans, E. (2001) Modulation of the neuronal dopamine transporter activity by the metabotropic glutamate receptor mGluR5 in rat striatal synaptosomes through phosphorylation mediated processes. *J. Neurochem.* **76,** 1282–1290.
23. Wang, J., and Johnson, K. M. (1995) Regulation of striatal cyclic-3,5-adenosine monophosphate accumulation and GABA release by glutamate metabotropic and dopaminergic receptors. *J. Pharmacol. Exp. Ther.* **275,** 877–884.
24. Paolillo, M., Montecucco, A., Zanassi, P., and Schinelli, S. (1998) Potentiation of dopamine-induced cAMP formation by group I metabotropic glutamate receptors via protein kinase C in cultured striatal neurons. *Eur. J. Neurosci.* **10,** 1937–1945.
25. Liu, F., Ma, X-H., Ule, J., et al. (2001) Regulation of cyclin-dependent kinase 5 and casein kinase 1 by metabotropic glutamate receptors. *Proc. Nat. Acad. Sci. USA* **98,** 11062–11068.

# III Glutamate and Opiate Drugs (Heroin) of Abuse

*Section Editors*

Barbara H. Herman

Jerry Frankenheim

# 19
# Role of the Glutamatergic System in Opioid Tolerance and Dependence
*Effects of NMDA Receptor Antagonists*

### Jianren Mao, MD, PhD

## 1. NMDA RECEPTORS AND OPIOID TOLERANCE/DEPENDENCE

The *N*-methyl-D-asparate (NMDA) receptor is a complex subtype of the glutamatergic receptor system. Glutamate and aspartate are endogenous ligands binding to NMDA receptors. Activation of NMDA receptors can be blocked at either the glutamate-binding site or the regulatory sites. Although the role of NMDA receptors in opioid tolerance and dependence was initially examined using µ-opioid agonists, studies have been carried out to investigate the effects of NMDA receptor antagonists on tolerance and dependence induced by κ- and σ-opioid agonists.

### 1.1. µ-Opioids

#### 1.1.1. Tolerance

Morphine has been used primarily as a µ-opioid agonist in a number of studies investigating µ-opioid-induced tolerance and dependence. Coadministration of morphine with MK-801, a noncompetitive NMDA receptor antagonist, was initially shown to be effective in preventing the development of antinociceptive tolerance in rats, mice, and guinea pigs *(1–4)*. Since these initial reports, a large number of studies have consistently shown the same results *(5–24)*. NMDA receptor antagonists and $Zn^{2+}$ also have been shown to block the development of acute tolerance to morphine *(25)*. The methodologies employed in these studies are rather diversified with regard to the route of drug administration, tolerance-inducing regimens, and the choice of NMDA receptor antagonists (see above-cited references):

1. Morphine has been given via the route of subcutaneous (sc), intravenous (iv), intrathecal (it), intracerebroventricular (icv), intraperitoneal (ip), and oral administration. The treatment regimens have included daily boluses for 5–8 d or continuous infusion through an osmotic pump for 3–7 d.
2. The behavioral tests employed in these studies have included the hot-plate, tail-flick, and formalin tests. Different species of laboratory animals (mice, rats, guinea pigs, etc.) have been used in these studies and no reliable differences in species were observed with regard to the effect of NMDA receptor antagonists on morphine tolerance.
3. The NMDA receptor antagonists used in these studies have included the competitive agents AP-5, LY235959, CGP 39551, and LY274614, the noncompetitive channel blockers MK-801, dextromethorphan (DM), MRZ 2/579, ketamine, *d*-methadone, and memantine, and the glycine-binding-site NMDA receptor antagonists MRZ 2/576, ACEA-1328, and HA966. NMDA receptor antagonists were given via the sc, it,

and ip routes. There were no reliable differences between competitive and noncompetitive NMDA receptor antagonists in preventing the development of morphine tolerance. Collectively, these data indicate a broad generality of the NMDA receptor involvement in the development of morphine tolerance across species, behavioral tests, and tolerance-inducing regimens.

It is of significance to point out that low-affinity, clinically available NMDA receptor antagonists such as DM, *d*-methadone, and memantine may have a role in preventing the development of tolerance to and dependence on opioid analgesics in clinical settings *(6,12,19,26,27)*. For instance, DM, commonly known as an antitussive drug, has been shown to be effective in preventing the development of opioid tolerance *(6,12,27)*. In a rat model of morphine tolerance, DM prevented or attenuated the development of tolerance to the antinociceptive effects of morphine (15,24, or 32 mg/kg) when DM was coadministered orally with morphine (ratios from 4 : 1 to 1 : 2). This combined oral treatment regimen also reduced naloxone-precipitated signs (teeth chattering, wet-dog shaking, or jumping) of physical dependence on morphine in the same rats. The data reveal a constant ratio range of the morphine/DM combination effective for preventing the development of morphine tolerance and dependence *(12)*. These results indicate that the combined treatment with clinically available NMDA receptor antagonists and morphine may be a useful approach for preventing morphine tolerance and dependence in humans.

*1.1.2. Selective μ-Opioid Agonists*

Although a prototypical μ-agonist, morphine does have interactions with other opioid receptor subtypes. In a recent study, the role of NMDA receptors in the antinociceptive tolerance induced by highly selective μ-opioids was examined *(5)*. It was reported that 0.1 mg/kg MK-801 given intraperitoncally did not prevent tolerance induced by the highly selective μ-opioid agonists DAMGO and fentanyl in mice, although the same dose of MK-801 prevented the development of tolerance to morphine in the same study. Because autoradiographic studies have shown that morphine also binds to δ- and κ-opioid receptors *(28–30)*, these results could imply that interactions among subtypes of opioid receptors would be important in determining the involvement of NMDA receptors in mechanisms of opioid tolerance. Although in another study the development of tolerance resulting from repeated it administration of either 6 μg or 1.5 μg DAMGO was prevented dose dependently by the it coadministration of DAMGO and MK801 *(31)*, it would be of interest to interest to further elucidate similarities and differences between the NMDA receptor-mediated mechanisms of tolerance induced by highly selective μ-opioids or morphine.

*1.1.3. Dependence*

The noncompetitive NMDA receptor antagonists (MK-801, memantine, DM, HA966) have been shown to prevent the development of dependence on morphine as assessed by naloxone-precipitated withdrawal signs (jumping, teeth chattering, diarrhea, wet-dog shakes, vocalization) in both mice and rats *(6,9,12–14,32,33)*. Consistent with these findings, it has been shown that the occurrence of naloxone-precipitated withdrawal signs in tolerant rats was associated with spinal cord release of glutamate *(9)*. In addition, daily transient blockade of morphine with naloxone resulted in greater tolerance than that from continuous chronic morphine administration, presumably due to an increased release of glutamate in the naloxone treatment group *(34)*. It would be of interest to examine differences in glutamate release, in the absence of naloxone precipitation, in animal models of opioid tolerance using daily bolus treatment versus continuous infusion with osmotic pumps.

*1.1.4. Other Issues*

Tolerance may involve both associative (constant presence of a cue) and nonassociative (lack of a cue) components. The role of NMDA receptors in associative versus nonassociative tolerance to morphine was examined *(4,8)*. The blockade of NMDA receptors with MK-801 was particularly effective for preventing nonassociative tolerance as opposed to associative tolerance. In contrast, spinal cord neurotensin appeared to be contributory to the development of associative tolerance *(8)*. Although mechanisms of such a distinction remain to be determined, these data indicate the importance of distinguishing these two processes in investigating the NMDA receptor-mediated mechanisms of opioid tolerance.

Although there is ample evidence indicating a critical role of NMDA receptors in the development of μ-opioid tolerance, the expression of tolerance as assessed by a behavioral test is not determined by the activity of NMDA receptors. The acute blockade of NMDA receptors with either MK-801 or LY274614 in animals already made tolerant to morphine did not reverse the tolerance status *(1,16)*. The shifted dose-response relationship in morphine-tolerant rats appears to be directly related to the status of opioid receptors, because naloxone, but not MK-801, altered the antinociceptive response in these rats *(11)*. However, NMDA receptors are contributory to the maintenance of an established tolerance status, because coadministration of morphine with a competitive NMDA receptor antagonist, LY274614, gradually (over days) reversed morphine tolerance in mice *(16)*. These concepts support a combined use of opioids and NMDA receptor antagonists in clinical settings even in those patients who are already tolerant to opioid analgesics.

### *1.2. δ-Opioids*

The data on the effects of NMDA receptor activation on δ-opioid tolerance remain inconclusive. In a mouse model of tolerance induced by repeated icv administration of 20 nmol DELT II (δ-2 agonist) twice daily for 3 d, neither MK-801 (0.1 mg/kg, ip) nor LY235959 (3 mg/kg) pretreatment before each DELT II dose prevented the development of tolerance *(5)*. MK-801 in the dose used in that study also failed to prevent tolerance to antinociception induced by cold-water swim stress, a process presumably mediated by endogenous δ-opioids *(5)*.

In direct contrast, in a similar experimental paradigm in which DELT II (20 μg) was given intracerebroventricularly twice daily for 4 d, MK-801 (0.03 and 0.1 mg/kg, ip) or LY235959 (4 mg/kg, ip) pretreatment effectively inhibited the development of antinociceptive tolerance to DELT II *(35)*. In addition, MK-801 (0.1 mg/kg, ip) or LY235959 (1, 2, or 4 mg/kg, ip) prevented the antinociceptive tolerance induced by twice daily icv administration of the δ-1 agonist DPDPE (20 μg) in mice *(36)*. The inhibition of DPDPE-induced tolerance also was observed when 1-aminocyclopropane carboxylic acid (ACPC) (150 mg/kg, sc, a competitive NMDA receptor antagonist) was coadministered with DPDPE (0.5 μg, it) in mice *(37)*. Collectively, it appears that antagonism of NMDA receptors may prevent the development of tolerance to selective δ-opioid agonists.

### *1.3. κ-Opioids*

Similar to δ-opioid-induced tolerance, mixed results have been reported concerning the role of NMDA receptors in κ-opioid tolerance. Coadministration with the κ-agonist U-50488H (5 mg/kg, sc, once daily for 5 d) of MK-801 (0.3 mg/kg, ip) or LY274614 (6 mg/kg, ip, or 24 mg/kg/24 h pump infusion) did not prevent the development of antinociceptive tolerance to U-50488H as assessed by the tail-flick test *(38)*. Likewise, ACPC (150 mg/kg, sc) also failed to prevent tolerance induced by repeated administration of U-50488H (5 mg/kg, sc) *(37)*. In addition, MK-801, LY274614, or ACPC in the doses described was ineffective in preventing the development of tolerance to the κ-3 agonist naloxone benzoylhydrazone (50 mg/kg, sc) given once daily for 5 d *(37,38)*.

In separate studies, MK-801 (0.01–0.3 mg/kg, ip) prevented the development of the antinociceptive tolerance induced by twice daily ip injection of 25 mg/kg U-50488H (4 d for rats and 9 d for mice) using the tail-flick test in both mice and rats *(39)*. One confounding factor to the assessment of these data is that κ-opioids may interact with both NMDA and κ-opioid receptors *(40,41)*. It is possible that the mixed results from these studies may be, in part, the dual functions of κ-opioids interacting with both NMDA and κ-opioid receptors. Little has been known with regard to the role of NMDA receptors in the development of dependence on δ- or κ-opioids.

## 1.4. Summary

It has been demonstrated in numerous studies that NMDA receptors play a significant role in tolerance to and dependence on opioid antinociceptive effects. Several important points may be drawn from these studies: (1) Coadministration of NMDA receptor antagonists with morphine prevents tolerance and dependence, indicating that NMDA receptors are crucial for the development of both tolerance to and dependence on morphine. (2) Although repeated application of NMDA receptor antagonists reverses tolerance over time, a single treatment with an NMDA receptor antagonist does not restore the antinociceptive effects of opioids in tolerant animals. Thus, activation of NMDA receptors is not required for the expression of opioid tolerance. (3) Except for morphine-induced tolerance and dependence, in which studies consistently show the effectiveness of NMDA receptor antagonists, the role of NMDA receptors in tolerance induced by selective opioid agonists (particularly δ- and κ-opioids) remains controversial. (4) Clinically available agents with NMDA receptor antagonist properties (such as DM, D-methadone, memantine) are generally as effective as MK-801 in preventing tolerance and dependence in preclinical trials.

## 2. ROLE OF INTRACELLULAR PROTEIN KINASES

Activation of NMDA receptors initiates intracellular processes that may lead to neuroplastic changes. A common feature of NMDA receptor activation is an increase in intracellular $Ca^{2+}$ concentration, which, in turn, initiates a number of second/third-messenger-mediated intracellular processes. One such process is the redistribution (translocation from cytosolic to membrane-bound form) and activation of the $Ca^{2+}$-sensitive protein kinase C (PKC). It is known that $Ca^{2+}$-regulated PKC translocation and activation are associated with a variety of central nervous system (CNS) functional changes that occur by means of protein phosphorylation. The following sections will briefly discuss the role of NMDA receptor-mediated activation of protein kinases, particularly PKC, in the development of opioid tolerance.

### 2.1. Behavioral Evidence

A growing body of evidence indicates an important role of protein kinases, particularly PKC, in the development of antinociceptive tolerance *(42–46)*. It has been shown that antinociceptive tolerance to morphine, butorphanol, DAMGO, or DELT II can be prevented by coadministration with opioids of the nonselective protein kinase inhibitor H7 (morphine, butorphanol) *(46,47)*, the PKC translocation blocker GM1 ganglioside (morphine) *(42,44)*, or the selective PKC inhibitor calphostin C (DELT II, DAMGO) *(45)*. Inhibition of protein kinase A (PKA) with KT5720 has not been effective in preventing the antinociceptive tolerance to DAMGO or DELT II *(45)*. Additional evidence for the involvement of PKC in DELT II-mediated antinociception is that it administration of phorbol 12,13-dibutyrate (PDBu), a PKC activator, produced calphostin C-reversible attenuation of antinociception induced by it DELT II, whereas PDBu alone had no effect on the nociceptive threshold in the mouse tail-flick test *(45)*.

### 2.2. Electorphysiological Evidence

Further evidence for the intracellular modulation of μ-opioid receptor desensitization through the calcium/calmodulin-dependent kinase and PKC was elegantly demonstrated in a recent study using a human μ-opioid receptor cDNA *(48)*. In that study, both μ-opioid receptor (encoded by a cDNA from human μ-opioid receptors) and a cloned G-protein-activated $K^+$ channel (displaying coupling to the μ-opioid receptor) were coexpressed in *Xenopus* oocytes. The verification of μ-opioid receptors was confirmed by binding to selective μ-opioid agonists and antagonists. Under these conditions, functional desensitization of μ-opioid receptors, as reflected by reduced $K^+$ currents following repeated exposure to μ-opioids, was potentiated by both the calcium/calmodulin-dependent kinase and PKC *(48)*. These results provide convincing evidence for the protein-kinase-mediated modulation of functional μ-opioid receptors. Interestingly, a recent study has shown that activation of PKC also decreases μ-opioid receptor mRNA levels *(49)*, suggesting that PKC may also have a role in regulating the μ-opioid receptor turnover. The functional importance of the PKC-mediated μ-opioid receptor turnover in mechanisms of opioid tolerance is yet to be determined.

### 2.3. Autoradiographic and Immunocytochemical Evidence

Spinal cord levels of membrane-bound PKC increase reliably as morphine tolerance develops *(42)*. This increase in membrane-bound PKC occurs mainly within laminae I–II of the spinal cord dorsal horn, a region showing increased levels of PKC translocation in nerve-injured animals with demonstrable thermal hyperalgesia *(50,51)*. This tolerance-associated increase in membrane-bound PKC was reduced by it treatment with the PKC translocation blocker GM1 ganglioside *(42)*. In another study, however, only cytosolic PKC, but not membrane-bound PKC, was upregulated in the pons and medulla following chronic morphine administration *(52)*, although the affinity of PDBu binding to membrane-bound PKC was increased in the brain under the same condition *(47)*.

More evidence for PKC changes in morphine-tolerant rats was provided by a recent study utilizing an immunocytochemical method. In that study, the development of morphine tolerance was shown to be associated with increases in immunoreactivity of a PKC isoform (PKCγ) in laminae I–II dorsal horn neurons *(43)*. Such increases in PKCγ immunoreactivity along with behavioral manifestations of morphine tolerance were prevented by it administration of MK-801 *(43)*. Increases in PKCγ immunoreactivity in similar spinal cord dorsal horn regions also were seen in rats with nerve-injury-induced thermal hyperalgesia, further suggesting the involvement of common regions of the spinal cord dorsal horn in mechanisms of hyperalgesia and morphine tolerance. In contrast, spinal cord PKC levels did not change in saline control rats receiving a single injection of morphine *(43)*, suggesting that activation of NMDA receptors and subsequent intracellular PKC within the spinal cord reflects neuroplastic changes following repeated exposure to opioids.

It should be noted that a different mechanism of μ-opioid receptor desensitization involving G-protein-coupled receptor kinases, but not PKC, also has been proposed *(53–55)*. It has been shown that β-adrenergic receptor kinase 2 and β-arrestin 2, both of which are G-protein-coupled receptor kinases, synergistically desensitize homologous μ-opioid as well as δ-opioid receptors *(54)*. It will be of interest to investigate how different intracellular processes converge to influence the cellular and molecular mechanismas of opioid tolerance.

## 3. MECHANISMS OF NMDA RECEPTOR-MEDIATED OPIOID TOLERANCE

### 3.1. Critical Issues Regarding NMDA Receptors and Opioid Tolerance

A critical issue concerning NMDA receptors and the development of opioid tolerance is how exogenous opioid administration leads to the activation of NMDA receptors. This remains the core issue for understanding the involvement of NMDA receptors in the cellular and molecular mechanisms of opioid tolerance and dependence. Several points should be considered in view of the data discussed above. First, in a well-controlled experimental model of opioid tolerance, the primary variable is the prolonged exposure of opioid receptors to an opioid agonist. Yet, NMDA receptors are critically recruited during the process, leading to the development of opioid tolerance, and there is no reliable evidence suggesting that opioids may directly activate the NMDA receptor. Second, NMDA receptor activation is not required for the expression of tolerance, nor does the acute blockade of NMDA receptors reverse an established tolerance status. Third, the NMDA receptor is a unique voltage/ligand-gated receptor and its activation requires the removal of the $Mg^{2+}$ blockade following a partial membrane depolarization. It appears difficult to envision that the NMDA receptor would be activated following exposure to opioids given the overwhelming hyperpolarization induced by opioids under most circumstances. Fourth, NMDA receptor-mediated intracellular events such as PKC activation are crucial for the development of opioid tolerance, as inhibition of PKC activation effectively prevents the development of opioid tolerance. Thus, a cellular model of NMDA receptor-mediated opioid tolerance should include these important experimental observations.

Although in an intact system the involvement of a neural circuitry cannot be ruled out in mechanisms of NMDA receptor-mediated opioid tolerance, studies do support interactions between NMDA and opioid receptors at a single-cell level *(56,57)*. In a study utilizing an in vitro trigeminal dorsal horn neuron preparation, NMDA receptor-mediated inward membrane current (depolarization) was initially induced by glutamate *(56)*. The magnitude of this inward membrane current was paradoxically enhanced by the addition of exogenous μ-opioid agonists. A critical intracellular component that mediates this μ-opioid effect must be PKC, because activating and inhibiting PKC, respectively, facilitated and blocked the enhancement by μ-opioid agonists of this NMDA-mediated inward membrane current *(56)*. This PKC effect was subsequently shown to result from a PKC-mediated removal of the $Mg^{2+}$ blockade of the NMDA receptor *(57)*, making it possible that the NMDA receptor could be activated even under overwhelming inhibitory opioid actions. A necessary condition for this to happen is that NMDA and (μ)-opioid receptors are colocalized in a single neuron. Indeed, such a colocalization has been shown in both supraspinal regions *(58–60)* and the spinal cord dorsal horn *(61)*. These lines of evidence are the building blocks for a cellular model of μ-opioid tolerance shown next.

### 3.2. A Spinal Cord Model of μ-Opioid Tolerance

Given the interactions between μ-opioid and NMDA receptors and the behavioral evidence indicating a role of NMDA receptors, PKC, and nitric oxide (NO) in μ-opioid tolerance, we previously proposed that the development of tolerance to the analgesic effects of μ-opioids is a consequence of a series of cellular and intracellular events initiated by opioid administration, and at least one central locus of such action is in the superficial laminae of the spinal cord dorsal horn *(42–44,62,63)*. The following discussion is based on the frame of a cellular model involving interactions between NMDA and μ-opioid receptors *(42,44,62)*.

As shown in Fig. 1, μ-opioid receptor occupation by an exogenous ligand such as morphine may initiate second-messenger (G-protein)-mediated PKC translocation and activation. The involvement of a second messenger is suggested by the observation that PKC-mediated NMDA receptor sensitization takes about 2–4 min to occur after the addition of exogenous μ-opioid agonists in an in vitro preparation *(56)*. PKC activation removes the $Mg^{2+}$ blockade from the NMDA receptor–channel complex via PKC-mediated phosphorylation of receptor–channel sites (57). With this blockade removed, even the physiological level of endogenous NMDA receptor ligands may activate the NMDA receptor *(64)* and allow a localized NMDA receptor/$Ca^{2+}$ channel opening leading to an increase in intracellular $Ca^{2+}$ concentration. An elevation of the intracellular $Ca^{2+}$ level may result in the activation of additional PKC, production of NO via $Ca^{2+}$ calmodulin-mediated activation of NO synthase, and/or regulation of gene expression via MAP kinases (MAPKs). Although PKC can be activated following repeated μ-opioid activation, the NMDA receptor activation is required in the process to ensure sufficient and enduring activation of PKC, via $Ca^{2+}$ actions and/or *de novo* PKC production and other intracellular molecules.

PKC may then modulate the μ-opioid-activated G-protein-coupled $K^+$ channel or uncouple G-proteins from the μ-opioid receptor, resulting in decreased responsiveness of μ-opioid receptors and behavioral manifestations of opioid tolerance. Changes in the responsiveness of opioid receptors and the dissociation between G-proteins and opioid receptors have already been suggested to be a contributory factor to the cellular mechanisms of opioid tolerance *(26,65–69)*. This hypothesis is in agreement with previous autoradiographic data showing that morphine tolerance develops without concurrent downregulation (decreases in receptor numbers and/or binding affinity) of μ-opioid receptors *(70,71)* and is supported by observations that protein kinases can indeed modulate the μ-opioid

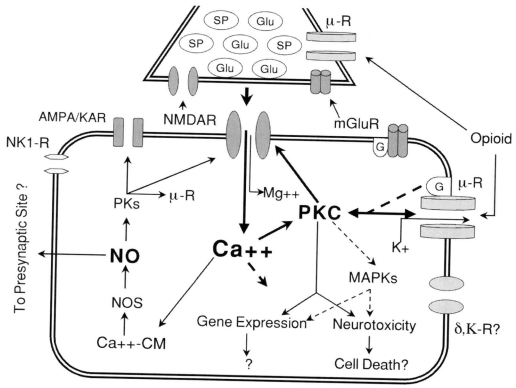

**Fig. 1.** A spinal cord model of μ-opioid tolerance. Postsynaptic opioid (μ) receptor occupation by an exogenous ligand such as morphine may initiate G-protein-mediated PKC translocation and activation. PKC translocation/activation facilitates the removal of the $Mg^{2+}$ blockade from NMDA receptors. With this blockade removed, the NMDA receptor could be activated in the absence of excessive release of glutamate from presynaptic sources. An elevation of the intracellular $Ca^{2+}$ level leads to the activation of additional PKC and production of NO via $Ca^{2+}$ calmodulin-mediated activation of NO synthase. PKC may modulate the μ-opioid-activated G-protein-coupled $K^+$ channel or attenuate G-protein activation. PKC may also regulate gene expression and contribute to the NMDA receptor-mediated neurotoxic process, either directly or indirectly via MAPKs. In addition, NO may activate various protein kinases by means of increasing cGMP and thus participate in the modulation of the μ-opioid-activated G-protein-coupled $K^+$ channel within the same cell. NO could also diffuse out of the neuron that produces it, thereby enhancing presynaptic release of glutamate/aspartate resulting in a positive feedback; that is, opioids may increase the basal level of presynaptic glutamate release via the NO mechanism initiated by the postsynaptic opioid action. The role of presynaptic μ-opioid, NMDA and metabotropic glutamate receptors as well as postsynaptic NK-1, non-NMDA receptors, and metabotropic glutanmate receptors in the development of opioid tolerance remains to be elucidated. Because many of the intracellular steps following activation of the NMDA receptor in this proposed model are similar to those that occur following nerve-injury-induced hyperalgesia, it has been shown that NMDA receptor-mediated intracellular changes initiated by peripheral nerve injury may lead to both the development of hyperalgesia and the reduced effectiveness of opioid antinociception, mimicking the status of pharmacological tolerance. Collectively, this model shows interactions between the cellular and molecular mechanisms of opioid tolerance and hyperalgesia.

receptor responsiveness *(45,46,48,72)*. In particular, PKC has been shown to facilitate the *desensitization* of μ-opioid receptor coupling through a G-protein to an inwardly rectifying $K^+$ channel *(73)*. More recently, a PKC isoform, PKCγ, has been shown convincingly to attenuate μ-opioid receptor-mediated G-protein activation following chronic administration of DAMGO, further indicating a critical role of the NMDA receptor/PKC pathway in the cellular mechanisms of μ-opioid tolerance *(74)*.

It is conceivable that NO may participate in the modulation of the μ-opioid-activated G-protein-coupled K⁺ channel within the same cell by activating various cGMP-associated protein kinases *(75)*. NO may also diffuse out of the neuron that produces it and influence the presynaptic glutamate/aspartate release *(76)*, resulting in a positive feedback. Such NO actions may then counteract presynaptic inhibitory effects of opioids on neurotransmitter release, further diminishing the analgesic effects of μ-opioids. In addition, spinal cord endogenous opioids and adrenergic agonists could be released via the action of extracellular NO *(77)* and their role in mechanisms of μ-opioid tolerance remains to be determined. Further, the downstream activation of intracellular pathways such as MAPKs may result in translational and posttranscriptional modifications leading to prolonged neuroplastic changes. Perhaps more importantly, neurotoxic consequences may occur following prolonged and excessive activation of NMDA receptors and the associated intracellular pathways, causing potentially irreversible changes within the central nervous system *(78–80)*.

### 3.3. Limitations of the Proposed Cellular Model

Although much effort has been made to formulate a cellular model that may incorporate the experimental data accumulated over the last decade *(44,62,81–83)*, it becomes clear that the cellular and molecular mechanisms of NMDA receptor-mediated opioid tolerance and dependence are much more complex across subtypes of opioid receptors. It should be noted that although interactions between postsynaptic μ-opioid and NMDA receptors are emphasized in this model, this consideration does not exclude a possible role of presynaptic glutamate receptors in the spinal cord mechanisms of μ-opioid tolerance. Activation of presynaptic glutamate receptors (the metabotropic type) has been shown to enhance glutamate release from cerebrocortical nerve terminals *(84)*. Indeed, metabotropic glutamate receptors (particularly type II/III) and their interactions with δ-opioid receptors have been suggested to be involved in the mechanisms of NMDA receptor-mediated μ-opioid tolerance primarily for the initiation of PKC activation following opioid administration *(81)*.

In addition, several studies utilizing a cellular model (in the locus coeruleus) of morphine tolerance have implicated the upregulation of the cAMP–PKA system in mechanisms of morphine tolerance *(68,85)*. It will be of interest to determine whether similar PKA upregulation and their interactions with PKC occur at the spinal cord level of tolerant animals. Similarly, investigations on the role of NMDA receptors in supraspinal mechanisms of μ-opioid tolerance would be expected to provide valuable information regarding the generality of the NMDA receptor-mediated cellular and intracellular mechanisms in μ-opioid tolerance. Finally, the involvement of NMDA receptors in mechanisms of δ- and κ-opioid tolerance remains to be elucidated.

## 4. INTERACTIONS BETWEEN NMDA RECEPTOR-MEDIATED MECHANISMS OF OPIOID TOLERANCE AND NEUROPATHIC PAIN

### 4.1. Opioid Tolerance and Neuropathic Pain

As discussed above, NMDA receptor-mediated cellular and intracellular changes occur within the rat's spinal cord following repeated exposure to opioids. Similar cellular and intracellular changes have been observed at the spinal cord level in animal models of neuropathic pain. Several lines of evidence suggest that these NMDA receptor-mediated changes are the neural basis of spinal cord neuroplastic changes responsible for the behavioral manifestations of both opioid tolerance and hyperalgesia (a sign of neuropathic pain) *(62)*. Because these changes occur at the same spinal cord loci, it is possible that interactions may occur between NMDA receptor-mediated changes following repeated opioid exposure and neuropathic pain; that is, opioids may exacerbate hyperalgesia associated with at least certain types of neuropathic pain and lead to hyperalgesia even in the absence of pre-existing neuropathic pain. A corollary of this is that neuroplastic changes from neuropathic pain may, under some circumstances, reduce the response to opioid analgesics.

Because NMDA receptors are critical for the development of both hyperalgesia and μ-opioid tolerance, conditions such as peripheral nerve injury leading to the development of hyperalgesia have been shown to decrease the antinociceptive effectiveness of opioids in the absence of repeated exposure to opioids *(86–88)*. The reduced opioid analgesia mimics pharmacological tolerance, thereby significantly hampering the ability of opioids to treat neuropathic pain. Conversely, because repeated exposure to opioids, a process leading to the development of pharmacological tolerance, involves activation of NMDA receptors, NMDA receptor-mediated hyperalgesia does, indeed, occur in association with the development of μ-opioid tolerance *(44,89,90)*, further reducing the effectiveness of opioid analgesics. In this regard, repeated opioid administration would not only result in the development of tolerance (a desensitization process) and also lead to activation of the pain facilitatory process (a sensitization process). Thus, interactions between opioid and NMDA receptors may play a significant role in neuroplastic changes following repeated opioid exposure and neuropathic pain.

## *4.2. Clinical Implications*

The interactions between neuroplastic changes following repeated opioid exposure and neuropathic pain have clear clinical implications. For example, it is a common clinical experience that opioids are less effective and often unreliable for treating neuropathic pain such as that resulting from peripheral nerve injury. It is conceivable that the diversity of clinical response patterns to opioid therapy in neuropathic pain patients may result from various degrees of CNS neuroplastic changes initiated by conditions of neuropathic pain. Such neuroplastic changes may underlie the development of neuropathic pain syndromes and also result in a reduction of the antinociceptive effects of opioids.

On the other hand, repeated treatment with opioids could lead to changes within the spinal cord through interactions between opioid and NMDA receptors, mimicking the condition of neuropathic pain following nerve injury. Apparently, a common factor in both directions is the activation of NMDA receptors. This concept is the basis for recommending a combined use of opioids and clinically available NMDA receptor antagonists, because these two classes of agents would complement each other in a well-balanced treatment regimen *(62)*. Importantly, such a strategy should be integrated into treatment regimens both for managing chronic pain syndromes and preventing an evolving pain condition such as that after nerve injury *(83,91)*.

Our understanding of the cellular and molecular mechanisms of opioid tolerance has been significantly advanced over the last decade. Evidence suggests that mechanisms of opioid tolerance have much in common with NMDA receptor-mediated neuroplastic changes that are associated with many types of substance abuse and neurological disorder. Although details of such mechanisms remain to be elaborated, it can be anticipated that studies on the NMDA receptor-mediated cellular and molecular mechanisms of opioid tolerance and their interactions with those of neuropathic pain will lead to improvement in opioid therapy for pain management.

## ACKNOWLEDGMENTS

This work is supported by NIH RO1 grant DA08835. The author thanks Dr. David J. Mayer and Dr. Donald D. Price for their invaluable input in a number of studies described in this chapter.

## REFERENCES

1. Trujillo, K. A. and Akil, H. (1991) Inhibition of morphine tolerance and dependence by the NMDA receptor antagonist MK-801. *Science* **251**, 85–87.
2. Marek, P., Ben Eliyahu, S., Vaccarino, A. L., and Liebeskind, J. C. (1991) Delayed application of MK-801 attenuates development of morphine tolerance in rats. *Brain Res.* **558**, 163–165.
3. Marek, P., Ben Eliyahu, S., Gold, M., and Liebeskind, J. C. (1991) Excitatory amino acid antagonists (kynurenic acid and MK-801) attenuate the development of morphine tolerance in the rat. *Brain Res.* **547**, 77–81.
4. Ben Eliyahu, S., Marek, P., Vaccarino, A. L., Mogil, J. S., Sternberg, W. F., and Liebeskind, J. C. (1992) The NMDA receptor antagonist MK-801 prevents long-lasting non-associative morphine tolerance in the rat. *Brain Res.* **575**, 304–308.

5. Bilsky, E. J., Inturrisi, C. E., Sadee, W., Hruby, V. J., and Porreca, F. (1996) Competitive and non-competitive NMDA antagonist block the development of antinociceptive tolerance to morphine, but not to selective mu or delta opioid agonists in mice. *Pain* **68,** 229–237.
6. Elliott, K. J., Hynansky, A., and Inturrisi, C. E. (1994) Dextromethorphan attenuates and reverses morphine tolerance. *Pain* **59,** 361–368.
7. Wong, C. S., Hsu, M. M., Chou, Y. Y., Tao, P. L., and Tung, C. S. (2000) Morphine tolerance increases [3H]MK-801 binding affinity and constitutive neuronal nitric oxide synthase expression in the rat spinal cord. *Br. J. Anaesth.* **85,** 587–591.
8. Grisel, J. E., Watkins, L. R., and Maier, S. F. (1996) Associative and non-associative mechanisms of morphine analgesic tolerance are neurochemically distinct in the rat spinal cord. *Psychopharmacology (Berl.)* **128,** 248–255.
9. Jhamandas, K. H., Marsala, M., Ibuki, T., and Yaksh, T. L. (1996) Spinal amino acid release and precipitated withdrawal in rats chronically infused with spinal morphine. *J. Neurosci.* **16,** 2758–2766.
10. Lutfy, K., Shen, K. Z., Woodward, R. M., and Weber, E. (1996) Inhibition of morphine tolerance by NMDA receptor antagonists in the formalin test. *Brain Res.* **731,** 171–181.
11. Manning, B. H., Mao, J., Frenk, H., Price, D. D., and Mayer, D. J. (1996) Continuous co-administration of dextromethorphan or MK-801 with morphine: attenuation of morphine dependence and nalixone-reversible attenuation of morphine tolerance. *Pain* **67,** 79–80.
12. Mao, J., Price, D. D., Caruso, F. S., and Mayer, D. J. (1996) Oral administration of dextromethorphan prevents the development of morphine tolerance and dependence in rats. *Pain* **67,** 361–368.
13. Popik, P. and Skolnick, P. (1996) The NMDA antagonist memantine blocks the expression and maintenance of morphine dependence. *Pharmacol. Biochem. Behav.* **53,** 791–797.
14. Tokuyama, S., Wakabayashi, H., and Ho, I. K. (1996) Direct evidence for a role of glutamate in the expression of the opioid withdrawal syndrome. *Eur. J. Pharamacol.* **295,** 123–129.
15. Wong, C. S., Cherng, C. H., Luk, H. N., Ho, S. T., and Tung, C. S. (1996) Effects of NMDA receptor antagonists on inhibition of morphine tolerance in rats: binding at mu-opioid receptors. *Eur. J. Pharmacol.* **297,** 27–33.
16. Tiseo, P. J. and Inturrisi, C. E. (1993) Attenuation and reversal of morphine tolerance by the competitive $N$-methyl-D-aspartate receptor antagonist, LY274614. *J. Pharmacol. Exp. Ther.* **264,** 1090–1096.
17. Lufty, K., Shen, K. Z., Kwon, I. S., Cai, S. X., Woodward, R. M., Keana, J. F., et al. (1995) Blockade of morphine tolerance by ACEA-1328, a novel NMDA receptor/glycine site antagonist. *Eur. J. Pharmacol.* **273,** 187–189.
18. Gonzalez, P., Cabello, P., Germany, A., Norris, B., and Contreras, E. (1997) Decrease of tolerance to, and physical dependence on morphine by, glutamate receptor antagonists. *Eur. J. Pharmacol.* **332,** 257–262.
19. Davis, A. M., and Inturrisi, C. E. (1999) $d$-Methadone blocks morphine tolerance and $N$-methyl-D-aspartate-induced hyperalgesia. *J. Pharmacol. Exp. Ther.* **289,** 1048–1053.
20. Lutfy, K., Doan, P., and Weber E. (1999) ACEA-128, an NMDA receptor/glycine site antagonist, acutely potentiates antinociciption and chronically attenuates tolerance induced by morphine. *Pharmacol. Res.* **40,** 435–442.
21. Christensen, D., Guilbaud, G., and Kayser, V. (2000) Complete prevention but stimulus-dependent reversion of morphine tolerance by the glycine/NMDA receptor antagonist (+)-HA966 in neuropathic rats. *Anesthesiology* **92,** 786–794.
22. Belozertseva, I. V., Danysz, W., and Bespalov, A. Y. (2000) Effects of a short-acting NMDA receptor antagonist MRZ 2/576 on morphine tolerance and development in mice. *Naunyn-Schmiedebergs Arch. Pharmacol.* **361,** 573–577.
23. Houghton, A. K., Parsons, C. G., and Headley, P. M. (2001) Mrz 2/579, a fast kinetic NMDA channel blocker, reduces the development of morphine tolerance in awake rats. *Pain* **91,** 201–207.
24. Allen, R. M. and Dykstra, L. A. (2000) Attenuation of mu-opioid tolerance and cross-tolerance by the competitive $N$-methyl-D-aspartate receptor antagonist LY235959 is related to tolerance and cross-tolerance magnitude. *J. Pharmacol. Exp. Ther.* **295,** 1012–1021.
25. Larson, A. A., Kovacs, K. J., and Spartz, A. K. (2000) Intrathecal $Zn^{2+}$ attenuates morphine antinociception and the development of acute tolerance. *Eur. J. Pharmacol.* **407,** 267–272.
26. Popik, P. and Kozela, E. (1999) Clinically available NMDA antagonist, memantine, attenuates tolerance to analgesic effects of morphine in a mouse tail flick test. *Pol. J. Pharmacol.* **51,** 223–231.
27. Popik, P., Kozela, E., and Danysz, W. (2000) Clinically available NMDA receptor antagonists memantine and dextromethorphan reverse existing tolerance to the antinociceptive effects of morphine in mice. *Naunyn-Schmiedebergs Arch. Pharmacol.* **361,** 425–432.
28. Chen, Z. R., Irvine, R. J., Somogyi, A. A., and Bochner, F. (1991) Mu receptor binding of some commonly used opioids and their metabolites. *Life Sci.* **48,** 2165–2171.
29. Emmerson, P. J., Liu, M. R., Woods, J. H., and Medzihradsky, F. (1994) Binding affinity and selectivity of opioids at mu, delta and kappa receptors in monkey brain membranes. *J. Pharmacol. Exp. Ther.* **271,** 1630–1637.
30. Mignat, C., Wille, U., and Ziegler, A. (1995) Affinity profiles of morphine, codeine, dihydrocodeine and their glucuronides at opioid receptor subtypes. *Life Sci.* **56,** 793–799.
31. Mao, J., Price, D. D., Lu, J., and Mayer, D. J. (1998) Antinociceptive tolerance to the mu-opioid agonist DAMGO is dose-dependently reduced by MK-801 in rats. *Neurosci. Lett.* **250,** 193–196.

32. Tanganelli, S., Antonelli, T., Morari, M., Bianchi, C., and Beani, L. (1991) Glutamate antagonists prevent morphine withdrawal in mice and guinea pigs. *Neurosci. Lett.* **122,** 270–272.
33. Christensen, D., Guilbaud, G., and Kayser, V. (2000) The effect of the glycine/NMDA receptor antagonist, (+)-HA966, on morphine dependence in neuropathic rats. *Neuropharmacology* **39,** 1589–1595.
34. Ibuki, T., Dunbar, S. A., and Yaksh, T. L. (1997) Effect of transient naloxone antagonism on tolerance development in rats receiving continous spinal morphine infusion. *Pain* **70,** 125–132.
35. Bhargava, H. N. and Zhao, G. M. (1996) Effects of *N*-methyl-D-aspartate receptor antagonists on the analgesia and tolerance to D-Ala2, Glu4 deltorphin II, a delta 2-opioid receptor agonist in mice. *Brain Res.* **719,** 56–61.
36. Zhao, G. M. and Bhargava, H. N. (1996) Effect of antagonism of the NMDA receptor on tolerance to <D-Pen2,D-Pen5> enkephalin, a delta 1-opioid receptor agonist. *Peptides* **17,** 233–236.
37. Kolesnikov, Y. A., Maccechini, M. L., and Pasternak, G. W. (1994) 1-Aminocyclopropane carboxylic acid (ACPC) prevents mu and delta opioid tolerance. *Life Sci.* **55,** 1393–1398.
38. Elliott, K., Minami, N., Kolesnikov, Y. A., Pasternak, G. W., and Inturrisi, C. E. (1994) The NMDA receptor antagonists, LY274614 and MK-801, and the nitric oxide synthase inhibitor, $N^G$-nitro-L-arginine, attenuate analgesic tolerance to the mu-opioid morphine but not to kappa opioids. *Pain* **56,** 69–75.
39. Bhargava, H. N. and Thorat, S. N. (1994) Effect of dizocilpine (MK-801) on analgesia and tolerance induced by U-50, 488H, a kappa-opioid receptor agonist, in the mouse. *Brain Res.* **649,** 111–116.
40. Chen, L., Gu, Y., and Huang, L. Y. (1995) The mechanism of action for the block of NMDA receptor channels by the opioid peptide dynorphin. *J. Neurosci.* **15,** 4602–4611.
41. Randic, M., Cheng, G., and Kojic, L. (1995) kappa-opioid receptor agonists modulate excitatory transmission in substantia gelatinosa neurons of the rat spinal cord. *J. Neurosci.* **15,** 6809–6826.
42. Mayer, D. J., Mao, J., and Price, D. D. (1995) The development of morphine tolerance and dependence in associated with translocation of protein kinase C. *Pain* **61,** 365–374.
43. Mao, J., Price, D. D., Phillips, L. L., Lu, J., and Mayer, D. J. (1995) Increases in protein kinase C gamma immunoreactivity in the spinal cord of rats associated with tolerance to the analgesic effects of morphine. *Brain Res.* **677,** 257–267.
44. Mao, J., Price, D. D., and Mayer, D. J. (1994) Thermal hyperalgesia in association with the development of morphine tolerance in rats: Roles of excitatory amino acid receptors and protein kinase C. *J. Neurosci.* **14,** 2301–2312.
45. Narita, M., Mizoguchi, H., Kampine, J. P., and Tseng, L. F. (1996) Role of protein kinase C in desensitization of spinal delta-opioid-mediated antinociception in the mouse. *Br. J. Pharmacol.* **118,** 1829–1835.
46. Narita, M., Mizoguchi, H., and Tseng, L. F. (1995) Inhibition of protein kinase C, but not of protein kinase A, blocks the development of acute antinociceptive tolerance to an intrathecally administered mu-opioid receptor agonist in the mouse. *Eur. J. Pharmacol.* **280,** R1–R3.
47. Bilsky, E. J., Bernstein, R., Wang, Z., Sadee, W., and Porreca, F. (1996) Effects of naloxone and D-phe-Cys-TyrD-Trp-Arg-Thr-Phe-Thr-NH2 and the protein kinase inhibitors H7 and H8 on acute morphine dependence and antinociceptive tolerance in mice. *J. Pharmacol. Exp. Ther.* **277,** 484–490.
48. Mestek, A., Hurley, J. H., Bye, L. S., Campbell, A. D., Chen, Y., Tian, M., et al. (1995) The human mu opioid receptor: modulation of functional desensitization by calcium/calmodulin-dependent protein kinase and protein kinase C. *J. Neurosci.* **15,** 2396–2406.
49. Gies, E. K., Peters, D. M., Gelb, C. R., Knag, K. M., and Peterfreund, R. A. (1998) Regulation of mu opioid receptor mRNA levels by activation of protein kinase C in human SH-SY5Y neuroblastoma cells. *Anesthesiology* **87,** 1127–1138.
50. Mao, J., Price, D. D., Mayer, D. J., and Hayes, R. L. (1992) Pain-related increases in spinal cord membrane-bound protein kinase C following peripheral nerve injury. *Brain Res.* **588,** 144–149.
51. Mao, J., Mayer, D. J., Hayes, R. L., and Price, D. D. (1993) Spatial patterns of increased spinal cord membrane-bound protein kinase C and their relation to increases in 14C-2-deoxyglucose metabolic activity in rats with painful peripheral mononeuropathy. *J. Neurophysiol.* **70,** 470–481.
52. Narita, M., Makimura, M., Feng, Y., Hoskins, B., and Ho, I. K. (1994) Influence of chronic morphine treatment on protein kinase C activity: comparison with butorphanol and implication for opioid tolerance. *Brain Res.* **650,** 175–179.
53. Kovoor, A., Celver, J. P., Wu, A., and Chavkin, C. (1998) Agonist induced homologous desensitization of μ-opioid receptors mediated by G protein-coupled receptor kinases is dependent on agonist efficacy. *Mol. Pharmacol.* **54,** 704–711.
54. Kovoor, A., Nappey, V., Kieffer, B. L., and Chavkin, C. (1997) Mu and delta opioid receptors are differentially desensitized by the coexpression of beta-adrenergic receptor kinase 2 and beta-arrestin 2 in *Xenopus oocytes*. *J. Biol. Chem.* **272,** 27,605–27,611.
55. Kovoor, A., Celver, J., Abdryashitov, R. I., Chavkin, C., and Gurevich, V. V. (1999) Targeted construction of phosphorylation-independent beta-arrestin mutants with constitutive activity in cells. *J. Biol. Chem.* **274,** 6831–6834.
56. Chen, L. and Huang, L. Y. M. (1991) Sustained potentiation of NMDA receptor-mediated glutamate responses through activation of protein kinase C by a μ-opioid. *Neuron* **7,** 319–326.
57. Chen, L. and Huang, L. Y. M. (1992) Protein kinase C reduces $Mg^{2+}$ block of NMDA-receptor channels as a mechanism of modulation. *Nature* **356,** 521–523.

58. Gracy, K. N., Svingos, A. L., and Pickel, V. M. (1997) Dual ultrastructural localization of mu-opioid receptors and NMDA-type glutamate receptors in the shell of the rat nucleus accumbens. *J. Neurosci.* **17,** 4839–4848.
59. Wang, H., Gracy, K. N., and Pickel, V. M. (1999) μ-Opioid and NMDA-type glutamate receptors are often colocalized in spiny neurons within patches of the caudata–putamen nucleus. *J. Comp. Neurol.* **412,** 132–146.
60. Commons, K. G., Van Bockstaele, E. J., and Pfaff, N. W. (1999) Frequent colocalization of mu opioid and NMDA-typer glutamate receptors at postsynaptic sites in periaquaductal gray neurons. *J. Comp. Neurol.* **408,** 549–559.
61. Keniston, L., Mao, J., Price, D. D., Lu, J., and Mayer, D. J. (1998) *Soc. Neurosci. Abst.* **24,** 390.
62. Mao, J., Price, D. D., and Mayer, D. J. (1995) Mechanisms of hyperalgesia and opiate tolerance: A current view of their possible interactions. *Pain* **62,** 259–274.
63. Granados-Soto, V., Kalcheva, I., Hua, X., Newton, A., and Yaksh, T. L. (2000) Spinal PKC activity and expression: role in tolerance produced by continuous spinal morphine infusion. *Pain* **85,** 395–404.
64. Fox, K. and Daw, N. W. (1993) Do NMDA receptors have a critical function in visual cortical plasticity? *Trends Neurosci.* **16,** 116–122.
65. Christie, M. J., Williams, J. T., and North, R. A. (1987) Cellular mechanisms of opioid tolerance: studies in single brain neurons. *Mol. Pharmacol.* **32,** 633–638.
66. Collin, E. and Cesselin, F. (1991) Neurobiological mechanisms of opioid tolerance and dependence. *Clin. Neuropharmacol.* **14,** 465–488.
67. Trujillo, K. A. and Akil, H. (1991) Opiate tolerance and dependence: recent findings and synthesis. *New Biologist* **3,** 915–923.
68. Nestler, E. J. (1992) Molecular Mechanisms of drug addiction. *J. Neurosci.* **12,** 2439–2450.
69. Pasternak, G. W. (1993) Pharmacological mechanisms of opioid analgesics. *Clin. Neuropharmacol.* **16,** 1–18.
70. Loh, H. H., Tao, P. L., and Smith, A. P. (1988) Role of receptor regulation in opioid tolerance mechanisms. *Synapse* **2,** 457–462.
71. Nishino, K., Su, Y. F., Wong, C. S., Watkins, W. D., and Chang, K. J. (1990) Dissociation of mu opioid tolerance from receptor down-regulation in rat spinal cord. *J. Pharmacol. Exp. Ther.* **253,** 67–72.
72. Harada, H., Ueda, H., Katada, T., Ui, M., and Satoh, M. (1990) Phosphorylated μ-opiate receptor purified from rat brains lacks functional coupling with Gil, a GTP-binding protein in reconstituted lipid vesicles. *Neurosci. Lett.* **113,** 47–49.
73. Chen, Y. and Yu, L. (1994) Differential regulation by cAMP-dependent protein kinase and protein kinase C of the μ opioid receptor coupling to a G protein activated K channel. *J. Biol. Chem.* **269,** 7839–7842.
74. Narita, M., Mizoguchi, H., Nagase, H., Suzuki, T., and Tseng, L. F. (2001) Involvement of spinal protein kinase C-gama in the attenuation of opioid-mu-receptor-mediated G-protein activation after chronic intrathecal administration of [D-Ala2,N-MePhe4,Gly-Ol5]enkephalin. *J. Neurosci.* **21,** 3715–3720.
75. Meller, S. T. and Gebhart, G. F. (1993) Nitric oxide (NO) and nociceptive processing in the spinal cord. *Pain* **52,** 127–136.
76. Sorkin, L. S. (1993) NMDA evokes an L-NAME sensitive spinal release of glutamate and citrulline. *NeuroReport* **4,** 479–482.
77. Montague, P. R., Gancayco, C. D., Winn, M. J., Marchase, R. B., and Friedlander, M. J. (1994) Role of NO production in NMDA receptor-mediated neurotransmitter release in cerebral cortex. *Science* **263,** 973–977.
78. Mao, J., Price, D. D., Zhu, J., Lu, J., and Mayer D. J. (1997) The inhibition of nitric oxide-activated poly(ADP-ribose) synthetase attenuates transsynaptic alteration of spinal cord dorsal horn neurons and neuropathic pain in the rat. *Pain* **72,** 355–366.
79. Azaryan, A. V., Coughlin, L. J., Buzas, B., Clock, B. J., and Cox, B. M. (1996) Effect of chronic cocaine treatment on mu- and delta-opioid receptor mRNA levels in dopaminergically innervated brain regions. *J. Neurochem.* **66,** 443–448.
80. Mayer, D. J., Mao, J., Holt, J., and Price, D. D. (1999) Cellular mechanisms of neuropathic pain, morphine tolerance, and their interactions. *Proc. Natl. Acad. Sci. USA* **96,** 7731–7736.
81. Fundytus, M. E. and Coderre, T. J. (1999) Opioid tolerance and dependence: a new model highlighting the role of metabotropic glutamate receptors. *Pain Forum* **8,** 3–13.
82. Trujillo, K. A. (1999) Cellular and molecular mechanisms of opioid tolerance and dependence. *Pain Forum* **8,** 29–33.
83. Herman, B. H., Vocci, F., and Bridge, P. (1995) The effects of NMDA receptor antagonists and nitric oxide synthase inhibitors on opioid tolerance and withdrawal: medicationdevelopment issues for opiate addiction. *Neuropsychopharmacology* **13,** 269–294.
84. Herrero, I., Miras-Potugal, M. T., and Sanche-Prieto, J. (1992) Positive feedback of glutamate exocytosis by metabotropic presynaptic receptor stimulation. *Nature* **360,** 163–166.
85. Mackler, S. A. and Eberwine, J. H. (1991) The molecular biology of addictive drugs. *Mol. Neurobiol.* **5,** 45–58.
86. Mao, J., Price, D. D., and Mayer, D. J. (1995) Experimental mononeuropathy reduces the antinociceptive effects of morphine: implications for common intracellular mechanisms involved in morphine tolerance and neuropathic pain. *Pain* **61,** 353–364.
87. Ossipov, M. H., Lopez, Y., Nichols, M. L., Bian, D., and Porreca, F. (1995) The loss of antinociceptive efficacy of spinal morphine in rats with nerve ligation injury is prevented by reducing spinal afferent drive. *Neurosci. Lett.* **199,** 87–90.

88. Wegert, S., Ossipov, M. H., Nichols, M. L., Bian, D., Vanderah, T. W., Malan, T. P., Jr., et al. (1997) Differential activities of intrathecal MK-801 or morphine to alter responses to thermal and mechanical stimuli in normal and nerve-injured rats. *Pain* **71,** 57–64.
89. Dunbar, S. A. and Pulai, I. J. (1998) Repetitive opioid abstinence causes progressive hyperalgesia sensitive to *N*-methyl-D-aspartate receptor blockade in the rat. *J. Pharmacol. Exp. Ther.* **284,** 678–686.
90. McNally, G. P. and Westbrook, R. F. (1998) Effects of systemic, intracerebral, or intrathecal administration of an *N*-methyl-D-aspartate receptor antagonist on associative morphine analgesic tolerance and hyperalgesia in rats. *Behav. Neurosci.* **112,** 966–978.
91. Herman, B. H. and O'Brien, C. P. (1997) Clinical medications development for opiate addiction: focus on nonopioids and opioid antagonists for the amelioration of opiate withdrawal symptoms and relapse prevention. *Semin. Neurosci.* **9,** 158–172.

# 20

## The Role of NMDA Receptors in Opiate Tolerance, Sensitization, and Physical Dependence

*A Review of the Research, A Cellular Model, and Implications for the Treatment of Pain and Addiction*

### Keith A. Trujillo, PhD

## 1. INTRODUCTION

When administered acutely, opiates such as morphine produce characteristic behavioral effects, including a decrease in pain responsiveness (analgesia) and an increase in pleasure (euphoria). Because of their profound ability to produce analgesia, opiates are the drugs of choice for the treatment of severe or chronic pain. Because of their ability to produce positive reinforcement and pleasure, these drugs are widely self-administered and, therefore, represent an important class of abused drugs. Long-term treatment with opiates, as well as other drugs of abuse, leads to three well-known consequences: *tolerance,* which is a decrease in an effect of a drug with chronic use; *sensitization,* which is an increase in an effect of drug with chronic use; and *physical dependence,* which is a physiological change produced by chronic use, such that the absence of the drug results in an unpleasant withdrawal syndrome* *(1–3)*. Tolerance, sensitization, and physical dependence are important in both the clinical use of opiates and in their self-administration. For example, the development of tolerance to the analgesic effect of opiates may lead to the need to escalate the dose during the treatment of chronic pain, whereas tolerance to the euphorigenic effect may be a factor in the escalation of drug use in addicts *(4)*. Conversely, tolerance to dose-limiting side effects may allow addicts to escalate drug intake and achieve greater euphorigenic effects. The development of sensitization is thought to be involved in the craving that occurs following chronic use of drugs of abuse and, therefore, critical to addiction *(5,6)*. Finally, the avoidance of withdrawal in physically dependent individuals is considered to be an important factor in maintaining self-administration *(2,4)*.

Although tolerance, sensitization, and physical dependence have been widely studied, there is still an incomplete understanding of the neural mechanisms involved in their development and expression. Recent experiments suggest that excitatory amino acid systems and, in particular, *N*-methyl-D-aspartate (NMDA) receptors may have an important role in these phenomena. This review will explore the role of NMDA receptors in the behavioral changes that occur following long-term opiate administration, including tolerance, sensitization, and physical dependence. I will focus on opiates that act at the

---

*It is important to note that tolerance and sensitization occur to selected behavioral effects of a drug, rather than to all effects. Because of this, tolerance may occur to some effects, sensitization to others, and yet other effects may show no change with chronic use.

From: *Contemporary Clinical Neuroscience: Glutamate and Addiction*
Edited by: Barbara H. Herman et al. © Humana Press Inc., Totowa, NJ

μ-opioid receptor, because these drugs have been more widely studied than drugs acting on the δ or κ receptors and are more relevant to both addiction and the treatment of pain. The evidence suggests that NMDA receptors are widely involved in opiate-induced neural and behavioral plasticity, including the development of tolerance, sensitization, and physical dependence. Key research findings are discussed, as well as controversies in the field, a potential cellular model, and clinical relevance.

## 2. NMDA RECEPTORS

Although NMDA receptors are described in detail in other chapters in this volume, it will be helpful to offer a brief description here before entering into a detailed discussion of their involvement in opiate tolerance, sensitization, and physical dependence. NMDA receptors are a type of ionotropic excitatory amino acid receptor—large protein complexes, with a central ion channel and several sites to which neurotransmitters and drugs can bind and affect receptor activity. Key sites on the receptor include the competitive site, the glycine site, the noncompetitive site, and the polyamine site. Binding of an excitatory amino acid to the competitive site on the receptor complex opens the ion channel and allows calcium to flow into the neuron. When calcium enters the neuron, it can activate a variety of calcium-dependent enzymes and thereby modify neuronal function. Activation of NMDA receptors by competitive site agonists requires coactivation of the glycine site on the complex. Because glycine appears to be required for receptor function, it has been referred to as a coagonist of the NMDA receptor. The noncompetitive or phencyclidine (PCP) site is located within the ion channel. Drugs acting at this site block the open ion channel and prevent the influx of calcium, thereby inhibiting receptor function (These drugs are sometimes referred to as uncompetitive antagonists because their ability to block the receptor is dependent on receptor activation.) The final key site is the polyamine site; drugs acting at this site noncompetitively affect receptor activity (Fig. 1). The number of modulatory sites on the NMDA receptor complex allows for numerous pharmacological tools to explore the role of this receptor in physiology and behavior *(7–13)*.

N-Methyl-D-aspartate receptors have been suggested to have a general role in neural and behavioral plasticity. Drugs that block these receptors have been found to interfere with several different types of neural and behavioral plasticity, including learning, long-term potentiation (LTP), long-term depression (LTD), neural development, kindling, and sensitization to pain *(7,10–12,14–24)*. A notable observation in these studies is that NMDA receptor antagonists interfere with the development or acquisition of these phenomena, but not their expression. For example, NMDA receptor antagonists interfere with certain types of learning if administered during training, but they do not abolish learned responses if administered during testing. Similarly, these drugs block LTP if administered during its induction, but do not reverse this phenomenon after it is established. This pattern of results suggests that NMDA receptors are involved in the cellular changes that underlie the development of these forms of neural and behavioral plasticity. Evidence suggests that calcium influx and subsequent intracellular events associated with this influx mediate the role of NMDA receptors in neural and behavioral plasticity *(7,10–12,19,20)*.

## 3. NMDA RECEPTORS AND THE DEVELOPMENT OF TOLERANCE TO OPIATE ANALGESIA

Over a decade ago, we began studies on the potential role of NMDA receptors in opiate tolerance and physical dependence. Reasoning that opiate tolerance and dependence were good examples of neural and behavioral plasticity, we hypothesized that NMDA receptors might be involved in these phenomena, in a manner similar to their involvement in learning and other forms of neural and behavioral plasticity. This hypothesis led us to predict that NMDA receptor antagonists would inhibit the development but not the expression of opiate tolerance and physical dependence. Initial studies on these phenomena suggested that this was, indeed, the case. In 1991, we reported that the potent

## 4.3. Suppression of Operant Responding

Another robust effect of opiates, when administered acutely, is the ability to suppress operant responding. This effect is likely related to opiate-induced locomotor depression, discussed above—opiates produce suppression of a variety of behaviors, including locomotor activity and operant responding, when administered acutely in high doses. Recently, Bespalov and co-workers *(77)* obtained evidence that NMDA receptor antagonists inhibit the development of tolerance to morphine-induced suppression of operant responding. Although the study was designed to explore the development of tolerance to the discriminitive stimulus effects of morphine, the investigators noted a potent reduction in the rate of responding with morphine and the development of tolerance to this effect. Each of the NMDA receptor antagonists used in this experiment (a noncompetitive antagonist, a competitive antagonist, a glycine-site antagonist, and a polyamine-site antagonist) inhibited tolerance to morphine-induced suppression of operant responding. The fact that drugs acting on all four sites on the NMDA receptor complex produced similar effects provides a powerful demonstration of the potential role of NMDA receptors in this phenomenon. Although a different behavioral end point was used than typically seen in studies of locomotor depression, we believe that these results are consistent with the idea that NMDA receptors are involved in tolerance to the locomotor-depressant effects of opiates. Alternatively, these results may suggest involvement of NMDA receptors in the development of tolerance to yet another opiate-induced behavior.

## 4.4. Discriminitive Stimulus Effects

As noted above, Bespalov and co-workers *(77)* explored the potential role of NMDA receptors in tolerance to the discriminitive stimulus effects of morphine. In this study, morphine produced a potent stimulus effect, and tolerance developed to this effect with repeated administration. Interestingly, the development of tolerance to the stimulus effect of morphine was inhibited by the competitive antagonist D-CPPene and the polyamine-site antagonist eliprodil, but not by MK-801 or the glycine-site partial agonist (+)-HA-966. Although it is nuclear why differences were seen across these compounds, the results offer intriguing preliminary evidence that NMDA receptors may be involved in tolerance to the discriminitive stimulus effects of opiates.

## 4.5. Summary

It is striking that NMDA receptor antagonists have the ability to inhibit tolerance to a variety of different opiate effects, including analgesia, locomotor depression, suppression of operant responding, and discriminitive stimulus effects. These results suggest that NMDA receptors may have a widespread role in the development of opiate tolerance. This role, however, is apparently not universal, as MK-801 did not inhibit the development of tolerance to morphine-induced hyperthermia. In a similar demonstration, Rauhala and co-workers *(87)* reported that the nitric oxide synthesis inhibitor *N*-nitro-L-arginine inhibited tolerance to the analgesic effect of morphine, but not the hormonal effects. These results raise the intriguing but not surprising possibility that NMDA receptors are involved in some forms of opiate-induced neural and behavioral plasticity, but not others. Opiate tolerance is not a singular phenomenon. Tolerance to different effects of opiates differs across a variety of parameters, including the time-course of development, threshold doses, degree of tolerance, and disappearance *(77,83,88–90)*. Additionally, the development of tolerance to different opiate effects involves different neural systems and likely involves different cellular events *(3,91–94)*.

As research progresses on the potential role of NMDA receptors in tolerance to different effects of opiates, studies should be designed in a manner to allow comparisons to be made both within and across experiments. Dose responses should be explored and different antagonists should be utilized before it is concluded that NMDA receptors are or are not involved in a particular effect. The most powerful data will come from those studies in which the effects of NMDA receptor antagonists are

**Table 4**
**Inhibition of the Development of Opiate Sensitization by NMDA Receptor Antagonists**

| NMDA receptor antagonist[a] | Species | Opiate effect | Inhibition of sensitization? | Citation |
|---|---|---|---|---|
| MK-801 (noncomp) | Rat | Locomotor stimulation | Yes | Wolf and Jeziorski, 1993 (95) |
| MK-801 (noncomp) CGS 19755 (comp) | Rat | Locomotor stimulation | Yes | Jeziorski et al., 1994 (74) |
| MK-801 (noncomp) | Rat | Stereotyped biting | Yes (but only high doses examined) | Livezey et al., 1995 (96) |
| MK-801 (noncomp) | Mouse | Locomotor stimulation | Yes (but only at high doses) | Iijima et al., 1996 (97) |
| MK-801 (noncomp) | Rat | Locomotor stimulation | Yes (but only at high doses) | Vanderschuren et al., 1996 (98) |
| MK-801 (noncomp) | Rat | Locomotor stimulation | Yes | Swadley-Lewellen and Trujillo, 1998 (76) |

[a] Noncomp = noncompetitive antagonist; comp = competitive antagonist. The study of Tzschentke and Schmidt (75) is not included in this table. As discussed previously (70) and noted in Table 2, this study appears to have been misinterpreted, as sensitization was not observed in the course of the experiments.

explored across different opiate effects in the same experimental animals, as this will allow for direct comparisons to be made across behaviors (25,29,76,77) (see ref. 70 for further discussion).

## 5. NMDA RECEPTORS AND THE DEVELOPMENT OF OPIATE SENSITIZATION

Wolf and co-workers were the first to report on the ability of NMDA receptor antagonists to inhibit opiate sensitization (74,95). These investigators noted that both MK-801 and the competitive NMDA receptor antagonist CGS 19755, at relatively low and selective doses, inhibited the development of sensitization to the locomotor-stimulant effect of morphine. The pattern observed was strikingly similar to that seen with morphine tolerance, in that the antagonists inhibited the development but not the expression of morphine sensitization (Table 4).

Recent research from others appears to present a more complicated picture. Vanderschuren et al. (98), for example, reported that only high doses of MK-801, which produced considerable lethality in combination with morphine, had the ability to inhibit sensitization to the locomotor-stimulant effect of morphine in rats. Lower doses, which did not produce lethality, did not affect sensitization. These authors therefore suggest that the effect may be a nonspecific drug interaction, rather than an inhibition of neural and behavioral plasticity.

Iijima and co-workers (97), in studies on mice, also reported that high doses of MK-801 were necessary to inhibit morphine sensitization. Although lethality was not seen in this study, the doses of MK-801 required to inhibit sensitization were very high and produced significant locomotor stimulation on their own. It should be noted that the locomotor effects of morphine in mice are quite different than those in rats. Whereas high doses of morphine in rats produce a biphasic locomotor response, with an initial locomotor depression followed by locomotor stimulation, this drug in mice produces dose-dependent stimulation of locomotion without the locomotor depression. The effects in mice may, therefore, be difficult to directly compare to those in rats.

In exploring an alternative behavior in rats, stereotyped biting, Livezey et al. (96) reported that a high dose of MK-801 inhibited both the development and expression of sensitization to this effect. Lower doses of MK-801 were not examined in this experiment so it is not clear if the high dose was required.

As mentioned earlier, we have begun studies exploring the role of NMDA receptors in tolerance and sensitization to the locomotor effects of morphine. In these studies, we observed that MK-801, at a relatively low and selective dose, inhibited the development but not the expression of sensitization to the locomotor-stimulant effect of morphine *(76,86)*. Adding strength to this finding, this inhibition paralleled inhibition of tolerance to the locomotor-depressant effect and tolerance to the analgesic effect of morphine in the same experimental animals. In more recent experiments, we found that the low-affinity noncompetitive NMDA receptor antagonist memantine also inhibited sensitization to the locomotor-stimulant effect of morphine (Peterson and Trujillo, unpublished observations). Results from our laboratory therefore support the findings of Wolf and co-workers that the development of opiate sensitization is inhibited by low selective doses of NMDA receptor antagonists.

Taken together, there appears to be general agreement that NMDA receptor antagonists inhibit the development of morphine sensitization; however, there is disagreement about the doses necessary to achieve this effect. At first glance, it is unclear why some studies report inhibition of this effect at low doses of NMDA receptor antagonists, whereas others require higher doses. However, Wolf *(6)* has suggested a potential resolution to this discrepancy. She suggested that context-dependent (or associative) sensitization may be less sensitive to inhibition by NMDA receptor antagonists than context-independent (or nonassociative) sensitization. Context-dependent sensitization arises when a drug is administered repeatedly in the same environment and the effects of the drug become intimately associated with environmental cues. This form of sensitization requires learning and is expressed only in the presence of the environmental cues in which drug administration occurred. In contrast, context-independent sensitization arises in the absence of environmental cues, does not involve learned associations between drug effects and environmental cues, and is, therefore, a more direct form of sensitization. If Wolf is correct, then this may help to explain why some studies find inhibition of morphine sensitization at relatively low doses, whereas others require higher doses—those that required high doses may have examined context-dependent sensitization, rather than context-independent sensitization. The studies in our laboratory were performed in a manner that promoted the development of context-independent sensitization, which may explain why lower doses were able to inhibit this phenomenon.

Although the results on sensitization are less clear than those on opiate tolerance, it appears that most investigators agree that sensitization to the locomotor-stimulant effect of morphine is inhibited by NMDA receptor antagonists, albeit at different doses and with different interpretations of the results. Taken together, we believe the data offers tentative support of a role for NMDA receptors in the development of opiate sensitization. Further studies should help to resolve this issue. As mentioned earlier in the discussion on tolerance, studies should be designed to facilitate comparisons with other behaviors and other forms of opiate-induced neural and behavioral plasticity.

It should be added that NMDA receptor antagonists have been found to inhibit the development of sensitization to the stimulant effects of several other drugs of abuse, including amphetamine, cocaine, and nicotine (see refs. *2,6,99,* and *100,* for review). These findings are very intriguing, suggesting that NMDA receptors may be involved in sensitization to a variety of different drugs of abuse. Understanding the mechanisms underlying sensitization is of particular interest to the field of substance abuse because this process may be involved in the craving that arises from repeated drug use *(5,6)*.

## 6. NMDA RECEPTORS AND THE DEVELOPMENT OF OPIATE PHYSICAL DEPENDENCE

At the same time that we first reported that MK-801 inhibited tolerance to morphine analgesia, we reported that this drug also inhibited physical dependence on morphine *(25)*. The effects were identical to those on tolerance in that MK-801 inhibited the development but not the expression of physical dependence. In other words, MK-801 inhibited withdrawal-related behaviors when administered with morphine during the acquisition of physical dependence; but the drug did not affect these behaviors when administered immediately prior to naloxone-precipitated withdrawal. A similar inhibition of the development of physical dependence by NMDA receptor antagonists has been observed more recently by others using a variety of NMDA receptor antagonists and a variety of behavioral measures of withdrawal (Table 5).

One of the most intriguing recent findings relevant to this topic is that of Zhu et al. *(108)*, demonstrating that treatment with antisense oligonucleotides directed against the NMDA receptor (NMDA-R1 subunit) can inhibit the development of opiate physical dependence. Intraventricular administration of antisense oligonucleotides, but not sense (control) oligonucleotides, decreased levels of brain NMDA receptors and inhibited the development of morphine physical dependence, as assessed by several different behavioral signs of withdrawal. This finding together with the pharmacological results provide strong evidence that activation of NMDA receptors is necessary for the normal development of opiate physical dependence.

The opitate-withdrawal syndrome is a complex phenomenon, characterized by several different signs and symptoms. It is presently unclear if NMDA receptor antagonists inhibit the development of the syndrome as a whole or a subset of signs and symptoms. Given the variety of physiological and behavioral effects expressed during withdrawal and their mediation by different physiological systems, it would be surprising if the development of all of these would be inhibited by NMDA receptor antagonists. Thus far, different studies have explored different signs and symptoms and utilized different methods of quantification. It is therefore difficult to make comparisons across experiments. NMDA receptor antagonists have been reported to interfere with "active" signs such as jumping, wet-dog shakes, and teeth chattering, "passive" signs such as diarrhea, weight loss, and piloerection, and "emotional" symptoms such as vocalizations and irritability (*see* Table 5). It will be useful to clarify the effects of NMDA receptor antagonists on these effects, both individually and collectively, in order to better understand the potential role of NMDA receptors in the development of these different signs and symptoms and their relationships to one another.

Although the effects of NMDA receptor antagonists on the development of physical dependence are relatively clear, the effects of these drugs on the expression of withdrawal are more problematic. Several groups have reported that NMDA receptor antagonists will inhibit the expression of opiate physical dependence *(45,109–123)*. However, this may be a nonspecific effect related to behavioral competition rather than a specific blockade of the neural processes responsible for opiate withdrawal *(2,124,125)*. An important clue to this possibility is found in the doses required to inhibit the development of physical dependence relative to those necessary to inhibit the expression of withdrawal. As we have discussed previously *(2,125,126)*, doses required to inhibit the expression of opiate withdrawal are typically greater than those necessary to inhibit the development of physical dependence. Recent studies reinforce this idea. Gonzalez and co-workers *(45)*, for example, found that doses of MK-801 necessary to inhibit the expression of withdrawal were *considerably* greater than those necessary to inhibit the development of physical dependence. Because doses of NMDA receptor antagonists required to inhibit the expression of opiate physical dependence are often high enough to produce significant locomotor effects, it has been suggested that these drugs may nonspecifically interfere with the expression of the opiate-withdrawal syndrome *(2,125,126)*. On the other hand, some studies have shown inhibition of the expression of certain aspects of withdrawal at relatively low doses of NMDA receptor antagonists *(45,114,118–120)*. It may be that NMDA receptors have a role not only in the development of opiate physical dependence but also in certain aspects of the expression of withdrawal. Future studies should focus on low doses of NMDA receptor antagonists, comparing their effects on the development versus the expression of opiate physical dependence, in order to help clarify this issue.

## 7. NMDA RECEPTORS AND THE SLOW REVERSAL OF OPIATE TOLERANCE AND PHYSICAL DEPENDENCE

As discussed above, there is widespread agreement that NMDA receptor antagonists inhibit the development of opiate tolerance, sensitization, and physical dependence. However, intriguing evidence

**Table 5**
**Inhibition of the Development of Opiate Physical Dependence by NMDA Receptor Antagonists**

| NMDA receptor antagonist | Species | Withdrawal behaviors inhibited | Citation |
|---|---|---|---|
| **Non-competitive Antagonists** | | | |
| MK-801 | Rat | Jumping | Trujillo and Akil, 1991 (25) |
| | Rat | Teeth chattering | Ben-Eliyahu et al., 1992 (101) |
| | Rat | Teeth chattering, writhing | Fundytus and Coderre, 1994 (102) |
| | Rat (intrathecal) | Hyperalgesia | Mao et al., 1994 (35) |
| | Mouse | Jumping | Verma and Kulkarni, 1995 (38) |
| | Rat (intrathecal) | Elicited vocalization, spontaneous vocalization, abnormal posture, ejaculation, jumping | Dunbar and Yaksh, 1996 (40) |
| | Rat | Rearing, teeth chattering, hippocampal norepinephrine release | Makimura et al., 1996 (103) |
| | Rat | Hyperalgesia, teeth chattering, jumping, wet-dog shakes | Manning et al., 1996 (42) |
| | Mouse | Micturition, running, diarrhea, piloerection, paw tremors, body shakes, jumping, convulsions | Gonzalez et al., 1997 (45) |
| | Mouse | Jumping (acute dependence) | McLemore et al., 1997 (46) |
| | Rat | Chewing, irritibility, stretching, tremor | Shoemaker et al., 1997 (104) |
| | Rat | Hyperalgesia (acute dependence) | Larcher et al., 1998 (49) |
| | Rat | Hyperalgesia | Laulin et al., 1998 (105) |
| | Rat (intrathecal) | Hyperalgesia | Mao et al., 1998 (51) |
| | Rat | Suppression of operant responding | Bespalov et al., 1999 (77) |
| | Rat | Hyperalgesia (acute dependence) | Celerier et al., 1999 (106) |
| | Rat | Weight loss | Koyuncuoglu et al., 1999 (53) |
| | Rat | Hyperalgesia | Laulin et al., 1999 (54) |
| ketamine | Mouse | Micturition, running, diarrhea, piloerection, paw tremors, body shakes, jumping, convulsions | Gonzalez et al., 1997 (45) |
| | Rat | Hyperalgesia | Celerier et al., 2000 (107) |
| Dextromethorphan | Rat | Hyperalgesia, teeth chattering, jumping, wet-dog shakes | Manning et al., 1996 (42) |
| | Rat | Teeth chattering, jumping | Mao et al., 1996 (58) |
| Ethanol | Rat | Chewing, irritibility, stretching, tremor | Shoemaker et al., 1997 (104) |
| **Competitive Antagonists** | | | |
| CGP 39551 | Mouse | Micturition, running, diarrhea, piloerection, paw tremors, body shakes, jumping, convulsions | Gonzalez et al., 1997 (45) |
| LY235959 | Mouse | Jumping (acute, not chronic dependence) | McLemore et al., 1997 (46) |
| D-CPPene | Rat | Suppression of operant responding | Bespalov et al., 1999 (77) |
| **Others** | | | |
| NMDA-R1 antisense | Rat | Jumping, rearing, teeth chattering, stretching | Zhu and Ho, 1998 (108) |

Table 6
Reversal of Opiate Tolerance and Dependence by NMDA Receptor Antagonists

| NMDA receptor antagonist[a] | Species | Observed effect | Citation |
|---|---|---|---|
| LY274614 (comp) | Rat | Reversal of tolerance to morphine analgesia | Tiseo and Inturrisi, 1993 (33) |
| Dextromethorphan (noncomp) | Mouse | Reversal of tolerance to morphine analgesia | Elliott et al., 1994 (34) |
| ACPC (glycine) | Mouse | Reversal of tolerance to morphine analgesia | Kolesnikov et al., 1994 (66) |
| LY274614 (comp) | Rat | Reversal of tolerance to morphine analgesia | Tiseo et al., 1994 (63) |
| Memantine (noncomp) | Mouse | Reversal of morphine physical dependence (jumping) | Popik et al., 1996 (118) |
| NPC 17742 (comp) | | Reversal of tolerance to morphine analgesia | Shimoyama et al., 1996 (55) |
| Ketamine (noncomp) | Rat | | |
| MRZ 2/579 (noncomp) | Mouse | Reversal of morphine physical dependence (jumping) | Popik et al., 1998 (120) |
| MRZ 2/570 (glycine) | | | |
| L-701,324 (glycine) | | | |
| LY235959 (comp) | Rat | Reversal of tolerance to morphine analgesia | Allen and Dykstra, 1999 (64) |
| MK-801 (noncomp) | Mouse | Reversal of tolerance to morphine analgesia | Kolesnikov and Pasternak, 1999a (52) |
| Ketamine (noncomp) | Mouse | Reversal of tolerance to morphine analgesia | Kolesnikov and Pasternak, 1999b (56) |
| ACEA-1328 (glycine) | Mouse | Reversal of tolerance to morphine analgesia | Lutfy et al., 1999 (68) |
| Memantine (noncomp) | Mouse | Reversal of tolerance to morphine analgesia | Popik et al., 2000 (127) |
| MRZ 2/579 (noncomp) | | | |

[a] noncomp = noncompetitive antagonist; comp = competitive antagonist; glycine = glycine-site antagonist. For the present purposes, "reversal" refers to the ability of NMDA receptor antagonists to slowly diminish tolerance and/or physical dependence with repeated administration.

suggests that these drugs may be able to slowly reverse tolerance and physical dependence under certain experimental circumstances. This effect is different than the acute blockade of the expression of tolerance or dependence, in that it does not occur immediately but takes days of administration to develop. The phenomenon was first reported by Tiseo and Inturrisi (33) in studies on morphine tolerance. These investigators found that the competitive NMDA receptor antagonist LY274614 had the ability to restore morphine analgesia if administered over several days to tolerant animals. Interestingly, this effect occurred whether LY274614 was administered in the presence or the absence of continued administration of morphine. These findings led the authors to suggest that NMDA receptors may be involved, not only in the development of morphine tolerance but also in the maintenance of this phenomenon. The ability of NMDA receptor antagonists to slowly reverse tolerance has since been replicated in several different studies, using several different NMDA receptor antagonists and different experimental approaches (Table 6). This ability does not appear to be isolated to opiate tolerance, as similar findings have been obtained with opiate physical dependence. Popik and co-workers (118,120) have demonstrated that competitive, noncompetitive, and glycine-site NMDA receptor antagonists will reverse physical dependence when administered over several days to physically dependent mice (Table 6).

The implications of the ability to reverse tolerance and dependence are self-evident. If NMDA receptor antagonists had the ability to simply prevent the development of these phenomena, they would be of little use in the treatment of addiction. Individuals seeking treatment for opiate addiction would presumably already be tolerant and dependent, and NMDA receptor antagonists would, therefore, be of little use (*see* refs. 2 and 126). However, the ability to reverse these phenomena suggests that NMDA receptor antagonists may be useful in detoxification, perhaps speeding up the process and lessening the impact of withdrawal.

## 8. QUESTIONS AND CONTROVERSIES

### 8.1. Potential Alternative Explanations

The hypothesis that led to studies in this area is relatively simple and straightforward: Because NMDA receptors are involved in many forms of neural and behavioral plasticity, these receptors may also be involved in the neural and behavioral plasticity arising from long-term administration of drugs of abuse. As discussed earlier, studies on the development of opiate tolerance, sensitization, and physical dependence are consistent with this hypothesis. However, there have been suggestions that the ability of NMDA receptor antagonists to inhibit the development of these phenomena may be due to other factors. Three of the most popular alternative explanations are drug side effects, inhibition of associative learning, and state dependency (70). Although these factors may indeed contribute to some of the results obtained in this area, the abundance of evidence suggests that they are not responsible for the majority of results (see ref. 70 for detailed discussion). Perhaps some of the most powerful evidence that these factors are not responsible is that inhibition of opiate-induced neural and behavioral plasticity is obtained with a variety of different NMDA receptor antagonists, with different behavioral and interoceptive effects, and typically at doses that produce few significant effects on ongoing behavior (see Tables 1–6 and ref. 70). Moreover, similar effects are obtained across different forms of plasticity, sometimes in the same experimental animals. These observations, together with others, suggest that the most parsimonious explanation for the results is that NMDA receptors are involved in the neural changes that underlie the development of tolerance, sensitization, and physical dependence to several different effects of opiates (70).

### 8.2. Morphine Versus Other Mu Agonists

Because of its widespread clinical use, its long history in psychopharmacological research, and its relative selectivity for μ-opiate receptors, morphine is the prototypical μ-opiate agonist, and is, therefore, used as the drug of choice when examining μ-opioid receptor function. This has been the case in the study of the role of NMDA receptors in opiate-induced neural and behavioral plasticity. It has been assumed, based on studies with morphine, that NMDA receptor antagonists would inhibit tolerance, sensitization, and physical dependence to other μ agonists in a similar manner. However, Bilsky and co-workers (39) have obtained evidence that tolerance to μ agonists other than morphine may *not* be affected by NMDA receptor antagonists. In studies on tolerance to opiate analgesia in mice, these authors reported that whereas MK-801 and the competitive NMDA receptor antagonist LY235959 inhibited development of tolerance to morphine, these drugs did not affect tolerance to the peptide μ agonists DAMGO and PL017 and the nonpeptide opioid fentanyl. It is possible, as suggested by the authors (39) that the development of tolerance to μ agonists that are more selective and/or more efficacious than morphine involves mechanisms that are independent of NMDA receptors. However, more recent studies question these findings. Mao et al. (51), for example, demonstrated that MK-801 can, indeed, inhibit the development of tolerance to the analgesic effect of the peptide μ agonist DAMGO, as well as the development of physical dependence as measured by hyperalgesia (this study used a different experimental approach than the Bilsky et al. study, which may help to explain the contrasting findings). Allen and Dykstra (65) have suggested that the ineffectiveness of the NMDA receptor antagonists in the Bilsky et al. study may be related to differences in the magnitude of tolerance produced by the different opioids. In studies on morphine tolerance in rats, Allen and Dykstra found that higher levels of tolerance produced by higher doses of morphine were more resistant to NMDA receptor inhibition than lower levels of tolerance. Thus, NMDA receptors may, indeed, be involved in opiate-induced neural and behavioral plasticity produced by a variety of μ agonists, but the involvement of NMDA receptors may diminish at higher levels of tolerance. As the magnitude of tolerance increases, other neural mechanisms, such as changes in receptor number, may contribute to the development of this phenomenon.

## 9. A CELLULAR MODEL

The ability of NMDA receptor antagonists to inhibit different forms of opiate-induced neural and behavioral plasticity suggests that there may be widespread interactions between NMDA receptors and opioid receptors in the body and that these interactions may be critical in the development of opiate tolerance, sensitization, and physical dependence. Because the majority of evidence in this area has been obtained from behavioral studies, it is unclear whether NMDA receptors and μ-opioids located together on individual cells are responsible for the interactions, or if the interactions may be the result of more distributed circuits. However, increasing evidence suggests that NMDA receptors and μ-opioid receptors may indeed interact at the level of single cells—anatomical studies suggest that NMDA receptors and μ-opioid receptors are colocalized on neurons in the central nervous system (CNS) *(128–130)*, biochemical studies suggest intracellular pathways through which these receptors may interact *(131–133)*, and physiological evidence demonstrates that cellular interactions between these receptors occur within the brain *(134,135)*. Therefore, there is reason to believe that cellular interactions between μ receptors and NMDA receptors may be involved in at least some of these forms of opiate-induced neural and behavioral plasticity. Cellular models for such interactions have previously been proposed for opiate tolerance and dependence *(72,93,136–139)*. The present model, based in part on these previous models, illustrates intracellular pathways by which NMDA receptors and μ-opioids may interact to produce opiate-induced changes in the brain and behavior.

The model is depicted in Figure 2. The cascade is initiated by stimulation of μ receptors, which causes activation of protein kinase C (PKC). PKC has the ability to phosphorylate NMDA receptors and relieve the magnesium ($Mg^{2+}$) block that keeps these receptors quiescent under normal physiological conditions. Relief of the magnesium block (likely together with activation of NMDA receptors by tonic synaptic levels of glutamate) leads to influx of calcium, which can activate several different calcium-dependent processes that may be involved in altering the cellular response to morphine. The first of these is activation of PKC. This may represent further activation of the original pool of PKC stimulated through μ receptor activation, or a second, distinct pool. PKC may phosphorylate μ receptors, μ-receptor-coupled G-proteins, or second-messenger enzymes, thereby modifying the proteins and producing changes in the coupling of μ receptors to key second-messenger cascades. In addition to PKC, calcium entering through NMDA receptor ion channels activates nitric oxide synthase (NOS), leading to production of the diffusible messenger nitric oxide. Nitric oxide (NO) can then activate guanylyl cyclase and cyclic GMP (cGMP) within the neuron, which stimulates cGMP-dependent protein kinases, which, in turn, may phosphorylate μ receptors or other key molecules in the second-messenger cascade, causing further changes in the coupling of μ receptors to second-messenger systems. Nitric oxide can also diffuse to the presynaptic cell and activate cGMP within this cell, causing release of glutamate and further activation of NMDA receptors, thereby increasing activation of the postsynaptic processes that inhibit μ receptor function. Another enzyme that may be activated by calcium entering the cell via NMDA receptor ion channels is calcium/calmodulin-dependent protein kinase II (CaMKII). Like PKC and the NO/cGMP pathway, this enzyme may contribute to the changes in coupling of μ receptors to critical second-messenger cascades. A functional decoupling of μ receptors from second-messenger cascades produced by PKC, NO/cGMP, and CaMKII would help to explain the decreased responsiveness to opiates that is characteristic of tolerance.

There are several key findings that support this model. First are observations that μ receptor stimulation leads to activation of PKC *(131,136,140–146)*. The activation of PKC by μ receptors may be mediated by phospholipase C(PLC) *(146–150)*. PKC activated by μ receptor stimulation potentiates

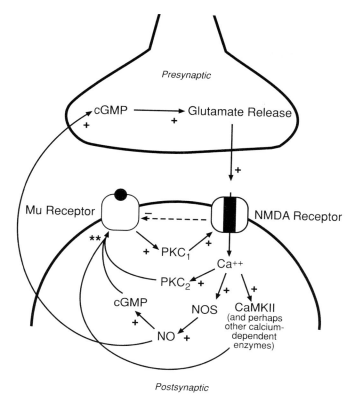

**Fig. 2.** Schematic diagram of cellular processes that may mediate interactions between μ-opioid receptors and NMDA receptors in opiate-induced neural and behavioral plasticity + refers to activation, − refers to inhibition, and ** refers to changes in receptor coupling between μ receptors and second messengers. The dashed line refers to the general idea that NMDA receptor activation is functionally antagonistic to μ receptor function. The solid arrows indicate identified pathways through which NMDA receptors and μ receptors may interact. The sequence of events depicted in the model is as follows: μ Receptor activation leads to activation of protein kinase C ($PKC_1$), which relieves the magnesium ($Mg^{2+}$) block on NMDA receptors. Relief of the $Mg^{2+}$ block leads to the influx of calcium ($Ca^{2+}$) and activation of at least three different calcium-dependent intracellular cascades: (1) $PKC_2$ (either greater activation of the initial PKC pool, or a distinct pool); (2) nitric oxide synthase (NOS), which leads to production of nitric oxide (NO) and activation of cyclic GMP (cGMP); and (3) calcium/calmodulin-dependent protein kinase II (CaMKII). Each of these cascades may produce changes in coupling between the μ receptor and its second-messenger systems (**), leading to the manifestations of tolerance, sensitization, and physical dependence. In addition, NO activation of presynaptic cGMP may lead to the release of glutamate, which further stimulates the postsynaptic processes that modify coupling between μ receptors and second messengers. Some intermediates in the cascades are omitted for clarity. Please see text for further details.

NMDA receptor function by removal of the $Mg^{2+}$ block on NMDA receptors *(131,132,135,151)*. Calcium influx resulting from NMDA receptor stimulation can then activate a variety of intracellular pathways, including PKC, NO/cGMP, and CaMKII. Studies using enzyme inhibitors have been instrumental in demonstrating a role for these pathways in opiate tolerance and dependence—inhibitors of PKC *(35,37,140,152–155)*, PLC *(147)*, NOS (*see* refs. *2, 71,73,* and *156*), guanylyl cylase (the intermediate enzyme between NO and cGMP activation) *(156)*, and CaMKII *(157)*, have each been found to inhibit the development of opiate tolerance and/or physical dependence. These enzymes, or others in their respective cascades, may modify μ receptors, G-proteins, or second messengers, producing changes in μ receptor second-messenger coupling. Additional biochemical evidence in support of this model is that PKC, activated through NMDA receptor stimulation, has the ability to interfere with coupling between opioid receptors and G-proteins *(133,158–160)*.

The classical view of opioid tolerance and physical dependence has been that these phenomena are mediated, at least in part, by the functional decoupling of opioid receptors from second messengers. Opiates produce primarily inhibitory effects on cellular responsiveness, via G-protein-mediated inhibition of adenylyl cyclase, enhanced potassium efflux, and decreased calcium influx (*see* refs. *161–163* for review). Decreases in coupling would, therefore, lead to a diminished ability of opiates to inhibit cell function, characteristic of tolerance, and perhaps producing a hyperresponsiveness upon withdrawal, characterisitic of physical dependence. However, more recent studies suggest a more complex and interesting picture—in addition to diminished inhibitory coupling, chronic opioid receptor activation may produce a switch whereby stimulation of opioid receptors produces activation of certain second-messenger cascades *(164–169)*, or an alternative switch whereby opioid receptors become constitutively active *(170,171)*.

A further complexity to interactions between NMDA receptors and μ-opioid receptors is suggested by recent electrophysiological data obtained by Martin and colleagues *(134,172)*. Following chronic treatment with morphine, these investigators observed a functional downregulation of postsynaptic NMDA receptors and a functional upregulation of presynaptic metabotropic glutamate receptors (mGluR) in the nucleus accumbens. In other words, postsynaptic NMDA receptors were less active, and glutamatergic autoreceptors were more active following chronic morphine treatment, leading to an overall NMDA receptor hypofunction (*see* Fig. 3). This pattern is suggestive of a compensatory response to overactivation of glutamatergic signaling in nucleus accumbens neurons containing NMDA receptors and μ-opioid receptors. Thus, these data are consistent with the above-described model whereby opioid receptor activation leads to NMDA receptor activation and, in a feed-forward process, causes increasingly greater NMDA receptor activation (Fig. 2).

The changes in glutamatergic signaling produced by chronic opiate administration raises a very important question: What is the consequences of these changes for opiate tolerance, sensitization, and physical dependence? It is possible that they are simply a compensatory response to NMDA receptor hyperactivity produced by μ receptor activation, dampening the effects of the NMDA receptor activation and helping the system to stabilize. Alternatively, however, these changes in glutamatergic signaling may have a more direct role in the behavioral changes produced by chronic morphine treatment, perhaps being involved in the expression of tolerance, sensitization, or physical dependence. In fact, we have recently obtained evidence that acute NMDA receptor hypofunction, in combination with acute morphine treatment, mimics the behavioral effects of chronic morphine treatment *(28)*. Together with the Martin et al. data *(134,172)*, the results suggest that decreases in NMDA receptor function produced by chronic morphine treatment may be, at least in part, responsible for the development of opiate tolerance. The cascade illustrated in Fig. 2 may represent the initiation of events leading to the development of tolerance, sensitization, and physical dependence, whereas the longer-term receptor modifications illustrated in Fig. 3 may represent some of the key changes responsible for the behavioral expression of these phenomena. If this is indeed the case, then it may help to explain both the inhibition of the development of tolerance and dependence and the reversal of these phenomena by NMDA receptor antagonists. During the development of tolerance and dependence, chronic treatment with NMDA receptor antagonists, by blocking the feed-forward NMDA receptor activation, should prevent the changes in NMDA receptor function produced by chronic opiate administration. After tolerance and dependence are established, chronic treatment with NMDA receptor antagonists should produce compensatory increases in NMDA receptor function, thereby reversing the chronic opiate-induced NMDA receptor hypofunction. Although further studies are necessary to confirm these possibilities, it is apparent that

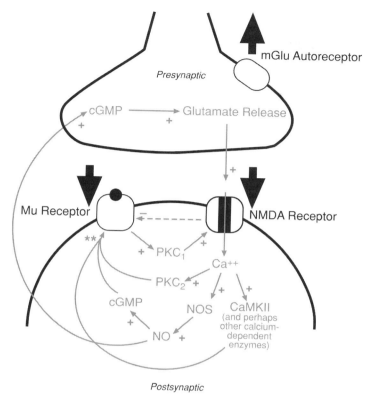

**Fig. 3.** Schematic diagram illustrating compensatory changes in μ-opioid receptors, NMDA receptors, and metabotropic glutamate receptors (mGluRs) following chronic opiate administration. Evidence suggests chronic opiate treatment leads to (1) functional decoupling of μ receptors from key second-messenger systems (large downward arrow), (2) functional downregulation of postsynaptic NMDA receptors (large downward arrow), and (3) functional upregulation of mGlu autoreceptors (large upward arrow). For comparison, the cellular events illustrated in Fig. 2 are shown in light gray. Please see legend for Figure 2 and text for further details.

dynamic interactions between NMDA receptors and μ-opioid receptors are important in opiate-induced neural and behavioral plasticity.

It needs to be acknowledged that there are many other mechanisms involved in opiate-induced neural and behavioral plasticity that have not been addressed in this model. For example, other intracellular pathways, such as adenylyl cyclase/cyclic AMP (AC/cAMP), G-protein receptor kinases (GRKs), and mitogen-activated protein kinases (MAPKs) may be involved, as well as regulation of opioid receptor internalization, sequestration, and biosynthesis *(94,163)*. Considering broader circuit interactions, regulation of opioid peptides and other neurotransmitters and receptors that modulate opiate function may also have a role *(3,27,173,174)*. Finally, learning processes involved in the development of context-dependent plasticity are also likely to be involved *(1,3,175)*. Given the complexity of intracellular, intercellular and circuit processes involved in opiate-induced plasticity, it will be a considerable challenge to place NMDA receptors and NMDA receptor-mediated second-messenger events in context with all of the other mechanisms involved in opiate tolerance, sensitization, and physical dependence (see ref. *93* for further discussion).

## 10. CLINICAL IMPLICATIONS

### 10.1. Chronic Pain

One of the most obvious possibilities for therapeutic intervention with NMDA receptor drugs is in the treatment of chronic pain. Because NMDA receptor antagonists have the ability to prevent the development of tolerance, sensitization, and physical dependence, these drugs may be useful adjuncts to opiates in situations in which long-term administration of opiates is necessary. Administration of an NMDA receptor antagonist together with an opiate should inhibit the development of tolerance, thereby allowing the opiate to maintain effectiveness for a greater period of time. The development of physical dependence is not typically considered to be a major problem in the treatment of chronic pain, as it can easily be managed by a well-trained physician. However, decreased physical dependence should offer reassurance to both patients and physicians overly concerned about "addiction" and decrease the time necessary to withdraw patients from opiates when necessary. Sensitization is not typically discussed in relation to chronic pain management. However, in theory, it is possible that sensitization to certain side effects of opiates may lead to their escalation during chronic administration. Inhibition of sensitization may, therefore, help to diminish the impact of such side effects in pain patients.

Other research, not discussed in the present review, suggests further uses for NMDA receptor antagonists in the treatment of pain. First, considerable evidence suggests that NMDA receptors may be involved in the development of sensitization to pain, in which pain increases in a pathological manner *(17,22,136,137,176)*. NMDA receptor antagonists have the ability to inhibit such sensitization, providing for their potential use in situations in which pain sensitization may be expected to occur. Second, although the majority of evidence suggests that NMDA receptor antagonists do not affect the acute analgesic actions of opiates at doses that inhibit the development of tolerance (discussed earlier), some studies suggest that these drugs may, in fact, potentiate opiate analgesia *(42,49,84,101,176–180)*. This potentiation is typically seen as a prolongation of opiate analgesia and is interpreted by most researchers as inhibition of the earliest phase of opiate tolerance *(42,49,101,106,178–180)*. Regardless of the interpretation, however, if NMDA antagonists have the ability to prolong opiate analgesia when administered acutely with an opiate, they may be useful adjuncts in the treatment of acute, as well as chronic pain. The potential for combination medications including an NMDA receptor antagonist and an opiate in the treatment of pain appears to be well on its way to being realized, as clinical testing of such combinations is currently underway *(181–184)*.

### 10.2. Addiction

The use of NMDA receptor antagonists in the treatment of addiction is less obvious if one views these drugs as inhibiting the development but not the expression of tolerance, sensitization, and physical dependence *(2,3,25,126)*. Because these processes should already be well established in an addict, it appears that the drugs would be of little use. On the other hand, under these conditions, there still might be a role for NMDA receptor antagonists as adjuncts in opiate maintenance therapy. Coadministration of these drugs during methadone treatment, for example, might lead to less tolerance and dependence arising from the maintenance therapy and easier cessation of treatment for the motivated individual.

The potential role of NMDA receptor antagonists in addiction becomes more evident when considering the findings, discussed earlier, that NMDA receptor antagonists may reverse opiate tolerance and dependence when administered over days. The ability to accelerate the extinction of tolerance and physical dependence (and perhaps sensitization) would offer a means to speed up the detoxification process and perhaps reverse some of the physiological processes responsible for addiction. Although very intriguing, further research is necessary in order to more firmly establish the conditions under which these drugs may reverse these phenomena and to demonstrate whether or not NMDA receptor antagonists will act in a similar manner in opiate addicts.

Evidence demonstrating that NMDA receptor antagonists may inhibit the expression of physical dependence under certain circumstances suggests that these drugs may be useful in acute detoxification. Preliminary clinical research offers conflicting views on this possibility. Whereas two studies suggest that NMDA receptor antagonists may block some symptoms of withdrawal in opiate addicts *(110,185)*, two other studies demonstrate no significant effects *(186,187)*. Perhaps in support of the use of these drugs, ibogaine, which is used in some clinics for acute detoxification, is thought to alleviate withdrawal, at least in part, through blockade of NMDA receptors *(188–194)*. Clearly, more research is necessary in order to determine the potential for NMDA receptor antagonists in alleviating the signs and symptoms of acute opiate withdrawal.

One final area related to opiate addiction where NMDA receptors may have a role is in the rewarding effects of these drugs. Opiate reward is thought to play a critical role in drug-seeking behavior in addicts. NMDA receptor antagonists have been found, in several studies, to block the ability of opiates to establish a conditioned place preference, an experimental procedure commonly used to study drug reward *(119,120,195–201)*. In support of the their ability to inhibit opiate reward, preliminary evidence suggests that some NMDA receptor antagonists may also inhibit the acquisition of opiate self-administration *(202)*. However, these drugs appear to increase the rewarding effects of opiates as measured by intracranial self-stimulation *(203,204)*. Thus, the role of NMDA receptors in opiate reward is presently unclear. However, the results on conditioned place preference and self-administration indicates that future studies should investigate the ability of NMDA receptor antagonists to modify opiate reward, and thereby impact opiate abuse.

## 10.3. Side Effects and the Therapeutic Potential of NMDA Receptor Antagonists

In discussions of the therapeutic potential of NMDA receptor antagonists, the issue of adverse side effects inevitably arises. High-affinity noncompetitive NMDA receptor antagonists, such as phencyclidine, produce significant effects on behavior that may limit their therapeutic use. In rodents, these effects are expressed as dose-dependent changes in locomotion and coordination—at relatively low doses, no significant effects are evident; at higher doses, increases in locomotor behavior are seen; at still higher doses, the increases in locomotor behavior are accompanied by significant ataxia; at the highest doses, animals become limp and show a complete lack of voluntary movement *(25,205–208)*. In addition, high-affinity noncompetitive NMDA receptor antagonists can show disruptive effects on learning and cognition *(209,210)* and show evidence of neurotoxicity *(211,212)*. Historically, these drugs (primarily ketamine and phencyclidine) have been used at relatively high doses as dissociative anesthetics *(213–218)*. Phencyclidine was withdrawn for use in humans because of problematic side effects—when individuals awakened from phencyclidine anesthesia, they sometimes reported disturbing hallucinations, thoughts, and dreams, and these were occasionally accompanied by behavioral disturbances *(213,214,216,218)*. Despite some evidence of the same symptoms, ketamine is still used clinically in humans. In addition to these effects, phencyclidine and ketamine are considered drugs of abuse. They show abuse potential in animals models and, according to news reports, are becoming increasingly popular. Given these effects, it is not surprising that caution is advised when considering clinical use of these compounds.

On the other hand, it appears that many of the concerning side effects of NMDA receptor blockers may be restricted to high-affinity noncompetitive antagonists. In preclinical studies, low-affinity noncompetitive antagonists, competitive antagonists, glycine-site antagonists, and polyamine-site antagonists produce fewer problematic side effects *(205,206,219–227)*. Moreover, the doses at which high-affinity noncompetitive antagonists inhibit opiate-induced neural and behavioral plasticity are often quite low, well below doses that cause significant side effects. Further research on these compounds will help to determine the therapeutic potential for NMDA receptor antagonists not only in pain and addiction but also in other clinical problems.

## ACKNOWLEDGMENTS

As always, I would like to thank Huda Akil for her generous support and contributions, intellectual and otherwise, to early portions of the work described in this chapter. I need to acknowledge the hard work and expert technical assistance of several students who contributed to these research efforts, including Ivan Sanchez, Terrance Conner, and Myrna Caballero at the University of Michigan, and Michelle Swadley-Lewellen, Kathleen Warmoth, Karen Watorski, Eric Ruzek, Dawn Albertson, David Peterson, and Ian Mendez at California State University San Marcos. The writing of this review was supported by the National Institute of General Medical Sciences (GM 59833). This chapter is dedicated to the memories of Virginia Fox and Terrance Conner.

## REFERENCES

1. Stewart, J. and Badiani, A. (1993) Tolerance and sensitization to the behavioral effects of drugs. *Behav. Pharmacol.* **4,** 289–312.
2. Trujillo, K. A. and Akil, H. (1995) Excitatory amino acids and drugs of abuse: a role for N-methyl-D-aspartate receptors in drug tolerance, sensitization and physical dependence. *Drug Alcohol Depend.* **38,** 139–154.
3. Trujillo, K. A. and Akil, H. (1991) Opiate tolerance and dependence: recent findings and synthesis. *New Biol.* **3,** 915–923.
4. Schulteis, G. and Koob, G. F. (1996) Reinforcement processes in opiate addiction: a homeostatic model. *Neurochem. Res.* **21,** 1437–1454.
5. Robinson, T. E. and Berridge, K. C. (1993) The neural basis of drug craving: an incentive-sensitization theory of drug craving. *Brain Res. Rev.* **18,** 247–291.
6. Wolf, M. E. (1998) The role of excitatory amino acids in behavioral sensitization to psychomotor stimulants. *Prog. Neurobiol.* **54,** 679–720.
7. Collingridge, G. L. and Watkins, J. C. (1994) *The NMDA receptor* 2nd ed., Oxford University Press, New York.
8. Meldrum, B. S. (1991) Excitatory amino acid antagonists, in *Frontiers in Pharmacology and Therapeutics,* Blackwell Scientific, Boston.
9. Seeburg, P. H. (1993) The TINS/TIPS Lecture: the molecular biology of mammalian glutamate receptor channels. *Trend Neurosci.* **16,** 359–365.
10. Collingridge, G. L. and Lester, R. A. J. (1989) Excitatory amino acid receptors in the vertebrate central nervous system. *Pharmacol. Rev.* **41,** 143–210.
11. Michaelis, E. K. (1998) Molecular biology of glutamate receptors in the central nervous system and their role in excitotoxicity, oxidative stress and aging. *Prog. Neurobiol.* **54,** 369–415.
12. Yamakura, T. and Shimoji, K. (1999) Subunit- and site-specific pharmacology of the NMDA receptor channel. *Prog. Neurobiol.* **59,** 279–298.
13. Dingledine, R., Borges, K., Bowie, D., and Traynelis, S. F. (1999) The glutamate receptor ion channels. *Pharmacol. Rev.* **51,** 7–61.
14. Artola, A. and Singer, W. (1993) Long-term depression of excitatory synaptic transmission and its relationship to long-term depression. *Trends Neurosci.* **16,** 480–487.
15. McMahon, S. B., Lewin. G. R., and Wall, P. D. (1993) Central hyperexicitability triggered by noxious inputs. *Curr. Opin. Neurobiol.* **3,** 602–610.
16. Malenka, R. C. and Nicoll, R. A. (1993) NMDA-receptor-dependent synaptic plasticity: multiple forms and mechanisms. *Trends Neurosci.* **16,** 521–527.
17. Coderre, T. J., Katz, J., Vaccarino, A. L., and Melzack, R. (1993) Contribution of central neuroplasticity to pathological pain: review of clinical and experimental evidence. *Pain* **53,** 259–285.
18. McDonald, J. W. and Johnston, M. V. (1990) Physiological and pathophysiological roles of excitatory amino acids during central nervous system development. *Brain Res. Rev.* **15,** 41–70.
19. Crepel, V., Congar, P., Aniksztejn, L., et al. (1998) Synaptic plasticity in ischemia: role of NMDA receptors. *Prog. Brain Res.* **116,** 273–285.
20. Abraham, W. C. and Bear, M. F. (1996) Metaplasticity: the plasticity of synaptic plasticity *Trends Neurosci.* **19,** 126–130.
21. Bear, M. F. (1996) NMDA-receptor-dependent synaptic plasticity in the visual cortex. *Prog. Brain Res.* **108,** 205–218.
22. Dickenson, A. H. (1995) Central acute pain mechanisms. *Ann. Med.* **27,** 223–227.

23. Bear, M. F. and Malenka, R. C. (1994) Synaptic plasticity: LTP and LTD. *Curr. Opin. Neurobiol.* **4,** 389–399.
24. Davis, M., Falls, W. A., Campeau, S., and Kim, M. (1993) Fear-potentiated startle: a neural and pharmacological analysis. *Behav. Brain Res.* **58,** 175–198.
25. Trujillo, K. A. and Akil, H. (1991) Inhibition of morphine tolerance and dependence by the NMDA receptor antagonist MK-801. *Science* **251,** 85–87.
26. Marek, P., Ben-Eliyahu, S., Gold, M., and Liebeskind, J. C. (1991) Excitatory amino acid antagonists (kynurenic acid and MK-801) attenuate the development of morphine tolerance in the rat. *Brain Res.* **547,** 77–81.
27. Marek, P., Ben-Eliyahu, S., Vaccarino, A. L., and Liebeskind, J. C. (1991) Delayed application of MK-801 attenuates development of morphine tolerance in rats. *Brain Res.* **558,** 163–165.
28. Trujillo, K. A., Warmoth, K. P., Peterson, D. J., Albertson, D. N., and Swadley-Lewellen, R. M. (2000) NMDA receptor blockade mimics tolerance to the locomotor effects of morphine. International Narcotics Research Conference, Seattle, WA.
29. Bhargava, H. N. and Matwyshyn, G. A. (1993) Dizocilpine (MK-801) blocks tolerance to the analgesic but not to the hyperthermic effect of morphine in the rat. *Pharmacology.* **47,** 344–350.
30. Gutstein, H. B. and Trujillo, K. A. (1993) MK-801 inhibits the development of morphine tolerance at spinal sites. *Brain Res.* **626,** 332–334.
31. Kest, B., Mogil, J. S., Shamgar, B.-E., et al. (1993) The NMDA receptor antagonist MK-801 protects against the development of morphine tolerance after intrathecal administration. *Proc. West. Pharmacol. Soc.* **36,** 307–310.
32. Lutfy, K., Hurlbut, D. E., and Weber, E. (1993) Blockade of morphine-induced analgesia and tolerance in mice by MK-801. *Brain Res.* **616,** 83–88.
33. Tiseo, P. J. and Inturrisi, C. E. (1993) Attenuation and reversal of morphine tolerance by the competitive *N*-methyl-D-aspartate receptor antagonist LY274614. *J. Pharmacol. Exp. Ther.* **264,** 1090–1096.
34. Elliott, K., Minami, N., Kolesnikov, Y. A., Pasternak, G. W., and Inturrisi, C. E. (1994) The NMDA receptor antagonists, LY274614 and MK-801, and the nitric oxide synthase inhibitor, $N^G$-nitro-L-arginine, attenuate analgesic tolerance to the mu-opioid morphine but not to kappa opioids. *Pain* **56,** 69–75.
35. Mao, J., Price, D. D., and Mayer, D. J. (1994) Thermal hyperalgesia in association with the development of morphine tolerance in rats: roles of excitatory amino acid receptors and protein kinase C. *J. Neurosci.* **14,** 2301–2312.
36. Trujillo, K. A. and Akil, H. (1994) Inhibition of opiate tolerance by non-competitive *N*-methyl-D-aspartate receptor antagonists. *Brain Res.* **633,** 178–188.
37. Mao, J., Price, D. D., Phillips, L. L., Lu, J., and Mayer, D. J. (1995) Increases in protein kinase C immunoreactivity in the spinal cord of rats associated with tolerance to the analgesic effects of morphine. *Brain Res.* **677,** 257–267.
38. Verma, A. and Kulkarni, S. K. (1995) Role of D1/D2 dopamine and *N*-methyl-D-aspartate (NMDA) receptors in morphine tolerance and dependence in mice. *Eur. Neuropsychopharmacol* **5,** 81–87.
39. Bilsky, E. J., Inturrisi, C. E., Sadee, W., Hruby, V. J., and Porreca, F. (1996) Competitive and non-competitive NMDA antagonists block the development of antinociceptive tolerance to morphine, but not to selective μ or δ opioid agonists in mice. *Pain* **68,** 229–237.
40. Dunbar, S. and Yaksh, T. L. (1996) Concurrent spinal infusion of MK801 blocks spinal tolerance and dependence induced by chronic intrathecal morphine in the rat. *Anesthesiology* **84,** 1177–1188.
41. Lutfy, K., Shen, K. Z., Woodward, R. M., and Weber, E. (1996) Inhibition of morphine tolerance by NMDA receptor antagonists in the formalin test. *Brain Res.* **731,** 171–181.
42. Manning, B. H., Mao, J., Frenk, H., Price, D. D., and Mayer, D. J. (1996) Continuous co-administration of dextromethorphan or MK-801 with morphine: attenuation of morphine dependence and naloxone-reversible attenuation of morphine tolerance. *Pain* **67,** 79–88.
43. Wong, C.-S., Cherng, C.-H., Luk, H. N., Ho, S.-T., and Tung, C.-S. (1996) Effects of NMDA receptor antagonists on inhibition of morphine tolerance in rats: binding at μ-opioid receptors. *Eur. J. Pharmacol.* **297,** 27–33.
44. Fairbanks, C. A. and Wilcox, G. L. (1997) Acute tolerance to spinally administered morphine compares mechanistically with chronically induced morphine tolerance. *J. Pharmacol. Exp. Ther.* **282,** 1408–1417.
45. Gonzalez, P., Cabello, P., Germany, A., Norris, B., and Contreras, E. (1997) Decrease of tolerance to, and dependence on morphine by, glutamate receptor antagonists. *Eur. J. Pharmacol.* **332,** 257–262.
46. McLemore, G. L., Kest, B., and Inturrisi, C. E. (1997) The effects of LY293558, and AMPA receptor antagonist, on acute and chronic morphine dependence. *Brain Res.* **778,** 120–126.
47. Vaccarino, A. L. and Clavier, M. C. (1997) Blockade of tolerance to stress-induced analgesia by MK-801 in mice. *Pharmacol. Biochem. Behav.* **56,** 435–439.
48. Belozertseva, I. V. and Bespalov, A. (1998) Effects of NMDA receptor channel blockers, dizocilpine and memantine, on the development of opiate analgesic tolerance induced by repeated morphine exposures or social defeats in mice. *Naunyn-Schmiedebergs Arch. Pharmacol.* **358,** 270–274.
49. Larcher, A., Laulin, J. P., Celerier, E., Le Moal, M., and Simonnet, G. (1998) Acute tolerance associated with a single opiate administration: involvement of *N*-methyl-D-aspartate-dependent pain facilitatory systems. *Neuroscience* **84,** 583–589.

50. McCarthy, R. J., Kroin, J. S., Tuman, K. J., Penn, R. D., and Ivankovich, A. D. (1998) Antinociceptive potentiation and attenuation of tolerance by intrathecal co-infusion of magnesium sulfate and morphine in rats. *Anesth. Analg.* **86,** 830–836.
51. Mao, J., Price, D. D., Lu, J., and Mayer, D. J. (1998) Antinociceptive tolerance to the mu-opioid agonist DAMGO is dose-dependently reduced by MK-801 in rats. *Neurosci. Lett.* **250,** 193–196.
52. Kolesnikov, Y. A. and Pasternak, G. W. (1999) Topical opioids in mice: analgesia and reversal of tolerance by a topical *N*-methyl-D-aspartate antagonist. *J. Pharmacol. Exp. Ther.* **290,** 247–252.
53. Koyuncuoglu, H., Nurten, A., Yamanturk, P., and Nurten, R. (1999). The importance of the number of NMDA receptors in the develpment of supersensitivity or tolerance to and dependence on morphine. *Pharmacol. Res.* **39,** 311–319.
54. Laulin, J. P., Celerier, E., Larcher, A., Le Moal, M., and Simonnet, G. (1999) Opiate tolerance to daily heroin administration: an apparent phenomenon associated with enhanced pain sensitivity. *Neuroscience* **89,** 631–636.
55. Shimoyama, N., Shimoyama, M., Inturrisi, C. E., and Elliott, K. J. (1996) Ketamine attenuates and reverses morphine tolerance in rats. *Anesthesiology* **85,** 1357–1366.
56. Kolesnikov, Y. A. and Pasternak, G. W. (1999) Peripheral blockade of topical morphine tolerance by ketamine. *Eur. J. Pharmacol.* **374,** R1–R2.
57. Elliott, K., Hynansky, A., and Inturrisi, C. E. (1994) Dextromethorphan attenuates and reverses analgesic tolerance to morphine. *Pain* **59,** 361–368.
58. Mao, J., Price, D. D., Caruso, F. S., and Mayer, D. J. (1996) Oral administration of dextromethorphan prevents the development of morphine tolerance and dependence in rats. *Pain* **67,** 361–368.
59. Popik, P. and Kozela, E. (1999) Clinically available NMDA antagonist, memantine, attenuates tolerance to analgesic effects of morphine in a mouse tail flick test. *Pol. J. Pharmacol.* **51,** 223–231.
60. Cao, Y. J. and Bhargava, H. N. (1997) Effects of ibogaine on the development of tolerance to antinociceptive action of mu-, delta- and kappa-opioid receptor agonists in mice. *Brain Res.* **752,** 250–254.
61. Davis, A. M. and Inturrisi, C. E. (1999) *d*-Methadone blocks morphine tolerance and *N*-methyl-D-aspartate-induced hyperalgesia. *J. Pharmacol. Exp. Ther.* **289,** 1048–1053.
62. Kolesnikov, Y. A., Ferkany, J., and Pasternak, G. W. (1993) Blockade of mu and kappa$_1$ opioid analgesic tolerance by NPC17742, a novel NMDA antagonist. *Life Sci.* **53,** 1489–1494.
63. Tiseo, P. J., Cheng, J., Pasternak, G. W., and Inturrisi, C. E. (1994) Modulation of morphine tolerance by the competitive *N*-methyl-D-aspartate receptor antagonist LY274614: assessment of opioid receptor changes. *J. Pharmacol. Exp. Ther.* **268,** 195–201.
64. Allen, R. M. and Dykstra, L. A. (1999) The competitive NMDA receptor antagonist LY235959 modulates the progression of morphine tolerance in rats. *Psychopharmacology* **142,** 209–214.
65. Allen, R. M., and Dykstra, L. A. (2000) Role of morphine maintenance dose in the development of tolerance and its attenuation by an NMDA receptor antagonist. *Psychopharmacology* **148,** 59–65.
66. Kolesnikov, Y., Maccechini, M. L., and Pasternak, G. W. (1994) 1-Aminocyclopropane carboxylic acid (ACPC) prevents mu and delta opioid tolerance. *Life Sci.* **55,** 1393–1398.
67. Lutfy, K., Shen, K. Z., Kwon, I. S., et al. (1995) Blockade of morphine tolerance by ACEA-1328, a novel NMDA receptor/glycine site antagonist. *Eur. J. Pharmacol.* **273,** 187–189.
68. Lutfy, K., Doan, P., and Weber, E. (1999) ACEA-1328 a NMDA receptor/glycine site antagonist, acutely potentiates antinociception and chronically attenuates tolerance induced by morphine. *Pharmacol. Res.* **40,** 435–442.
69. Bhargava, H. N. (1994) Diversity of agents that modify opioid tolerance, physical dependence, abstinence syndrome, and self-administrative behavior. *Pharmacol. Rev.* **46,** 293–324.
70. Trujillo, K. A. (2000) Are NMDA receptors involved in opiate-induced neural and behavioral plasticity? A review of preclinical findings. *Psychopharmacology* **151,** 121–141.
71. Herman, B. H., Vocci, F., and Bridge, P. (1995) The effects of NMDA receptor antagonists and nitric oxide synthase inhibitors on opioid tolerance and withdrawal. Medication development issues for opiate addiction. *Neuropsychopharmacology* **13,** 269–293.
72. Inturrisi, C. E. (1997) Preclinical evidence for a role of glutamatergic systems in opioid tolerance and dependence. *Semin. Neurosci.* **9,** 110–119.
73. Pasternak, G. W., Kolesnikov, Y. A., and Babey, A. M. (1995) Perspectives on the *N*-methyl-D-aspartate/nitric oxide cascade and opioid tolerance. *Neuropsychopharmacology* **13,** 309–313.
74. Jeziorski, M., White, F. J., and Wolf, M. E. (1994) MK-801 prevents the development of behavioral senstization during repeated morphine administration. *Synapse* **16,** 137–147.
75. Tzschentke, T. M. and Schmidt, W. J. (1996) Procedural examination of behavioral sensitisation to morphine: lack of blockade by MK-801, occurrence of sensitised sniffing, and evidence for cross-sensitisation between morphine and MK-801. *Behav. Pharmacol.* **7,** 169–184.
76. Swadley-Lewellen, R. M. and Trujillo, K. A. (1998) Are NMDA receptors involved in tolerance and sensitization to the locomotor effects of morphine? *Soc. Neurosci. Abstr.* **24,** 1968.

77. Bespalov, A. Y., Balster, R. L., and Beardsley, P. M. (1999) N-Methyl-D-aspartate receptor antagonists and the development of tolerance to the discriminitive stimulus effects of morphine in rats. *J. Pharmacol. Exp. Ther.* **290**, 20–27.
78. Babbini, M. and Davis, W. M. (1972) Time–dose relationships for locomotor activity effects of morphine after acute or repeated treatment. *Br. J. Pharmacol.* **46**, 213–224.
79. Brady, L. S. and Holtzman, S. G. (1981) Locomotor behavior in morphine-dependent and post-dependent rats. *Pharmacol. Biochem. Behav.* **14**, 361–370.
80. Domino, E. F., Vaski, M. R., and Wilson, A. E. (1976) Mixed depressant and stimulant actions of morphine and their relationship to brain acetylcholine. *Life Sci.* **18**, 361–376.
81. Bartoletti, M., Gairdi, M., Gubellini, C., Bacchi, A., and Babbini, M. (1987) Previous treatment with morphine and sensitization to the excitatory actions of opiates: dose–effect relationship. *Neuropharmacology* **26**, 115–119.
82. Babbini, M., Gaiardi, M., and Bartoletti, M. (1975) Persistence of chronic morphine effects upon activity in rats 8 months after ceasing the treatment. *Neuropharmacology.* **14**, 611–614.
83. Vasko, M. R. and Domino, E. F. (1978) Tolerance development to the biphasic effects of morphine on locomotor activity and brain acetylcholine in the rat. *J. Pharmacol. Exp. Ther.* **207**, 848–858.
84. Swadley-Lewellen, R. M., Watorski, K., Warmoth, K., and Trujillo, K. A. (1998) Acute behavioral interactions between opiates and NMDA receptor antagonists in rats. *NIDA Res. Monogr.* **179**, 95.
85. Trujillo, K. A. and Swadley-Lewellen, R. M. (1997) Tolerance, sensitization and physical dependence to the locomotor effects of morphine in rats. *Soc. Neurosci. Abstr.* **23**, 2413.
86. Trujillo, K. A. and Watorski, K. (1999) Does MK-801 affect the expression of opiate tolerance, sensitization or physical dependence. *Soc. Neurosci. Abstr.* **25**, 2076.
87. Rauhala, P., Idanpaan-Heikkila, J. J., Tuominen, R. K., and Mannisto, P. T. (1994) N-Nitro-L-arginine attenuates development of tolerance to antinociceptive but not to hormonal effects of morphine. *Eur. J. Pharmacol.* **259**, 57–64.
88. Fernandes, M., Kluwe, S., and Coper, H. (1982) Development and loss of tolerance to morphine in the rat. *Psychopharmacology* **78**, 234–238.
89. Rauhala, P., Idanpaan-Heikkila, J. J., Tuominen, R. K., and Mannisto, P. T. (1995) Differential disappearance of tolerance to thermal, hormonal and locomotor effects of morphine in the male rat. *Eur. J. Pharmacol.* **285**, 69–77.
90. Mannisto, P. T., Borisenko, S. A., Rauhala, P., Tuomainen, P., and Tuominen, R. K. (1994) Variation in tolerance to the antinociceptive, hormonal and thermal effects of morphine after a 5-day pre-treatment of male rats with increasing doses of morphine. *Naunyn-Schmiedebergs Arch. Pharmacol.* **349**, 161–169.
91. Cox, B. M. (1991) Molecular and cellular mechanisms in opioid tolerance, in *Towards a New Pharmacotherapy of Pain* (Basbaum, A. I. and Besson, J.-M., eds.), Wiley, New York, pp. 137–156.
92. Cox, B. M. and Werling, L. L. (1991) Opioid tolerance and dependence, in *The Biological Basis of Drug Tolerance and Dependence*, (Pratt, J. A., ed.), Academic, San Diego, CA, pp. 199–229.
93. Trujillo, K. A. (1999) Cellular and molecular and mechanisms of opioid tolerance and dependence: progress and pitfalls. *Pain Forum* **8**, 29–33.
94. Nestler, E. J. and Aghajanian, E. J. (1997) Molecular and cellular basis of addiction. *Science* **278**, 58–63.
95. Wolf, M. E., and Jeziorski, M. (1993) Coadministration of MK-801 with amphetamine, cocaine or morphine prevents rather than transiently masks the development of behavioral sensitization. *Brain Res.* **613**, 291–294.
98. Vanderschuren, L. J., Schoffelmeer, A. N., and De Vries, T. J. (1997) Does dizocilpine (MK-801) inhibit the development of morphine-induced behavioral sensitization in rats? *Life Sci.* **61**, 427–433.
97. Iijima, Y., Asami, T., and Kuribara, H. (1996) Modification by MK-801 (dizocilpine), a noncompetitive NMDA receptor antagonist, of morphine sensitization: evaluation by ambulation in mice. *Jpn. J. Psychopharmacol.* **16**, 11–18.
96. Livezey, R. T., Pearce, L. B., and Kornetsky, C. (1995) The effect of MK-801 and SCH23390 on the expression and sensitization of morphine-induced oral stereotypy. *Brain Res.* **692**, 93–98.
99. Stephens, D. N. (1995) A glutamatergic hypothesis of drug dependence: extrapolations from benzodiazepine receptor ligands. *Behav. Pharmacol.* **6**, 425–446.
100. Kalivas, P. W. (1995) Interactions between dopamine and excitatory amino acids in behavioral sensitization to psychostimulants. *Drug Alcohol Depend.* **37**, 95–100.
101. Ben-Eliyahu, S., Marek, P., Vaccarino, A. L., et al. (1992) The NMDA receptor antagonist MK-801 prevents long-lasting non-associative morphine tolerance in the rat. *Brain Res.* **575**, 304–308.
102. Fundytus, M. E. and Coderre, T. J. (1994) Effects of activity at metabotropic, as well as ionotropic (NMDA), glutamate recepotrs on morphine dependence. *Br. J. Pharmacol.* **113**, 1215–1220.
103. Makimura, M., Sugimoto, H., Shinomiya, K., Kabasawa, Y., and Fukuda, H. (1996) Inhibitory effect of the NMDA receptor antagonist, dizocilpine (MK-801), on the development of morphine dependence. *J. Toxicol. Sci.* **21**, 135–141.
104. Shoemaker, W. J., Kosten, T. A., and Muly, S. M. (1997) Ethanol attenuation of morphine dependence: comparison to dizocilpine. *Psychopharmacology* **134**, 83–87.
105. Laulin, J.-P., Larher, A., Celerier, E., Le Moal, M., and Simonnet, G. (1998) Long-lasting increased pain sensitivity in rat following exposure to heroin for the first time. *Eur. J. Neurosci.* **10**, 782–785.

106. Celerier, E., Laulin, J., Larcher, A., Le Moal, M., and Simonnet, G. (1999) Evidence for opiate-activated NMDA processes masking opiate analgesia in rats. *Brain Res.* **847,** 18–25.
107. Celerier, E., Rivat, C., Jun, Y., et al. (2000) Long-lasting hyperalgesia induced by fentanyl in rats: preventive effect of ketamine [see comments]. *Anesthesiology* **92,** 465–472.
108. Zhu, H. and Ho, I. K. (1998) NMDA-R1 antisense oligonucleotide attenuates withdrawal signs from morphine. *Eur. J. Pharmacol.* **352,** 151–156.
109. Koyuncuoglu, H., Gungor, M., Sagduyu, H., and Aricioglu, F. (1990) Suppression by ketamine and dextromethorphan of precipitated abstinence syndrome in rats. *Pharmacol. Biochem. Behav.* **35,** 829–832.
110. Koyuncuoglu, H. and Saydam, B. (1990) The treatment of heroin addicts with dextromethorphan. A double-blind comparison of dextromethorphan with chlorpromazine. *Int. J. Clin. Pharmacol. Ther. Toxicol.* **28,** 147–452.
111. Rasmussen, K., Krystal, J. H., and Aghajanian, G. K. (1991) Excitatory amino acids and morphine withdrawal: differential effects of central and peripheral kynurenic acid administration. *Psychopharmacology* **105,** 508–512.
112. Rasmussen, K., Fuller, R. W., Stockton, M. E., et al. (1991) NMDA receptor antagonists suppress behaviors but not norepinephrine turnover or locus coeruleus unit activity induced by opiate withdrawal. *Eur. J. Pharmacol.* **197,** 9–16.
113. Tanganelli, S., Antonelli, T., Morari, M., Bianchi, C., and Beani, L. (1991) Glutamate antagonists prevent morphine withdrawal in mice and guinea pigs. *Neurosci. Lett.* **122,** 270–272.
114. Koyuncuoglu, H., Dizdar, Y., Aricioglu, F., and Sayin, U. (1992) Effects of MK-801 on morphine physical dependence: attenuation and intensification. *Pharmacol. Biochem. Behav.* **43,** 487–490.
115. Higgins, G. A., Nguyen, P., and Sellers, E. M. (1992) The NMDA antagonist dizocilpine (MK-801) attenuates motivational as well as somatic aspects of naloxone precipitated opioid withdrawal. *Life Sci.* **50,** PL167–PL172.
116. Brent, P. J., and Chahl, L. A. (1993) Enhancement of the opiate withdrawal response by antipsychotic drugs in guinea pigs is not mediated by sigma binding sites. *Eur. Neuropsychopharmacol.* **3,** 23–32.
117. Cappendijk, S. L. T., de Vries, R., and Dzoljic, M. R. (1993) Excitatory amino acid receptor antagonists and naloxone-precipitated withdrawal syndrome in morphine-dependent mice. *Eur. Neuropsychopharmacol.* **3,** 111–116.
118. Popik, P. and Skolnick, P. (1996) The NMDA antagonist memantine blocks the expression and maintenance of morphine dependence. *Pharmacol. Biochem. Behav.* **53,** 791–797.
119. Popik, P. and Danysz, W. (1997) Inhibition of reinforcing effects of morphine and motivational aspects of naloxone-precipitated opioid withdrawal by N-methyl-D-aspartate receptor antagonist, memantine. *J. Pharmacol. Exp. Ther.* **280,** 854–865.
120. Popik, P., Mamczarz, J., Fraczek, M., et al. (1998) Inhibition of reinforcing effects of morphine and naloxone-precipitated opioid withdrawal by novel glycine site and uncompetitive NMDA receptor antagonists. *Neuropharmacology* **37,** 1033–1042.
121. Bristow, L. J., Hogg, J. E., and Hutson, P. H. (1997) Competitive and glycine/NMDA receptor antagonists attenuate withdrawal-induced behaviors and increased hippocampal acetylcholine efflux in morphine-dependent rats. *Neuropharmacology* **36,** 241–250.
122. Medvedev, I. O., Dravolina, O. A., and Bespalov, A. Y. 1998. Effects of N-methyl-D-aspartate receptor antagonists on discriminative stimulus effects of naloxone in morphine-dependent rats using the Y-maze drug discrimination paradigm. *J. Pharmacol. Exp. Ther.* **286,** 1260–1268.
123. Belozertseva, I. V., Danysz, W., and Bespalov, A. Y. (2000) Short-acting NMDA receptor antagonist MRZ 2/576 produces prolonged suppression of morphine withdrawal in mice. *Naunyn-Schmiedebergs Arch. Pharmacol.* **361,** 279–282.
124. Bläsig, J., Herz, A., Reinhold, K., and Zieglgänsberger, S. (1973) Development of physical dependence on morphine in respect to time and dosage and quantification of the precipitated withdrawal syndrome in rats. *Psychopharmacology* **33,** 19–38.
125. Trujillo, K. A. and Akil, H. (1993) Does MK-801 block naloxone-precipitated opiate withdrawal? *Soc. Neurosci. Abstr.* **19,** 1246.
126. Trujillo, K. A. (1995) Effects of non-competitive N-methyl-D-aspartate receptor antagonists on opiate tolerance and physical dependence. *Neuropsychopharmacology* **13,** 301–307.
127. Popik, P., Kozela, E., and Danysz, W. (2000) Clinically available NMDA receptor antagonists memantine and dextromethorphan reverse existing tolerance to the antinociceptive effects of morphine in mice. *Naunyn-Schmiedebergs Arch. Pharmacol.* **361,** 425–432.
128. Commons, K. G., van Bockstaele, E. J., and Pfaff, D. W. (1999) Frequent colocalization of mu opioid and NMDA-type glutamate receptors at postsynaptic sites in periaqueductal gray neurons. *J. Comp. Neurol.* **408,** 549–559.
129. Gracy, K. N., Svingos, A. L., and Pickel, V. M. 1997. Dual ultrastructural localization of mu-opioid receptors and NMDA-type glutamate receptors in the shell of the rat nucleus accumbens. *J. Neurosci.* **17,** 4839–4848.
130. Wang, H., Gracy, K. N., and Pickel, V. M. (1999) Mu-opioid and NMDA-type glutamate receptors are often colocalized in spiny neurons within patches of the caudate-putamen nucleus. *J. Comp. Neurol.* **412,** 132–146.
131. Chen, L. and Huang, L. Y. (1991) Sustained potentiation of NMDA receptor-mediated glutamate responses through activation of protein kinase C by a mu opioid. *Neuron* **7,** 319–326.
132. Chen, L. and Huang, L. Y. (1992) Protein kinase C reduces $Mg^{++}$ block of NMDA-receptor channels as a mechanism of modulation. *Nature* **356,** 521–523.

133. Fan, G.-H., Zhao, J., Wu, Y. L., et al. (1998) N-Methyl-D-aspartate attenuates opioid receptor-mediated G protein activation and this process involves protein kinase C. *Mol. Pharmacol.* **53,** 684–690.
134. Martin, G., Przewlocki, R., and Siggins, G. R. (1999) Chronic morphine treatment selectively augments metabotropic glutamate receptor-induced inhibition of N-methyl-D-aspartate receptor-mediated neurotransmission in nucleus accumbens. *J. Pharmacol. Exp. Ther.* **288,** 30–35.
135. Martin, G., Nie, Z., and Siggins, G. R. (1997) Mu-Opioid receptors modulate NMDA receptor-mediated responses in nucleus accumbens neurons. *J. Neurosci.* **17,** 11–22.
136. Mao, J., Price, D. D., and Mayer, D. J. (1995) Mechanisms of hyperalgesia and morphine tolerance: a current view of their possible interactions. *Pain* **62,** 259–274.
137. Mayer, D. J., Mao, J., and Price, D. D. (1995) The association of neuropathic pain, morphine tolerance and dependence, and the translocation of protein kinase C. *NIDA Res. Monogr.* **147,** 269–298.
138. Mayer, D. J. and Mao, J. (1999) Mechanisms of opioid tolerance: a current view of cellular mechanisms. *Pain Forum* **8,** 14–18.
139. Fundytus, M. E. and Coderre, T. J. (1999) Opioid tolerance and dependence: a new model highlighting the role of metabotropic glutamate receptors. *Pain Forum* **8,** 3–13.
140. Mayer, D. J., Mao, J., and Price, D. D. (1995) The development of morphine tolerance and dependence is associated with translocation of protein kinase C. *Pain* **61,** 365–374.
141. Mestek, A., Hurley, J. H., Bye, L. S., et al. (1995) The human mu opioid receptor: modulation of functional desensitization by calcium/calmodulin-dependent protein kinase and protein kinase C. *J. Neurosci.* **15,** 2396–2406.
142. Narita, M., Makimura, M., Feng, Y., Hoskins, B., and Ho, I. K. (1994) Influence of chronic morphine treatment on protein kinase C activity: comparison with butorphanol and implication for opioid tolerance. *Brain Res.* **650,** 175–179.
143. Makimura, M., Iwai, M., Sugimoto, H., and Fukuda, H. (1997) Effect of NMDA receptor antagonists on protein kinase activated by chronic morphine treatment. *J. Toxicol. Sci.* **22,** 59–66.
144. Ventayol, P., Busquets, X., and Garcia-Sevilla, J. A. (1997) Modulation of immunoreactive protein kinase C-alpha and beta isoforms and G proteins by acute and chronic treatments with morphine and other opiate drugs in rat brain. *Naunyn-Schmiedebergs Arch. Pharmacol.* **355,** 491–500.
145. Li, Y., and Roerig, S. C. (1999) Alteration of spinal protein kinase C expression and kinetics in morphine, but not clonidine, tolerance. *Biochem. Pharmacol.* **58,** 493–501.
146. Smart, D., and Lambert, D. G. (1996) The stimulatory effects of opioids and their possible role in the development of tolerance. *Trends Pharmacol. Sci.* **17,** 264–269.
147. Smith, F. L., Lohmann, A. B., and Dewey, W. L. (1999) Involvement of phospholipid signal transduction in morphine tolerance in mice. *Br. J. Pharmacol.* **128,** 220–226.
148. Smart, D., Hirst, R. A., Hirota, K., Grandy, D. K., and Lambert, D. G. (1997) The effects of recominant rat μ-opioid receptor activation in CHO cell on phospholipase C, $[Ca^{2+}]_i$ and adenylyl cyclase. *Br. J. Pharmacol.* **120,** 1165–1171.
149. Ueda, H., Miyamae, T., Fukushima, N., et al. (1995) Opioid μ- and κ-receptor mediate phospholipase C activation through $G_{i1}$ in *Xenopus* oocytes. *Mol. Brain Res.* **32,** 166–170.
150. Xie, W., Samoriski, G. M., McLaughlin, J. P., et al. (1999) Genetic alterations of phospholipase C β3 expression modulates behavioral and cellular responses to μ opioids. *Proc. Natl. Acad. Sci. USA* **96,** 10,385–10,390.
151. Przewlocki, R., Parsons, K. L., Sweeney, D. D., et al. (1999) Opioid enhancement of calcium oscillations and burst events involving NMDA receptors and L-type calcium channels in cultured hippocampal neurons. *J. Neurosci.* **19,** 9705–9715.
152. Narita, M., Narita, M., Mizoguchi, H., and Tseng, L. F. (1995) Inhibition of protein kinase C, but not of protein kinase A, blocks the development of acute antinociceptive tolerance to an intrathecally administered μ-opioid receptor agonist in the mouse. *Eur. J. Pharmacol.* **280,** R1–R3.
153. Granados-Soto, V., Kalcheva, I., Hua, X., Newton, A., and Yaksh, T. L. (2000) Spinal PKC activity and expression: role in tolerance produced by continuous spinal morphine infusion. *Pain* **85,** 395–404.
154. Inoue, M., and Ueda, H. (2000) Protein kinase C-mediated acute tolerance to peripheral mu-opioid analgesia in the bradykinin-nociception test in mice. *J. Pharmacol. Exp. Ther.* **293,** 662–669.
155. Fundytus, M. E., and Coderre, T. J. (1996) Chronic inhibition of intracellular $Ca^{2+}$ release or protein kinase C activation significantly reduces the development of morphine dependence. *Eur. J. Pharmacol.* **300,** 173–181.
156. Xu, J. Y., Hill, K. P., and Bidlack, J. M. (1998) The nitric oxide/cyclic GMP system at the supraspinal site is involved in the development of acute morphine antinociceptive tolerance. *J. Pharmacol. Exp. Ther.* **284,** 196–201.
157. Fan, G.-H., Wang, L.-Z., Qiu, H.-C., Ma, L., and Pei, G. (1999) Inhibition of calcium/calmodulin-dependent protein kinase II in rat hippocampus attenuates morphine tolerance and dependence. *Mol. Pharm.* **56,** 39–45.
158. Cai, Y.-C., Ma, L., Fan, G.-H., et al. (1997) Activation of N-methyl-D-aspartate receptor attenuates acute responsiveness of δ-opioid receptors. *Mol. Pharm.* **51,** 583–587.
159. Chen, Y. and Yu, L. (1994) Differential regulation by cAMP-dependent protein kinase and protein kinase C of the μ opioid receptor coupling to a G protein-activated $K^+$ channel. *J. Biol. Chem.* **269,** 7839–7842.
160. Zhang, L., Yu, Y., Mackin, S., et al. (1996) Differential μ opiate receptor phosphorylation and desensitization induced by agonists and phorbol esters. *J. Biol. Chem.* **271,** 11,449–11,454.

161. Childers, S. (1993) Opioid receptor-coupled second messenger systems, in *Opioids I* Akil, H. and Simon, E. J., eds.), Springer-Verlag, New York, pp. 189–216.
162. Jordan, B. and Devi, L. A. (1998) Molecular mechanisms of opioid receptor signal transduction. *Br. J. Anaesth.* **81**, 12–19.
163. Law, P.-Y., Wong, Y. H., and Loh, H. H. (2000) Molecular mechanisms and regulation of opioid receptor signalling. *Annu. Rev. Pharmacol. Toxicol.* **40**, 389–430.
164. Wang, L. and Gintzler, A. R. (1995) Morphine tolerance and physical dependence: reversal of opioid inhibition to enhancement of cyclic AMP formation. *J. Neurochem.* **64**, 1102–1106.
165. Crain, S. M., and Shen, K.-F. (1990) Opioids can evoke direct receptor-mediated excitatory effects on sensory neurons. *Trends Pharmacol. Sci.* **11**, 77–81.
166. Crain, S. M., and Shen, K.-F. (1998) Modulation of opioid analgesia, tolerance and dependence by Gs-coupled, GM1 ganglioside-regulated opioid receptor functions. *Trends Pharmacol. Sci.* **19**, 358–365.
167. Crain, S. M. and Shen, K.-F. (1996) Modulatory effects of Gs-coupled excitatory opioid receptor functions on opioid analgesia, tolerance and dependence. *Neurochem. Res.* **21**, 1347–1351.
168. Wang, L., Medina, V. M., Rivera, M., and Gintzler, A. R. (1996) Relevance of phosphorylation state to opioid responsiveness in opiate naive and tolerant/dependent tissue. *Brain Res.* **723**, 61–69.
169. Wang, L., and Gintzler, A. R. 1997. Altered mu-opiate receptor-G protein signal transduction following chronic morphine exposure. *J. Neurochem.* **68**, 248–254.
170. Sadee, W. and Wang, Z. (1995) Agonist induced constitutive receptor activation as a novel regulatory mechanism. Mu receptor regulation. *Adv. Exp. Med. Biol.* **373**, 85–90.
171. Wang, Z., Bilsky, E. J., Porreca, F., and Sadee, W. (1994) Constituitive µ opioid receptor activation as a regulatory mechanism underlying narcotic tolerance and dependence. *Life Sci.* **54**, PL339–PL350.
172. Martin, G., Ahmed, S. H., Blank, T., et al. (1999) Chronic morphine treatment alters NMDA receptor-mediated synaptic transmission in the nucleus accumbens. *J. Neurosci.* **19**, 9081–9089.
173. Rothman, R. B. (1992) A review of the role of anti-opioid peptides in morphine tolerance and dependence. *Synapse* **12**, 129–138.
174. Trujillo, K. A., Bronstein, D. M., and Akil, H. (1993) Regulation of opioid peptides by self-administered drugs, in *The Neurobiology of Opiates* (Hammer, R. P., Jr., ed.), CRC, Boca Raton, FL, pp. 223–256.
175. Tiffany, S. T., and Maude-Griffith, P. M. (1988) Tolerance to morphine in the rat: associative and non-associative effects. *Behav. Neurosci.* **102**, 534–543.
176. Dray, A. and Urban, L. (1996) New pharmacological strategies for pain relief. *Annu. Rev. Pharmacol. Toxicol.* **36**, 253–280.
176a. Grass, S., Hoffmann, O., Xu, X. J., and Wiesenfeld, Z. (1996) $N$-methyl-D-aspartate receptor antagonists potentiate morphine's antinociceptive effect in the rat. *Acta Physiol. Scand.* **158**, 269–273.
177. Hoffmann, O. and Wiesenfeld, Z. (1996) Dextromethorphan potentiates morphine antinociception, but does not reverse tolerance in rats. *NeuroReport* **7**, 838–840.
178. Bespalov, A., Kudryashova, M., and Zvartau, E. (1998) Prolongation of morphine analgesia by competitive NMDA receptor antagonist D-CPPene (SDZ EAA 494) in rats. *Eur. J. Pharmacol.* **351**, 299–305.
179. Advokat, C. and Rhein, F. Q. (1995) Potentiation of morphine-induced antinociception in acute spinal rats by the NMDA antagonist dextrorphan. *Brain Res.* **699**, 157–160.
180. Belozertseva, I. V., Dravolina, O. A., Neznanova, O. N., Danysz, W., and Bespalov, A. Y. 2000. Antinociceptive activity of combination of morphine and NMDA receptor antagonists depends on the inter-injection interval. *Eur. J. Pharmacol.* **396**, 77–83.
181. Goldblum, R. (2000) Long-term safety of MorphiDex. *J. Pain Symptom Manage.* **19**, S50–S56.
182. Chevlen, E. (2000) Morphine with dextromethorphan: conversion from other opioid analgesics. *J. Pain Symptom Manage.* **19**, S42–S49.
183. Katz, N. P. (2000) MorphiDex (MS: DM) double-blind, multiple-dose studies in chronic pain patients. *J. Pain Symptom Manage.* **19**, S37–S41.
184. Caruso, F. S. (2000) MorphiDex pharmacokinetic studies and single-dose analgesic efficacy studies in patients with postoperative pain. *J. Pain Symptom Manage.* **19**, S31–S36.
185. Koyuncuoglu, H. (1995) The combination of tizanidine markedly improves the treatment with dextromethorphan of heroin addicted outpatients. *Int. J. Clin. Pharmacol. Ther.* **33**, 13–19.
186. Rosen, M. I., Pearsall, H. R., and Kosten, T. R. (1998) The effect of lamotrigine on naloxone-precipitated opiate withdrawal. *Drug Alcohol Depend.* **52**, 173–176.
187. Rosen, M. I., McMahon, T. J., Woods, S. W., Pearsall, H. R., and Kosten, T. R. (1996) A pilot study of dextromethorphan in naloxone-precipitated opiate withdrawal. *Eur. J. Pharmacol.* **307**, 251–257.
188. Alper, K. R., Lotsof, H. S., Frenken, G. M., Luciano, D. J., and Bastiaans, J. (1999) Treatment of acute opioid withdrawal with ibogaine. *Am. J. Addict.* **8**, 234–242.

189. Mash, D. C., Kovera, C. A., Buck, B. E., et al. (1998) Medication development of ibogaine as a pharmacotherapy for drug dependence. *Ann. N Y Acad. Sci.* **844,** 274–292.
190. Itzhak, Y. and Ali, S. F. (1998) Effect of ibogaine on the various sites of the NMDA receptor complex and sigma binding sites in rat brain. *Ann. NY Acad. Sci.* **844,** 245–251.
191. Layer, R. T., Skolnick, P., Bertha, C. M., et al. (1996) Structurally modified ibogaine analogs exhibit differing affinities for NMDA receptors. *Eur. J. Pharmacol.* **309,** 159–165.
192. Popik, P., Layer, R. T., Fossom, L. H., et al. (1995) NMDA antagonist properties of the putative antiaddictive drug, ibogaine. *J. Pharmacol. Exp. Ther.* **275,** 753–760.
193. Popik, P., Layer, R. T., and Skolnick, P. (1995) 100 years of ibogaine: neurochemical and pharmacological actions of a putative anti-addictive drug. *Pharmacol. Rev.* **47,** 235–253.
194. Sheppard, S. G. (1994) A preliminary investigation of ibogaine: case reports and recommendations for further study. *J. Subst Abuse Treat.* **11,** 379–385.
195. Tzschentke, T. M. and Schmidt, W. J. (1995) *N*-methyl-D-aspartic acid-receptor antagonists block morphine-induced conditioned place preference in rats. *Neurosci. Lett.* **193,** 37–40.
196. Kim, H.-S., Jang, C.-G., and Park, W.-K. (1996) Inhibition by MK-801 of morphine-induced conditioned place preference and postsynaptic dopamine receptor supersensitivity in mice. *Pharmacol. Biochem. Behav.* **55,** 11–17.
197. Del Pozo, E., Barrios, M., and Baeyens, J. M. (1996) The NMDA receptor antagonist dizocilpine (MK-801) stereoselectively inhibits morphine-induced place preference conditioning in mice. *Psychopharmacology* **125,** 209–213.
198. Tzschentke, T. M. and Schmidt, W. J. (1997) Interactions of MK-801 and GYKI 52466 with morphine and amphetamine in place preference conditioning and behavioural sensitization. *Behav. Brain Res.* **84,** 99–107.
199. Tzschentke, T. M. and Schmidt, W. J. (1998) Blockade of morphine- and amphetamine-induced conditioned place preference in the rat by riluzole. *Neurosci. Lett.* **242,** 114–116.
200. Kotlinska, J. and Biala, G. (1999) Effects of the NMDA/glycine receptor antagonist, L-701,324, on morphine- and cocaine-induced place preference. *Pol. J. Pharmacol.* **51,** 323–330.
201. Suzuki, T., Kato, H., Tsuda, M., Suzuki, H., and Misawa, M. (1999) Effects of the non-competitive NMDA receptor antagonist ifenprodil on the morphine-induced place preference in mice. *Life Sci.* **64,** L151–L156.
202. Semenova, S., Danysz, W., and Bespalov, A. (1999) Low-affinity NMDA receptor channel blockers inhibit acquisition of intravenous morphine self-administration in naive mice. *Eur. J. Pharmacol.* **378,** 1–8.
203. Tzschentke, T. M. and Schmidt, W. J. (2000) Effects of the non-competitive NMDA-receptor antagonist memantine on morphine- and cocaine-induced potentiation of lateral hypothalamic brain stimulation reward. *Psychopharmacology* **149,** 225–234.
204. Carlezon, W. A. and Wise, R. A. (1993) Morphine-induced potentiation of brain stimulation reward is enhanced by MK-801. *Brain Res.* **620,** 339–342.
205. Bubser, M., Keseberg, U., Notz, P. K., and Schmidt, W. J. (1992) Differential behavioral and neurochemical effects of competitive and non-competitive NMDA receptor antagonists in rats. *Eur. J. Pharmacol.* **229,** 75–82.
206. Danysz, W., Essman, U., Bresink, I., and Wilke, R. (1994) Glutamate antagonists have different effects on spontaneous locomotor activity in rats. *Pharmacol. Biochem. Behav.* **48,** 111–118.
207. Hiramatsu, M., Cho, A. K., and Nabeshima, T. (1989) Comparison of the behavioral and biochemical effects of the NMDA receptor antagonists, MK-801 and phencyclidine. *Eur. J. Pharmacol.* **166,** 359–366.
208. Tricklebank, M. D., Singh, L., Oles, R. J., Preston, C., and Iversen, S. D. (1989) The behavioral effects of MK-801: a comparison with antagonists acting non-competitively and competitively at the NMDA receptor. *Eur. J. Pharmacol.* **167,** 127–135.
209. Newcomer, J. W., Farber, N. B., Jevtovic-Todorovic, V., et al. (1999) Ketamine-induced NMDA receptor hypofunction as a model of memory impairment and psychosis. *Neuropsychopharmacology* **20,** 106–118.
210. Jentsch, J. D. and Roth, R. H. (1999) The neuropsychopharmacology of phencyclidine: from NMDA receptor hypofunction to the dopamine hypothesis of schizophrenia. *Neuropsychopharmacology* **20,** 201–225.
211. Olney, J. W. (1994) Neurotoxicity of NMDA receptor antagonists: an overview. *Psychopharmacol. Bull.* **30,** 533–540.
212. Olney, J. W. and Farber, N. B. (1995) NMDA antagonists as neurotherapeutic drugs, psychotogens, neurotoxins, and research tools for studying schizophrenia. *Neuropsychopharmacology* **13,** 335–345.
213. Domino, E. F. (1990) *Status of Ketamine in Anesthesiology,* NPP Books, Ann Arbor, MI.
214. Gutstein, H. B., Johnson, K. L., Heard, M. B., and Gregory, G. A. (1992) Oral ketamine preanesthetic medication in children. *Anesthesiology* **76,** 28–33.
215. Krystal, J. H., Karper, L. P., Seibyl, J. P., et al. (1994) Subanesthetic effects of the non-competitive NMDA antagonist, ketamine, in humans: psychotomimetic, perceptual, cognitive and neuroendocrine responses. *Arch. Gen. Psychiatry* **51,** 199–214.
216. Marshall, B. E. and Longnecker, D. E. (1996) General anesthetics, in *Goodman & Gilman's The Pharmacological Basis of Therapeutics.* (Hardman, J. G., Limbird, L. E., Molinoff, P. B., Ruddon, K. W., and Gilman, A. G., eds.), McGraw-Hill, New York, pp. 331–348.

217. O'Brien, C. P. (1996) Drug addiction and drug abuse, in *Goodman & Gilman's The Pharmacological Basis of Therapeutics* (Hardman, J. G., Limbird, L. E., Molinoff, P. B., Ruddon, R. W., and Gilman, A. G., eds.), McGraw-Hill, New York, pp. 557–578.
218. Gorelick, D. A. and Balster, R. L. (1995) Phencyclidine (PCP), in *Psychopharmacology: The Fourth Generation of Progress* (Bloom, F. E. and Kupfer, D. J., eds.), Raven, New York, pp. 1767–1776.
220. Danysz, W. and Parsons, C. G. (1998) Glycine and N-methyl-d-aspartate receptors: physiological significance and possible therapeutic applications. *Pharmacol. Rev.* **50,** 597–664.
221. Kemp, J. A. and Leeson, P. D. (1993) The glycine site of the NMDA receptor—five years on. *Trends Pharmacol. Sci.* **14,** 20–25.
222. Lipton, S. A. (1993) Prospects for clinically tolerated NMDA antagonists: open channel blockers and alternative redox states of nitric oxide. *Trends Neurosci.* **16,** 527–532.
223. Rogawski, M. A. and Porter, R. J. (1990) Antiepileptic drugs: pharmacological mechanisms and clinical efficacy with consideration of promising developmental stage compounds. *Pharmacol. Rev.* **42,** 223–286.
224. Rogawski, M. A. (1993) Therapeutic potential of excitatory amino acid antagonists: channel blockers and 2,3-benzodiazepines. *Trends Pharmacol. Sci.* **14,** 325–331.
225. Parsons, C. G., Danysz, W., and Quack, G. (1999) Memantine is a clinically well-tolerated N-methyl-d-aspartate (NMDA) receptor antagonists—a review of preclinical data. *Neuropharmacology* **38,** 735–767.
226. Willets, J., Balster, R. L., and Leander, J. D. (1990) The behavioral pharmacology of NMDA receptor antagonists. *Trends Pharmacol. Sci.* **11,** 423–428.
227. Leeson, P. D. and Iversen, L. L. (1994) The glycine site on the NMDA receptor: structure–activity relationships and therapeutic potential. *J. Med. Chem.* **37,** 4053–4065.

# 21
# Modification of Conditioned Reward by N-Methyl-D-aspartate Receptor Antagonists

## Piotr Popik, MD, PhD

## 1. INTRODUCTION

Alcohol and other drugs of abuse produce dependence, a chronic and relapsing disorder that is an enormously destructive public health problem. In the United States, mortality that can be attributed to these disorders is greater than mortality attributable to all other factors combined (1). Currently, there are a limited number of available medications to treat alcohol, opioid, and nicotine dependence, and no medication to treat cocaine, other stimulants, or cannabis dependence. Success rate of available treatments is relatively low, only 30–50% of patients remain abstinent at 1 yr after completion of the most successful treatment programs (2,3). Despite recent advances in the understanding of the neurobiological basis for these disorders and the development of new psychotherapeutic approaches, there is a lack of viable pharmacological treatments. At the same time, societal and political pressures continue for the development of effective and inexpensive treatments.

Substances that are abused by humans, including opiates, psychostimulants, marijuana, alcohol, and sedatives, differ in chemical structures and acute pharmacological effects. However, the pathological, compulsive pattern of their intake share many common features. This suggests that a limited number of neural pathways may mediate the "addictive" qualities of drugs of abuse. If there is a common neural substrate that is central to the development and the maintenance of drugs of abuse dependence, it might be a target for pharmacological treatments. It has previously been suggested (4–6) that N-methyl-D-aspartate (NMDA) receptors constitute the target component of this pathway. Preclinical research suggests that antagonists of this receptor modulate pathophysiological processes common to the development, maintenance, and expression of drugs of abuse dependence and have a potential to be developed as pharmacotherapeutics. Preliminary results from clinical trials are encouraging further clinical development.

## 2. CONDITIONED REWARD

Animals and people associate internal drive states with the knowledge about the external environment in which these states were experienced. The use of such associations provide means for a better quality of life. This is because learning that some environmental stimuli may be predictive of the occurrence of primary reinforcers (unconditioned stimuli, UCS) enhance chances of survival. For example, the knowledge that certain stimuli (conditioned stimuli, CS) are signaling the availability of food, sexual partner, or another potent reinforcer, such as a drug of abuse, increases its availability, and thus enhances the (subjective) well-being of an individual (7).

The laboratory model to test the development and expression of such associations is the conditioned place preference (CPP) test (8,9). It allows investigating associations of biologically relevant

From: *Contemporary Clinical Neuroscience: Glutamate and Addiction*
Edited by: Barbara H. Herman et al. © Humana Press Inc., Totowa, NJ

stimuli (internal drive states produced by primary reinforcers) with previously neutral contextual cues; such associations develop following contingent pairings. In its simplest form, the CPP apparatus is made of two or more distinctive environments (compartments) that differ in visual, olfactory, and tactile cues. During conditioning, a drug effect is consistently paired with one chamber, whereas the other chamber is paired with the effect of the vehicle. After several pairings, the testing phase is carried out, during which the animal (now drug-free and having a free choice) demonstrates a preference to a drug-paired environment. This is measured by the time that the animal spends in drug-associated chamber. All of the numerous variants of this procedure are based on the assumption that an animal learns to approach stimuli paired with rewarding properties of the drug and that, through repetitive pairing, an initially neutral environment gains incentive salience. These procedures are related to operant second-order schedule studies, which have shown that exteroceptive stimuli can exert powerful control over drug-seeking and drug-taking *(10)*. CPP is now among the most commonly used methods for the assessment of rewarding effects of pharmacological and nonpharmacological reinforcers *(11)* (see refs. *12* and *13* for a review). This paradigm represents an indirect measure of drug reinforcement (as opposed to, e.g., the drug self-administration procedure). There are a number of advantages of the CPP technique over those techniques based on other measures of drug reward, like drug self-administration procedure. These include extreme sensitivity to the low doses of drugs of abuse [e.g., 0.08 mg/kg of morphine *(11)*] and the ability to produce CPP response after a single drug-environment pairing *(11,14)*. The single pairing is important in light of human research indicating that the effect of the initial drug experience may be the best predictor of later drug abuse *(15)* and in light of animal data demonstrating the development of tolerance *(16)* accompanying administration of several doses of an opiate that could be a confounding factor in studying drug reward. Moreover, in CPP studies, both the rewarding as well as aversive properties of drugs may be measured at the same time. In this paradigm, the testing is conducted under drug-free conditions (eliminating, e.g., motor effects of drugs that may obscure the measurement of its reinforcing effects in other tests). In addition, most often, drugs are given parenterally, which is impossible in self-administration studies. The CPP paradigm allows administration of known and planned drug doses (which is difficult in paradigms where animals are themselves administering the drugs and, therefore, the dose depends on the animal's rate of responding). Finally, it is possible to study nonpharmacological rewards as well; these include the rewarding properties of food and that of social and sexual interactions.

## 3. EFFECT OF NMDA RECEPTOR ANTAGONISTS ON THE ACQUISITION OF CPP PRODUCED BY DRUGS OF ABUSE

To determine the putative effects of NMDA receptor antagonists on the acquisition of drug-induced CPP, one can give injections of NMDA receptor antagonists during conditioning. For morphine reward, such inhibitory effects have been demonstrated for the nonselective glutamate antagonist, kynurenic acid *(17)*, MK-801 (dizocilpine, NMDA receptor channel blocker) *(18–20)*, MRZ 2/570 and L-701,324 (glycine/NMDA receptor antagonists) *(21,22)*, MRZ 2/579 (NMDA receptor channel blocker) *(21)*, and ifenprodil (polyamine-site antagonist selective for NR2B-containing NMDA receptors) *(23)*. Similar inhibitory effects on the acquisition of morphine-induced CPP have been reported for CGP 37849 and NPC 17742 (competitive NMDA receptor antagonists) *(19,24)*, the glutamate release inhibitor riluzole *(25)*, and ACPC (a partial agonist at strychnine-insensitive glycine/NMDA sites) *(26)*. Data from our laboratory demonstrate similar inhibitory effects of memantine (NMDA receptor channel blocker and clinically available NMDA receptor antagonist) *(27)*. Memantine given at the doses that inhibited morphine reward appear not to alter food reward or the interoceptive properties of morphine in the drug-discrimination paradigm *(27)*.

With regard to other drugs of abuse, the data are less abundant. Pretreatment with dextromethorphan (NMDA receptor channel blocker and clinically available NMDA receptor antagonist) dose-dependently decreased the CPP for 10 and 20 mg cocaine/kg but increased the CPP for 2.5–5 mg/kg of cocaine, suggesting a biphasic effect on cocaine-induced CPP *(28)*. The development of cocaine-induced CPP was also attenuated by MK-801 *(29,30)* and ACPC *(26)* (which, in addition, blocked the development of CPP induced by amphetamine, nomifensine, nicotine, and diazepam). Furthermore, metamphetamine-induced *(31)* but not amphetamine-induced *(32)* CPP was blocked by MK-801. MK-801 and L-701,324 attenuated also the development of CPP induced by alcohol *(33)*.

Regarding the specificity of inhibitory effects on conditioned drug reward, the effects of MK-801 on morphine-induced CPP appear not to result from the state-dependent learning *(20)* and the data reported by Del Pozo et al. *(34)* demonstrate that this inhibition is stereoselective. The data on the rewarding effects of NMDA receptor antagonists itself remain controversial. Thus, memantine *(27)*, kynurenic acid *(17)*, dextrorphan (NMDA receptor channel blocker) *(11)*, L-701,324 *(33)*, or ifenprodil *(23)* produced neither place preference nor aversion. On the other hand, MK-801 produced CPP in most *(32–35)* but not all *(19)* studies, and such effects appear unrelated to the dose and are not stereoselective. CGP37849 produced a small but significant CPP *(19)* and phencyclidine (NMDA receptor channel blocker and abused substance) produced place aversion *(36)*.

The effects of NMDA receptor antagonists on the acquisition of drug-induced conditioned place preference appear to be extremely consistent across the literature and provide substantial insight to the mechanisms of the development of drug addiction. However, by definition, their outcome is not that significant to the clinical applications.

## 4. EFFECT OF NMDA RECEPTOR ANTAGONISTS ON THE EXPRESSION OF CPP PRODUCED BY DRUGS OF ABUSE

To investigate the putative effects of NMDA receptor antagonists on the expression of drug-induced CPP, one can give a single injection of NMDA antagonist before the final measurement (posttest). Studying the inhibitory effects on the *expression* of conditioned reward produced by drugs (rather than on the *acquisition* of drug–environment associations) approaches the therapeutic standpoint much closer. It may be hypothesized that the treatment able to diminish the secondarily rewarding effects of drugs would likely diminish the incentive properties of drug-related cues and environments. It has been well documented that individuals recovering from drug and alcohol addiction remain highly susceptible to the environmental stimuli associated with previous exposure to drugs *(37,38)*. This abnormal reactivity of the neural systems, altered by repetitive pairing of drug effects and specific environments or situations, represents a form of learning and is postulated to underlie the vulnerability to relapse for a long time after the termination of drug use *(39)*. The main feature of the ideal pharmacological treatment for addictions would thus be the protective effect against the conditioned and other factors that are known to increase risk of the relapse.

The inhibitory effects of NMDA receptor antagonists on the expression of morphine-induced CPP have been shown for kynurenic acid *(17)*, ACPC *(26)*, NPC 17742 *(24)*, and MK-801 *(20)*. Data from this laboratory showed similar inhibitory effects of memantine, MRZ 2/570, MRZ 2/579, and L-701,324 on the expression of morphine-induced CPP *(21,27)*. Memantine appeared to act selectively, because given at the doses that inhibited morphine reward, it did not affect the expression of food-induced CPP.

The effects of intracerebral injections of NMDA receptor antagonist on the expression of morphine-induced CPP were investigated by Popik and Kolasiewicz *(24)*. These data provide some insight into the mechanism by which NMDA receptor antagonists inhibit the expression of conditioned reward produced by drugs of abuse. Several lines of evidence indicate the essential involvement of mesolimbic areas as the major neural substrate of the reward produced by drugs of abuse *(40)*. Among other elements, this system consists of dopaminergic cells in the ventral tegmental area and their projections to the regions such as nucleus accumbens. Opiates potently activate dopaminergic mesolimbic neurons through the inhibition of inhibitory GABAergic interneurons *(41)* that subsequently increases dopaminergic transmission to the nucleus accumbens *(42)*. Such activation is thought to be associated with a reward or the feeling of euphoria that—through a complicated and not completely understood mechanism(s)—may ultimately lead to compulsive drug-seeking and drug-taking.

NPC 17742 has been found to significantly reduce the expression of morphine-induced CPP when injected into the nucleus accumbens and ventral tegmental area *(24)*. Although the effects of the higher dose of intra-accumbens NPC 17742 produced behavioral stimulation, intrategmental injection did not change it. These findings demonstrate that stimulation of NMDA receptors in the nucleus accumbens and ventral tegmental area are necessary for the expression of morphine-induced CPP and suggest that alteration of locomotor activity produced by NMDA receptor antagonists are not essential for these agents to be effective attenuating reinforcing effects of morphine.

With regard to other drugs of abuse, the data, again, are less abundant. The expression of cocaine-induced CPP was unaffected by MK-801 *(29)* or L-701,324 *(22)*. Gruca and Papp *(26)* reported that ACPC blocked the expression of CPP induced by nicotine and diazepam, but not that induced by cocaine, amphetamine, and nomifensine.

## 5. THE EFFECT OF NMDA RECEPTOR ANTAGONISTS ON THE MAINTENANCE OF CPP PRODUCED BY DRUGS OF ABUSE

The inhibitory effects of NMDA receptor antagonists on the acquisition and expression of conditioned reward produced by drugs of abuse discussed in the preceding sections, if applied within clinical settings, would perhaps facilitate the process of detoxification, although they may not lead to the elimination of drug addiction *per se*. Therefore, it may be hypothesized that the ideal "antiaddictive" medication should be able to inhibit the process of ongoing addiction/dependence (i.e., to inhibit and/or eliminate the maintenance of the addictive state). In fact, CPP produced by heroin *(43)* and morphine *(21)* lasts for several days or even weeks, indicating that the state of association between the unconditioned drug stimulus and conditioned environmental cues is maintained for a substantial time. Similar observations have been made in human addicts and show that a presentation of a stimulus associated with heroin taking may evoke heroin craving long after detoxification *(39)*.

It was hypothesized *(21)* that the treatment with a NMDA receptor antagonist should inhibit the maintenance of conditioned morphine reward (i.e., the abnormal reactivity of the neural systems altered by repetitive pairing of drug effects and specific environments), as it inhibits the maintenance of morphine dependence *(44)*. Thus, groups of rats were conditioned to morphine-rewarding effects, and at the end of conditioning, their preference to the morphine-associated environment has been found, as expected. During the subsequent 3 d (i.e., after conditioning), these subjects received twice a day injections of the glycine/NMDA antagonist L-701,324 or MRZ 2/570, or the channel blocker MRZ 2/579. The preference to the morphine-associated environment investigated in drug-free rats on the next day indicated that the treatment with NMDA receptor antagonists did not influence the maintenance of the conditioned response to morphine *(21)*. These data suggest that NMDA receptor antagonists may not influence the maintenance of CPP induced by morphine.

## 6. EFFECTS OF NMDA RECEPTOR ANTAGONISTS IN OTHER MODELS OF CONDITIONED REWARD

The above- summarized findings are supported by the still limited evidence generated using other types of drug conditioning paradigm such as conditioned facilitation of intracranial self-stimulation *(45)* and conditioned reinstatement of intravenous cocaine self-administration *(46)*. Interestingly, NMDA receptor blockade may effectively counteract the ability of abused drugs (amphetamine) to potentiate the responding for conditioned reward *(47)* (see, however, ref. *48* for opposite effect with memantine used at a fairly high dose).

In conclusion, antagonists of NMDA receptor appear to inhibit a conditioned reward produced by drugs of abuse (particularly opiates and psychostimulants). Preliminary clinical findings described in other chapters of this volume strongly suggest that the same effects can be observed in humans. It could be expected that a medication that has NMDA receptor antagonist properties might be a useful pharmacotherapy for disorders related to abuse of a variety of substances. The most promising candidates for such development in the near future include low-affinity NMDA receptor channel blockers like memantine, dextromethorphan, and MRZ 2/579 (Nerimexane®). Others include glycine-site antagonists and, perhaps, other noncompetitive antagonists.

## ACKNOWLEDGMENTS

The preparation of this chapter was supported by KBN grant no. P05A 04217.

## REFERENCES

1. McGinnis, J. L. and Foege, W. H. (1993) Actual causes of death in the United States. *JAMA* **270**, 2207–2212.
2. Ball, J. C. and Ross, A. (1991) Follow-up study of 105 patients who left treatment, in *The Effectiveness of Methadone Maintenance Treatment. Patients, Programs, Services, and Outcome* (Ball, J. C. and Ross, A., eds.), Springer-Verlag, New York, pp. 176–187.
3. Gonzalez, J. P. and Brogden, R. N. (1988) Naltrexone. A review of its pharmacodynamic and pharmacokinetic properties and therapeutic efficacy in the management of opioid dependence. *Drugs* **35**, 192–213.
4. Bisaga, A., Popik, P., Bespalov, A. Y., and Danysz, W. (2000) Therapeutic potential of NMDA receptor antagonists in the treatment of alcohol and substance use disorders. *Expert Opin. Invest. Drugs* **9**, 2233–2248.
5. Bisaga, A. and Popik, P. (2000) In search of a new pharmacological treatment of drug addiction: $N$-methyl-D-aspartate (NMDA) antagonists. *Drug Alcohol Depend.* **59**, 1–15.
6. Popik, P. (2001) Glutamate antagonists inhibit tolerance, dependence and reward produced by morphine, in *Excitatory Amino Acids—Ten Years Later* (Turski, L., Schoepp, D. D., and Cavalheiro, E. A., eds.), IOS Amsterdam, pp. 231–259.
7. Childress, A. R., Ehrman, R., Rohsenow, D. J., Robbins, S. J., and O'Brien, C. P. (1992) Classically conditioned factors in drug dependence, in *Substance Abuse: A Comprehensive Textbook.* (Lowinson, J. H., Ruiz, P., Milliman, R. B., and Langrod, J. G., eds.), Williams & Wilkins, Baltimore, MD, pp. 56–69.
8. Beach, H. D. (1957) Morphine addiction in rats. *Can J. Psychol.* **11**, 104–112.
9. Garcia, J., Kimeldorf, D. J., and Hunt, E. L. (1957) Spatial avoidance in the rat as a result of exposure to ionizing radiation. *Br. J. Radiat.* **30**, 318–321.
10. Davis, W. M. and Smith, S. G. (1987) Conditioned reinforcement as a measure of the rewarding properties of drug, in *Methods of Assessing the Reinforcing Properties of Abused Drugs* (Bozarth, M. A., ed.), Springer-Verlag, New York, pp. 199–210.
11. Mucha, R. F., Van der Kooy, D., O'Shaughnessy, M., and Bucenieks, P. (1982) Drug reinforcement studied by the use of place conditioning in the rat. *Brain Res.* **243**, 91–105.
12. Carr, G. F., Fibiger, H. C. and Phillips, A. G. (1989) Conditioned place preference as a measure of drug reward, in *The Neuropharmacological Basis of Reward* (Liebman, J. and Cooper, S. J., eds.), Clarendon Ps, Oxford, pp. 264–319.
13. Schulteis, G., Gold, L. H. and Koob, G. F. (1997) Preclinical behavioral models for addressing unmet needs in opiate addiction. *Semin. Neurosci.* **9**, 94–109.
14. Bardo, M. T. and Neisewander, J. L. (1986) Single-trial conditioned place preference using intravenous morphine. *Pharmacol. Biochem. Behav.* **25**, 1101–1105.
15. Haertzen, C. A., Kocher, T. R. and Miyasato, K. (1983) Reinforcements from the first drug experience can predict later drug habits and/or addiction: results with coffee, cigarettes, alcohol, barbiturates, minor and major tranquilizers, stimulants, marijuana, hallucinogens, heroin, opiates and cocaine. *Drug Alcohol Depend.* **11**, 147–165.
16. Shippenberg, T. S., Emmett Oglesby, M. W., Ayesta, F. J., and Herz, A. (1988) Tolerance and selective cross-tolerance to the motivational effects of opioids. *Psychopharmacology* **96**, 110–115.
17. Bespalov, A., Dumpis, M., Piotrovsky, L., and Zvartau, E. (1994) Excitatory amino acid receptor antagonist kynurenic acid attenuates rewarding potential of morphine. *Eur. J. Pharmacol.* **264**, 233–239.
18. Kim, H. S., Jang, C. G., and Park, W. K. (1996) Inhibition by MK-801 of morphine-induced conditioned place preference and postsynaptic dopamine receptor supersensitivity in mice. *Pharmacol. Biochem. Behav.* **55**, 11–17.
19. Tzschentke, T. M. and Schmidt, W. J. (1995) $N$-Methyl-D-aspartic acid-receptor antagonists block morphine- induced conditioned place preference in rats. *Neurosci. Lett.* **193**, 37–40.
20. Tzschentke, T. M. and Schmidt, W. J. (1997) Interactions of MK-801 and GYKI 52466 with morphine and amphetamine in place preference conditioning and behavioural sensitization. *Behav. Brain Res.* **84**, 99–107.

21. Popik, P., Mamczarz, J., Fraczek, M., Widla, M., Hesselink, M., and Danysz, W. (1998) Inhibition of reinforcing effects of morphine and naloxone-precipitated opioid withdrawal by novel glycine site and uncompetitive NMDA receptor antagonists. *Neuropharmacology* **37**, 1033–1042.
22. Kotlinska, J. and Biala, G. (1999) Effects of the NMDA/glycine receptor antagonist, L-701, 324, on morphine- and cocaine-induced place preference. *Pol. J. Pharmacol.* **51**, 323–330.
23. Suzuki, T., Kato, H., Tsuda, M., Suzuki, H., and Misawa, M. (1999) Effects of the non-competitive NMDA receptor antagonist ifenprodil on the morphine-induced place preference in mice. *Life Sci.* **64**, L151–L156.
24. Popik, P. and Kolasiewicz, W. (1999) Mesolimbic NMDA receptors are implicated in the expression of conditioned morphine reward. *Naunyn-Schmiedebergs Arch. Pharmacol.* **359**, 288–294.
25. Tzschentke, T. M. and Schmidt, W. J. (1998) Blockade of morphine-and amphetamine- induced conditioned place preference in the rat by riluzole. *Neurosci. Lett.* **242**, 114–116.
26. Gruca, P. and Papp, M. (1999) ACPC, a partial agonist of glycine receptors, attenuates reinforcing properties of drugs of abuse but not of natural rewards and aversive agents. *Eur. Neuropsychopharmacol.* **8**, S18.
27. Popik, P. and Danysz, W. (1997) Inhibition of reinforcing effects of morphine and motivational aspects of naloxone-precipitated opioid withdrawal by NMDA receptor antagonist, memantine. *J. Pharmacol. Exp. Ther.* **280**, 854–865.
28. Jhoo, W. K., Shin, E. J., Lee, Y. H., et al. (2000) Dual effects of dextromethorphan on cocaine-induced conditioned place preference in mice. *Neurosci. Lett.* **288**, 76–80.
29. Cervo, L. and Samanin, R. (1995) Effects of dopaminergic and glutamatergic receptor antagonists on the acquisition and expression of cocaine conditioning place preference. *Brain Res.* **673**, 242–250.
30. Kim, H. S., Park, W. K., Jang, C. G., and Oh, S. (1996) Inhibition by MK-801 of cocaine-induced sensitization, conditioned place preference, and dopamine-receptor supersensitivity in mice. *Brain Res. Bull.* **40**, 201–207.
31. Kim, H. S. and Jang, C. G. (1997) MK-801 inhibits methamphetamine-induced conditioned place preference and behavioral sensitization to apomorphine in mice. *Brain Res. Bull.* **44**, 221–227.
32. Hoffman, D. C. (1994) The noncompetitive NMDA antagonist MK-801 fails to block amphetamine-induced place conditioning in rats. *Pharmacol. Biochem. Behav.* **47**, 907–912.
33. Biala, G. and Kotlinska, J. (1999) Blockade of the acquisition of ethanol-induced conditioned place preference by N-methyl-D-aspartate receptor antagonists. *Alcohol Alcohol.* **34**, 175–182.
34. Del Pozo, E., Barrios, M. and Baeyens, J. M. (1996) The NMDA receptor antagonist dizocilpine (MK-801) stereoselectively inhibits morphine induced place preference conditioning in mice. *Psychopharmacology* **125**, 209–213.
35. Layer, R. T., Kaddis, F. G., and Wallace, L. J. (1993) The NMDA receptor antagonist MK-801 elicits conditioned place preference in rats. *Pharmacol. Biochem. Behav.* **44**, 245–247.
36. Barr, G. A., Paredes, W., and Bridger, W. H. (1985) Place conditioning with morphine and phencyclidine: dose dependent effects. *Life Sci.* **36**, 363–368.
37. O'Brien, C. P., Testa, T., O'Brien, T. J., Brady, J. P., and Wells, B. (1977) Conditioned narcotic withdrawal in humans. *Science* **195**, 1000–1002.
38. Childress, A. R., McLellan, A. T., and O'Brien, C. P. (1986) Role of conditioning factors in the development of drug dependence. *Psychiatr. Clin. North Am.* **9**, 413–425.
39. O'Brien, C. P., Childress, A. R., Ehrman, R., and Robbins, S. J. (1998) Conditioning factors in drug abuse: can they explain compulsion? *J. Psychopharmacol.* **12**, 15–22.
40. Wise, R. A. (1989) Opiate reward: sites and substrates. *Neurosci. Biobehav. Rev.* **13**, 129–133.
41. Johnson, S. W. and North, R. A. (1992) Opioids excite dopamine neurons by hyperpolarization of local interneurons. *J. Neurosci.* **12**, 483–488.
42. Leone, P., Pocock, D., and Wise, R. A. (1991) Morphine-dopamine interaction: ventral tegmental morphine increases nucleus accumbens dopamine release. *Pharmacol. Biochem. Behav.* **39**, 469–472.
43. Hand, T. H., Stinus, L., and Le Moal, M. (1989) Differential mechanisms in the acquisition and expression of heroin-induced place preference. *Psychopharmacology (Berl.)* **98**, 61–67.
44. Popik, P. and Skolnick, P. (1996) The NMDA antagonist memantine blocks the expression and maintenance of morphine dependence. *Pharmacol. Biochem. Behav.* **53**, 791–798.
45. Bespalov, A. and Zvartau, E. (1997) NMDA receptor antagonists prevent conditioned activation of intracranial self-stimulation in rats. *Eur. J. Pharmacol.* **326**, 109–112.
46. Bespalov, A. Y., Zvartau, E. E., Beardsley, P. M., and Balster, R. L. (2000) Effects of NMDA receptor antagonists on reinstatement of cocaine self-administration behavior by priming injections of cocaine or exposures to cocaine-associated cues in rats. *Behav. Pharmacol.* **11**, 37–44.
47. Kelley, A. E. and Throne, L. C. (1992) NMDA receptors mediate the behavioral effects of amphetamine infused into the nucleus accumbens. *Brain Res. Bull.* **29**, 247–254.
48. Tzschentke, T. M. and Schmidt, W. J. (2000) Effects of the non-competitive NMDA-receptor antagonist memantine on morphine- and cocaine-induced potentiation of lateral hypothalamic brain stimulation reward. *Psychopharmacology* **149**, 225–234.

# 22
# Morphine Withdrawal as a State of Glutamate Hyperactivity
*The Effects of Glutamate Receptor Subtype Ligands on Morphine-Withdrawal Symptoms*

### Kurt Rasmussen, PhD

## 1. INTRODUCTION

Cessation of the repeated administration of opiates results in a characteristic morbidity in humans, including anxiety, nausea, insomnia, hot and cold flashes, muscle aches, perspiration, and diarrhea *(1)*. Great strides have been made in understanding the neurophysiology underlying these opiate-withdrawal symptoms. Several neurotransmitter systems have been shown to play an important role in opiate withdrawal, including the dopaminergic *(2–4)* and cholinergic *(5–7)* systems. This chapter will discuss evidence for a role of the glutamate system in morphine withdrawal. Specifically, the idea that morphine withdrawal is a state of glutamate hyperactivity in defined brain regions will be discussed. One of those brain regions is the locus coeruleus.

## 2. THE LOCUS COERULEUS AND OPIATE WITHDRAWAL

The locus coeruleus (LC) is the largest cluster of noradrenergic neurons in the mammalian brain *(8,9)*. Although the cell bodies are confined to a small area near the forth ventricle in the anterior pons, LC neurons send projections to most of the central nervous system, including the cerebral cortex, hippocampus, cerebellum, and spinal cord *(10,11)*. Owing to these wide-ranging projections, the LC is in a position to influence the activity of many parts of the neuraxis and has been hypothesized to play a role in many behaviors, physiological processes, and disease states.

The LC receives numerous afferent inputs. Sites sending projections to the LC include the nucleus paragigantocellularis (PGi), the prepositus hypoglossi, subregions of the hypothalamus, the Kolliker–Fuse nucleus, the periaquaductal gray, Barrington's nucleus in the brainstem, the nucleus of the solitary tract, and the central nucleus of the amygdala *(12–16)*. The projection to the LC from the PGi has both inhibitory and excitatory components and has a strong influence on the activity of LC neurons. The excitatory input from the PGi is mediated, at least in part, via glutamatergic projections *(17)* and has been shown to play an important role in the activation of the LC observed during morphine withdrawal *(18)*.

In opiate-dependent rats, the activity of LC neurons increases dramatically during antagonist-precipitated withdrawal *(19–22)*. This increased activity of the LC has been hypothesized to play an important role in opiate-withdrawal symptoms. This hypothesis is supported by several lines of evidence. First, the increased activity of LC neurons correlates temporally with withdrawal behaviors

*(21)*. Second, administration of clonidine, an $\alpha_2$-adrenergic receptor agonist (either systemically or locally infused into the LC), suppresses the increased LC unit activity *(19)*, the increase in norepinephrine turnover and release in LC projection areas *(23–25)*, and many behavioral symptoms *(26–28)* seen during opiate withdrawal. Third, destruction of the LC decreases physical signs of opiate withdrawal *(29)*. Fourth, the LC is the most sensitive site for the induction of withdrawal signs following the local injection of an opiate antagonist *(30)*.

However, the role of the LC in opiate withdrawal has also been questioned. In one study, a neurochemical lesion of the LC did not alter opiate-withdrawal symptoms or the ability of clonidine to reverse opiate withdrawal *(31)*. In addition, lesions of LC noradrenergic projections did not alter opiate-withdrawal-induced conditioned place aversion *(32)*. Indeed, some investigators have suggested that brain structures which are independent of the LC-noradrenergic system play a more important role in the expression of opiate-withdrawal symptoms *(33)*. Whatever the precise role of the LC in the production of opiate-withdrawal symptoms, there is clearly a strong activation of LC neurons during opiate withdrawal. This strong activation of LC neurons can serve as a model of glutamate hyperactivity during opiate withdrawal.

## 3. THE ROLE OF GLUTAMATE RECEPTOR SUBTYPES IN OPIATE WITHDRAWAL

Glutamate receptors have been divided into two broad categories: iontotropic and metabotropic. Iontotropic glutamate receptors contain cation-specific ion channels as a component of their protein complex, whereas metabotropic glutamate receptors are coupled to G-proteins and modulate intracellular second-messenger systems. Iontotropic receptors are divided into three main subtypes: *N*-methyl-D-aspartate (NMDA), D-2-amino-3-hydroxy-5-methyl-4-isoxazole-propionic acid (AMPA), and kainate. Eight different clones for metabotropic glutamate (mGlu) receptors have been isolated (mGlu1–8). Based on agonist interactions, sequence homology, and second-messenger coupling, the eight mGlu receptors have been grouped into three large families *(34)*. Group I mGlu receptors include mGlu1 and mGlu5, group II mGlu receptors include mGlu2 and mGlu3, and group III mGlu receptors include mGlu4, 6, 7, and 8. mGlu receptors can differentially modulate synaptic function through both presynaptic and postsynaptic sites *(35)*. Group I mGlu receptors are primarily located postsynaptically and typically regulate neuronal excitability. Group II and III mGlu receptors are primarily located presynaptically and affect the release of glutamate and other neurotransmitters *(36–39)*.

A role for glutamate receptors in opiate withdrawal was demonstrated through the use of kynurenic acid. A naturally occurring metabolite of tryptophan, kynurenic acid is a nonselective excitatory amino acid antagonist that does not readily cross the blood-brain barrier. Intraventricular administration of kynurenic acid dose-dependently attenuated the behavioral signs of naltrexone-precipitated withdrawal in morphine-dependent rats *(40)*. Intraventricular administration does not allow precise localization of the excitatory amino acid receptors involved in the attenuation of the morphine-withdrawal symptoms. However, glutamate receptors in the LC are implicated as morphine-withdrawal-induced activation of the LC was also blocked by intraventricular kynurenic acid administration *(18,40,41)*. The role of glutamate in the opiate-withdrawal-induced activation of the LC was also supported by microdialysis experiments showing increases of glutamate and aspartate release in the LC during morphine withdrawal *(42,43)*, butorphanol withdrawal *(44)*, and U-69,593 (a selective κ-opioid agonist) withdrawal *(45)*.

As mentioned above, other brain areas beside the LC play a role in opiate withdrawal. Indeed, in the studies with kynurenic acid, the block of the morphine-withdrawal-induced activation of the LC was nearly complete, whereas the suppression of withdrawal symptoms was only partial *(40)*. Although anesthesia could account for some of this difference, another possibility is the participation of glutamate receptors in other brain areas in the genesis of morphine-withdrawal symptoms. One such area may be the nucleus accumbens. The nucleus accumbens has been hypothesized to play an important role in the effects of drugs of abuse, including a prominent role in opioid addiction and withdrawal *(46)*. In particular, the nucleus accumbens has been hypothesized to play an important role in the aversive stimulus properties of opiate withdrawal *(47,48)*. While decreased dopamine and serotonin release have been hypothesized to play a role in the effects of the nucleus accumbens during opiate withdrawal *(49,50)*, glutamate may also play a role as glutamate and aspartate release increase by 300% during morphine withdrawal *(51)*.

Another brain region that displays increased release of glutamate during opiate withdrawal is the spinal cord. In animals that received repeated spinal infusions of morphine, naloxone administration evoked a 300% increase of glutamate release *(52)*. In addition, mRNA for the glutamate transporter GLT-1 has been shown to significantly increase in the striatum during morphine withdrawal, an effect most likely explained by enhanced glutamate release *(53)*. However, not all brain areas will have an increased release of glutamate during morphine withdrawal. For example, in the ventral tegmental area, there may be a decreased release of glutamate during morphine withdrawal *(54)*.

## 3.1. NMDA Receptors

The role of different subtypes of glutamate receptors in morphine withdrawal has been examined with the use of selective pharmacological tools *(55,56)*. Coadministration of competitive (e.g., MK-801) and noncompetitive (e.g., LY274614) NMDA antagonists can attenuate the development of morphine tolerance *(57–60)*. In addition, administration of NMDA antagonists blocked the behavioral signs of withdrawal in morphine-dependent rats *(61–63)*. However, the same doses of MK801 that blocked morphine withdrawal also simultaneously produced phencyclidine (PCP)-like behavioral effects (i.e., head-weaving, falls, and increased locomotor activity). The competitive NMDA antagonist LY274614 blocked the behavioral signs of withdrawal in morphine-dependent rats but did not produce any PCP-like behavioral effects (although sedation occurred at the higher doses tested). It is important to note that not all withdrawal symptoms were blocked by the NMDA antagonists. For example, both MK-801 and LY274614 produced reductions in the occurrence of teeth chatter, erections, ptosis, chews, diarrhea, and weight loss, whereas neither affected lacrimation or salivation. Thus, although NMDA receptors may play a role in the occurrence of many withdrawal signs, they are unlikely to play a role in all.

The effects of one NMDA antagonist on opiate-withdrawal symptoms has been examined in humans. Dextromethorphan is an over-the-counter antitussive agent that is (along with its metabolite dextrophan) a moderately potent NMDA antagonist *(64,65)*. Administration of dextromethorphan has been shown to decrease opiate-withdrawal signs in rats and mice *(66,67)*. In humans, dextromethorphan showed some positive effects on opiate-withdrawal symptoms in one study *(68)* but not in another *(69)*. Recently, higher doses of dextromethorphan than used in previous studies *(68,69)* have been shown to be well tolerated in methadone-maintained opiate-dependent subjects *(70)*. Thus, it is possible that higher doses of dextromethorphan (i.e., 480 mg/d) may show efficacy in reducing opiate-withdrawal symptoms in man. In addition, more potent NMDA antagonists may have stronger effects on opiate withdrawal in humans; however, PCP-like side effects may preclude their routine use.

Electrophysiological recordings from LC neurons in morphine-dependent animals showed that neither MK801 nor LY274614 blocked the withdrawal-induced activation of these neurons *(61)*. In addition, neither NMDA antagonist blocked the withdrawal-induced increase in norepinephrine turnover in the cortex, hippocampus, or hypothalamus. Thus, the LC-noradrenergic system appears to be fully activated in animals that are showing few overt signs of withdrawal because of pretreatment with an NMDA antagonist. It is important to note that these results do not necessarily indicate that the LC does not play a role in opiate withdrawal, as the NMDA antagonists may be blocking the effects of LC activation at a site distal to the LC.

*N*-Methyl-D-aspartate antagonists attenuate many signs of morphine withdrawal without blocking the withdrawal-induced increase of LC unit activity. Therefore, these studies indicate that the glutamate-induced activation of the LC during opiate withdrawal is not mediated primarily through NMDA receptors. Another study reported that direct injections of an NMDA antagonist into the LC produced a modest (approx 20%), but significant, reduction in withdrawal-induced activation of LC neurons (22). Results indicating that the excitatory amino acid projection to the LC from the PGi is mediated by non-NMDA receptors (17) are consistent with a relatively minor role of NMDA receptors in the morphine-withdrawal-induced activation of the LC.

Other brain regions that may be important for the effects of NMDA antagonists during morphine withdrawal have been suggested by studies of the induction of c-*fos*. The nuclear protein Fos is a product of the c-*fos* proto-oncogene that can regulate the transcription of cellular genes (71–74). The expression of c-fos mRNA and protein is rapidly stimulated in response to increases in neuronal activity. Thus, the presence of c-fos mRNA or protein can be used as a measure of neuronal activation (72–74). Opiate withdrawal leads to an induction of the c-*fos* proto-oncogene, Fos-like immunoreactivity, and Fos-related antigens (FRAs) in several regions of the rat and guinea pig brain (49, 75–77).

One study examined the effects of MK-801 and LY274614 on naltrexone-precipitated morphine withdrawal increased c-fos mRNA levels in the nucleus accumbens, frontal cortex, amygdala, and hippocampus (78). Pretreatment with MK-801 blocked the withdrawal-induced increased c-fos expression in the amygdala, but not in the nucleus accumbens, frontal cortex, or hippocampus, whereas pretreatment with LY274614 (or the $\alpha_2$-adrenergic agonist clonidine) blocked the withdrawal-induced increased c-fos expression in the amygdala and nucleus accumbens, but not in the frontal cortex or hippocampus. Because NMDA receptor sites are present in all four of these areas (79), these results indicate that the increased c-fos expression seen in the amygdala and nucleus accumbens during morphine withdrawal is mediated, at least in part, by activation of NMDA receptors. Conversely, the increased c-fos expression during morphine-withdrawal seen in the hippocampus and frontal cortex does not seem to be mediated primarily by activation of NMDA receptors. The nucleus accumbens and the amygdala have been suggested to play an important role in the aversive effects of opiate withdrawal (48) and MK-801 and clonidine have been shown to attenuate at least some aversive effects of opiate withdrawal (80,81). Thus, the effects of MK-801 and clonidine in the amygdala may play an important role in its ability to attenuate aversive effects of opiate withdrawal.

### 3.2. AMPA Receptors

Coadministration of the AMPA antagonist LY293558 has been shown to attenuate analgesic tolerance and behavioral sensitization to morphine (82–84). Pretreatment with this same AMPA antagonist (or its racemate LY215490) will also block many morphine-withdrawal signs (85,86). Significant decreases in the occurrence of writhes, wet-dog shakes, stereotyped head movements, ptosis, lacrimation, salivation, diarrhea, and chews were observed following pretreatment with LY293558. No significant change in the occurrence of teeth chatter, irritability, erections, or the amount of weight loss was observed. Thus, the morphine-withdrawal symptoms attenuated by antagonism of AMPA receptors are similar, but not the same as those attenuated by NMDA receptors. Furthermore, AMPA antagonists do not produce PCP-like side effects and, thus, may be useful for treating opiate-withdrawal symptoms in humans.

The site of action of AMPA antagonists for the suppression of opiate-withdrawal signs has also been studied. Administration of LY293558 antagonized the morphine-withdrawal-induced activation of LC neurons in a dose-dependent manner. Thus, the morphine-withdrawal-induced activation of the LC appears to be mediated primarily by glutamate acting through AMPA receptors. These findings agree with an earlier study showing that intra-LC application of the nonselective AMPA/kainate antagonist CNQX can block most of the withdrawal-induced activation of LC neurons (22). Further support for the role of LC activation in the genesis of morphine-withdrawal signs was seen by a study showing that intra-LC infusions of CNQX significantly attenuated many signs of naloxone-precipitated morphine withdrawal (87).

The dose of LY293558 that was able to suppress most of the withdrawal-induced activation of LC neurons in anesthetized animals only suppressed the physical signs of opiate withdrawal by about 50%. Although anesthesia could account for some of this difference, another possibility is the participation of AMPA receptors in other brain areas in the genesis of morphine-withdrawal symptoms. Other potential brain sites include those with the highest density of AMPA receptors [(i.e., hippocampus, layers I–III of the cortex, dorsal lateral septum, striatum, and the molecular layer of cerebellum *(88,89)*]. Another area may be the central nucleus of the amygdala. The central nucleus of the amygdala has been hypothesized to play a role in aversive states *(90)* and morphine withdrawal *(30)*. Indeed, local infusion of CNQX into the central nucleus of the amygdala also significantly attenuated many morphine-withdrawal symptoms, including irritability, ptosis, lacrimation, penile erections, wet-dog shakes, teeth chattering, and weight loss, but not diarrhea, rhinorrhea, abnormal posture, rearing, and grooming *(87)*.

Although LY293558 is more selective for AMPA (iGluR1-4) receptors than CNQX (and is systemically available), it has been shown to also have high affinity for one type of kainate receptor (iGluR5) in addition to AMPA receptors. Therefore, we examined the effects of a selective iGluR1–4 noncompetitive antagonist LY300168 [GYKI 53655 *(91,92)*] and a selective iGluR5 antagonist, LY382884 *(93)*, on the morphine-withdrawal-induced activation of LC neurons and behavioral signs of morphine withdrawal *(94)*. Administration of LY300168, but not LY382884, significantly attenuated the occurrence of morphine-withdrawal signs. LY300168 attenuated the occurrence of writhes, wet-dog shakes, ptosis, digging, salivation, irritability, diarrhea, chews, and weight gain, but not teeth chatter, jumps, or erections. LY382884 attenuated only lacrimation. The effect of LY382884 on lacrimation confirms that the compound is having biological activity under the present conditions and implicates iGluR5 receptors in morphine-withdrawal-induced lacrimation. Administration of LY300168 also completely attenuated the morphine-withdrawal-induced activation of LC neurons in a dose-dependent manner. However, administration of LY382884 did not affect the morphine-withdrawal-induced activation of LC neurons. LY382884 has previously been shown to have activity in vivo in rats following systemic administration at doses at and below those used in this study [i.e., 5– 100 mg/kg *(95)*]. Therefore, these results support the conclusion that the morphine-withdrawal-induced activation of LC neurons is mediated by glutamate acting at AMPA (iGluR 1–4) receptors and they indicate that iGluR5 receptors play little, if any, role.

### 3.3. Metabotropic Glutamate Receptors

Several studies have supported a role of metabotropic glutamate (mGlu) receptors in morphine dependence *(96,97)*. Based on these results, a model in which the effects of mGlu receptors on intracellular second messengers influence opiate tolerance and dependence has been proposed *(98)*. In addition, an mGlu receptor group II agonist decreased the severity of some morphine withdrawal signs *(97)*. However, the interpretation of these experiments is clouded by the nonselective nature and lack of central penetration of some of the compounds employed (e.g., ACPD and DCG-IV). The selective, centrally penetrant mGlu receptor group II agonist LY354740 has also been studied in morphine dependence and withdrawal. Administration of LY354740 blocked morphine, but not fentanyl (a selective μ-opiate agonist), tolerance *(99)*. LY354740 also has been shown to block opiate-withdrawal symptoms in the mouse *(100)* and rat *(101)*. In the rat, pretreatment with LY354740 decreased the occurrence of writhes, digging, salivation, diarrhea, chews, wet-dog shakes, and ptosis, whereas teeth chatter, lacrimation, irritability, erections, and weight loss were not affected. Thus, LY354740 had similar effects on individual morphine-withdrawal symptoms as AMPA antagonists *(86)*. The studies suggest that mGlu group II receptor agonists may be a novel treatment for opiate withdrawal in humans.

LY354740, but not its inactive isomer LY317207, significantly reduced morphine-withdrawal-induced acativation of LC neurons *(101)*. This finding is consistent with reports that presynaptic mGlu receptors function as glutamate autoreceptors to inhibit activation of LC neurons *(102)*. The release of glutamte in the LC during morphine withdrawal, as mentioned above is elevated *(42,43)* and LY354740 reduces veratridine-stimulated release of glutamate in vivo *(103)*. Thus, it seems likely that LY354740 attenuates the morphine-withdrawal-induced activation of LC neurons, at least in part, by decreasing the release of glutamate. However, indirect effects of presynaptic mGlu receptors on the release of other neurotransmitter in the LC [e.g., GABA *(104)*] may also play a role in the effects of LY354740. Further studies are needed to explore the affects of LY354740 on the release of glutamate, GABA, and/or other neurotransmitter in the LC during morphine withdrawal.

In addition to the presynaptic actions of LY354740, the activation of postsynaptic mGlu2/3 receptors in the LC may also be involved in the suppression of morphine-withdrawal symptoms. LY354740 and other mGlu2/3 agonists can act via postsynaptic receptors to inhibit cAMP formation and adenylate cyclase (AC) activity *(105)*. In the LC, upregulation of cAMP and AC pathways plays an important role in the development and expression of morphine dependence *(106–108)*. For example, chronic morphine administration increases levels of AC and cAMP-dependent protein kinases activity in the LC, and intra-LC administration of cAMP-dependent protein kinase inhibitors attenuates opiate withdrawal *(109,110)*. Thus, LY354740 may attenuate the morphine-withdrawal-induced activation of LC neurons by reducing the production of cAMP in addition to inhibiting the release of glutamate.

The actions of LY354740 in other brain areas, in addition to the LC, may play a role in its suppression of morphine-withdrawal symptoms. Other potential sites of action of LY354740 include those with the highest densities of mGlu2/3 receptors, including the cerebral cortex, hippocampus, substantia nigra, habenula, and spinal cord *(111)*. Importantly, mGlu2/3 receptors are located in some areas thought to be involved in opiate-withdrawal behaviors such as the amygdala, periaquaductal grey area, and spinal cord *(30,111)*. One region that may especially be important for the actions of LY354740 during morphine withdrawal is the nucleus paragigantocellularis (PGi). The PGi contains mGlu2/3 receptors, sends a major glutamatergic afferent to the LC, and lesions of the PGi reduce morphine withdrawal symptoms *(17,18,111)*. Thus, activation of mGlu2/3 receptors in the PGi may reduce subsequent release of glutamate in the LC and attenuate activation of LC neurons.

The relative contribution of mGlu2 versus mGlu3 receptors to the action of LY354740 is not clear. However, LY354740 has a higher affinity for mGlu2 than mGlu3 receptors *(112)* and mGlu2 and mGlu3 receptors have a differential distribution in the brain *(113,114)*. Thus, activation of mGlu2 and mGlu3 receptors may have different effects on morphine withdrawal. Additional studies with compounds selective for mGlu2 or mGlu3 receptors will help shed light on the role of these receptor subtypes in opiate withdrawal.

## 4. CONCLUSION

A great deal of evidence supports an important role for the glutamatergic system in morphine withdrawal. The nonselective glutamate antagonist kynurenic acid can attenuate many symptoms of morphine withdrawal. Antagonists selective for NMDA and AMPA receptors can also attenuate many symptoms of morphine withdrawal. In addition, mGlu receptor group II receptor agonists, which can suppress the release of glutamate, can also attenuate many symptoms of opiate withdrawal. As NMDA antagonists may produce PCP-like side effects, AMPA receptor antagonists and mGlu group II receptor agonists may be novel pharmacotherapies in the treatment of opiate withdrawal in humans. There is also evidence for enhanced glutamate activity in select brain regions during morphine withdrawal. There is an increased release of glutamate in several brain regions during morphine withdrawal (e.g., LC, nucleus accumbens, spinal cord) and NMDA antagonists can block c-*fos* activation in several brain regions observed during morphine withdrawal (e.g., nucleus accumbens, amygdala). Activation of the LC during opiate withdrawal has been particularly well studied and is mediated primarily by increased release of glutamate acting at AMPA receptors. Taken together, these studies support the idea of morphine withdrawal as a state of glutamate hyperactivity in selected brain regions.

## REFERENCES

1. Kolb, L. and Himmelsbach, C. K. (1938) Clinical studies of drug addiction, III. A critical review of the withdrawal treatments with method of evaluating abstinence syndromes. *Am. J. Psychiatry* **94**, 759–764.
2. Pathos, E., Rada, P., Mark, G. P., and Hoebel, B. G. (1991) Dopamine microdialysis in the nucleus accumbens during acute and chronic, naloxone-precipitated withdrawal and clonidine treatment. *Brain Res.* **566**, 348–350.
3. Harris, G. C. and Aston-Jones, G. A. (1994) Involvement of D2 dopamine receptors in the nucleus accumbens in the opiate withdrawal syndrome. *Nature* **371**, 155–157.
4. Druhan, J. P., Walters, C. L., and Aston-Jones, G. (2000) Behavioral activation induced by $D_2$-like receptor stimulation during opiate withdrawal. *J. Pharmcol. Exp. Ther.* **294**, 531–538.
5. Buccafusco, J. J. (1991) Inhibition of the morphine withdrawal syndrome by a novel muscarinic antagonist (4-DAMP). *Life Sci.* **48**, 749–756.
6. Buccafusco, J. J. (1992) Neuropharmacologic and behavioral actions of clonidine: interactions with central neurotransmitters. *Int. Rev. Neurobiol.* **33**, 55–107.
7. Zhang, L. C. and Buccafusco, J. J. (2000) Adaptive changes in M1 muscarinic receptors localized to specific rostral brain regions during and after morphine withdrawal. *Neuropharmacology* **39**, 1720–1734.
8. Dahlstrom, A. and Fuxe, K. (1965) Evidence for the existence of monoamine-containing neurons in the central nervous system. I. Demonstration of monoamines in the cell bodies of brainstem neurons. *Acta Physiol. Scand.* **232 (Suppl.)**, 1–55.
9. Foote, S. L., Bloom, F. E., and Aston-Jones, G. (1983) Nucleus locus coeruleus: new evidence of anatomical and physiological specificity. *Physiol. Rev.* **63**, 844–914.
10. Jones, B. E. and Moore, R. Y. (1977) Ascending projections of the locus coeruleus in the rat. II. Autoradiographic study. *Brain Res.* **127**, 23–53.
11. Nygren, L. G. and Olson, L. (1977) A new major projection from locus coeruleus: the main source of noradrenergic nerve terminals in the ventral and dorsal columns of the spinal cord. *Brain Res.* **132**, 85–93.
12. Cedarbaum, J. M. and Aghajanian, G. K. (1978) Activation of locus coeruleus neurons by peripheral stimuli: modulation by a collateral inhibitory mechanism. *Life Sci.* **23**, 1383–1392.
13. Aston-Jones, G., Ennis, M., Pieribone, V. A., Nickell, W. T., and Shipley, M. T. (1986) The brain nucleus locus coeruleus: restricted afferent control of a broad efferent network. *Science* **234**, 734–737.
14. Luppi, P. H., Aston-Jones, G., Akaoka, H., Chouvet, G., and Jouvet, M. (1995) Afferent projections to the rat locus coeruleus demonstrated by retrograde and anterograde tracing with cholera-toxin B subunit and phaseolus vulgaris leucoagglutinin. *Neuroscience* **65**, 119–160.
15. Valentino, R. J., Curtis, A. L., Page, M. E., Pavcovich, L. A., and Florin-Lechner, S. M. (1998) Activation of the locus coeruleus brain noradrenergic system during stress: circuitry, consequences, and regulation. *Adv. Pharmacol.* **42**, 781–784.
16. Van Bockstaele, E. J., Bajic, D., Proudfit, H., and Valentino, R. J. (2001) Topographic architecture of stress-related pathways targeting the noradrenergic locus coeruleus. *Physiol. Behav.* **73**, 273–283.
17. Ennis, M. and Aston-Jones, G. (1988) Activation of locus coeruleus from nucleus paragigantocellularis: a new excitatory amino acid pathway in brain. *J. Neurosci.* **8**, 3644–3657.
18. Rasmussen, K. and Aghajanian, G. K. (1989) Withdrawal-induced activation of locus coeruleus neurons in opiate-dependent rats: attenuation by lesions of the nucleus paragigantocellularis. *Brain Res.* **505**, 346–350.
19. Aghajanian, G. K. (1978) Tolerance of locus coeruleus neurones to morphine and suppression of withdrawal response by clonidine. *Nature* **276**, 186–187.
20. Valentino, R. J. and Wehby, R. G. (1989) Locus coeruleus discharge characteristics of morphine-dependent rats: Effects of naltrexone. *Brain Res.* **488**, 126–134.
21. Rasmussen, K., Beitner, D. B., Krystal, J. H., Aghajanian, G. K., and Nestler, E. J. (1990) Opiate withdrawal and the rat locus coeruleus: behavioral, electrophysiological and biochemical correlates. *J. Neurosci.* **10**, 2308–2317.
22. Akaoka, H. and Aston-Jones, G. A. (1991) Opiate withdrawal-induced hyperactivity of locus coeruleus neurons is substantially mediated by augmented excitatory amino acid input. *J. Neurosci.* **11**, 3830–3839.
23. Crawley, J. N., Laverty, R., and Roth, R. (1979) Clonidine reversal of increased norepinephrine metabolite levels during morphine withdrawal. *Eur. J. Pharm.* **57**, 247–250.
24. Laverty, R. and Roth, R. H. (1980) Clonidine reverses the increased norepinephrine turnover during morphine withdrawal in rats. *Brain Res.* **182**, 482.
25. Done, C., Silverstone, P., and Sharp, T. (1992) Effect of naloxone-precipitated morphine withdrawal on noradrenaline release in rat hippocampus in vivo. *Eur. J. Pharm.* **215**, 333–336.
26. Tseng, L. F., Loh, H. H., and Wei, E. T. (1975) Effects of clonidine on morphine withdrawal signs in the rat. *Eur. J. Pharmacol.* **30**, 93–99.
27. Gold, M. S., Redmond, D. E., Jr., and Kleber, H. D. (1978) Clonidine blocks acute opiate-withdrawal symptoms. *Lancet* **2**, 599–602.

28. Taylor, J. R., Elsworth, J. D., Garcia, E. J., Grant, S. J., Roth, R. H., and Redmond, D. E., Jr. (1988) Clonidine infusion into the locus coeruleus attenuates behavioral and neurochemical changes associated with naloxone-precipitated withdrawal. *Psychopharmacology* **96,** 121–134.
29. Maldonado, R. and Koob, G. F. (1993) Destruction of the locus coeruleus decreases physical signs of opiate withdrawal. *Brain Res.* **605,** 128–138.
30. Maldonado, R., Stinus, L., Gold, L. H., and Koob, G. F., (1992) Role of different brain structures in the expression of the physical morphine withdrawal syndrome. *J. Pharmacol. Exp. Ther.* **261,** 669–677.
31. Caille, S., Espejo, E. F., Reneric, J., Cador, M., Koob, G. F., and Stinus, L. (1999) Total neurochemical lesion of noradrenergic neurons of the locus coeruleus does not alter either naloxone-precipitated or spontaneous opiate withdrawal nor does it influence ability of clonidine to reverse opiate withdrawal. *J. Pharmacol. Exp. Ther.* **290,** 881–892.
32. Delfs, J. M., Zhu, Y., Druhan, J. P., and Aston-Jones, G. (2000) Noradrenaline in the ventral forebrain is critical for opiate withdrawal-induced aversion. *Nature* **403,** 430–434.
33. MacDonald, J. C., Williams, J. T., Osborne, P. B., and Bellchambers, C. E. (1997) Where is the locus in opioid withdrawal? *TIPS* **18,** 134–140.
34. Conn, P. J. and Pin, J. P. (1997) Pharmacology and functions of metabotropic glutamate receptors. *Annu. Rev. Pharmacol.* **37,** 205–237.
35. Schoepp, D. D. and Conn, J. P. (1993) Metabotropic glutamate receptors in brain function and pathology. *Trends Pharm. Sci.* **14,** 13–20.
36. Pin, J. P. and Duvoisin, R. (1994) The metabotropic glutamate receptors: structure and functions. *Neuropharmacology* **34,** 1–26.
37. Herrero, I., Miras-Portugal, M. T., and Sanches-Prieto, J. (1992) Positive feedback of glutamate exocytosis by metabotropic presynaptic receptor stimulation. *Nature* **360,** 163–166.
38. Gereau, R. W. and Conn, P. J. (1995) Multiple presynaptic metabotropic glutamate receptors modulate excitatory and inhibitory synaptic transmission in hippocampal area CA1. *J. Neurosci.* **15,** 6879–6889.
39. Cartmell, J. and Schoepp, D. D. (2000) Regulation of neurotransmitter release by metabotropic glutamate receptors. *J. Neurochem.* **75,** 889–907.
40. Rasmussen, K., Krystal, J. H., and Aghajanian, G. K. (1991) Excitatory amino acids and morphine withdrawal: diffferential effects of central and peripheral kynurenic acid administration. *Psychopharmacology* **105,** 508–512.
41. Tung, C. S., Grenhoff, J., Svensson, T. H. (1990) Morphine withdrawal responses of rat locus coeruleus neurons are blocked by an excitatory amino-acid antagonist. *Acta Phys. Scand.* **138,** 581–582.
42. Aghajanian, G. K., Kogan, J. H., and Moghaddam, B. (1994) Opiate withdrawal increases glutamate and aspartate efflux in the locus coeruleus: an in vivo microdialysis study. *Brain Res.* **636,** 126–130.
43. Zhang, T., Feng, Y., Rockhold, R. W., and Ho, I. K. (1994) Naloxone-precipitated morphine withdrawal increases pontine glutamate levels in the rat. *Life Sci.* **55,** PL25–PL31.
44. Feng, Y. Z., Zhang, T., Rockhold, R. W., and Ho I. K. (1995) Increased locus coeruleus glutamate levels are associated with naloxone-precipitated withdrawal from butorphanol in the rat. *Neurochem. Res.* **20,** 745–751.
45. Hoshi, K., Ma, T, and Ho, I. K. (1996) Precipitated κ-opioid receptor agonist withdrawal increases glutamate in rat locus coeruleus. *Eur. J. Pharmacol.* **314,** 301–30.
46. Koob, G. F. and Bloom, F. E., (1988) Cellular and molecular mechanisms of drug dependence. *Science* **242,** 715–723.
47. Koob, G. F., Wall, T. L., and Bloom F. E. (1989) Nucleus accumbens as a substrate for the aversive stimulus effects of opiate withdrawal. *Psychopharmacology* **98,** 530–534.
48. Stinus, L., Le Moal, M., and Koob, G. F. (1990) Nucleus accumbens and amygdala are possible substrates for the aversive stimulus effects of opiate withdrawal. *Neuroscience* **37,** 767–773.
49. Walters, C. L., Aston-Jones, G., and Druhan, J. P. (2000) Expression of Fos-related antigens in the nucleus accumbens during opiate withdrawal and their attenuation by a D2 dopamine receptor agonist. *Neuropsychopharmacology* **23,** 307–315.
50. Harris, G. and Aston-Jones, G. (2001) Augmented accumbal serotonin levels decrease the preference for a morphine associated environment during withdrawal. *Neuropsychopharmacology* **22,** 75–85.
51. Sepulveda, M. J., Hernandez, L., Rada, P., Tucci, S., and Contreras, E. (1998) Effect of precipitated withdrawal on extracellular glutamate and aspartate in the nucleus accumbens of chronically morphine-treated rats: an in vivo microdialysis study. *Pharmacol. Biochem. Behav.* **60,** 255–262.
52. Jhamandas, K. H., Marsala, M., Ibuki, T., and Yaksh, T. L. (1996) Spinal amino acid release and precipitated withdrawal in rats chronically infused with spinal morphine. *J. Neurosci.* **16,** 2758–2766.
53. Ozawa, T., Nakagawa, T., Shige, K., Minami, M., and Satoh, M. (2001) Changes in the expression of glial glutamate transporters in the rat brain accompanied with morphine dependence and naloxone-precipitated withdrawal. *Brain Res.* **905,** 245–258.
54. Manzoni, O. J. and Williams, J. T. (1999) Presynaptic regulation of glutamate release in the ventral tegmental area during morphine withdrawal. *J. Neurosci.* **19,** 6629–6636.
55. Herman, B. H., Vocci, F., and Bridge, P. (1995) The effects of NMDA receptor antagonists and nitric oxide synthase inhibitors on opioid tolerance and withdrawal: medication development issues for opiate addiction. *Neuropsychopharmacology* **13,** 269–294.

56. Herman, B. H. and O'Brien, C. P. (1997) Clinical medications development for opiate addiction: focus on nonopioids and opioid antagonists for the amelioration of opiate withdrawal symptoms and relapse prevention. *Sem. Neuroscience* **9**, 158–172.
57. Trujillo, K. A. and Akil, H. (1991) Inhibition of morphine tolerance and dependence by the NMDA receptor antagonist MK-801. *Science* **251**, 85–87.
58. Tiseo, P. J. and Inturrisi, C. E. (1993) Attenuation and reversal of morphine tolerance by the competitive N-Methyl-D-Aspartate receptor antagonist, LY274614. *J. Pharmacol. Exp. Ther.* **264**, 1090–1096.
59. Tiseo, P. J., Cheng, J., Pasternak, G. W., and Inturrisi, C. E. (1994) Modulation of morphine tolerance by the competitive N-methyl-D-aspartate receptor antagonist LY274614: assesment of opioid receptor changes. *J. Pharmacol. Exp. Ther.* **268**, 195–201.
60. Elliott, K., Minami, N., Kolesnikov, Y. A., Pasternak, G. W., and Inturrisi, C. E. (1994) The NMDA receptor antagonists, LY274614 and MK-801, and the nitric oxide synthase inhibitor, $N^G$-nitro-L-arginine, attenuate analgesic tolerance to the muopioid morphine but not to kappa opioids. *Pain* **56**, 69–74.
61. Rasmussen, K, Fuller, R. W., Stockton, M. E., Perry, K. W., Swinford, R. M., and Ornstein, P. L. (1991b) NMDA receptor antagonists suppress behaviors but not norepinephrine turnover or locus coeruleus unit activity induced by opiate withdrawal. *Eur. J. Pharmacol.* **117**, 9–16.
62. Popik, P. and Danysz, W. (1997) Inhibition of reinforcing effects of morphine and motivational aspects of naloxone-precipitated opioid withdrawal by N-methyl-D-aspartate receptor antagonist, memantine. *J. Pharmacol. Exp. Ther.* **80**, 854–865.
63. Popik, P., Mamczarz, J., Fraczek, M., Widla, M., Hesselink, M., and Danysz, W. (1998) Inhibition of reinforcing effects of morphine and naloxone-precipitated opioid withdrawal by novel glycine site and uncompetitive NMDA receptor antagonists. *Neuropharmacology* **37**, 1033–1042.
64. Wong, B. Y., Coulter, D. A., Choi, D. W., and Prince, D. A. (1988) Dextrophan and dextromethorphan, common antiussives, are antiepileptic and antagonize N-methyl-D-aspartate in brain slices. *Neurosci. Lett.* **85**, 261–266.
65. Franklin, P. H., and Murray, T. F. (1992) High affinity [$^3$H]dextrorphan binding in rat brain is localized to a noncompetitive antagonist site of the activated N-methyl-D-aspartate receptor-cation channel. *Mol. Pharmacol.* **41**, 134.
66. Koyuncouglu, H., Gungor, M., Sagduyu, H., and Aricioglu, F. (1990) Suppression by ketamine and dextromoethorphan of precipitated abstinence syndrome in rats. *Pharmacol. Biochem. Behav.* **35**, 829.
67. Farzin, D. (1999) Modification of naloxone-induced withdrawal signs by dextromethorphan in morphine-dependent mice. *Eur. J. Pharmacol.* **377**, 35–42.
68. Koyuncouglu, H. and Saydam, B. (1990) The treatment of heroin addicts with dextromethorphan: a double-blind comparison of dextromethorphan with chlorpromazine. *Int. J. Clin. Pharmacol. Ther. Toxicol.* **28**, 147.
69. Rosen, M. I., McMahon, T. J., Woods, S. W., Pearsall H. R., and Kosten, T. R. (1996) A pilot study of dextromethorphan in naloxone-precipitated opiate withdrawal. *Eur. J. Pharmacol.* **307**, 251–257.
70. Cornish, J. W., Herman, B. H., Ehrman, R. N., Robbins, S. J., Childress, A. R., Bead, V., et al. (2001) A randomized, double-blind, placebo-controlled safety study of high-dose dextromethorphan in methadone-maintained male inpatients. *Drug Alcohol Depend.*, **61**, 183–189.
71. Curran, T., Abate, C., Cohen, D. R., Macgregor, P. F., Rauscher, F. J. 3d, Sonnenberg, J. L., et al. (1990) Inducible proto-oncogene transcription factors: third messengers in the brain. *Cold Spring Harb. Syml. Quant. Biol.* **55**, 225–234.
72. Morgan, J. I. and Curran, T. (1991) Stimulus-transcription coupling in the nervous system: involvement of the inducible proto-oncogenes fos and jun. *Annu. Rev. Neurosci.* **14**, 421–451.
73. Sheng, M. and Greenberg, M. E. (1990) The regulation and function of c-fos and other immediate early genes in the nervous system. *Neuron* **4**, 477–485.
74. Sonnenberg, J. L., Macgregor-Leon, P. F., Curran, T., and Morgan, J. I. (1989) Dynamic alterations occur in the levels and composition of transcription factor AP-1 complexes after seizure. *Neuron*, **3**, 359–365.
75. Hayward, M. D, Duman, R. S., and Nestler, E. J. (1990): Induction of the c-fos proto-oncogene during opiate withdrawal in the locus coeruleus and other regions of rat brain. *Brain Res.* **525**, 256–266.
76. Stornetta, R. L., Norton, F. E., and Guyenet, P. G. (1993) Autonomic areas of rat brain exhibit increased Fos-like immunoreactivity during opiate withdrawal in rats. *Brain Res.* **624**, 19–28.
77. Chahl, L. A., Leah, J., Herdegen, T., Trueman, L., and Lynch-Frame, A. M. (1996) Distribution of c-Fos in guinea-pig brain following morphine withdrawal. *Brain Res.* **717**, 127–134.
78. Rasmussen, K., Brodsky, M., and Inturrisi, C. E. (1995) NMDA antagonists and clonidine block c-fos expression during morphine withdrawal. *Synapse* **20**, 68–74.
79. Monaghan, D. T. and Cotman, C. W. (1985) Distribution of N-methyl-D-aspartate-sensitive L-[$^3$H]glutamate-binding sites in rat brain. *J. Neurosci.* **5**, 2909–2919.
80. Higgins, G. A., Nguyen, P., and Sellers, E. M. (1992) The NMDA antagonist dizocilpine (MK801) attenuates motivational as well as somatic aspects of naloxone precipitated opioid withdrawal. *Life Sci.* **50**, PL167–PL172.
81. Kosten, T. A. (1994) Clonidine attenuates conditioned aversion produced by naloxone-precipitated opiate withdrawal. *Eur. J. Pharm.* **254**, 59–63.

82. Kest, K., McLemore, G., Kao, B., and Inturrisi, C. E. (1997) The competitive a-amino-3-hydroxy-5-methylisoxazole-4-propoinate receptor antagonist LY293558 attenuates and reverses analgesic tolerance to morphine but not to delta or kappa opioids. *J. Pharmacol. Exp. Ther.* **283**, 1249–1255.
83. McLemore, G. L., Kest, B., and Inturrisi, C. E. (1997) The effects of LY293558, an AMPA receptor antagonist, on acute and chronic morphine dependence. *Brain Res.* **778**, 120–126.
84. Carlezon, W. A., Rasmussen, K., and Nestler, E. J. (1999) AMPA antagonist LY293558 blocks the development, without blocking the expression, of behavioral sensitization to morphine. *Synapse* **31**, 256–262.
85. Rasmussen, K. (1995) The role of the locus coeruleus and N-methyl-D-aspartic acid (NMDA) and AMPA receptors in opiate withdrawal. *Neuropsychopharmacology* **13**, 295–300.
86. Rasmussen, K., Kendrick, W. T., Kogan, J. H., and Aghajanian, G. K., (1996) A selective AMPA antagonist, LY293558, antagonizes morphine-withdrawal-induced activation of locus coeruleus neurons and behavioral signs of morphine withdrawal. *Neuropsychopharmacology* **15**, 497–505.
87. Taylor, J. R., Punch, L. J, and Elsworth, J. D. (1998) A comparison of the effects of clonidine and CNQX infusion into the locus coeruleus and the amygdala on naloxone-precipitated opiate withdrawal in the rat. *Psychopharmacology* **138**, 133–142.
88. Monaghan, D. T., Yao, D., and Cotman, C. W. (1984) Distribution of [$^3$H] AMPA binding sites in rat brain as determined by quantitative autoradiography. *Brain Res.* **324**, 160–164.
89. Young, A. B. and Fagg, G. E. (1990) Excitatory amino acid receptors in the brain: membrane binding and receptor autoradiographic approaches. *Trends Pharm. Sci.* **11**, 126–133.
90. Davis, M. (1992) The role of the amygdala in fear and anxiety. *Annu. Rev. Neurosci.* **15**, 353–373.
91. Vizi, E. S., Mike, A., and Tarnawa, I. (1996) 2,3-Benzodiazepines (GYKI 52466 and analogs): negative allosteric modulators of AMPA receptors. *CNS Drug Rev.* **2**, 91–126.
92. Bleakman, D., Ballyk, B. A., Schoepp, D. D., Palmer, A. J., Bath, C. P, Sharpe, E. F., et al. (1996) Activity of 2,3-benzodiazepines at native rat and recombinant human glutamate receptors in vitro: stereospecificity and selectivity profiles. *Neuropharmacology* **35**, 1689–1702.
93. Bortolotto, Z. A., Clarke, V. R., Delany, C. M., Parry, M. C., Smolders, I., Vignes, M., et al. (1999) Kainate receptors are involved in synaptic plasticity. *Nature* **402**, 297–301.
94. Rasmussen, K. and Vandergriff, J. L. (1997) The selective AMPA antagonist LY300168 suppresses morphine-withdrawal-induced activation of locus coeruleus neurons and behavioral signs of morphine withdrawal, *Soc. Neurosci. Abstr.* **23**, 1201.
95. Simmons, R. A., Li, D. L., Hoo, K. H., Deverill, M., Ornstein, P. L., and Iyengar, S. (1998) Kainate GluR5 receptor subtype mediates the nociceptive response to formalin in the rat. *Neuropharmacology* **37**, 25–36.
96. Fundytus, M. E. and Coderre, T. J. (1994) Effect of activity at metabotropic, as well as ionotropic (NMDA), glutamate receptors on morphine dependence. *Br. J. Pharmacol.* **113**, 1215–1220.
97. Fundytus, M. E., Ritchie, J., and Coderre, T. J. (1997) Attenuation of morphine withdrawal symptoms by subtype-selective metabotropic glutamate receptor antagonists. *Br. J. Pharmacol.* **120**, 1015–1020.
98. Fundytus, M. E. and Coderre, T. J. (1999) Opioid tolerance and dependence: a new model highlighting the role of metabotropic glutamate receptors. *Pain Forum* **8**, 3–13.
99. Popik, P., Kozela, E., and Pilc, A. (2000) Selective agonist of group II glutamate metabotropic receptors, LY354740, inhibits tolerance to analgesic effects of morphine in mice. *Br. J. Pharmacol.* **130**, 1425–1431.
100. Klodzinska, A., Chojnacka, W. E., Palucha, A., Branski, P., Popik, P., and Pilc, A. (1999) Potential anti-anxiety and anti-addictive effects of LY354740. A selective group II glutamate metabotropic receptors agonist in animal models. *Neuropharmacology* **38**, 1831–1839.
101. Vandergriff, J. and Rasmussen, K. (1999) The selective mGlu2/3 receptor agonist LY354740 attenuates morphine-withdrawal-induced activation of locus coeruleus neurons and behavioral signs of morphine withdrawal. *Neuropharmacology* **38**, 217–222.
102. Dube, G. R. and Marshall, K. C. (1997) Modulation of excitatory synaptic transmission in locus coeruleus by multiple presynaptic metabotropic glutamate receptors. *Neuroscience* **80**, 511–521.
103. Battaglia, G., Monn, J. A., and Schoepp, D. D. (1997) In vivo inhibition of veratridine-evoked release of striatal excitatory amino acids by the group II metabotropic glutamate receptor agonist LY354740 in rats. *Neurosci. Lett.* **229**, 161–164.
104. Salt, T. E. and Eaton, S. A. (1995) Distinct presynaptic metabotropic receptors for L-AP4 and CCG1 on gabaergic terminals: pharmacological evidence using novel alpha-methyl derivative mGluR antagonists. MAP4 and MCCG, in the rat thalamus in vivo. *Neuroscience* **65**, 5–13.
105. Schaffhauser, H., Cartmell, J., Jakob-Rotne, R., and Mutel, V. (1997) Pharmacological characterization of metabotropic glutamate receptors linked to the inhibition of adenylate cyclase activity in rat striatal slices. *Neuropharmacology* **36**, 933–940.
106. Nestler, E. J. (1996) Under siege: the brain on opiates. *Neuron* **16**, 897–900.
107. Nestler, E. J. and Aghajanian, G. K. (1997) Molecular and cellular basis of addiction. *Science* **278**, 58–63.
108. Ivanov, A. and Aston-Jones, G. (2001) Local opiate withdrawal in locus coeruleus neurons in vitro. *J. Neurophysiol.* **85**, 2388–2397.

109. Lane-Ladd, S. B., Pineda, J., Boundy, V. A., Pfeuffer, T., Krupinski, J., Aghajanian, G. K., et al. (1997) CREB (cAMP response element-binding protein) in the locus coeruleus: biochemical, physiological, and behavioral evidence for a role in opiate dependence. *J. Neurosci.* **17,** 7890–7901.
110. Punch, L. J., Self, D. W., Nestler, E. J., and Taylor J. R. (1997) Opposite modulation of opiate withdrawal behaviors on microinfusion of a protein kinase A inhibitor versus activator into the locus coeruleus or periaqueductal gray. *J. Neurosci.* **17,** 8520–8527.
111. Petralia, R. S., Wang, Y. X., Niedzielski, A. S., and Wenthold, R. J. (1995) The metabotropic glutamate receptors, mGluR2 and mGluR3, show unique postsynaptic, presynaptic and glial localizations. *Neuroscience* **71,** 949–976.
112. Schoepp, D. D., Johnson, B. G., Wright, R. A., Salhoff, C. R., Mayne, N. G., Wu, S., et al. (1997) Ly354740 is a potent and highly selective group II metabotropic glutamate receptor agonist in cells expressing human glutamate receptors. *Neuropharmacology* **36,** 1–11.
113. Ohishi, H., Shigemoto, R., Nakanishi, S., and Mizuno, N. (1993a) Distribution of the mRNA for a metabotropic glutamate receptor (mGluR3) in the rat brain: an in situ hybridization study. *J. Comp. Neurol.* **335,** 252–266.
114. Ohishi, H., Shigemoto, R., Nakanishi, S., and Mizuno, N. (1993b) Distribution of the mRNA for a metabotropic glutamate receptor, mGluR2, in the central nervous system of the rat. *Neuroscience* **53,** 1009–1018.

# IV Glutamate and Alcohol Abuse and Alcoholism

*Section Editors*

Forrest F. Weight
Raye Z. Litten

# 23
# Alcohol Actions on Glutamate Receptors

## Robert W. Peoples, PhD

## 1. INTRODUCTION

Alcohol is arguably the oldest drug known to man, its use dating back at least 10,000 yr to the dawn of human civilization *(1)*. Although illicit drug use often receives more attention in contemporary society, alcohol abuse exacts a devastating toll: In the United States at present, over 7% of the population meet diagnostic criteria for alcohol abuse or alcoholism *(2)*, over 28% of children under 18 yr of age are exposed to alcohol abuse or dependence in the home *(3)*, and the overall economic cost to society of alcohol abuse has been estimated at $ 185 billion *(4)*. Despite intensive research since the latter part of the previous century, it is clear that the biological actions that are responsible for the characteristic effects of ethyl alcohol, or ethanol, on human physiology and behavior are still incompletely understood. Because of the simple chemical structure of ethanol (it differs from water only by two methylene groups) and its low potency (it produces most of its biological effects at millimolar concentrations), alcohol undoubtedly interacts with multiple sites in the central nervous system. The biological effects of alcohol almost certainly reflect its concerted actions at a number of these sites. Of the many possible targets of alcohol actions, neurotransmitter receptors, and in particular, neurotransmitter-gated receptor-ion channels are currently believed to be among the most important *(5)*. Because the neurotransmitter glutamate mediates the majority of fast excitatory neurotransmission in the central nervous system via actions on glutamate-gated receptor-ion channels *(6)*, effects of alcohol on these ion channels could profoundly alter central nervous system function.

In addition to postsynaptic effects on glutamate receptors, alcohol could also influence glutamatergic neurotransmission presynaptically by altering release of glutamate or its clearance from the synaptic cleft. Few studies have addressed this, however *(7–9)*, and results of studies in brain slices suggest that presynaptic effects of alcohol on glutamatergic transmission are likely to be of lesser physiological importance relative to postsynaptic effects *(10,11)*. The content of this chapter will be restricted to the effects of acute exposure to alcohol on pharmacologically isolated or recombinant glutamate receptors. Although effects of repeated or chronic exposure to alcohol on glutamate receptors are of great interest because of their relevance to human alcohol abuse and alcoholism, this topic is addressed in Chapters 24–26. For a more detailed discussion of the effects of alcohol on glutamatergic synaptic transmission, including studies performed prior to glutamate receptor cloning, the reader is referred to the review by Weight *(12)*.

## 2. ACTIONS OF ALCOHOL ON NMDA RECEPTORS

Glutamate-gated membrane ion channels are broadly divided into *N*-methyl-D-aspartate (NMDA) receptors and non-NMDA receptors [*(6)*; see Chapters 1 and 2].) NMDA receptor-ion channels are involved in nervous system excitability, cognitive function, forms of neural plasticity believed to

From: *Contemporary Clinical Neuroscience: Glutamate and Addiction*
Edited by: Barbara H. Herman et al. © Humana Press Inc., Totowa, NJ

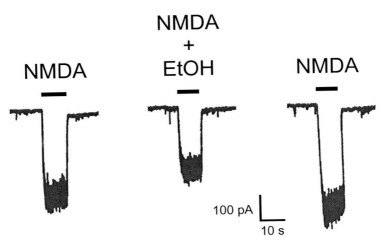

**Fig. 1.** Ethanol inhibits NMDA receptors at physiologically relevant concentrations. Traces are current activated by 25 μ$M$ NMDA and 10 μ$M$ glycine and its inhibition by 50 m$M$ ethanol (EtOH) in a rat hippocampal neuron. The bars over the traces correspond to the duration of agonist and ethanol application. (Data from ref. 24.)

underlie learning and memory, and motor coordination (13–16), all of which have obvious relevance to the intoxicating effects of alcohol. Perhaps the first evidence of an effect of alcohol on NMDA receptors was the finding that ethanol, as well as the alcohols methanol, 1-propanol, and 1-butanol, inhibited NMDA-evoked $^{22}Na^+$ efflux from rat striatal slices (17). The first direct evidence for alcohol inhibition of NMDA receptors in neurons was reported in a study in mouse hippocampal neurons in culture, in which alcohols from methanol to isopentanol inhibited NMDA-, kainate-, and quisqualate-activated ion current (18). Other studies in the same year also demonstrated ethanol inhibition of NMDA receptor single-channel currents (19) and ethanol inhibition of NMDA-stimulated $^{45}Ca^{2+}$ uptake (20,21), cyclic GMP production (21,22), and neurotransmitter release (23). Importantly, ethanol inhibits NMDA receptors at physiologically relevant concentrations (Fig. 1). Although the potency of ethanol may vary depending on experimental conditions, the large number of studies to date that have reported inhibition of NMDA receptor-mediated responses by concentrations of ethanol in the intoxicating range in many different tissues and preparations using various experimental techniques testifies to the robustness of this effect.

## 2.1. Effects of Alcohol on NMDA Receptor Subunits

*N*-methyl-D-aspartate receptors are heteromeric assemblies containing NR1 subunits, of which there are eight variants due to alternate RNA splicing of three cassettes (N1, C1, and C2), and NR2 subunits, of which there are four subtypes, NR2A–NR2D (6). Because the distribution of these subunits varies among brain regions (25), any differences in ethanol sensitivity among subunits could result in brain region-specific effects of ethanol (26–28). Such differences in ethanol sensitivity among NR1 subunit splice variants and NR2 subunits have been observed in some, but not all, studies. In *Xenopus laevis* oocytes expressing recombinant NMDA receptor subunits, ethanol inhibition of NMDA receptors was greatest when the NR1 subunit contained the N1, C1, and C2 cassettes, appeared to decrease in NR1 subunits lacking either the N1 or C1 cassettes and was lowest in NR1 subunits containing only the C2 cassette (29). Interestingly, these differences in ethanol potency were not observed when calcium in the extracellular bathing solution was replaced with barium. In a later study, ethanol sensitivity of native NMDA receptors in rat striatal neurons or recombinant NMDA

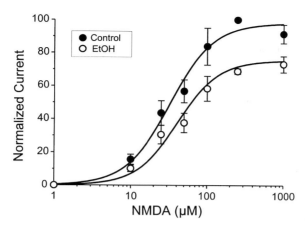

**Fig. 2.** Ethanol inhibition of NMDA receptors is not competitive with NMDA. The graph plots the percentage of current activated by 250 μ$M$ NMDA as a function of NMDA concentration in the absence (filled circles) and presence (open circles) of 100 m$M$ ethanol. Solutions of NMDA also contained 10 μ$M$ glycine. Each data point is the mean ± S.E. of six to seven neurons. The curves shown are the best fits of the data to the equation $y = E_{max}/[1 + (X/EC_{50})^n]$, where $x$ and $y$ are concentration and response, respectively, $EC_{50}$ is the half-maximal concentration, $n$ is the slope factor (Hill coefficient), and $E_{max}$ is the maximal response. Ethanol decreased the maximal response of the NMDA concentration–response curve without changing the $EC_{50}$ of NMDA. (Data from ref. *40*.)

receptors in transfected cells did not differ depending on the presence or absence of the NR1 N-terminal cassette *(30)*. In some studies, the NR2 subunit type was reported to alter ethanol sensitivity of NMDA receptors, with NR2A and NR2B subunit-containing receptors generally being the most sensitive to ethanol inhibition *(31–36)*, whereas other studies reported little if any influence of the NR2 subunit on ethanol sensitivity *(30, 37–39)*. If there is any consensus on this point at present, it is that NR2A and NR2B subunits under some conditions confer the highest ethanol sensitivity, with NR2B being perhaps the most sensitive, whereas NR2C and NR2D subunits are less sensitive. The contribution of NR1 splice variants to ethanol sensitivity in neurons under normal physiological conditions is probably of lesser importance. It should be appreciated, however, that any differences in ethanol sensitivity among NMDA receptor subunits are subtle at best. Thus, brain region-specific ethanol sensitivity of NMDA receptors is most probably not attributable to simple differences in ethanol sensitivity of subunit combinations, but instead is more likely to arise primarily from factors such as phosphorylation state or intracellular modulatory proteins (*see* below). Differences in regional expression of NMDA receptor subunits may still be important, however, in that modulation of ethanol sensitivity by phosphorylation or intracellular proteins may differ among subunits.

## 2.2. Mechanisms of Alcohol Action on NMDA Receptors

Studies designed to identify the mechanism of alcohol action on the NMDA receptor have been inconclusive to date. The agonist-binding site of the NMDA receptor is clearly not the site of ethanol action, based on observations of noncompetitive inhibition in experiments using electrophysiological recording of NMDA-activated current in hippocampal neurons *(40)* (Fig. 2), NMDA-evoked release of [$^3$H]norepinephrine from cerebral cortical slices *(23,41)* or of [$^3$H]dopamine from striatal slices *(42)*, NMDA-stimulated $Ca^{2+}$ influx in cerebellar granule cells *(43)* or dissociated whole brain cells *(44)*, and NMDA-activated current in *Xenopus* oocytes injected with rat hippocampal mRNA *(28)*. In addition, radiolabeled ligand-binding experiments in membranes from mouse cerebral cortex or hippocampus indicate that ethanol does not alter the binding affinity of [$^3$H]L-glutamate or of the NMDA

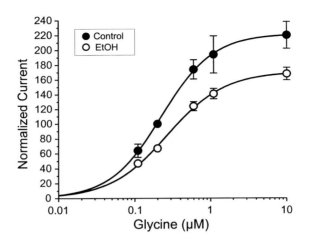

**Fig. 3.** Ethanol inhibition of NMDA receptors is not competitive with glycine. The graph plots the percentage of current activated by 50 μ$M$ NMDA as a function of glycine concentration in the absence (filled circles) and presence (open circles) of 50 m$M$ ethanol. Each data point is the mean ± S.E. of at least seven neurons. The curves shown are the best fits of the data to the equation described in the legend to Fig. 1. Ethanol decreased the maximal response of the glycine concentration–response curve without changing the $EC_{50}$ of glycine. (Data from ref. 40.)

competitive antagonist [$^3$H]CGS 19755 (45). Similarly, alcohols do not interact with the modulatory sites for dissociative anesthetics, oxidation–reduction reagents, polyamines, protons, and $Zn^{2+}$ on the NMDA receptor channel (32,40,46). In contrast, results from several of the early studies suggested that alcohol interacts with the NMDA receptor glycine coagonist site (21,42–44,47) and that $Mg^{2+}$ could enhance the inhibitory effect of ethanol (48–50), whereas in other studies, interactions of alcohols with the glycine (24,32,34,40,51,52) or $Mg^{2+}$ sites were not observed (32,40,41,44,46) (Fig. 3). The discrepant results regarding ethanol interaction with the glycine site have not yet been entirely resolved, but may, in part, arise from differences in experimental techniques, differences in NMDA receptor subunit composition (53), or differences in intracellular modulators that vary with cell type and experimental protocol. For example, ethanol inhibition and its reversal by high concentrations of glycine differ in cerebral cortical and cerebellar granule neurons (46,54). In one study using rat cerebellar granule neurons in culture, high concentrations of glycine reversed ethanol inhibition of NMDA-activated steady-state current when the perforated-patch recording mode was used, which largely preserves the intracellular milieu, but not when the intracellular environment was altered by using the whole-cell patch-clamp recording mode (55). Furthermore, pre-exposure to ethanol in the absence of NMDA enhanced ethanol inhibition of peak NMDA-activated current in these neurons. Thus, one or more intracellular factors, such as a protein kinase (54), rather than a site on the NMDA receptor itself, may contribute significantly to glycine reversal of ethanol inhibition of NMDA receptors in this cell type. With regard to $Mg^{2+}$ enhancement of ethanol inhibition of NMDA receptors, the difference in the results obtained among the various studies most probably arises from the different techniques used. Studies reporting an enhancement of ethanol inhibition by $Mg^{2+}$ did not measure NMDA receptor function directly, but, instead, measured changes in membrane potential or second-messenger levels. Because $Mg^{2+}$ has effects on proteins other than the NMDA receptor, actions of $Mg^{2+}$ on these additional sites could influence or obscure effects mediated by the NMDA receptor, which could account for the increases in the slopes of the NMDA concentration–response curves in the presence of $Mg^{2+}$ observed in those studies (48–50), as well as for the observation of apparent competitive inhibition of NMDA receptor-mediated responses by $Mg^{2+}$ (50), a known ion-channel blocker (56,57). Finally, although $Mg^{2+}$ was observed to alter the half-maximal inhibitory concentration ($IC_{50}$) of ethanol in these studies, results of experiments in which ethanol and $Mg^{2+}$ concentration were covaried indicated separate sites of action of ethanol and $Mg^{2+}$ *(48,49)*.

As would be predicted from the observations of noncompetitive inhibition by alcohol of NMDA receptors, results obtained using single-channel recording have shown that ethanol and related alcohols inhibit NMDA receptors via effects on gating of the ion channel. In outside-out membrane patches from rat hippocampal neurons, ethanol inhibited NMDA-activated single-channel current primarily by decreasing the mean open time of the channel *(19)*. Similarly, in outside-out membrane patches from mouse cerebral cortical and hippocampal neurons, ethanol inhibited NMDA receptor single-channel current by reducing both the mean open time and the frequency of opening of the ion channels by approximately the same extent *(58)*. The observation in this study that there were no changes in fast closed-state kinetics or open-channel conductance in the presence of ethanol also provides evidence that ethanol does not produce open-channel block of the receptor.

## 2.3. Molecular Sites of Alcohol Action on NMDA Receptors

The identity of the precise molecular sites of action of alcohol on NMDA receptors has remained elusive. One approach used to determine the characteristics of such a site was based on the observation that alcohols exhibit a "cutoff" effect *(59–64)*. As the molecular size of a series of analogous alcohols is increased, a point is reached at which the biological potency attains a maximum; potency then levels off or declines with further increases in size. In mouse hippocampal neurons, the NMDA receptor inhibitory potency of a series of primary straight-chain alcohols exhibited a distinct cutoff: potency increased with increases in carbon chain length up to seven carbon atoms, and decreased precipitously above this point *(65)*. These observations were originally interpreted to result from exclusion of the larger alcohols from an amphiphilic alcohol-binding site of fixed dimensions, which would allow estimation of the molecular volume of this site. Results of a subsequent study, however, indicate that this effect appears to be attributable primarily to an inability of the higher alcohols to achieve adequate aqueous concentrations, rather than to the inability of these alcohols to bind to a site of action because of their physical dimensions *(66)*. In a recent study *(67)*, two alternative approaches were used in an attempt to localize the site of alcohol action on the NMDA receptor. The first approach involved truncation of the intracellular C-terminal regions of the NR1 and NR2 subunits to determine whether these regions contained the site of action of alcohol. Results of this study indicated that the C-terminal domain of the NMDA receptor is highly unlikely to contain the site of ethanol action, because removal of virtually all of the C-termini of the NR1 and NR2B subunits did not abolish the inhibitory effect of ethanol *(67)* (Fig. 4). The second approach involved selective application of alcohols to either the extracellular or intracellular side of the membrane in cells transfected with NMDA receptor subunits and cell-free membrane patches from these cells. In these experiments, alcohols inhibited NMDA-activated current only when applied to the extracellular side of the membrane (Fig. 5). Under the conditions used in these experiments, the alcohols would have access to residues in the transmembrane regions of the receptor protein that are exposed to the surrounding membrane lipids. Thus, the site of alcohol action is located either in a region of the receptor protein that is directly exposed to the extracellular environment, or in a transmembrane domain in a region that is accessible only from the extracellular environment (Fig. 6).

## 2.4. Modulation of Alcohol Sensitivity of NMDA Receptors

A number of sites on the NMDA receptor protein have been identified that modulate its alcohol sensitivity, but are unlikely to be alcohol-binding sites. In rat cerebellar granule neurons, alcohol sensitivity of NMDA receptor-mediated $Ca^{2+}$ influx was altered by phosphorylation of the receptor or an associated protein by protein kinase C *(47)*. The protein tyrosine kinase Fyn was also suggested to regulate NMDA receptor alcohol sensitivity following the initial observations that the duration of

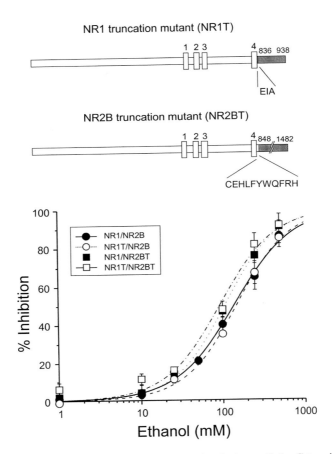

**Fig. 4.** Ethanol inhibition of NMDA receptors does not require the intracellular C-terminal regions. **(Top)** Diagrams of the C-terminal truncation mutants of NR1 and NR2B. Regions deleted from wild-type subunits are shown in dark shading, the positions of the C-terminal residues in the truncated and wild-type subunits are shown above the diagrams, and the remaining sequence following TM4 is shown below the diagrams. **(Bottom)** Concentration–response curves for inhibition of NMDA-activated ion current by ethanol in each combination of wild-type and truncation mutant NR1 and NR2B subunits. Concentrations of NMDA and glycine used were 25 and 10 μ$M$, respectively. Data points are means ± S.E. of six to seven cells, and lines shown are least-squares fits to the equation in the legend to Fig. 1. Truncation of the NR1 subunit did not significantly alter the ethanol IC$_{50}$ (149 ± 9.76 m$M$ vs a control value of 138 ± 8.71 m$M$; $p > 0.05$), whereas truncation of the NR2B subunit produced a slight decrease in the ethanol IC$_{50}$ [107 ± 7.25 m$M$ vs a control value of 138 ± 8.71 m$M$; $p < 0.01$, ALLFIT *(68)* analysis]. (Data from ref. *67*.)

alcohol hypnosis was increased twofold to three-fold in genetically engineered mice lacking the Fyn protein (Fyn "knockout" mice) and that acute tolerance to ethanol inhibition of NMDA receptor-mediated postsynaptic potentials developed in hippocampal slices from control, but not Fyn knockout, mice *(69)*. Because the acute tolerance in hippocampal slices from control mice in this study was blocked by the NR2B subunit-selective selective antagonist ifenprodil and because NR2B subunit phosphorylation was enhanced by ethanol administration, phosphorylation of the NR2B subunit was thought to decrease the alcohol sensitivity of hippocampal NMDA receptors. A subsequent study, however, found that although Fyn kinase phosphorylated both NR2A and NR2B subunits expressed in mammalian cells, it reduced ethanol inhibition of NMDA receptors that contained the NR2A, but not

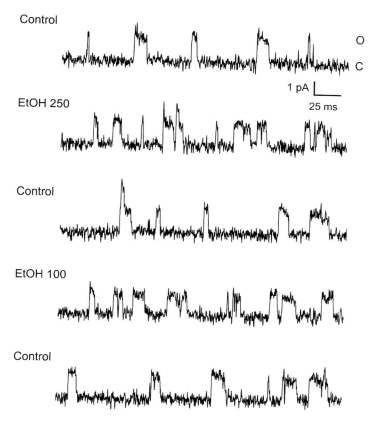

**Fig. 5.** Ethanol does not inhibit NMDA receptors when applied to the cytoplasmic face of the membrane. Traces are single-channel currents activated by 100 μM NMDA and 10 μM glycine in an inside-out patch from a single cell. The patch was voltage clamped at +50 mV; ion-channel openings are upward. Traces are representative segments of 30-s records and are sequential from top to bottom. The cytoplasmic face of each patch was continuously bathed in extracellular solution, either in the absence or the presence of ethanol (EtOH) at 100 or 250 mM. The total charge conducted by NMDA receptors in this patch was 103% and 104% of control in the presence of 100 and 250 mM ethanol, respectively; similar results were obtained in four patches tested. (Data from ref. 67.)

the NR2B, subunit (70). Although the results of the latter study do not confirm the regulation of alcohol sensitivity of the NR2B NMDA receptor subunit by Fyn kinase, the observation that ethanol sensitivity of NR2A subunit-containing NMDA receptors was altered by Fyn kinase is consistent with results from other studies suggesting that the intracellular C-terminal domain of the NMDA receptor channel, which contains multiple phosphorylation sites for a number of serine/threonine and tyrosine protein kinases (6), might be involved in the regulation of alcohol sensitivity. In one study, high extracellular calcium increased the ethanol sensitivity of receptors containing NR2A, but not NR2B or NR2C, subunits expressed in *Xenopus laevis* oocytes, and this increase was dependent on the presence of the initial segment (C0 domain) of the C-terminus of the NR1 subunit (71). Another study from the same laboratory reported that the enhanced ethanol sensitivity in the presence of high calcium concentrations was reduced by mutating residues in the NR1 C0 domain responsible for calcium-dependent inactivation in NMDA receptors containing NR2A, but not NR2B or NR2C, subunits in transfected cells (72). This latter study also demonstrated that coexpressing the anchoring/scaffold protein α-actinin 2, which binds to the C0 domain of NR1 (73), reduced the inhibitory effect of

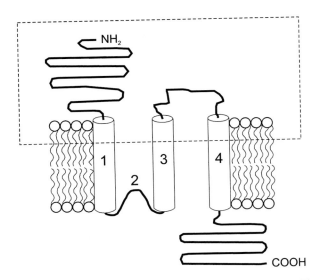

**Fig. 6.** The region of the NMDA receptor containing the site of alcohol action. The diagram shows a single NMDA receptor subunit in a membrane bilayer. The part of the figure above the membrane corresponds to the extracellular environment, and the regions numbered 1–4 are the membrane-associated domains. The probable region containing the site of alcohol action is located within the box (dashed lines).

ethanol. Thus, proteins that interact with the intracellular C-terminal region of NMDA receptor subunits may also regulate ethanol sensitivity. As discussed earlier, however, the C-termini of the NR1 and NR2B subunits do not contain the site of action of ethanol. Interestingly, truncating the intracellular C-terminal domain of the NR2B subunit slightly increased NMDA receptor ethanol sensitivity (67) (Fig. 4), suggesting that the regulatory influence of the C-terminal domain upon alcohol sensitivity differs among subunits and depends on experimental conditions.

Recent studies using site-directed mutagenesis in recombinant NMDA receptor subunits have reported that a residue in the third transmembrane domain of the NR1 subunit can influence ethanol sensitivity (74,75). As this amino acid is located in the region proposed to contain the site of alcohol action, it may constitute or form part of the alcohol-binding site.

## 3. ACTIONS OF ALCOHOL ON AMPA/KAINATE RECEPTORS

Non-NMDA glutamate-gated ion channels are further divided into two groups based on their relative sensitivity to α-amino-3-hydroxy-5-methyl-4-isoxazole propionic acid (AMPA) or kainate (6) (Chapters 1 and 2). AMPA receptors mediate the great majority of fast excitatory neurotransmission in the central nervous system (6,76). Kainate receptors also mediate fast excitatory neurotransmission in the central nervous system and may also modulate neurotransmitter release through a presynaptic action (76). Because of the widespread distribution of AMPA/kainate receptors and their importance in neurotransmission, the effects of alcohols on these receptors could profoundly influence the excitatory tone and function of the central nervous system. The first observation of alcohol modulation of AMPA/kainate receptor function was that efflux of $^{22}Na^+$ from slices of rat striatum evoked by the non-NMDA receptor agonists kainate or quisqualate was inhibited by alcohols from methanol to butanol (17). In this study, alcohols inhibited responses to kainate or quisqualate more potently than responses to NMDA, which is in contrast to results obtained in the majority of subsequent studies. For example, alcohols from methanol to isopentanol inhibited kainate- and quisqualate-activated ion current less potently than NMDA-activated current in mouse hippocampal neurons in culture (18).

Preferential inhibition of NMDA versus non-NMDA receptors was also observed in a study using extracellular recording in rat hippocampal slices, in which 100 m*M* ethanol inhibited NMDA receptor-mediated postsynaptic potentials by approx 45% and AMPA/kainate receptor-mediated postsynaptic potentials by approx 10% *(10)*. A number of other groups have also reported more potent inhibition of NMDA versus non-NMDA receptor-mediated responses, including agonist-evoked norepinephrine release in human cerebral cortical slices *(77)*, postsynaptic potentials in rat spinal cord slices *(78)*, depolarization in rat brain slices *(79)*, agonist-activated current in rat cerebellar granule neurons *(80)*, and cortical neurons *(81)*. It is likely that the responses in the above-described studies were mediated predominantly by AMPA receptors rather than kainate receptors, even in cases where kainate was used as the agonist, because AMPA receptor expression is much greater than kainate receptor expression at the majority of synapses and kainate can activate AMPA receptors as well as kainate receptors *(76)*. Intriguing results obtained in a recent study suggest that kainate receptors may be more sensitive than AMPA receptors to ethanol inhibition. In this study, kainate receptor-mediated postsynaptic potentials and kainate-activated current in rat hippocampal CA3 pyramidal neurons were inhibited by relatively low concentrations of ethanol (e.g., 20 m*M*), whereas AMPA receptor-mediated postsynaptic potentials were not affected by ethanol at these concentrations *(82)*.

### 3.1. Effects of Alcohol on AMPA/Kainate Receptor Subunits

The AMPA receptors are composed of one or more types of the non-NMDA glutamate receptor subunits GluR1–4, each of which occurs in forms termed "flip" and "flop" because of alternative RNA splicing at a 38-amino-acid segment preceding the fourth transmembrane domain, and kainate receptors are composed of one or more types of the subunits GluR5–7 and KA1–2 *(6)*. Despite initial reports of lower ethanol potency for inhibition of AMPA/kainate receptors in native neurons and tissues relative to NMDA receptors, ethanol inhibited recombinant AMPA receptors formed from various combinations of GluR1, GluR2, and GluR4 subunits transfected in mammalian cells with potencies comparable to those observed for NMDA receptor inhibition *(83)*. Although the flip splice variants of these subunits were used in this study, the ethanol sensitivity reportedly did not differ in the flop forms; however, a recent study in *Xenopus* oocytes reported lower potencies for ethanol inhibition of the flop forms of GluR1 and GluR3 *(84)*. In addition, ethanol inhibition was similar among various combinations of the AMPA receptor subunits GluR1, GluR2, and GluR3 expressed in *Xenopus* oocytes *(85)*. Results obtained to date suggest that potency for ethanol inhibition of kainate receptors also does not highly depend on subunit composition. Ethanol inhibition was reported to be similar in receptors containing GluR5, GluR6, KA1, and KA2 subunits expressed homomerically (in the case of GluR5 and GluR6) and in various heteromeric combinations in both *Xenopus* oocytes and transfected mammalian cells *(86)*.

### 3.2. Mechanism of Alcohol Action on AMPA/Kainate Receptors

As is true of NMDA receptors, ethanol inhibition appears to be noncompetitive with respect to agonist in both AMPA *(28)* and kainate *(86)* receptors; beyond this, little is known at present about the mechanism of alcohol action on AMPA/kainate receptors. A site in the fourth membrane-associated domain of the GluR6 subunit that is involved in volatile anesthetic sensitivity apparently has no influence on the alcohol sensitivity of this subunit *(87)*. The cutoff for inhibition of AMPA/kainate receptors by a series of alcohols (*see* above) appears to differ little among subunits *(84,88,89)*, and such observations have not revealed specific structural information about putative alcohol-binding sites.

### 3.3. Modulation of Alcohol Sensitivity of AMPA/Kainate Receptors

Studies on the influence of phosphorylation on the alcohol sensitivity of AMPA/kainate receptors have not yielded consistent results, perhaps because of differences in subunit composition of the receptors, experimental methods, and tissue type. In *Xenopus* oocytes, high extracellular calcium

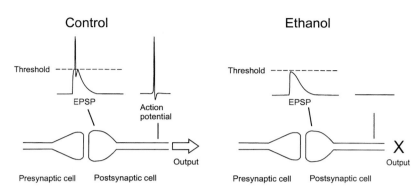

**Fig. 7.** Moderate inhibition of excitatory postsynaptic currents by ethanol can result in pronounced inhibition of neuronal signaling. The diagram on the left shows an excitatory glutamatergic synapse under normal conditions. Glutamate released from the presynaptic terminal (in response to its invasion by an action potential) causes an excitatory potential in the postsynaptic cell via activation of AMPA/kainate and NMDA receptors. Because this excitatory potential exceeds the threshold of the postsynaptic cell, it results in the firing of an action potential (shown superimposed on the postsynaptic potential as well as in the axon) and thus propagates the signal to the next synapse(s) in the pathway. The diagram on the right illustrates that a slight inhibition of the postsynaptic potential that drops its amplitude below the threshold can prevent action potential firing and thus prevent the further propagation of the signal.

concentrations have been reported to enhance ethanol inhibition via stimulation of protein kinase C in coexpressed GluR1 and GluR3 AMPA receptor subunits *(85)*, whereas in rat cerebellar granule neurons in culture, ethanol inhibition of kainate-evoked increases in intracellular calcium was reversed by stimulation of protein kinase C *(47)*. In homomeric GluR6 kainate receptors however, ethanol inhibition was not altered by activation of protein kinases A and C, calcium/calmodulin-dependent protein kinase II, protein tyrosine kinases, and serine–threonine protein phosphatases *(90)*.

## 4. SUMMARY AND CONCLUSIONS

Inhibition of glutamate-gated membrane ion channels by alcohols at concentrations in the intoxicating range has been extensively documented in a large number of studies performed over more than a decade of research. Although it is sometimes argued that the extent of inhibition of these channels by physiological concentrations of alcohols is too small to be responsible for the behavioral effects of alcohols, this argument fails to appreciate the physiology of neuronal signaling in the brain and spinal cord. Figure 7 illustrates that a small reduction in the amplitude of an excitatory postsynaptic potential (EPSP) mediated by glutamate receptors could prevent the membrane potential of the postsynaptic neuron from reaching the threshold for action potential firing, resulting in a complete inhibition of signal transmission at this synapse. In a neuronal pathway containing many neurons and synapses, it is not difficult to envision how this phenomenon could result in the amplification of the inhibitory effect of alcohol. Such an effect has, in fact, been observed in rat hippocampal slices, in which 50 m$M$ ethanol inhibited NMDA receptor-mediated EPSPs by approx 30%, but inhibited the resultant population spike (as a result of action potentials at many synapses) by nearly 70% *(10)*.

In addition to the observations that concentrations of alcohols in the intoxicating range inhibit NMDA receptors, other lines of evidence are also consistent with the involvement of NMDA receptors in the behavioral effects of alcohols. For example, NMDA receptor inhibitory potency of alcohols has been shown to correlate with intoxication potency in vivo *(18,65)*, and the potency of ethanol for producing anesthesia in rats is similar to its potency for modulation of NMDA, but not GABA$_A$, receptors in hippocampal neurons *(91)*. In drug-discrimination studies, in which animals are trained to produce an operant behavior in response to ethanol administration, NMDA antagonists can substitute for ethanol in rats and pigeons *(92–96)*, suggesting that the animals perceive the subjective effects of ethanol and NMDA antagonists to be similar. Furthermore, microinjection of NMDA receptor antagonists into the nucleus accumbens or hippocampus was able to substitute for ethanol administered systemically in rats *(97)*. Observations that NMDA receptor antagonists reduce self-administration of ethanol in rats are also consistent with similar subjective effects of NMDA receptor antagonists and alcohol *(98,99)*. Finally, human alcoholics perceive the subjective effects of ethanol to be similar to those of an NMDA antagonist [*(100)*; Chapter 26]. Thus, NMDA receptors appear to play an important role in mediating the intoxicating actions of alcohols; AMPA/kainate receptors may be involved in the effects of higher concentrations of alcohols, such as anesthesia *(101)*, because of their generally lower potencies for inhibition by alcohols.

A more complete understanding of the molecular mechanisms that underlie the actions of alcohol on glutamate receptors and the modulation of the alcohol sensitivity of these receptors by interactions with other proteins and biochemical modifications such as phosphorylation should lead not only to a better understanding of the physiology and pharmacology of glutamate receptors but also to the development of novel therapeutic agents and more effective therapeutic regimens for the treatment of alcohol abuse and alcoholism.

## REFERENCES

1. Vallee, B. L. (1994) *Toward a Molecular Basis of Alcohol Use and Abuse,* Birkhauser, Berlin.
2. Grant, B. F., Harford, T. C., Dawson, D. A., Chou, P., DuFour, M., and Pickering, R. (1994) Prevalence of DSM-IV alcohol abuse and dependence: United States, 1992. *Alcohol Health Res. World* **18,** 243–248.
3. Grant, B. F. (2000) Estimates of U. S. children exposed to alcohol abuse and dependence in the family. *Am. J. Public Health* **90,** 112–115.
4. Harwood, H. (2000) Updating estimates of the economic costs of alcohol abuse in the United States: estimates, update methods and data. Report prepared by the Lewin Group for the National Institute on Alcohol Abuse and Alcoholism.
5. Diamond, I. and Gordon, A. S. (1997) Cellular and molecular neuroscience of alcoholism. *Physiol. Rev.* **77,** 1–20.
6. Dingledine, R., Borges, K., Bowie, D., and Traynelis, S. F. (1999) The glutamate receptor ion channels. *Pharmacol. Rev.* **51,** 7–61.
7. Nelson, T. E., Ur, C. L., and Gruol, D. L. (1999) Chronic intermittent ethanol exposure alters CA1 synaptic transmission in rat hippocampal slices. *Neuroscience* **94,** 431–442.
8. Martin, D. and Swartzwelder, H. S. (1992) Ethanol inhibits release of excitatory amino acids from slices of hippocampal area CA1. *Eur. J. Pharmacol.* **219,** 469–472.
9. Clark, M. and Dar, M. S. (1989) Release of endogenous glutamate from rat cerebellar synaptosomes: interactions with adenosine and ethanol. *Life Sci.* **44,** 1625–1635.
10. Lovinger, D. M., White, G., and Weight, F. F. (1990) NMDA receptor-mediated synaptic excitation selectively inhibited by ethanol in hippocampal slice from adult rat. *J. Neurosci.* **10,** 1372–1379.
11. Morrisett, R. A. and Swartzwelder, H. S. (1993) Attenuation of hippocampal long-term potentiation by ethanol: a patch-clamp analysis of glutamatergic and GABAergic mechanisms. *J. Neurosci.* **13,** 2264–2272.
12. Weight, F. F. (1992) Cellular and molecular physiology of alcohol actions in the nervous system. *Int. Rev. Neurobiol.* **33,** 289–348.
13. Bliss, T. V. P. and Collingridge, G. L. (1993) A synaptic model of memory: long-term potentiation in the hippocampus. *Nature* **361,** 31–39.
14. Morris, R. G. M., Anderson, E., Lynch, G. S., and Baudry, M. (1986) Selective impairment of learning and blockade of long-term potentiation by an *N*-methyl-D-aspartate antagonist, AP5. *Nature* **319,** 774–776.
15. Willetts, J., Balster, R. L., and Leander, J. D. (1990) The behavioral pharmacology of NMDA receptor antagonists. *Trends Pharmacol. Sci.* **11,** 423–428.
16. Malenka, R. C. and Nicoll, R. A. (1999) Long-term potentiation—a decade of progress? *Science* **285,** 1870–1874.
17. Teichberg, V. I., Tal, N., Goldberg, O., and Luini, A. (1984) Barbiturates, alcohols and the CNS excitatory neurotransmission: specific effects on the kainate and quisqualate receptors. *Brain Res.* **291,** 285–292.
18. Lovinger, D. M., White, G., and Weight, F. F. (1989) Ethanol inhibits NMDA-activated ion current in hippocampal neurons. *Science* **243,** 1721–1724.

19. Lima-Landman, M. T. R. and Albuquerque, E. X. (1989) Ethanol potentiates and blocks NMDA-activated single-channel currents in rat hippocampal pyramidal cells. *FEBS Lett.* **247**, 61–67.
20. Dildy, J. E. and Leslie, S. W. (1989) Ethanol inhibits NMDA-induced increases in free intracellular $Ca^{2+}$ in dissociated brain cells. *Brain Res.* **499**, 383–387.
21. Hoffman, P. L., Rabe, C. S., Moses, F., and Tabakoff, B. (1989) N-methyl-D-aspartate receptors and ethanol: inhibition of calcium flux and cyclic GMP production. *J. Neurochem.* **52**, 1937–1940.
22. Hoffman, P. L., Moses, F., and Tabakoff, B. (1989) Selective inhibition by ethanol of glutamate-stimulated cyclic GMP production in primary cultures of cerebellar granule cells. *Neuropharmacology* **28**, 1239–1243.
23. Göthert, M. and Fink, K. (1989) Inhibition of N-methyl-D-aspartate (NMDA)- and L-glutamate-induced noradrenaline and acetylcholine release in the rat brain by ethanol. *Naunyn-Schmiedebergs Arch. Pharmacol.* **340**, 516–521.
24. Peoples, R. W. and Weight, F. F. (1992) Ethanol inhibition of N-methyl-D-aspartate-activated ion current in rat hippocampal neurons is not competitive with glycine. *Brain Res.* **571**, 342–344.
25. McBain, C. J. and Mayer, M. L. (1994) N-methyl-D-aspartic acid receptor structure and function. *Physiol. Rev.* **74**, 723–723.
26. McCown, T. J., Frye, G. D., and Breese, G. R. (1985) Evidence for site specific ethanol actions in the CNS. *Alcohol Drug Res.* **6**, 423–429.
27. Givens, B. S. and Breese, G. R. (1990) Electrophysiological evidence that ethanol alters function of medial septal area without affecting lateral septal function. *J. Pharmacol. Exp. Ther.* **253**, 95–103.
28. Dildy-Mayfield, J. E. and Harris, R. A. (1992) Comparison of ethanol sensitivity of rat brain kainate, DL-α-amino-3-hydroxy-5-methyl-4-isoxalone proprionic acid and N-methyl-D-aspartate receptors expressed in *Xenopus* oocytes. *J. Pharmacol. Exp. Ther.* **262**, 487–494.
29. Koltchine, V., Anantharam, V., Wilson, A., Bayley, H., and Treistman, S. N. (1993) Homomeric assemblies of NMDAR1 splice variants are sensitive to ethanol. *Neurosci. Lett.* **152**, 13–16.
30. Popp, R. L., Lickteig, R., Browning, M. D., and Lovinger, D. M. (1998) Ethanol sensitivity and subunit composition of NMDA receptors in cultured striatal neurons. *Neuropharmacology* **37**, 45–56.
31. Masood, K., Wu, C., Brauneis, U., and Weight, F. F. (1994) Differential ethanol sensitivity of recombinant N-methyl-D-aspartate receptor subunits. *Mol. Pharmacol.* **45**, 324–329.
32. Chu, B., Anantharam, V., and Treistman, S. N. (1995) Ethanol inhibition of recombinant heteromeric NMDA channels in the presence and absence of modulators. *J. Neurochem.* **65**, 140–148.
33. Lovinger, D. M. (1995) Developmental decrease in ethanol inhibition of N-methyl-D-aspartate receptors in rat neocortical neurons: relation to the actions of ifenprodil. *J. Pharmacol. Exp. Ther.* **274**, 164–172.
34. Mirshahi, T. and Woodward, J. J. (1995) Ethanol sensitivity of heteromeric NMDA receptors: effects of subunit assembly, glycine and NMDAR1 $Mg^{2+}$-insensitive mutants. *Neuropharmacology* **34**, 347–355.
35. Yang, X. H., Criswell, H. E., Simson, P., Moy, S., and Breese, G. R. (1996) Evidence for a selective effect of ethanol on N-methyl-D-aspartate responses: ethanol affects a subtype of the ifenprodil-sensitive N-methyl-D-aspartate receptors. *J. Pharmacol. Exp. Ther.* **278**, 114–124.
36. Wirkner, K., Poelchen, W., Köles, L., et al. (1999) Ethanol-induced inhibition of NMDA receptor channels. *Neurochem. Int.* **35**, 153–162.
37. Kuner, T., Schoepfer, R., and Korpi, E. R. (1993) Ethanol inhibits glutamate-induced currents in heteromeric NMDA receptor subtypes. *NeuroReport* **5**, 297–300.
38. Blevins, T., Mirshahi, T., Chandler, L. J., and Woodward, J. J. (1997) Effects of acute and chronic ethanol exposure on heteromeric N-methyl-D-aspartate receptors expressed in HEK 293 cells. *J. Neurochem.* **69**, 2345–2354.
39. Engblom, A. C., Courtney, M. J., Kukkonen, J. P., and Akerman, K. E. (1997) Ethanol specifically inhibits NMDA receptors with affinity for ifenprodil in the low micromolar range in cultured cerebellar granule cells. *J. Neurochem.* **69**, 2162–2168.
40. Peoples, R. W., White, G., Lovinger, D. M., and Weight, F. F. (1997) Ethanol inhibition of N-methyl-D-aspartate-activated current in mouse hippocampal neurones: whole-cell patch-clamp analysis. *Br. J. Pharmacol.* **122**, 1035–1042.
41. Gonzales, R. A. and Woodward, J. J. (1990) Ethanol inhibits N-methyl-D-aspartate-stimulated [$^3$H] norepinephrine release from rat cortical slices. *J. Pharmacol. Exp. Ther.* **253**, 1138–1144.
42. Woodward, J. J. and Gonzales, R. A. (1990) Ethanol inhibition of N-methyl-D-aspartate-stimulated endogenous dopamine release from rat striatal slices: reversal by glycine. *J. Neurochem.* **54**, 712–715.
43. Rabe, C. S. and Tabakoff, B. (1990) Glycine site-directed agonists reverse the actions of ethanol at the N-methyl-D-aspartate receptor. *Mol. Pharmacol.* **38**, 753–757.
44. Dildy-Mayfield, J. E. and Leslie, S. W. (1991) Mechanism of inhibition of N-methyl-D-aspartate-stimulated increases in free intracellular $Ca^{2+}$ concentration by ethanol. *J. Neurochem.* **56**, 1536–1543.
45. Snell, L. D., Tabakoff, B., and Hoffman, P. L. (1993) Radioligand binding to the N-methyl-D-aspartate receptor/ionophore complex: alterations by ethanol in vitro and by chronic in vivo ethanol ingestion. *Brain Res.* **602**, 91–98.
46. Bhave, S. V., Snell, L. D., Tabakoff, B., and Hoffman, P. L. (1996) Mechanism of ethanol inhibition of NMDA receptor function in primary cultures of cerebral cortical cells. *Alcohol. Clin. Exp. Res.* **20**, 934–941.

47. Snell, L. D., Tabakoff, B., and Hoffman, P. L. (1994) Involvement of protein kinase C in ethanol-induced inhibition of NMDA receptor function in cerebellar granule cells. *Alcohol. Clin. Exp. Res.* **18**, 81–85.
48. Morrisett, R. A., Martin, D., Oetting, T. A., Lewis, D. V., Wilson, W. A., and Swartzwelder, H. S. (1991) Ethanol and magnesium ions inhibit N-methyl-D-aspartate-mediated synaptic potentials in an interactive manner. *Neuropharmacology* **30**, 1173–1178.
49. Martin, D., Morrisett, R. A., Bian, X. P., Wilson, W. A., and Swartzwelder, H. S. (1991) Ethanol inhibition of NMDA mediated depolarizations is increased in the presence of $Mg^{2+}$. *Brain Res.* **546**, 227–234.
50. Chandler, L. J., Guzman, N. J., Sumners, C., and Crews, F. T. (1994) Magnesium and zinc potentiate ethanol inhibition of N-methyl-D-aspartate-stimulated nitric oxide synthase in cortical neurons. *J. Pharmacol. Exp. Ther.* **271**, 67–75.
51. Woodward, J. J. (1994) A comparison of the effects of ethanol and the competitive glycine antagonist 7-chlorokynurenic acid on N-methyl-D-aspartic acid-induced neurotransmitter release from rat hippocampal slices. *J. Neurochem.* **62**, 987–991.
52. Cebers, G., Cebere, A., Zharkovsky, A., and Liljequist, S. (1996) Glycine does not reverse the inhibitory actions of ethanol on NMDA receptor functions in cerebellar granule cells. *Naunyn-Schmiedebergs Arch. Pharmacol.* **354**, 736–745.
53. Buller, A. L., Larson, H. C., Morrisett, R. A., and Monaghan, D. T. (1995) Glycine modulates ethanol inhibition of heteromeric N-methyl-D-aspartate receptors expressed in *Xenopus* oocytes. *Mol. Pharmacol.* **48**, 717–723.
54. Hoffman, P. L., Snell, L. D., Bhave, S. V., and Tabakoff, B. (1994) Ethanol inhibition of NMDA receptor function in primary cultures of rat cerebellar granule cells and cerebral cortical cells. *Alcohol Alcohol.* **29** (Suppl.), 199–204.
55. Popp, R. L., Lickteig, R. L., and Lovinger, D. M. (1999) Factors that enhance ethanol inhibition of N-methyl-D-aspartate receptors in cerebellar granule cells. *J. Pharmacol. Exp. Ther.* **289**, 1564–1574.
56. Mayer, M. L., Westbrook, G. L., and Guthrie, P. B. (1984) Voltage-dependent block by $Mg^{2+}$ of NMDA responses in spinal cord neurones. *Nature* **309** 261–263.
57. Nowak, L., Bregestovski, P., Ascher, P., Herbet, A., and Prochiantz, A. (1984) Magnesium gates glutamate-activated channels in mouse central neurones. *Nature* **307**, 462–465.
58. Wright, J. M., Peoples, R. W., and Weight, F. F. (1996) Single-channel and whole-cell analysis of ethanol inhibition of NMDA-activated currents in cultured mouse cortical and hippocampal neurons. *Brain Res.* **738**, 249–256.
59. Mullins, L. J. (1954) Some physical mechanisms in narcosis. *Chem. Rev.* **54**, 289–323.
60. Rang, H. P. (1960) Unspecific drug action. The effects of a homologous series of primary alcohols. *Br. J. Pharmacol.* **15**, 185–200.
61. McCreery, M. J. and Hunt, W. A. (1978) Physico-chemical correlates of alcohol intoxication. *Neuropharmacology* **17**, 451–461.
62. Lyon, R. C., McComb, J. A., Schreurs, J., and Goldstein, D. B. (1981) A relationship between alcohol intoxication and the disordering of brain membranes by a series of short-chain alcohols. *J. Pharmacol. Exp. Ther.* **218**, 669–675.
63. Franks, N. P. and Lieb, W. R. (1985) Mapping of general anaesthetic target sites provides a molecular basis for cutoff effects. *Nature* **316**, 349–351.
64. Alifimoff, J. K., Firestone, L. L., and Miller, K. W. (1989) Anaesthetic potencies of primary alkanols: implications for the molecular dimensions of the anaesthetic site. *Br. J. Pharmacol.* **96**, 9–16.
65. Peoples, R. W. and Weight, F. F. (1995) Cutoff in potency implicates alcohol inhibition of N-methyl-D-aspartate receptors in alcohol intoxication. *Proc. Natl. Acad. Sci. USA* **92**, 2825–2829.
66. Peoples, R. W. and Ren, H. (2002) Inhibition of N-methyl-D-aspartate receptors by straight-chain diols: implications for the mechanism of the alcohol cutoff effect. *Mol. Pharmacol.*, **61**, 169–176.
67. Peoples, R. W. and Stewart, R. R. (2000) Alcohols inhibit N-methyl-D-aspartate receptors via a site exposed to the extracellular environment. *Neuropharmacology* **39**, 1681–1691.
68. DeLean, A., Munson, P. J., and Rodbard, D. (1978) Simultaneous analysis of families of sigmoidal curves: application to bioassay, radioligand assay, and physiological dose-response curves. *Am. J. Physiol.* **235**, E97–E102.
69. Miyakawa, T., Yagi, T., Kitazawa, H., et al. (1997) Fyn-kinase as a determinant of ethanol sensitivity: relation to NMDA-receptor function. *Science* **278**, 698–701.
70. Anders, D. L., Blevins, T., Sutton, G., Swope, S., Chandler, L. J., and Woodward, J. J. (1999) Fyn tyrosine kinase reduces the ethanol inhibition of recombinant NR1/NR2A but not NR1/NR2B NMDA receptors expressed in HEK 293 cells. *J. Neurochem.* **72**, 1389–1393.
71. Mirshahi, T., Anders, D. L., Ronald, K. M., and Woodward, J. J. (1998) Intracellular calcium enhances the ethanol sensitivity of NMDA receptors through an interaction with the CO domain of the NR1 subunit. *J. Neurochem.* **71**, 1095–1107.
72. Anders, D. L., Blevins, T., Smothers, C. T., and Woodward, J. J. (2000) Reduced ethanol inhibition of N-methyl-D-aspartate receptors by deletion of the NR1 CO domain or overexpression of α-actinin-2 proteins. *J. Biol. Chem.* **275**, 15,019–15,024.
73. Wyszynski, M., Lin, J., Rao, A., et al. (1997) Competitive binding of α-actinin and calmodulin to the NMDA receptor. *Nature* **385**, 439–442.
74. Ronald, K. M., Anders, D. L., Blevins, T., and Woodward, J. J. (2000) Effect of transmembrane domain mutations on the ethanol sensitivity of NMDA receptors. *Alcohol Clin. Exp. Res.* **24**, 9A (abstract).

75. Ronald, K. M., Anders, D. L., Blevins, T., and Woodward, J. J. (2000) Effect of transmembrane domain three mutations on the ethanol sensitivity of NMDA receptors. *Alcohol Clin. Exp. Res.* **25,** 9A (abstract).
76. Frerking, M. and Nicoll, R. A. (2000) Synaptic kainate receptors. *Curr. Opin. Neurobiol.* **10,** 342–351.
77. Fink, K., Schultheiss, R., and Göthert, M. (1992) Inhibition of *N*-methyl-D-aspartate-and kainate-evoked noradrenaline release in human cerebral cortex slices by ethanol. *Naunyn-Schmiedebergs Arch. Pharmacol.* **345,** 700–703.
78. Wong, S. M. E., Fong, E., Tauck, D. L., and Kendig, J. J. (1997) Ethanol as a general anesthetic: actions in spinal cord. *Eur. J. Pharmacol.* **329,** 121–127.
79. Martin, D., Tayyeb, M. I., and Swartzwelder, H. S. (1995) Ethanol inhibition of AMPA and kainate receptor-mediated depolarizations of hippocampal area CA1. *Alcoholism (NY)* **19,** 1312–1316.
80. Valenzuela, C. F., Bhave, S., Hoffman, A., and Harris, R. A. (1998) Acute effects of ethanol on pharmacologically isolated kainate receptors in cerebellar granule neurons: comparison with NMDA and AMPA receptors. *J. Neurochem.* **71,** 1777–1780.
81. Wirkner, K., Eberts, C., Poelchen, W., Allgaier, C., and Illes, P. (2000) Mechanism of inhibition by ethanol of NMDA and AMPA receptor channel functions in cultured rat cortical neurons. *Naunyn Schmiedebergs Arch. Pharmacol.* **362,** 568–576.
82. Weiner, J. L., Dunwiddie, T. V., and Valenzuela, C. F. (1999) Ethanol inhibition of synaptically evoked kainate responses in rat hippocampal CA3 pyramidal neurons. *Mol. Pharmacol.* **56,** 85–90.
83. Lovinger, D. M. (1993) High ethanol sensitivity of recombinant AMPA-type glutamate receptors expressed in mammalian cells. *Neurosci. Lett.* **159,** 83–87.
84. Akinshola, B. E. (2001) Straight-chain alcohols exhibit a cutoff in potency for the inhibition of recombinant glutamate receptor subunits. *Br. J. Pharmacol.* **133,** 651–658.
85. Dildy-Mayfield, J. E. and Harris, R. A. (1995) Ethanol inhibits kainate responses of glutamate receptors expressed in *Xenopus* oocytes: role of calcium and protein kinase C. *J. Neurosci.* **15,** 3162–3171.
86. Valenzuela, C. F. and Cardoso, R. A. (1999) Acute effects of ethanol on kainate receptors with different subunit compositions. *J. Pharmacol. Exp. Ther.* **288,** 1199–1206.
87. Minami, K., Wick, M. J., Stern-Bach, Y., et al. (1998) Sites of volatile anesthetic action on kainate (glutamate receptor 6) receptors. *J. Biol. Chem.* **273,** 8248–8255.
88. Dildy-Mayfield, J. E. and Harris, R. A. (1995) Inhibition of NMDA and kainate currents by a series of alcohols: studies of receptor composition and the cut-off phenomenon. *Alcohol. Clin. Exp. Res.* **19,** 7A (abstract).
89. Dildy-Mayfield, J. E., Mihic, S. J., Liu, Y., Deitrich, R. A., and Harris, R. A. (1996) Actions of long chain alcohols on GABA$_A$ and glutamate receptors: relation to *in vivo* effects. *Br. J. Pharmacol.* **118,** 378–384.
90. Valenzuela, C. F., Cardoso, R. A., Lickteig, R., Browning, M. D., and Nixon, K. M. (1998) Acute effects of ethanol on recombinant kainate receptors: Lack of role of protein phosphorylation. *Alcohol. Clin. Exp. Res.* **22,** 1292–1299.
91. Peoples, R. W. and Weight, F. F. (1999) Differential alcohol modulation of GABA$_A$ and NMDA receptors. *NeuroReport* **10,** 97–101.
92. Grant, K. A., Knisely, J. S., Tabakoff, B., Barrett, J. E., and Balster, R. L. (1991) Ethanol-like discriminative stimulus effects of non-competitive *N*-methyl-D-aspartate antagonists. *Behav. Pharmacol.* **2,** 87–95.
93. Grant, K. A. and Colombo, G. (1993) Discriminative stimulus effects of ethanol: effect of training dose on the substitution of *N*-methyl-D-aspartate antagonists. *J. Pharmacol. Exp. Ther.* **264,** 1241–1247.
94. Sanger, D. J. (1993) Substitution by NMDA antagonists and other drugs in rats trained to discriminate ethanol. *Behav. Pharmacol.* **4,** 523–528.
95. Schechter, M. D., Meehan, S. M., Gordon, T. L., and McBurney, D. M. (1993) The NMDA receptor antagonist MK-801 produces ethanol-like discrimination in the rat. *Alcohol* **10,** 197–201.
96. Shelton, K. L. and Balster, R. L. (1994) Ethanol drug discrimination in rats: substitution with GABA agonists and NMDA antagonists. *Behav. Pharmacol.* **5,** 441–450.
97. Hodge, C. W. and Cox, A. A. (1998) The discriminative stimulus effects of ethanol are mediated by NMDA and GABA$_A$ receptors in specific limbic brain regions. *Psychopharmacology (Berl.)* **139,** 95–107.
98. Bienkowski, P., Koros, E., Kostowski, W., and Danysz, W. (1999) Effects of *N*-methyl-D-aspartate receptor antagonists on reinforced and nonreinforced responding for ethanol in rats. *Alcohol* **18,** 131–137.
99. Holter, S. M., Danysz, W., and Spanagel, R. (2000) Novel uncompetitive *N*-methyl-D-aspartate (NMDA)-receptor antagonist MRZ 2/579 suppresses ethanol intake in long-term ethanol-experienced rats and generalizes to ethanol cue in drug discrimination procedure. *J. Pharmacol. Exp. Ther.* **292,** 545–552.
100. Krystal, J. H., Petrakis, I. L., Webb, E., et al. (1998) Dose-related ethanol-like effects of the NMDA antagonist, ketamine, in recently detoxified alcoholics. *Arch. Gen. Psychiatry* **55,** 354–360.
101. Peoples, R. W. and Weight, F. F. (1997) Anesthetic actions on excitatory amino acid receptors, in *Anesthesia: Biologic Foundations* (Biebuyck, J. F., Lynch, C., Maze, M., Saidman, L. J., Yaksh, T. L., and Zapol, W. M., eds.), Lippincott–Raven, New York, pp. 239–258.

# 24
# Glutamate and Alcohol-Induced Neurotoxicity

### Fulton T. Crews, PhD, Joseph G. Rudolph, PhD, and L. Judson Chandler, PhD

## 1. INTRODUCTION

Alcohol is a major drug of use and abuse in the United States. An estimated 15 million Americans (1988 NHIS study) are alcohol abusers or alcohol dependent. Lifetime prevalence of alcohol dependence is estimated at 13% and 4% for American men and women over 18 yr of age, respectively (1). It is well established that chronic excessive ethanol consumption produces marked deficits in cognitive and motor abilities (2,3). Alcohol is a leading cause of adult dementia in the United States, accounting for approx 10% of cases (Alzheimer's disease is the leading cause, accounting for 40–60% of cases). Although there is evidence of reversibility of deficits with sobriety (4), a variety of studies report that 50–75% of sober, detoxified, long term alcohol-dependent individuals suffer from some degree of cognitive impairment, and approx 10% of these are seriously demented (5). The damaging effects of alcohol appear to lie on a continuum, with moderate deficits in the majority of long-term alcoholics, progressing to the much more severe deficits of Wernicke's disease and Wernicke's encephalopathy with Korsakoff's amnestic syndrome (6,7). A variety of lifestyle factors, including nutrition, are implicated in the more severe cases. However, all deficiencies on the continuum appear to be related to alcohol consumption and to the amount of alcohol regularly consumed; that is, the most severe damage is associated with chronic long-term alcoholism (6,7).

## 2. MORPHOLOGICAL CHANGES

Alcohol-induced changes in the brain have been studied in both humans and rodents. A variety of postmortem histological analyses as well as supporting imaging analyses suggest that chronic alcohol exposure changes brain structure (2,3). Computed tomography (CT) and magnetic resonance imaging (MRI) studies of the human brain have repeatedly shown enlargement of the cerebral ventricles and sulci in most alcoholics. The enlargement of the ventricles and sulci essentially reflect a shrinking of the brain mass. This is consistent with studies on postmortem brain tissue, which show that alcoholics have a reduction in total brain weight relative to controls. Particularly severe alcoholics have significant reductions in global cerebral hemisphere and cerebellar mass compared to controls and moderate drinkers (8). Some of this loss of brain mass is likely the result of actual loss of neurons and the resulting loss of myelin sheath white matter, which normally envelops neuronal extensions. However, a portion of this loss in brain mass is likely not to be the result of the actual loss of cells, but to a reduction in the brain parenchyma (i.e., the size of the cells and their processes). Recent studies have indicated that within 1–5 mo of recovery from alcoholism and with sustained abstinence, the size of the brain returns toward normal levels. It is likely that this return involves an increase in neuronal cell size, arborization, and density of the neuronal processes that make up cellular brain mass, as well as

From: *Contemporary Clinical Neuroscience: Glutamate and Addiction*
Edited by: Barbara H. Herman et al. © Humana Press Inc., Totowa, NJ

increases in the number and size of glial cells (9). Although it is not clear exactly how alcoholism leads to a reduction in brain weight and volume, it is clear that this does occur during active alcohol abuse and that some recovery of brain mass does occur during abstinence. More studies are needed to more clearly understand how chronic alcohol abuse leads to a reduction in brain mass and what occurs during recovery of brain mass in abstinence.

The frontal lobes appear to be particularly affected in persons with chronic alcoholism, as first observed in early neuropathological studies (10) and confirmed more recently with neuropathological (11) and in vivo neuroradiological studies (12–14). Quantitative morphometry suggests that the frontal lobes of the human alcoholic brain show the greatest loss and account for much of the associated ventricular enlargement (15). Specific types of brain cells appear to be affected. Both gray matter, which is composed largely of neurons, and white matter, which involves neuronal tracks surrounded by myelin sheaths, appear to decrease. Studies have found that neuronal density in the superior frontal cortex is reduced by 22% in alcoholics compared to nonalcoholic controls, in contrast to other areas of the cortex, in which there was no difference between the groups (16). Further, the complexity of the basal dendritic arborization of layer III pyramidal cells in both superior frontal and motor cortices was significantly reduced in alcoholics compared to controls. A reduction in dendritic arborization of Purkinje cells in the anterior superior vermis of the cerebellum is also found in alcoholics. Taken together, these data demonstrate selective neuronal loss, dendritic simplification, and reduction of synaptic complexity in specific brain regions of alcoholics. It remains uncertain how these cellular lesions relate to the selective loss of white matter that appears to occur in frontal lobes. One reason that these frontal lobe changes are more evident is the greater proportion of white matter to cortical gray matter in the frontal regions.

Frontal lobe shrinkage has been reported with and without seizures, but recent studies suggest that temporal lobe shrinkage occurs particularly in individuals with alcohol withdrawal seizure history (17). Decreases in the amounts of N-acetyl aspartate in the frontal lobe, a measure of neuronal volume or density, also illustrate frontal lobe degeneration in alcoholics (18). Alcoholics with more severe brain disorders, such as Wernicke's and/or Korsakoff's syndrome, show more reduction in white matter and more extensive brain-region degeneration, consistent with their greater alcohol consumption.

In addition to the global shrinkage of brain regions, certain key neuronal nuclei that have broad-ranging functions on brain activity are selectively lost with chronic alcohol abuse. Important regions altered in alcoholism are the cholinergic basal forebrain nuclei, which are also lost in Alzheimer's disease. Animal studies and some human studies have suggested that this region is particularly damaged in alcoholic subjects. Arendt found a significant loss of neurons in this region in alcoholic Korsakoff's psychosis patients (19).

Additional brain nuclei that appear to be particularly sensitive are the locus coeruleus and raphe nuclei. These two nuclei contain many of the noradrenergic and serotonergic neurons within the brain, respectively. Although these nuclei are small in size, they are particularly important because their neuronal processes project throughout the brain and modulate global aspects of brain activity. Chemical studies have shown abnormally low levels of serotonergic metabolites in the cerebral spinal fluid of alcoholics with Wernicke–Korsakoff syndrome, and more recent morphological studies have found significant reductions (e.g., 50%) in the number of serotonergic neurons from the raphe nuclei of the brains of alcoholics compared to controls. Thus, the serotonergic system appears to be disrupted, particularly in severe alcoholics (20,21). Several studies have also reported significant noradrenergic cell loss in the locus coeruleus (22–24), although not all studies have found this loss (8). Recent studies have also indicated that certain neurons that contain the peptide vasopressin may be sensitive to chronic ethanol-induced neurotoxicity in both rats and humans (25,26). Damage to hypothalamic vasopressin and other peptide-containing neurons could disrupt a variety of hormone functions as well as daily rhythms that are important for healthy living. Additional studies are needed to determine which specific cell groups within the brain might be particularly damaged. Specific neuronal loss in

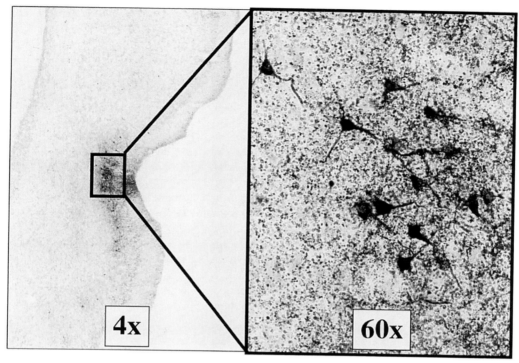

**Fig. 1.** Binge-drinking-induced brain damage in perirhinal cortex as revealed by amino cupric silver staining. Shown are two different magnifications of a rat brain stained with amino cupric silver stain to identify damaged neurons following a 4-d binge ethanol treatment *(28)*. Rats were treated intragastricly with ethanol in a high dose model of ethanol binge drinking. Brains were sectioned and stained to reveal neuronal damage using the amino cupric stain of de Olmos *(29,30)*. Shown on the left is a 4× magnification illustrating the edge of the brain and the perirhinal fissure with significant silver staining in pyramidal cells in that region. On the right at 60× magnification, one can clearly see neuronal-like structures revealing that short-term ethanol administration can significantly damage the brain *(28)*.

small but functionally significant brain areas could result in global changes in attention, mood, and personality that are difficult to quantify but have a great impact on brain function and overall behavior.

Recent studies have found that long-term ethanol intoxication is not necessary to cause brain damage. Studies in rats show that as little as a few days of intoxication can lead to neuronal loss in several brain areas, including dentate gyrus, entorhinal, piriform, insular, orbital and, perirhinal cortices and in the olfactory bulb *(27,28)* (Fig. 1). These structures are involved in frontal cortical neuronal circuits, including the limbic and association cortex. These findings are consistent with recent human studies reporting damage to the entorhinal cortex *(31)* and significant hippocampal shrinkage *(32)* in alcoholics. Hippocampal damage during chronic ethanol treatment has been correlated with deficits in spatial learning and memory *(9)*. Thus, cortical and hippocampal damage occur with chronic ethanol treatment, and relatively short durations of alcohol abuse may cause some form of damage. Additional studies are needed to understand the molecular mechanisms involved in selective neuronal death and the factors that regulate brain-regional sensitivity to ethanol neurotoxicity.

Alcoholics who do not have Korsakoff's amnestic syndrome show decreased neuropsychological performance compared to per nonalcoholics on tests of learning, memory, abstracting, problem-solving, visuospatial and perceptual motor functioning, and information processing *(33)*. Alcoholics are not only less accurate, but take considerably longer to complete tasks and are differentially vulnerable to these deficits. Many of the deficits appear to recover to age-appropriate levels of performance over a 4- to 5-yr period of abstinence *(33)*. Although global cerebral mass returns toward normal levels with extended abstinence, not all cognitive functions return. Some abstinent alcoholics appear to have permanent cognitive impairments, particularly in memory and visual–spatial–motor skills *(34)*. Other studies show a loss of logical memory and diminished paired association learning in alcoholics that may be long-lasting *(35)*. This suggests long-term changes in brain function following chronic alcohol abuse that likely relate to permanent changes in neuronal plasticity and circuitry.

Electrophysiological studies using brain electroencephalogram (EEG) and early receptor potential (ERP) have revealed that alcoholics have difficulty differentiating relevant and irrelevant, easy and difficult, and familiar and unfamiliar stimuli *(36)*. These deficits may be related to frontal cortical function. One hypothesis regarding the development of alcoholism is that an initial state of disinhibition/hyperexcitability, perhaps because of low frontal cortical impulse inhibition, is a predisposition for alcohol abuse leading to dependence *(37)*. Interestingly, these are the same areas damaged by chronic alcohol abuse, suggesting a cycle where frontal cortical hypofunction promotes further alcohol consumption, contributing to the progression to addiction *(2)*. Both clinical and experimental studies show frontal cortical involvement in the neuropsychological dysfunction of alcoholics, particularly those with Korsakoff's syndrome *(38)*. Functions affected include emotional abilities, disinhibition, perseverative responding, problem-solving abilities, and attention. It is of note that the prefrontal cortex, which sends and receives projections from the basal ganglion, may play an important role in addiction and drug-seeking behavior. Prefrontal damage is typically associated with changes in personality and elusive cognitive abnormalities. Recent studies have emphasized the role of the prefrontal cortex in executive cognitive functions *(39)*. Executive cognitive functions are those higher mental abilities such as attention, planning, organization, sequencing, abstract reasoning, and the ability to utilize external and internal feedback to adaptively modulate future behavior *(40)*. These abilities are diminished in alcoholism and other diseases showing prefrontal damage *(41)*, and disruption of these functions has been implicated in the aggression associated with substance abuse *(42)*. Although these types of change in brain function are difficult to assess, they are consistent with the morphological changes found in the frontal cortex, as well as damage to the association cortex in animals shown in recent studies.

## 3. ETHANOL, NMDA, AND GABA$_A$ RECEPTORS

There is evidence that alcohol-induced neurotoxicity is related to changes in glutamate and GABA transmission. The major excitatory and inhibitory neurotransmitters in the brain are glutamate and GABA, respectively. It follows that the ionotrophic glutamate receptors $N$-methyl-D-aspartate (NMDA), $(RS)$-$\alpha$-amino-3-hydroxy-5-methyl-4-isoxazole propionic acid (AMPA), and kainate, and the GABA$_A$ receptors are among the most widely distributed and abundant receptor-operated ion channels in the central nervous system (CNS). Both NMDA and GABA$_A$ receptors are composed of multiple subunit proteins, which are thought to assemble as heteropentameric structures that exhibit distinct properties depending upon the particular subunit composition. Identified subunits include $\alpha 1$–$\alpha 6$, $\beta 1$–$\beta 4$, $\gamma 1$–$\gamma 4$, $\delta$, and $\rho 1$–$2$ for GABA$_A$ receptors, and NR1, NR2A–NR2D, and NR3A for NMDA receptors. Additional variants of both GABA$_A$ and NMDA receptors are generated by alternative splicing *(43)*. There is convincing evidence that excitatory NMDA and inhibitory GABA$_A$ receptors are important sites of action of ethanol. Studies using a variety of tissue preparations (i.e., heterogeneous neuronal preparations and cells transfected with receptor subunits) demonstrate that pharmacologically relevant concentrations of ethanol can potentiate GABA$_A$ receptor currents and antagonize NMDA receptor currents *(43,44)*. It is likely that the combination of reduced excitatory glutamate-mediated activity and enhanced inhibitory GABA-mediated activity contributes to ethanol intoxication. Brain-regional differences in the ethanol sensitivity of NMDA and GABA$_A$ receptors have been noted *(43,45)*, leading to the hypothesis that the ethanol sensitivity of native NMDA and $GABA_A$ receptors is determined, at least in part, by the subunit composition of the receptor. This is supported by studies using recombinant expression systems showing that the ethanol sensitivity of these receptors varies with the particular subunits expressed *(43,46)*, there is considerable evidence that specific hydrophobic sites of the receptor polypeptides are crucial for modulation of ionotrophic glutamate and $GABA_A$ receptor function by ethanol and volatile anesthetics *(47,48)*. In addition, phosphorylation and dephosphorylation play an important role in regulating NMDA and $GABA_A$ receptors, and second-messenger phosphorylating systems that directly or indirectly modulate receptor function can be disrupted by ethanol.

## 4. CHRONIC ETHANOL TOLERANCE

Hyperexcitability of the CNS is a characteristic component of ethanol withdrawal, and there is good evidence for both a reduction in GABA-mediated inhibitory neurotransmission and an increase in glutamate-mediated excitatory neurotransmission following chronic ethanol exposure. Studies using primary neuronal cultures have shown that prolonged exposure to ethanol leads to a supersensitization of NMDA receptor-mediated events, such as $Ca^{2+}$ influx *(49)* and $Ca^{2+}$-dependent processes, including glutamate excitotoxicity *(50)*, and glutamate–NMDA receptor-stimulated nitric oxide (NO) formation *(51)*. Similarly, studies with isolated brain preparations have reported that ethanol-mediated enhancement of $GABA_A$ receptor-coupled $Cl^-$ flux is decreased following chronic ethanol exposure *(52)*. Thus, chronic ethanol-induced NMDA supersensitivity and $GABA_A$ receptor desensitization occur with chronic ethanol exposure and likely contribute to the hyperexcitability of the alcohol-withdrawal syndrome and alcohol neurotoxicity.

There are likely to be multiple mechanisms of chronic ethanol-induced changes in glutamate–NMDA receptor sensitivity. Several studies have reported increased NMDA receptor binding and/or subunit protein levels following chronic ethanol exposure both in vitro and in vivo *(53–55)*, whereas other studies have failed to find any such changes *(51,56–58)*. It has been shown that exposure of cultured cortical neurons to ethanol leads to enhancement of NMDA–stimulated NO formation (but not that stimulated by kainate, AMPA or ionomycin) without a change in receptor density, suggesting that the main factor in chronic ethanol-induced NMDA supersensitivity is related to posttranslational modification of NMDA receptors *(51)*. Similarly, in cells stably transfected with $GABA_A$ receptor subunits, ethanol was shown to cause changes in $GABA_A$ receptor function similar to those observed in vivo, but no change in surface receptor density *(59)*. This finding is consistent with previous observations in rats *(60)* and mice *(61)* chronically exposed to ethanol. Because the transfected cells contain defined $GABA_A$ receptor subunits with expression controlled by the dexamethasone-sensitive promoter, it is unlikely that ethanol affected subunit expression or produced subunit substitution. Thus, it is possible that posttranslational modification(s) underlie these functional changes. No change in $GABA_A$ receptor density following chronic ethanol exposure has been observed in the majority of studies. However, changes in $GABA_A$ receptor subunit expression in the brain have been reported, suggesting that subunit changes could play an important role in vivo *(62)*.

Phosphorylation is important in direct and indirect modulation of NMDA receptors and might play a role in synaptic modifications underlying ethanol tolerance. Tyrosine phosphorylation within the C-terminal region of NR2 subunits enhances NMDA currents *(63–66)* and tyrosine phosphatases reduce NMDA currents *(67–69)* It is thought that tyrosine kinases and phosphatases participate in a dynamic process that regulates channel activity. Phosphorylation of tyrosine residues within the C-terminal region of the NR2 subunit appears to potentiate NMDA currents by reducing tonic inhibition of the receptor by zinc *(70)*. Calmodulin kinase II has also been reported to phosphorylate the C-terminal region of NR2 *(71)*, but it is not known how this affects channel function. Phosphorylation of the Cl-cassette by protein kinase A (PKA) can enhance NMDA receptor function. In the absence of synaptic activity, it appears that NMDA receptors are phosphorylated by basally active PKA, thus enhancing the activity of quiescent receptors *(72,73)*. Ca$^{2+}$ influx during receptor activation leads to calcineurin-mediated dephosphorylation and receptor downregulation, which can be overcome by β-adrenoceptor-mediated stimulation of PKA activity *(72)*. As noted above, protein kinase C (PKC) phosphorylation of the Cl-cassette can also regulate channel function. Thus, phosphorylation by a variety of kinases can regulate NMDA receptor sensitivity in ways consistent with ethanol tolerance.

Whether the phosphorylation state of NMDA and GABA$_A$ receptors is altered by chronic ethanol exposure is not yet known. Chronic ethanol can increase both PKC levels and activity *(74–78)* as well as induce heterologous desensitization of cAMP signaling with decreased PKA activity *(79–81)*. Some of these wide-ranging effects of ethanol exposure could relate to changes in subcellular translocation and localization. Ethanol has been shown to stimulate translocation to the nucleus of the catalytic subunit of PKA, where it remains sequestered for as long as ethanol is present *(82)*, and to stimulate translocation of PKC-δ and PKC-ε to new intracellular sites *(78)*. Translocation of PKC and PKA isozymes to subcellular anchoring proteins is thought to be important in targeting specific signaling events. Furthermore, PKA and calcineurin (protein phosphatase 2B) are concentrated in postsynaptic densities via a common A-kinase anchoring protein (AKAP79), putting them in position to regulate phosphorylation and/or dephosphorylation of key postsynaptic proteins *(83)*. Clearly, changes in PKA and/or PKC activity and subcellular targeting could play an important role in ethanol-induced changes in synaptic function, including modulation of NMDA and GABA$_A$ receptors.

It is estimated that approximately 90% of excitatory glutamatergic synapses in the mammalian brain occur on dendritic spines. Dendritic spines, proposed to be the primary sites of synaptic plasticity in the brain, contain a pronounced postsynaptic density (PSD) enriched in neurotransmitter receptors and associated signal-transducing proteins. Recent studies have demonstrated the importance of cytoskeletal elements and scaffolding proteins in anchoring the molecular components within the PSD. Changes in dendritic spine shape have been correlated with behavioral alterations, such as learning and memory, and may provide a structural basis for plasticity in the brain. Shape changes can occur within seconds and are coupled to changes in synaptic activity *(84)*. In particular, AMPA and NMDA receptor activity is associated with spine dynamics *(85)*. Ethanol may disrupt structural plasticity through alterations in AMPA/NMDA receptor function via direct action on the channels themselves or indirectly through changes in synaptic activity (e.g., enhanced GABAergic and decreased glutamatergic activity). Signaling through Rho GTPases, which regulate the organization of the actin cytoskeleton, appear to play a key role in regulation of actin-based plasticity of dendritic spines *(86,87)*. (Fig. 2). Thus, ethanol-induced changes in NMDA responses could be related to changes in dendritic localization.

Another potentially important process in NMDA and GABA$_A$ receptor adaptation during ethanol exposure is receptor–cytoskeletal interaction *(88)*. NMDA receptor NR2 subunits bind to the PDZ domains of a family of closely related postsynaptic density proteins (PSDs) [ PSD-95/synapse-associated protein (SAP)-90, Chapsyn-110/PSD-93, and SAP-102] *(89)*. These proteins appear to function by transporting receptor proteins to the synapse as well as anchoring them at synaptic sites. PSD-95 proteins can undergo head-to-head disulfide linkage resulting in a multimodular scaffold for clustering receptors and/or ion channels and coupling receptor–enzyme complexes and receptor–downstream signaling molecules *(90)*. NMDA receptors are required for activity-dependent synaptic remodeling during development, and studies in hippocampal cultures have shown that the subcellular distribution of NMDA receptors is modulated by receptor activity. Chronic treatment with an NMDA receptor antagonist leads to increased NMDA receptor clustering at synaptic sites. Conversely, spontaneous activity leads to decreased synaptic NMDA receptor clustering *(91)*. Because studies in primary neuronal cell cultures might more closely model developmental processes, an important question to be addressed is whether this activity-dependent redistribution of NMDA receptors also occurs in mature neurons. Whether the functional property of the NMDA receptor itself is altered by clustering and redistribution (i.e., synaptic versus nonsynaptic) is unknown. Receptor redistribution could represent a

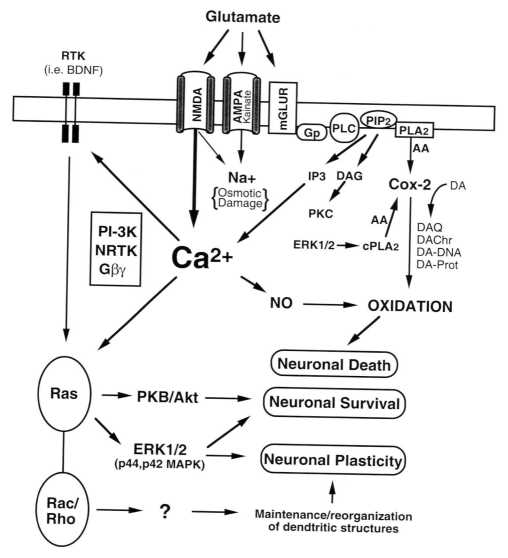

**Fig. 2.** A schematic diagram of glutamate-induced signal transduction associated with brain damage. It is known that excessive glutamate stimulation of NMDA receptors can lead to neuronal death in a form known as excitotoxicity. Rapid death can occur through osmotic damage. Delayed neuronal death is associated with excessive NMDA activation, neuronal depolarization, and other glutamate receptor activation, leading to excessively high levels of calcium that can activate a series of cascades, including oxidative stress and other mechanisms yet to be revealed, that lead to neuronal death. In addition, less dramatic activation of NMDA receptors and/or other factors activate a trophic pathway involving tyrosine kinases including the extracellular-signal regulated kinase–mitogen-activated protein kinase (ERK–MAPKinase) pathway that leads to activation of neuronal survival genes. Thus, alcohol—by changing NMDA receptor sensitivity—can alter the ratios between cell-death pathways activated through glutamate NMDA receptors and cell-survival pathways activated through glutamate NMDA receptors.

novel form of activity-dependent synaptic modification (plasticity), and prolonged inhibition of the NMDA receptor during chronic ethanol exposure might also lead to an increase in NMDA receptor clustering at synaptic sites. This is an intriguing hypothesis that needs testing.

The mitogen-activated protein kinase (MAPK) signaling cascade is one of the most highly conserved signal-transduction systems in eukaryotes. MAPK comprises three signaling modules that include the extracllular-signal regulated kinases (ERKs), stress-activated protein (SAP) kinase/jun N-terminal kinase (JNK), and the p38 kinase pathway. These signaling pathways utilize a small G-protein that couples to activation of a downstream signaling cassette of sequentially acting kinases (92). Ras proteins belong to the superfamily of small GTPases that cycle between inactive GDP-bound states and active GTP-bound states and represent a point of convergence for the transduction and integration of many extracellular signals (93). Ras activity is regulated through activation of guanine nucleotide exchange. Ras–GTP activates the serine/threonine kinase Raf-1 by a complex and poorly understood mechanism that involves its recruitment to the membrane. In addition, Raf-1 has multiple phosphorylation sites that can promote both activation and inhibition. Raf-1 activation initiates a kinase cascade that involves activation of the dual-specificity kinase MAPK/ERK (MEK) that, in turn, activates ERK1 and ERK2 (ERK1/2). Activated ERKs phosphorylate cellular substrates and/or translocate to the nucleus where they regulate transcription of genes critical in proliferation, differentiation, and survival in non-neuronal cells. In neurons, increases in intracellular calcium can activate Ras–ERK signaling, and calcium-dependent NMDA receptor-mediated ERK activation appears to play an important role in NMDA-associated synaptic plasticity and survival (94). Chronic ethanol treatment alters both NMDA receptor sensitivity and activation of ERK cascades through phosphorylation of ERK (Fig. 3). In other examples, NMDA-dependent hippocampal long-term potentiation (LTP) is associated with activation of ERK and is blocked by compounds that inhibit the ability of MEK to activate ERK (95). ERK activation has also been shown to be required for hippocampal dependent-associative learning, and Ras–GRF knockout mice display impaired amygdala-dependent memory consolidation (96,97). Ras–ERK signaling is also associated with regulation of dendrite outgrowth and refinement of neuronal processes. Thus, ERK and other tyrosine kinases can be activated by NMDA receptor-stimulated calcium flux.

Other signaling molecules that have gained attention of late are the lipid kinase phosphoinositol 3 kinase (PI3K), and its downstream target, protein kinase B (PKB). The PI3K pathway consists of a heterodimer of a regulatory subunit (P85) and a catalytic subunit (p110). Activation of PI3K protects cells from apoptosis and is thus considered a survival signal (98). Although the lipid products produced by PI3K can stimulate multiple kinases, its antiapoptotic signal is thought to be mediated through activation of PKB. A number of substrates have been identified for PKB, including the apoptotic protein Bad, the forkhead transcription factor FKHRL1 (a transcriptional regulator of many proapoptotic proteins), glycogen synthase kinase-3 (Gsk3), caspase-9, CREB, IκB-kinase-α (IKKα), and nitric oxide synthase (NOS). Phosphorylation by PKB is often associated with negative regulation by promotion of binding to the protein 14-3-3. Phosphorylation of Bad and FKHRL1 creates a recognition binding site for 14-3-3, which leads to their sequestration upon binding 14-3-3. However, some PKB-mediated inactivations do not involve sequestration by 14-3-3 (such as caspase-9 and Gsk3), and still other substrates are not inactivated by PKB (such as CREB). A number of recent studies have shed light on the fact that a great deal of crosstalk exists between ERK and PKB signaling systems. PI3K–PKB activation is frequently observed to be dependent on Ras activation, and Ras may represent a point of convergence for upstream signals that couple to activation of Raf/ERK and PI3K/PKB. In addition, recent studies have shown that PKB can regulate Raf/ERK signaling, apparently through PKB-mediated phosphorylation of Raf-1 on Ser$^{259}$ (99). Interestingly, recent evidence suggests that PI3K–PKB activation may also be coupled to NMDA receptor activation.

As discussed earlier, NMDA receptors are major targets of ethanol and appear to play a central role in the effects of acute and chronic ethanol on brain function. Acute ethanol exposure inhibits NMDA

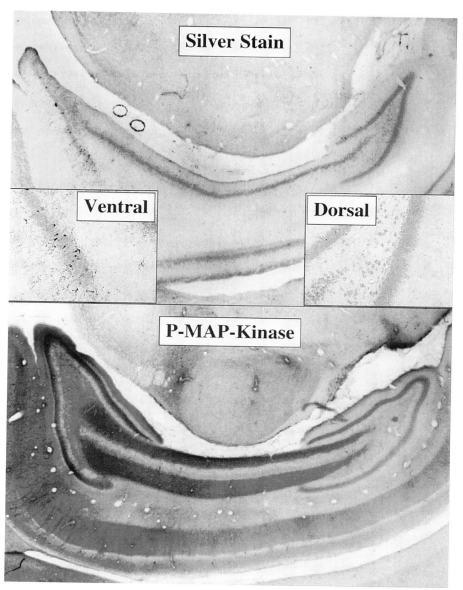

**Fig. 3.** Hippocampal damage and MAPKinase phosphorylation following binge ethanol treatment of rats (see ref. *28* for details). Shown are sections of hippocampus revealing neuronal damage through amino cupric silver staining and phospho-MAPKinase. (e.g., ERK1/2) through immunohistochemistry using phospho-MAPKinase specific antibodies. Phosphorylation of MAPK–ERK1/2 leads to activation of tyrosine-kinase-mediated neuronal survival–plasticity signals. This figure illustrates how chronic ethanol dependence leads to activation of cell-death pathways and activation of phospho-MAPKinase pathways in similar portions of the hippocampus. Shown is a section of hippocampus with dorsal and ventral (temporal hippocampus) areas. The ventral section here (temporal hippocampus) shows significant brain damage as visualized by silver stain in the upper image (see insets with higher magnification). Less significant amino cupric silver staining cell death is seen in the dorsal hippocampal regions. In the lower image, immunohistochemistry from phospho-MAPKinase also shows a gradient of increased phospho-MAPKinase presence in the ventral dentate gyrus, temporal hippocampal dentate gyrus, as opposed to the more dorsal or upper regions of the hippocampus. Thus, with chronic ethanol treatment, cellular death pathways and cellular survival pathways appear to be activated in similar areas consistent with both signals being related to ethanol-induced NMDA supersensitivity.

stimulation of ERK and PKB in vitro *(100,101)*, whereas chronic ethanol exposure has been reported to enhance ERK and PKB activation in vitro (100) and ERK in vivo (Fig. 3). In addition, as discussed below, expression of several growth factors and their receptor tyrosine kinases that couple to ERK and PKB activation are altered by prolonged ethanol exposure. Thus, disruption of ERK and PKB signaling by ethanol may play a key role in ethanol-related neuropathology and ethanol-induced alterations in neurocircuitry and neuroplasticity of the developing and adult brain.

## 5. GLUTAMATE RELEASE AND ETHANOL DEPENDENCE

In addition to the NMDA supersensitivity that occurs during chronic ethanol treatment and likely contributes to the hyperexcitability of ethanol withdrawal, ethanol also alters extracellular glutamate levels. Acutely ethanol tends to lower CNS extracellular glutamate levels consistent with NMDA receptor antagonism by ethanol *(102,103)*. In alcohol-dependent rats, extracellular levels of glutamate return to control levels. However, during withdrawal, there is a threefold increase in the levels of extracellular glutamate that corresponds in time with the progression of the ethanol-withdrawal syndrome *(104–106)*. Thus, it is likely that the hyperexcitability of the ethanol-withdrawal syndrome involves contributions from supersensitive NMDA receptor responses, increases glutamate release, and blunted GABA inhibitory responses.

## 6. NEUROTROPHIC FACTORS

Neurotrophins are small protein growth factors that have profound influences upon the development, survival, regulation of function, and plasticity of neurons. The neurotrophin family includes nerve growth factor (NGF), brain-derived neurotrophic factor (BDNF), neurotrophin-3 (NT-3), and neurotrophin-4/5. Although the members of the neurotrophin family are 50–55% homologous, the different neurotrophins promote survival of distinct sets of neurons through distinct receptors. For example, sympathetic neurons respond to NGF and NT-3 but not to BDNF *(107)*, whereas dopamine neurons respond to BDNF and NT-3, but not to NGF *(108)*. All members of the neurotrophin family bind to a low-affinity receptor, p75, and each member binds to a high-affinity *trk* tyrosine kinase receptor. Signal transduction involves both tyrosine kinase signals as well as internalization and transport of the neurotrophin–receptor complex. Because neurotrophins contribute both to the survival of neurons as well as resistance to toxicity, ethanol-induced changes in neurotrophin levels and/or function could contribute to ethanol-induced neurotoxicity.

Much of the present research on chronic ethanol-induced toxicity and the protective effects of trophic factors have focused on the neurons in the hippocampus and septohippocampal pathways, of which the cholinergic and GABAergic neurons have been shown to be susceptible to neuronal loss and atrophy *(19,109–113)*. These neurons are known to require neurotrophins for normal function and survival. Arendt and colleagues found that 28 wk of ethanol liquid diet (20% v/v) decreased choline acetyltransferase and other cholinergic-specific neuronal markers by 60–80%. However, the reduction in the number of neurons expressing the nerve growth factor receptor component p75 was only reduced 20–30% *(22)*. In contrast to reduced neurotrophin receptors, NGF mRNA was significantly increased throughout the brain, with pronounced increases in the hippocampus, where chronic ethanol treatment increased NGF mRNA approximately twofold *(114)*. Increased NGF expression is known to occur in response to traumatic brain injury *(115,116)*. The increased NGF mRNA levels following chronic ethanol treatment may be the result of neuronal damage induced by ethanol exposure. Arendt et al. *(114)* observed increased NGF mRNA levels up to 4 wk after being removed from the ethanol diet, which is consistent with a long-term alteration in NGF expression in response to ethanol neurotoxicity. Although increased NGF mRNA levels were associated with dendritic remodeling, the expression of choline acetyltransferase remained decreased. Assuming increased NGF translation, this suggests that the increased NGF level is not robust enough to completely restore choline acetyltransferase expression in chronic ethanol-damaged rat brain *(114)*. These studies suggest that chronic ethanol-induced damage induced NGF expression. Walker's group *(117)* reported that 28 wk of chronic ethanol diet causing septohippocampal damage does not change immunoreactive NGF levels or NGF mRNA when animals are sacrificed just after ethanol has cleared from the blood. Interestingly, a third study has found that chronic ethanol treatment of rats leads to elevated NGF content in the hippocampus after 2- or 4-wk exposures, but not after 12-wk exposures *(118)*. Nine months of ethanol liquid diet treatment decreased sciatic nerve NGF by 54% but had no effect on NGF content of the iris, superior cervical ganglion, trigeminal ganglion, or submandibular ganglion *(119)*. Taken together, these studies suggest that NGF levels increase under certain conditions in response to tissue injury, but not sufficiently to correct chronic ethanol-induced damage to the septohippocampal pathways.

Although NGF levels were either unchanged or increased following chronic ethanol treatment, chronic ethanol exposure decreased both neuronal survival (–25%) and neurite-outgrowth (–50%) activities of hippocampal extracts relative to controls, assessed using neuronal culture bioassays *(120)*. At least 21 wk of chronic ethanol treatment were required to reduce neurotrophic activity within the hippocampus. After 28 wk of chronic ethanol treatment, both neurotrophic activity levels and morphological changes resulting from ethanol were found in the septal and hippocampal areas *(111,112)*, suggesting that there may be a relationship between damage and loss of trophic activity. Measurement of mRNA for NT-3, bFGF, and BDNF indicated that only BDNF mRNA appeared to be reduced by 21–28 wk of chronic ethanol treatment *(121,122)*. The loss of BDNF during chronic ethanol treatment likely plays a role in chronic ethanol-induced damage to the septohippocampal pathway *(121,122)*.

Growth factors and NMDA receptors are linked in a number of ways. Growth factors have been found to reduce neuronal sensitivity to NMDA excitotoxicity and oxidative radical formation *(123)*. In contrast, sublethal NMDA stimulation has been shown to have a protective effect on survival of cerebellar granule cells by inducing BDNF *(124,125)*. BDNF, through a tyrosine kinase mechanism, can actually enhance NMDA receptor responses *(126)*. Ethanol-induced blockade of NMDA receptors during chronic administration could contribute to decreased BDNF expression and increased neurotoxicity *(124)*. BDNF has been shown to increase the survival of both dopamine neurons *(127,128)* and serotonin neurons *(129)*; therefore, loss of BDNF would have a deleterious effect on these neuronal populations. For example, in neuronal cultures treatment with NMDA for a short period of time actually increased the density of dopamine neurons, consistent with sublethal NMDA stimulation having a trophic effect *(130)*. However, pretreatment of neuronal cultures with ethanol for 2 d followed by ethanol removal, simulating ethanol withdrawal, resulted in an NMDA neurotoxic response to dopaminergic and serotonergic neurons *(131)*. These results are consistent with other studies showing that chronic ethanol can cause supersensitive NMDA excitotoxicity *(50,132)*. The addition of BDNF to the culture protected against ethanol-induced sensitization to NMDA excitotoxicity in dopamine neurons *(131)*. Chronic ethanol exposure has been found to reduce brain levels of BDNF *(121,122)*. Thus, a mechanism of ethanol neurotoxicity includes reductions in BDNF production sensitizing neurons to insults, including NMDA excitotoxicity.

Receptors for the growth factors are also modified by chronic ethanol treatment. Studies have found that chronic ethanol exposure can decrease levels of p75 neurotrophin receptors *(22,133)*. In contrast, chronic ethanol treatment of rats for 28 wk has been found to increase *trk* B-like protein expression, suggesting an upregulation of BDNF receptor response elements, perhaps as a compensatory reaction to the decrease in BDNF levels *(121)*. Other studies have found ethanol disruption of neuronal calcium that can be reversed by NGF *(134)*. Ethanol has also been reported to disrupt tyrosine kinase signaling of insulin growth factor receptors *(135)*. Thus, the interaction of ethanol with growth factors could play a key role in ethanol neurotoxicity. The study of growth factor action and its role in ethanol-induced brain damage represents an exciting new area of research with tremendous potential to provide new approaches to the treatment of neurodegeneration.

## 7. SUMMARY AND CONCLUSIONS

Human and animal studies have clearly established that chronic alcohol intake causes a loss of brain mass as a result of ethanol-induced brain damage. Although human studies show loss of brain tissue and the loss of specific brain nuclei after years of alcohol consumption, recent animal studies have indicated that neuronal loss can occur after only a few days of binge drinking. The neuropathology in humans involves both gray and white matter, whereas the animal studies have focused on neuronal loss (e.g., gray matter). It is possible that long-term animal studies could confirm the human findings. Alcohol-induced neuropathology is associated with the loss of some cognitive functions, including the executive functions associated with the frontal cortex. Neuropathology and the loss of executive functions, including impulse inhibition and goal setting, could contribute to the development of alcohol dependence.

Ethanol-induced brain damage appears to be secondary to changes in glutamate and GABA transmission. Acute ethanol exposure inhibits NMDA receptors and reduces extracellular glutamate, which could reduce NMDA trophic signaling through tyrosine kinases. Chronic ethanol exposure leads to NMDA supersensitivity because of changes in the receptor response. Although some reports have found an increase in NMDA receptor number, more studies have shown supersensitive NMDA responses in the absence of more receptors. Nor do models of alcohol dependence show increased NMDA receptor density. This suggests that other mechanisms of NMDA supersensitivity are involved in this adaptive response to ethanol. The supersensitive NMDA response includes chronic ethanol-induced supersensitivity to NMDA excitotoxicity. NMDA supersensitivity combined with increased extracellular glutamate during ethanol withdrawal are likely contributors to ethanol neurotoxicity. Blunting of GABA inhibitory responses would be expected to enhance neurotoxicity, as would the loss of trophic signals. The loss of BDNF and other trophic signals during chronic ethanol treatment can sensitize cells to NMDA toxic insults. Thus, the combination of NMDA supersensitivity, increased extracellular glutamate, GABA blunting, and loss of trophic signals are likely factors in ethanol-induced neurotoxicity. Unique aspects of these factors in combination probably contribute to specific brain-region sensitivity. Although additional studies are needed to determine the exact role of brain-region-specific ethanol neurotoxicity and its relationship to the development of alcohol dependence, it is clear that glutamate plays a significant role in ethanol neurotoxicity.

## REFERENCES

1. Grant, B. F. and Dawson, D. A. (1997) Age of onset of alcohol use and its association with DSM-IV alcohol abuse and dependence: results from the National Longitudinal Alcohol Epidemiologic Survey. *J. Subst. Abuse* **9**, 103–110.
2. Crews, F. T. (1999) Alcohol and neurodegeneration. *CNS Drug Rev.* **5**, 379–394.
3. Sullivan, E. V., Rosenbloom, M. J., and Pfefferbaum, A. (2000) Pattern of motor and cognitive deficits in detoxified alcoholic men. *Alcohol. Clin. Exp. Res.* **24**, 611–621.
4. Sullivan, E. V., Rosenbloom, M. J., Lim, K. O., and Pfefferbaum, A. (2000) Longitudinal changes in cognition, gait, and balance in abstinent and relapsed alcoholic men: relationships to changes in brain structure. *Neuropsychology* **14**, 178–188.
5. Martin, P. R., Adinoff, B., Weingartner, H., Mukherjee, A. B., and Edhardt, M. J. (1986) Alcoholic organic brain disease: nosology and pathophysiologic mechanisms. *Prog. Neuropsychopharmacol. Biol. Psychiatry* **10**, 147–164.
6. Butterworth, R. F. (1995) Pathophysiology of alcoholic brain damage: synergistic effects of ethanol, thiamine deficiency and alcoholic liver disease. *Metab. Brain Dis.* **10**, 1–8.
7. Pfefferbaum, A., Lim, K. O., Desmond, J. E., and Sullivan, E. V. (1996) Thinning of the corpus callosum in older alcoholic men: A magnetic resonance imaging study. *Alcohol. Clin. Exp. Res.* **20**, 752–757.
8. Harper, C. G. and Kril, J. J. (1993) Neuropathological changes in alcoholics, in *Alcohol-Induced Brain Damage* (Hunt, W. A. and Nixon, S. J., eds.), NIAAA/NIH, Rockville, MD, pp. 39–70.
9. Franke, H., Kittner, H., Berger, P., Wirkner, K., and Schramek, J. (1997) The reaction of astrocytes and neurons in the hippocampus of adult rats during chronic ethanol treatment and correlations to behavioral impairments. *Alcohol* **14**, 445–454.
10. Courville, C. B. (1955) *Effects of Alcohol on the Nervous System in Man*, San Lucas, Los Angeles.
11. Harper, C. B. and Krill, J. J. (1990) Neuropathology of alcoholism. *Alcohol* **25**, 207–216.

12. Nicolas, J. M., Estruch, R., Salamero, M., Orteu, N., Fernandez-Sola, J., Sacanella, E., et al. (1997) Brain impairment in well-nourished chronic alcoholics is related to ethanol intake. *Ann. Neurol.* **41,** 590–598.
13. Pfefferbaum, A., Sullivan, E. V., Mathalon, D. H., and Lim, K. O. (1997) Frontal lobe volume loss observed with magnetic resonance imaging in oldler chronic alcoholics. *Alcohol. Clin. Exp. Res.* **21,** 521–529.
14. Ron, M. A., Acker, R. W., Shaw, G. K., and Lishman, W. A. (1982) Computerized tomography of the brain in chronic alcoholism: a survey and follow-up study. *Brain* **105,** 497–514.
15. Jernigan, T. L., Butters, N., DiTraglia, G., Schafer, K., Smith, T., Irwin, M., et al. (1991) Reduced cerebral grey matter observed in alcoholics using magnetic resonance imaging. *Alcohol. Clin. Exp. Res.* **15,** 418–427.
16. Harper, C. G., Kril, J. J., and Daly, J. (1987) Are we drinking our neurons away? *Br. Med. J.* **294,** 534–536.
17. Sullivan, E. V., Marsh, L., Mathalon, D. H., Lim, K. O., and Pfefferbaum, A. (1996) Relationship between alcohol withdrawal seizures and temporal lobe white matter volume deficits. *Alcohol. Clin. Exp. Res.* **20,** 348–354.
18. Jagannathan, M. R., Desai, M. G., and Raghunathan, P. (1996) Brain metabolite changes in alcoholism: an *in vivo* proton magnetic resonance spectroscopy. *Magn. Reson. Imaging* **14,** 553–557.
19. Arendt, T. (1993) The cholinergic differentiation of the cerebral cortex induced by chronic consumption of alcohol: reversal by cholinergic drugs and transplantation, in *Alcohol-Induced Brain Damage* (Hunt, W. A. and Nixon, S. J., eds.), NIAAA/NIH, Rockville, MD, pp. 431–460.
20. Halliday, G., Baker, K., and Harper, C. (1995) Serotonin and alcohol-related brain damage. *Metab. Brain. Dis.* **10,** 25–30.
21. Baker, K. G., Halliday, G. M., Kril, J. J., and Harper, C. G. (1996) Chronic alcoholism in the absence of Wernicke–Korsakoff syndrome and cirrhosis does not result in the loss of serotonergic neurons from the median raphe nucleus. *Metab. Brain Dis.* **11,** 217–227.
22. Arendt, T., Bruckner, M. K., Magliusi, S., and Krell, T. (1995) Degeneration of rat cholinergic basal forebrain neurons and reactive changes in nerve growth factor expression after chronic neurotoxic injury-I. Degeneration and plastic response of basal forebrain neurons. *Neuroscience* **65,** 633–645.
23. Arango, V., Underwood, M. D., Pauler, D. K., Kass, R. E., and Mann, J. J. (1996) Differential age-related loss of pigmented locus coeruleus neurons in suicides, alcholics, and alcoholic suicides. *Alcohol. Clin. Exp. Res.* **20,** 1141–1147.
24. Lu, W., Jaatinen, P., Rintala, J., Sarviharju, M., Kiianmaa, K., and Hervonen, A. (1997) Effects of life-long ethanol consumption on rat locus coeruleus. *Alcohol Alcohol.* **32,** 463–470.
25. Harding, A. J., Halliday, G. M., Ng, J. L., Harper, C. G., and Kril, J. J. (1996) Loss of vasopressin-immunoreactive neurons in alcoholics is dose-related and time-dependent. *Neuroscience* **72,** 699–708.
26. Madeira, M. D., Andrade, J. P., Lieberman, A. R., Sousa, N., Almeida, O. F., and Paula-Barbosa, M. M. (1997) Chronic alcohol consumption and withdrawal do not induce cell death in the suprachiasmatic nucleus, but lead to irreversible depression of peptide immunoreactivity and mRNA levels. *J. Neurosci.* **17,** 1302–1319.
27. Collins, M., Corso, T., and Neafsey, E. (1996) Neuronal degeneration in rat cerebrocortical olfactory regions during subchronic "binge" intoxication with ethanol: possible explanation for olfactory deficits in alcoholics. *Alcohol. Clin. Exp. Res.* **20,** 284–292.
28. Crews, F. T., Braun, C. J., Hoplight, B., Switzer, R. C., and Knapp, D. J. (2000) Binge ethanol consumption causes differential brain damage in young-adolescent compared to adult rats. *Alcohol. Clin. Exp. Res.* **24,** 1712–1723.
29. Switzer R. C., III (2000) Application of silver degeneration stains for neurotoxicity testing. *Toxicol. Pathol.* **28,** 70–83.
30. de Olmos, J. S., Beltramino, C. A., and de Olmos de Lorenzo, S. (1994) Use of an amino-cupric-silver technique for the detection of early and semiacute neuronal degeneration caused by neurotoxicants, hypoxia, and physical trauma. *Neurotoxicol. Teratol.* **16,** 545–561.
31. Ibanez, J., Herrero, M. T., Insausti, R., Balzunegui, T., Tunon, T., Garcia-Bragado, F., et al. (1995) Chronic alcoholism decreases neuronal nuclear size in the human entorhinal cortex. *Neurosci. Lett.* **183,** 71–74.
32. Harding, A. J., Wong, A., Svoboda, M., Kril, J. J., and Halliday, G. M. (1997) Chronic alcohol consumption does not cause hippocampal neuron loss in humans. *Hippocampus* **7,** 78–87.
33. Parsons, O. A. (1993) Impaired neuropsychological cognitive functioning in sober alcoholics, in *Alcohol-Induced Brain Damage* (Hunt, W. A. and Nixon, S. J., eds.), NIAAA/NIH, Rockville, MD, pp. 173–194.
34. Di Sclafani, V., Ezekiel, F., Meyerhoff, D. J., Mackay, S., Dillon, W. P., and Weiner, M. W. (1995) Brain atrophy and cognitive function in older abstinent alcoholic men. *Alcohol. Clin. Exp. Res.* **19,** 1121–1126.
35. Eckardt, M. J., Rohrbaugh, J. W., Stapleton, J. M., Davis, E. Z., Martin, P. R., and Weingartner, H. J. (1996) Attention-related brain potential and cognition in alcoholism-associated organic brain disorders. *Biol. Psychiatry* **39,** 143–146.
36. Porjesz, B. and Begleiter, H. (1993) Neurophysiological factors associated with alcoholism, in *Alcohol-Induced Brain Damage* (Hunt, W. A. and Nixon, S. J., eds.), NIAAA/NIH, Rockville, MD, pp. 89–120.
37. Begleiter, H. and Porjesz, B. (1999) What is inherited in the predisposition toward alcoholism? A proposed model. *Alcohol. Clin. Exp. Res.* **23,** 1125–1135.
38. Oscar-Berman, M. and Hutner, N. (1993) Frontal lobe changes after chronic alcohol ingestion, in *Alcohol-Induced Brain Damage* (Hunt, W. A. and Nixon, S. J., eds.), NIAAA/NIH, Rockville, MD, pp. 121–156.
39. Giancola, P. R. and Zeichner, A. (1995) Alcohol-related aggression in males and females: effects of blood alcohol concentration, subjective intoxication, personality, and provocation. *Alcohol. Clin. Exp. Res.* **19,** 130–134.

40. Foster, J., Eskes, G., and Stuss, D. (1994) The cognitive neuropsychology of attention: a frontal lobe perspective. *Cogn. Neuropsychol.* **11,** 133–147.
41. Boller, F., Traykov, L., Dao-Castellana, M. H., and Fontaine-Dabernard, A. (1995) Cognitive functioning in "diffuse pathology": Role of prefrontal and limbic structures. *Ann. NY Acad. Sci.* **769,** 23–39.
42. Hoaken, P. N. S., Giancola, P. R., and Pihl, R. O. (1998) Executive cognitive functions as mediators of alcohol-related aggression. *Alcohol Alcohol.* **33,** 47–54.
43. Crews, F. T., Morrow, A. L., Criswell, H., and Breese, G. (1996) Effects of ethanol on ion channels, in *International Review of Neurobiology* (Bradley, R., Harris, R., and Jenner, P., eds.), Academic P, San Diego, CA, pp. 283–367.
44. Diamond, I. and Gordon, A. S. (1997) Cellular and molecular neuroscience of alcoholism. *Physiol. Rev.* **77,** 1–120.
45. Criswell, H. E., Simson, P. E., Duncan, G. E., McCown, T. J., Herbert, J. S., Morrow, A. L., et al. (1993) Molecular basis for regionally specific action of ethanol on γ-aminobutyric acid$_A$ receptors: generalization to other ligand-gated ion channels. *J. Pharmacol. Exp. Ther.* **267,** 522–537.
46. Masood, K., Wu, C., Brauneis, U., and Weight, F. F. (1994) Differential ethanol sensitivity of recombinant $N$-methyl-D-aspartate receptor subunits. *Mol. Pharmacol.* **45,** 317–323.
47. Minami, K., Wick, M. J., Stern-Bach, Y., Dildy-Mayfield, J. E., Brozowiski, S. J., Gonzales, E. L., et al. (1998) Sites of volatile anesthetic action on kainate (glutamate receptor 6) receptors. *J. Biol. Chem.* **273,** 8248–8255.
48. Wick, M., Mihic, S. J., Ueno, S., Mascia, M. P., Trudell, J. R., Brozowski, S. J., et al. (1998) Mutations of gamma-aminobutyric acid and glycine receptors change alcohol cutoff: evidence for an alcohol receptor? *Proc. Natl. Acad. Sci. USA* **95,** 6504–6509.
49. Iorio, K. R., Reinlib, L., Tabakoff, B., and Hoffman, P. L. (1992) Chronic exposure of cerebellar granule cells to ethanol results in increased $N$-methyl-D-aspartate receptor function. *Mol. Pharmacol.* **41,** 1142–1148.
50. Chandler, L. J., Newson, H., Sumners, C., and Crews, F. T. (1993) Chronic ethanol exposure potentiates NMDA excitotoxicity in cerebral cortical neurons. *J. Neurochem.* **60,** 1578–1581.
51. Chandler, L. J., Sutton, G., Norwood, D., Sumners, C., and Crews, F. T. (1997) Chronic ethanol increases NMDA-stimulated nitric oxide formation but not receptor density in cultured cortical neurons. *Mol. Pharmacol.* **51,** 733–740.
52. Morrow, A. L., Suzdak, P. D., and Paul, S. M. (1988) Chronic ethanol administration alters GABA, pentobarbital and ethanol-mediated 36CL-uptake in cerebral cortical synaptoneurosomes. *J. Pharmacol. Exp. Ther.* **246,** 158–164.
53. Grant, K. A., Valverius, P., Hudspith, M., and Tabakoff, B. (1990) Ethanol withdrawal seizures and the NMDA receptor complex. *Eur. J. Pharmacol.* **176,** 289–296.
54. Hu, X. J. and Ticku, M. (1995) Chronic ethanol treatment upregulates the NMDA receptor function and binding in mammalian cortical neurons. *Brain Res. Mol. Brain Res.* **30,** 347–356.
55. Snell, L. D., Tabakoff, B., and Hoffman, P. L. (1993) Radioligand binding to the $N$-methyl-D-aspartate receptor/ionophore complex: alterations by ethanol *in vitro* and by chronic *in vivo* ethanol ingestion. *Brain Res.* **602,** 91–98.
56. Tremwel, M. F., Anderson, K. J., and Hunter, B. E. (1994) Stability of [³H]MK-801 binding sites following chronic ethanol consumption. *Alcohol. Clin. Exp. Res.* **18,** 1004–1008.
57. Carter, L. A., Belknap, J. K., Crabbe, J. C., and Janoksky, A. (1995) Allosteric regulation of the $N$-methyl-D-aspartate receptor-linked ion channel complex and effects of ethanol in ethanol withdrawal seizure-prone and -resistant mice. *J. Neurochem.* **64,** 213–219.
58. Rudolph, J. G., Walker, D. W., Iimuro, Y., Thurman, R. G., and Crews, F. T. (1997) NMDA receptor binding in adult rat brain after several chronic ethanol treatment protocols. *Alcohol. Clin. Exp. Res.* **21,** 1508–1519.
59. Harris, R. A., Valenzuela, C. F., Brozowski, S., Chuang, L., Hadingham, K., and Whiting, P. J. (1998) Adaptation of γ-aminobutyric acid type A receptors to alcohol exposure: studies with stably transfected cells. *J. Pharmacol. Exp. Ther.* **284,** 180–188.
60. Rastogi, S. K. and Ticku, M. K. (1986) Anticonvulsant profile of drugs which facilitate GABAergic transmission on convulsions mediated by a GABAergic mechanism. *Neuropharmacology* **25,** 175–185.
61. Buck, K. J. and Harris, R. A. (1990) Benzodiazepine agonist and inverse agonist actions on GABAA receptor-operated chloride channels. II. Chronic effects of ethanol. *J. Pharmacol. Exp. Ther.* **253,** 713–719.
62. Morrow, A. L. (1995) Regulation of GABA$_A$ receptor function and gene expression in the central nervous system, in *International Review of Neurobiology* (Bradley, R. J. and Harris, R. A., eds.), Academic P, San Diego, CA, pp. 1–41.
63. Rosenblum, K., Dudai, Y., and Richter-Levin, G. (1996) Long-term potentiation increases tyrosine phosphorylation of the $N$-methyl-D-aspartate receptor subunit 2β in rat dentate gyrus in vivo. *Proc. Natl. Acad. Sci. USA* **93,** 10,457–10,460.
64. Suzuki, T. and Okumura-Noji, K. (1995) NMDA receptor subunits epsilon 1 (NR2A) and epsilon 2 (NR2B) are substrates for Fyn in the postsynaptic density fraction isolated from the rat brain. *Biochem. Biophys. Res. Commun.* **216,** 582–588.
65. Rostas, J. A., Brent, V. A., Voss, K., Errington, M. L., Bliss, T. V., and Gurd, J. W. (1996) Enhanced tyrosine phosphorylation of the 2β subunit of the $N$-methyl-D-aspartate receptor in long-term potentiation. *Proc. Natl. Acad. Sci. USA* **93,** 10,452–10,456.
66. Chen, C. and Leonard, J. P. (1996) Protein tyrosine kinase-mediated potentiation of currents from cloned NMDA receptors. *J. Neurochem.* **67,** 194–200.

67. Wang, Y. T., Yu, X. M., and Salter, M. W. (1996) Ca(2+)-independent reduction of *N*-methyl-D-aspartate channel activity by protein tyrosine phosphatase. *Proc. Natl. Acad. Sci. USA* **93,** 1721–1725.
68. Wang, Y. T. and Salter, M. W. (1994) Regulation of NMDA receptors by tyrosine kinases and phosphatases. *Nature* **369,** 233–235.
69. Hall, R. A. and Soderling, T. R. (1997) Differential surface expression and phosphorylation of the *N*-methyl-D-aspartate receptor subunits NR1 and NR2 in cultured hippocampal neurons. *J. Biol. Chem.* **272,** 4135–4140.
70. Zheng, F., Gingrich, M. B., Traynelis, S. F., and Conn, P. J. (1998) Tyrosine kinase potentiates NMDA receptor currents by reducing tonic zinc inhibition. *Nature Neurosci.* **1,** 185–191.
71. Omkumar, R. V., Kiely, M. J., Rosenstein, A. J., Min, K. T., and Kennedy, M. B. (1996) Identification of a phosphorylation site for calcium/calmodulin dependent protein kinase II in the NR2B subunit of the *N*-methyl-D-aspartate receptor. *J. Biol. Chem.* **271,** 31,670–31,678.
72. Tong, G., Shepherd, D., and Jahr, C. E. (1995) Synaptic desensitization of NMDA receptors by calcineurin. *Science* **267,** 1510–1512.
73. Raman, I. M., Tong, G., and Jahr, C. E. (1996) β-Adrenergic regulation of synaptic NMDA receptors by cAMP-dependent protein kinase. *Neuron* **16,** 415–421.
74. Messing, R. O., Petersen, P. J., and Henrich, C. J. (1991) Chronic ethanol exposure increases levels of protein kinase C delta and epsilon and protein kinase C-mediated phosphorylation in cultured neural cells. *J. Biol. Chem.* **266,** 23,428–23,432.
75. Roivainen, R., McMahon, T., and Messing, R. O. (1993) Protein kinase C isozymes that mediate enhancement of neurite outgrowth by ethanol and phorbol esters in PC12 cells. *Brain Res.* **624,** 85–93.
76. DePetrillo, P. B. and Liou, C. S. (1993) Ethanol exposure increases total protein kinase C activity in human lumphocytes. *Alcohol. Clin. Exp. Res.* **17,** 351–354.
77. Coe, I. R., Yao, L., Diamond, I., and Gordon, A. S. (1996) The role of protein kinase C in cellular tolerance to ethanol. *J. Biol. Chem.* **271,** 29,468–29,472.
78. Gordon, A. L., Yao, K., Wu, Z. L., Coe, I. R., and Diamond, I. (1997) Ethanol alters the subcellular localization of delta and epsilon protein kinase C in NG 108-15 cells. *Mol. Pharmacol.* **52,** 554–559.
79. Gordon, A. S., Collier, K., and Diamond, I. (1986) Ethanol regulation of adenosine receptor-stimulated cAMP levels in a clonal neural cell line: an in vitro model of cellular tolerance to ethanol. *Proc. Natl. Acad. Sci. USA* **83,** 2105–2108.
80. Rabin, R. A., Edelman, A. M., and Wagner, J. A. (1992) Activation of protein kinase A is necessary but not sufficient for ethanol-induced desensitization of cyclic AMP production. *J. Pharmacol. Exp. Ther.* **262,** 257–262.
81. Coe, I. R., Dohrman, D. P., Constantinescu, A., Diamond, I., and Gordon, A. S. (1996) Activation of cyclic AMP-dependent protein kinase reverses tolerance of a nucleoside transporter to ethanol. *J. Pharmacol. Exp. Ther.* **276,** 365–369.
82. Dohrman, D. P., Diamond, I., and Gordon, A. S. (1996) Ethanol causes translocation of cAMP-dependent protein kinase catalytic subunit to the nucleus. *Proc. Natl. Acad. Sci. USA* **93,** 10,217–10,221.
83. Coghlan, V. M., Perrino, B. A., Howard, M., Langeberg, L. K., Hicks, J. B., Gallatin, W. M., et al. (1995) Association of protein kinase A and protein phosphatase 2B with a common anchoring protein. *Science* **267,** 108–111.
84. Fischer, M., Kaech, S., Knutti, D., and Matus, A. (1998) Rapid actin-based plasticity in dendritic spines. *Neuron* **20,** 847–854.
85. Fischer, M., Kaech, S., Wagner, U., Brinkhaus, H., and Matus, A. (2000) Glutamate receptors regulate actin-based plasticity in dendritic spines. *Nature Neurosci.* **3,** 887–890.
86. Mackay, D. J. G. and Hall, A. (1998) Rho GTPases. *J. Biol. Chem.* **273,** 20,685–20,688.
87. Nakayama, A. Y., Harms, M. B., and Luo, L. (2000) Small GTPases Rac and Rho in the maintenance of dendritic spines and branches in hippocampal pyramidal neurons. *J. Neurosci.* **20,** 5329–5338.
88. Chandler, L. J., Harris, R. A., and Crews, F. T. (1998) Ethanol tolerance and synaptic plasticity. *Trends Pharmacol. Sci.* **19,** 491–495.
89. Sheng, M. (1996) PDZs and receptor/channel custering: rounding up the latest suspects. *Neuron* **17,** 575–578.
90. Hsueh, Y.-P., Kim, E., and Sheng, M. (1997) Disulfide-linked head-to-head multimerization in the mechanism of ion channel clustering by PSD-95. *Neuron* **18,** 803–814.
91. Rao, A. and Craig, A. M. (1997) Activity regulates the synaptic localization of the NMDA receptor in hippocampal neurons. *Neuron* **19,** 801–812.
92. Widmann, C., Gibson, S., Jarpe, M. B., and Johnson, G. L. (1999) Mitogen-activated protein kinase: conservation in a three-kinase module from yeast to human. *Physiol. Rev.* **79,** 143–180.
93. Campbell, S. L., Khosravi-Far, R., Rossman, K. L., Clark, G. J., and Der, C. J. (1998) Increasing complexity of Ras signaling. *Oncogene* **17,** 1395–1413.
94. Kornhauser, J. M. and Greenberg, M. E. (1997) A kinase to remember: dual roles for MAP kinase in long-term memory. *Neuron* **18,** 839–842.
95. English, J. D. and Sweatt, J. D. (1997) A requirement for the mitogen-activated protein kinase cascade in hippocampal long-term potentiation. *J. Biol. Chem.* **272,** 19,103–19,106.
96. Brambilla, R., Gnesutta, N., Minichiello, L., White, G., Roylance, A. J., Herron, C. E., et al. (1997) A role for the Ras signaling pathway in synaptic transmission and long-term memory. *Nature* **390,** 281–286.

97. Atkins, C. M., Selcher, J. C., Petraitis, J. J., Tzaskos, J. M., and Sweatt, J. D. (1998) The MAPK cascade is required for mammalian associative learning. *Nature Neurosci.* **1**, 602–609.
98. Datta, S. R., Brunet, A., and Greenberg, M. E. (1999) Cellular survival: a play in three Akts. *Genes Dev.* **13**, 2905–2927.
99. Zimmermann, S. and Moelling, D. (1999) Phosphorylation and regulation of Raf by Akt (protein kinase b). *Science* **286**, 1741–1744.
100. Chandler, L. J., Sutton, G., and Norwood, D. (1998) NMDA receptors regulate phosphorylation/dephosphorylation of ERK2 (p42$^{MAPK}$): modulation by ethanol. *Alcohol. Clin. Exp. Res.* **22**, 498.
101. Sutton, G., Dorairaj, N., and Chandler, L. J. (1999) Effects of ethanol on NMDA receptor modulation of ERK and AKT/PKB activation in cortical cultures. *Alcohol. Clin. Exp. Res.* **23**, 93A.
102. Moghaddam, B. and Bolinao, M. (1994) Biphasic effects of ethanol on extracellular accumulation of glutamate in the hippocampus and the nucleus accumbens. *Neurosci. Lett.* **178**, 99–102.
103. Carboni, S., Isola, R., Gessa, G. L., and Rossetti, Z. L. (1993) Ethanol prevents the glutamate release induced by *N*-methyl-D-aspartate in the rat striatum. *Neurosci. Lett.* **152**, 133–136.
104. Gonzales, R., Bungay, P. M., Kiianmaa, K., Samson, H. H., and Rossetti, Z. L. (1996) In vivo links between neurochemistry and behavioral effects of ethanol. *Alcohol. Clin. Exp. Res.* **20(Suppl.)**, 203A–205A.
105. Rossetti, Z. L. and Carboni, S. (1995) Ethanol withdrawal is associated with increased extracellular glutamate in the rat striatum. *Eur. J. Pharmacol.* **283**, 177–183.
106. Fadda, F. and Rossetti, Z. (1998) Chronic ethanol consumption: from neuroadaptation to neurodegeneration. *Prog. Neurobiol.* **56**, 385–431.
107. Maisonpierre, P. C., Belluscio, L., Squinto, S., Ip, N. Y., Furth, M. E., Lindsay, R. M., et al. (1990) Neurotrophin-3: a neurotrophic factor related to NGF and BDNF. *Science* **247**, 1446–1451.
108. Hyman, C., Juhasz, M., Jackson, C., Wright, P., Ip, N. Y., and Lindsay, R. M. (1994) Overlapping and distinct actions of the neurotrophins BDNF, NT-3, and NT-4/5 on cultured dopaminergic and GABAergic neurons of the ventral mesencephalon. *J. Neurosci.* **14**, 335–347.
109. Walker, D. W., Barnes, D. E., Zornetzer, S. F., Hunter, B. E., and Kubanis, P. (1980) Neuronal loss in hippocampus induced by prolonged ethanol consumption in rats. *Science* **209**, 711–713.
110. Walker, D. W., Hunter, B. E., and Abraham, W. C. (1981) Neuroanatomical and functional deficits subsequent to chronic ethanol administration in animals. *Alcohol. Clin. Exp. Res.* **5**, 267–282.
111. Walker, D. W., Heaton, M. B., Lee, N., King, M. A., and Hunter, B. E. (1993) Effect of chronic ethanol on the septohippocampal system: a role for neurotrophic factors? *Alcohol. Clin. Exp. Res.* **17**, 12–18.
112. Walker, D. W., King, M. A., and Hunter, B. E. (1993) Alterations in the structure of the hippocampus after long-term ethanol consumption, in *Alcohol-Induced Brain Damage* (Hunt, W. A. and Nixon, S. J., eds.), NIAAA, Washington, DC, pp. 231–247.
113. Nordberg, A., Larsson, C., Perdahl, E., and Winblad, B. (1983) Changes in cholinergic activity in human hippocampus following chronic alcohol abuse. *Pharmacol. Biochem. Behav.* **(Suppl.) 18**, 397–400.
114. Arendt, T., Bruckner, M. K., Magliusi, S., and Krell, T. (1995) Degeneration of rat cholinergic basal forebrain neurons and reactive changes in nerve growth factor expression after chronic neurotoxic injury—II. Reactive expression of the nerve growth factor gene in astrocytes. *Neuroscience* **65**, 647–659.
115. Korsching, S., Heumann, R., Thoenen, H., and Hefti, F. (1986) Cholinergic denervation of the rat hippocampus by fimbrial transection leads to a transient accumulation of nerve growth factor (NGF) without change in mRNANGF content. *Neurosci. Lett.* **66**, 175–180.
116. Gasser, U. E., Weskamp, G., Otten, U., and Dravid, A. R. (1986) Time course of the elevation of nerve growth factor (NGF) content in the hippocampus and septum following lesions of the septohippocampal pathway in rats. *Brain Res.* **376**, 351–356.
117. Baek, J.-K., Heaton, M. B., and Walker, D. W. (1994) Chronic alcohol ingestion: nerve growth factor gene expression and neurotrophic activity in rat hippocampus. *Alcohol. Clin. Exp. Res.* **18**, 1368–1376.
118. Nakano, T., Fujimoto, T., Shimooki, S., Fukudome, T., Uchida, T., Tsuji, T., et al. (1996) Transient elevation of nerve growth factor content in the rat hippocampus and frontal cortex by chronic ethanol treatment. *Psychiatry Clin. Neurosci.* **50**, 157–160.
119. Hellweg, R., Baethge, C., Hartung, H. D., Bruckner, M. K., and Arendt, T. (1996) NGF level in the rat sciatic nerve is decreased after long-term consumption of ethanol. *NeuroReport* **7**, 777–780.
120. Walker, D. W., Lee, N., Heaton, M. B., King, M. A., and Hunter, B. E. (1992) Chronic ethanol consumption reduced the neurotrophic activity in rat hippocampus. *Neurosci. Lett.* **147**, 77–80.
121. Baek, J. K., Heaton, M. B., and Walker, D. W. (1996) Up-regulation of high-affinity neurotrophin receptor, trk B-like protein on western blots of rat cortex after chronic ethanol. *Brain Res.* **40**, 161–164.
122. MacLennan, A. J., Lee, N., and Walker, D. W. (1995) Chronic ethanol administration decreases brain-derived neurotrophic factor gene expression in the rat hippocampus. *Neurosci. Lett.* **197**, 105–108.
123. Mattson, M. P., Lovell, M. A., Furukawa, K., and Markesbery, W. R. (1995) Neurotrophic factors attenuate glutamate-induced accumulation of peroxides, elevation of intracellular calcium concentration, and neurotoxicity and increase antioxidant enzyme activities in hippocampal neurons. *J. Neurochem.* **65**, 1740–1751.

124. Bhave, S. V., Ghoda, L., and Hoffman, P. L. (1999) Brain-derived neurotrophic factor mediates the anti-apoptotic effect of NMDA in cerebellar granule neurons: signal transduction cascades and site of ethanol action. *J. Neurosci.* **19,** 3277–3286.
125. Marini, A. M., Rabin, S. J., Lipsky, R. H., and Mocchetti, I. (1998) Activity-dependent release of brain-derived neurotrophic factor underlies the neuroprotective effect of *N*-methyl-D-aspartate. *J. Biol. Chem.* **273,** 29,394–29,399.
126. Levine, E. S., Crozier, R. A., Black, I. B., and Plummer, M. R. (1998) Brain-derived neurotrophic factor modulates hippocampal synaptic transmission by increasing *N*-methyl-D-aspartic acid receptor activity. *Proc. Natl. Acad. Sci. USA* **95,** 10,235–10,239.
127. Studer, L., Spenger, C., Seiler, R. W., Altar, C. A., Lindsay, R. M., and Hyman, C. (1995) Comparison of the effects of the neurotrophins on the morphological structure of dopaminergic neurons in cultures of rat substantia nigra. *Eur. J. Neurosci.* **7,** 223–233.
128. Hyman, C., Hofer, M., Barde, Y. A., Juhasz, M., Yancopoulos, G. D., Squinto, S. P., et al. (1991) BDNF is a neurotrophic factor for dopaminergic neurons of the substantia nigra. *Nature* **350,** 230–232.
129. Mamounas, L. A., Blue, M. E., Siuciak, J. A., and Altar, C. A. (1995) Brain-derived neurotrophic factor promotes the survival and sprouting of serotonergic axons in rat brain. *J. Neurosci.* **15,** 7929–7939.
130. Castoldi, A. F., Barni, S., Randine, G., Costa, L. G., and Manzo, L. (1998) Ethanol selectively interferes with the trophic action of NMDA and carbachol on cultured cerebellar granule neurons undergoing apoptosis. *Brain Res. Dev. Res.* **111,** 279–289.
131. Crews, F. T., Waage, H. G., Wilkie, M. B., and Lauder, J. M. (1999) Ethanol pretreatment enhances NMDA excitotoxicity in biogenic amine neurons: protection by brain derived neurotrophic factor. *Alcohol. Clin. Exp. Res.* **23,** 1834–1842.
132. Crews, F. T., Newsom, H., Gerber, M., Sumners, C., Chandler, L. J., and Freund, G. (1993) Molecular mechanisms of alcohol neurotoxicity, in *Alcohol, Cell Membranes, and Signal Transduction in Brain* (Alling, C. and Sun, G., eds.), Plenum P, Lund, Sweden, pp. 123–138.
133. Seabold, G. K., Luo, J., and Miller, M. W. (1998) Effect of ethanol on neurotrophin-mediated cell survival and receptor expression in cultures of cortical neurons. *Brain Res. Dev. Brain Res.* **108,** 139–145.
134. Webb, B., Suarez, S. S., Heaton, M. B., and Walker, D. W. (1997) Cultured postnatal rat septohippocampal neurons change intracellular calcium in response to ethanol and nerve growth factor. *Brain Res.* **778,** 354–366.
135. Xu, Y. Y., Bhavani, K., Wands, J. R., and de la Monte, S. M. (1995) Ethanol inhibits insulin receptor substrate-1 tyrosine phosphorylation and insulin-stimulated neuronal thread protein gene expression. *Biochem. J.* **310,** 125–132.

# 25
# Role of Glutamate in Alcohol Withdrawal Kindling

## Howard C. Becker, PhD and Nicole Redmond, PhD

## 1. INTRODUCTION

It is well known that continued excessive alcohol consumption can lead to the development of physiological dependence. When drinking is abruptly terminated or substantially reduced in the dependent individual, a characteristic withdrawal syndrome ensues. As with other central nervous system (CNS) depressants, withdrawal symptoms associated with cessation of chronic alcohol use are opposite in nature to the effects of intoxication. Thus, clinical features of alcohol withdrawal include signs of heightened autonomic nervous system activation (e.g., tachycardia, elevated blood pressure, diaphoresis, tremor), CNS hyperexcitability that may culminate in motor seizures, and, in its most severe form, hallucinosis and delerium tremens *(1–3)*. In addition to physical signs of withdrawal, a constellation of symptoms contributing to psychological discomfort (e.g., irritability, agitation, anxiety, dysphoria) constitute a significant component of the withdrawal syndrome *(4–7)*. The overall intensity of the withdrawal syndrome is presumed to reflect the degree of physiological dependence developed during the course of chronic alcohol use/abuse.

Although a number of factors have been shown to influence the development and expression of alcohol dependence, the amount, duration, and pattern of alcohol consumption appear to play a critical role in influencing the intensity of withdrawal symptoms *(8)*. Alcoholism and alcohol abuse often are characterized by frequent bouts of heavy drinking interspersed with periods of abstinence. Thus, it is not uncommon for many alcoholics to experience multiple episodes of withdrawal during the course of the disease *(9)*. It has been suggested that repeated experience with alcohol may render individuals more vulnerable and susceptible to more complicated and severe withdrawal episodes in the future. Furthermore, the progressive intensification of withdrawal symptoms that results from repeated withdrawal experience has been postulated to represent the manifestations of a "kindlinglike" phenomenon. This chapter will provide an overview of this "kindling" hypothesis of alcohol withdrawal, as well as review experimental evidence suggestive of an important role for glutamate neurotransmission in mediating this phenomenon and related consequences.

## 2. THE KINDLING HYPOTHESIS OF ALCOHOL DEPENDENCE AND WITHDRAWAL

The term "kindling" was first introduced by Goddard et al. *(10)* and refers to the phenomenon wherein subthreshold electrical stimulation of discrete brain regions that initially produces no overt behavioral effects comes to produce, upon repeated periodic application, full motor seizures. It subsequently has been demonstrated that the stimulus may be chemical in nature as well (i.e., repeated systemic or central administration of subthreshold doses of various chemoconvulsants will come to produce full motor seizures). Enhanced brain excitability and susceptibility to behavioral convulsions

From: *Contemporary Clinical Neuroscience: Glutamate and Addiction*
Edited by: Barbara H. Herman et al. © Humana Press Inc., Totowa, NJ

resultant from this kindling process has been shown to be long-lasting and thought to be most likely reflective of long-term changes in neuronal circuitry and function *(11)*.

Extending this kindling phenomenon to alcohol withdrawal, it has been postulated that each episode of CNS hyperexcitability that normally accompanies alcohol withdrawal may serve as a stimulus supportive of a "kindling" process. Thus, although episodes of heavy drinking may not initially result in serious, or even noticeable, withdrawal symptoms, repeated experience with this pattern of excessive alcohol intoxication followed by periods of interrupted drinking (abstinence) may lead to a worsening of future withdrawal-related symptoms. This kindling or sensitization process then may underlie the commonly observed progression of withdrawal symptoms, from relatively mild responses characteristic of initial withdrawal episodes (irritability, tremors) to more severe symptoms associated with subsequent withdrawal syndromes, such as seizures and delirium tremens *(12)*.

A substantial body of clinical and experimental evidence has accumulated in support of the "kindling" hypothesis of alcohol withdrawal *(13–15)*. Clinical studies have revealed that alcoholics with a history of multiple previous detoxifications are more likely to experience a seizure during detoxification than patients without such detoxification histories *(16–20)*. Additionally, a history of previous detoxifications was found to be associated not only with more severe and medically complicated withdrawal syndromes but also with an increased likelihood of hospital readmission for alcohol-related problems *(16)*. More recently, clinical studies have revealed that a history of multiple detoxifications is associated with abnormal regional brain activity *(21)*, as well as greater resistance to treatment and enhanced susceptibility to relapse *(22,23)*.

Animal models have provided critical support for these findings, where multiple detoxifications are studied in models of repeated alcohol "withdrawals." For example, we have established a mouse model of alcohol dependence that is sensitive to the effects of a prior withdrawal experience *(24)*. Mice experiencing repeated cycles of alcohol intoxication and withdrawal exhibit a significantly more severe withdrawal seizure response in comparison to animals tested following a single-withdrawal episode. Both intensity and duration of exacerbated withdrawal seizures were found to be positively correlated with the number of previously experienced withdrawal episodes *(25,26)*. This sensitized withdrawal seizure response was observed even when the total amount of alcohol exposure was equated across groups *(24,27)*. Importantly, the differential withdrawal response among groups with different histories of withdrawal experience does not appear to be related to an alteration in alcohol pharmacokinetics; that is, both peak blood alcohol levels and rate of alcohol elimination did not differ between animals following single or multiple cycles of alcohol intoxication and withdrawal *(25)*.

Other studies employing different experimental procedures have similarly demonstrated an increase in the severity of alcohol-withdrawal symptoms following prior withdrawal experience *(28–32)*. Although this "kindling" or sensitization of alcohol withdrawal has primarily focused on withdrawal-related CNS hyperexcitability (e.g., seizures), there is some evidence that other aspects of the withdrawal syndrome (e.g., anxiety) may be susceptible to this kindling phenomenon as well *(33)*.

Exacerbated behavioral symptoms (i.e., indices of CNS hyperexcitability) observed in animals that experienced repeated episodes of withdrawal have been shown to be accompanied by progressively greater changes in EEG, including bursts of spike activity that reverberate among several brain regions, as well as abnormal patterns of epileptiform activity in specific brain regions *(34–37)*. In fact, electrical activity in some brain regions (as measured by electroencepalography [EEG]) may be particularly sensitive to repeated withdrawal experience, whereas activity of other brain sites may be more responsive to total amount of alcohol exposure prior to withdrawal *(35)*. In addition, multiple-withdrawal experience has been shown to result in more intense changes in brain local metabolic activity *(38)* and neuroendocrine responses *(15)*, as well as cognitive/memory impairments *(39)* and susceptibility to neurotoxicity *(40)*. There is also some evidence suggesting cross-sensitization between alcohol withdrawal and other forms of kindling. For example, animal studies have shown that prior electrical or chemical kindling potentiates the symptoms of subsequent alcohol withdrawal *(30,41)*. Conversely, repeated experience with alcohol withdrawal has been reported to subsequently alter the development of electrical kindling in various brain structures *(42–45)*.

Taken together, both preclinical and clinical studies have provided corroborating evidence supportive of the kindling hypothesis of alcohol withdrawal. The significance of this collective body of evidence is that, in addition to the amount (dose) and duration of alcohol consumption prior to withdrawal, a history of previous withdrawal experience appears to represent a critical factor that contributes to the severity of a given withdrawal episode.

## 3. ROLE OF GLUTAMATE IN ALCOHOL WITHDRAWAL KINDLING

Mechanisms underlying alcohol withdrawal sensitization or "kindling" are not well understood. In fact, the terms "sensitization" and "kindling" are used interchangeably here to describe observed exacerbation of withdrawal signs (noted in both clinical and preclinical studies) rather than infer a particular mechanism, *per se*. Both neuroadaptive changes in response to alcohol exposure as well as changes unique to withdrawal from alcohol most likely play a joint role in the final expression of the phenomenon *(15,46)*. Furthermore, mechanisms underlying the *development* of withdrawal kindling or sensitization may be distinct from those critical for *expression* of the phenomenon. Experimental work in recent years is beginning to elucidate neural substrates involved in the complex and dynamic changes in brain function associated with multiple-withdrawal experience.

It is well documented that excessive alcohol consumption results in neuroadaptive changes in many neurochemical systems. These compensatory neurochemical alterations are thought to mediate, to varying extents, the myriad of withdrawal symptoms *(47,48)*. A hallmark feature of alcohol withdrawal is CNS hyperexcitability, which reflects a general imbalance in brain function characterized by reduced inhibitory neurotransmission along with enhanced excitatory neurotransmission *(49)*. Adaptive changes in numerous neurochemical systems most likely contribute to this resultant general state of CNS hyperexcitability. Presumably, changes in any number of these systems that progressively intensify with each successive withdrawal occurrence may contribute to a persistent hyperexcitable state that is manifested as an augmented withdrawal response. The magnification or accrual of neuroadaptive changes over several withdrawal episodes may reflect a kindling process. Thus, through this kindling-like mechanism, it is postulated that the brain is rendered hyperexcitable, such that subsequent bouts of intoxication may result in exaggerated withdrawal reactions.

### 3.1. NMDA Glutamate Receptors and Alcohol Withdrawal

Among many candidate neurochemical systems that are responsive to chronic alcohol exposure, a great deal of attention has focused on the role of glutamate-mediated excitatory neurotransmission. Much of this work has focused on the ionotropic *N*-methyl-D-aspartate (NMDA) glutamate receptor subtype. Whereas acute alcohol inhibits NMDA receptor function, chronic alcohol exposure has been shown to result in glutamatergic hyperfunction, possibly a result of an upregulation of NMDA receptor density and/or functional activity (for reviews, see refs. *48* and *50–53*). Several studies have reported an increased number of NMDA receptor-binding sites following chronic in vivo or in vitro alcohol exposure *(54–58)*, although this outcome has not been universally observed *(59,60)*. In some cases *(61,62)*, but not others *(63,64)*, increased glutamate binding has been reported in brain from alcoholic patients, as well. Amount of alcohol exposure, the radioligand employed, and brain-regional differences most likely account for the reported discrepancies.

Chronic alcohol exposure also produces changes in the assembly of subunits that define various isoforms of NMDA receptors. Both in vivo and in vitro chronic alcohol treatments have been shown to increase mRNA and polypeptide levels for the NR1, NR2A, and NR2B NMDA receptor subunits. However, results have been inconsistent regarding selective changes, and the pattern of results may depend on regional differences as well as gender differences *(65–75)*. Congruent with these biochemical and molecular findings, electrophysiological studies involving chronic alcohol exposure and ex vivo or in vitro analysis of hippocampal slices provide further support for a role of NMDA receptors in neural hyperexcitability associated with alcohol withdrawal *(76,77)*. Finally, additional support for a role of NMDA receptors in alcohol withdrawal comes from studies showing that competitive *(78,79)*, uncompetitive *(54,80,81)*, and noncompetitive *(82)* NMDA receptor antagonists are effective in reducing the severity of alcohol-withdrawal seizures. Conversely, systemic as well as direct (intrahippocampal) administration of NMDA itself during alcohol withdrawal increased the severity of withdrawal-related seizures *(54,83)*. Thus, there is ample evidence to suggest that changes in excitatory neurotransmission mediated through NMDA receptors play a prominent role in alcohol dependence and the expression of withdrawal symptoms.

### 3.2. Role of NMDA Glutamate Receptors in Alcohol Withdrawal Kindling

Aside from the aforementioned neuroadaptive changes exhibited by NMDA receptors following chronic alcohol exposure, there is additional reason to suspect that alterations in NMDA receptor function may represent and important neural substrate underlying sensitization or kindling of alcohol withdrawal. It is well documented that NMDA receptors, and glutamate neurotransmission in general, play a prominent role in various forms of neural plasticity *(84)*. This includes synaptic plasticity events such as long-term potentiation/depression *(85,86)* and epileptiform activity *(87,88)*. Of particular significance is that NMDA receptors are involved in various sensitization and chemical kindling models of epilepsy *(89–92)*. There is also some evidence for changes in NMDA receptor function following repeated alcohol withdrawal exprience.

Several studies have reported enhanced behavioral sensitivity to NMDA in animals with multiple alcohol withdrawal experience. For example, our laboratory examined whether sensitivity to the convulsant properties of NMDA is altered as a function of withdrawal history. Mice were chronically exposed to alcohol vapor in inhalation chambers, where stable blood alcohol levels were maintained during the course of intoxication (165–185 mg/dL). One group received four cycles of 16 h alcohol exposure separated by 8 h periods of withdrawal, a second group was tested after a single 16 h bout of exposure, and a third group was maintained in control (air) chambers, serving as ethanol-naive controls. The mice were assessed for handling-induced convulsions after being injected with NMDA (15 mg/kg; ip) at various time-points following either a single or fourth withdrawal from chronic alcohol exposure. As illustrated in Fig. 1, NMDA-induced seizures were significantly more severe in the multiple-withdrawal group compared with controls as well as animals tested after a single withdrawal. Enhanced sensitivity to the convulsant effects of NMDA was apparent at the time of withdrawal (when mice were still intoxicated), at peak withdrawal (8 h postwithdrawal), and at a more remote time point (24 h) when behavioral expression of seizure activity is typically similar among single- and multiple-withdrawal groups.

Similar results have been obtained in a separate study in which seizure threshold dosage after intravenous NMDA infusion was determined in mice with single- or multiple-withdrawal experience *(93)*. At both 8 and 24 h postwithdrawal, the NMDA seizure threshold dosage was significantly lower in animals with multiple-withdrawal experience in comparison to those that were tested following a single-withdrawal episode. Furthermore, relative to single-withdrawal and control groups, decreased latency and reduced NMDA dosage were required for multiple-withdrawal mice to transition from initial signs of seizure activity (myoclonus) to more severe end-stage (tonic/clonic) convulsions *(93)*. In another study involving chronic administration of alcohol in liquid diets, direct injection of NMDA in the inferior colliculus reduced the amount of electrical stimulation necessary to elicit a seizure in rats that experienced multiple (10) prior withdrawal episodes compared to controls *(94)*. Thus, there is some evidence to suggest that enhanced sensitivity to the neuroexcitatory/convulsant properties of NMDA results from multiple-alcohol-withdrawal experience.

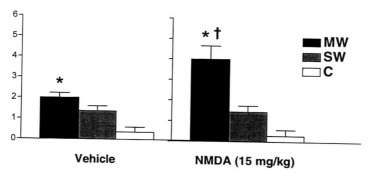
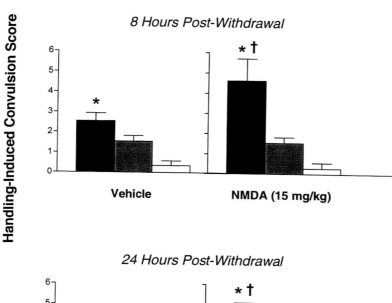
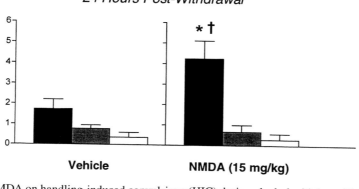

**Fig. 1.** Effects of NMDA on handling-induced convulsions (HIC) during alcohol withdrawal in mice with different withdrawal histories. Data are presented as mean ± S.E.M. HIC score for each group. MW: multiple-withdrawal group; SW: single-withdrawal group; C: alcohol-naive controls (see text for details of treatment). All mice were injected (ip) with vehicle (saline) or NMDA (15 mg/kg) at a time corresponding to 4 h (top panel), 8 h (middle panel), and 24 h (bottom panel) following the final (fourth) withdrawal cycle for the MW condition. HIC severity was determined by scoring the responses 5 min after injections on a scale ranging from 0 to 7. *Significantly differs from SW and C groups ($p < 0.05$); †significantly differs from corrsponding vehicle-injected group ($p < 0.05$).

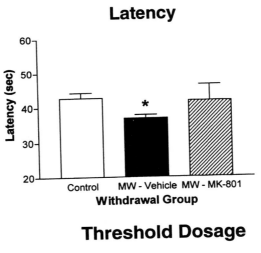

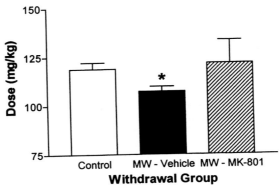

**Fig. 2.** Effects of treating early alcohol withdrawals with MK-801 on sensitivity to NMDA-induced seizures during a subsequent untreated withdrawal episode. Data are presented as mean ± S.E.M. latency **(top)** and threshold dosage **(bottom)** for NMDA-induced clonic seizures. Multiple-withdrawal (MW) groups were injected (ip) with either vehicle (saline) or MK-801 (0.3 mg/kg) at 1 h into each of the first three successive withdrawal cycles. Alcohol-naive controls were given saline injections at the equivalent times. No injections were given on the last (fourth) withdrawal episode. At 8 h following final withdrawal, latency and threshold dosage required for eliciting clonic seizures were determined following iv infusion of NMDA (15 mg/mL; 0.28 mL/min). *Significantly differs from both Control and MW/MK-801 groups ($p < 0.05$).

Additional support for the notion that altered NMDA receptor function may underlie sensitization or kindling of alcohol withdrawal comes from studies demonstrating that pharmacological antagonism of NMDA receptors during early withdrawals may influence the severity of subsequent withdrawal episodes. In our model of repeated withdrawals, dizocilpine (MK-801; 0.1–0.6 mg/kg) was found to significantly reduce alcohol-withdrawal seizures in a dose-dependent manner when administered at the beginning of each of three successive withdrawal episodes. This resulted in a dose-dependent attenuation of seizure activity during a fourth untreated withdrawal cycle *(95)*. Similarly, as shown in Fig. 2, MK-801 treatment of early withdrawals was found to reverse the reduction in NMDA seizure threshold dosage exhibited during a subsequent (untreated) withdrawal episode in mice with multiple-withdrawal experience. However, as a note of caution, although repeated MK-801 treatment for multiple alcohol withdrawals attenuated the early phase of a later untreated withdrawal episode (1–10 h postwithdrawal), this treatment also resulted in an exacerbation of seizure activity at later time-points (10–72 h postwithdrawal) *(95)*. A similar pattern of results was reported when the competitive NMDA receptor antagonis CGP-39551 was administered chronically during the course of alcohol exposure *(78)*. This suggests that repeated (chronic) use of these antagonists for alcohol dependence may increase adaptive changes in NMDA receptors that ultimately contribute to withdrawal-related hyperexcitability.

Evidence for a role of increased NMDA receptor density in alcohol withdrawal sensitization has been mixed. Repeated bouts of heavy alcohol intoxication interspersed by periods of withdrawal did not significantly alter the number of [$^3$H]MK-801-binding sites in the various regions of rat brain *(96)*. However, a small increase in [$^3$H]MK-801-binding sites in the hippocampus and entorhinal cortex was observed in the multiple-withdrawal group that evidenced seizures during prior withdrawal episodes in comparison to multiple-withdrawal rats that did not exhibit such spontaneous seizure activity. In our mouse model of repeated alcohol withdrawals, the density of [$^3$H]MK-801-binding sites did not significantly differ from controls in hippocampus or cortex (unpublished data). In an in vitro model of chronic intermittent alcohol exposure, an upregulation in [$^3$H]MK-801 binding was reported 1 wk following final withdrawal *(97)*. This persistent increase in number of NMDA receptors was accompanied by an increase in the ability of NMDA to stimulate calcium influx in cultured cortical neurons. Different results regarding changes in NMDA receptor density following repeated alcohol withdrawal may depend on the method and intensity of chronic alcohol exposure. Alcohol withdrawal sensitization also may be associated with changes in the subunit composition of NMDA receptors, which may not be revealed by radioligand-binding studies but, rather, impact on receptor function. Such investigations are currently be conducted by a number of laboratories.

### 3.3. Non-NMDA Glutamate Receptors and Alcohol Withdrawal Kindling

Although studies investigating the role of metabotropic glutamate receptors in alcohol actions are rather limited, an emerging body of evidence suggests that non-NMDA ionotropic glutamate receptors (α-amino-3-hydroxy-5-methyl-4-isoxazole propronic acid [AMPA]/kainate receptors) may contribute to both acute and chronic effects of alcohol *(50,98)*. In addition, AMPA/kainate receptors are known to play a significant role in various forms of neuroplasticity *(85,87,99)*. Thus, there is some basis upon which to suspect that these non-NMDA glutamate receptor systems may play a role in mediating kindling or sensitization of alcohol withdrawal.

Alcohol inhibits glutamate-mediated excitatory transmission through AMPA/kainate receptors in a complex fashion, dependent on numerous factors *(98)*. Whereas in vitro studies have demonstrated neuroadaptation to these effects, the effects of chronic in vivo alcohol treatment on AMPA/kainate receptors have been mixed. For example, chronic alcohol exposure did not alter [$^3$H]kainate binding in hippocampus or cortex *(58)*, and expression of GluR1 and GluR2 levels was not altered in various brain regions *(75)*. On the other hand, rats undergoing alcohol withdrawal were reported to exhibit enhanced sensitivity to the convulsant properties of kainate *(57)*.

Few studies have examined whether changes in non-NMDA glutamate receptors accompany enhanced CNS hyperexcitability associated with repeated alcohol withdrawals. In one study, repeated cycles of heavy intoxication and withdrawal in rats resulted in no changes in [$^3$H]kainate binding, but the density of [$^3$H]AMPA-binding sites was decreased in several brain regions *(96)*. This latter effect was suggested to reflect a possible compensatory response to augmented CNS hyperexcitability. In another study, although mice with multiple-withdrawal experience were found to exhibit enhanced sensitivity to NMDA (reduced seizure threshold dosage), the seizure threshold dosage for kainate was significantly elevated *(93)*. Whether this blunted sensitivity to the convulsant properties of kainate represents a compensatory response in these mice is unclear. However, because multiple alcohol withdrawal experience does not appear to uniformly alter sensitivity to all chemoconvulsants, this suggests that exacerbated seizure activity exhibited by animals with such repeated withdrawal experience may not reflect a global non-specific heightening of neural excitation. Additional studies will be needed to more clearly define the role of AMPA/kainate receptors in the alcohol-withdrawal kindling phenomenon.

## 4. GLUTAMATE AND ALCOHOL WITHDRAWAL KINDLING: IMPLICATIONS FOR NEUROTOXICITY

A dangerous consequence of alcohol withdrawal kindling relates to alcohol-induced neurotoxcity. It has been suggested that multiple alcohol withdrawal experience may render individuals more vulnerable to neurological damage and associated congnitive impairments *(13,15)*. There is both preclinical and clinical evidence in support of this notion. For example, chronic intermittent alcohol exposure (i.e., coupled with numerous intervening periods of abstinence) was reported to result in hippocampal cell loss in rats, an effect not observed in animals exposed to alcohol in a continuous fashion *(40,100)*. In a study involving postmortem analysis of alcoholic brains, temporal lobe pathology was reported to be associated with a history of alcohol withdrawal seizures *(101)*. Enhanced vulnerability and susceptibility to alcohol-related neuropathology as a result of multiple-withdrawal experience may, in turn, underlie cognitive deficits associated with chronic bingelike drinking *(102)*. Indeed, there is both clinical *(103)* and experimental *(39)* evidence indicating that a history of repeated alcohol withdrawals is associated with greater cognitive dysfunction.

Given the prominent role of glutamate in neurotoxicity and neurodegeneration *(104,105)*, enhanced vulnerability to neuropathologic insult following repeated alcohol withdrawals may be related to an associated upregulation in NMDA receptor function *(106,107)*. In in vitro studies, although acute alcohol has been shown to inhibit NMDA-mediated cell death *(108,109)*, chronic alcohol exposure results in an exacerbation of NMDA excitotoxicity in cortical *(110–112)* and cerebellar granule cells *(113,114)*. Similar results have been reported following chronic alcohol exposure in hippocampal slice explants *(115–117)*. Moreover, chronic alcohol exposure resulted in enhanced NMDA-mediated, but not AMPA- or kainate-mediated excitotoxicity in cultured hippocampal cells *(118)*.

Studies employing in vivo microdialysis techniques have demonstrated increased extracellular glutamate levels in striatum *(119,120)*, nucleus accumbens *(121,122)*, and hippocampus *(123)* during withdrawal from chronic alcohol exposure in rats. Similarly, elevated levels of glutamate were reported in cerebrospinal fluid (CSF) from alcoholic patients for as long as 1 mon following abstinence *(107)*. Withdrawal-related increases in glutamate release presumably result in increased activation of NMDA receptors, which may relate to reported increased in vivo sensitivity to NMDA-mediated excitotoxic damage in the hippocampus of rats undergoing alcohol withdrawal *(83,124)*.

Whether sensitivity to glutamate-related excitotoxicity is enhanced as a function of repeated alcohol withdrawals has not been extensively explored. In one study, repeated alcohol withdrawals resulted in a decrease (toward baseline), rather than a further increase, in extracellular glutamate levels in the hippocampus *(123)*. Because there was a robust increase in glutamate release following the first withdrawal period, it is not clear if this represents some protective response that serves to dampen further neural damage during subsequent withdrawals. In contrast, repeated in vitro exposure and withdrawal from alcohol in rat hippocampal slices was reported to enhance sensitivity to NMDA excitotoxicity *(125,126)*. Furthermore, this enhanced sensitivity to the excitotoxic effects of NMDA challenge was greatest in hippocampal cultures exposed to the greatest number of withdrawal (washout) periods interspersed between alcohol exposures *(125)*. Given the serious clinical implications of alcohol-related neuropathologic damage and cognitive impairment, the potential for multiple-withdrawal experience to exacerbate these effects, possibly through enhancement of glutamate function, remains an area of active investigation.

## 5. SUMMARY

A growing body of clinical and experimental evidence indicates that multiple alcohol withdrawal experiences may result in more severe and complicated future withdrawal episodes. It has been postulated that this progressive intensification of withdrawal symptoms may reflect a kindling-like phenomenon. Thus, CNS hyperexcitability that normally accompanies alcohol withdrawal (but does not necessarily result in severe or even noticeable symptoms) may progressively magnify over successive withdrawal episodes, ultimately being manifested as an exaggerated ("kindled") withdrawal response. Although the pathophysiological mechanisms underlying withdrawal sensitization or kindling are not well understood, there is evidence that neuroadaptive changes in glutamate-mediated excitatory neurotransmission may play a significant role in the phenomenon. This notion is further bolstered by the fact that glutamate is known to play a prominent role in various forms of neuroplasticity, many of which bear great resemblance to a kindlinglike process.

A convergent body of evidence from behavioral, biochemical, electrophysiological, and molecular studies has demonstrated that enhanced glutamatergic activity, primarily through increased density and/or function of NMDA receptors, underlies CNS hyperexcitability associated with alcohol withdrawal. There is some evidence indicating that these neuroadaptive changes (enhanced NMDA receptor function) may become further magnified as a consequence of repeated alcohol-withdrawal experience. Although non-NMDA receptors (AMPA/kainate receptors) have been shown to be responsive to acute and chronic alcohol exposure, their role in withdrawal kindling has been less extensively studied. Of course, compensatory changes in other excitatory systems (e.g., voltage-gated calcium channels) as well as inhibitory systems (e.g., GABA, adenosine) undoubtedly contribute to exaggerated CNS hyperexcitability and the expression of sensitized or kindled withdrawal symptoms. These changes may occur either upstream or downstream from upregulated glutamate function. Elucidation of the complex changes in brain function following chronic alcohol exposure as well as the dynamic alterations that are associated with repeated cycles of alcohol intoxication and withdrawal is an area of active investigation.

Aside from the potentially life-threatening consequences of augmented withdrawal-related CNS hyperexcitability (seizures), another serious concern regarding abusive alcohol consumption and a history of multiple-withdrawal experience is enhanced potential for neurotoxicity. Indeed, there is both clinical and preclinical evidence that suggests neurological damage and associated cognitive deficits are more extensive as a result of multiple-withdrawal experience. Glutamate-mediated neurotransmission, particularly through NMDA receptor activation, has been shown to play a prominent role in neurotoxicity and neurodegradation. Several studies have demonstrated chronic alcohol exposure and withdrawal result in enhanced NMDA-mediated excitotoxicity, and there is some evidence to suggest that this effect is further enhanced following intermittent alcohol exposure that involves repeated withdrawals. The clinical significance of alcohol-induced neuropathology and cognitive impairment, as well as the manifestations of withdrawal-related CNS hyperexcitability add further relevance to understanding the potential for repeated withdrawals to exacerbate these deleterious consequences through alterations in glutamate-mediated neurotransmission. Further elucidation of the role of glutamate in mediating the alcohol-withdrawal kindling phenomenon could provide the necessary insight needed for the development of more targeted and effective pharmacotherapy treatment strategies for alcohol detoxification, as well as the long-term management of alcohol dependence and alcoholism.

# REFERENCES

1. Anton, R. F. and Becker, H. C. (1995) Pharmacotherapy and pathophysiology of alcohol withdrawal, in *The Pharmacology of Alcohol Abuse* (Kranzler, H. R. ed.), Springer-Verlag, Berlin, pp. 315–367.
2. Saitz, R. (1998) Introduction to alcohol withdrawal. *Alcohol Health Res. World* **22(1),** 5–12.
3. Yost, D. A. (1996) Alcohol withdrawal syndrome [published erratum appears in American Family Physician 1996; **54(8):**2377]. *Am. Fam. Physician* **54(2),** 657–64, 669.
4. De Soto, C. B., et al. (1985) Symptomatology in alcoholics at various stages of abstinence. *Alcohol. Clin. Exp. Res.* **9(6),** 505–512.
5. Roelofs, S. (1985) Hyperventilation, anxiety, craving for alcohol: a subacute alcohol withdrawal syndrome. *Alcohol* **2,(3),** 501–505.

6. Schuckit, M. A., et al. (1998) Clinical relevance of the distinction between alcohol dependence with and without a physiological component. *Am. J. Psychiatry* **155(6)**, 733–740.
7. Stockwell, T. (1994) Alcohol withdrawal: an adaptation to heavy drinking of no practical significance? *Addiction* **89**, 1447–1453.
8. Finn, D. A. and Crabbe, J. C. (1997) Exploring alcohol withdrawal syndrome. *Alcohol Health Res. World* **21(2)**, 149–156.
9. Hillbom, M. (1990) Alcohol withdrawal seizures and binge versus chronic drinking, in *Alcohol and Seizures: Basic Mechanisms and Clinical Conceapts* (Porter, R. J., et al., eds.). F. A. Davis, Philadelphia, pp. 206–215.
10. Goddard, G. V., McIntyre, D. C., and Leech, C. K. (1969) A permanent change in brain function resulting from daily electrical stimulation. *Exp. Neurol.* **25**, 295–330.
11. McNamara, J. O. and Wada, J. A. (1997) Kindling Model, in *Epilepsy: A Comprehensive Textbook* (Engel, J. and Pedley, T. A., eds.), Lippincott–Raven, Philadelphia, pp. 419–425.
12. Ballenger, J. C. and Post, R. M. (1978) Kindling as a model for alcohol withdrawal syndromes. *Br. J. Psychiatry.* **133**, 1–14.
13. Becker, H. C. and Littleton, J. M. (1996) The alcohol withdrawal "kindling" phenomenon: clinical and experimental findings. *Alcohol. Clin. Exp. Res.* **20**, 121A–124A.
14. Becker, H. C. (1998) Kindling in alcohol withdrawal. *Alcohol Health Res. World* **22(1)**, 25–33.
15. Becker, H. C. (1999) Alcohol withdrawal: neuroadaptation and sensitization. *CNS Spectrums* **4(1)**, 38–65.
16. Booth, B. M. and Blow, F. C. (1993) The kindling hypothesis: further evidence from a U.S. national study of alcoholic men. *Alcohol Alcohol.* **28**, 593–598.
17. Daryanani, H. E., et al. (1994) Alcoholic withdrawal syndrome and seizures. *Alcohol Alcohol.* **29**, 323–328.
18. Brown, M. E., et al. (1988) Alcohol detoxification and withdrawal seizures: clinical support for a kindling hypothesis. *Biol. Psychiatry* **23**, 507–514.
19. Worner, T. M. (1996) Relative kindling effect of readmissions in alcoholics. *Alcohol Alcohol.* **31**, 375–380.
20. Moak, D. H. and Anton, R. F. (1996) Alcohol-related seizures and the kindling effect of repeated detoxifications: the influence of cocaine. *Alcohol Alcohol.* **31**, 135–143.
21. George, M. S., et al. (1999) Multiple previous alcohol detoxifications are associated with decreased medial temporal and paralimbic function in the postwithdrawal period. *Alcohol. Clin. Exp. Res.* **23(6)**, 1077–1084.
22. Malcolm, R., et al. (2000) Recurrent detoxification may elevate alcohol craving as measured by the Obsessive Compulsive Drinking scale. *Alcohol* **20(2)**, 181–185.
23. Malcolm, R., et al. (2000) Multiple previous detoxifications are associated with less responsive treatment and heavier drinking during an index outpatient detoxification. *Alcohol* **22**, 159–164.
24. Becker, H. C. and Hale, R. L. (1993) Repeated episodes of ethanol withdrawal potentiate the severity of subsequent withdrawal seizures: an animal model of alcohol withdrawal "kindling." *Alcohol. Clin. Exp. Res.* **17(1)**, 94–98.
25. Becker, H. C. (1994) Positive relationship between the number of prior ethanol withdrawal episodes and the severity of subsequent withdrawal seizures. *Psychopharmacology* **116**, 26–32.
26. Becker, H. C., Diaz-Granados, J. L., and Weathersby, R. T. (1997) Repeated ethanol withdrawal experience increases the severity and duration of subsequent withdrawal seizures in mice. *Alcohol* **14(4)**, 319–326.
27. Becker, H. C., Diaz-Granados, J. L., and Hale, R. L. (1997) Exacerbation of ethanol withdrawal seizures in mice with a history of multiple withdrawal experience. *Pharmacol. Biochem. Behav.* **57(1–2)**, 179–183.
28. Maier, D. M. and Pohorecky, L. A. (1989) The effect of repeated withdrawal episodes on subsequent withdrawal severity in ethanol-treated rats. *Drug Alcohol Depend.* **23**, 103–110.
29. Clemmesen, L. and Hemmingsen, R. (1984) Physical dependence on ethanol during multiple intoxication and withdrawal episodes in the rat: evidence of a potentiation. *Acta Pharmacol. Toxicol.* **55**, 345–350.
30. Kokka, N., et al. (1993) The kindling model of alcohol dependence: similar persistent reduction in seizure threshold to pentylenetetrazol in animals receiving chronic ethanol or chronic pentylenetetrazol. *Alcohol. Clin. Exp. Res.* **17(3)**, 525–531.
31. Ulrichsen, J., Clemmesen, L., and Hemmingsen, R. (1992) Convulsive behavior during alcohol dependence: discrimination between the role of intoxication and withdrawal. *Psychopharmacology* **107**, 97–102.
32. Pohorecky, L. A. and Roberts, P. (1991) Development of tolerance to and physical dependence on ethanol: daily versus repeated cycles treatment with ethanol. *Alcohol. Clin. Exp. Res.* **15**, 824–833.
33. Becker, H. C., Fernandes, K. G., and Weathersby, R. T. (1994) Repeated ethanol withdrawal experience differentially influences withdrawal-related seizures and "anxiety" responses in mice. *Soc. Neurosci. Abstr.* **20(2)**, 1618.
34. Poldrugo, F. and Snead, 3rd, O. C. (1984) Electroencephalographic and behavioral correlates in rats during repeated ethanol withdrawal syndromes. *Psychopharmacology* **83(2)**, 140–146.
35. Veatch, L. M. and Gonzalez, L. P. (1996) Repeated ethanol withdrawal produces site-dependent increases in EEG spiking. *Alcohol. Clin. Exp. Res.* **20(2)**, 262–267.
36. Veatch, L. M., Coykendall, D. S., and Becker, H. C. (1999) Sensitization of electrographic (EEG) activity and behavioral seizures following repeated ethanol withdrawals in different mouse strains. *Soc. Neurosci. Abstr.* **25**, 1077.
37. Walker, D. W. and Zornetzer, S. F. (1974) Alcohol withdrawal in mice: electroencephalographic and behavioral correlates. *Electroencephalogr. Clin. Neurophysiol.* **36(3)**, 233–243.

38. Clemmesen, L., et al. (1988) Local cerebral glucose consumption during ethanol withdrawal in the rat: effects of single and multiple episodes and previous convulsive seizures. *Brain Res.* **453**, 204–214.
39. Bond, N. W. (1979) Impairment of shuttlebox avoidance learning following repeated alcohol withdrawal episodes in rats. *Pharmacol. Biochem. Behav.* **11**, 589–591.
40. Lundqvist, C., et al. (1995) Intermittent ethanol exposure of adult rats: hippocampal cell loss after one month of treatment. *Alcohol Alcohol.* **30**, 737–748.
41. Pinel, J. P. J. (1980) Alcohol withdrawal seizures: implications of kindling. *Pharmacol. Biochem. Behav.* **13(Suppl. 1)**, 225–231.
42. McCown, T. J. and Breese, G. R. (1990) Multiple withdrawals from chronic ethanol "kindles" inferior collicular seizure activity: evidence for kindling of seizures associated with alcoholism. *Alcohol. Clin. Exp. Res.* **14(3)**, 394–399.
43. Veatch, L. M. and Gonzalez, L. P. (1997) Chronic ethanol retards kindling of hippocampal area $CA_3$. *NeuroReport* **8**, 1903–1906.
44. Ulrichsen, J., et al. (1998) Electrical amygdala kindling in alcohol-withdrawal kindled rats. *Alcohol Alcohol.* **33(3)**, 244–254.
45. Veatch, L. M. and Gonzalez, L. P. (1999) Repeated ethanol withdrawal delays development of focal seizures in hippocampal kindling. *Alcohol. Clin. Exp. Res.* **23(7)**, 1145–1150.
46. Becker, H. C. (1996) The alcohol withdrawal "kindling" phenomenon: clinical and experimental findings. *Alcohol. Clin. Exp. Res.* **20(8 Suppl.)**, 121A–124A.
47. Deitrich, R. A., Radcliffe, R., and Erwin, V. G. (1996) Pharmacological effects in the development of physiological tolerance and physical dependence, in *The Pharmacology of Alcohol and Alcohol Dependence* (Begleiter, H. and Kissin, B., eds.), Oxford. University Press, New York, pp. 431–476.
48. Tabakoff, B. and Hoffman, P. L. (1996) Effect of alcohol on neurotransmitters and their receptors and enzymes, in *The Pharmacology of Alcohol and Alcohol Dependence* (Begleiter, H. and Kissin, B., eds.), Oxford University Press, New York.
49. Littleton, J. (1998) Neurochemical mechanisms underlying alcohol withdrawal. *Alcohol Health Res. World* **21(1)**, 13–24.
50. Dodd, P. R., et al. (2000) Glutamate-mediated transmission, alcohol, and alcoholism. *Neurochem. Int.* **37(5–6)**, 509–533.
51. Crews, F. T., et al. (1996) Effects of ethanol on ion channels. *Int. Rev. Neurobiol.* **39**, 283–367.
52. Fadda, F. and Rossetti, Z. L. (1998) Chronic ethanol consumption: from neuroadaptation to neurodegeneration. *Prog. Neurobiol.* **56**, 385–431.
53. Lovinger, D. M. (1997) Alcohols and neurotransmitter gated ion channels: past, present and future. *Naunyn-Schmiedebergs Arch. Pharmacol.* **356**, 267–282.
54. Grant, K. A., et al. (1990) Ethanol withdrawal seizures and the NMDA receptor complex. *Eur. J. Pharmacol.* **176(3)**, 289–296.
55. Gulya, K., et al., (1991) Brain regional specificity and time-course of changes in the NMDA receptor-ionophore complex during ethanol withdrawal. *Brain Res.* **547(1)**, 129–134.
56. Hu, X. J. and Ticku, M. K. (1995) Chronic ethanol treatment upregulates the NMDA receptor function and binding in mammalian cortical neurons. *Mol. Brain Res.* **30**, 347–356.
57. Sanna, E., et al. (1993) Chronic ethanol intoxication induces differential effects on GABAA and NMDA receptor function in the rat brain. *Alcohol. Clin. Exp. Res.* **17(1)**, 115–123.
58. Snell, L. D., Tabakoff, B., and Hoffman, P. L. (1993) Radioligand binding to the *N*-methyl-D-aspartate receptor/ionophore complex: alterations by ethanol in vitro and by chronic in vivo ethanol ingestion. *Brain Res.* **602(1)**, 91–98.
59. Carter, L. A., et al. (1995) Allosteric regulation of the *N*-methyl-D-aspartate receptor-linked ion channel complex and effects of ethanol in ethanol-withdrawal seizure-prone and -resistant mice. *J. Neurochem.* **64(1)**, 213–219.
60. Rudolph, J. G., et al. (1997) NMDA receptor binding in adult rat brain after several chronic ethanol treatment protocols. *Alcohol. Clin. Exp. Res.* **21(8)**, 1508–1519.
61. Freund, G. and Anderson, K. J. (1996) Glutamate receptors in the frontal cortex of alcoholics. *Alcohol. Clin. Exp. Res.* **20**, 1165–1172.
62. Michaelis, E. K., et al. (1993) Glutamate receptor changes in brain synaptic membranes during chronic alcohol intake. *Alcohol Alcohol.* **2(Suppl.)**, 377–381.
63. Dodd, P. R., et al. (1992) Amino acid neurotransmitter receptor changes in cerebral cortex in alcoholism: effect of cirrhosis of the liver. *J. Neurochem.* **59(4)**, 1506–1515.
64. Freund, G. and Anderson K. J. (1999) Glutamate receptors in the cingulate cortex, hippocampus, and cerebellar vermis of alcoholics. *Alcohol. Clin. Exp. Res.* **23(1)**, 1–6.
65. Devaud, L. L. and Morrow, A. L. (1999) Gender-selective effects of ethanol dependence on NMDA receptor subunit expression in cerebral cortex, hippocampus and hypothalamus. *Eur. J. Pharmacol.* **369(3)**, 331–334.
66. Follesa, P. and Ticku, M. K. (1995) Chronic ethanol treatment differentially regulates NMDA receptor subunit mRNA expression in rat brain. *Mol. Brain Res.* **29**, 99–106.
67. Follesa, P. and Ticku, M. K. (1996) Chronic ethanol-mediated up-regulation of the *N*-methyl-D-aspartate receptor polypeptide subunits in mouse cortical neurons in culture. *J. Biol. Chem.* **271(23)**, 13,297–13,299.

68. Hardy, P. A., Chen, W., and Wilce, P. (1999) Chronic ethanol exposure and withdrawal influence NMDA receptor subunit and splice variant mRNA expression in the rat cerebral cortex. *Brain Res.* **819(1–2),** 33–39.
69. Hu, X.-J., Follesa, P., and Ticku, M. K. (1996) Chronic ethanol treatment produces a selective upregulation of the NMDA receptor subunit gene expression in mammalian cultures cortical neurons. *Mol. Brain Res.* **36,** 211–218.
70. Kalluri, H. S., Mehta, A. K., and Ticku, M. K. (1998) Up-regulation of NMDA receptor subunits in rat brain following chronic ethanol treatment. *Brain Res. Mol. Brain Res.* **58(1–2),** 221–224.
71. Kumari, M. and Ticku, M. K. (1998) Ethanol and regulation of the NMDA receptor subunits in fetal cortical neurons. *J. Neurochem.* **70(4),** 1467–1473.
72. Kumari, M. and Ticku, M. K. (2000) Regulation of NMDA receptors by ethanol. *Prog. Drug Res.* **54,** 152–189.
73. Ortiz, J., et al. (1995) Biochemical actions of chronic ethanol exposure in the mesolimbic dopamine system. *Synapse* **21(4),** 289–298.
74. Snell, L. D., et al. (1996) Regional and subunit specific changes in NMDA receptor mRNA and immunoreactivity in mouse brain following chronic ethanol ingestion. *Mol. Brain Res.* **40,** 71–78.
75. Trevisan, L., et al. (1994) Chronic ingestion of ethanol up-regulates NMDAR1 receptor subunit immunoreactivity in rat hippocampus. *J. Neurochem.* **62,** 1635–1638.
76. Thomas, M. P., Monaghan, D. T., and Morrisett, R. A. (1998) Evidence for a causative role of *N*-methyl-D-aspartate receptors in an in vitro model of alcohol withdrawal hyperexcitability. *J. Pharmacol. Exp. Ther.* **287(1),** 87–97.
77. Whittington, M. A., Lambert, J. D. C., and Little, H. J. (1995) Increased NMDA receptor and calcium channel activity underlying ethanol withdrawal hyperexcitability. *Alcohol Alcohol.* **30(1),** 105–114.
78. Ripley, T. L. and Little, H. J. (1995) Effects on ethanol withdrawal hyperexcitability of chronic treatment with a competitive n-methyl-d-aspartate receptor antagonist. *J. Pharmacol. Exp. Ther.* **272,** 112–118.
79. Liljequist, S. (1991) The competitive NMDA receptor antagonist, CGP 39551, inhibits ethanol withdrawal seizures. *Eur. J. Pharmacol.* **192(1),** 197–198.
80. Morrisett, R. A., et al. (1990) MK-801 potently inhibits alcohol withdrawal seizures in rats. *Eur. J. Pharmacol.* **176(1),** 103–105.
81. Grant, K. A., et al. (1992) Comparison of the effects of the uncompetitive *N*-methyl-D-aspartate antagonist (+–)-5-aminocarbonyl-10,11-dihydro-5*H*-dibenzo[*a,d*] cyclohepten-5, 10-imine (ADCI) with its structural analogs dizocilpine (MK-801) and carbamazepine on ethanol withdrawal seizures. *J. Pharmacol. Exp. Ther.* **260(3),** 1017–1022.
82. Kotlinska, J. and Liljequist, S. (1996) Oral administration of glycine and polyamine receptor antagonists blocks ethanol withdrawal seizures. *Psychopharmacology* **127,** 238–244.
83. Davidson, M. D., Wilce, P., and Chanley, B. C. (1993) Increased sensitivity of the hippocampus in ethanol-dependent rats to toxic effect of *N*-methyl-D-aspartic acid in vivo. *Brain Res.* **606,** 5–9.
84. Collingridge, G. L. and Singer, W. (1990) Excitatory amino acid receptors and synaptic plasticity. *Trends Pharmacol. Sci.* **11(7),** 290–296.
85. Kullmann, D. M., Asztely, F., and Walker, M. C. (2000) The role of mammalian ionotropic receptors in synaptic plasticity: LTP, LTD and epilepsy [in process citation]. *Cell. Mol. Life Sci.* **57(11),** 1551–1561.
86. Nicoll, R. A. and Malenka, R. C. (1999) Expression mechanisms underlying NMDA receptor-dependent long-term potentiation. *Ann. NY Acad. Sci.* **868,** 515–525.
87. Chapman, A. G. (1998) Glutamate receptors in epilepsy. *Prog. Brain Res.* **116,** 371–383.
88. Meldrum, B. S., Akbar, M. T., and Chapman, A. G. (1999) Glutamate receptors and transporters in genetic and acquired models of epilepsy. *Epilepsy Res.* **36(2–3),** 189–204.
89. Camarini, R., et al. (2000) MK-801 blocks the development of behavioral sensitization to the ethanol. *Alcohol. Clin. Exp. Res.* **24(3),** 285–290.
90. Ekonomou, A. and Angelatou, F. (1999) Upregulation of NMDA receptors in hippocampus and cortex in the pentylenetrazol-induced "kindling" model of epilepsy. *Neurochem. Res.* **24(12),** 1515–1522.
91. Itzhak, Y. and Martin, J. L. (2000) Cocaine-induced kindling is associated with elevated NMDA receptor binding in discrete mouse brain regions. *Neuropharmacology* **39(1),** 32–39.
92. Vanderschuren, L. J. and Kalivas, P. W. (2000) Alterations in dopaminergic and glutamatergic transmission in the induction and expression of behavioral sensitization: a critical review of preclinical studies. *Psychopharmacology (Berl.)* **151(2–3),** 99–120.
93. Becker, H. C., Veatch, L. M., and Diaz-Granados, J. L. (1998) Repeated ethanol withdrawal experience selectively alters sensitivity to different chemoconvulsant drugs in mice. *Psychopharmacology* **139,** 145–153.
94. McCown, T. J. and Breese, G. R. (1993) A potential contribution to ethanol withdrawal kindling—reduced GABA function in the inferior collicular cortex. *Alcohol. Clin. Exp. Res.* **17,** 1290–1294.
95. Becker, H. C., Diaz-Granados, J. L., and Reich, R. R. (1997) MK-801 treatment of early ethanol withdrawals reduces the severity of seizures during a subsequent untreated withdrawal episode. *Alcohol. Clin. Exp. Res.* **21(3),** 235A.
96. Ulrichsen, J., et al. (1996) Glutamate and benzodiazepine receptor autoradiography in rat brain after repetition of alcohol dependence. *Psychopharmacology (Berl.)* **126(1),** 31–41.

97. Hu, X. J. and Ticku, M. K. (1997) Functional characterization of a kindling-like model of ethanol withdrawal in cortical cultured neurons after chronic intermittent ethanol exposure. *Brain Res.* **767(2)**, 228–234.
98. Woodward, J. J. (1999) Ionotropic glutamate receptors as sites of action for ethanol in the brain. *Neurochem. Int.* **35(2)**, 107–113.
99. Bortolotto, Z. A. et al. (1999) Kainate receptors are involved in synaptic plasticity. *Nature* **402(6759)**, 297–301.
100. Lundqvist, C., et al. (1994) Long-term effects of intermittent versus continuous ethanol exposure on hippocampal synapses of the rat. *Acta Neuropathology* **87**, 242–249.
101. Sullivan, E. V., et al. (1996) Relationship between alcohol withdrawal seizures and temporal lobe white matter volume deficits. *Alcohol. Clin. Exp. Res.* **20(2)**, 348–354.
102. Hunt, W. A. (1993) Are binge drinkers more at risk of developing brain damage? *Alcohol* **10(6)**, 559–561.
103. Glenn, S. W., et al. (1988) The effects of repeated withdrawals from alcohol on the memory of male and female alcoholics. *Alcohol Alcohol.* **23(5)**, 337–342.
104. Choi, D. W. (1988) Glutamate neurotoxicity and diseases of the nervous system. *Neuron* **1(8)**, 623–634.
105. Coyle, J. T. and Puttfarcken, P. (1993) Oxidative stress, glutamate, and neurodegenerative disorders. *Science* **262(5134)**, 689–695.
106. Lovinger, D. M. (1993) Excitotoxicity and alcohol-related brain damage. *Alcohol. Clin. Exp. Res.* **17(1)**, 19–27.
107. Tsai, G. E., et al. (1998) Increased glutamatergic neurotransmission and oxidative stress after alcohol withdrawal. *Am. J. Psychiatry* **155**, 726–732.
108. Chandler, L. J., Summners, C., and Crews, F. T. (1993) Ethanol inhibits NMDA receptor-mediated excitotoxicity in rat primary neuronal cultures. *Alcohol. Clin. Exp. Res.* **17(1)**, 54–60.
109. Lustig, H. S., Chan, J., and Greenberg, D. A. (1992) Ethanol inhibits excitotoxicity in cerebral cortical cultures. *Neurosci. Lett.* **135(2)**, 259–261.
110. Ahern, K. B., Lustig, H. S., and Greenberg, D. A. (1994) Enhancement of NMDA toxicity and calcium responses by chronic exposure of cultured cortical neurons to ethanol. *Neurosci. Lett.* **165(1–2)**, 211–214.
111. Chandler, L. J., et al. (1993) Chronic ethanol exposure potentiates NMDA excitotoxicity in cerebral cortical neurons. *J. Neurochem.* **60(4)**, 1578–1581.
112. Nagy, J., Muller, F., and Laszlo, L. (2001) Cytotoxic effect of alcohol-withdrawal on primary cultures of cortical neurones. *Drug. Alcohol Depend.* **61(2)**, 155–162.
113. Hoffman, P. L., et al. (1995) Attenuation of glutamate-induced neurotoxicity in chronically ethanol-exposed cerebellar granule cells by NMDA receptor antagonists and ganglioside GM1. *Alcohol. Clin. Exp. Res.* **19(3)**, 721–726.
114. Iorio, K. R., Tabakoff, B., and Hoffman, P. L. (1993) Glutamate-induced neurotoxicity is increased in cerebellar granule cells exposed chronically to ethanol. *Eur. J. Pharmacol.* **248(2)**, 209–212.
115. Prendergast, M. A., et al. (2000) In vitro effects of ethanol withdrawal and spermidine on viability of hippocampus from male and female rat. *Alcohol. Clin. Exp. Res.* **24(12)**, 1855–1861.
116. Prendergast, M. A., et al. (2000) Chronic, but not acute, nicotine exposure attenuates ethanol withdrawal-induced hippocampal damage in vitro. *Alcohol. Clin. Exp. Res.* **24(10)**, 1583–1592.
117. Thomas, M. P. and Morrisett, R. A. (2000) Dynamics of NMDAR-mediated neurotoxicity during chronic ethanol exposure and withdrawal. *Neuropharmacology* **39(2)**, 218–226.
118. Smothers, C. T., Mrotek, J. J., and Lovinger, D. M. (1997) Chronic ethanol exposure leads to a selective enhancement of *N*-methyl-D-aspartate receptor function in cultured hippocampal neurons. *J. Pharmacol. Exp. Ther.* **283(3)**, 1214–1222.
119. Rossetti, Z. L. and Carboni, S. (1995) Ethanol withdrawal is associated with increased extracellular glutamate in the rat striatum. *Eur. J. Pharmacol.* **283(1–3)**, 177–183.
120. Rossetti, Z. L., Carboni, S., and Fadda, F. (1999) Glutamate-induced increase of extracellular glutamate through *N*-methyl-D-aspartate receptors in ethanol withdrawal. *Neuroscience* **93(3)**, 1135–1140.
121. Dahchour, A., et al. (1998) Central effects of acamprosate: part 1. Acamprosate blocks the glutamate increase in the nucleus accumbens microdialysate in ethanol withdrawan rats. *Psychiatry Res.* **82(2)**, 107–114.
122. Dahchour, A. and De Witte, P. (2000) Taurine blocks the glutamate increase in the nucleus accumbens microdialysate of ethanol-dependent rats. *Pharmacol. Biochem. Behav.* **65(2)**, 345–350.
123. Dahchour, A. and De Witte, P. (1999) Effect of repeated ethanol withdrawal on glutamate microdialysate in the hippocampus. *Alcohol. Clin. Exp. Res.* **23(10)**, 1698–1703.
124. Davidson, M., Shanley, B., and Wilce, P. (1995) Increased NMDA-induced excitability during ethanol withdrawal: a behavioural and histological study. *Brain Res.* **674**, 91–96.
125. Mayer, S. and Littleton, J. (1996) Increased excititoxicity associated with repeated ethanol withdrawal in rat hippocampus slice cultures. *Alcohol. Clin. Exp. Res.* **20(2)**, 77A.
126. Mayer, S., et al. (1999) Regional neurotoxicity due to ethanol withdrawal hyperexcitotoxicity and neuroprotection by nicotine in organotypic hippocampal slices. *Alcohol. Clin. Exp. Res.* **23(5)**, 69A.

# 26
# Alcohol and Glutamate Neurotransmission in Humans
*Implications for Reward, Dependence, and Treatment*

### John H. Krystal, MD, Ismene L. Petrakis, MD, D. Cyril D'Souza, MD, Graeme Mason, PhD, and Louis Trevisan, MD

## 1. INTRODUCTION

Ethanol has multiple specific targets in the brain that combine to yield a complexly nuanced psychoactive agent *(1)*. However, the study of glutamatergic targets of ethanol have been a recent development *(2)*. The recency of these clinical studies may be surprising. Glutamate is the most prevalent excitatory neurotransmitter in the cerebral cortex and it mediates most output of the cortex and limbic system *(3)*. Also, the *N*-methyl-D-aspartate (NMDA) subtype of glutamate receptor is among the highest-affinity targets of ethanol inthe brain *(4)*. This chapter will provide an introduction to studies indicating that NMDA receptor blockade contributes to the behavioral effects of ethanol in humans. In doing so, it will provide clinical insights into the neurobiology of the rewarding and dysphoric effects associated with the blockade of NMDA receptors by ethanol. This chapter will then describe evidence of glutamatergic dysregulation in ethanol-dependent patients. In doing so, it will emphasize the hypothesis that ethanol tolerance may be associated with alterations in the reward valence of the NMDA antagonist component of ethanol action. It will also describe evidence that the familial vulnerability to alcoholism may be associated with alterations in NMDA receptor function that promote its use.

## 2. NMDA GLUTAMATE RECEPTOR ANTAGONISM AND THE BEHAVIORAL EFFECTS OF ETHANOL

The NMDA receptor is among the most potent targets for ethanol in the brain. As described in Fig. 1, this glutamate receptor includes a calcium channel that is activated by the joint binding of glutamate and glycine under conditions where limited membrane depolarizaition has first removed a magnesium-ion blockade *(6)*. Ethanol produces a dose-related capacity blockade of NMDA receptor function between 5 and 100 m$M$ *(4,7,8)*. In animals, ethanol shares many properties with other NMDA receptor antagonists. With acute administration, NMDA antagonist have ethanollike discriminative stimulus properties, with the effects of NMDA antagonist being most similar to higher training doses of ethanol *(9)*.

There is a growing body of literature suggesting that NMDA antagonists have ethanollike effects in healthy human subjects and recovering ethanol-dependent patients. Healthy subjects spontaneously reported that the uncompetitive NMDA antagonists phencyclidine *(10)* and ketamine *(11)* produced a sense of intoxication that resembled that produced by ethanol. In a subsequent studies, ketamine *(12)* and dextromethorphan *(13)* produced ethanollike effects in recently detoxified alcoholic patients. As

From: *Contemporary Clinical Neuroscience: Glutamate and Addiction*
Edited by: Barbara H. Herman et al. © Humana Press Inc., Totowa, NJ

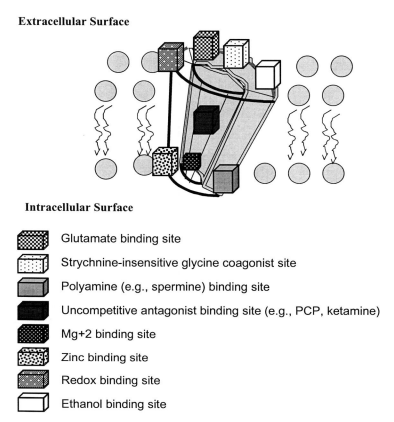

**Fig. 1.** Schematic of *N*-methyl-D-aspartate glutamate receptor illustrating multiple binding sites. (From ref. *5.*)

with the preclinical studies *(9)*, both the intensity and the degree of similarity of the ethanollike effects of ketamine were greater at 0.5 mg/kg than at 0.1 mg/kg (*see* Fig. 2). The higher dose was judged to be similar to approximately eight to nine standard ethanol drinks while the lower dose was similar to one to two standard ethanol drinks. Typically, 0.5 mg/kg ketamine has more pronounced perceptual effects than 8–10 ethanol drinks in healthy subjects *(11)*. However, the perceptual effects of ketamine were blunted in recovering ethanol-dependent dependent patients and these individuals also had more insight into perceptual effects associated with very high levels of ethanol intoxication. As one subject in this study put it, "I have these (perceptual) effects when I drink, then I pass out" (Krystal, unpublished observation). Ketamine effects were judged more similar to ethanol effects than the effects of marijuana or cocaine. Ketamine did not stimulate ethanol-craving in patients, although craving was associated with the ethanollike effects of another NMDA antagonist, dextromethorphan *(13)*.

Clinical studies examining the interactive effects of ketamine and other drugs may provide insights into the NMDA antagonist component of ethanol effects in the brain. The NMDA antagonist-induced euphoria does not yet appear dopamine dependent. For example, the euphoric properties of ketamine are not blocked by haloperidol pretreatment *(15)* or markedly potentiated by amphetamine pretreatment *(16)*. These findings parallel clinical findings describing the lack of interaction of ethanol and amphetamine *(17)*. In contrast, the euphoric effects of ketamine *(18)*, like ethanol *(19)*, are attenuated by pretreatment with the µ-opiate receptor antagonist naltrexone. Ethanol may possess actions at other brain targets that attenuate the dysphoric properties arising from its blockade of NMDA receptors. For

# Alcohol and Glutamate Neurotransmission in Humans

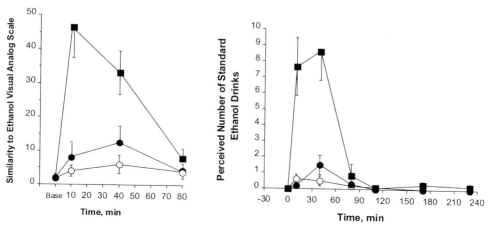

**Fig. 2.** Evidence that doses of ketamine that are judged to be more similar to ethanol are also thought to be similar to the consumption of greater amounts of ethanol. The left figure presents ratings made by recovering ethanol-dependent patients ($n = 20$) on a visual analog scale measuring similarity of a given subjective state to ethanol intoxication. This figure illustrates that the 0.5-mg/kg dose of ketamine was substantially more "ethanol-like" in these patients relative to 0.1 mg/kg, ketamine or a saline placebo. The right figure presents the subjective reports from this same patient group regarding the intensity of ethanollike effects reflected by the "Number of Drinks Scale," which measures the number of standard ethanol drinks (30 cm$^3$ absolute ethanol) thought to produce comparable levels of intoxication to a given subjective state. The effects of 0.5 mg/kg ketamine were judged to produce significantly more intense ethanollike effects than either 0.1 mg/kg ketamine or a saline placebo. (Adapted from ref. *14*.) ○, placebo; ●, 0.1 mg/kg ketamine hydrochloride; ■, 0.5 mg/kg ketamine hydrochloride.

example, ethanol facilitates GABA$_A$ receptor function and blocks voltage-gate ion calcium channels *(2)*. The combination of ketamine and the benzodiazepine agonist lorazepam is more tolerable than ketamine alone because lorazepam attenuates anxiogenic and perceptual ketamine effects *(20)*. Similarly, pretreatment with the voltage-gated cation channel antagonists lamotrigine or nimodipine also attenuated the behavioral effects of ketamine and improved the tolerability of ketamine *(21,22)*.

Drugs acting at the glycine-B modulatory site of the NMDA receptor modulate ethanol intoxication (*see* Fig. 1). In cultured cerebellar granule cells, ethanol lowered the NMDA receptor affinity for glycine and ethanol effects were partially reversed by raising glycine levels *(2)*. In humans, intravenous 0.1–0.2g/kg glycine raised cerebrospinal fluid glycine levels several-fold *(23)*. These doses of glycine also show other psychoactive properties in humans consistent with enhanced stimulation of the glycine-B site: attenuation of the cognitive and perceptual effects of ketamine *(24)* and enhancement of the euphoric and sedative effects of the glycine-B partial agonist D-cycloserine (1000 mg., po) *(25)*. The latter finding supports the hypothesis that exogenous glycine raises synaptic glycine levels because the "NMDA antagonistlike" properties of glycine-B partial agonists would be predicted to become more prominent as synaptic glycine levels rise *(26)*. A preliminary report also suggests that 500 mg D-cycloserine at its peak blood levels may mildly increase ethanol intoxication without increasing ethanol levels *(27)*.

Together, these studies suggest that the NMDA antagonist actions of ethanol contribute to its behavioral effects in humans. These studies also suggest that drugs that modulate the rewarding or dysphoric effects of NMDA antagonists in humans also may modulate ethanol intoxication in manners that may promote or attenuate the abuse of ethanol. From this perspective, drugs that block the rewarding effects or promote the dysphoric effects of NMDA antagonists are possible pharmacotherapies for alcoholism. To date, this approach suceeded in detecting a therapeutic effect of an approved alcoholism treatment, naltrexone. It will be important, then, to determine whether high-dose glycine has therapeutic value related to a capacity to attenuate ethanol intoxication. Alternatively, the capacity of high-dose D-cycloserine to potentiate dysphoric effects of ethanol may also signal a potential therapeutic role for this drug as well.

Other binding sites of the NMDA receptor may contribute to the behavioral pharmacology of ethanol and its treatment. Acamprosate, for example, shows efficacy in the treatment of alcoholism *(28,29)*. Its site of action is not clear at this time, although it is known to modulate NMDA receptor function. Recent data suggest that it may attenuate the stimulatory effects of spermine on NMDA receptor function via the polyamine-binding site of the NMDA receptor complex *(30)* (*see* Fig. 1). This action may be consistent with the capacity of acamprosate to produce mild impairments in delayed recall *(31)* without producing ethanollike behavioral effects in animals *(32)*. Other NMDA receptor-binding sites have been probed indirectly in clinical research studies. For example, desipramine is an antagonist of NMDA receptor function, perhaps via the zinc-binding site *(33)*. This medication has also shown efficacy in the treatment of alcoholism *(34)*. Overall, the psychopharmacology of regulatory sites of the NMDA receptor complex other than the glycine, glutamate, and channel (PCP, ketamine) sites are poorly understood.

## 3. GLUTAMATERGIC DYSREGULATION IN ETHANOL-DEPENDENT PATIENTS

There is compelling preclinical evidence suggesting that adaptations within the brain to chronic ethanol administration contribute to the phenomena of tolerance and withdrawal. In response to the chronic blockade of NMDA receptors associated with sustained ethanol administration, there are increases the levels of mRNA or protein for NMDA receptor subunits and increased NMDA receptor function *(2)*. These changes, combined with increased glutamate release, are temporally related to the expression of withdrawal seizures in animals *(35)*. The importance of NMDA receptor adaptations associated with ethanol dependence for withdrawal is supported by evidence that NMDA antagonists suppress ethanol withdrawal *(36)* and that inbred rodent strains with increased expression of withdrawal seizures show elevated levels of NMDA receptors *(37)*.

Preclinical research suggests that acute withdrawal is associated with a paradoxical convergence: upregulation of postsynaptic NMDA receptor function and increased glutamate release. The cause of increased glutamate release during ethanol withdrawal is unknown, but it has many hypothesized etiologies. For example, chronic ethanol administration depresses GABA synthesis and reduces GABA receptor function *(38,39)*. Inhibitory deficits might contribute to enhanced glutamate release. Also, chronic ethanol administration increases the levels of several subtypes of voltage-gated calcium channels that are involved in neurotransmitter release *(2)*. In addition to promoting glutamate release, upregulation of voltage-gated calcium channels could also serve to recruit NMDA receptors by reducing membrane potential, removing the $Mg^{2+}$ blockade and shifting NMDA receptors to the "active state." This process would be facilitated because of the common colocalization of NMDA and voltage-gated cation channels *(40)*. The potential clinical consequences of increased presynaptic and postsynaptic glutamatergic activation during acute withdrawal are serious because preclinical data suggest that this convergence contributes to seizures and neurotoxicity *(41,42)*.

Clinical research studies have begun to document the enhancement in glutamatergic function during withdrawal. Postmortem studies of ethanol-dependent individuals suggest that the $B_{max}$ or $K_D$ of NMDA receptors are increased in cortical structures alcoholics *(43,44)*. In vivo, ethanol withdrawal increases cerebrospinal fluid glutamate levels *(45)*, consistent with preclinical evidence of enhanced glutamate release *(46,47)*. Repeated episodes of withdrawal may promote the initiation of forms of neural sensitization that may contribute to increased startle magnitude *(48)* and enhanced seizure risk *(49,50)*. It is possible that withdrawal-related neuroplasticity contributes to associative learning, as might be reflected in drug-craving *(51)*.

Although the most severe consequences of withdrawal appear during the initial week of sobriety, protracted components of withdrawal persist for many months and may contribute to relapse *(52)*. Protracted withdrawal symptoms include insomnia, anergia, and depressed mood. During the period of protracted withdrawal, patients may show reduced glutamatergic activity. This hypothesis is primarily supported by preliminary indirect evidence from a proton magnetic resonance spectroscopy study that occipital cortex glutamine levels, an amino acid metabolite associated with glutamatergic neurotransmission, may be reduced in the weeks following detoxification *(53)*. This finding may be consistent with positron emission computerized tomography (PET) studies indicating that cortical metabolism is reduced during the recovering from ethanol dependence *(54–56)*.

*N*-Methyl-D-aspartate receptor functional alterations in recovering ethanol-dependent patients also may shift the reward valence of the NMDA antagoinst component of ethanol response, promoting further alcohol use. Recently detoxified ethanol-dependent patients show marked reductions in their sensitivity to the perceptual, mood, and cognitive effects of ketamine *(57)* and the glycine-B partial agonist D-cycloserine *(58)*. In contrast, preliminary data suggest that these patients exhibited a trend for increased euphoric responses to ketamine (Krystal, unpublished data). Thus, NMDA receptor alterations associated with ethanol dependence might contribute to relapse to ethanol use in two ways:(1) by contributing to the signs and symptoms of ethanol withdrawal and (2) by enhancing the rewarding properties or reducing the dysphoric properties of ethanol during the early phases of relapse to ethanol use. In light of these possibilities, there may be a unique role for agents that directly address glutamatergic alterations in the treatment of ethanol withdrawal and the prevention of relapse.

An additional possible role for glutamatergic agents in the treatment of alcoholism may be in the restoration of cognitive function, normalizing cortical metabolism, or in promoting the recovery of cortical connectivity. Studies suggest that recovering ethanol-dependent patients show reductions in cortical volume on magnetic resonance imaging studies that are more pronounced with heavier drinking and with advancing age *(59,60)*. More recent research also suggests that heavy drinking during adolescence may impair brain development *(61)*. These studies are consistent with behavioral studies suggesting that heavy sustained drinking impairs cognitive functions, particularly those associated with frontal and temporo-hippocampal networks *(62,63)*. These cognitive deficits are associated with reduced prefrontal cortical metabolism *(54)*. Reductions in cortical glutamate turnover described previously during protracted withdrawal may reflect a need to enhance glutamatergic neurotransmission to normalize regional brain activity. Drugs that facilitate NMDA receptor function without independently stimulating neurotoxicity, such as the glycine-B agonists *(26)*, might play a role in helping to restore functional coritcal connectivity during recovery. AMPAkines *(64)* and other glutamatergic nootropic strategies may also have a role in directly ameliorating cognitive deficits in recovering ethanol-dependent patients.

However, a component of the cognitive changes in patients reflect toxic or atrophic consequences of alcoholism *(2)*. As reviewed earlier, withdrawal-related neurotoxicity may reflect the increase in glutamate release during withdrawal and the upregulation in NMDA receptor function. However, long-term ethanol administration may also contribute to brain atrophy by interfering with neurotrophic functions directly or indirectly associated with intact NMDA receptor activity *(65–68)*. Because neurotrophic functions generally decline with age *(69,70)*, it might not be surprising that the ethanol-related effects on these functions might compound the effects of aging. Further research will be needed to determine whether glutamatergic or other treatment strategies can contribute to the recovery from ethanol-related neurotoxicity or atrophy.

## 4. GLUTAMATERGIC DYSREGULATION AND THE VULNERABILITY TO ALCOHOLISM

Healthy individuals at increased familial risk for developing alcoholism, relative to a "family history negative" group, show reductions in the dysphoric effects of NMDA receptor antagonists resembling the changes seen in ethanol-dependent patients *(71).* Thus, inherited differences related to NMDA receptor function may contribute to alterations in the set point for sensitivity to ethanol effects that promote the development of the abuse of ethanol. Further research will be needed to clarify the impact of ethanol dependence and alcoholism vulnerability on glutamatergic function.

The mechanism through which ketamine sensitivity is altered in individuals at increased familial risk for alcoholism is not yet clear. In individuals without a family history of alcoholism, antagonism of voltage-gated cation channels reduces the dysphoric effects of ketamine and enhances its euphoric effects *(21)* (i.e., produces changes in the reward valence of ketamine effects that are similar to the alterations associated with a family history of ethanol dependence). The genes underlying altered ketamine response in individuals at risk for alcoholism are not yet known.

## 5. SUMMARY AND TREATMENT IMPLICATIONS

There is now growing evidence in humans that (1) glutamate receptors are an important target for ethanol in the brain, (2) ethanol actions at NMDA receptors contribute to its behavioral effects, (3) ethanol dependence may be associated with upregulation of NMDA receptors, (4) acute ethanol withdrawal may be associated with increased glutamate release and protracted withdrawal may be associated with reduced brain glutamate turnover, and (5) the familial risk for developing alcoholism may be associated with alterations in NMDA receptor function. Genetic variation that might link glutamatergic systems to the vulnerability to alcoholism and its treatment have yet to be explicated. Acamprosate is the first agent developed for the treatment of alcoholism with a mechanism of action that may be related to glutamate function. Future pharmacotherapy research glutamatergic pharmacotherapy research may explore medications designed to suppress withdrawal, reduce ethanol consumption, attenuate alcoholism-related cognitive deficits, protect neuronal viability, and reverse cerebral atrophy.

## ACKNOWLEDGMENTS

This work was supported by NIAAA (KO2 AA 00261-01, RO1 AA11321-01A1) and the Department of Veterans Affairs (Alcohol Research Center, Clinical Neurosciences Division, National Center for Post-traumatic Stress Disorder).

## REFERENCES

1. Green, K. L. and Grant, K. A. (1998) Evidence for overshadowing by components of the heterogeneous discriminative stimulus effects of ethanol. *Drug Alcohol Depend.* **52,** 49–59.
2. Krystal, J. H. and Tabakoff, B. (2001) Ethanol abuse, dependence, and withdrawal: neurobiology and clinical implications, in *Psychopharmacology: A Fifth Generation of Progress.*
3. Charney, D. S., Nestler, E., and Bunney, B. S. (1999) *Neurobiology of Mental Illness,* Oxford University Press, New York.
4. Grant, K. A. and Lovinger, D. M. (1995) Cellular and behavioral neurobiology of alcohol: receptor-mediated neuronal processes. *Clin. Neurosci.* **3,** 155–164.
5. Krystal, J. H., Belger, A., Abi-Saab, W., Moghaddam, B., Charney, D. S., Anand, A., et al. (2000). Glutamatergic contributions to cognitive dysfunction in schizophrenia, in *Cognitive Functioning in Schizophrenia* (Harvey, P. D. and Sharma, T., eds.), Oxford University Press, London.
6. Sucher, N. J., Awobuluyi, M., Choi, Y. B., and Lipton, S. A. (1996) NMDA receptors: from genes to channels. *Trends Pharmacol. Sci.* **17,** 348–355.
7. Lovinger, D. M., White, G., and Weight, F. F. (1989) Ethanol inhibits NMDA-activated ion current in hippocampal neurons. *Science* **243,** 1721–1724.
8. Hoffman, P. L., Rabe, C. S., Grant, K. A., Valverius, P., Hudspith, M., and Tabakoff, B. (1990) Ethanol and the NMDA receptor. *Alcohol* **7,** 229–231.
9. Grant, K. A. and Colombo, G. (1993) Discriminative stimulus effects of ethanol: effect of training dose on the substitution of N-methyl-D-aspartate antagonists. *Pharmacol. Exp. Ther.* **264,** 1241–1247.
10. Luby, E. D., Cohen, B. D., Rosenbaum, G., Gottlieb, J. S., and Kelley, R. (1959) Study of a new schizophrenomimetic drug—sernyl. *Arch. Neurol. Psychiatry* **81,** 363–369.

11. Krystal, J. H., Karper, L. P., Seibyl, J. P., Freeman, G. K., Delaney, R., Bremner, J. D., et al. (1994) Subanesthetic effects of the noncompetitive NMDA antagonist, ketamine, in humans. Psychotomimetic, perceptual, cognitive, and neuroendocrine responses. *Arch. Gen. Psychiatry.* **51**, 199–214.
12. Krystal, J. H., Petrakis, I. L., Webb, E., Cooney, N. L., Karper, L. P., Namanworth, S., et al. (1998) Dose-related ethanol-like effects of the NMDA antagonist, ketamine, in recently detoxified alcoholics. *Arch. Gen. Psychiatry* **55**, 354–360.
13. Schutz, C. G. and Soyka, M. (2000) Dextromethorphan challenge in alcohol-dependent patients and controls [letter; comment.] *Arch. Gen. Psychiatry* **57**, 291–292.
14. Krystal, J. H., Petrakis, I. L., Webb, E., Cooney, N. L., Karper, L. P., Namanworth, S., et al. (1998). Dose-related ethanol-like effects of the NMDA antagonist, ketamine, in recently detoxified alcoholics. *Arch. Gen. Psychiatry* **55**, 354–360.
15. Krystal, J. H., D'Souza, D. C., Karper, L. P., Bennett, A., Abi-Dargham, A., Abi-Saab, D., et al. (1999) Interactive effects of subanesthetic ketamine and haloperidol. *Psychopharmacology* **145**, 193–204.
16. Krystal, J. H. and Petrakis, I. L. (2000) Dextromethorphan challenge in alcohol dependent patients and controls—reply. *Arch. Gen. Psychiatry* **57**, 292.
17. Perez-Reyes, M., White, W. R., McDonald, S. A., and Hicks, R. E. (1992) Interaction between ethanol and dextroamphetamine: effects on psychomotor performance. *Alcohol. Clin. Exp. Res.* **16**, 75–81.
18. Madonick, S. M., D'Souza, D. C., Brush, L., Cassello, K., Wray, Y., Larvey, K., et al. (1999) Naltrexone blockade of ketamine intoxication in healthy humans. *Alcohol. Clin. Exp. Res.* **23(Suppl.)** 103A (abstract).
19. Swift, R. M., Whelihan, W., Kuznetsov, O., Buongiorno, G., and Hsuing, H. (1994) Naltrexone-induced alterations in human ethanol intoxication. *Am. J. Psychiatry.* **151**, 1463–1467.
20. Krystal, J. H., Karper, L. P., Bennett, A., D'Souza, D. C., Abi-Dargham, A., Morrissey, K., et al. (1998) Interactive effects of subanesthetic ketamine and subhypnotic lorazepam in humans. *Psychopharmacology* **135**, 213–229.
21. Anand, A., Charney, D. S., Cappiello, A., Berman, R. M., Oren, D. A., and Krystal, J. H. (2000) Lamotrigine attenuates ketamine effects in humans: support for hyperglutamatergic effects of NMDA antagonists. *Arch. Gen. Psychiatry* **57**, 270–276.
22. Krupitsky, E., Burakov, A., Romanova, T., Grinenko, N., Grinenko, A., Vegso, S., et al. (2000) Interactive ethanol-like cognitive and behavioral effects of nimodipine and ketamine in abstinent alcoholics. *Alcohol. Clin. Exp. Res.* **24**, 12A.
23. D'Souza, D. C., Gil, R., Cassello, K., Morrissey, K., Abi-Saab, D., White, J., et al. (2000) IV glycine and oral D-cycloserine effects on plasma and CSF amino acids in healty humans. *Biol. Psychiatry* **47**, 450–462.
24. D'Souza, D. C., Gil, R., Zuzarte, E., Abi-Saab, D., Damon, D., White, et al. (1997) Glycine ketamine interactions in healthy humans. *Schizophr. Res.* **24**, 213.
25. Krystal, J., Petrakis, I., Krasnicki, S., Trevisan, L., Boutros, N., and D'Souza, D. C. (1998) Altered responses to agonists of the strychnine-insensitive glycine NMDA coagonist (SIGLY) site in recently detoxified alcoholics. *Alcohol. Clin. Exp. Res.* **22**, 94A.
26. D'Souza, D. C., Charney, D. S., and Krystal, J. H. (1995) Glycine site agonists of the NMDA receptor: a review. *CNS Drug Rev.* **1**, 227–260.
27. Trevisan, L., Nammanworth, S., Petrakis, I., Randall, P., Charney, D. S., and Krystal, J. H. (1995) Cycloserine modulation of human ethanol intoxication: implications for NMDA receptor contributions to human ethanol intoxication. *Alcohol. Clin. Exp. Res.* **19**, 4A(abstract).
28. Mason, B. J. (2001) Results of the multicenter study of acamprosate in the treatment of alcoholism. *Biol. Psychiatry* **49**, 77S.
29. Sass, H., Soyka, M., Mann, K., and Zieglgansberger, W. (1996) Relapse prevention by acamprosate. Results from a placebo-controlled study on alcohol dependence [published erratum appears in *Arch. Gen. Psychiatry* **53(12)**:1097.] *Arch. Gen. Psychiatry* **53**, 673–680.
30. Popp, R. L. and Lovinger, D. M. (2000) Interaction of acamprosate with ethanol and spermine on NMDA receptors in primary cultured neurons. *Eur. J. Pharmacol.* **394**, 221–231.
31. Schneider, U., Wohlfarth, K., Schulze-Bonhage, A., Haacker, T., Muller-Vahl, K. R., Zedler, M., et al. (1999) Effects of acamprosate on memory in healthy young subjects. *J. Studies Alcohol* **60**, 172–175.
32. Grant, K. A. and Woolverton, W. L. (1989) Reinforcing and discriminative stimulus effects of Ca-acetyl homotaurine in animals. *Pharmacol. Biochem. Behav.* **32**, 607–611.
33. Reynolds, I. J., and Miller, R. J. (1988) Trcyclic antidepressants block *N*-methyl-D-aspartate receptors: similarities to the action of zinc. *Br. J. Pharmacol.* **95**, 95–102.
34. Mason, B. J., Kocsis, J. H., Ritvo, E. C., and Cutler, R. B. (1996) A double-blind, placebo-controlled trial of desipramine for primary alcohol dependence stratified on the presence or absence of major depression [see comments.] *JAMA* **275**, 761–767.
35. Tsai, G. and Coyle, J. T. (1998) The role of glutamatergic neurotransmission in the pathophysiology of alcoholism. *Ann. Rev. Med.* **49**, 173–184.
36. Grant, K. A., Snell, L. D., Rogawski, M. A., Thurkauf, A., and Tabakoff, B. (1992) Comparison of the effects of the uncompetitive *N*-methyl-D-aspartate antagonist (+–)-5-aminocarbonyl-10, 11-dihydro-5*H*-dibenzo[*a*,*d*] cyclohepten-5, 10-imine (ADCI) with its structural analogs dizocilpine (MK-801) and carbamazepine on ethanol withdrawal seizures. *J. Pharmacol. Exp. Ther.* **260**, 1017–1022.
37. Valverius, P., Crabbe, J., Hoffman, P., and Tabakoff, B. (1990) NMDA receptors in mice bred to be prone or resistant to ethanol withdrawal seizures. *Eur. J. Pharmacol.* **184**, 185–189.

38. Hemmingsen, R., Braestrup, C., Nielsen, M., and Barry, D. L. (1982) The benzodiazepine/GABA receptor complex during severe ethanol intoxication and withdrawal in the rat. *Acta Psychiatr. Scand.* **65,** 120–126.
39. Harris, R. A., Valenzuela, C. F., Brozowski, S., Chuang, L., Hadingham, K., and Whiting, P. J. (1998) Adaptation of gamma-aminobutyric acid type A receptors to alcohol exposure: studies with stably transfected cells. *J. Pharmacol. Exp. Ther.* **284,** 180–188.
40. Schiller, J., Schiller, Y., and Clapham, D. E. (1998) NMDA receptors amplify calcium influx into dendritic spines during associative pre-and postsynaptic activation. *Nature Neurosci.* **1,** 114–118.
41. Hoffman, P. L., Grant, K. A., Snell, L. D., Reinlib, L., Iorio, K. and Tabakoff, B. (1992) NMDA receptors: role in ethanol withdrawal seizures. *Ann. NY Acad. Sci.* **654,** 52–60.
42. Hoffman, P. L. (1995) Glutamate receptors in alcohol withdrawal-induced neurotoxicity. *Metab. Brain Dis.* **10,** 73–79.
43. Michaelis, E. K., Freed, W. J., Galton, N., Foye, J., Michaelis, M. L., Phillips, I., et al. (1990) Glutamate receptor changes in brain synaptic membranes from human alcoholics. *Neurochem. Res.* **15,** 1055–1063.
44. Freund, G. and Anderson, K. J. (1996) Glutamate receptors in the frontal cortex of alcoholics. *Alcohol. Clin. Exp. Res.* **20,** 1165–1172.
45. Tsai, G. E., Ragan, P., Chang, R., Chen, S., Linnoila, V. M., and Coyle, J. T., (1998) Increased glutamatergic neurotransmission and oxidative stress after alcohol withdrawal. *Am. J. Psychiatry* **155,** 726–732.
46. Keller, E., Cummins, J. T., and von Hungen, K. (1983) Regional effects of ethanol on glutamate levels, uptake and release in slice and synaptosome preparations from rat brain. *Subst. Alcohol Actions/Misuse* **4,** 383–392.
47. Tsai, G., Gastfriend, D. R., and Coyle, J. T. (1995) The glutamatergic basis of human alcoholism. *Am. J. Psychiatry* **15,** 332–340.
48. Krystal, J. H., Webb, E., Grillon, C., Cooney, N., Casal, L., Morgan, C. A., 3rd, et al. (1997) Evidence of acoustic startle hyperreflexia in recently detoxified early onset male alcoholics: modulation by yohimbine and *m*-chlorophenylpiperazine (mCPP). *Psychopharmacology* **131,** 207–215.
49. Brown, M. E., Anton, R. F., Malcolm, R., and Ballenger, J. C. (1988) Alcohol detoxification and withdrawal seizures: clinical support for a kindling hypothesis. *Biol. Psychiatry* **23,** 507–514.
50. Lechtenberg, R. and Worner, T. M. (1992) Total ethanol consumption as a seizure risk factor in alcoholics. *Acta Neurol. Scand.* **85,** 90–94.
51. Malcolm, R., Herron, J. E., Anton, R. F., Roberts, J., and Moore, J. (2000) Recurrent detoxification may elevate alcohol craving as measured by the Obsessive Compulsive Drinking scale. *Alcohol* **20,** 181–185.
52. De Soto, C. B., O'Donnell, W. E., Allred, L. J., and Lopes, C. E. (1985) Symptomatology in alcoholics at various stages of abstinence. *Alcohol. Clin. Exp. Res.* **9,** 505–512.
53. Behar, K., Rothman, D., Petersen, K., Hooten, M., Namanworth, S., Delaney, R., et al. (1999) Preliminary evidence of reduced cortical GABA levels in localized $^1$H NMR spectra of alcohol dependent and hepatic encephalopathy patients. *Am. J. Psychiatry* **156,** 952–954.
54. Adams, K. M., Gilman, S., Koeppe, R. A., Kluin, K. J., Brunberg, J. A., Dede, D., et al. (1993) Neuropsychological deficits are correlated with frontal hypometabolism in positron emission tomography studies of older alcoholic patients. *Alcohol. Clin. Exp. Res.* **17,** 205–210.
55. Volkow, N. D., Wang, G. J., Hitzemann, R., Fowler, J. S., Overall, J. E., Burr, G., et al. (1994) Recovery of brain glucose metabolism in detoxified alcoholics. *Am. J. Psychiatry* **15,** 178–183.
56. Volkow, N. D., Hitzemann, R., Wang, G. J., Fowler, J. S., Burr, G., Pascani, K., et al. (1992) Decreased brain metabolism in neurologically intact healthy alcoholics. *Am. J. Psychiatry* **149,** 1016–1022.
57. Krystal, J., Karper, L., D'Souza, D. C., Webb, E., Bennett, A., Abi-Dargham, A., et al. (1995) Differentaiating NMDA dysregulation in schizophrenia and alcoholism using ketamine. *Schizophr. Res.* **15,** 156.
58. Krystal, J., Petrakis, I., Krasnicki, S., Trevisan, L., Boustros, N., and D'Souza, D. C. (1999) Altered responses to agonists of the strychnine-insensitive glycine NMDA coagonist (SIGLY) site in recently detoxified alcoholics. *Alcohol. Clin. Exp. Res.* **22,** 94A.
59. Pfefferbaum, A., Sullivan, E. V., Rosenbloom, M. J., Mathalon, D. H., and Lim, K. O. (1998) A controlled study of cortical gray matter and ventricular changes in alcoholic men over a 5-year interval. *Arch. Gen. Psychiatry* **55,** 905–912.
60. Pfefferbaum, A., Sullivan, E. V., Mathalon, D. H., and Lim, K. O. (1997) Frontal lobe volume loss observed with magnetic resonance imaging in older chronic alcoholics. *Alcohol. Clin. Exp. Res.* **21,** 521–529.
61. De Bellis, M. D., Clark, D. B., Beers, S. R., Soloff, P. H., Boring, A. M., Hall, J., et al. (2000) Hippocampal volume in adolescent-onset alcohol use disorders. *Am. J. Psychiatry* **157,** 737–744.
62. Parsons, O. A. and NIxon, S. J. (1993) Neurobehavioral sequelae of alcoholism. *Neurol. Clin.* **11,** 205–218.
63. Schafer, K., Butters, N., Smith, T., Irwin, M., Brown, S., Hanger, P., et al. (1991) Cognitive performance of alcoholics: a longitudinal evaluation of the role of drinking history, depression, liver function, nutrition, and family history. *Alcohol. Clin. Exp. Res.* **15,** 653–660.
64. Goff, D., Berman, I., Posever, T., Leahy, L., and Lynch, G. (1999). A preliminary dose-escalation trial of CX-516 (ampakine) added to clozapine in schizophrenia. *Schizophr. Res.* **36,** 280.

65. Springer, J. E., Gwag, B. J., and Sessler, F. M. (1994), Central administration of the excitotoxin N-methyl-D-aspartate increases nerve growth factor mRNA in vivo. *Neurotoxicology* **15,** 483–489.
66. Heisenberg, C. P., Cooper, J. D., Berke, J., and Sofroniew, M. V. (1994). NMDA potentiates NGF-induced sprouting of septal cholinergic fibres. *NeuroReport* **5,** 413–416.
67. Favaron, M., Manev, R. M., Rimland, J. M., Candeo, P., Beccaro, M., and Manev, H. (1993) NMDA-stimulated expression of BDNF mRNA in cultured cerebellar granule neurones. *NeuroReport* **4,** 1171–1174.
68. Lindefors, N., Ballarin, M., Ernfors, P., Falkenberg, T., and Persson, H. (1992) Stimulation of glutamate receptors increases expression of brain-derived neurotrophic factor mRNA in rat hippocampus. *Ann. NY Acad. Sci.* **648,** 296–299.
69. McLay, R. N., Freeman, S. M., Harlan, R. E., Ide, C. F., Kastin, A. J., and Zadina, J. E. (1997) Aging in the hippocampus: interrelated actions of neurotrophins and glucocorticoids. *Neurosci. Biobehav. Rev.* **21,** 615–629.
70. Rylett, R. J. and Williams, L. R. (1994). Role of neurotrophins in cholinergic-neurone function in the adult and aged CNS. *Trends Neurosci.* **17,** 486–490.
71. Petrakis, I. L., Boutros, N. N., O'Malley, S. M., Gelernter, J., and Krystal, J. H. (2000). NMDA dysregulation in individuals with a family vulnerability to alcoholism. *Alcohol. Clin. Exp. Res.* **24,** 12A.

# 27
# Mechanism of Action of Acamprosate Focusing on the Glutamatergic System

### W. Zieglgänsberger, G. Rammes, R. Spanagel, W. Danysz, and Ch. Parsons

## 1. INTRODUCTION

The taurine analog acamprosate (calcium acetylhomotaurinate) has received considerable attention in Europe for its ability to prevent relapse in abstained alcoholics [(1); Chapter 28] and has been suggested to act by reducing craving associated with conditioned withdrawal (2,3).

## 2. ANTICRAVING AND ANTIRELAPSE PROPERTIES OF ACAMPROSATE

Novel aspects of addictive behavior to alcohol (craving, relapse, and sensitization processes) are uncovered by a new animal model of long-term, free-choice, alcohol self-administration followed by alcohol-deprivation phases. After several months of voluntary alcohol consumption, the drug-taking behavior following a deprivation (withdrawal) phase is characterized by increased alcohol intake and preference. During this so-called alcohol-deprivation effect (relapselike behavior) rats exhibit a high motivation for alcohol (4). This behavior is interpreted as craving, and this model has been used to investigate the potential of new anticraving agents (5). In this model, acamprosate (50–200 mg/kg, ip) administered twice daily during the alcohol-deprivation phase dose-dependently reduced the subsequent alcohol deprivation effect (6). The effects of acamprosate on drinking behavior under operant conditions, both during normal training conditions (i.e., at baseline) and during the alcohol-deprivation effect, were also studied (7). Under baseline conditions, acamprosate reduced operant responding in long-term alcohol-drinking rats. At maximal acamprosate levels in the blood and the brain, however, the agent reduced alcohol consumption more effectively during the alcohol-deprivation effect than during baseline drinking. Because the intensity of the alcohol-deprivation effect can serve as a measure of craving, these findings suggest that acamprosate indeed has anticraving properties. Acute administration of acamprosate (400 mg/kg) reduced oral ethanol consumption under limited access where rats were trained to respond for ethanol (10% w/v) or water in a two-lever free-choice operant paradigm (8). Repeated administration of lower daily doses of acamprosate (100 and 200 mg/kg) selectively blocked increased ethanol consumption typically observed in the same model after an imposed abstinece period (8). In a model in which ethanol administration was repeatedly paired with plus-maze exposure, the opioid antagonist naltrexone had no significant effect on alcohol-conditioned abstinence behavior in the plus-maze, but acamprosate reduced the incidence of stretched-attend postures (9). This difference was attributed to effects of acamprosate on conditioned negative reinforcement, whereas naltrexone is thought to have effects on positive reinforcement for ethanol.

From: *Contemporary Clinical Neuroscience: Glutamate and Addiction*
Edited by: Barbara H. Herman et al. © Humana Press Inc., Totowa, NJ

The mechanism of action of acamprosate in the central nervous system (CNS) is still unclear. A better understanding of this is important to increase our knowledge of the fundamental processes governing alcohol abuse that would, in turn, allow the development of better drugs to prevent relapse in weaned alcoholics. Although early studies indirectly suggested an action at GABA receptors *(10–13)*, more recent data suggest that acamprosate rather or also interacts with the *N*-methyl-D-aspartate (NMDA) subclass of ionotropic glutamate receptor. These $Ca^{2+}$-permeable channels have, in turn, been implicated in the induction of alcohol dependence.

## 3. THE GLUTAMATERGIC/NMDA RECEPTOR SYSTEM

*N*-Methyl-D-aspartate receptors consist of tetrameric and heteromeric subunit assemblies that have different physiological and pharmacological properties and are differentially distributed throughout the CNS *(14–19)*. So far, three major subunit families designated NR1, NR2, and NR3 have been cloned. Functional receptors in the mammalian CNS are almost certainly only formed by combination of NR1 and NR2 subunits, which express the glycine and glutamate recognition sites, respectively *(20,21)*. NR3 subunits seem to inhibit receptor function and are expressed at higher levels during development *(22)*.

Alternative splicing generates eight isoforms for the NR1 subfamily *(23–25)*. The variants arise from splicing at three exons; one encodes a 21-amino acid insert in the N-terminal domain (N1, exon 5) and two encode adjacent sequences of 37 and 38 amino acids in the C-terminal domain (C1, exon 21, and C2, exon 22). NR1 variants are sometimes denoted by the presence or absence of these three alternatively spliced exons (from N to C1 to C2); $NR1_{111}$ has all three exons, $NR1_{000}$ has none, and $NR1_{100}$ has only the N-terminal exon *(23)*. The variants from $NR1_{000}$ to $NR1_{111}$ are alternatively denoted as NMDAR1 -4a, -2a, -3a, -1a, -4b, -2b, -3b, and -1b. The NR2 subfamily consists of four individual subunits, NR2A to NR2D *(14–19)*. Various heteromeric NMDA receptor channels formed by combinations of NR1 and NR2 subunits are known to differ in gating properties, magnesium sensitivity, and pharmacological profile (e.g., see Table 1 of ref. *15*). Only the heteromeric assembly of NR1 and NR2B subunits for instance are potentiated in a glycine-independent manner by the polyamines spermine and spermidine and are selectively blocked by ifenprodil and related compounds. *In situ* hybridization has revealed overlapping but different expression for NR2 mRNA; for example, NR2A mRNA is distributed ubiquitously like NR1 with highest densities occuring in hippocampal regions and NR2B is expressed predominantly in the forebrain but not in the cerebellum, where NR2C predominates.

Glycine is a coagonist at NMDA receptors at a strychnine-insensitive recognition site ($glycine_B$) and its presence at moderate nanomolar concentrations is a prerequisite for channel activation by glutamate or NMDA *(26–28)* (for review, see ref. *16*). Physiological concentrations of glycine reduce one form of relatively rapid NMDA receptor desensitization *(29–31)*. Recently, it has been suggested that D-serine may be more important than glycine as an endogenous coagonist at NMDA receptors in the telencephalon and developing cerebellum *(32)*.

The endogenous polyamines spermine and spermidine have multiple effects on the activity of NMDA receptors. These include an increase in the magnitude of NMDA-induced whole-cell currents seen in the presence of saturating concentrations of glycine, an increase in glycine affinity, a decrease in glutamate affinity, and voltage-dependent inhibition at higher concentrations *(33,34)*. Glycine-independent stimulation requires the presence of NR1 variants that lack an amino-terminal insert such as NR1a ($NR1_{011}$) but not NR1b ($NR1_{111}$). The stimulatory effect is also controlled by NR2 subunits in heteromeric complexes—it is observed at heteromeric NR1a/NR2B receptors but not at heteromeric NR1a/NR2A or NR1a/NR2C receptors *(35,36)*. Glycine-dependent stimulation is mediated via an increase in glycine affinity and probably involves a second binding site, as it is also seen at NR1a/NR2A receptors *(36,37)*. Spermine also induces a small decrease in the affinity of NR1a/NR2B but not NR1a/2A receptors for NMDA and glutamate *(37)*. The voltage-dependent inhibitory effect of higher concentrations of spermine is similar for NR1A/NR2A and NR1A/NR2B receptors but is apparently absent at NR1A/NR2C receptors *(36)*. This effect seems to be mediated at the $Mg^{2+}$ channel site. Endogenous polyamines could therefore act as a bidirectional gain control of NMDA receptors, by dampening toxic chronic activation by low concentrations of glutamate—through changes in glutamate affinity and voltage-dependent blockade—but enhancing transient synaptic responses to millimolar concentrations of glutamate *(33)*.

The polyamine modulatory site of the NMDA receptor was discovered in the late 1980s and then ifenprodil and its analog eliprodil were found to block NMDA receptors in a spermine-sensitive manner and proposed to be polymaine antagonists *(38–40)*. Initial patch-clamp evidence for NMDA receptor-subtype selectivity of ifenprodil and eliprodil was provided by Legendre and Westbrook *(41)*, followed by conclusive evidence for NR2B subtype selectivity by Williams *(42)*.

## 4. ACTION OF ALCOHOL ON THE GLUTAMATERGIC/NMDA RECEPTOR SYSTEM

Ethanol can be seen as an NMDA receptor antagonist *(43–45)* at concentrations reached in the brains of alcohol abusers. There is some in vitro evidence that the effects of ethanol may be related to selective actions at NR2B receptors *(46,47)*. NR2B selective actions are supported by the finding that chronic ethanol treatment of cultured cortical neurons has been reported to increase NR2B mRNA *(48)* and receptor expression *(49–51)*. Moreover, chronic exposure of cultured cortical neurons or NR2B-transfected HEK-293 cells to ethanol increases sensitivity to NMDA measured by $Ca^{2+}$ influx and neurotoxicity and causes a 10-fold increased sensitivity to the blockade by ifenprodil *(49)*. In vivo data based on comparing sensitivity to ifenprodil and alcohol also suggest that both interact with the same subtype of NMDA receptors [i.e., containing the NR2B subunit *(47)*]. In contrast, Mirshahi and Woodward *(45)* reported that ethanol is somewhat more potent against NR2A receptors in *Xenopus* oocytes. Furthermore, no selective change in NR2B was observed in two other studies on chronic exposure of cultures to ethanol *(52–53)*. It should also be noted that, apart from interactions with both NMDA and GABA receptors, at low millimolar concentrations (reached in the brain during intoxication), actions on nicotinic receptor micromolar levels were also detected recently ($IC_{50} = 89\ \mu M$) *(54)*.

In rats, at 9 h after withdrawal from chronic treatment with ethanol, an increase in NR2A and NR2B but no change in NR1 was seen *(55)*. A recent postmortem study on the brains of alcoholics showed a modest (but significant) increased binding for [$^3$H]glutamate and [$^3$H]CGP-39653—competitive NMDA receptor antagonists *(56)*. In humans with a history of alcohol abuse, an increase in immunoreactivity toward AMPA GluR2 and GluR3 subunits was also found *(57)*.

It has been shown that upon withdrawal from ethanol in dependent rats, an increase in glutamate release is seen in the striatum that temporally corresponds to the duration of withdrawal syndrome (hyperactivity, treading, shakes, jerks, twitches) *(58)*. (+)MK-801 normalized both effects, whereas diazepam only affected the behavioral aspects. Treatment of mice with ethanol increases fast α-amino-3-hydroxy-5-methyl-4-isoxazole propionic acid (AMPA) receptor-mediated excitatory postsynapthic currents (EPSPs) in the CA1 region in hippocampal slices ex vivo at 4–6 h but not 2 h after withdrawal *(59)*. Under similar conditions, $Ca^{2+}$ channel activity and NMDA receptor-mediated EPSPs were reported to be potentiated in the hippocampus *(60)*. NMDA receptor antagonists (MK-801 and CGP-39551) inhibit the development of tolerance to ethanol (measured as a decrease of sleeping time) even if given 120 min after the daily alcohol injection *(61)*. This indicates that NMDA receptor activation is probably involved in plastic changes following chronic ethanol-induced receptor adaptation. Consistent with the role of the NMDA receptor in alcohol tolerance is the finding that the partial agonist D-cycloserince enhances tolerance to the motor-impairing effect of ethanol in rats *(62)*.

In the alcohol "craving"/relapse model discussed above for acamprosate, memantine given via osmotic mini-pumps at a dose leading to steady-state serum levels similar to those seen in clinical practice also greatly inhibited alcohol consumption during the relapse phase *(63)*. This suggests that The blockade of NMDA receptors inhibits some aspects of alcohol dependence. There are several possibilities as to how NMDA receptor antagonists exert this effect: (1) produce alcohol-like effects because memantine and other NMDA receptor antagonists show partial or full generalization to the ethanol cue in rats trained to discriminate ethanol *(64,65)*; (2) block recognition of the alcohol cue; (3) inhibit association of environmental cues with alcohol use (4) directly interfere with the reinforcing properties of ethanol. Another study examined the effects of the NMDA receptor antagonist AP5 on alcohol preference. Following intracerebroventral (icv) infusion of AP5, ethanol preference was reduced in untrained but not trained rats *(66)*. This indicates that NMDA receptor antagonists may interact with the association but not the recognition of the alcohol cue or inhibition of its hedonic effects.

There is also increasing evidence that NMDA receptors may participate in the execution of pathological changes in Wernicke–Korsakoff syndrome. In animals, thiamine deficiency produces a pattern on neuronal damage resembling that found in Wernicke–Korsakoff syndrome (e.g., in the thalamus). In such animals, an increase in extracellular glutamate in the brain is observed and NMDA receptor antagonists prevent both neurodegeneration and the increase in glutamate *(67)*. In a goat version of this model, there is a decrease in the NMDA receptors in the motor cortex, probably reflecting damage to neurons rich in this receptor type *(68)*. Interestingly, in traumatic brain injury in rats, alcohol exerts potent neuroprotective potential at moderate doses (1–2.5 g/kg), whereas at higher doses, aggravation of injury was observed *(69)*. This later effect is possibly related to ethanol-induced hemodynamic and respiratory depression.

## 5. INTERACTION OF ACAMPROSATE WITH THE GLUTAMATERGIC/NMDA RECEPTOR SYSTEM

Acamprosate has been reported to bind to a specific, spermidine-sensitive site ($K_d$ of 120 µ$M$ and a $B_{max}$ of 450 pmol/mg of protein) and modulate NMDA receptor function by acting as a partial coagonist. Thus, low concentrations enhanced functional [$^3$H]-MK-801 binding (under nonequilibrium conditions) when receptor activity was low (i.e., in the absence of added agonists), whereas higher concentrations (>100 µ$M$) were inhibitory under high levels of receptor activation (i.e., in the presence of 100 µ$M$ glutamate and 30 µ$M$ glycine) *(70)*. Interestingly, only the inhibitory effects of acamprosate were seen in rats made dependent on alcohol following 10 d of ethanol inhalation *(71)*. Similar effects were seen in rats that had received 400 mg/kg/d of acamprosate in their drinking water with or without concurrent ethanol inhalation for 10 d *(71)*. These results suggest that the inhibitory effects of acamprosate are more important for its therapeutic effects. However, a recent patch-clamp study showed no effect of acamprosate (0.1–300 µ$M$) on NMDA- or glutamate-induced currents in primary cultured cerebellar granule cells under control conditions or in the presence of spermine and no modification of the potency of ethanol as an NMDA receptor antagonist in these cells *(72)*. However, this study did report reversal of polyamine potentiation in a subset of cultured striatal neurons, although, again, no influence on the potency of ethanol was seen in these cells. Similarly, another recent study reported extremely weak effects of acamprosate on NMDA receptor in cultured hippocampal neurons and NR1a/2A and NR1a/2B receptors expressed in *Xenopus* oocytes or HEK-293 cells (all IC$_{50}$'s > 100 µ$M$) and no increase in potency of acamprosate following in vitro exposure of neurons to ethanol for 48 h *(73)*. This study was also unable to show any interaction of acamprosate with the polyamine site or influence on agonist affinity. However, in this same study, acamprosate produced similar increases in NR1 and NR2B receptor expression in vivo to those seen following acute treatment with (+)MK-801 and memantine, indicating that acamprosate may produce changes in the CNS that are similar to those seen following NMDA receptor antagonists and that these changes may, in turn, underlie the effects of both kinds of drugs in the treatment of alcohol abuse.

The nucleus accumbens (NAc) is a brain region thought to mediate ethanol reinforcement. Acamprosate (300 µ$M$) has been reported to selectively increase the NMDA receptor-mediated component of EPSCs recorded from NAc core neurons in vitro with no effect on resting membrane potential or the AMPA receptor-mediated component *(74)*. In the same preparation, acamprosate had little effect on postsynaptic GABA$_A$ receptors (i.e., monosynaptic IPSCs) but significantly decreased paired-pulse inhibition (PPI) in the presence of D-APV and CNQX. This latter finding was taken to imply that acamprosate may concomitantly enhance NMDA receptor-mediated excitatory transmission and disinhibit NAc core neurons by blocking presynaptic GABA$_B$ receptors *(74)*. A similar selective enhancement of NMDA receptor-mediated transmission was reported for the Schaffer collateral input onto CA1 neurons recorded in hippocampal slices *(75)*. In this study, the effects of acamprosate seemed to be mediated directly via actions at postsynaptic NMDA receptors because acamprosate (100–1000 μ$M$) dramatically increased inward current responses in most CA1 neurons to exogenous NMDA applied in the presence of TTX to block synaptic transmission. In contrast, higher concentrations of acamprosate (100–1000 μ$M$) inhibited both inhibitory and excitatory transmission in the neocortex in vitro and blocked responses to iontophoretic excitatory amino acids both in vitro and in vivo *(76)*.

Cotreatment with acamprosate (400 mg/kg/d) for 4 wk. has also been reported to block the withdrawal-induced increase in glutamate increase in the nucleus accumbens microdialysate in rats made dependent on ethanol by inhalation *(77)*. One of the known behavioral actions of acamprosate is to decrease hypermotility during alcohol withdrawal. It therefore, seems plausible that glutamate release in the nucleus accumbens could underlie hyperexcitability during ethanol withdrawal. More recent results from the same group indicate that this effect could be secondary due to an increase in the levels of taurine or GABA *(78,79)*. Although acamprosate (200 mg/kg, ip) caused an increase in c-Fos expression in the hippocampus (CA1) and the cerebellum in drug-naive animals, the same dose of acamprosate reduced elevated c-Fos mRNA levels in these structures following 24 h of ethanol withdrawal in alcohol-dependent rats or ip administration of the convulsant pentylenetetrazole *(80)*. This finding also supports the notion that acamprosate elicits its preventive effect on relapse by reducing the hyperexcitability of central neurons during withdrawal, following long-term ethanol consumption.

## 6. EFFECTS OF ACAMPROSATE ON MORPHINE DEPENDENCE

Acamprosate attenuated the expression of sensitized locomotor activity and dopamine release in the nucleus accumbens following daily injection of morphine (10 mg/kg, sc) for 14 d; however, it did not have any consistent effect on either iv heroin self-administration during the maintenance phase or the relapse to heroin seeking in a drug-free state induced by priming injections of heroin or a foot-shock stressor after a 5- to 8-d period of extinction *(81)*. This, in turn, implies that acamprosate may not have similar beneficial effects to NMDA receptor antagonists in other forms of drug-seeking behavior such as heroin or cocaine abuse (see ref. *15*). This is supported by the finding that although acamprosate has similar effects to NMDA receptor antagonists in animal models of alcohol abuse, it does not cross-discriminate for the high-affinity uncompetitive NMDA receptor antagonist (+)MK-801 and, in contrast to this agent, also does not cross-discriminate for the ethanol cue *(6,82,87)*. Moreover, it is important to note that acamprosate lacked both reinforcing properties and discriminative stimulus properties similar to D-amphetamine or pentobarbital in rhesus monkeys, suggesting that it has little or no abuse potential in it's own regard *(83,86)*. Similarly, acamprosate (170 and 320 mg/kg, ip) did not cross-discriminate for morphine or amphetamine in rats *(84)*; however, these data are questionable because the sodium salt of acamprosate was used in this study, which barely crosses the blood-brain barrier. In contrast, acquisition of conditioned place aversion by naloxone 5–6 d after the subcutaneous implantation of a 75-mg morphine pellet was completely inhibited by the pretreatment with acamprosate (200 mg/kg, ip) prior to conditioning, indicating that ethanol and opiates share similar properties in the neuronal mechanisms of conditioned withdrawal and craving *(85)*.

## 7. CONCLUSIONS AND PERSPECTIVES

In summary, recent studies have shown that acamprosate primarily interacts with the glutamatergic system. Chronic alcohol use leads to an upregulation of this system (i.e., enhanced glutamate release, less glutamate reuptake, and alterations at NMDA receptors). Although acamprosate probably has not direct antagonistic effect at NMDA receptors, it counteracts alcohol-induced alterations in the glutamatergic system. It is important to note, however, that not all patients benefit from acamprosate. The task for the future will be to identify, prior to medication, those who will respond to acamprosate and those who will not respond to this treatment. Interdisciplinary research in animals and humans has shown that relapse for alcohol involves multiple pathways with different neurobiological mechanisms. The first pathway may trigger relapse as a result of the mood-enhancing and appetitive effects of alcohol intake (alcohol-associated positive-mood states). The second pathway may solicit relapse by negative motivational states, including conditioned withdrawal and stress (alcohol-associated negative-mood states). Because acamprosate interacts primarily with the glutamatergic system and might affect alcohol-associated negative-mood states, it is hypothesized that acamprosate is most effective among individuals with alcohol-associated negative-mood states. In this respect, preclinical studies will be useful, and for this purpose, reinstatement of alcohol-seeking behavior in long-term alcohol-experienced rats will be studied—a procedure recognized as a model of craving and relapse. Reinstatement will be tested in response to different classes of craving and relapse inducing stimuli. In particular, alcohol-associated positive- and negative-mood states will be induced and acamprosate treatment will be matched to specific relapse-inducing stimuli. This procedure will allow us to identify responders to acamprosate treatment. This knowledge can subsequently be transferred to the human situation and will lead to better treatment success.

## REFERENCES

1. Garbutt, J. C., West, S. L., Carey, T. S., Lohr, K. N., and Crews, F. T. (1999) Pharmacological treatment of alcohol dependence: a review of the evidence. *J. Am. Med. Assoc.* **281,** 1318–1325.
2. Littleton, J. (1995) Acamprosate in alcohol dependence: how does it work? *Addiction* **90,** 1179–1188.
3. Spanagel, R. and Zieglgänsberger, W. (1997) Anti-craving compounds for ethanol: new pharmacological tools to study addictive processes. *Trends Pharmacol. Sci.* **18,** 54–59.
4. Spanagel, R. and Holter, S. M. (1999) Long-term alcohol self-administration with repeated alcohol deprivation phases: an animal model of alcoholism? *Alcohol Alcohol.* **34,** 231–243.
5. Spanagel, R. and Holter, S. M. (2000) Pharmacological validation of a new animal model of alcoholism, *J. Neural Transm.* **107,** 669–680.
6. Spanagel, R., Holter, S. M., Allingham, K., Landgraf, R., and Zieglgansberger, W. (1996) Acamprosate and alcohol: I. Effects on alcohol intake following alcohol deprivation in the rat. *Eur. J. Pharmacol.* **305,** 39–44.
7. Holter, S. M., Landgraf, R., Zieglgänsberger, W., and Spanagel, R. (1997) Time course of acamprosate action on operant ethanol self-administration after ethanol deprivation. *Alcohol. Clin. Exp. Res.* **21,** 862–868.
8. Heyser, C. J., Schulteis, G., Durbin, P., and Koob, G. F. (1998) Chronic acamprosate eliminates the alcohol deprivation effect while having limited effects on baseline responding for ethanol in rats. *Neuropsychopharmacology.* **18,** 125–133.
9. Cole, J. C., Littleton, J. M., and Little, H. J. (2000) Acamprosate, but not naltrexone, inhibits conditioned abstinence behaviour associated with repeated ethanol administration and exposure to a plus-maze. *Psychopharmacology* **147,** 403–411.
10. Boismare, F., Daoust, M., Moore, N., et al. (1984) A homotaurine derivative reduces the voluntary intake of ethanol by rats: are cerebral GABA receptors involved? *Pharmacol. Biochem. Behav.* **21,** 787–789.
11. Daoust, M., Lhuintre, J. P., Saligaut, C., Moore, N., Flipo, J. L., Boismare, F. (1987) Noradrenaline and GABA brain receptors are co-involved in the voluntary intake of ethanol by rats. *Alcohol Alcohol.* **1(Suppl.),** 319–322.
12. Daoust, M., Legrand, E., Gewiss, M., et al. (1992) Acamprosate modulates synaptosomal GABA transmission in chronically alcoholised rats. *Pharmacol. Biochem. Behav.* **41,** 669–674.
13. Rassnick, S., D'Amico, E., Riley, E., Pulvirenti, L., Zieglgänsberger, W., and Koob, G. F. (1992) GABA and nucleus accumbens glutamate neurotransmission modulate ethanol self-administration in rats. *Ann. NY Acad. Sci.* **654,** 502–505.
14. Danysz, W., Parsons, C. G., Bresink, I., and Quack, G. (1995) Glutamate in CNS disorders: a revived target for drug development? *Drug News Perspect.* **8,** 261–277.
15. Parsons, C. G., Danysz, W., and Quack, G. (1998) Glutamate in CNS disorders as a target for drug development: an update. *Drug News Perspect.* **11,** 523–569.
16. Danysz, W. and Parsons, C. G. (1998) Glycine and N-methyl-D-aspartate receptors: physiological significance and possible therapeutic applications. *Pharmacol. Rev.* **50,** 597–664.
17. Hollmann, M. and Heinemann, S. (1994) Cloned glutamate receptors. *Annu. Rev. Neurosci.* **17,** 31–108.
18. McBain, C. J. and Mayer, M. L. (1994) N-Methyl-D-aspartic acid receptor structure and function. *Physiol. Rev.* **74,** 723–760.

19. Seeburg, P. H. (1993) The molecular biology of mammalian glutamate receptor channels. *Trends Pharmacol. Sci.* **14**, 297–303.
20. Laube, B., Hirai, H., Sturges, M., Betz, H., and Kuhse, J. (1997) Molecular determinants of agonist discrimination by NMDA receptor subunits: analysis of the glutamate binding site on the NR2B subunit. *Neuron* **18**, 493–503.
21. Hirai, H., Kirsch, J., Laube, B., Betz, H., and Kuhse, J., (1996) The glycine binding site of the *N*-methyl-D-aspartate receptor subunit NR1: identification of novel determinants of co-agonist potentiation in the extracellular M3-M4 loop region. *Proc. Natl. Acad. Sci. USA* **93**, 6031–6036.
22. Das, S., Sasaki, Y. F., Rothe, T., et al. (1998) Increased NMDA current and spine density in mice lacking the NMDA receptor subunit NR3A. *Nature* **393**, 377–381.
23. Durand, G. M., Bennett, M. V., and Zukin, R. S. (1993) Splice variants of the *N*-methyl-D-aspartate receptor NR1 identify domains involved in regulation by polyamines and protein kinase C. [published erratum appeared in *Proc. Natl. Acad. Sci. USA* 1993, **90**:9739] *Proc. Natl. Acad. Sci. USA* **90**, 6731–6735.
24. Zukin, R. S. and Bennett, M. V. L. (1995) Alternatively spliced isoforms of the NMDARI receptor subunit. *Trends Neurosci.* **18**, 306–313.
25. Barnard, E. A. (1997). Ionotropic glutamate receptors—new types and new concepts. *Trends Pharmacol. Sci.* **18**, 141–148.
26. Fadda, E., Danysz, W., Wroblewski, J. T., and Costa, E. (1988) Glycine and D-serine increase the affinity of the *N*-methyl-D-aspartate sensitive glutamate binding sites in rat brain synaptic membranes. *Neuropharmacology* **27**, 1183–1185.
27. Kleckner, N. W. and Dingledine, R. (1988) Requirement for glycine in activation of NMDA receptors expressed in *Xenopus* oocytes. *Science* **214**, 835–837.
28. Johnson, J. W. and Ascher, P. (1987) Glycine potentiates the NMDA response in cultured mouse brain neurones. *Nature* **325**, 529–531.
29. Mayer, M. L., Vyklicky, L. J., and Sernagor, E. (1989) A physiologist's view of the *N*-methyl-D-aspartate receptor: an allosteric ion channel with multiple regulatory sites. *Drug Dev. Res.* **17**, 263–280.
30. Mayer, M. L., Vyklicky, L. J., and Clements, J. (1989) Regulation of NMDA receptor desensitization in mouse hippocampal neurons by glycine. *Nature* **338**, 425–427.
31. Parsons, C. G., Zong, X. G., and Lux, H. D., (1993) Whole cell and single channel analysis of the kinetics of glycine-sensitive *N*-methyl-D-aspartate receptor desensitization. *Br. J. Pharmacol.* **109**, 213–221.
32. Hashimoto, A. and Oka, T. (1997) Free D-aspartate and D-serine in the mammalian brain and periphery. *Prog. Neurobiol.* **52**, 325–353.
33. Williams, K. (1997) Modulation and block of ion channels: a new biology of polyamines. *Cell Signal* **9**, 1–13.
34. Johnson, T. D. (1996) Modulation of channel function by polyamines. *Trends Pharmacol. Sci.* **17**, 22–27.
35. Kashiwagi, K., Pahk, A. J., Masuko, T., Igarashi, K., and Williams, K. (1997) Block and modulation of *N*-methyl-D-aspartate receptors by polyamines and protons: role of amino acid residues in the transmembrane and pore-forming regions of NR1 and NR2 subunits. *Mol. Pharmacol.* **52**, 701–713.
36. Williams, K., Zappia, A. M., Pritchett, D. B., Shen, Y. M., and Molinoff, P. B. (1994) Sensitivity of the *N*-methyl-D-aspartate receptor to polyamines is controlled by NR2 subunits. *Mol. Pharmacol.* **45**, 803–809.
37. Williams, K. (1994) Mechanisms influencing stimulatory effects of spermine at recombinant *N*-methyl-D-aspartate receptors. *Mol. Pharmacol.* **46**, 161–168.
38. Carter, C., Benavides, J., Legendre, P., et al. (1998) Ifenprodil and SL 82.0715 as cerebral anti-ischemic agents. II. Evidence for *N*-methyl-D-aspartate receptor antagonist properties. *J. Phramacol. Exp. Ther.* **247**, 1222–1232.
39. Gotti, B., Duverger, D., Bertin, J., et al. (1988) Ifenprodil and SL 82.0715 as cerebral anti-ischemic agents. I. Evidence for efficacy in models of focal cerebral ischemia. *J. Pharmacol. Exp. Ther.* **247**, 1211–1221.
40. Carter, C., Rivy, J. P., and Scatton, B. (1989) Ifenprodil and SL 82.0715 are antagonists at the polyamine site of the *N*-methyl-D-aspartate (NMDA) receptor. *Eur. J. Pharmacol.* **164**, 611–612.
41. Legendre, P. and Westbrook, G. L. (1991) Ifenprodil blocks *N*-methyl-D-aspartate receptors by a two-component mechanism. *Mol. Pharmacol.* **40**, 289–298.
42. Williams, K. (1993) Ifenprodil discriminates subtypes of the *N*-methyl-D-aspartate receptor—selectivity and mechanisms at recombinant heteromeric receptors. *Mol. Pharmacol.* **44**, 851–859.
43. Buller, A. L., Larson, H. C., Morrisett, R. A., and Monaghan, D. T. (1995) Glycine modulates ethanol inhibition of heteromeric *N*-methyl-D-aspartate receptors expressed in *Xenopus* oocytes. *Mol. Pharmacol.* **48**, 717–723.
44. Morrisett, R. A. and Swartzwelder, H. S. (1993) Attenuation of hippocampal long-term potentiation by ethanol: a patch-clamp analysis of glutamatergic and GABAergic mechanisms. *J. Neurosci.* **13**, 2264–2272.
45. Mirshahi, T. and Woodward, J. J. (1995) Ethanol sensitivity of heteromeric NMDA receptors: effects of subunit assembly, glycine and NMDAR1 $Mg^{2+}$-insensitive mutants. *Neuropharmacology* **34**, 347–355.
46. Lovinger, D. M. and Zieglgänsberger, W. (1996) Interactions between ethanol and agents that act on the NMDA-type glutamate receptor. *Alcohol. Clin. Exp. Res.* **20**, A187–A191.
47. Yang, X. H., Criswell, H. E., Simson, P., Moy, S., and Breese, G. R. (1996) Evidence for a selective effect of ethanol on *N*-methyl-D-aspartate responses: ethanol affects a subtype of the ifenprodil-sensitive *N*-methyl-D-aspartate receptors. *J. Pharmacol. Exp. Ther.* **278**, 114–124.

48. Hu, X. J., Follesa, P., and Ticku, M. K. (1996) Chronic ethanol treatment produces a selective upregulation of the NMDA receptor subunit gene expression in mammalian cultured cortical neurons. *Mol. Brain Res.* **36,** 211–218.
49. Blevins, T., Mirshahi, T., and Woodward, J. J. (1995) Increased agonist and antagonist sensitivity of *N*-methyl-D-aspartate stimulated calcium flux in cultured neurons following chronic ethanol exposure. *Neurosci. Lett.* **200,** 214–218.
50. Follesa, P. and Ticku, M. K. (1996) Chronic ethanol-mediated up-regulation of the *N*-methyl-D-aspartate receptor polypeptide subunits in mouse cortical neurons in culture. *J. Biol. Chem.* **271,** 13,297–13,299.
51. Follesa, P. and Ticku, M. K. (1996) NMDA receptor upregulation: molecular studies in cultured mouse cortical neurons after chronic antagonist exposure. *J. Neurosci.* **16,** 2172–2178.
52. Hoffman, P. L., Bhave, S. V., Kumar, K. N., Iorio, K. R., Snell, L. D., Tabakoff, B., et al. (1996) The 71 kDa glutamate-binding protein is increased in cerebellar granule cells after chronic ethanol treatment. *Mol. Brain Res.* **39,** 167–176.
53. Chandler, L. J., Sutton, G., Norwood, D., Sumners, C., and Crews, F. T. (1997) Chronic ethanol increases *N*-methyl-D-aspartate-stimulated nitric oxide formation but not receptor density in cultured cortical neurons. *Mol. Pharmacol.* **51,** 733–740.
54. Nagata, K., Aistrup, G. L., Huang, C. S., et al. (1996) Potent modulation of neuronal nicotinic acetylcholine receptor-channel by ethanol. *Neurosci. Lett.* **217,** 189–193.
55. Follesa, P. and Ticku, M. K. (1995) Chronic ethanol treatment differentially regulates NMDA receptor subunit mRNA expression in rat brain. *Mol. Brain Res.* **29,** 99–106.
56. Freund, G. and Anderson, K. J. (1996) Glutamate receptors in the frontal cortex of alcoholics. *Alcohol. Clin. Exp. Res.* **20,** 1165–1172.
57. Breese, C. R., Freedman, R., and Leonard, S. S. (1995) Glutamate receptor subtype expression in human postmortem brain tissue from schizophrenics and alcohol abusers. *Brain Res.* **674,** 82–90.
58. Rossetti, Z. L. and Carboni, S. (1995) Ethanol withdrawal is associated with increased extracellular glutamate in the rat striatum. *Eur. J. Pharmacol.* **283,** 177–183.
59. Molleman, A. and Little, H. J. (1995) Increases in non-*N*-methyl-D-aspartate glutamatergic transmission, but no change in gamma-aminobutyric acid (B) transmission, in CA1 neurons during withdrawal from in vivo chronic ethanol treatment. *J. Pharmacol. Exp. Ther.* **274,** 1035–1041.
60. Whittington, M. A., Lambert, J. D. C., and Little, H. J. (1995) Increased NMDA receptor and calcium channel activity underlying ethanol withdrawal hyperexcitability. *Alcohol Alcohol.* **30,** 105–114.
61. Karcz-Kubicha, M. and Liljequist, S. (1995) Effects of post ethanol administration of nmda and non-NMDA receptor antagonists on the development of ethanol tolerance in c57bi mice. *Psychopharmacology.* **120,** 49–56.
62. Khanna, J. M., Morato, G. S., Chau, A., and Shah, G. (1995) D-Cycloserine enhances rapid tolerance to ethanol motor incoordination. *Pharmacol. Biochem. Behav.* **52,** 609–614.
63. Holter, S. M., Danysz, W., and Spanagel, R. (1996) Evidence for alcohol anti-craving properties of memantine. *Eur. J. Pharmacol.* **314,** R1–R2.
64. Hundt, W., Danysz, W., Holter, S. M., and Spanagel, R. (1998) Ethanol and *N*-methyl-D-aspartate receptor complex interactions: a detailed drug discrimination study in the rat. *Psychopharmacology* **135,** 44–51.
65. Bienkowski, P., Stefanski, R., and Kostowski, W. (1997) Discriminative stimulus effects of ethanol: lack of antagonism with *N*-methyl-D-aspartate and D-cycloserine. *Alcohol* **14,** 345–350.
66. Lin, N. and Hubbard, J.I. (1995) An NMDA receptor antagonist reduces ethanol preference in untrained but not trained rats. *Brain Res. Bull.* **36,** 421–424.
67. Hazell, A. S., Butterworth, R. F., and Hakim, A. M. (1993) Cerebral vulnerability is associated with selective increase in extracellular glutamate concentration in experimental thiamine deficiency. *J. Neurochem.* **61,** 1155–1158.
68. Dodd, P. R., Thomas, G. J., McCloskey, A., Crane, D. I., and Smith, I.D. (1996) The neurochemical pathology of thiamine deficiency: GABA(A) and glutamate (NMDA) receptor binding sites in a goat model. *Metab. Brain. Dis.* **11,** 39–54.
69. Kelly, D. F., Lee, S. M., Pinanong, P. A., and Hovda, D. A. (1997) Paradoxical effects of acute ethanolism in experimental brain injury. *J. Neurosurg.* **86,** 876–882.
70. Naassila, M., Hammoumi, S., Legrand, E., Durbin, P., and Daoust, M. (1998) Mechanism of action of acamprosate. Part I. Characterization of spermidine-sensitive acamprosate binding site in rat brain. *Alcohol. Clin. Exp. Res.* **22,** 802–809.
71. al Qatari, M., Bouchenafa, O., and Littleton, J. (1998) Mechanism of action of acamprosate. Part II. Ethanol dependence modifies effects of acamprosate on NMDA receptor binding in membranes from rat cerebral cortex. *Alcohol. Clin. Exp. Res.* **22,** 810–814.
72. Popp, R. L. and Lovinger, D. M. (2000) Interaction of acamprosate with ethanol and spermine on NMDA receptors in primary cultured neurons. *Eur. J. Pharmacol.* **394,** 221–231.
73. Rammes G., Mahal B., Putzke J., et al. (2001) The anti-craving compound acamprosate acts as a weak NMDA-receptor antagonist, but modulates NMDA-receptor subunit expression similar to memantine and MK-801. Neuropharmacology **40,** 749–760.
74. Berton, F., Francesconi, W. G., Madamba, S. G., Zieglgänsberger, W., and Siggins, G. R. (1998) Acamprosate enhances *N*-methyl-D-apartate receptor-mediated neurotransmission but inhibits presynaptic GABA(B) receptors in nucleus accumbens neurons. *Alcohol. Clin. Exp. Res.* **22,** 183–191.

75. Madamba, S. G., Schweitzer, P., Zieglgänsberger, W., and Siggins, G. R. (1996) Acamprosate (calcium acetylhomotaurinate) enhances the N-methyl-D-aspartate component of excitatory neurotransmission in rat hippocampal CA1 neurons in vitro. *Alcohol. Clin. Exp. Res.* **20,** 651–658.
76. Zeise, M. L., Kasparov, S., Capogna, M. and Zieglgansberger, W. (1993) Acamprosate (calciumacetylhomotaurinate) decreases postsynaptic potentials in the rat neocortex: possible involvement of excitatory amino acid receptors. *Eur. J. Pharmacol.* **231,** 47–52.
77. Dahchour, A., DeWitte, P., Bolo, N., et al. (1998) Central effects of acamprosate: Part 1. Acamprosate blocks the glutamate increase in the nucleus accumbens microdialysate in ethanol withdrawn rats. *Psychiat. Res. Neuroimag.* **82,** 107–114.
78. Dahchour, A., and De Witte, P. (1999) Acamprosate decreases the hypermotility during repeated ethanol withdrawal. *Alcohol* **18,** 77–81.
79. Dahchour, A. and De Witte, P. (2000) Ethanol and amino acids in the central nervous system: assessment of the pharmacological actions of acamprosate. *Prog. Neurobiol.* **60,** 343–362.
80. Putzke, J., Spanagel, R., Tolle, T. R., and Zieglgänsberger, W. (1996) The anti-craving drug acamprosate reduces c-fos expression in rats undergoing ethanol withdrawal. *Eur. J. Pharmacol.* **317,** 39–48.
81. Spanagel, R., Sillaber, I., Zieglgänsberger, W., Corrigall, W. A., Stewart, J., and Shaham, Y. (1998) Acamprosate suppresses the expression of morphine-induced sensitization in rats but does not affect heroin self-administration or relapse induced by heroin or stress. *Psychopharmacology* **139,** 391–401.
82. Spanagel, R., Putzke, J., Stefferl, A., Schobitz, B., and Zieglgansberger, W. (1996) Acamprosate and alcohol: II. Effects on alcohol withdrawal in the rat. *Eur. J. Pharmacol.* **305,** 45–50.
83. Grant, K. A. and Woolverton, W. L. (1989) Reinforcing and discriminative stimulus effects of Ca-acetyl homotaurine in animals. *Pharmacol. Biochem. Behav.* **32,** 607–611.
84. Pascucci, T., Cioli, I., Pisetzky, F., Dupre, S., Spirito, A., and Nencini, P. (1999) Acamprosate does not antagonise the discriminative stimulus properties of amphetamine and morphine in rats. *Pharmacol. Res.* **40,** 333–338.
85. Kratzer, U. and Schmidt, W. J. (1998) The anti-craving drug acamprosate inhibits the conditioned place aversion induced by naloxone-precipitated morphine withdrawal in rats. *Neurosci. Lett.* **252,** 53–56.
86. Schneider, U., Wohlfahrt, K., SchulzeBonhage, A., Haacker, T., Caspary, A., Zedler, M., et al. (1998) Lack of psychotomimetic or impairing effects on psychomotor performance of acamprosate. *Pharmacopsychiatry* **31,** 110–113.
87. Spanagel, R., Zieglgänsberger, W., and Hundt, W. (1996) Acamprosate and alcohol: III. Effects on alcohol discrimination in the rat. *Eur. J. Pharmacol.* **305,** 51–56.

# 28
# The NMDA/Nitric Oxide Synthase Cascade in Opioid Analgesia and Tolerance

## Gavril W. Pasternak, MD, PhD and Yuri Kolesnikov, MD, PhD

## 1. INTRODUCTION

Despite their widespread use in the management of pain, opioids have a number of issues that limit their overall utility. Chronic use of opioids is associated with a progressive decline in their analgesic efficacy, an effect commonly termed tolerance. Although analgesic responses can usually be regained with escalation of the dose, the therapeutic index of these drugs decreases with tolerance because tolerance develops to various opioid actions at different rates, making the clinical management of patients more difficult. Thus, understanding the ways in which tolerance can either be eliminated or diminished would be a significant advantage in the therapeutic use of these agents.

## 2. NMDA/ NITRIC OXIDE CASCADE IN OPIOID TOLERANCE AND DEPENDENCE

A number of years ago, several laboratories reported that opioid tolerance could be impeded by the blockade of *N*-methyl-D-aspartate (NMDA) receptors *(1,2)*, findings that were quickly confirmed by a large number of laboratories *(3–8)*. In brief, the blockade of NMDA receptors prevents tolerance without interfering with opioid analgesia. This is a remarkable observation because it clearly dissociates the mechanisms responsible for the two. A wide range of NMDA antagonists are effective, including those that are both competitive and noncompetitive and even agents working on the glycine regulatory site. In general, we have found marked similarities between µ and δ systems, with δ tolerance showing the same reversal with NMDA *(4,8–11)*.

*N*-Methyl-D-aspartate receptors are closely associated with nitric oxide synthase (NOS) *(12,13)*, leading to the question of whether nitric oxide (NO) also is involved in modulation of opioid tolerance. Shortly after the initial reports with NMDA antagonists, we explored the involvement of nitric oxide using inhibitors of the enzyme nitric oxide synthase (NOS), including $N^G$-nitroarginine (NOArg). Whereas the response to a fixed morphine dose alone progressively declined and was not observed after 5 d, coadministration of NOArg maintained significant analgesia for 4 wk (Fig. 1). Chronic administration of morphine alone in this paradigm led to a twofold shift of the dose-response curve by 5 d (Table 1). Even though we did not see any analgesia at this morphine dose after 5 d, continued administration of morphine increased the levels of tolerance even further, as shown by the progressive increase in the median effective dose ($ED_{50}$) values at 10 and 28 d to fourfold and eightfold, respectively (Table 1). Yet, after 10 d of morphine coadministered with NOArg, the $ED_{50}$ value was virtually the same as in naive mice. Tolerance to either the $\kappa_1$ drug U50, 488H or the $\kappa_3$ agent naloxone benzoylhydrazone (NalBzoH) were not affected by NOArg in this paradigm, implying that κ tolerance

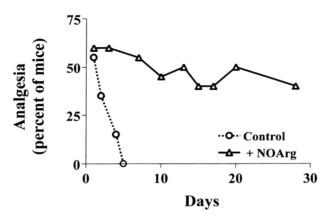

**Fig. 1.** Effect of NOArg on morphine analgesia. Mice received morphine (5 mg/kg, sc) alone or in combination with NOArg. (2 mg/kg, sc) once daily. By d 10, the morphine-alone group differed significantly from the morphine + NOArg group ($p < 0.001$). (Data from ref. *14*).

Table 1
**Effects of Modulators of NOS on Morphine Analgesia**

| Treatment | Morphine $ED_{50}$ value (mg/Kg, sc) | |
|---|---|---|
| | Without NOArg | With NOArg |
| Morphine | | |
| Acute | 4.4 (3.6, 5.3) | 3.9 (2.8, 5.4) |
| 5 d | 8.8 (6.2, 11) | 4.6 (3.1, 6.7) |
| 10 d | 17 (12, 24) | 5.8 (4.0, 8.7) |
| 28 d | 38 (27, 51) | |
| L-Arginine | | |
| Acute | 6.6 (4.8, 8.5) | |
| 3 d | 9.9 (7.5, 13) | |
| 5 d | 10 (7.0, 14) | 4.0 (3.6, 5.7) |
| 10 d | 8.7 (6.1, 12) | |
| D-Arginine | | |
| Acute | 4.6 (2.7, 5.9) | |
| 5 d | 4.4 (2.4, 6.1) | |

*Source:* Data from refs. *14* and *15*.

involved different mechanisms. As noted earlier, δ tolerance showed the same sensitivity to these treatments as morphine *(4,8–11)*.

Because blockade of NOS potentiated opioid action over time, eliminating tolerance, we also explored whether the opposite was true. L-Arginine is the natural substrate for the production of NO and its administration can induce NO formation. A single dose of L-arginine lowered morphine analgesic activity, but the effect was more pronounced after L-arginine was given alone over a few days (Table 1). Coadminsitration of NOArg, however, prevented the effects of L-arginine, confirming its actions through NOS. As expected, D-arginine, which is not a substrate for NOS, was ineffective. Again, we observed differences between μ and κ systems, with κ systems insensitive to these treatments.

Can tolerance be reversed once it develops? If NOArg is added to the chronic morphine treatment after tolerance is established, the animals regained their sensitivity toward morphine despite

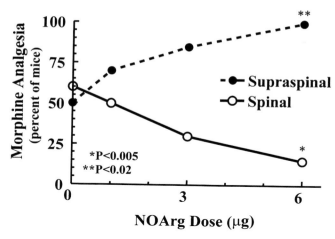

**Fig. 2.** Effect of acute spinal and supraspinal NOArg on morphine analgesia. Groups of mice received the indicated dose of NOArg either spinally or supraspinally, followed by morphine (5 mg/kg, sc) 30 min later. Results are from ref. 20.)

its continued administration. Thus, the blockade of NOS receptor blockade restored analgesic sensitivity in morphine-tolerant animals. In addition to opening a novel therapeutic approach toward ameliorating tolerance, these observations also have implications with regard to the mechanism of tolerance. It suggests that tolerance is not a static state, but rather a dynamic one balancing factors that are inducing it and others acting to restore analgesic sensitivity.

Physical dependence has long been associated with tolerance, although there are suggestions that their mechanisms may differ (16,17). NOArg also attenuated signs of morphine dependence in mice, although higher doses of drug seemed to be needed (14). Others reported similar findings (18,19).

## 3. ANTISENSE MAPPING NOS AND OPIOID TOLERANCE

Although NOArg and other NOS inhibitors clearly implied a role for nitric oxide in morphine tolerance, some observations were not easily reconciled. Given supraspinally, NOArg potentiated systemic morphine analgesia. However, spinal NOArg had the opposite effect, lowering systemic morphine analgesia in a dose-dependent manner (Fig. 2) (20). Earlier work from our laboratory had found no effect of NOArg given systemically on morphine analgesia, presumably the result of the simultaneous inhibition of the opposing spinal and supraspinal systems. How could these opposite actions between supraspinal and spinal systems be reconciled? Supraspinally, NO impedes morphine analgesia, as shown by the acute effects of supraspinal NOArg and its role in inducing tolerance. Yet, at the spinal level, NO is integrally related to the production of morphine analgesia because its blockade impairs morphine's analgesic response.

These differences between spinal and supraspinal NOS actions may be simply because of the circuitry of the region. However, it also was possible that these actions might be mediated through different enzymes, particularly because a number of nNOS splice variants have been reported (21–23). By designing antisense probes against individual exons, it is possible to selectively downregulate individual splice variants, a technique termed antisense mapping (24–25). We compared the pharmacology of a major isoform of neuronal nitric oxide synthase (nNOS-1) with a variant lacking exons 9 and 10 (nNOS-2). By specifically targeting exon 10 with an antisense probe (Fig. 3A), we were able to downregulate the major isoform (nNOS-1), as determined by reverse transcription – polymerase chain reaction (RT-PCR). Furthermore, this did not interfere with the levels of the other splice variant (nNOS-2)

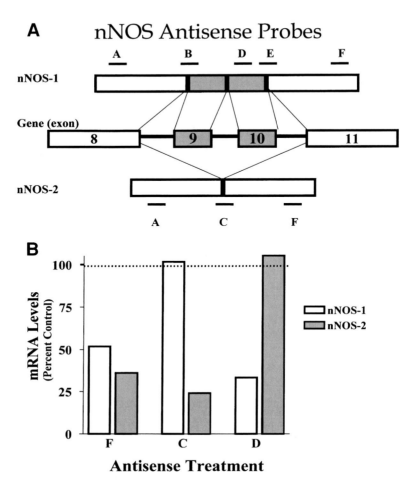

**Fig. 3.** Antisense mapping nNOS-1 and nNOS-2. **(A)** A schematic of the nNOS gene is presented that illustrates the portion of the gene that differs between nNOS-1 and nNOS-2. The sites targeted by the antisense probes are indicated by the heavy lines. Note: Antisense C spans the splice site between exons 8 and 11. **(B)** RT-PCR was performed on extracts of the periventricular gray following antisense treatment with the indicated antisense probe. The bands were digitized and plotted as percent of control. (From ref. 20.)

that lacked exons 9 and 10 (Fig. 3B). Selectively downregulating nNOS-2 proved more difficult because all of the sequences in this variant were also contained within the sequence of nNOS-1. However, we were able to accomplish this by designing a probe across the exon 8/11 splice junction. The antisense probe effectively downregulated nNOS-2, but not nNOS-1, mRNA (Fig. 3B). Presumably, the length of sequence on either side of the splice site alone capable of annealing to the antisense oligodeoxynucleotide was insufficient for activity and the probe required the full sequence, which was only available in nNOS-2.

Antisense mapping defined interesting, and distinct, pharmacological profiles for the two variants. Downregulation of nNOS-2 either spinally or supraspinally markedly impaired morphine analgesia acutely (Fig. 4), implying that nNOS-2 was important in the production of morphine analgesia. In contrast, the antisense probes selective for nNOS-1 all blocked morphine tolerance without interfering with its analgesic actions (Fig. 5). Thus, antisense mapping has defined two distinct

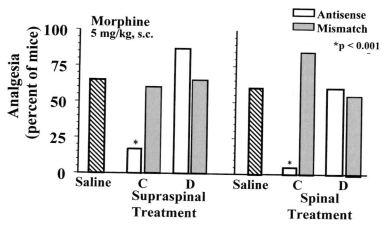

**Fig. 4.** Effects of nNOS antisense probes on morphine analgesia. Groups of animals were treated with the indicated antisense or mismatch (5 μg) treatment either supraspinally or spinally on d 1, 3, and 5 and then tested with morphine (5 mg/kg, sc). Antisense C, selective for nNOS-2, lowered morphine analgesia in both regions ($p < 0.001$). Results are from ref. 20.)

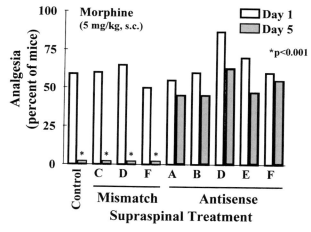

**Fig. 5.** Effects of nNOS antisense probes on morphine tolerance. Groups of animals were treated with the indicated antisense or mismatch (5 μg) treatment supraspinally on d 1, 3, and 5. On the last day of oligodeoxynucleotide treatment (d 1), the mice received daily morphine injections (5 mg/kg, sc). Analgesic responses following the d 1 and the d 5 morphine doses are presented. Tolerance developed in the mismatch animals ($p < 0.001$), but not in the antisense animals. Note that antisense C was not examined because it lowered morphine analgesia acutely, making its interpretation quite difficult. (Data from ref. 20.)

nitric oxide systems. Activation of nNOS-2 was important in producing morphine analgesia, whereas activation of nNOS-1 acted in an opposite manner by diminishing morphine's analgesic responses. Thus, the effects of NOS inhibitors such as NOArg are due to the simultaneous inactivation of both systems, with the pharmacological responses dependent on the relative importance of each in the region examined. Presumably, the actions of nNOS-2 predominate spinally because the NOS inhibitor given spinally impairs morphine analgesia, whereas nNOS-1 predominates supraspinally because the NOS inhibitor given supraspinally enhances morphine responses.

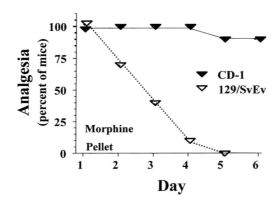

**Fig. 6.** Effects of chronic morphine in CD-1 and 129/SvEv mice. Groups of mice received morphine pellets (75 mg free base) on d 1 and analgesia was assessed at the indicated time. Results are from ref. 26.

## 4. LACK OF TOLERANCE IN 129/SvEv MICE

The earlier studies showed the importance of the NMDA receptor/nitric oxide synthase cascade when they were dissected pharmacologically. Recently we demonstratd that defects in this cascade occurred naturally (26), results that have since been confirmed (27). Most knockout mice strains are generated from ES cells from the 129/SvEv strain and we wanted to confirm its genetic background pharmacologically before embarking upon detailed tolerance studies in knockout mice. We were quite surprised to see that chronic administration of morphine in the 129/SvEv animals did not result in tolerance, regardless of whether morphine was given by pellet (Fig. 6), systemic injection, or supraspinally (26). Furthermore, tolerance to the δ-selective drug [D-Pen$^2$,D-Pen$^5$]enkephalin (DPDPE) also was lost in the 129/SvEv mice. In our earlier study, blockade of NOS prevented tolerance to morphine, but not κ drugs. Similarly, 129/SvEv mice developed tolerance to the $\kappa_1$ agent U50, 488H and the $\kappa_3$ drug NalBzoH at a rate indistinguishable from the traditional strains. Thus, the 129/SvEv mice show the same pharmacological profile as previously seen with NMDA antagonists and NOS inhibitors.

We then further examined the site of the defect along the NMDA/NO cascade. In CD-1 mice, NMDA given for 5 d shifted the dose-response curve fourfold. In contrast, the NMDA treatment had little demonstrable effect in the 129/SvEv mice, with only a 1.1-fold shift. We explored the actions of nitric oxide by administering L-arginine, the precursor that enhances the generation of nitric oxide. After 3 d, L-arginine shifted the morphine dose-response curve 2.7-fold to the right in the 129/SvEv mice, a response quite similar to that of the CD-1 mice (Table 2). Together, these findings imply that the nitric oxide synthase system is intact and the defect exists at either the NMDA receptor itself or a site further downstream, but before nitric oxide synthase. Thus, defects in the NMDA/NO cascade exist naturally, which raises interesting questions regarding genetic differences in the development of tolerance that may extend into the clinical arena.

## 5. CONCLUSIONS

The NMDA receptor/nitric oxide synthase cascade has been implicated in many functions, including opioid tolerance. A variety of NMDA antagonists effectively block tolerance in animal models, including both competitive and noncompetitive antagonists and agents acting through the glycine site. Similarly, studies also have implicated the major variant of nNOS, nNOS-1, in the development of tolerance. It was interesting to find that a different nNOS variant, nNOS-2, had opposing actions.

### Table 2
### Effects of NMDA and L-Arginine on Morphine Analgesia in CD-1 and 129/SvEv Mice

| | Morphine $ED_{50}$ (mg/kg, sc) | | | |
|---|---|---|---|---|
| | CD-1 | | 129/SvEv | |
| | $ED_{50}$ value | Shift | $ED_{50}$ value | Shift |
| Naive | 3.1 (1.6, 4.4) | | 2.0 (1.4, 2.8) | |
| NMDA | 12.4 (6.1, 14.8) | 4 | 2.2 (1.2, 3.6) | 1.1 |
| L-Arginine | 9.9 (7.5, 12.5) | 3.2 | 5.2 (2.1, 9.1) | 2.7 |

*Note:* Mice were treated with either nothing (naive), NMDA (1 mg/kg for 5 d), or L-arginine (50 mg/kg, ip for 3 d) and then examined for morphine analgesia.

*Source:* Data from ref. 26.

Because NOS inhibitors do not distinguish between these variants, their actions reflect the blockade of both systems and the pharmacological activity is likely the result of the net difference between the two systems. Clearly, selective inhibitors of these variants would be interesting pharmacologically. A major question is whether this cascade is a therapeutic target clinically. A number of NMDA receptor antagonists have been developed, but they have not been widely used because of psychotomimetic side effects. Nitric oxide synthase inhibitors have the potential of serious vascular adverse effects. Yet, the more basic question is the overall importance of the NMDA/NOS cascade in opioid tolerance. It is important to note that most studies documenting the role of the cascade in tolerance have examined low levels of opioid tolerance. It is not entirely clear that the same mechanisms are important in more profound tolerance, as is often seen in clinical situations. Opioid tolerance is likely to involve a vast array of mechanisms, ranging from the receptor itself, to transduction systems within the cell, to circuitry. Thus, the question of the therapeutic potential of blockers of this cascade remains open. However, it is tantalizing to consider the possibility that analgesic tolerance can be prevented through inhibition of this cascade while tolerance still develops to the side effects associated with opioid use. For example, inhibition of gastrointestinal motility and respiratory depression undergo tolerance more slowing than analgesia, implying that they may involve different mechanisms of tolerance.

## ACKNOWLEDGMENTS

This work was supported by grants from the National Institute on Drug Abuse (DA06241 and DA07242) to GWP, who is supported by a Senior Scientist Award (DA00220) and a Mentored Scientist Award (DA00405) to YAK, as well as a Core Grant from the National Cancer Institute (CA8748) to Memorial Sloan-Kettering Cancer Center.

## REFERENCES

1. Ben-Eliyahu, S., Marek, P., Vaccarino, A. L., Mogil, J. S., Sternberg, W. F., and Liebeskind, J. C. (1992) The NMDA receptor antagonist MK-801 prevents long-lasting non-associative morphine tolerance in the rat. *Brain Res.* **575,** 304–308.
2. Trujillo, K. A. and Akil, H. (1991) Inhibition of morphine tolerance and dependence by the NMDA receptor anagonist MK-801. *Science* **251,** 85–87.
3. Gutstein, H. B. and Trujillo, K. A. (1993) MK-801 inhibits the development of morphine tolerance at spinal sites. *Brain Res.* **626,** 332–334.
4. Kolesnikov, Y. A., Maccechini, M.-L., and Pasternak, G. W. (1994) 1-Aminocyclopropane carboxylic acid (ACPC) prevents mu and delta opioid tolerance. *Life Sci.* **55,** 1393–1398.
5. Kolesnikov, Y. A., Ferkany, J., and Pasternak, G. W. (1993) Blockade of mu and kappa$_1$ opioid analgesic tolerance by NPC17742, a novel NMDA antagonist. *Life Sci.* **53,** 1489–1494.
6. Tiseo, P. and Inturrisi, C. E. (1993) Attenuation and reversal of morphine tolerance by the competitive *N*-methyl-D-aspartate antagonist LY274614. *J. Pharmacol. Exp. Ther.* **264,** 1090–1096.

7. Bhargava, H. N. and Thorat, S. N. (1994) Effect of dizocilpine (MK-801) on analgesia and tolerance induced by U-50,488H, a kappa-opioid receptor agonist, in the mouse. *Brain Res.* **649,** 111–116.
8. Elliott, K., Minami, N., Kolesnikov, Y. A., Pasternak, G. W., and Inturrisi, C. E. (1994) The NMDA receptor antagonists, LY274614 and MK-801, and the nitric oxide synthase inhibitor, NG-nitro-L-arginine, attenuate analgesic tolerance to the mu-opioid morphine but not to kappa opioids. *Pain* **56,** 69–75.
9. Pasternak, G. W. and Inturrisi, C. (1995) Pharmacological modulation of opioid tolerance. *Exp. Opin. Invest. Drugs* **4,** 271–281.
10. Pasternak, G. W., Kolesnikov, Y. A., and Babey, A. M. (1995) Perspectives on the *N*-methyl-D-aspartate nitric oxide cascade and opioid tolerance. *Neuropsychopharmacology* **13,** 309–313.
11. Kolesnikov, Y., Jain, S., and Pasternak, G. W. (1996) Modulation of opioid analgesia by agmatine. *Eur. J. Pharmacol.* **296,** 17–22.
12. Dawson, V. L., Dawson, T. M., London, E. D., Bredt, D. S., and Snyder, S. H. (1991) Nitric oxide mediates glutamate neurotoxicity in primary cortical cultures. *Proc. Natl. Acad. Sci. USA* **88,** 6368–6371.
13. Dawson, V. L., Dawson, T. M., Bartley, D. A., Uhl, G. R., and Snyder, S. H. (1993) Mechanisms of nitric oxide-mediated neurotoxicity in primary brain cultures. *J. Neurosci.* **13,** 2651–2661.
14. Kolesnikov, Y. A., Pick, C. G., Ciszewska, G., and Pasternak, G. W. (1993) Blockade of tolerance to morphine but not to kappa opioids by a nitric oxide synthase inhibitor. *Proc. Natl. Acad. Sci. USA* **90,** 5162–5166.
15. Babey, A. M., Kolesnikov, Y., Cheng, J., Inturrisi, C. E., Trifilletti, R. R., and Pasternak, G. W. (1994) Nitric oxide and opioid tolerance. *Neuropharmacology* **33,** 1463–1470.
16. Bohn, L. M., Gainetdinov, R. R., Lin, F. T., Lefkowitz, R. J., and Caron, M. G. (2000) Mu-opioid receptor desensitization by beta-arrestin-2 determines morphine tolerance but not dependence. *Nature* **408,** 720–723.
17. Aley, K. O. and Levine, J. D. (1997) Dissociation of tolerance and dependence for opioid peripheral antinociception in rats. *J. Neurosci.* **17,** 3907–3912.
18. London, E. D., Vaupel, D. B., and Kimes, A. S. (1994) Inhibitors of nitric oxide synthase as potential treatments for opioid withdrawal. *Regul. Pept.* **54,** 165–166.
19. Bhargava, H. N. (1995) Attenuation of tolerance to, and physical dependence on, morphine in the rat by inhibition of nitric oxide synthase. *Gen. Pharmacol.* **26,** 1049–1053.
20. Kolesnikov, Y. A., Pan, Y. X., Babey, A. M., Jain, S., Wilson, R., and Pasternak, G. W. (1997) Functionally differentiating two neuronal nitric oxide synthase isoforms through antisense mapping: Evidence for opposing NO actions on morphine analgesia and tolerance. *Proc. Natl. Acad. Sci. USA* **94,** 8220–8225.
21. Brenman, J. E., Chao, D. S., Gee, S. H., McGee, A. W., Craven, S. E., Santillano, D. R., et al. (1996) Interaction of nitric oxide synthase with the postsynaptic density protein PSD-95 and α1-syntrophin mediated by PDZ domains. *Cell* **84,** 757–767.
22. Ogura, T., Yokoyama, T., Fujisawa, H., Kurashima, Y., and Esumi, H. (1993) Structural diversity of neuronal nitric oxide synthase mRNA in the nervous system. *Biochem. Biophys. Res. Commun.* **193,** 1014–1022.
23. Hall A. V., Antoniou, H., Wang, Y., Cheung, A. H., Arbus, A. M., Olson, S. L., et al. (1994) Structural organization of the human neuronal nitric oxide synthase gene *(NOSI). J. Biol. Chem.* **269,** 33,082–33,090.
24. Pasternak, G. W. and Standifer, K. M. (1995) Mapping of opioid receptors using antisense oligodeoxynucleotides: correlating their molecular biology and pharmacology. *Trends Pharmacol. Sci.* **16,** 344–350.
25. Pasternak, G. W. and Pan, Y.-X. (2000) Antisense mapping: assessing the functional significance of genes and splice variants, in *Antisense Techniques* (Phillips, M.I. ed.), Academic, Orlando, FL, pp. 51–60.
26. Kolesnikov, Y., Jain, S., Wilson, R., and Pasternak, G. W. (1998) Lack of morphine and enkephalin tolerance in 129/SvEv mice: evidence for a NMDA receptor defect. *J. Pharmacol. Exp. Ther.* **284,** 455–459.
27. Crain, S. M. and Shen, K. (2000) Enhanced analgesic potency and reduced tolerance of morphine in 129/SvEv mice: evidence for a deficiency in GM1 ganglioside-regulated excitatory opioid receptor functions. *Brain Res.* **856,** 227–235.

# 29
# Overview of Clinical Studies for Acamprosate

## Adriaan S. Potgieter, MD

## 1. INTRODUCTION

The first double-blind, placebo-controlled study to test the efficacy of acamprosate in alcohol-dependent patients took place in France in 1982, and results were published by Lhuintre et al. in 1985 *(1)*. Since then, more than 25 clinical studies have been performed, assessing various efficacy and safety criteria within the context of varying clinical settings and psychosocial support systems. Seventeen of these studies, which included a total of 4523 patients (2371 on acamprosate, 2152 on placebo), were double-blind studies primarily designed to establish efficacy and to evaluate drug safety. The first market authorization was granted in France in 1989. At present, acamprosate is registered in Europe (19 countries), Latin America (14 countries), Mauritius, Australia, and Hong Kong. An application in the United States is pending. This chapter will summarize clinical evidence and other information obtained through the trials on acamprosate. Information was obtained mainly from published data, but when these were incomplete, additional data were obtained through examination of original study reports *(2)*.

## 2. PHARMACODYNAMICS AND KINETICS

Acamprosate is available as gastro-resistant 333-mg tablets at a recommended daily dosage of four to six tablets per day, to be taken three times daily. It is poorly and slowly absorbed from the gastrointestinal tract via the paracellular route *(3)*. The absolute bioavailability is 11.1%. Steady-state plasma levels are reached after 5–7 d of administration. Interaction studies have confirmed that after a single administration with food, absorption is decreased by approx 20%.

## 3. DOUBLE-BLIND STUDIES

Eighteen double-blind, placebo-controlled clinical studies have been completed since 1982. Of these, 17 were performed in Europe (*see* Table 1) and 1 in the United States. The results of the latter project have not been released by the sponsoring company. Sixteen of the 18 studies employed DSM criteria for diagnosis of alcohol dependence. Patients were between the ages of 18 and 65 and treatment periods varied from 3 to 12 mo. Other drug dependencies were always excluded.

In all but two of the studies, patients had to have already withdrawn from alcohol and started the study medication immediately *after* the acute period of detoxification (which, on average, lasted between 3 and 14 d). The reason for not starting during the acute withdrawal period was to isolate the drug effect from possible confounding effects of concomitant psychotropic and anticonvulsive detoxification medications. Exceptions to this policy were in the trial performed in Spain by Gual and collaborators *(4)*, who introduced the study medication from the first day of weaning, and in the UK study by Chick and collaborators *(5)*, who allowed a period of "stabilization" or washout of up to 5 wk after the

From: *Contemporary Clinical Neuroscience: Glutamate and Addiction*
Edited by: Barbara H. Herman et al. © Humana Press Inc., Totowa, NJ

**Table 1**
**Double-Blind Studies with Acamprosate**

| Study | No. of patients A: P | Treatment duration (mo) | Follow-up duration (mo) | CAD (d) | Time to first drink (d) | Abstinence rate at last visit on treatment (%) |
|---|---|---|---|---|---|---|
| Three Month Studies | | | | | | |
| Lhuintre et al. (France 1982–1983) | 85 A: 42 P: 43 | 3 | 0 | N/A | N/A | N/A |
| Lhuintre et al. (France 1984–1990) | 569 A: 279 P: 290 | 3 | 3 | N/A | N/A | N/A |
| Rousseaux et al (Belgium 1987–1989) | 127 A: 63 P: 64 | 3 | 0 | N/A | N/A | A: 29 P: 33 |
| Pelc et al. (Belgium 1990–1992) | 188 A1998: 63 A1332: 63 P: 62 | 3 | 0 | A1998: 56.6 A1332: 51.9 P: 34.3 | A1998: 56 A1332: 55 P: 15 | A1998: 51 A1332: 44 P: 26 |

| Six Month Studies | No. of Patients A: P | Treatment duration (mo) | Follow-up duration (mo) | Mean CAD (%) | Mean time to first drink (d) | Abstinence rate at last visit on treatment (%) |
|---|---|---|---|---|---|---|
| Pelc et al. (Belgium 1988–1990) | 102 A: 55 P: 47 | 6 | 6 | A: 33% P: 27% | N/A | A: 33 P: 9 |
| Ladewig et al. (Switzerland 1989–1991) | 61 A: 29 P: 32 | 6 | 6 | A: 43% P: 24% | N/A | A: 43 P: 23 |
| Poldrugo et al. (Italy 1989–1993) | 246 A: 122 P: 124 | 6 | 6 | A: 72% P: 59% | A: 151 P: 61 | A: 48 P: 32 |
| Tempesta et al. (Italy 1989–1993) | 330 A: 164 P: 166 | 6 | 3 | A: 66% P: 54% | A: 135 P: 58% | A: 58 P: 45 |
| Geerlings et al. (Benelux countries 1990–1992) | 262 A: 128 P: 134 | 6 | 6 | A: 34% P: 24% | A: 45 P: 15 | A: 25 P: 13 |
| Chick et al. (United Kingdom 1990–1993) | 581 A: 289 P: 292 | 6 | 1.5 | A: 43% P: 45% | A: 37 P: 40 | A: 23 P: 24 |
| Borg et al. (Sweden 1991–1993) | 14 A: 7 P: 7 | 6 | 0 | | | |
| Gual et al. (Spain) | 288 A: 141 P: 147 | 6 | 0 | A: 52% P: 41% | | |

*(continues)*

**Table 1**
*(Continued)*

| Twelve Month Studies | No. of patients A: P | Treatment duration (mo) | Follow-up duration (mo) | CAD (%) | Time to first drink (d) | Abstinence rate at last visit on treatment (%) |
|---|---|---|---|---|---|---|
| Paille et al (France 1989–1992) (Published 1995) | 538 A1998: 173 A1332: 188 P: 177 | 12 | 6 | A1998: 62 A1332: 55 P: 48 | A1998: 153 A1332: 55 P: 102 | A1998: 35 A1332: 28 A1332: 136 P: 19 |
| Barrias et al. (Portugal 1989–1992) | 302 A: 150 P: 152 | 12 | 6 | A: 49% P: 36% | A: 111 P: 55 | A: 39 P: 26 |
| Whitworth et al. (Austria 1989–1993) | 448 A: 224 P: 224 | 12 | 12 | A: 39% P: 30% | A: 55 P: 43 | A: 30 P: 21 |
| Besson et al. (Switzerland 1989–1993) | 118 A: 31 A+D: 24 P: 33 P+D: 22 | 12 | 12 | A: 40% A+D: 55% P: 21% P+D: 31% | A: 55 P: 43 | A: 30 P: 21 |
| Sass et al. (Germany 1990–1992) | 272 A: 136 P: 136 | 12 | 12 | A: 62% P: 45% | A: 165 P: 112 | A: 43 P: 21 |

*Note:* A = acamprosate; P = placebo; CAD = cumulative abstinence duration.

acute weaning treatment. The latter study resulted in some patients (32%) relapsing between the acute weaning phase and inclusion into the acamprosate study.

Most earlier studies differentiated medication dosage according to body weight (1998 mg/d for patients over 60 kg and 1332 mg/d for patient under 60 kg), but this practice proved to be nonessential and has been altered in some countries to dosage regimens irrespective of body weight.

Psychosocial support and treatment protocols in the European studies were not manual driven, and participating centers were allowed to follow local clinical practices as usual with the simple addition of the double-blinded test medication. The exception was the US study, where a manual for the psychosocial therapy was used. Several open-label studies were also conducted that provided additional information on the usage of psychosocial programs with acamprosate, including the NEAT studies *(6)* and the Micado study *(7)*.

In all investigations, the study medication was withdrawn at the end of the study period without tapering dosages. All of the projects except for two of them (*see* Table 1) continued to evaluate patients after study drug withdrawal for periods varying between 6 wk and 1 yr.

Outcome assessment in all studies was primarily considered total abstinence, rather than controlled or reduced drinking. Outcome measures included trials to first drink, proportion of patients drinking, abstinent throughout the study period, and cumulative number of abstinent days.

### 3.1. General Results

A total of 4523 patients were included in the 17 European double-blind studies; 2371 received acamprosate and 2152 received placebo. Between 16% and 31% of patients were females, and the mean age was 42.79 yr. No race differentiation was recorded.

## 3.2. Dose Ranging Study Summaries

### 3.2.1. Paille et al. (France 1989–1992) (Published 1995) (8).

Although preliminary open-dose ranging studies had been performed earlier (Poinso et al. unpublished data, from Merck/Lipha, France 1986), the first double-blind dose ranging study was initiated in 1989 by Paille et al. (8). Five hundred thirty-eight alcohol-dependent patients were administered either 1332 mg or 1998 mg of acamprosate or placebo daily for 12 mo and were then followed for an additional period of 6 mo during which all patients received placebo. The study demonstrated enhanced efficacy with no increase in adverse events for patients treated at the higher dose of acamprosate. The mean cumulative abstinence duration (CAD) of the patients receiving 1998 mg acamprosate was 223 d; for 1332 mg acamprosate, 198 d; and for placebo, 173 d. The difference between the 1998-mg acamprosate group and placebo reached statistical significance ($p = 0.0005$), whereas differences between the 1332-mg acamprosate and placebo groups failed to achieve significance ($p = 0.055$). Another criteria, continuous abstinence, showed greatest efficacy for the 1998-mg dose of acamprosate and the lowest for placebo. Although many patients dropped out during the last 6 mo of placebo follow-up, it nevertheless appeared that the treatment advantage observed during the first 12 mo was maintained. Interestingly, craving and $\gamma$-glutamyl transferase (GGT) levels did not significantly differ between the two active medication groups.

### 3.2.2. Pelc et al. (Belgium 1990–1992) (9)

This study of 188 alcohol-dependent patients tested the same dose range as in the study by Paille et al. in France, but covered a period of 3 mo and included a placebo follow-up period following the active treatment period. The differences between the acamprosate and placebo treatments significantly favored the active medication. Although trends identical to these found in the Paille et al. study were noted (i.e., better efficacy and equal numbers of adverse regardless of dose), differences between the two acamprosate dosages were less pronounced. The study also found fewer dropouts in the acamprosate groups than in the placebo-treated group.

### 3.2.3. Mason et al. (United States 1997–1999)

This study included one group of patients receiving 2 g acamprosate per day and a smaller group receiving 3 g acamprosate per day. The results of this study are not released yet.

## 3.3. Single-Dose Studies

### 3.3.1. Studies with 3-mo Active Treatment Period

Four studies covering treatment periods of 3 mo were performed between 1982 and 1992. Three reported significantly better outcomes on acamprosate than on placebo.

#### 3.3.1.1. LHUINTRE ET AL. (FRANCE 1982–1983) (10)

This was the first double-blind, placebo-controlled study using acamprosate with alcohol-dependnet patients. Eight-five patients were tested on a dosage of 25 mg acamprosate per kg body weight, but not less than 1500 mg nor more than 2500 mg per day. Psychosocial supportive treatment was permitted. Of the 70 patients who completed the study, 61% of those on acamprosate maintained abstinence and had normalization of GGT and mean corpuscular volume (MCV) versus 32% on placebo ($p = 0.02$). The success of this study stimulated the subsequent extensive development program with the drug.

#### 3.3.1.2. LHUINTRE ET AL. (FRANCE 1984–1986) (11)

This study in 569 patients from 31 centers in France had as a primary outcome criterion normalization of GGT in alcohol-dependent patients. All patients received 1332 mg acamprosate or matching placebo. Psychosocial supportive treatment was permitted and patients were followed for 3 mo without treatment and after medication was ceased. At the end of the active treatment period, GGT levels in patients treated with acamprosate were significantly lower than those of patients on placebo ($p < 0.016$). In addition, tongue trembling significantly improved in more patients on acamprosate than on placebo. Data on drinking behavior were not reported. The authors noted that patients of lower body weight experienced more favorable outcome on acamprosate. This was viewed as evidence of a possible dose-related effect with the study drug. The results were considered to be compelling significant confirmation of efficacy, and authorization was granted to market the drug in France in 1989.

### 3.3.1.3. ROUSSEAUX ET AL. (BELGIUM 1987–1989) *(12)*

This single-center study included 127 patients—45 alcohol abusing and 82 alcohol dependent—distributed equally between treatment groups, with 90 patients completing the 3-mo treatment period. Outcome criteria included absolute abstinence and objective markers such as GGT and red cell MCV, but not duration of abstinence. The study medication was dose-adjusted according to the following regimen: Patients below 60 kg body weight received 1332 mg acamprosate daily and those above 60 kg received 1998 mg acamprosate daily or matching placebo. At the end of the study, 21 patients on placebo and 18 on acamprosate were abstinent and improvements in GGT and MCV were comparable in both treatment groups. The study therefore failed to demonstrate any therapeutic advantage of acamprosate.

### 3.3.1.4. PELC ET AL. (BELGIUM 1990–1992) *(9)*

This dose ranging study comparing outcomes among 1998 mg and 1332 mg acamprosate and matching placebo confirmed the efficacy of acamprosate and is discussed in detail in the above section on dose ranging studies.

## 3.3.2. Studies with 6 mo Active Treatment Period

Between 1988 and 2000, nine of the placebo-controlled studies performed with acamprosate tested the drug over a 6-mo active treatment period. The European studies were conducted in eight different countries, whereas the other study was conducted in the United States Despite some differences in methodology, primary outcome criteria were reasonably consistent across projects and evaluated the time to first drink, proportion of patients relapsing, abstinence period throughout the study period, and CAD. European study protocols permitted a liberal approach for concomitant psychosocial intervention, allowing institutional practices of the different participating centres, whereas the study in the United States used a manual-driven psychosocial intervention for all patients. Eight studies had follow-up periods (varying between 1.5 and 6 mo) after termination of active study medication.

Five of the eight European studies reported significantly better outcome in the main criteria of efficacy in favor of acamprosate.

### 3.3.2.1. PELC ET AL. (BELGIUM 1988–1990) *(13)*

This study reported on outcome of 102 alcohol-dependent patients from 5 treatment centers. As in the Rousseaux study *(12)*, the daily dose of acamprosate was adjusted as follows: Patients below 60 kg body weight received 1332 mg acamprosate, whereas those above 60 kg received 1998 mg acamprosate. Attrition rates were high, with 79% of patients on placebo and 56% of those on acamprosate dropping out before the end of the 6-mo period. Patients on acamprosate demonstrated significantly longer CAD (60 vs 40 d) and a higher rate of complete abstinence (24% vs 4%). Measures of clinical global impression, depression, anxiety, craving, and psychological and physiological dependence did not reveal differences between treatment groups.

### 3.3.2.2. LADEWIG ET AL. (SWITZERLAND 1989–1991) *(14,15)*

During 1989, four acamprosate studies with treatment periods of 6 mo (and five studies with treatment periods of 12 mo—see below) were started in different European countries. The study by Ladewig et al. *(14)* in Switzerland was the smallest of these and reported results in 61 alcohol-dependent patients from 3 centers. As in the 1987 Rousseaux *(12)* and the 1988 Pelc *(9)* studies, the study medication was dose-adjusted according to body weight below or above 60 kg for 1332 mg or 1998 mg acamprosate, respectively. The 6-mo treatment period was followed by a 6-mo drug-free follow-up period. Patients on acamprosate had significantly longer CAD (121 vs 78 d). Thirty-eight percent of acamprosate patients and 24% of placebo patients were abstinent at the end of the treatment period. Although the difference in proportions remained during the follow-up period, the statistical difference disappeared.

### 3.3.2.3. POLDRUGO ET AL. (ITALY 1989–1993) *(16)*

This study in northern Italy included alcohol-dependent patients who participated in a postwithdrawal community-based alcohol-treatment program. Acamprosate dose were adjusted according to body weight and use of concomitant disulfiram was allowed. Two hundred forty-six patients were analyzed, including 112 patients completing the 6-mo treatment period and 101 finishing the 6 mo of drug-free follow-up. Patients treated with acamprosate experienced a higher rate of abstinence at the 3- and 6-mo treatment periods, longer CAD at 6 mo (99 vs 70 d) and 12 mo follow-up (168 vs 120 d), and longer time to first relapse (150 vs 61 d). Disulfiram did not appear to influence the outcome.

### 3.3.2.4. TEMPESTA ET AL. (ITALY 1989–1993) *(17)*

Three hundred thirty alcohol-dependent patients from 18 centers in southern Italy participated in this study of 6-mo treatment with 3 mo follow-up. After 6 mo of treatment, 25% dropped out. Patients on acamprosate had significantly higher continuous-abstinence rates, longer CADs, and longer periods before the first relapse occurred. Whereas most studies on acamprosate assessed outcome according to measures of abstinence (proportion and duration), this study also reported outcome in terms of alcohol consumption during periods of relapse. Some significant, ableit limited, differences in the quantity and frequency of drinking in those patients who relapsed were reported suggesting that patients on acamprosate consumed less alcohol during relapse periods. As with most other studies in the acamprosate development program, differences between treatment groups could be not detected for depression and anxiety levels or measurements associated with craving or desire to drink.

### 3.3.2.5. GEERLINGS ET AL. (THE BENELUX COUNTRIES 1990–1992) *(18)*

Two hundred sixty-two patients from Belgium, the Netherlands, and Luxembourg participated in this study, which was characterized by a 6-mo follow-up period after the 6-mo active treatment phase. The drug dose was adjusted to body weight and, as with all other studies, abruptly terminated at the end of the active study period, (i.e., without any tapering down of dosage). Drug compliance based on pill count was 86%. The study suffered from high attrition rates (64% over the first 6 mo and another 16% over the second 6-mo follow-up period), but, nevertheless, demonstrated significant treatment effects in favor of acamprosate. The therapeutic advantage seemed to have been maintained during the drug-free follow-up period, although the number of subjects at the end of the period were too small to statistically conclude this.

### 3.3.2.6. CHICK ET AL. (UNITED KINGDOM 1990–1993) *(5)*

This study comprised patients up to 7 wk after an acute withdrawal treatment, a longer period than any of the other studies. Hence, 32% of the 581 patients included in the study had already relapsed after the required withdrawal treatment and before their inclusion into the study. All patients on acamprosate received 1998 mg/d. Sixty-five percent had dropped out after 6 mo and general compliance to drug-taking was considered low by the authors, with 57% of patients taking 90% of the total pills. No therapeutic advantage of acamprosate over placebo to reduce relapse was observed. Interestingly, patients treated with acamprosate had significantly lower craving scores after 2 and 4 wk of treatment and lower levels of anxiety after 4 wk. The authors noted that this study differed from other acamprosate studies in that patients suffered more social problems, received less psychosocial support, and were more likely to be episodic drinkers. Although none of the patients who relapsed during the washout period attained complete abstinence, no specific responder profile could be identified.

### 3.3.2.7. BORG ET AL. (SWEDEN 1991–1993) (UNPUBLISHED DATA, MERCK/LIPHA, FRANCE)

This small study of only 14 patients was designed for close daily monitoring of patients during 6 mo of treatment, rather than intended as a comparative study with the objective to demonstrate treatment differences. Over the 6-mo treatment period, alcohol consumption was measured three times per week, serum analyses for alcohol markers conducted weekly, and urine analysis for ethanol and 5-hydroxytryptophol performed daily. Ten patients completed the study. Abstinence varied between 73% and 100% of treatment days, but the sample was too small to detect possible differences between treatment groups.

### 3.3.2.8. GUAL ET AL. (SPAIN 1993–1994) *(4)*

The 288 patients in this study began daily acamprosate (1998 mg) or matching placebo concomitantly with the start of acute withdrawal treatment. This earlier introduction of acamprosate in the treatment regimen was hypothesized to potentially enhance drug efficacy, because plasma steady state for acamprosate takes approximately 5 d to develop. The study confirmed significantly longer CAD and longer periods of abstinence after the last relapse with acamprosate. In addition, the measure of continuous abstinence revealed some trend in favor of acamprosate, although this was not statistically significant. Nonetheless, the study failed to suggest some clinical advantage of starting acamprosate earlier, although this possibility merits a direct study. Finally, no adverse interaction occurred between acamprosate and the early-withdrawal treatment regimens.

## *3.3.3. Studies with a 12-mo Active Treatment Period*

### 3.3.3.1. PAILLE ET AL. (FRANCE 1989–1992) (PUBLISHED 1995) *(6)*

This dose ranging study was discussed in Section 3.2.1.

### 3.3.3.2. BARRIAS ET AL. (PORTUGAL 1989–1992) *(19)*

Three hundred two alcohol-dependent patients from 14 centers participated in this study of 12 mo treatment and 6 mo follow-up. Study medication was dose adjusted for body weight. Patients on benzodiazepines for more than 1 mo prior to the study were permitted to continue this practice during the study, and new prescriptions of concomitant oxazepam and temazepam were allowed for periods not exceeding 2 wk. Thirty percent of patients dropped out during the first 6 mo, 14% during the second 6 mo, and 9% during the last 6 mo. The proportion of abstinent patients, time to first relapse, and CAD were all showed a significant advantage for acamprosate during the active treatment period. During the follow-up period, the difference between the two groups became less marked.

### 3.3.3.3. WHITWORTH ET AL. (AUSTRIA 1989–1993) *(20)*

Four hundred forty-eight alcohol-dependent patients from five treatment centers with similar psychosocial treatment programs participated in this study of 12 mo treatment and 12 mo follow up. As with most studies, acamprosate dosage was adjusted for body weight. Sixty percent of patients dropped out during the treatment year (per trimester: 34%, 15%, 6%, and 5%, respectively) and a further 7% dropped out during the follow-up year. Following the "intent to treat" principle, patients who received concomitant neuroleptics (65), antidepressants (34), and benzodiazepines (23) were distributed equally across groups and were maintained in the study despite this protocol violation. The study confirmed the efficacy of acamprosate. Those treated with acamprosate had a greater rate of abstinence, a longer time to first relapse, and a longer CAD than placebo-treated subjects.

### 3.3.3.4. BESSON ET AL. (SWITZERLAND 1989–1993) *(21)*

The particular interest of this study lies in the fact that both the acamprosate and placebo treatment groups were stratified into two subgroups: patients who chose to have concomitant open-label disulfiram treatment and patients who did not. Patients on disulfiram had daily visits and observed disulfiram administration, suggesting more medical contact than patients not opting for disulfiram. One hundred eighteen patients participated in the study (12 mo treatment, 12 mo follow up), with 67% dropping out the first year and 17% in the second year. No adverse interaction between acamprosate and disulfiram was noted. The study outcome confirmed significantly better results on acamprosate than placebo. Analysis of the strata indicated that acamprosate and disulfiram produced the best outcome, disulfiram or acamprosate the second most positive, and no medication the worst outcome.

### 3.3.3.5. Sass et al. (Germany 1990–1992) *(22)*

This intensively documented study of 272 alcohol-dependent patients recruited from 12 centers was considered a pivotal study in the official European application for marketing authorization for acamprosate. Patients received 12 mo of active treatment and 12 mo follow-up. Counseling and psychotherapy differed among sites but, in general, was rather intensive, consisting of 1 h sessions every week for 18 wk, followed by two weekly group sessions. Fifty-one percent dropped out during the first year and 11% in the second year. Similar outcome criteria for drinking behavior as mentioned in the above studies clearly demonstrated the efficacy of acamprosate. No differential effect on craving, however, could be detected, and extensive psychological assessments did not indicate differences between treatment groups. Interestingly, analysis of complete abstinence demonstrated an advantage for acamprosate-treated patients during the drug-free follow-up year. This may suggest a stabilization of abstinent behavior in the acamprosate patients. Psychological assessments at the end of the double-blind period did not suggest any rebound phenomena after the abrupt termination of study medication. Subsequent analyses of this study data by the authors failed to identify a clear responder profile for acamprosate.

## 4. OPEN STUDIES

Several open-label studies have been performed for acamprosate. These include the following:

- A 2-wk study in 591 patients to investigate tolerance of coprescription with psychotropic medications (tetrabamate, meprobamate, and oxazepam) commonly used in France during acute detoxification. No adverse drug interactions were recorded *(23)*.
- Six multicenter studies in six countries to compare the efficacy of acamprosate with different types of psychosocial programmes and determine tolerance to acamprosate. One of these studies was recently published and the data confirmed the good safety profile of acamprosate *(6)*.
- A study with 248 patients in the Netherlands *(7)* to contrast the outcomes of acamprosate treatment in three groups: one without any psychosocial support, another with minimal intervention, and the last with a brief intervention. No significant differences were found among the groups.

## 5. SAFETY DATA

Side effects reported with acamprosate are rare and transient. These include minor gastrointestinal effects such as diarrhea in 10% of patients, nausea and abdominal discomfort in fewer than 10% of patients, and minor skin irritations in fewer than 10% of patients. Decreases as well as increases in libido have been observed in a few patients. The only symptom consistently associated with overdose has been diarrhea. No interactions between acamprosate and any other medications (e.g., disulfiram, antidepressants, anxiolytics, neuroleptics, or hypnotics) have been identified so far. Acamprosate is contraindicated in patients with renal insufficiency and those with known hypersensitivity against the drug.

## 6. REVIEW

The global interpretation of the evidence from the double-blind, placebo-controlled studies with acamprosate confirms that acamprosate has a moderate but fairly consistent effect in reducing relapse in alcohol dependence. Most studies reported that acamprosate increased abstinent rates, increased cumulated abstinence periods, and prolonged periods between withdrawal and first relapse. Acamprosate was also associated with lower patient dropout from treatment. Furthermore, no physical or psychological dependence was observed. Craving or desire to drink was occasionally reported to be less on acamprosate, although more studies are needed. Most studies that measured mood (anxiety and depression) did not report any significant medication effects. None of the analyses published to date has succeeded in identifying a particular responder profile. Data from most studies showed a perceivable drug effect within 1 mo of treatment. However, the minimum required duration of acamprosate treatment has not been clearly established. The more consistent or lasting effects may be achieved when acamprosate is administered for 6–12 mo. All double-blind study projects employed some form of psychosocial support, but no consistent evidence has emerged on any particular requirements of such support during acamprosate treatment. In one open study, results suggested that patients without significant psychosocial support may do as well as those receiving manual-driven minimal or brief interventions. The majority of studies incorporated used regimens of dose adjustment according to body weight. Although official authorization to prescribe the medication in several countries follow this treatment schedule, there seems to be no clear evidence that this is indeed necessary. The documented safety and tolerance profile of acamprosate is exceptionally good. Dose ranging studies indicated better outcome with 1998 mg acamprosate when two tablets are taken three times a day. Although taking six tablets daily may be a practical problem to some patients, high levels of compliance have been recorded in most studies.

# REFERENCES

1. Lhuintre, J. P., Moore, N. D., Saligaut, C., Boismare, F., Daoust, M., Chretien, P., et al. (1985) Ability of calcium bis acetyl homotaurine, a GABA agonist, to prevent relapse in weaned alcoholics. *Lancet* **1(8436)**, 1014–1016.
2. Merck/Lipha (1994) Acamprosate Gastro-Resistant Tablets. Part IV, Clinical documentation. Merck Lipha, France.
3. Saivin, S., Hulot, T., Chabac, S., Potgieter, A., Durbin, P. H., and Houin, G. (1998) Clinical pharmacokinetics of acamprosate. *Clin. Pharmacokinet* **35(5)**, 331–345.
4. Gual, A. (1997) Acamprosate versus placebo in alcoholics. The Spanish study. *Alcohol Alcohol.* **32(3)**, 325.
5. Chick, J., Howlett, H., Morgan, M. Y., and Ritson, B. (2000) United Kingdom Multicentre Acamprosate Study (UKMAS): a 6 month prospective study of acamprosate versus placebo in preventing relapse after withdrawal from alcohol. *Alcohol Alcohol.* **35**, 176–187.
6. Ansoms, C., Deckers, F., Lehert, P., Pelc, I., and Potgieter, A. (2000). An open study with Campral in Belgium and Luxemburg: results on sociodemographics, supportive treatment and outcome. *Eur. Addict. Res.* **6**, 132–140.
7. De Wildt, et al. (2000), submitted.
8. Paille, F., Guelfi, J. D., Perkins, A., Royer, R. J., Steru, L., and Parot, P. (1995) Double-blind randomized multicentre trial of acamprosate in maintaining abstinence from alcohol. *Alcohol Alcohol.* **30(2)**, 239–247.
9. Pelc, L., Verbanck, P., Le Bon, O., Gavrilovic, M., Lion, K., and Lehert, P. (1997) Efficacy and safety of acamprosate in the treatment of detoxified alcohol-dependent patients. *Br. J. Psychiatry* **171**, 73–77.
10. Lhuintre, J. P., Moore, N. D., Saligaut, C., et al. (1985) Ability of calcium bis acetyl homotaurine, a GABA agonist, to prevent relapse in weaned alcoholics. *Lancet* **1(8436)**, 1014–1016.
11. Lhuintre, J. P., Moore, N. D., Tran, G., et al. (1990) Acamprosate appears to decrease alcohol intake in weaned alcoholics. *Alcohol Alcohol.* **25(6)**, 613–622.
12. Rousseaux, J.-P., Hers, D., and Ferauge, M. (1996) L'acamprosatediminue-t-il l'appétence pourl'alcool chez l'alcoolique sevré? *J. Pharm. Belg.* **51(2)**, 65–68.
13. Pelc, I., Le Bon, O., Verbanck, P., Lehert, P., and Opsomer, L. (1992) Calcium-acetylhomotaurinate for maintaining abstinence in weaned alcoholic patients: a placebo-controlled double-blind multicentre study, in *Novel Pharmacological Interventions in Alcoholism* (Narango, C. A. and Sellers, E., eds.), Springer-Verlag New York, pp. 348–352.
14. Ladewig, D., Knecht, T., Lehert, P., and Fendl, A. (1993) Acamprosat-cia Stabilisierungsfaktor in der Langzeitentwohnung von Alkoholabhangigen. *Ther. Umsch.* **50(3)**, 182–198.
15. Merck/Lipha: Acamprosate European Marketing Autorisation Dossier, Vol. 38.
16. Poldrugo, F. (1997) Acamprosate treatment in a long-term community-based alcohol rehabilitation. *Addiction* **92(II)**, 1537–1546.
17. Tempesta, E., Janiri, L., Bignamini, A. Chabac, S., and Potgieter, A. (2000) Acamprosate and relapse prevention in the treatment of alcohol dependence: a placebo controlled study. *Alcohol Alcohol.* **35(2)**, 202–209.
18. Geerlings, P., Ansoms, C., and Van Den Brink, W. (1997) Acamprosate and prevention of relapse in alcoholics. Results from a randomized, placebo-controlled double-blind study in out-patients alcoholics in the Netherlands, Belgium and Luxembourg. *Eur. Addict. Res.* **3**, 129–137.

19. Barrias, J. A., Chabac, S., Ferreira, L., Fonte, A., Potgieter, A. S., and Teixeira de Sousa, E. (1997) Acamprosate: estudo portugues, multicentrico de avaliaqao da eficaciae *Psiquiatr. Clin.* **18(2),** 149–160.
20. Whitworth, A. B., Fisher, F., Lesch, O. M., et al. (1996) Comparison of acamprosate and placebo in long-term treatment of alcohol dependence. *Lancet* **347(9013),** 1438–1442.
21. Besson, J., Aeby, F., Kasas, A., Lehert, P., and Potgieter, A. S. (1998) Combined efficacy of acamprosate and disulfiram in the treatment of alcoholism: a controlled study. *Alcohol. Clin. Exp. Res.* **22(3),** 573–579.
22. Sass, H., Soyka, M., Mann, K., and Zieglgansberger, W. (1996) Relapse prevention by Acamprosate. Results from a placebo-controlled study of alcohol dependence *Arch. Gen. Psychiatry.* **53(8),** 673–680.
23. Aubin, H. J., Lehert, P., Beaupere, B., Parot, P., and Barrucand, D. (1995) Tolerability of the combination of acamprosate with drugs used to prevent alcohol withdrawal syndrome. Alcoholism **31(1–2),** 25–38.

# INDEX

**A**

Acamprosate,
  alcoholism treatment,
    anticraving and antirelapse properties, 407, 408, 424, 425
    dose ranging studies, 420
    double-blind studies, overview, 417–419
    mechanism of action,
      $N$-methyl-$\text{D}$-aspartate receptor interactions, 408–412
      overview, 392, 407, 408
    open-label studies, 424
    pharmacodynamics, 417
    safety, 424, 425
    single-dose studies,
      3-mo active treatment period, 420, 421
      6-mo active treatment period, 421–423
      12-mo active treatment period, 423, 424
  morphine dependence treatment, 411
  nucleus accumbens effects, 410, 411
ACEA-1021,
  cocaine convulsant effect treatment, 249
  cocaine lethality treatment, 253
ACEA-1031,
  cocaine convulsant effect treatment, 249
  cocaine lethality treatment, 253
ACEA-1328,
  cocaine convulsant effect treatment, 249
  cocaine lethality treatment, 253
$N$-Acetylated $\alpha$-linked-acidic dipeptidase (NAALASase), inhibitors, 41
ACPC, $N$-methyl-$\text{D}$-aspartate receptor binding, 31
ADCI,
  $N$-methyl-$\text{D}$-aspartate receptor binding, 26, 27, 33
  receptor specificity, 30
Addiction, *see also* Behavioral sensitization; Motive circuit,
    definition, 127
    psychomotor stimulant sensitization and self-administration, glutamate–dopamine interactions, 188–190, 193, 194
    stable drug-induced changes in brain, candidate mechanisms, 129–131
Alcohol,
  abuse prevalence and impact, 343
  acamprosate treatment,
    anticraving and antirelapse properties, 407, 408, 424, 425
    dose ranging studies, 420
    double-blind studies, overview, 417–419
    mechanism of action,
      $N$-methyl-$\text{D}$-aspartate receptor interactions, 408–412
      overview, 392, 407, 408
    open-label studies, 424
    pharmacodynamics, 417
    safety, 424, 425
    single-dose studies,
      3-mo active treatment period, 420, 421
      6-mo active treatment period, 421–423
      12-mo active treatment period, 423, 424
  $\alpha$-amino-3-hydroxy-5-methyl-4-isoxazole propionic acid receptor effects,
    mechanism of action, 351
    overview, 350, 351
    sensitivity modulation, 351, 352
    subunit sensitivity, 351
  $\text{GABA}_\text{A}$ effects,
    chloride flux decrease, 361
    phosphorylation and kinases, 361, 362
    toxicity role, 366–368
  glutamate release,
    dependence, 366
    withdrawal, 392, 409

kainate receptor effects,
  mechanism of action, 351
  overview, 350, 351
  sensitivity modulation, 351, 352
  subunit sensitivity, 351
$N$-methyl-D-aspartate receptor effects,
  antagonist studies,
    glycine-B partial agonists, 391, 392
    intoxication effects, 389, 390
    mechanisms of action, 409, 410
    self-administration effects, 353
  behavioral modulation, 389–392
  cytoskeletal interactions and clustering, 362, 364
  familial risk, 393, 394
  inhibition potency, 344, 352
  magnesium enhancement, 346, 347
  mechanisms of action, 345–347
  mitogen-activated protein kinase cascade, 364, 366
  molecular sites of action, 347
  nitric oxide induction, 361
  phospholipid signaling, 364
  phosphorylation and kinases, 361, 362
  sensitivity modulation, 347–350
  subunit sensitivity and composition, 344, 345, 377, 408, 409
  toxicity role, 366–368
  withdrawal changes, 377, 378, 392, 393, 409
morphological changes in brain, 357–360
neuropsychological effects in alcoholics, 359, 360, 393
neurotrophin depletion in toxicity, 366
nitric oxide role in conditioned place preference, 233, 234, 239
withdrawal,
  kindling,
    α-amino-3-hydroxy-5-methyl-4-isoxazole propionic acid receptor role, 381–383
    animal models, 376, 377
    hypothesis of dependence and withdrawal, 375–377
    kainate receptor role, 381–383

    $N$-methyl-D-aspartate receptor role, 378, 380, 381, 383
    neurotoxicity implications, 382, 383
  syndrome, 375
$N$-Allyl-normetazocine, $N$-methyl-D-aspartate receptor binding, 24
Amantadine, $N$-methyl-D-aspartate receptor binding, 28
γ-Aminobutyric acid (GABA) receptors, GABA$_A$,
  alcohol effects,
    chloride flux decrease, 361
    phosphorylation and kinases, 361, 362
    toxicity role, 366–368
  subunits, 360
  psychomotor stimulants, behavioral sensitization and induced stereotypy mediation by dopamine, glutamate, and γ-aminobutyric acid system interactions,
  intracortical drug effects,
    agonist–antagonist interactions, 120–122
    agonists, 119, 120
    antagonists, 118, 119
  intrastriatal drug effects,
    agonist–antagonist interactions, 117, 118
    antagonists, 115–117
  overview, 122–124
  study design, 107, 108
  systemic antagonist effects,
    qualitative data, 108–111
    quantitative data, 111–115
α-Amino-3-hydroxy-5-methyl-4-isoxazole propionic acid (AMPA) receptor,
  activation, 13–14
  activity-dependent synaptic plasticity and behavioral sensitization induction in ventral tegmental area, 131–134
  alcohol effects,
    mechanism of action, 351
    overview, 350, 351
    sensitivity modulation, 351, 352
    subunit sensitivity, 351
    withdrawal kindling role, 381–383

# Index

calcium permeability, 65
cocaine-induced conditioned increases in locomotor activity role, 97, 100, 101
desensitization modulators, 15–16
G protein coupling, 12
ligands,
  allosteric modulators,
    negative, 36–37
    positive, 37
  competitive antagonists, 34–36
long-term potentiation role, 130, 131
maturational regulation and neuroplasticity, 65
opiate withdrawal, antagonist effects, 332, 333
phosphorylation, 11, 130
protein–protein interactions, 11–12
splice and editing variants, 8–9
subunits, 6–7, 65, 130
AMPA receptor, see α-Amino-3-hydroxy-5-methyl-4-isoxazole propionic acid receptor
Amphetamine,
  behavioral sensitization and induced stereotypy mediation by dopamine, glutamate, and γ-aminobutyric acid system interactions,
    intracortical drug effects,
      agonist–antagonist interactions, 120–122
      agonists, 119, 120
      antagonists, 118, 119
    intrastriatal drug effects,
      agonist–antagonist interactions, 117, 118
      antagonists, 115–117
      overview, 122–124
    study design, 107, 108
    systemic antagonist effects,
      qualitative data, 108–111
      quantitative data, 111–115
  locomotor sensitization,
    dopamine role, 184, 186
    glutamate–dopamine interactions,
      induction, 186, 187
      metabotropic glutamate receptor expression, 187, 188
      nucleus accumbens, 186
      ventral tegmental area, 186
    motive circuit neuroadaptation and glutamate and dopamine interactions,
      acute drug administration, 149
      animal models of addiction, 149
      craving implications, 150, 151
      repeated drug administration, 150
    sensitization and self-administration, glutamate–dopamine interactions, 188–190, 193, 194
Amygdala, motive circuit,
  neuronal electrophysiology, 147
  neurotransmitter release, 148
Apoptosis,
  glutamate neurotoxicity, 52–55
  methamphetamine neurotoxicity,
    Bcl-2, 204, 205
    p53, 205, 206
Aptiganel, N-methyl-D-aspartate receptor binding, 26

# B

Basic fibroblast growth factor (bFGF), addiction role, 129
BDNF, see Brain-derived neurotrophic factor
Behavioral sensitization, see also Conditioned place preference,
  cocaine, nitric oxide role,
    conditioned increases in locomotor activity role, 90, 91
    conditioned place preference role, 233, 234, 239
    neuronal nitric oxide synthase inhibition studies,
      knockout mouse studies, 231, 232, 234
      7-nitroindazole effects, 230, 233, 234
    consequences of psychostimulant sensitization in humans, 230
    history of glutamate antagonist studies, 127, 128

psychomotor stimulants, behavioral sensitization and induced stereotypy mediation by dopamine, glutamate, and γ-aminobutyric acid system interactions,
  intracortical drug effects,
    agonist–antagonist interactions, 120–122
    agonists, 119, 120
    antagonists, 118, 119
  intrastriatal drug effects,
    agonist–antagonist interactions, 117, 118
    antagonists, 115–117
  overview, 122–124
  study design, 107, 108
  systemic antagonist effects,
    qualitative data, 108–111
    quantitative data, 111–115
  ventral tegmental area and activity-dependent synaptic plasticity, 131–134
bFGF, see Basic fibroblast growth factor
Brain-derived neurotrophic factor (BDNF),
  alcohol effects on levels, 367, 368
  strategies for increasing of levels, 41, 42

**C**

Calcium flux,
  glutamate neurotoxicity,
    calcium set point, 54, 55
    neuron culture studies, 52
  methamphetamine neurotoxicity, 216
  mitochondrial effects, 73
  p38 activation, 73
  striatal neuron expression regulation by metabotropic glutamate receptors, 161–163
CNQX,
  α-amino-3-hydroxy-5-methyl-4-isoxazole propionic acid receptor binding, 34
  kainate receptor binding, 37
Cocaine,
  behavioral sensitization and induced stereotypy mediation by dopamine, glutamate, and γ-aminobutyric acid system interactions,
    intracortical drug effects,
      agonist–antagonist interactions, 120–122
      agonists, 119, 120
      antagonists, 118, 119
    intrastriatal drug effects,
      agonist–antagonist interactions, 117, 118
      antagonists, 115–117
    overview, 122–124
    study design, 107, 108
    systemic antagonist effects,
      qualitative data, 108–111
      quantitative data, 111–115
  conditioned increases in locomotor activity,
    behavioral sensitization, nonassociative versus associative factors, 83, 84
    drug addiction relevance, 102
    glutamate role,
      α-amino-3-hydroxy-5-methyl-4-isoxazole propionic acid receptor role, 97, 100, 101
      long-term potentiation parallels, 101
      $N$-methyl-D-aspartate receptor role, 85, 86, 100
      MK-801 blockade, 85–88, 90–94, 96–106
      neuroanatomy, 86, 87, 100, 101
      nitric oxide role, 90, 91
    neurobiology and dopamine role, 84, 85, 101
  conditioned place preference, $N$-methyl-D-aspartate receptor antagonist effects,
    acquisition, 324, 325
    expression, 326
  convulsant effects,
    glutamate release inhibitor treatment, 252
    incidence, 248
    metabotropic glutamate receptor antagonist blocking, 251, 252
    $N$-methyl-D-aspartate receptor antagonist blocking,
      allosteric modulators, 250, 251

# Index

AMPA/kainate receptor antagonist combination therapy, 251
channel blockers, 250
clinical implications, 255, 256
competitive antagonists, 250
NMDA/glycine antagonists, 248–250
dopamine transporter binding, 229
glutamate-dependent synaptic plasticity and addiction, 176–178
glutamate neurotransmission in limbic system, role in abuse, 172–176
lethality,
   glutamate release inhibitor treatment, 252, 254
   N-methyl-D-aspartate receptor antagonist blocking,
     allosteric modulators, 254
     AMPA/kainate receptor antagonist combination therapy, 254
     channel blockers, 254
     clinical implications, 255, 256
     competitive antagonists, 254
     NMDA/glycine antagonist pretreatment and posttreatment, 253
   onset, 252, 253
mGluR5 modulation studies,
   dopamine signaling, 273, 274, 276
   learning tasks in mice, 273
   locomotor and reinforcing effects, 273, 274
   transgenic mice, 272, 273
motive circuit neuroadaptation and glutamate and dopamine interactions,
   acute drug administration, 149
   animal models of addiction, 149
   craving implications, 150, 151
   repeated drug administration, 150
N-methyl-D-aspartate receptor antagonists, withdrawal and abstinence studies, 266
neural substrates of abuse, 171, 172
overdose adverse effects, 243
sensitization and self-administration, glutamate–dopamine interactions, 188–190, 193, 194
treatment stages for withdrawal, 261
Conditioned place preference (CPP), *see also* Behavioral sensitization,
conditioned reward model, 323, 324
N-methyl-D-aspartate receptor antagonist effects,
   acquisition,
     cocaine, 324, 325
     morphine, 324, 325
   expression,
     cocaine, 326
     morphine, 325, 326
   maintenance effects with morphine, 326
nitric oxide role,
   alcohol, 233, 234
   cocaine, 233, 234, 239
   nicotine, 233, 234
CPP,
conditioned place preference, *see* Conditioned place preference
drug psychomotor stimulant studies, behavioral sensitization and induced stereotypy mediation by dopamine, glutamate, and γ-aminobutyric acid system interactions, 116, 117, 120–122
CREB, addiction role, 129
CX516, α-amino-3-hydroxy-5-methyl-4-isoxazole propionic acid receptor binding, 37
D-Cycloserine,
   alcohol intoxication blocking, 391, 392
   N-methyl-D-aspartate receptor binding, 249, 250

# D

D-CPP-ene, N-methyl-D-aspartate receptor binding, 30
DCQX,
   cocaine convulsant effect treatment, 249
   cocaine lethality treatment, 253
Dependence, definition, 295
$\delta 1$, subunits, 6
$\delta 2$, subunits, 6
Dextromethorphan,
   cocaine addiction studies, 175

opioid dependence, phase II trials, 263, 265
Dizocipline, *see* MK-801
DNQX,
    cocaine addiction studies, 175
    kainate receptor binding, 37
    psychomotor stimulant studies, behavioral sensitization and induced stereotypy mediation by dopamine, glutamate, and γ-aminobutyric acid system interactions, 113–115
Dopamine,
    amphetamine locomotor sensitization role, 184, 186
    cocaine-induced conditioned increases in locomotor activity, 84, 85, 101
    meso-accumbens pathway for addiction, 143, 144
    methamphetamine neurotoxicity and glutamate mediation, 211, 212, 214, 215, 220, 221
    phosphorylation of glutamate receptors, mediation, 130
    psychomotor stimulant sensitization and self-administration, glutamate–dopamine interactions, 188–190, 193, 194
Dopamine receptor,
    motor activity regulation by striatal metabotropic glutamate receptors, 159, 160
    psychomotor stimulants, behavioral sensitization and induced stereotypy mediation by dopamine, glutamate, and g-aminobutyric acid system interactions,
        intracortical drug effects,
            agonist–antagonist interactions, 120–122
            agonists, 119, 120
            antagonists, 118, 119
        intrastriatal drug effects,
            agonist–antagonist interactions, 117, 118
            antagonists, 115–117
        overview, 122–124

        study design, 107, 108
        systemic antagonist effects,
            qualitative data, 108–111
            quantitative data, 111–115
    types, 145

**E**
Eliprodil, $N$-methyl-D-aspartate receptor binding, 33
Ethanol, *see* Alcohol
Excitoxicity, *see* Glutamate excitoxicity

**F**
Felbamate,
    $N$-methyl-D-aspartate receptor binding, 30, 33
    receptor specificity, 30
FosB, addiction role, 129
Frontal lobes, chronic alcoholism effects, 358
Fyn, $N$-methyl-D-aspartate receptor alcohol sensitivity modulation, 347–349

**G**
GABA receptors, *see* γ-Aminobutyric acid receptors
Glial cell, glutamate receptors, 14–15, 74
GluR1, phosphorylation, 130
Glutamate excitotoxicity, *see also specific drugs*,
    animal models, 53, 54
    calcium set point, 54, 55
    cultured neuron studies,
        apoptosis studies, 52, 53
        calcium flux, 52
        potassium efflux, 53
        zinc flux, 53
    excitatory amino acid receptors, 71, 72
    history of study, 51, 72
    methamphetamine neurotoxicity, 203, 204, 212–217
Glutamate receptors, *see specific receptors*
Glutamate release,
    alcohol,
        dependence, 366
        withdrawal, 392, 409
    inhibitor types, 40, 41

# Index

GYKI52466, α-amino-3-hydroxy-5-methyl-4-isoxazole propionic acid receptor binding, 36

## H

HIV, *see* Human immunodeficiency virus
Homer, metabotropic glutamate receptor interactions, 5
Human immunodeficiency virus (HIV),
  brain cell infection, 71
  chemokine receptors, 71, 74, 75
  dementia,
    features, 74
    $N$-methyl-D-aspartate receptor stimulation, 75–77
    neurotoxicity of viral components, 75
  pathogenesis, 74, 75
  risk factors, 74

## I

Ibogaine,
  $N$-methyl-D-aspartate receptor binding, 30
  opioid dependence management, 265
IEGs, *see* Immediate early genes
Ifendopril, $N$-methyl-D-aspartate receptor binding, 33
Immediate early genes (IEGs), striatal neuron expression regulation by metabotropic glutamate receptors, 160, 161
Ionotropic glutamate receptors, *see also specific receptors*,
  classes, 6, 330
  phosphorylation, 11
  protein–protein interactions, 11–12
  psychomotor stimulants, behavioral sensitization and induced stereotypy mediation by dopamine, glutamate, and γ-aminobutyric acid system interactions,
    intracortical drug effects,
      agonist–antagonist interactions, 120–122
      agonists, 119, 120
      antagonists, 118, 119
    intrastriatal drug effects,
      agonist–antagonist interactions, 117, 118
      antagonists, 115–117
      overview, 122–124
      study design, 107, 108
      systemic antagonist effects,
        qualitative data, 108–111
        quantitative data, 111–115
  splice and editing variants, 8–9, 11
  subunits,
    GluR2 crystal structure, 8
    modulatory sites, 15–16
    stoichiometry, 7–8
    types, 6–7
  topology, 8

## K

Kainate receptor,
  activation, 14
  alcohol effects,
    mechanism of action, 351
    overview, 350, 351
    sensitivity modulation, 351, 352
    subunit sensitivity, 351
    withdrawal kindling role, 381–383
  desensitization modulators, 15–16
  ligands, 37, 38
  maturational regulation and neuroplasticity, 66
  morphological changes in brain, 357–360
  neuropsychological effects in alcoholics, 359, 360, 393
  phosphorylation, 11
  splice and editing variants, 8–9
  subunits, 6–7, 66, 145
  types, 37
Ketamine,
  ethanol-like effects, 389, 390
  $N$-methyl-D-aspartate receptor binding, 24, 26
Kynurenic acid, opiate withdrawal studies, 330

## L

LC, *see* Locus coeruleus
Lithium chloride, conditioned place aversion, 233, 234

Locus coeruleus (LC),
    activity in opioid withdrawal, 329–334
    alcohol effects, 358
Long-term depression (LTD),
    corticostriatal pathway, 134, 135
    drug alterations, 134–136
    mechanisms, 132, 133, 135
    *N*-methyl-D-aspartate receptor role, 296
Long-term potentiation (LTP),
    addiction parallels, 101, 129–131
    drug alterations, 134–136
    glutamate dependence, 176
    *N*-methyl-D-aspartate receptor role, 296
LTD, *see* Long-term depression
LTP, *see* Long-term potentiation
LY 235959, *N*-methyl-D-aspartate receptor binding, 31
LY 274614, *N*-methyl-D-aspartate receptor binding, 31
LY 293558,
    α-amino-3-hydroxy-5-methyl-4-isoxazole propionic acid receptor binding, 36
    kainate receptor binding, 38
LY 300164, α-amino-3-hydroxy-5-methyl-4-isoxazole propionic acid receptor binding, 36
LY 300168, α-amino-3-hydroxy-5-methyl-4-isoxazole propionic acid receptor binding, 36
LY 303070, α-amino-3-hydroxy-5-methyl-4-isoxazole propionic acid receptor binding, 36
LY 354740, metabotropic glutamate receptor binding, 39, 40
LY 377770, kainate receptor binding, 38
LY 382884, kainate receptor binding, 38

## M

MAPK, *see* Mitogen-activated protein kinase
Memantine,
    cocaine convulsant effect treatment, 250
    *N*-methyl-D-aspartate receptor binding, 26–28
    opioid dependence, phase II trials, 263–265
Metabotropic glutamate receptors (mGluRs),
    amphetamine locomotor sensitization, glutamate–dopamine interactions,
        expression of receptors, 187, 188
        induction, 186, 187
        nucleus accumbens, 186
        ventral tegmental area, 186
    classification, 3, 4, 38, 66, 67, 145, 271, 330
    cocaine convulsant effect blocking by antagonists, 251, 252
    G protein coupling, 3, 6
    Homer interactions, 5
    ligands, 38–40
    maturational regulation and neuroplasticity, 66, 67
    mGluR5,
        brain expression and distribution, 271, 272
        cocaine modulation studies,
            dopamine signaling, 273, 274, 276
            learning tasks in mice, 273
            locomotor and reinforcing effects, 273, 274
        transgenic mice, 272, 273
    opiate withdrawal, antagonist effects, 333, 334
    psychomotor stimulants, behavioral sensitization and induced stereotypy mediation by dopamine, glutamate, and γ-aminobutyric acid system interactions,
        intracortical drug effects,
            agonist–antagonist interactions, 120–122
            agonists, 119, 120
            antagonists, 118, 119
        intrastriatal drug effects,
            agonist–antagonist interactions, 117, 118
            antagonists, 115–117
        overview, 122–124
        study design, 107, 108
        systemic antagonist effects,
            qualitative data, 108–111
            quantitative data, 111–115
    signaling, 3, 6, 38, 157
    splice variants, 5

striatal neuron glutamate cascade,
  gene expression regulation,
    constitutive genes, 160–163
    dopamine-dependent genes, 163, 164
    overview, 157, 158
  localization of receptors, 158, 159
  motor activity regulation, 159, 160
  therapeutic targeting of receptors, 164, 165
  synaptic functions, 14
  targeting, 5–6
Methamphetamine,
  abuse rates, 211
  behavioral sensitization, *see* Behavioral sensitization
  clinical toxicity, 202, 203, 211
  neurotoxicity mechanism,
    apoptosis,
      Bcl-2, 204, 205
      p53, 205, 206
    ATP loss in metabolic stress, 218–220
    dopamine and glutamate mediation, 211, 212, 214, 215, 220, 221
    glutamate excitotoxicity, 203, 204, 212–217
    hydroxy radicals, 217
    hyperthermia, 235
    nitric oxide, 203, 204, 235, 236
    peroxynitrite, 218, 235
    poly(ADP-ribose) polymerase activation, 204
    serotonin and glutamate mediation, 220, 221
    superoxide radicals, 203, 217
$N$-Methyl-D-aspartate (NMDA) receptor,
  acamprosate interactions, 408–412
  activation, 13–14
  alcohol effects,
    antagonist studies,
      glycine-B partial agonists, 391, 392
      intoxication effects, 389, 390
      mechanisms of action, 409, 410
      self-administration effects, 353
    behavioral modulation, 389–392
    cytoskeletal interactions and clustering, 362, 364
  familial risk, 393, 394
  inhibition potency, 344, 352
  magnesium enhancement, 346, 347
  mechanisms of action, 345–347
  mitogen-activated protein kinase cascade, 364, 366
  molecular sites of action, 347
  nitric oxide induction, 361
  phospholipid signaling, 364
  phosphorylation and kinases, 361, 362
  sensitivity modulation, 347–350
  subunit sensitivity and composition, 344, 345, 377, 408, 409
  toxicity role, 366–368
  withdrawal changes, 377, 378, 392, 393, 409
  withdrawal kindling role, 378, 380, 381, 383
calcium flux and excitotoxicity, 63, 73
cocaine-induced conditioned increases in locomotor activity and MK-801 blockade, 85–88, 90–94, 96–106
conditioned place preference, antagonist effects,
  acquisition,
    cocaine, 324, 325
    morphine, 324, 325
  expression,
    cocaine, 326
    morphine, 325, 326
  maintenance effects with morphine, 326
dependence, phase II trials of antagonist treatment,
  cocaine withdrawal and abstinence studies, 266
  drug self-administration models, 264
  opioid dependence,
    abstinence promotion, 266
    dextromethorphan, 263, 265
    human models, 262, 263
    ibogaine, 265
    memantine, 263–265
  prospects, 266, 267
  study design, 262

human immunodeficiency virus stimulation in dementia, 75–77
ligands,
  agonists, 24
  binding site overview, 389, 390, 408, 409
  competitive antagonists, 30, 31
  glycine-site ligands, 31, 32
  magnesium, 63, 64
  NMDA/glycine antagonists,
    clinical implications, 255, 256
    cocaine convulsion treatment, 248–252
    cocaine lethality prevention, 252–255
    development, 248
  noncompetitive antagonists, 15
  NR2B-selective ligands, 33–34
  overview, 245–247
  side effects, 245–247
  uncompetitive antagonists, 24–30
maturational regulation and neuroplasticity, 63–65
NR1 splice variants, 408
nucleus accumbens blockade and cocaine addiction effects, 174, 175
opioid tolerance and dependence role,
  activation, 285, 286
  antagonist prevention,
    addiction treatment, 312, 313
    antagonist types and effects, 298, 299, 305, 306
    chronic pain treatment, 312
    dependence, 303, 304
    discriminative stimulus effects, 301
    history of study, 296–299
    locomotor depression, 299, 300
    operant responding, 301
    δ-opioids, 283
    κ-opioids, 283
    μ-opioids, 281–283, 307, 308
    reversal of tolerance and dependence, 304, 306, 307
    sensitization, 302, 303
    side effects and therapeutic potential, 313, 314
    temperature regulation, 299
    withdrawal effects, 331, 332
  neuropathic pain and tolerance mechanism interactions, 288, 289
  protein kinase C role,
    autoradiographical and immunocytochemical evidence, 285
    behavioral evidence, 284, 286
    electrophysiological evidence, 284, 285, 310
    signaling model, 308–310
    spinal cord model of μ-opioid tolerance studies, 286–288
  signal transduction, 308–311
phosphorylation and kinases, 11
protein–protein interactions, 12
subunits, 6–7, 33, 64, 243, 244, 408
mGluRs, see Metabotropic glutamate receptors
Mitogen-activated protein kinase (MAPK),
  alcohol effects on activation, 364
  cascade, 364
MK-801,
  cocaine addiction studies, 174, 175
  cocaine-induced conditioned increases in locomotor activity, blockade, 85–88, 90–94, 96–106
  $N$-methyl-D-aspartate receptor binding, 24, 26
  opioid tolerance and dependence prevention,
    addiction treatment, 312, 313
    antagonist types and effects, 298, 299, 305, 306
    chronic pain treatment, 312
    dependence, 303, 304
    discriminative stimulus effects, 301
    history of study, 296–299
    locomotor depression, 299, 300
    operant responding, 301
    δ-opioids, 283
    κ-opioids, 283
    μ-opioids, 281–283, 307, 308
    reversal of tolerance and dependence, 304, 306, 307
    sensitization, 302, 303
    side effects and therapeutic potential, 313, 314

# Index

temperature regulation, 299
side effects, 245
Morphine,
    acamprosate treatment of dependence, 411
    conditioned place preference, N-methyl-D-aspartate receptor antagonist effects,
        acquisition, 324, 325
        expression, 325, 326
        maintenance, 326
    N-methyl-D-aspartate receptor antagonist prevention of tolerance,
        associative versus nonassociative tolerance, 282, 283
        dependence prevention, 282
        μ-opioid agonist studies, 282, 307, 308
        study design, 281, 282
Motive circuit,
    amygdala,
        neuronal electrophysiology, 147
        neurotransmitter release, 148
    nucleus accumbens,
        neuronal electrophysiology, 146
        neurotransmitter release, 147, 148
    prefrontal cortex,
        neuronal electrophysiology, 146, 147
        neurotransmitter release, 148
    psychostimulant-induced neuroadaptation in glutamate and dopamine interactions,
        acute drug administration, 149
        animal models of addiction, 149
        craving implications, 150, 151
        repeated drug administration, 150
    ventral tegmental area,
        neuronal electrophysiology, 146
        neurotransmitter release, 148
MPEP, metabotropic glutamate receptor binding, 40

# N

NAALASase, *see* N-Acetylated α-linked-acidic dipeptidase
NAc, *see* Nucleus accumbens
Narp, glutamate receptor binding, 12
NBQX,
    α-amino-3-hydroxy-5-methyl-4-isoxazole propionic acid receptor binding, 34, 35
    cocaine addiction studies, 176
    kainate receptor binding, 37, 38
Nerve growth factor (NGF), alcohol effects on levels, 366, 367
Neurexin, glutamate receptor binding, 12
Neuroligin, glutamate receptor binding, 12
NGF, *see* Nerve growth factor
Nicotine, nitric oxide role in conditioned place preference, 233, 234, 29
Nitric oxide (NO),
    cocaine and methamphetamine behavioral sensitization role,
        conditioned increases in locomotor activity role, 90, 91
        conditioned place preference role, 233, 234, 239
        neuronal nitric oxide synthase inhibition studies,
            knockout mouse studies, 231, 232, 234
            7-nitroindazole effects, 230, 233, 234
    methamphetamine neurotoxicity, 203, 204, 218, 234–236
Nitric oxide synthase (NOS),
    cocaine addiction studies with inhibitors, 177
    isoforms, 40, 229
    N-methyl-D-aspartate receptor modulation, 177, 178, 229
    neuronal enzyme,
        inhibitors, 40, 229
        knockout mice, 231, 232, 234, 236, 239
    opioid effect mediation,
        antisense mapping of splice variants, 401–403
        blockade inhibition of tolerance, 399, 400, 404
        reversal of tolerance, 400, 401
        therapeutic targeting, 404, 405
NMCQX, receptor specificity, 32
NMDA receptor, *see* N-Methyl-D-aspartate receptor

NMDX, receptor, specificity, 32
NO, see Nitric oxide
NOS, see Nitric oxide synthase
NS-102, kainate receptor binding, 38
NS-257, α-amino-3-hydroxy-5-methyl-4-isoxazole propionic acid receptor binding, 34
Nucleus accumbens (NAc),
  acamprosate effects, 410, 411
  amphetamine locomotor sensitization, glutamate–dopamine interactions, 186
  drug effects on synaptic plasticity, 134–136
  glutamate neurotransmission in limbic system, 172–174
  motive circuit,
    neuronal electrophysiology, 146
    neurotransmitter release, 147, 148
  opioid withdrawal role, 331

## O

Opioids,
  N-methyl-D-aspartate receptor in tolerance and dependence,
    activation, 285, 286
    antagonist prevention,
      addiction treatment, 312, 313
      antagonist types and effects, 298, 299, 305, 306
      chronic pain treatment, 312
      dependence, 303, 304
      discriminative stimulus effects, 301
      history of study, 296–299, 399
      locomotor depression, 299, 300
      operant responding, 301
      δ-opioids, 283
      κ-opioids, 283
      μ-opioids, 281–283, 307, 308
      reversal of tolerance and dependence, 304, 306, 307
      sensitization, 302, 303
      side effects and therapeutic potential, 313, 314
      temperature regulation, 299
    neuropathic pain and tolerance mechanism interactions, 288, 289
    phase II trials of antagonists,
      abstinence promotion, 266
      dextromethorphan, 263, 265
      human models, 262, 263
      ibogaine, 265
      memantine, 263–265
    protein kinase C role,
      autoradiographical and immunocytochemical evidence, 285
      behavioral evidence, 284, 286
      electrophysiological evidence, 284, 285, 310
      signaling model, 308–310
      spinal cord model of μ-opioid tolerance studies, 286–288
      signal transduction, 308–311
  nitric oxide synthase mediation of effects,
    antisense mapping of splice variants, 401–403
    blockade inhibition of tolerance, 399, 400, 404
    reversal of tolerance, 400, 401
    therapeutic targeting, 404, 405
  μ-opioids, see Morphine
  tolerance deficiency in 129/SvEv mice, 404
  withdrawal,
    glutamate receptor antagonist studies,
      α-amino-3-hydroxy-5-methyl-4-isoxazole propionic acid receptor, 332, 333
      metabotropic glutamate receptor, 333, 334
      N-methyl-D-aspartate receptor, 331, 332
      therapeutic prospects, 334
    locus coeruleus activity, 329–334
    nucleus accumbens role, 331
    spinal cord glutamate release, 331
    symptoms, 329
    treatment stages, 261
Oxidative stress,
  antioxidant defenses in brain, 202
  methamphetamine neurotoxicity,
    hydroxy radicals, 217

peroxynitrite, 218
superoxide radicals, 203, 217
reactive oxygen and nitrogen species in brain, 201, 202

## P

p75, alcohol effects on levels, 367
PARP, see Poly(ADP-ribose) polymerase
PCP, see Phencyclidine
PFC, see Prefrontal cortex
Phencyclidine (PCP),
  cocaine addiction studies, 174, 175
  N-methyl-D-aspartate receptor binding, 24, 26, 296
PKB, see Protein kinase B
PKC, see Protein kinase C
Poly(ADP-ribose) polymerase (PARP), methamphetamine activation, 204
PPD, see Preprodynorphin
PPE, see Preproenkephalin
Prefrontal cortex (PFC),
  glutamate neurotransmission in limbic system, 172, 173
  motive circuit,
    neuronal electrophysiology, 146, 147
    neurotransmitter release, 148
  metabotropic glutamate receptors, 157, 161, 163
Preproenkephalin (PPE), striatal neuron expression regulation by metabotropic glutamate receptors, 157, 161, 163
Protein kinase B (PKB), alcohol effects on phospholipid signaling, 364, 366
Protein kinase C (PKC),
  alcohol activation, 362
  opioid tolerance and dependence role,
    autoradiographical and immunocytochemical evidence, 285
    behavioral evidence, 284, 286
    electrophysiological evidence, 284, 285, 310
    signaling model, 308–310
    spinal cord model of μ-opioid tolerance studies, 286–288

## R

Reactive oxygen species, see Oxidative stress
Ro 48-8587, α-amino-3-hydroxy-5-methyl-4-isoxazole propionic acid receptor binding, 36

## S

SCH-23390, psychomotor stimulant studies, behavioral sensitization and induced stereotypy mediation by dopamine, glutamate, and γ-aminobutyric acid system interactions, 110, 112, 115, 122
Selfotel, N-methyl-D-aspartate receptor binding, 30–31
Sensitization, definition, 295
Serotonin, methamphetamine neurotoxicity and glutamate mediation, 220, 221
SIB 1893, metabotropic glutamate receptor binding, 40
SP, see Substance P
Spinal cord, glutamate release in opioid withdrawal, 331
Substance P (SP), striatal neuron expression regulation by metabotropic glutamate receptors, 157, 161
SYM 2189, α-amino-3-hydroxy-5-methyl-4-isoxazole propionic acid receptor binding, 36
SYM 2207, a-amino-3-hydroxy-5-methyl-4-isoxazole propionic acid receptor binding, 37

## T

Thiocyanate, α-amino-3-hydroxy-5-methyl-4-isoxazole propionic acid receptor binding, 37
THIP, psychomotor stimulant studies, behavioral sensitization and induced stereotypy mediation by dopamine, glutamate, and γ-aminobutyric acid system interactions, 117, 120
Tolerance, definition, 295

## V

Ventral tegmental area (VTA), behavioral sensitization and activity-dependent synaptic plasticity, 131–134
    amphetamine locomotor sensitization, glutamate–dopamine interactions, 186
    motive circuit,
        neuronal electrophysiology, 146
        neurotransmitter release, 148

VTA, *see* Ventral tegmental area

## Y

YM90K, α-amino-3-hydroxy-5-methyl-4-isoxazole propionic acid receptor binding, 34, 35

## Z

Zinc flux, glutamate neurotoxicity, 53